Geophysical Monograph Series

Including

IUGG Volumes
Maurice Ewing Volumes
Mineral Physics Volumes

Geophysical Monograph 125

Space Weather

Paul Song
Howard J. Singer
George L. Siscoe
Editors

American Geophysical Union
Washington, DC

Library of Congress Cataloging-in-Publication Data

Space weather / Paul Song, Howard J. Singer, George L. Siscoe, editors.
    p. cm. – (Geophysical monograph ; 125)
    ISBN 0-87590-984-1
    1. Space environment. I. Song, P. (Paul), 1954- II. Singer, Howard J. III. Siscoe, George L. IV. Series.

QB505 .S64 2001
629.4'16–dc21   2001033760

ISBN 0-87590-984-1
ISSN 0065-8448

Cover image: A composite of space weather images comprising a SOHO satellite image of an erupting coronal mass ejection, a global magnetohydrodynamic (MHD) rendition of the open magnetosphere intercepting the solar wind, magnetic field lines connecting Earth to the solar wind, and at Earth a simulation of ionospheric electron density at the F-region from a coupled thermosphere-ionosphere-plasmasphere-electrodynamic model. Photos courtesy of NASA, Mission Research Corporation, and NOAA Space Environment Center.

CONTENTS

CONTENTS

CONTENTS

SECTION IV: Specification and Prediction of the Ionosphere and Thermosphere

CONTENTS

PREFACE

This volume provides a comprehensive overview of our current observational knowledge, theoretical understanding, and numerical capability with regard to the phenomena known as space weather. Space weather refers to conditions on the Sun and in the solar wind, magnetosphere, ionosphere, and thermosphere that can influence the performance and reliability of space-borne and ground-based technological systems, and can endanger human life or health. The rapid advance in these technologies has provided us with unprecedented capability and convenience, and we have come to rely on them more and more. Technology has reduced society's risk to many kinds of natural disasters, but through its own vulnerability, it has actually increased society's risk to space weather. Adverse conditions in the space environment can cause disruption of satellite operations, communications, navigation, and electric power distribution grids, leading to a variety of socioeconomic losses.

In the last five years, largely as a result of the National Space Weather Program of the United States, research on space weather has grown intense. Research is transforming space weather from descriptive to predictive and from qualitative to quantitative. The aim of the research is ambitious: progressively and systematically develop our knowledge of the causally linked Sun-solar wind-magnetosphere-ionosphere/thermosphere system, so as to steadily improve our ability to specify and predict the environment where space-vulnerable systems operate. Consequently space weather research has become an anchor for space science programs in many countries. It highlights the importance of space physics and provides an outlet for contributions of the space physics commu-nity to society.

The content of the volume is as follows. First is a suite of articles covering broad scientific, technological, industrial, commercial, and programmatic topics on space weather. Nonexperts can readily read these articles. Then come three technical sections that trace space weather's causal links from the Sun to the Earth and the space in between. Several articles address the linkage among this energy chain and provide more integrated views on the coupling through different regions. Tutorials begin each major topic.

This book owes much to the advent of the United States' National Space Weather Program (NSWP). Accordingly the first article of the volume archives the major events associated with the birth of the NSWP. The NSWP was inaugurated in 1995 with the aim of addressing increasing needs to specify and predict conditions in space. To prepare for the technology challenges in the new millennium, the NSWP set out to achieve, within 10 years, an active, synergistic, interagency system to provide timely, accurate, and reliable space environment observations, specifications and forecasts. Since the advent of the NSWP, space weather research has expanded from a semi-private activity, pursued mostly to address operational concerns of industries and military branches affected by space weather, to a broad subject pursued in part for its capacity to integrate research across divisions in the Space Physics and Aeronomy community. Space weather programs have also been proposed and inaugurated in countries other than the United States.

Over the NSWP's first five years, progress on long-standing space weather projects has accelerated. An empirical model for predicting solar wind conditions at the Earth from solar magnetic observations became markedly more accurate. An empirical model for specifying the magnetospheric magnetic field, already a community workhorse five years ago, considerably extended its range of applicability. A model for specifying the radiation environment that solar storms induce along the orbit of the International Space Station reached maturity. A globe-spanning network of radars is now providing data on ionospheric electrodynamics virtually in real time. A program is now working to enable useful nowcasts and forecasts of ionospheric scintillations, which adversely affect communication and navigation. Reports on these advances are included in the volume.

Breakthroughs that might have seemed unfeasible five years ago have been made: "Halo events"—sudden brightenings of the solar corona—reveal storm clouds approaching from the Sun.

Magnetospheric models input with solar wind data demonstrate 80-90% accuracy in forecasting the onset and intensity of geo-magnetic storms and the occurrence of relativistic, "killer" electron events. A global magnetospheric magneto-hydrodynamic (MHD) code, which five years ago ran much slower than real time, has been streamlined and parallelized so that, given upstream solar wind condition, it can compute magnetospheric and geomagnetic conditions faster than real time, a prerequisite for an operational forecast code.

In the same half-decade, significant milestones and trends emerged. The first Sun-to-ionosphere numerical code was demon-strated. Hybrid global numerical codes combining the power of MHD for the outer magnetosphere and particle drift physics for the inner magnetosphere were initiated. Data assimilation moved from a little known topic in ionospheric and magnetospheric modeling to center stage. Powerful empirical prediction models, some of which use neural networks and techniques from nonlinear dy-namics, emerged to treat aspects from solar activity to the size of the auroral oval. The first commercial vendors of space weather services, complete with attractive Web sites, have come into being.

Looking ahead, while space weather programs investigate in depth many scientific aspects and draw on more extensive international collaborations, NASA has initiated a new program that stresses space weather. With this program, named "Living with a Star" (LWS), NASA will construct installations of spacecraft in strategic locations around the Sun and Earth to gather data that will give simultaneous views of aspects of the Sun-to-ionosphere system pertinent to space weather. The program's emphasis will be on system-wide data gathering and modeling. The impact of LWS on space weather research could be enormous. The volume contains an introduction to this new program.

Space weather is a cross-disciplinary scientific effort. Solar, heliosphere, magnetosphere physicists, and aeronomers have em-braced a common program partly out of necessity, since in space weather each discipline depends on the others. The current volume puts emphasis on the interfaces between scientific disciplines and integration of space weather programs over different disciplines. Engineering, user-interface, and education and outreach aspects of space weather have been omitted, however, for demands of space.

The predictive nature of space weather research distinguishes it from conventional space physics research. Space weather, despite often being viewed as applied science, is more accurately regarded as an enabling science defined by the intersection in the space arena of the spheres of pure and applied science. Accordingly this volume emphasizes research that leads to improvements in predictability, especially quantitative predictability.

The book benefits from material prepared for and presented at a Chapman Conference on Space Weather. Over 160 scientists and students from more than a dozen countries gathered March 20-24, 2000, in Clearwater, Florida, to attend the conference, which was made possible by support from the National Science Foundation, National Aeronautics and Space Administration, National Oceanic and Atmospheric Administration, and International Association of Geomagnetism and Aeronomy (NSF, NASA, NOAA and IAGA). The conference provided a forum for identifying, reviewing, and debating the issues presented in this volume, as well as for defining the next crucial phases of space weather research. Presentations at the conference provided a starting point for most of the articles contained in the book but in many cases the authors have expanded and considerably modified their written contributions in response to discussions at the conference, as well as comments from reviewers and editors. Some papers presented at the conference were not submitted for publication while a few papers in the volume were not presented at the meeting and were solicited by the editors in order to provide more complete coverage. Many new research results were contained in poster presentations at the conference; a selection of these papers will appear in the Journal of Geophysical Research, Space Physics, in a special section devoted to space weather and scheduled for publication later in 2001.

We express our gratitude to Richard Behnke and Robert Robinson of the National Science Foundation who perceived the scientific significance and social impact of space weather, and along with other leaders in the community formulated and implemented the National Space Weather Program. Without their vision, enthusiasm, and diligent efforts, the NSWP, and this book, would not have been possible. The editors wish to thank the following individuals who participated in the evaluation of papers submitted to the volume: J. M. Albert, B. J. Anderson, N. Arge, D. N. Baker, C. C. Balch, J. Barth, S. Basu, P. Bellaire, J. Berchem, J. Birn, W. J. Burke, A. G. Burns, A. Chan, J. K.Chao, J. Chen, M. W. Chen, C. R. Clauer, T. R. Detman, M. Dryer, D. Fairfield, J. F. Fennell, F. Fenrich, J. Feynman, T. Forbes, J. Foster, N. Fox, J. W. Freeman, B. E. Gilchrist, T. I. Gombosi, G. R. Heckman, E. Hildner, J. W. Hirman, M. K. Hudson, W. J. Hughes, J. Joselyn, S. Kahler, Y. Kamide, A. Klimas, J. Klimchuck, L. J. Lanzerotti, F. Li, X. Li, M. Liemohn, R. Lin, B. C. Low, H. Lundstedt, K. Macpherson, G. Mason, N. C. Maynard, R. L. McPherron, S. Milan, T. E. Moore, R. J. Niciejewski, K. Nishikawa, P. O'Brien, T. G. Onsager, V. O. Papitashvili, W. K. Peterson, R. Pirjola, V. Pizzo, J. Raeder, D. Reames, G. D. Reeves, J. Richardson, I. Richardson, A. Ridley, R. M. Robinson, C. T. Russell, D. Rust, K. Schatten, M. Schulz, R. Schunk, M. Shea, J.-H. Shue, D. Sibeck, D. Smart, O. C. St. Cyr, D. P. Stern, K. Takahashi, J. Thayer, M. Thomsen, D. Vassiliadis, W. Wang, D. Webb, W. W. White, M. Wiltberger, G. I. Withbroe, S. T. Wu, Q. Wu, L. Zelenye, X. P. Zhao, and L. Zhu.

Paul Song
Center for Atmospheric Research
Department of Environmental
Earth and Atmospheric Sciences
University of Massachusetts
Lowell, Massachusetts

Howard J. Singer
Space Environment Center
National Oceanic and Atmospheric Administration
Boulder, Colorado

George L. Siscoe
Center for Space Physics
Boston University
Boston, Massachusetts

The U. S. National Space Weather Program: A Retrospective

R. M. Robinson and R. A. Behnke

National Science Foundation, Arlington, Virginia

The National Space Weather Program began in 1994 in response to community efforts to highlight the strategic nature of space science research. In response to these efforts, several government agencies established an interagency working groups to coordinate space weather activities. Program elements include basic research, modeling, observations, technology transition, operational forecasting, and education. The goals and objectives of the program have been delineated in a strategic plan and an implementation plan that has been recently updated. Accomplishments during the past five years include the support of targeted basic research, various workshops and meetings focussing on the needs of space weather customers, the development of a modeling center for transitioning space weather models, and education and outreach efforts that have resulted in unprecedented awareness of space weather. The program emphasizes the treatment of the space environment as an integrated system encompassing the sun, solar wind, magnetosphere, ionosphere, and thermosphere. The history of the National Space Weather Program underscores the merits of interagency partnering and user-oriented goals in enhancing the quality and relevance of scientific research.

INTRODUCTION

The National Space Weather Program (NSWP) is an excellent example of interagency cooperation toward the achievement of both scientific and societal goals. Because interagency cooperation in program development is by no means a guarantee of success, it is instructive to examine the history of the NSWP as an example of a program that has gained favorable recognition within the government, as well as in the international scientific community.

This paper summarizes the steps leading up to the formation of the NSWP and the major milestones that have taken place since its inception. The intent is to provide insight into the factors responsible for the program's success and call particular attention to the agencies, institutions, and people that have contributed to its development. This monograph is a tribute to the extraordinary impacts of

Space Weather
Geophysical Monograph 125

the NSWP thus far, coming some seven years after the start of the program. Many will claim that the jury is still out regarding the overall contributions of the NSWP--that the most appropriate test of success is in the concrete improvements achieved in our ability to forecast the state of the space environment. While this is certainly true, the activities outlined below will demonstrate that the program's existence has removed major stumbling blocks in our ability to work together toward common goals.

EARLY DAYS

The first real step toward the creation of the NSWP came at a time when the National Science Foundation (NSF), along with other government agencies, was being strongly encouraged by the U. S. Congress to show the societal benefits of its activities. Among the sciences, this new paradigm for scientific research was met with considerable dismay, as some scientists felt hampered by requirements to demonstrate potential use or commercial value of scientific research. However, the newly defined standards actually resonated with many space scientists, who saw great opportunities in this change of focus. Accordingly, a small group of space scientists requested a meeting with the Assistant Director for Geosciences at NSF, Robert Corell. This meeting took place early in 1994 and included George Siscoe, Ernie Hildner, Bill Lotko, Lou Lanzerotti, and Tim Killeen. The idea presented to Corell was that space research contributes to our ability to forecast conditions in the space environment, and that this activity has had tremendous societal importance. Furthermore, space science research in this area, which is often referred to as space weather, was poised to achieve great leaps in understanding and predictive capability due to recent advances in model development, and to the impending deployment of powerful new instruments, both in space and on the ground. Whereas other sciences might have to significantly alter their approaches to scientific research, demonstrating societal relevance in space weather research would require only a small change in the way our research was conducted.

Corell immediately recognized the logic in this argument and encouraged the group to initiate efforts to develop an interagency program in space weather. He suggested that the first step was to bring all the stakeholders together: government agencies, members of the research community, and representatives from industry. This meeting took place March 2, 1994. Ironically, that day was the day of one of the largest snowstorms of the decade in the Washington area. Although some attendees were unable to get to the meeting because of the storm, the number of attendees that did show up was a testimony in itself to the level of interest in space weather.

It is significant to note that the outcome of the meeting was by no means a hue and cry for the merits of a space weather program. In fact, many of the speakers at the meeting were doubtful about the likelihood of success for such an effort. Indeed, perhaps all of the major pitfalls that would later plague the NSWP were identified during that meeting, such as, industry's reluctance to disclose problems with their technical systems, the small amount of money that industry and government agencies were willing to devote to space weather research, the lack of information about the economic benefits of a space weather program, the problems associated with transitioning research results to operational improvements, the need for educational activities to improve awareness of space weather effects, identifying potential customers for space weather services, and the preference for engineering solutions to space weather hazards.

Despite these challenges, a second interagency meeting was convened on June 16 for the purpose of developing the organizational structure for an interagency program in space weather. To prepare for the meeting, NSF worked with a small group of scientists to develop a draft strategic plan for a new space weather program. The plan was to be distributed in advance to interested agency representatives, and was to be the focal point for meeting discussions.

In soliciting agency representatives to attend the meeting, NSF contacted Colonel Tom Tascione of Air Force Weather, who had, in his presentation at the NSF meeting, characterized the likelihood of getting increased funding from the Air Force for space weather as virtually nil. Colonel Tascione was willing to work with NSF in the development of a program, but wished to meet separately prior to the June 16 meeting. The hazards of interagency cooperation hit home when Colonel Tascione suggested the meeting take place at 7:00 AM. As with subsequent disagreements between agencies, the exact meeting time was settled via a compromise—8:00 AM, and at that early morning meeting, the organizational structure for the future NSWP was laid out.

Colonel Tascione called attention to the fact that a group already existed to oversee space weather activities. It was called the Committee for Space Environment Forecasting (CSEF), organized under the Office of the Federal Coordinator for Meteorology (OFCM), an agency of the federal government loosely connected to the National Oceanic and Atmospheric Administration (NOAA), mandated with the oversight of interagency programs in meteorology. It was notable that several members of the CSEF did not even realize they belonged to the committee, partly because the group typically met only once per year. Its primary responsibility was to approve, on an annual basis, a report produced by OFCM on the status of space environmental services. Despite its passive approach to space weather planning, the organizational structure of the group, with members from all the relevant government agencies, represented the logical approach to formulating a more proactive program in space environment forecasting.

These ideas were presented to agency representatives at the June 16 meeting along with an outline of specific goals and recommendations for a national program. All attendees concurred with the program objectives and agreed that the next step in the development of the NSWP was to contact OFCM and request a restructuring of the CSEF, an update in its membership, and modifications to its charter. During the next few months, a small contingent of the CSEF met on a regular basis to begin developing a strategic plan for the NSWP and a new charter for the group giving it more extended responsibilities for coordinating interagency space weather activities. On November 22, 1994, the CSEF officially changed its name to the Working Group for the National Space Weather Program, and a new Terms of Reference for the group was put into effect. On December 14, the main oversight group for the OFCM, the Federal Committee for Meteorological Services and Supporting Research, reviewed the progress that had been made and approved the organization of the National Space Weather Program Council consisting of high ranking members of each of the government agencies. The Program Council was to provide high-level visibility within, and coordination among, the agencies. It would meet when significant milestones in the program had been reached, or when controversial issues required high-level attention.

It should be noted the decision was made early to include only representatives of federal agencies as members of the working group. This was done to create a body that would not appear biased toward any particular academic institution or threatening to industry stakeholders. However, the charter for the working group, which would eventually become the Committee for Space Weather (CSW), allowed for technical representatives in addition to regular members. These technical representatives are selected from non-government institutions and serve on the committee as required. Also, industry representatives and academic scientists are often invited as visitors to CSW meetings when technical input on specific issues is desired.

Immediately after the June 16 meeting at NSF, the Coupling, Energetics, and Dynamics of Atmospheric Regions (CEDAR) Program had its annual workshop in Boulder, Colorado. The official definition for space weather was created during that meeting at a late night get-together of NSF's Upper Atmosphere Research Section. The following definition was unanimously adopted: "Space weather refers to conditions on the sun and in the solar wind, magnetosphere, ionosphere, and thermosphere that can influence the performance and reliability of space-borne and ground-based technological systems and can endanger human life or health. Adverse conditions in the space environment can cause disruption of satellite operations, communications, navigation, and electric power distribution grids, leading to a variety of socioeconomic losses." This definition subsequently appeared in so many places that had it been copyrighted, its creators would no longer have to worry about socioeconomic losses.

Throughout the fall of 1994, when the organizational structure for the NSWP was being put in place, the Working Group continued efforts on the Strategic Plan. The Strategic Plan presented the motivation for an interagency program and identified the critical elements shown in Figure 1. It also delineated the role each agency would play in the efforts to improve operational space weather capabilities. As inclusive as the elements in Figure 1 may be in defining the scope of the NSWP, it fails to do justice to the extensive discussions that went into it. In particular, the committee was acutely aware that the weakest links in the chain of knowledge connecting research to operations was effective feedback between representatives of the two groups, and the process of transitioning technology from the research community to the operational environment. Fortunately, in the strategic plan, the solution to such problems could remain vague and left to the next stage in space weather planning which was development of an implementation plan.

While struggling with solutions to problems associated with carrying out the space weather program, the committee also had to contend with pervasive suspicions and

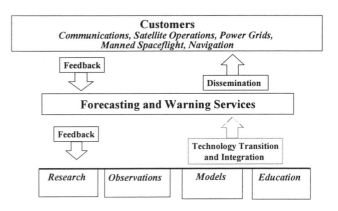

Figure 1. Strategic elements of the National Space Weather Program.

skepticism within communities that were to be best served by the program. Many presentations on space weather in these early days were necessarily defensive in nature. Anecdotal evidence about the impacts of space weather was offered as proof of the vital importance of the program. Space weather proponents often had to fend off challenges to produce a vociferous customer base that would justify the program's existence, and the research community had to be continually reminded that the NSWP did not threaten basic research in space sciences.

It is interesting to note that although the program was looked at with skepticism outside the federal government, within the government it was achieving a high degree of respectability. The best evidence for this is that space weather was accepted as an area of concern by the Subcommittee on Natural Disaster Relief (SNDR), a subcommittee of the Committee on Environment and Natural Resources of the National Science and Technology Council. The SNDR is made up of representatives from federal agencies who examine natural hazards from the points of view of assessment, mitigation, and warning. The goal of the SNDR is to create a sustainable society, resilient to natural hazards. Unlike the skeptics, this group recognized the importance of early preparation for potential threats, having learned from experience that disasters need not occur before preventive measures are put in place. The uniqueness of space weather is that, unlike other sources of natural disasters, civilization's susceptibility increases with advancing technology.

In a February 1995 issue of Newsweek a full-page advertisement for Martin Marietta showed an idyllic little town nestled in a valley bathed only in light from an auroral form which dominated the upper half of the image. It is obviously nighttime, yet there are no lights in any of the windows of the town. The ad proclaims the following statement: "Did the weatherman say anything about solar winds tonight?" The ad describes the famous electric power blackout in Quebec in March 1989 and explains how the solar wind creates the aurora. "It can also create havoc," the ad proclaims, and then goes on to boast about Martin Marietta's role in NASA's Wind mission. With Martin Marrieta's permission, this picture was adopted as the official cover page for the NSWP Strategic Plan and the two implementation plans. References to solar storms creating "havoc" have also been found in many subsequent space weather documents.

Though it now had an official cover page, the program had to wait four more years before adopting an official logo. The logo was designed by Kile Baker when he was the Program Manager for Magnetospheric Physics at NSF. The logo is oval shaped with color shading from yellow to black to blue horizontally across the length of the oval. The yellow represents the sun (on the left, of course), the black represents interplanetary space and the magnetosphere, and the blue represents Earth's atmosphere. The letters NSWP, written across the oval, link the three domains. The logo has become a useful adornment for space weather documents and presentations, just as the little town bathed in auroral light has become a familiar design for the covers of space weather planning documents.

NSWP PLANNING

The Strategic Plan for the NSWP was approved by the Program Council on August 4, 1995. At the approval meeting the Program Council was briefed on progress in the development of the first implementation plan. Stressing the importance of performance accountability, the Program Council recommended that the plan include a process for the development of metrics by which to gage the expected improvements in space weather forecasting.

Work on the Implementation Plan had actually started earlier in 1995 after NSF convened a workshop that took place in Arlington, Virginia, on June 15 and 16. The purpose of the meeting was to identify the scientific research that was required to achieve the desired improvement in operational forecasting. The operational requirements were developed primarily by Air Force Weather and NOAA's Space Environment Center (SEC), and were presented to the group of about 20 scientists attending the workshop. The scientists were divided into three working groups corresponding to the three domains relevant to space weather: the sun and solar wind, the magnetosphere, and the ionosphere/thermosphere. In a remarkably productive two days, these working groups developed a comprehensive set of science questions that needed to be addressed to meet the operational goals of the program. These findings appear in the appendices of both the first and second implementation plans.

Perhaps the most critical part of the implementation plans, and the part that consumed the most time and effort in development, were the road maps to guide space weather activities, shown in generalized form in Figure 2. Separate maps were developed for each of the three space

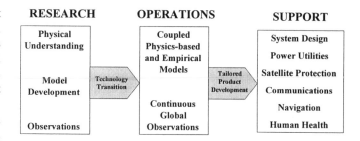

Figure 2. Generalized road map for the development of enhanced space weather products.

weather domains; each included activities related to research observations and models and operational observations and models. Modeling received particular attention in space weather planning because of the accepted philosophy that physics-based models are the best long-term means to achieve reliable and accurate forecasts necessary to meet operational requirements. Modeling considerations had a strong heritage in the planning that went into the development of the Global Geospace Circulation Model for the GEM program. Even earlier, Bob Carovillano led a small team of researchers in the development of the report "Quantitative Magnetospheric Predictions Program, A Magnetospheric Space Weather Research Initiative." This study, submitted to NASA Headquarters, laid the framework for the development of physics-based models of the magnetosphere and its response to solar wind conditions. The NSWP Implementation Plan road maps showed how such physics-based models for the entire space weather system could be constructed in a 10-year time frame by linking individual models.

Although the space weather road maps delineated the process for achieving long-term goals of the program, short-term advances were viewed as essential to maintain customer enthusiasm for the program. Therefore, the implementation plan contained a list of short-term goals as well. These short-term objectives were incorporated in the research announcements released for each of the NSF-sponsored space weather proposal competitions.

The first implementation plan was approved by the Space Weather Program Council on January 7, 1997. Because the Implementation Plan was meant to be a living document, to be updated as the program progressed, the Program Council recommended revisions be published on an annual basis. The OFCM has a term it uses for action items that are not closed, but in some way or other become outdated. The OFCM refers to such items as "overcome by events". The plan to update the Implementation Plan on an annual basis was one of those tasks "overcome by events". The second plan was not released until more than three years later. However, the Space Weather Working Group was by no means idle during the intervening period. In fact, one of the main reasons for the delay was the flurry of space weather activity that commenced soon after the first implementation plan was released.

PARALLEL SPACE WEATHER ACTIVITIES

The NSWP has been an excellent catalyst for a host of related activity initiated by government agencies concerned about space weather. Probably the earliest of these was a workshop convened by the Naval Research Laboratory in September 1995. This meeting highlighted the scientific endeavors currently underway related to space weather. The keynote speaker of the meeting was General Thomas Lennon of Air Force Weather, who argued for the importance of space weather in the future battlespace environment and emphasized the need for seamless modeling of the space weather system.

Another major activity made possible by the NSWP was the annual space weather proposal competition. Most of the funding for this program has come from NSF which received a budget increase of $1M for the NSWP. NSF funds have been supplemented by contributions from both the Air Force Office of Scientific Research (AFOSR) and the Office of Naval Research (ONR). Four such competitions have been conducted so far, and this monograph presents some of the results enabled by this funding. More than 90 of these small individual investigator proposals have been funded through the program to date with annual support that has increased to $3M.

The NSWP also inspired the National Security Space Architect study on space weather. The Space Architect, who heads a joint office under the National Reconnaissance Office and the Under-Secretary of the Department of Defense (DOD), is charged with developing a strategy for meeting warfighter requirements over the next 25 years. Space weather was recognized as an aspect of the future military environment that needed to be addressed. In this DOD-led effort, a small army of scientists and space weather operations experts was mobilized on short notice and invited to attend a series of intensive meetings. The outcome of these meetings was a plan, referred to as a "vector," which delineated steps necessary to meet the warfighter requirements of the 21st Century.

There are obvious differences between these plans and those put forth by the NSWP. First, the Space Architect plan extends out to 2025, while the NSWP plan covers only the next ten years. Second, the Space Architect study emphasizes those activities necessary to meet DOD requirements, while the NSWP includes commercial interests. But the differences between the two plans are negligible in comparison with the similarities. Both plans emphasize the need for comprehensive space weather research, observations and modeling, and the importance of effective procedures for transitioning research and technology into the operational environment. The Space Architect study was followed by the development of a

Table 1. Growth in agency participation in the National Space Weather Program since its inception.

Then	Now
NSF	NSF
NOAA/OAR	NOAA/OAR
NOAA/ERL	NOAA/SEC
	NOAA/NESDIS
	NOAA/National Weather Service
USAF/XOW	USAF XOW
USAF/ Space Command	USAF Space Command
	USAF 55th Space Weather Squadron
	USAF Office of Scientific Research
	USAF Research Laboratory
	USAF Space Division-NPOESS Office
	Office of Naval Research
	US Naval Observatory
	Naval Research Laboratory
	Naval Deputy to NOAA
	Naval Meteorology and Oceanography Command
	US Army Corps of Engineers
NASA	NASA
DOI/US Geologic Survey	DOI/US Geologic Survey
DOT/Coast Guard	DOT/Coast Guard
	DOT/Federal Aviation Administration
	DOE

transition plan that included a list of recommendations on how to achieve the desired space weather vector. The Committee for Space Weather has assumed responsibility for the implementation of this transition plan.

By 1997, agency participation in the NSWP had blossomed enormously. Table 1 shows a comparison of the agencies involved in the program from the time of the program's inception to the present time. Notable among the growing cast of characters are the Federal Aviation Agency (FAA) and the host of DOD agencies largely drawn in by the Space Architect study. Urged on by the growing attention to space weather and the huge success of solar missions such as the Solar and Heliospheric Observatory (SOHO) and Yohkoh, NASA participated in high level discussions with the Air Force Space Command (AFSPC) that led to a strategic partnership involving the two organizations. The purpose of the partnership was to prepare recommendations as to how the agencies could work together to address problems of mutual interest. In these discussions, George Withbroe of NASA Headquarters and Major Tom Frooninckx of Air Force Weather argued to include space weather as one of the areas of interest. As a result, an AFSPC/NASA/NOAA working group on the space environment examined the status of space weather modeling. The group recommended that the problem of transitioning research models to operations could best be addressed by developing a modeling center responsible for the testing and validation of research codes.

At the time, both the DOD and NOAA were attempting to set up Rapid Prototyping Centers (RPCs) that would

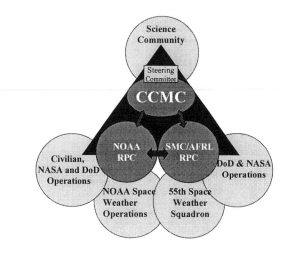

Figure 3. Conceptual design for the Community Coordinated Modeling Center for linking research results and operational requirements.

receive validated research models and mold them into software that could run efficiently on operational platforms. This strategy was originally suggested in the first Space Weather Implementation Plan. However, it was clear the RPCs could not handle the process of validation and testing of research models. The AFSPC/NASA partnership eventually inspired the creation of the Community Coordinated Modeling Center (CCMC). Figure 3 illustrates the strategy whereby the CCMC would provide the critical link between the research community and the RPCs.

In the true spirit of interagency cooperation, the CCMC uses high performance computer assets funded by the DOD, front-end computers provided by funding from NSF and AFOSR, and the management, personnel, and infrastructure support provided by NASA. The center is located at Goddard with high speed networking links to the computer and the Air Force Weather Agency in Omaha, as well as to other supercomputers within the DOD system. The ultimate goal is for the CCMC to be the clearinghouse for research models with operational potential. The CCMC will also be a valuable asset to the research community for model testing and validation.

In large part, the CCMC is built around the successful transitioning of the Magnetospheric Specification Model and the Magnetospheric Specification and Forecast Model undertaken by the Air Force during the 1980s. These efforts also contributed to defining the extensive array of space environment sensors to be flown on the current series of Defense Meteorological Satellite Program (DMSP) satellites. Also, the first real time solar wind measurements were made possible from the Wind satellite through Air Force support. Air Force efforts to develop models to take advantage of these new databases formed the foundation for the strategy behind the CCMC.

Space weather modeling efforts have also received tremendous new impetus through more recent support from federal agencies. The Office of Naval Research and the Air Force Office of Scientific Research have supported two Multi-University Research Initiatives (MURIs) focused on space weather modeling. NSF has supported a two-year program led by Boston University in collaboration with other institutions to begin development of an end-to-end space weather model. The University Partnership for Operational Support (UPOS) centered at the Applied Physics Laboratory of Johns Hopkins University and the Geophysical Institute at the University of Alaska is developing space weather products for near-term operational use. Over the next several years, these initiatives will enhance both specification and predictive capabilities within the operational space weather centers.

Another activity initiated during the course of space weather planning was a series of workshops aimed at assessing customer needs in space weather. The first such meeting was to examine the needs of the electrical power industry for space weather services. Held at Electric Power Research Institute (EPRI) headquarters in Washington DC in October 1996, the workshop featured open discussions between representatives of the electric power industry and space scientists on what was being done and what could be done to help the industry protect itself from potential space weather hazards. The success of the meeting is perhaps open to interpretation. On the one hand, one could argue that industry representatives did not come out with strong feelings that electric power grids were terribly susceptible to catastrophic failure due to space weather. On the other hand, the discussions demonstrated a willingness on the part of both industry representatives and scientists to learn more about each other's disciplines and to establish and maintain future avenues of communication. An example is the continued success of the Sunburst Program, an activity sponsored by EPRI involving electric power companies, which provides for direct measurement of induced currents in power lines over a wide grid. One of the important issues discussed at the meeting was that federal deregulation of the power industry would create a more competitive environment. In attempting to keep costs down, power companies would operate grids at levels near saturation, thus making them more susceptible to transient surges arising from geomagnetically-induced currents. Studies in this area remain a high priority, both within the science community and within some concerned representatives of the power industry.

A second workshop was held at COMSAT headquarters in Bethesda, Maryland in October 1997. This workshop was dedicated to communication and navigation customers of space weather services. Representatives from the academic community, government laboratories, the DOD, and industry, attended the meeting which featured talks, posters, and working group discussions. The working groups addressed such topics as ionospheric effects on Global Positioning System (GPS) navigation systems and radiowave propagation, and recommended observational systems that could help mitigate the potential impacts of space weather events.

A third workshop was planned to address space weather needs of the satellite industry. However, the difficulty in assembling a group of representatives from this industry willing to discuss potential weaknesses in their systems proved insurmountable. In lieu of a workshop, an NSF award was made to Sterling Software Inc. to interview industry executives on space weather issues. Although penetrating the "secrecy barrier" proved harder than anticipated, the study reveals three areas (science, observations, and funding) that limit the ability to meet the needs of the private sector. Translating customer requirements into statements meaningful to space science researchers was identified as a significant challenge in bridging the gap between scientists and space weather users.

The need for any further user workshops has been largely supplanted by the expansion of NOAA's Space Environment Center (SEC) Users Meeting. In the past, SEC had been sponsoring a meeting every three years. SEC invited its customers to its center in Boulder, Colorado, to discuss the status of space weather services and obtain feedback from customers on the quality of services. Aside from the scientists at SEC, members of the broader space science community were notably absent from these meetings. The only assembly of researchers to address space weather issues occurred in January 1998 in Boulder, cosponsored by NSF and the Air Force Research Laboratory. The meeting was called *Space Weather: Research to Operations*. After this meeting, it became apparent that it would be better to combine it with the SEC Users meeting. The combined group met first in 1999. The meeting was such a success that SEC and its co-sponsors decided to continue the joint assembly on an annual basis, at least through solar max. The joint meeting has subsequently been named Space Weather Week. This gathering represents the primary forum for exchange of ideas between space weather researchers, forecasters, and customers.

Perhaps the most significant evidence of the growing importance of space weather is the recent introduction of NASA's Living With a Star (LWS) program. The ambitious, $0.5B program will address not only space weather but also climatology and solar influences on global change. Currently, notional LWS flight elements include the Solar Dynamics Observatory, Solar Sentinels, Radiation Belt Mappers, and Ionospheric Mappers. A parallel theory and modeling effort will enhance the scientific output of the LWS observations. LWS represents an excellent opportunity to make solid advances in our understanding of the space weather system and our ability to predict and prepare for potential impacts on society.

One area of space weather that still needs work is metrics. In response to the Space Weather Program Council's recommendation that well-defined metrics be used to track progress, three working groups were organized to develop metrics in each of the three space weather domains. The working groups met first in September 1997 to establish common ground rules for the metrics that were to be developed. As with other space weather activities, the development of metrics proved to be more difficult than expected, but the first "research" metrics for the magnetosphere and ionosphere are now defined and we expect to plot our first point on the "metrics graph" sometime during the summer of 2001.

The list of activities supporting space weather shown in Figure 1 includes educational activities. NOAA's SEC has traditionally been the primary center for education and outreach efforts related to space weather, along with NASA's solar-terrestrial physics outreach program. Incremental funding for such efforts resulting from the NSWP has been rather limited thus far. NSF has made one award to the Space Science Institute in Boulder for the development of a web site and other space weather outreach efforts. The funding has been used to enhance *Electric Space*, a traveling museum exhibit which has been a highly successful endeavor involving many institutions and reaching thousands of people in cities nationwide. NSF has also supported the development of an IMAX film called *SolarMax*. The film uses solar images from the Yohkoh, SOHO, and TRACE satellites and auroral footage obtained by Robert Eather of KEO Consultants. The film is scheduled for distribution in theaters and museums world-wide.

Despite these activities, educational thrusts in space weather need more attention. The implementation plan calls for a broad-based approach involving both informal and formal education and encompassing customers, forecasters, and the general public. Obviously this area needs to be attacked more aggressively, but thus far education and outreach has yet to receive the infusion of funds that would do justice to its importance.

Another area of space weather that received little attention in the initial planning documents is coordination with the international space weather community. This was partially because of the DOD involvement in the program, but also because the Working Group for Space Weather had its hands full dealing with the intricacies of interagency cooperation within the U. S. let alone on a global scale. But this did not hinder other countries from embarking on their own organizational efforts to parallel those in the U. S. Among the countries initiating new programs or enhancing existing programs are Australia, Belgium, Canada, Finland, France, Indonesia, Japan, Korea, and Taiwan. Planning for international space weather activities has been initiated by the Scientific Committee on Solar Terrestrial Physics under S-RAMP (the Results, Applications, and

Space Weather Publications

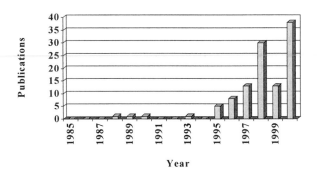

Figure 4. Growth in the usage of "space weather" in journal articles during the past 15 years.

Modeling Phase of STEP, the Solar Terrestrial Energy Program). The S-RAMP subcommittee has already selected a retrospective study of a magnetic storm in 1998 and coordinated observations for the space weather campaign conducted in September of 1999. The AGU Western Pacific meeting in Taipei in July 1998 featured a three-day session on space weather, a European Space Agency workshop on space weather took place in November of 1998, and a space weather session took place at the S-RAMP Symposium in Japan in October 2000. Meanwhile, the ten operational Regional Warning Centers of the International Space Environment Service (ISES) have been enhanced by adapting to regional customer needs and applying new scientific results and technology.

One question related to space weather that remains largely unanswered is the economic impacts of space weather hazards, and the quantitative benefits that might accompany improvements in forecasting capabilities. In a sense, it may be too late now to embark on such a study, because at this point it would be impossible to separate out the improvements in space weather preparedness that may have already been achieved owing to the existence of the program. Awareness of space weather is at an all-time high. This heightened awareness must have a positive effect on the nation's preparedness for space weather events. Thus, the fact that no space weather related catastrophe has occurred in the past seven years should not be construed as a failure of the program, or that the program is not necessary.

A vital element in the success of the NSWP has been its ability to encompass and place into context the space weather efforts of individual government agencies. The NSWP does not attempt to sculpt each agency program according to a grand master plan. Rather the program can be thought of as providing the clay from which each agency molds its own strategies. It is an endeavor that has been infused with the spirit of interagency cooperation and

the sincere desires of all stakeholders to work together toward a common goal.

SPACE WEATHER: AN APPLIED SCIENCE?

Evidence for the blossoming awareness in space weather is that the topic is increasingly finding its way into the popular media with newspaper articles, magazine articles, and TV specials appearing on a regular basis. On the science side, a study of journal articles during the past 15 years shows the tremendous growth of interest in space weather as a scientific discipline. A search through the Science Citation Index for articles with "space weather" in the title or the abstract reveals the sudden increase in attention to space weather. As shown in Figure 4, prior to 1995 there was virtually no specific mention of space weather in scientific journals. Apparently, the idea of linking progress in scientific understanding to applications with societal benefits was not emphasized in publications prior to 1995.

Why did it take so long for space weather to work its way into scientific thought? After all, the concept has been around for a long time. For example, in 1983, just after the maximum of solar cycle 21, a meeting dealing with predicting the space environment for societal benefit and "space weather" was held at what is now the Air Force Research Laboratory. Although this meeting led to many important breakthroughs in the development and transitioning of space weather models within the Air Force, few, if any, scientific papers on space weather followed. Similarly, in 1990, during the maximum of solar cycle 22, Juan Roederer said:

"We are going through some interesting times in Solar-Terrestrial research. The opportunities are tremendous. We are nearing a maximum of solar activity that is beginning to have important effects on man-made systems, alerting government officials and politicians to new facts about our global environment that go beyond the cliches of the global warming scare. Solar activity related space hazards have not killed people (yet), but they have caused very expensive, even irreparable damage to technological systems in space and on Earth, and they can present a serious threat to humans in space. We now have a chance to improve the predictability of these hazards related to weather and climate in space."

This description of space weather applications is prophetic, but it was a concept ahead of its time. To best understand what has changed, consider Figure 5 from *Pasteur's Quadrant* by Donald Stokes. The figure suggests two reasons for doing basic research: 1) for better understanding; and 2) for societal benefits or applications. During the last 10 years, space weather has moved up greatly on *both* the

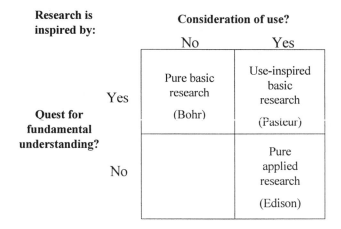

Research is inspired by:	Consideration of use?	
	No	Yes
Quest for fundamental understanding? **Yes**	Pure basic research (Bohr)	Use-inspired basic research (Pasteur)
No		Pure applied research (Edison)

Figure 5. Inspiration for scientific research (from Stokes, 1997).

"better understanding" axis, as well as the "societal benefit" axis.

With regard to better understanding, our community has seen the advent of powerful new models that can couple together different space domains such as the ionosphere and magnetosphere. With the advent of the Internet, there has been a revolution in information technology that has enabled data transfer and processing in ways unheard of in the last solar cycle. And, of course, we have seen powerful new instrumentation that has provided scientists with crucial new data. Foremost among these observational platforms was SOHO, which was launched in 1995, and provided stunning imagery of the sun and solar wind that has changed forever our ideas about our nearest star.

Also contributing to our better understanding of space weather is the NSWP's emphasis on treating the space environment as a strongly coupled system. Space weather is the intellectual glue that binds together the entire sun-Earth system. Lou Lanzerotti summed this up beautifully in an e-mail message to George Siscoe:

"While it is true that applications will result from the initiative, the science that will be accomplished will be first rate. To have applications result will be an added bonus. Indeed, the initiative provides a context in which much of solar terrestrial physics can be and should be done."

In addition to the exciting new possibilities for better understanding that space weather presents, it is just as important to stress its service to society. This is the second of Stokes' axes and is no less vital. In the past ten years, hundreds more satellites have been launched, most of which are vital to our everyday lives and our national security. Power lines have been increasingly webbed together, becoming more and more vulnerable to transient interruption

caused by induced magnetic fields at a distant part of the power grid. And, of course, astronauts spend more and more time in space as we move toward a permanent human presence in space. The nation has become truly vulnerable to space weather.

An important driver for the paradigm shift in the way solar-terrestrial science is being conducted is the recent ease in accessibility to real-time data. The communication revolution now allows us to drive predictive models with real-time data. An important example is the real-time access to solar wind data from the Advanced Composition Explorer (ACE) satellite inaugurated at a multi-agency (NOAA, NASA, and Air Force) ribbon-cutting ceremony at SEC in January 1998.

So to understand the power of space weather and the key to the success of the NSWP, we should not consider separately the importance of space weather in providing us with fundamental knowledge of the sun-Earth system nor the importance of space weather in providing needed societal benefits. It is the *combination* of these elements. It is the one-two punch of both science and applications that gives the concept its tremendous power.

CONCLUSION

This AGU monograph contains a sampling of space weather science accomplished during the last few years. As predicted by Lanzerotti, the program has indeed provided the context within which much excellent science has been and will be accomplished. As with all basic research, the resulting tangible benefits to society may not materialize for years to come, but without a doubt we are seeing improvements in our understanding of the space weather system.

In addition to the excellent science, the NSWP has had other benefits. It has gotten various government agencies to work together, thus eliminating redundant efforts and enabling better communication among stakeholders. It has stimulated the attention of additional U. S. and international government agencies and organizations that have an interest in space weather. The NSWP has provided a focus for many meetings, workshops, and special sessions highlighting all aspects of the space weather system. And it has motivated media attention that has heightened public awareness of space weather.

The success of the NSWP is due to a variety of factors. First and foremost is that it inspires good science, particularly because of its emphasis on prediction, which is the ultimate test of knowledge. It is multi-disciplinary, encompassing the entire space weather system from the sun to Earth's surface. The program brings together the research and operations communities, as well as a diverse customer base. It is multi-agency with participation from NSF,

NASA, and the Departments of Commerce, Defense, Energy, Interior, and Transportation. The program has societal relevance, being responsive to both civilian and DOD needs. It has public appeal because of its association with beautiful and dramatic natural phenomena such as solar storms and aurora, as well as its potential to cause multi-million dollar losses to high technology systems. It has international relevance, with many countries already actively involved in developing parallel programs. Last, but not least, like all good programs, the NSWP is community-driven, tracing its roots from an assembly of research scientists with a common vision.

The future looks extremely bright for the NSWP. Many on-going and planned activities have given the program real staying power. The entire space weather community of researchers, forecasters, customers, and educators has come to realize the benefits the program offers. Future activities of the Committee for Space Weather will concentrate on coordinating the different agency thrusts that have emerged recently. Metrics will be adopted and used to initiate tracking of improvements in space weather services. Efforts will continue to develop the Community Coordinated Modeling Center into a focal point for linking, validating, and testing space weather models for eventual transitioning to the Rapid Prototyping Centers operated by NOAA and the Air Force, and eventually to the operational centers of those agencies. The annual space weather competition will continue with hopefully increased funding levels and broader participation by other agencies. New observational platforms will be deployed and used for both operational and research purposes. Enhanced educational activities will increase public knowledge and national awareness of the space weather system. And efforts will continue to foster effective feedback between researchers, forecasters, and space weather customers, as well as to evaluate the cost benefits of space weather predictions.

Acknowledgments· Because of the number of people involved in the National Space Weather Program, it would be impossible to identify all the contributors from the international science community who have contributed to establishing, planning, and supporting the program.

At NSF, we acknowledge the contributions of Sunanda Basu who has been an indispensable participant in all the activities described here, but most notably in coordinating international space weather activities. Odile de la Beaujardiere, Bob Clauer, Kile Baker, David Sime, and Ken Schatten all contributed during their visiting scientist appointments at NSF.

The program has benefited from a long line of DOD representatives from the Air Force Office of Weather, including Lt. Col. Willow Cliffswallow who wrote large parts of the Strategic Plan. She was succeeded by Major Amanda Preble who assisted in the development of the first implementation plan, particularly the road maps which took a tremendous effort over many months. Later, Major Tom Frooninckx helped to ensure that space weather was included as an area of interest in the AFSPC/NASA partnership. More recently, the Air Force Weather representative has been Lt. Col. Michael Bonadonna, whose energy and enthusiasm helped launch the CCMC, and who was the primary NSWP point of contact during the Space Architect study. Colonel Michael Jamilkowski of the Office of the Assistant Secretary of Defense for C³I has led the effort to coordinate the NSWP with the Space Architect Transition Plan. The 55th Weather Squadron, the DOD's forecast center, was ably represented by Kevin Scro, who spearheaded the development of the CCMC.

Ernie Hildner coordinated space weather activities within NOAA's Space Environment Center. Other SEC contributors included Terry Onsager and Howard Singer, along with Gary Heckman, who represented civilian operational support services. At NOAA headquarters, Amy Holman provided extremely valuable interaction between the NSWP and Department of Commerce upper management levels. We are also grateful to Bill Hooke, formerly of NOAA and the former chair of SNDR, who recognized the importance of space weather to the nation's safety.

George Withbroe has coordinated NSWP activities at NASA, and Robert Carovillano and Larry Zanetti contributed to space weather planning while visiting scientists at NASA Headquarters.

The success of the NSWP is also a result of the excellent support from within the Office of the Federal Coordinator for Meteorology, beginning with James Harrison, who provided valuable advice during the early days of the program. He was followed by Colonel Jud Stailey whose diligence was critical to the successful development of the first Implementation Plan. After Colonel Stailey's departure from OFCM, the position was filled in succession by Majors Michael Hunsucker and Michael Babcock, both of whom carried on this outstanding support for the program. Major Babcock undertook the assembly and production of the second Implementation Plan almost single-handedly.

We extend our deep thanks and gratitude to these people, as well as the many others not named here, for their contributions to the program. We gratefully acknowledge the help provided by the reviewers of this manuscript, who kindly corrected several omissions, errors, and misconceptions in the original version.

REFERENCES

Stokes, Donald E., *Pasteur's Quadrant,* Brookings Institution Press, Washington, D. C., 1997.

R. A. Behnke and R. M. Robinson, National Science Foundation, 4201 Wilson Blvd., Arlington, VA 22230, U. S. A.

Space Weather Effects on Technologies

Louis J. Lanzerotti

Bell Laboratories, Lucent Technologies, Murray Hill, New Jersey

In the last about 150 years, the variety of human technologies that are embedded in space-affected environments have vastly increased. This paper presents some of the history of the subject of "space weather", beginning with early electric telegraph systems and continuing to today. An overview is presented of the present-day technologies that can be affected by solar-terrestrial phenomena such as galactic cosmic rays, solar-produced plasmas, and geomagnetic disturbances in the Earth's magnetosphere.

INTRODUCTION AND SOME HISTORY

While pursuing his very extensive and systematic program of observations and descriptions (by drawings and words) of sun spots, Richard Carrington, FRS, recorded an exceptionally large area of spots in the northern solar hemisphere at the end of August 1859. Figure 1 is a reproduction of Plate 80 from the comprehensive records of his studies, which were carried out over a more than seven year interval in the mid-nineteenth century [*Carrington*, 1863]. The large sunspot area at about 45° N solar latitude on August 31 is especially notable.

This observation of an extensive dark region on the solar face was to ultimately prove to be more out of the ordinary than Carrington's past research would have originally suggested to him. Quoting from his description of this region, "...at [the observatory at] Redhill [I] witnessed ... a singular outbreak of light which lasted about 5 minutes, and moved sensibly over the entire contour of the spot ...." Some hours following this white light event on the sun (the first ever reported), large disturbances were observed in magnetic measuring instruments on Earth, and the aurora borealis was seen as far south as Hawaii and Rome.

During the previous solar cycle (the 8th) in the year 1847, telegraph systems that were just beginning to be de-

ployed in common use were found to often exhibit "anomalous currents" in their wires [*Barlow*, 1848]. These anomalous electrical currents were not at all understood. Thus, the large and disruptive electrical disturbances that were recorded in numerous telegraph systems during the magnetic disturbances that followed Carrington's event were a great surprise. Indeed, during the same several day interval that the large auroral displays were widely seen, strange effects were measured in telegraph systems all across Europe – from Scandinavia to Tuscany. In the Eastern United States, it was reported [*Prescott*, 1860] that on the telegraph line from Boston to Portland (Maine) during "...Friday, September 2d, 1859 [the operators] continued to use the line [without batteries] for about two hours when, the aurora having subsided, the batteries were resumed."

In addition to the "anomalous" electrical currents flowing in the Earth, the early telegraph systems were also very vulnerable to atmospheric electrical disturbances in the form of thunderstorms. As noted by Silliman [1850], "One curious fact connected with the operation of the telegraph is the induction of atmospheric electricity upon the wires ... often to cause the machines at several stations to record the approach of a thunderstorm." While disturbances by thunderstorms on the "machines" could apparently be identified as to their source, the source(s) of the "anomalous currents" described by Barlow [1848] remained largely a mystery, even though he noted that "...in every case which has come under [his] observation the telegraph needles have been deflected whenever aurora has been visible".

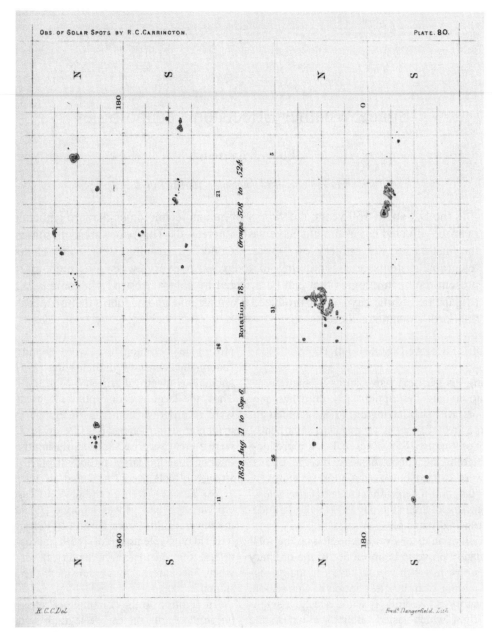

Figure 1. Plate 80 from Carrington [1863] showing his sunspot drawings for August 11 to September 6, 1859. The large spot area at about 45° N solar latitude on August 31 is especially notable.

The decades that followed the solar event of 1859 produced significant amounts of attention by telegraph engineers and operators to the effects on their systems of Earth electrical currents. Although little recognized for almost fifty years afterwards, the sun was indeed seriously affecting the first "modern" technology that was used for communications.

The invention of intercontinental wireless communications, with the long wavelength radio transmissions from Poldhu Station, Cornwall, to St. John's, Newfoundland, by Marconi in 1901, eliminated Earth currents as a source of disturbances on communications that were sent through the atmosphere. Marconi's achievement (for which he was awarded the Nobel Prize in Physics in 1909) was only possible because of the existence of the reflecting ionosphere. This reflecting layer was definitively identified some two decades later by Briet and Tuve [1925] and by Appleton and Barnett [1925]. Because wireless remained the only method for cross-oceanic voice communications for more than five decades following Marconi's feat, physical changes in the radio wave-reflecting layer (even before it was "discovered") were critical to the success (or failure)

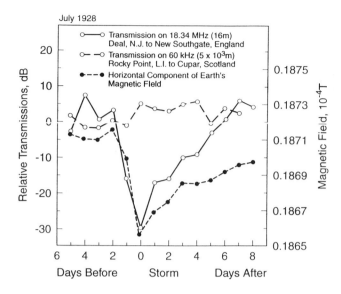

Figure 2. Yearly average daylight cross-Atlantic transmission signal strengths and monthly average sunspot numbers for the interval 1915 – 1932.

of reliable transmissions. (An excellent chronology and summary of events in the history of wireless from 1819 to 1976 is provided in the first chapter of Snyder and Bragaw [1986].)

The same ionosphere electrical currents that could produce "spontaneous" electrical currents within the Earth could also affect the reception and fidelity of the transmitted long-distance wireless signals. Indeed, Marconi [1928] commented on this phenomenon when he wrote that the "…times of bad fading [of radio signals] practically always coincide with the appearance of large sun-spots and intense aurora-boreali usually accompanied by magnetic storms …." These are "… the same periods when cables and land lines experience difficulties or are thrown out of action."

An example of the types of studies that were pursued in the early years of long-distance, very long wavelength, wireless is shown in Figure 2. Plotted in this Figure (reproduced from Fagen [1975], which contains historical notes on early wireless research in the old Bell Telephone System) are yearly average daylight cross-Atlantic transmission signal strengths for the years 1915 – 1932 (upper trace). The intensities in the signal strength curves were derived by averaging the values from about 10 European stations that were broadcasting in the ~15 to 23 kHz band, and after reducing them to a common base (the signal from Nauen, Germany, was used as the base). Plotted in the lower trace of the Figure are the monthly average sunspot numbers per year. Clearly, an association is seen between the two physical quantities. From the perspectives of space weather predictions, this relationship of the received electrical field strengths to the yearly solar activity as represented by the number of sunspots could be used by wireless engineers at that time to provide them some expectation as to transmission quality on a year to year basis.

The relationship of the "abnormal" propagation of long wavelength radio transmissions and solar disturbances was first identified in 1923 [Anderson, 1928]. The technical literature of the early wireless era showed clearly that solar-originating disturbances were serious assaults on the integrity of communications during the first several decades of the twentieth century. Communications engineers pursued a number of methodologies to alleviate or mitigate the assaults. One of these methodologies that sought more basic understanding was illustrated in the context of Figure 2. Another methodology utilized alternative communications "routes". As Figure 3 illustrates for the radio electric field strength data recorded during a solar and geomagnetic disturbance on July 8, 1928, the transmissions at long wave lengths were relatively undisturbed while those at the shorter wavelengths were seriously degraded [Anderson, 1929].

The practical effects of the technical conclusions of Figure 3 are well exemplified by a headline which appeared over a front page article in the Sunday, January 23, 1938, issue of The New York Times. This headline noted that "Violent magnetic storm disrupts short-wave radio communication." The sub-headline related that "Transoceanic services transfer phone and other traffic to long wave lengths as sunspot disturbance strikes". The engineering work-around that shifted the cross-Atlantic wireless traffic from short to longer wavelengths prevented the complete disruption of voice messages at that time.

The solar maximum interval that began near the end of the 1930's (the 17th solar cycle) saw the first significant ef-

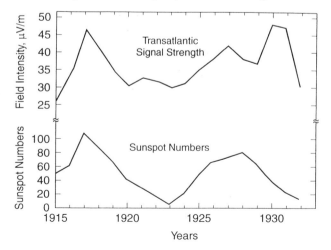

Figure 3. Trans-Atlantic wireless transmissions from the Eastern U. S. to the U. K. on two frequencies before and during a magnetic storm event in July 1928. Also shown are the values of the horizontal component of the Earth's magnetic field.

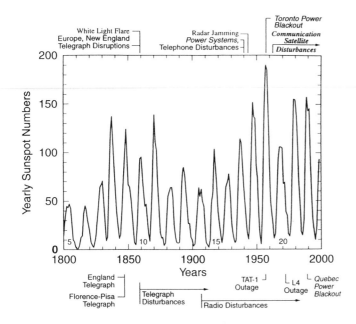

Figure 4. Yearly sunspot numbers with indicated times of selected major impacts of the solar-terrestrial environment on largely ground-based technical systems. The numbers just above the horizontal axis are the conventional numbers of the sunspot cycles.

fects of solar-initiated disturbances on technologies other than communications. During a large magnetic storm on March 24, 1940, some ten transformer banks were tripped in the region of the Ontario Hydro Commission. Numerous problems, including transformer tripping and reactive power surges, were encountered in electrical power systems in the northeast and northern United States [*Davidson*, 1940]. As might well be expected, there were widespread effects felt in both landline and radio telephony. Indeed, voltages as high as 2 V/km were recorded on a telephone line from Minneapolis to Fargo [*Germaine*, 1940].

THE SPACE ERA

From the earliest installations of the telegraph to the beginnings of wireless communications, from the establishment of inter-connected power grids to the flight in recent decades of humans in space, the role of the solar-terrestrial environment for successfully implementing and operating many new technologies has continued to increase in importance. Illustrated in Figure 4 are selected examples of the times of large solar-originating disturbances on ground-based communications and power systems, together with a plot of the sunspot number cycle to the rise in the present cycle (the 23rd).

Other than radio and satellites, all of the disturbances on the ground-based power and communications technologies

that are identified in Figure 4 have the same underlying physical cause [e.g., *Lanzerotti*, 1983; *Lanzerotti and Gregori*, 1986; *Boteler* et al., 1998]. Greatly increased electrical current systems in the magnetosphere and the ionosphere cause large variations in the time rate of change of the magnetic field at the Earth's surface. These time variations in the field in turn induce potential differences across large areas of the Earth's surface which are spanned by the communications and power systems. Electrical power lines and telecommunications lines that use the Earth itself as a ground return for their circuits provide the path for concentrating the electrical currents that flow between these newly established, but temporary, "batteries".

The precise effects of these "anomalous" electrical currents depend upon the technical system to which the long conductors are connected. In the case of long telecommunications lines, the Earth potentials can cause overruns of the compensating voltage swings that are designed into the power supplies [e.g., *Anderson et al.*, 1974]. There are several modes of degradation and failure that can occur in the grids of electrical power systems [e.g., *Albertson et al.*, 1973; 1974; *Kappenman et al.*1981; *Pirjola*, 1985; *Boteler et al.*, 1989]. Finally, the "anomalous" Earth currents can also degrade pipelines and cause disruptions to monitoring systems that are installed in such lines [e.g., *Pirjola and Lehtinen*, 1985; *Campbell*, 1986; *Viljanen*, 1989].

The advent of the space era has not led to any lessening of concern about the effects of solar-terrestrial processes on ground-based technologies. Indeed, four of the large disturbance effects indicated in Figure 4 occurred after the launch of the first Earth-encircling spacecraft in 1957. The magnetic storm of February 1958 not only disrupted voice communications on the first cross-Atlantic telecommunications cable (TAT-1 from Newfoundland to Scotland), but also plunged the Toronto area into darkness by the tripping of power company circuits. As recounted by Brooks [1959], the "… temporary darkness [in the Toronto area was] broken only by the strange light of the aurora overhead."

The outage for nearly an hour of a major continental telecommunications cable that stretched from near Chicago to the west coast was disrupted between its Illinois and Iowa power stations by the magnetic storm of August 1972 [*Anderson et al.*, 1974; *Boteler and Jansen van Beek*, 1999]. In March 1989, the entire province of Quebec suffered a power outage for nearly a day as major transformers failed under the onslaught of a large geomagnetic storm [*Czech et al.*, 1992]. At the same time, the first cross-Atlantic fiber voice cable was rendered nearly inoperative by the large potential difference that was established between the cable terminals on the coasts of New Jersey and England [*Medford et al.*, 1989].

Point-to-point high frequency (HF) wireless communications links, most often used today for some national defense communications and for civil emergency communications, continues to suffer the vagaries of the sun's interactions with the Earth's space environment. Those who are involved with, and/or users of, such systems are familiar with many anecdotes of solar-produced effects and disruptions. For example, near the peak of the 21st solar cycle in 1979, a distress signal from a downed commuter plane was received by an Orange County, California, fire department – which responded, only to discover that the signal had originated from an accident site in West Virginia.

The placing into space of ever-advancing technologies – for both civilian and national defense purposes – meant that ever more sophisticated understanding of the space environment was required in order to ensure reliability and survivability. Indeed, the first active civil telecommunications satellite, the low Earth-orbiting Telstar 1(launched June 10, 1962), carried solid state sensors to monitor the radiation environment encountered by the spacecraft [*Brown et al.*, 1963; *Buck et al.*, 1964]. As has been the case for many engineering investigations that have been made over the years in space-related research, these sensors also returned valuable new science information, such as measurements of the trapping lifetime of the radiation debris following high altitude nuclear explosions [*Brown*, 1966].

The operations and survivability of both ground- and space-based technical systems have often encountered unanticipated surprises because of natural space environmental effects. As technologies have increased in sophistication, as well as in miniaturization and in interconnectedness, more sophisticated understanding of the Earth's space environment continues to be required. In addition, the increasing diversity of technical systems that can be affected by space weather processes is accompanied by continual changes in the dominance of one technology over another for specific uses. For example, in 1988, satellites were the dominant carrier of transocean messages and data; only about two percent of this traffic was over ocean cables. By 1990, the advent of the wide bandwidths provided by fiber optics meant that 80% of the transocean traffic was now via ocean cable [*Mandell*, 2000]. Hence, any space weather effects on cables must now be considered more seriously than they may have been a decade ago.

SPACE ENVIRONMENTAL EFFECTS ON TECHNOLOGIES

Many present-day technologies that must include considerations of the solar-terrestrial environment in their designs and/or operations are listed in Table 1 [Lanzerotti et al., 1999]. The systems are grouped into broad categories that have similar physical origins. Plate 1 schematically illustrates these effects.

Ionosphere and Wireless

A century after Marconi's feat, the ionosphere remains both a facilitator and an intruder in numerous communications applications. As noted above, the military, as well as police and fire emergency agencies in many nations, continue to rely on wireless links that make extensive use of frequencies from kHz to hundreds of MHz that use the ionosphere as a reflector. Changes in the reflections produced by solar activity, both from solar UV and x-ray emissions, as well as by magnetic storms, can significantly alter the propagation of these wireless signals.

At higher, few GHz, frequencies the production of "bubbles" in ionosphere densities in equatorial regions of the Earth can be a prime source of scintillations in satellite to ground signals. Engineers at the COMSAT Corporation first discovered these effects after the initial deployment of the INTELSAT network at geosynchronous orbit (GEO)

Table 1 Impacts of Solar-Terrestrial Processes on Technologies

Ionosphere Variations
    Induction of electrical currents in the Earth
        Power distribution systems
        Long communications cables
        Pipelines
    Interference with geophysical prospecting
    Source for geophysical prospecting
    Wireless signal reflection, propagation, attenuation
    Communication satellite signal interference,
        scintillation
Magnetic Field Variations
    Attitude control of spacecraft
    Compasses
Solar Radio Bursts
    Excess noise in wireless communications systems
Radiation
    Solar cell damage
    Semiconductor device damage and failure
    Faulty of semiconductor devices
    Spacecraft charging, surface and interior materials
    Astronaut safety
    Airline passenger safety
Micrometeoroids and Artificial Space Debris
    Solar cell damage
    Damage to mirrors, surfaces, materials, complete
        vehicles
Atmosphere
    Low altitude satellite drag
    Attenuation and scatter of wireless signals

[*Taur*, 1973]. Plasma processes in the ionosphere are also the cause of considerable problems in the use of single frequency signals from the Global Positioning System (GPS) for precise position location. The intent to evolve the GPS system to a dual, and eventually a three, frequency system over the next five to ten years should eliminate this ionosphere nuisance.

There remain large uncertainties in the knowledge base of the processes that determine the initiation and scale sizes of the ionosphere irregularities that are responsible for the scintillation of radio signals that propagate through the ionosphere. Thus, it is very difficult to define diversity strategies for receivers and/or space-based transmitters that might be applicable under most conditions.

Ionosphere and Earth Currents

The basic physical chain of events behind the production of Earth potentials and the effects of the resulting enhanced Earth currents on technical systems with long conductors was outlined in the previous Section. The major issues that can arise in attempting to understand in detail (and in predicting) the effects of enhanced space electrical currents on these ground-based systems are several fold. At present, the time variations and spatial dependencies of the space electrical currents are not at all well understood or predicable from one geomagnetic storm to the next. This is of considerable importance since the Earth potentials that are induced are very much dependent upon the conductivity structure of the Earth underlying the affected ionosphere regions. Similar electrical current variations in the space environment can produce vastly different Earth potential drops depending upon the nature and orientation of underground Earth conductivity structures in relationship to the variable overhead currents.

Modeling of these effects is becoming quite advanced in many cases. However, the use of the model results for "predictions" for practical purposes is quite difficult at present even when accurate knowledge of interplanetary conditions close to Earth is available. This is an area of research that involves a close interplay between space plasma geophysics and solid Earth geophysics, and is one that is not often addressed collaboratively by these two very distinct research communities (except by the somewhat limited group of researchers who pursue electromagnetic investigations of the Earth). This is basically because the educational backgrounds and the community of colleagues are very different between the two research groups.

Solar Radio

Solar radio noise and bursts were discovered nearly six decades ago by Southworth [1945] and by Hey[1946] during the early research on radar at the time of the Second World War. Solar radio bursts produced unexpected (and unrecognized at first) jamming of this new technology that was under rapid development and deployment for war-time use [*Hey*, 1973]. Extensive post-war research established that solar radio emissions can exhibit a wide range of spectral shapes and intensity levels [e.g., *Kundu*, 1965; *Castelli et al.*, 1973;*Guidice and Castelli*, 1975; *Barron et al.*, 1985]. Research on solar radio phenomena remains an active and productive field of research today [e.g., *Bastian et al.*, 1998].

Some analysis of local noon time solar radio noise levels that are routinely taken by the U.S. Air Force and that are made available by the NOAA World Data Center have recently been made in order to assess the noise in the context of modern communications technologies. These analyses show that during 1991 (within the sunspot maximum interval of the 22nd cycle) the average noon fluxes measured at 1.145 GHz and at 15.4 GHz were −162.5 and −156 dBW/(m^2 4kHz), respectively [*Lanzerotti et al.*, 1999]. These values are only about 6 dB and 12 dB above the 273° K (Earth's surface temperature) thermal noise of −168.2 dBW/(m^2 4kHz). Further, these two values are only about 20 dB and 14 dB, respectively, below the maximum flux of −142 dBW/(m^2 4kHz) that is allowed for satellite downlinks by the ITU regulation RR2566.

Short term variations often occur within solar radio bursts, with time variations ranging from milliseconds to seconds and more [e.g., *Barron et al.*, 1980; *Benz*, 1986; *Isliker and Benz*, 1994]. Such short time variations can often be many tens of dB larger than the underlying solar burst intensities upon which they are superimposed. As one example, the large solar flare event of May 23, 1967, produced a radio flux level (as measured at Earth) of > 100,000 solar flux units (1 SFU = 10^{-22} W/(m^2 Hz) at 1 GHz, and perhaps much higher [*Castelli et al.*, 1973]. Such a sfu level corresponds to −129 dBW/(m^2 4kHz), or 13 dB above the maximum limit of −142 dBW/(m^2 4kHz) that was noted above.

Space Radiation Effects

The discovery by Van Allen in 1958 of the trapped radiation around Earth immediately implied that the space environment would not be benign for technologies placed within it. As noted by McCormac [1966], the high altitude "nuclear test Star Fish [on 9 July 1962] focused attention on the degradation of sensitive electronics by trapped electrons and protons." At that time, "...the basic processes occurring in the solar cells and transistors by the electrons and protons [were] poorly understood."

Some 200 or so in-use satellites now occupy the geosynchronous orbit. The charged particle radiation that permeates the Earth's space environment remains a diffi-

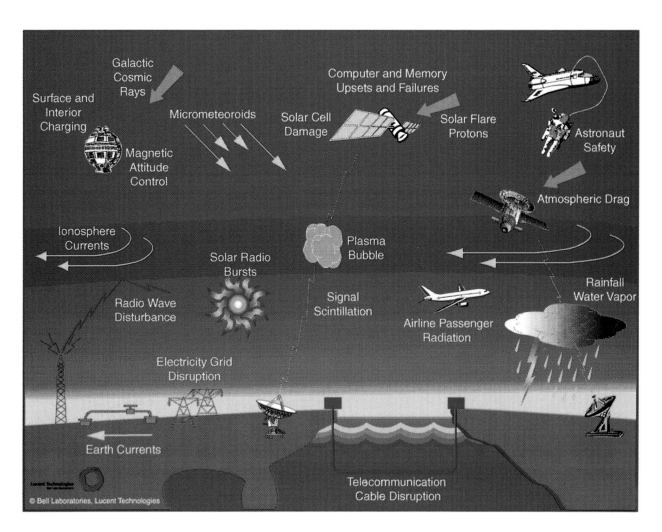

Plate 1. Some of the effects of space weather on technical systems that are deployed on the Earth's surface and in space, and/or whose signals propagate through the space environment.

Telstar 4 Charge Plates

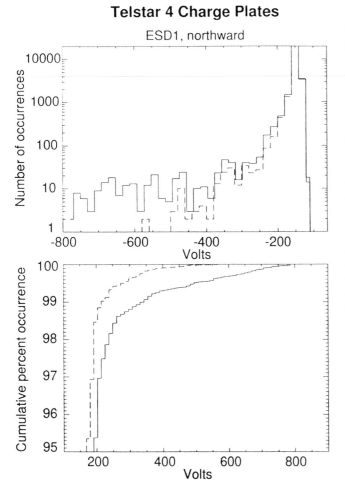

Figure 5. Statistical distribution of surface charging recorded on a charge plate sensor on the Telstar 4 spacecraft during the month of January 1997 (solid line) and for the same month with data from January 10th (the date of a large magnetic storm) removed. The upper panel records the number of voltage occurrences; the lower panel plots the cumulative percent voltage occurrences above 95% in order to show the extreme events.

cult problem for the design and operations of these and other space-based systems [e.g., *Shea and Smart*, 1998; *Koons et al.*, 1999]. In addition, space radiation places severe constraints on many aspects of human space flight, both within the Earth's magnetosphere as well as for scenarios for flights back to the moon and further into the solar system. It is likely, for example, that had any astronauts been enroute to the moon during the solar event of August 1972 – when the AT&T L4 cable was put out of service (Figure 4) – they would have suffered serious, and even life-threatening, radiation exposure [*National Research Council*, 2000]. It was a fortunate coincidence that the last two Apollo flights bracketed this event by several months.

A textbook discussion of the space environment and the implications for satellite design is contained in Tribble [1995]. The low energy (few eV to few keV) particles in the Earth's magnetosphere plasma population can produce different levels of surface charging on the materials (principally for thermal control) that encase a satellite [*Garrett*, 1981]. If good electrical connections are not established between the various surface materials, and between the materials and the solar arrays, differential charging on the surfaces can produce lightning-like breakdown discharges between the materials. These discharges can produce electromagnetic interference and serious damage to components and subsystems [e.g., *Vampola*, 1987; *Koons*, 1980; *Gussenhoven and Mullen*, 1983].

The plasma populations of the magnetosphere can be highly variable in time and in intensity levels. Under conditions of enhanced geomagnetic activity, the cross-magnetosphere electric field will convect earthward the plasma sheet in the Earth's magnetotail. When this occurs, the plasma sheet will extend earthward to within the geosynchronous (GEO) spacecraft orbit. When this occurs, on-board anomalies from surface charging effects will occur; these tend to be most prevalent in the local midnight to dawn sector of the orbit [*Mizera*, 1983].

While records of spacecraft anomalies exist, there is not a comprehensive body of published data on the statistical characteristics of charging on spacecraft surfaces, especially from commercial satellites that are used so extensively for communications. Two surface-mounted charge plate sensors were specifically flown on the AT&T (now Loral) Telstar 4 GEO satellite to monitor surface charging effects. Figure 5 shows the statistical distributions of charging on one of the sensors in January 1997 [*Lanzerotti et al.*, 1998]. The solid line in each panel corresponds to the charging statistics for the entire month, while the dashed lines omit data from a large magnetic storm event on January 10th (statistics shown by the solid lines). Charging voltages as large as –800 V were recorded on the charge plate sensor during the magnetic storm, an event during which a permanent failure of the Telstar 401 satellite occurred (although the failure has not been attributed specifically to the space conditions).

The intensities of higher energy particles in the magnetosphere (MeV energy electrons to tens of MeV energy protons) can change by many orders of magnitude over the course of minutes, hours, and days. These intensity increases occur through a variety of processes, including plasma physics energization processes in the magnetosphere and ready access of solar particles to GEO [*Lanzerotti*, 1968]. Generally it is prohibitively expensive to provide sufficient shielding of all interior spacecraft subsystems against high energy particles (in terms of including

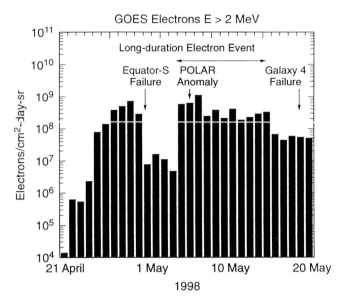

Figure 6. Daily flux values of electrons (> 2 MeV) measured on the NOAA geosynchronous GOES satellite for a one month interval, April – May, 1998. Dates of operational problems on three spacecraft are indicated.

additional spacecraft mass in lieu of additional transponders or orbit control gas, for example).

The range of a 100 MeV proton in aluminum (a typical spacecraft material) is ~ 40 mm. The range of a 3 MeV electron is ~ 6 mm. These particles can therefore penetrate deeply into the interior regions of a satellite. In addition to producing transient upsets in signal and control electronics, such particles can also cause electrical charges to build up in interior insulating materials such as those used in coax cables. If the charge buildup is sufficiently large, these interior materials will eventually suffer electrical breakdowns. There will be electromagnetic interference and damage to the electronics.

A number of spacecraft anomalies, and even failures, have been identified as to having occurred following many days of significantly elevated fluxes of > MeV energy electrons [*Baker et al.*, 1987; 1994; 1996; *Reeves et al.*, 1998]. Plotted as histograms in Figure 6 are the daily average fluxes of > 2 MeV electrons for one month in 1998 when two spacecraft failures occurred and one anomaly was noted on a NASA satellite (all indicated in the Figure; Baker et al. [1998]). While the association of the times of occurrence of the spacecraft problems with the increased electron fluxes is evident, it is still uncertain which, if any, of the failures can be ascribed specifically to space radiation effects [*Baker et al.*, 1998].

The significant uncertainties in placing, and retaining, a spacecraft in a revenue-returning orbital location has led to a large business in risk insurance and re-insurance for one

or more of the stages in a satellite's history. The loss of a spacecraft, or one or more transponders, from adverse space weather conditions is only one of many contingencies that can be insured against. In some years the space insurance industry is quite profitable, and in some years there are serious losses in net revenue after paying claims [e.g., *Todd*, 2000]. For example, Todd [2000] states that in 1998 there were claims totaling more than $1.71 billion after salvage, an amount just less than about twice that received in premiums. These numbers vary by large amounts from year to year.

No realistic shielding is possible for most technological systems in space or on Earth that are under bombardment by galactic cosmic rays (energies ~ 1 GeV and greater). These very energetic particles can produce upsets and errors in spacecraft electronics as well as in computer chips that are intended for use on Earth [*IBM*, 1996]. Often DRAMs and other logic devices that are destined for use on Earth are tested under cosmic ray fluxes at different altitudes on Earth's surface, as well as on aircraft, in order to better understand the sensitivities of these devices to realistic fluxes.

Magnetic Field Variations

The designs of those GEO spacecraft that use the Earth's magnetic field for attitude control must take into account the high probability that the satellite will on occasion, during a large magnetic disturbance, find itself outside the magnetosphere on the sunward side of the Earth. Enhanced solar wind flow velocities and densities, such as those that can occur in a coronal mass ejection event, can easily distort the dayside magnetopause and push it inside the geosynchronous orbit. The highly spatial and time varying magnetic fields that occur at the boundary and outside the magnetosphere can seriously disrupt the satellite stabilization if appropriate precautions have not been incorporated in the design of the control system. The magnetic field outside the magnetopause will have a polarity that is predominantly opposite to that in which the spacecraft is normally oriented, so a complete "flip" of the orientation could occur when the magnetopause is crossed.

Micrometeoroids and Space Debris

The impacts on spacecraft of solid objects, such as from mircrometeoroids and from debris left in orbit from space launches and from satellites that break up for whatever the reason, can seriously disorient a satellite and even cause a total loss. The U.S. Air Force systematically tracks thousands of space debris items that are circling the Earth, most of which are in low altitude orbits. Space debris is a problem that is taken seriously for human flight in the space shuttle and space station orbits. The design of the international space station includes debris-withstanding shielding.

Recent research related to micrometeoroid effects on satellites include overviews of the subject [e.g., *Beech et al.*, 1995; 1997; *McBride*, 1997], more targeted studies of meteor streams [e.g., *Jenniskens*, 1994; 1995], of electromagnetic interference from impacts [*Foschini*, 1998], and of laboratory-based hypervelocity impact experiments [*Gardner et al.*, 1997]. Many of these current works have been carried out in the context of the possibility that the recent appearances of the Leonids would be particularly intense [see also, for example, *McBride and McDonnell*, 1999; *Beech and Brown*, 1994; *Yeomans et al.*, 1996]. To date, few detrimental effects were reported from the Leonids, although perhaps a portion of the absence of serious collisional effects was due to the safeing procedures that were followed for many spacecraft, including the temporary changing of the orientations of solar panels.

Atmosphere: Low Altitude Spacecraft Drag

The ultraviolet emissions from the sun change by more than a factor of two at wavelengths ≤ 170 nm during a solar cycle [*Hunten et al*, 1991]. This is significantly more than the $\sim 0.1\%$ changes that are typical of the visible radiation. The heating of the atmosphere by the increased solar UV radiation causes the atmosphere to expand. The heating is sufficient to raise the "top" of the atmosphere by several hundred km during solar maximum. The greater densities at the higher altitudes result in increased drag on both space debris and on spacecraft in low Earth orbits (LEO), including the space shuttle, space stations (the international space station was most recently raised in orbit in May 2000 by a special shuttle flight), and telecommunications satellites.

Skylab, the first United States space station in the 1970's, was lost due to the effects of atmospheric drag in the solar maximum period of the late 1970's (the 21st solar cycle). The space shuttle was not flight-ready in time to be able to be used to boost Skylab to a higher orbit, as had been planned. Telecommunications spacecraft that fly in LEO have to plan to use some amount of their orbit control fuel to maintain orbit altitude during the buildup to, and in, solar maximum conditions [e.g., *Picholtz*, 1996].

Atmosphere Water Vapor

At frequencies in the Ka band that are planned for high bandwidth space-to-ground applications (as well as for point-to-point communications between ground terminals), water vapor in the neutral atmosphere is the most significant natural phenomenon that can serious affect the signals [*Gordon and Morgan*, 1993]. It would appear that, in general, the space environment can reasonably be ignored when designing around the limitations imposed by rain and water vapor in the atmosphere.

A caveat to this claim would arise if it were definitely to be proven that there are effects of magnetosphere and ionosphere processes (and thus the interplanetary medium) on terrestrial weather. It is well recognized that even at GHz frequencies the ionized channels caused by lightning strokes, and possibly even charge separations in clouds, can reflect radar signals. Lightning and cloud charging phenomena may produce as yet unrecognized noise sources for low-level wireless signals. Thus, if it were to be learned that ionosphere electrical fields influenced the production of weather disturbances in the troposphere, the space environment could be said to effect even those wireless signals that might be disturbed by lightning. Much further research is required in this area of speculation.

SUMMARY

In the last 150 years, the variety of technologies that are embedded within space-affected environments have vastly increased. The variety of underlying physical phenomena that affect these technologies are limited. However, the increasing sophistication of technologies, and how they relate to the environments in which they are embedded, means that ever more sophisticated understanding of the physical phenomena is needed.

At the same time, most present-day technologies that are affected by space phenomena are very dynamic (especially communications and deregulated electrical power). These technologies can not wait for optimum knowledge to be acquired before new embodiments are created, implemented, and marketed. Indeed, those companies that might seek perfectionist understanding can be left behind by the marketplace. A balance is needed between deeper understanding of physical phenomena and "engineering" solutions to current crises. The research community must try to understand, and operate in, this dynamic environment.

Acknowledgments. I thank numerous colleagues for their discussions on this topic over many years, including C. G. Maclennan, D. J. Thomson, G. Siscoe, J. B. Blake, G. Paulikas, A. Vampola, H. C. Koons, and L. J. Zanetti.

REFERENCES

Albertson, V. D., J. M. Thorson, R. E. Clayton, and R. E. Tripathy, Solar induced currents in power systems: cause and effects, *IEEE Trans, Power App. & Sys.*, PAS-92, 471, 1973.

Albertson, V. D., J. M. Thorson, and S. A. Miske, The effects of geomagnetic storms on electrical power systems, *IEEE Trans. Power App. & Sys.*, PAS-93, 1031, 1974.

Anderson, C. N., Correlation of radio transmission and solar activity, *Proc. I. R. E.*, 16, 297, 1928.

Anderson, C. N., Notes on the effects of solar disturbances on transatlantic radio transmissions, *Proc. I. R. E.*, 17, 1528, 1929.

Anderson, C. W., III, L. J. Lanzerotti, and C. G. Maclennan, Outage of the L4 system and the geomagnetic disturbance of 4 August 1972, *The Bell Sys. Tech. J.*, 53, 1817, 1974.

Appleton, E. V., and M. A. F. Barnett, Local reflection of wireless waves from the upper atmosphere, *Nature*, 115, 333, 1925.

Baker, D. N., R. D. Balian, P. R. Higbie, et al., Deep dielectric charging effects due to high energy electrons in Earth's outer magnetosphere, *J. Electrost.*, 20, 3, 1987.

Baker, D. N., S. Kanekal, J. B. Blake, et al., Satellite anomaly linked to electron increase in the magnetosphere, *Eos Trans. Am. Geophys. Union*, 75, 401, 1994.

Baker, D. N., An assessment of space environment conditions during the recent Anik E1 spacecraft operational failure, *ISTP Newsletter*, 6, 8, 1996.

Baker, D. N., J. H. Allen, S. G. Kanekal, and G. D. Reeves, Disturbed space environment may have been related to pager satellite failure, *Eos Trans. Am. Geophys. Union*, 79, 477, 1998.

Barlow, W. H., On the spontaneous electrical currents observed in the wires of the electric telegraph, *Phil. Trans. R. Soc.*, 61A, 61, 1849.

Barron, W. R., E. W. Cliver, D. A. Guidice, and V. L. Badillo, *An Atlas of Selected Multi-Frequency Radio Bursts from the Twentieth Solar Cycle*, Air Force Geophysics Laboratory, Space Physics Division, Project 4643, Hanscom AFB, MA, 1980.

Barron, W. R., E. W. Cliver, J. P. Cronin, and D. A. Guidice, Solar radio emission, in *Handbook of Geophysics and the Space Environment*, ed. A. S. Jura, Chap. 11, AFGL, USAF, 1985.

Bastian, T. S., A. O. Benz, and D. E. Gary, Radio emission from solar flares, *Ann. Rev. Astron. Astrophys.*, 36, 131, 1998.

Beech, M., and P. Brown, Space platform impact probabilities – the threat from the Leonids, *ESA J.*, 18, 63, 1994.

Beech, M., P. Brown, and J. Jones, The potential danger to satellites from meteor storm activity, *Q.J. R. Astr. Soc.*, 36, 127, 1995.

Benz, A. O., Millisecond radio spikes, *Solar Phys.*, 104, 99, 1986.

Beech, M., P. Brown, J. Jones, and A. R. Webster, The danger to satellites from meteor storms, *Adv. Space Res.*, 20, 1509, 1997.

Boteler, D. H., R. M. Shier, T. Watanabe, and R. E. Horita, Effects of geomagnetically induced currents in the B. C. Hydro 500 kV system, *IEEE Trans. Power Delivery*, 4, 818, 1989.

Boteler, D. H., R. J. Pirjola, and H. Nevanlinna, The effects of geomagnetic disturbances on electrical systems at the Earth's surface, *Adv. Space Res.*, 22, 17, 1998.

Boteler, D. H., and G. Jansen van Beek, August 4, 1972 revisited: A new look at the geomagnetic disturbance that caused the L4 cable system outage, *Geophys. Res. Lett.*, 26, 577, 1999.

Breit, M. A., and M. A. Tuve, A test of the existence of the conducting layer, *Nature*, 116, 357, 1925.

Brooks, J., A reporter at large; The subtle storm, *New Yorker*, February 19, 1959.

Brown, W. L., Observations of the transient behavior of electrons in the artificial radiation belts, in *Radiation Trapped in the Earth's Magnetic Field*, ed. B. M. McCormac, D. Reidel Pub. Co., Dordrecht, Holland, 610, 1966.

Brown, W. L., T. M. Buck, L. V. Medford, E. W. Thomas, H.

K. Gummel, G. L. Miller, and F. M. Smith, *Bell Sys. Tech. J.*, 42, 899, 1963.

Buck, T. M., H. G. Wheatley, and J. W. Rogers, *IEEE Trans. Nucl. Sci.*, 11, 294, 1964.

Campbell, W. H., An interpretation of induced electrical currents in long pipelines caused by natural geomagnetic sources of the upper atmosphere, *Surveys Geophys.*, 8, 239, 1986.

Carrington, R. C., *Observation of the Spots on the Sun from November 9, 1853, to March 24, 1863, Made at Redhill*, William and Norgate, London and Edinburgh, 167, 1863.

Castelli, J. P., J. Aarons, D. A.. Guidice, and R. M. Straka, The solar radio patrol network of the USAF and its application, *Proc. IEEE*, 61, 1307, 1973.

Czech, P., S. Chano, H. Huynh, and A. Dutil, The Hydro-Quebec system blackout of 13 March 1989: System response to geomagnetic disturbance, *Proc. EPRI Conf. Geomagnetically Induced Currents*, EPRI TR-100450, Burlingame, CA, 19, 1992.

Davidson, W. F., The magnetic storm of March 24, 1940 – Effects in the power system, *Edison Electric Inst. Bulletin*, 365, 1940.

Fagen, M. D., *A History of Science and Engineering in the Bell System*, Bell Tel. Labs., Inc., Murray Hill, NJ, 1975.

Foschini, L., Electromagnetic interference from plasmas generated in meteoroid impacts, *Europhys. Lett.*, 43, 226, 1998.

Gardner, D. J., J. A. M. McDonnell, and L. Collier, Hole growth characterization for hypervelocity impacts in thin targets, *Int. J. Impact Eng.*, 19, 589, 1997.

Garrett, H. B., The charging of spacecraft surfaces, *Revs. Geophys.*, 19, 577, 1981.

Germaine, L. W., The magnetic storm of March 24, 1940 – effects in the communication system, *Edison Electric Institute Bulletin*, July 1940.

Gordon, G. D., and W. L. Morgan, *Principals of Communications Satellites*, John Wiley, New York, 178-192, 1993.

Guidice, D. A., and J. P. Castelli, Spectral characteristics of microwave bursts, in *Proc. NASA Symp. High Energy Phenomena on the Sun*, Goddard Space Flight Center, Greenbelt, MD, 1972.

Gussenhoven, M. S., and E. G. Mullen, Geosynchronous environment for severe spacecraft charging, *J. Spacecraft Rockets*, 20, 26, 1983.

Hey, J. S., Solar radiations in the 4 – 6 metre radio wavelength band, Nature, 158, 234, 1946.

Hey, J. S., *The Evolution of Radio Astronomy*, Neale Watson Academic Pub. Inc., New York, 1973.

Hunten, D. M., J.-C. Gerard, and L. M. Francois, The atmosphere's response to solar irradiation, in *The Sun in Time*, ed. C. P. Sonnett, M. S. Giampapa, and M. S. Matthews, Univ. Arizona Press, Tucson, 463, 1991.

IBM Journal of Research and Development, 40, 1-136, 1996.

Isliker, H., and A. O. Benz, Catalogue of 1 – 3 GHz solar flare radio emission, *Astron. Astrophys. Suppl. Ser.*, 104, 145, 1994.

Jenniskens, P., Meteor stream activity, I. The annual streams, *Astron, Astrophys.*, 287, 990, 1994.

Jenniskens, P., Meteor stream activity, II. Meteor outbursts, *Astron. Astrophys.*, 295, 206, 1995.

Koons, H. C., Characteristics of electrical discharges on the

P78-2 satellite (SCATHA), *18th Aerospace Sciences Meeting*, AIAA 80-0334, Pasadena, CA, 1980.

Kappenman, J. G., V. D. Albertson, and N. Mohan, Current transformer and relay performance in the presence of geomagnetically-induced currents, *IEEE Trans. Power App. & Sys.*, PAS-100, 1078, 1981.

Koons, H., C., J. E. Mazur, R. S. Selesnick, J. B. Blake, J. F. Fennel, J. L. Roeder, and P. C. Anderson, *The Impact of the Space Environment on Space Systems*, Engineering and Technology Group, The Aerospace Corp., Report TR-99(1670), El Segundo, CA, 1999.

Kundu, M. R., *Solar Radio Astronomy*, Interscience, New York, 1965.

Lanzerotti, L. J., Penetration of solar protons and alphas to the geomagnetic equator, *Phys. Rev. Lett.*, 21, 929, 1968.

Lanzerotti, L. J., Geomagnetic induction effects in ground-based systems, *Space Sci. Rev.*, 34, 347, 1983.

Lanzerotti, L. J., and G. P. Gregori, Telluric currents: The natural environment and interactions with man-made systems, in *The Earth's Electrical Environment*, National Academies Press, Washington, D. C., 1986.

Lanzerotti, L. J., C. Breglia, D. W. Maurer, and C. G. Maclennan, Studies of spacecraft charging on a geosynchronous telecommunications satellite, *Adv. Space Res.*, 22, 79, 1998.

Lanzerotti, L. J., C. G. Maclennan, and D. J. Thomson, Engineering issues in space weather, in *Modern Radio Science*, ed. M. A. Stuchly, Oxford, 25, 1999.

Mandell, M., 120000 leagues under the sea, *IEEE Spectrum*, 50, April 2000.

Marconi, G., Radio communication, *Proc. IRE*, 16, 40, 1928.

McBride, N., The importance of the annual meteoroid streams to spacecraft and their detectors, *Adv. Space Res.*, 20, 1513, 1997.

McBride, N., and J. A. M. McDonnell, Meteoroid impacts on spacecraft: sporadics, streams, and the 1999 Leonids, *Plant. Space Sci.*, 47, 1005, 1999.

McCormac, B. M., Summary, in *Radiation Trapped in the Earth's Magnetic Field*, ed. B. M. McCormac, D. Reidel Pub. Co., Dordrecht, Holland, 887, 1966.

Medford, L. V., L. J. Lanzerotti, J. S. Kraus, and C. G. Maclennan, Trans-Atlantic earth potential variations during the March 1989 magnetic storms, *Geophys. Res. Lett.*, 16, 1145, 1989.

Mizera, P. F., A summary of spacecraft charging results, *J. Spacecraft Rockets*, 20, 438, 1983.

National Research Council, *Radiation and the International Space Station*, Space Studies Board and Board on Atmospheric Sciences and Climate, National Academy Press, Washington, DC, 2000.

Pickholtz, R. L., Communications by means of low Earth orbiting satellites, *in Modern Radio Science 1996*, ed. J. Hamlin, Oxford U. Press, 133, 1996.

Pirjola, R., On currents induced in power transmission systems during geomagnetic variations, *IEEE Trans, Power App. & Sys.*, PAS-104, 2825, 1985.

Pirjola, R., and M. Lehtinen, Currents produced in the Finnish 400 kV power transmission grid and in the Finnish natural gas pipeline by geomagnetically induced electric fields, *Annales Geophysicae*, 3, 485, 1985.

Prescott, G. B., *Theory and Practice of the Electric Telegraph*, IV ed., Tichnor and Fields, Boston, 1860.

Reeves, G. D., The relativistic electron response at geosynchronous orbit during January 1997 magnetic storm, *J. Geophys., Res.*,103, 17559, 1998.

Shea, M. A., and D. F. Smart, Space weather: The effects on operations in space, *Adv. Space Res.*, 22, 29, 1998.

Silliman, Jr., B., *First Principals of Chemistry*, Peck and Bliss, Philadelphia, 1850.

Snyder, W. F., and C. L. Bragaw, *Achievement in Radio*, U. S. Department of Commerce, U. S. Government Printing Office, 1986.

Southworth, G. C., Microwave radiation from the sun, *J. Franklin Inst.*, 239, 285, 1945.

Taur, R. R., Ionospheric scintillation at 4 and 6 GHz, *COMSAT Technical Review*, 3, 145, 1973.

Todd, D., Letter to *Space News*, pg. 12, March 6, 2000.

Tribble, A. C., *The Space Environment, Implications for Spacecraft Design*, Princeton Univ. Press, Princeton, NJ, 1995.

Vampola, A., The aerospace environment at high altitudes and its implications for spacecraft charging and communications, *J. Electrost.*, 20, 21, 1987.

Viljanen, A., Geomagnetically induced currents in the Finnish natural gas pipeline, *Geophysica*, **25**, 135, 1989.

Yeomans, D. K., K. K. Yau, and R. R. Weissman, The impending appearance of comet Tempel-Tuttle and the Leonid meteors, *Icarus*, 124, 407, 1996.

Bell Laboratories, Lucent Technologies, Murray Hill, NJ 07974 USA. Ljl@bell-labs.com

Space Weather Forecasting: A Grand Challenge

H.J. Singer, G.R Heckman, and J.W. Hirman

NOAA Space Environment Center

Space Environment Center (SEC) is the United States' official source of space weather alerts, warnings, and forecasts. Forecasts are used to support activities that are impacted by space weather such as electric power transmission, satellite operations, humans in space, navigation, and communication. This article presents a brief review of current space weather forecasting capabilities, and then focuses on the science, the models, the data, the new technologies, and the process for transitioning research into operations that is needed to meet the challenge to improve space weather forecasting in the new millennium. Forecasting critical parameters such as the interplanetary magnetic field at the magnetopause, and critical events such as coronal mass ejections are two examples of challenges to the research, observation, and modeling communities. Major improvements in space weather forecasting will be achieved when these, as well as other, challenges are met. The forecasting challenge is also discussed in the context of the goals of the US National Space Weather Program (NSWP) and other international activities.

1. INTRODUCTION

Space weather forecasting is a relatively new enterprise that combines space sciences research with the development of practical applications for the benefit of human activities affected by the space environment. While space weather forecasting is only decades old, forecasting for human benefit has been common throughout history and in many different areas of human endeavor. It seems to be a part of human nature to want to tell the future. Whether it is forecasting tropospheric weather, the stock market, earthquakes, or a company's business plans, there is one thing that all forecasting has in common - it is a grand challenge. As expressed by Yogi Berra, "It's tough to make predictions, especially about the future."

An article in the *New Yorker Magazine* in 1959 (Brooks, 1959) described the effects on technology of a

geomagnetic storm that occurred during the previous year. The article, "The Subtle Storm," was prophetic in suggesting that "In future years, it may be that the Weather Bureau or some Space Age equivalent will warn us of approaching magnetic storms, just as we are now warned of approaching hurricanes..." Today, as part of the U.S. National Oceanic and Atmospheric Administration's National Weather Service (NWS), the Space Environment Center (SEC) in Boulder, Colorado performs the forecasting functions foretold in that article. As well as being part of the NWS, SEC is also a part of NOAA's office of Oceanic and Atmospheric Research (OAR). As part of OAR, SEC carries out research and development that advances our understanding of the space environment and the ability of our Space Weather Operations to provide improved space weather alerts, warnings, and forecasts. This paper will discuss some of the current forecasting capabilities at SEC, and give a glimpse of forecasting tools on the horizon and those needed in the future.

Space weather has been defined as "...conditions on the sun and in the solar wind, magnetosphere, ionosphere, and thermosphere that can influence the performance and

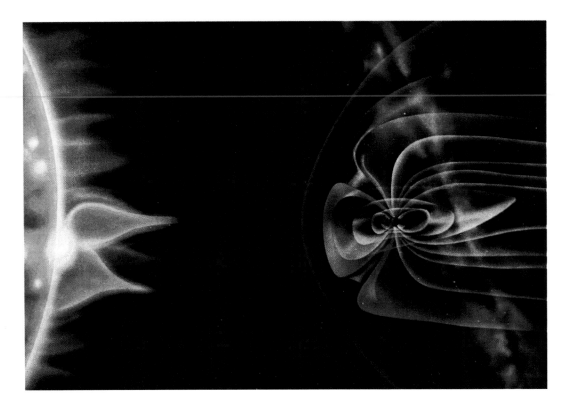

Plate 1. Artist concept of the solar-terrestrial system showing the active sun and Earth's magnetosphere. (NOAA/SEC).

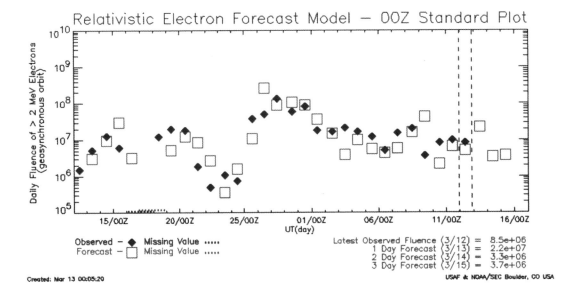

Plate 2. Relativistic electron forecast model (Smithtro and Onsager). Forecasts compared to observations above are 1-day forecasts. Forecasts beyond the observations are 1-, 2-, and 3-day forecasts.

reliability of space-borne and ground-based technological systems and can endanger human life or health. Adverse conditions in the space environment can cause disruption of satellite operations, communications, navigation, and electric power distribution grids, leading to a variety of socioeconomic losses" (NSWP Strategic Plan, 1995).

Other papers in this volume will discuss technologies affected by space weather conditions, while this paper will focus on forecasting those conditions. It is useful first to note the distinction between "prediction' and 'forecasting' as has been done in the "Metrics Report" under preparation by the NSWP. "A model 'prediction' can be any estimation issued by a model concerning a type of physical activity independently of whether the activity takes place after the prediction is made, or has taken place before the prediction. A prediction can be made in retrospective analyses (analyses of historical data). 'Forecasting' refers to predictions issued before the activity takes place. 'Nowcast' or specification is a model prediction of what is occurring at this moment (short-term forecast)."

The words forecasting and prediction are often used interchangeably. The point here is not to greatly influence our usage of these words, but to recognize that the space weather forecaster's challenge is different from that of the scientific researcher/model developer. The researcher often knows how an event turns out. The forecaster, on the other hand, uses minute-by-minute observations and model outputs to synthesize a forecast of future conditions, or conditions at locations where no observations are made, before the story unfolds.

2. THE GRAND CHALLENGE

The solar-terrestrial system is illustrated in Plate 1. Space weather originates on the sun and in its turbulent atmosphere that expands, sometimes explosively, into the background solar wind. The sun emits electromagnetic radiation extending from long-wavelength radio emissions, through the infrared and visible portions of the spectrum, to the ultraviolet, X-rays, and gamma radiation. It also emits energy in the form of energetic particles, mostly protons and electrons, but also heavier ions.

Nearly all of the solar emissions have consequences for space weather on Earth. The background solar wind, with speeds of several hundred km/sec and densities of several particles per cm^3, takes several days to reach Earth. At Earth it encounters Earth's dipole-like magnetic field. The solar wind compresses the field on the dayside of Earth and drags it out into a long geomagnetic tail on the nightside. The boundary between Earth's field and the solar wind is called the magnetopause. Just in front of the magnetopause, a bow shock slows the solar wind as it enters a

region between the magnetopause and the solar wind called the magnetosheath. When the solar wind's pressure increases dramatically, due primarily to velocity and/or density increases, Earth's field on the dayside is compressed and the magnetopause moves inward from its typical location about 10 Earth radii in front of Earth, sometimes to within geosynchronous orbit at 6.6 Earth radii. During large compression events, geosynchronous satellites on the dayside can become immersed in the magnetosheath's strong magnetic field, oriented oppositely to what is typically observed. This condition causes some satellites to experience attitude control difficulties.

The solar wind plasma carries with it an embedded magnetic field that is typically a few nT. The orientation of that magnetic field, when it encounters Earth's magneto-pause, is the key to opening the magnetic shield that surrounds Earth and to coupling solar wind energy into Earth's near-space environment. When the solar wind magnetic field is directed southward, opposite Earth's northward directed field at the magnetopause, the coupling of energy into the magnetosphere and ionosphere is most efficient and it is during these intervals that frequent substorms, or large geomagnetic storms occur.

Sometimes, superimposed on the background solar wind, transient solar events such as solar flares, coronal mass ejections (CME's), and solar proton events (SPE's) influence near-Earth space weather (Joselyn, 1995). The x-ray emissions associated with solar flares reach Earth at the speed of light and cause changes in the dayside ionosphere that affect high-frequency radio propagation. CME's can contain about 10^{16} g and those that are Earth directed reach Earth in a few days and are often associated with large geomagnetic storms. SPE's, associated with both flares and CME's, reach Earth in a few hours and can affect astronauts, satellite electronics, and radio propagation that depends on conditions in Earth's polar cap.

The grand challenge is to forecast conditions in this complex and dynamic coupled solar-terrestrial system with skill, reliability, and timeliness. Improved forecasting depends on new understanding through research, new advances in observation, and improved numerical models. But new research, observation, and modeling alone will not improve forecasts unless there is significant effort made to transition these capabilities to space weather operations and to give forecasters the tools to synthesize the knowledge, observations and model output into a useful forecast.

A framework to meet this challenge exists today within the National Space Weather Program (NSWP). The NSWP in the United States has a goal "to achieve, within the next 10 years, an active, synergistic, interagency system to provide timely, accurate, and reliable space environment observations, specifications, and forecasts." Participating

TABLE 1. Space Weather Forecasting Today

Parameter	Input or Technique (examples)	Effects or Users (examples)
F10.7 Solar Radio Flux	Persistence; recurrence; climatology; solar active region observations	Satellite drag; ionospheric models
Sunspot Number	Climatology; statistics; precursor methods; neural networks	Hubble telescope; International Space Station; orbital debris
Flare Probability and Proton Events	Persistence; magnetic structures; sunspot classification; solar images; conditional climatology	Radiation levels (NASA Space Radiation and Analysis Group); LORAN on dayside; single event upsets (SEU's); polar cap absorption (PCA)
A Index and Geomagnetic Storm Probability	Solar events; solar images (SOHO); radio type II, IV; X-ray and He 1083 nm images; solar wind (ACE); recurrence; climatology; persistence	Induced currents (power utilities); satellite operations; navigation systems
Equatorial Radiation Environment[a]	Kp forecast by neural net using solar wind data (ACE)	Satellite surface charging
Kp[a]	Solar wind data (ACE) and neural net	Models; power utilities; satellite drag
MeV Electrons	Solar wind velocity (ACE)	Satellite deep dielectric charging

[a] Test forecast products that are not yet considered operational.

agencies include: National Science Foundation, Department of Commerce, Department of Defense, National Aeronautics and Space Administration, Department of Energy and Department of Transportation. The NSWP rests on pedestals of research, observation, models, and education. These are just the themes that are critical for meeting the broad scope of the forecasting challenge. The remainder of the paper will discuss in more detail the forecasting that is done today and future forecasting needs.

3. FORECASTING TODAY

The NOAA Space Environment Center, Space Weather Operations (SWO), is the official civilian source of space weather alerts, warnings, and forecasts in the United States. The forecast center in Boulder, Colorado is staffed 24 hours per day, 7 days per week. Space weather products depend on global observations on Earth and in near-Earth space, as well as solar and interplanetary observations, model outputs, and scientific understanding of processes on the sun and the near-Earth space environment. The analysis of global effects involves global collaborations with partners at the US Air Force's 55th Space Weather Squadron, other U.S. Agencies, the International Space Environment

Services organization, the international scientific community, and other international space weather organizations.

To understand better the forecasting challenges, it is useful to provide an overview of today's forecasting capabilities. Table 1 identifies the principal parameters that are forecast today at the NOAA SEC Space Weather Operations, along with examples of the inputs that are considered to prepare a forecast, and the region of the environment, or user, that is affected. As one might expect, there are fewer forecasts than there are products that specify current environmental conditions (see SEC web pages for other products: http://www.sec.noaa.gov); however, new observations and numerical models are becoming part of the forecaster's arsenal.

A brief description of a few of the parameters that are forecast will illustrate the process, the progress, and the challenges. Beginning with the sun, F10.7 is the whole sun solar radio flux at 10.7 cm wavelength, useful as a proxy for solar extreme ultraviolet (EUV) radiation. EUV is responsible for the heating and ionization of neutrals in Earth's upper atmosphere. Consequently, increases in F10.7 are associated with increases in the height of the neutral atmosphere that produce additional drag on satellites and change ionospheric plasma properties that affect, for example, radio communications. Forecasts of F10.7 are made on time scales ranging from days to the 11-year solar cycle. Prior to NASA Space Shuttle flights, forecasts are prepared for the duration of the flight.

Preparation of F10.7 forecasts depends on knowledge of solar processes, observations, past experience, and historical trends. F10.7, and more importantly solar EUV output are correlated with solar active regions. Since active regions tend to persist from day-to-day and tend to recur from one solar rotation to the next, persistence and recurrence are two factors taken into consideration in preparing F10.7 forecasts. In addition, observations of the chromospheric plage network provide forecasters with an indication of the solar EUV output. Longer-term forecasts of F10.7, on solar cycle time scales, are needed for operations planning for the International Space Station, for the Hubble telescope and other similar space-based activities where reboosts may be needed to keep these structures at high enough altitude so that they do not re-enter prematurely. A recent study by Joselyn et al. (1997) used a variety of techniques, including climatology, statistics, and ideas about the recirculation of solar magnetic flux to forecast both F10.7 and sunspot number for the current Solar Cycle 23.

Another challenge to forecasters is to forecast the probability of a solar flare occurring in various time intervals. Solar flares are sudden eruptions of energy on the sun, lasting from minutes to hours. The largest flares can release as much as $\sim 10^{32}$ ergs in $\sim 10^3$ seconds. By comparison the total solar output is about 4×10^{33} erg/s. Flares are identified by a sudden intensification of X-ray radiation and are sometimes accompanied by Solar Energetic Particle events (SEP's); although SEP's are also associated with shocks in the solar wind produced by Coronal Mass Ejections (CME's). The X-ray radiation reaches Earth in about 8 minutes and causes changes in the dayside ionosphere that affect communications. SEP events contain energetic multi-MeV protons that can reach Earth in hours. These particles can be harmful to astronauts, especially those conducting extra-vehicular activity (EVA), or astronauts on future missions to the Moon or Mars. The particles also give rise to concern for the radiation dose to high flying aircraft, such as the SST and those flown by DoD. SEC Space Weather Operations provides daily forecasts to the US Federal Aviation Agency as well as to NASA during crewed space missions. Another effect of SEP events is to disturb satellite operations though Single Event Upsets (SEU's) that interfere with critical electronic components.

Flare forecasts are based on observations such as solar X-ray and H-alpha images, images of solar magnetic structures, and sunspot observations. Consideration is given to the persistence of solar active regions, as well as the complexity and growth/decline of magnetic structures. It is a major challenge to the reseach community to provide a better understanding of solar conditions that lead to the onset of individual flares. There is also promising work being done on the occurrence of flare associated SEP events. Using knowledge of flare location, maximum flare temperature, and maximum flare flux, an empirical relationship has been derived for proton event probability that is being tested at SEC (Garcia et al., 1999).

Forecasts of geomagnetic storms, the equatorial radiation environment, and geomagnetic activity indices, as noted in Table 1, have all been aided by the availability of solar wind data from the NASA ACE satellite upstream of Earth, solar observations from the ESA/NASA SOHO satellite, and new models. These are discussed further, albeit briefly, in the final section of the paper. However, as a final note on today's forecasting capabilities, it is worth mentioning relativistic electrons, often called MeV electrons, or even "killer electrons." These particles with millions of electron volt energy, populate the outer zone of Earth's radiation environment from about L = 2 - 8 where L is the equatorial, geocentric distance in Earth radii of a magntic field line. They cause charge build-up in spacecraft dielectrics and subsequent discharges can seriously damage spacecraft electronics. For this reason, and because the process by which these electrons are energized is unknown, research on forecasting MeV electrons is one of the outstanding challenges in magnetospheric physics today.

Even though we don't understand the mechanism(s) for energizing the particles, we know there is a clear relationship between enhanced solar wind velocity and the buildup of the flux of these electrons in the inner magnetosphere over the few days following the velocity increase. Using this empirical relationship, models (e.g., Baker et al., 1990) have been developed to forecast the enhancement of the electrons at geosynchronous orbit.

Plate 2 shows a modified version of the Baker et al. model (developed by Smithtro and Onsager at NOAA SEC) that has been implemented as a test forecast product at SEC. The figure compares the daily fluence of greater than 2 MeV electrons observed each day by the GOES satellites with 1-day, 2-day, and 3-day model forecasts that are driven by solar wind velocity measurements from the NASA ACE spacecraft. Improvements to the Baker model include: model coefficients for 2 and 3 day forecasts using

TABLE 2. Space Weather Forecasting Goals[a]

Space Weather Domain	Goal (Specify and Forecast:)
Solar Coronal Mass Ejections	Occurrence, magnitude, duration, and magnetic field structure
Solar Flares	Occurrence, magnitude, and duration
Solar and Galactic Energetic Particles	At satellite orbit, the temporal profile and spectral variation during a Solar Particle Event (SPE)
Solar Wind	Solar wind density, velocity, magnetic field strength and direction
Magnetospheric Particles and Fields	Global magnetic field, magnetospheric electrons and ions, and strength and location of field-aligned current systems; high-latitude electric fields, and electrojet current systems
Geomagnetic Disturbances	Geomagnetic indices and storm onset, intensity, and duration
Radiation Belts	Trapped ions and electrons (1 to 12 Earth Radii)
Ionosphere/Thermosphere	Neutral and electron density, variability, storm onset and storm recovery

[a]Adapted from National Research Council report on Radiation and the International Space Station (2000).

30 day intervals of data; and each day the prediction is modified by a factor that depends on the difference between the previous day's prediction and the actual measurement.

4. FORECASTING CHALLENGES AND CONCLUSIONS

Throughout this paper we have noted challenges to forecasting space weather conditions. However, it is worth summarizing the major challenges in terms of the specification and forecast goals for the various regions of space and types of space weather activity (Table 2).

Forecasting solar activity is key to providing long-lead time forecasts that will give various users time to react to impending disturbances. Some solar disturbances, such as increased X-rays from flares, or the initial increased flux of particles from solar proton events, reach Earth in minutes to hours. Others such as CME's that propagate through the background solar wind may take several days to reach Earth. It is important to note that even those solar disturbances that reach Earth within minutes can be forecast with long lead times if we understand the conditions that lead to those events. Understanding conditions on the sun that result in CME's, flare activity, or solar proton events depends on breakthroughs in scientific understanding of solar processes, new solar models, and improved solar observations.

In recent years, forecasting the effects of solar disturbances has improved with the use of observations from scientific missions such as the Japanese Yohkoh satellite and the ESA/NASA SOHO satellite. Yohkoh has made available spectacular X-ray images of the sun that show important solar features such as the magnetic structure of active regions, the location of flares, and coronal holes. The SOHO satellite has provided remarkable UV images of solar activity and coronagraph images of CME's as they leave the sun. These scientific instruments have provided forecasters the opportunity to test the value of new observational tools, and to determine which new observations should become part of future operational capabilities. For a tool to be operational, forecasters and customers must be able to count on long-term uninterrupted service.

New, planned solar missions include the NASA STEREO satellites and a Solar Dynamics Observatory that will be part of the proposed NASA Living With A Star Program (LWS). NASA LWS heliospheric missions also include observations of the far side of the sun and solar wind observations in the vicinity of 0.5 to 0.8 astronomical units (AU). STEREO will use two satellites orbiting the sun at 1 AU, with one ahead of Earth and the other behind. Solar images from the two vantage points will make it possible to construct a three-dimensional view of CME's as they travel towards Earth, providing a major advance in forecasting the CME arrival at Earth. Another new solar mission that will provide improved monitoring of coronal holes and solar flares is a Solar X-ray Imager (SXI). SXI will provide 1 image per minute of the sun, and be an operational instrument on the NOAA GOES satellites beginning with GOES M about 2001.

Forecasting geomagnetic activity depends critically upon our ability to forecast the properties of the solar wind and its imbedded magnetic field, particularly the Bz, or north-south, component of the interplanetary magnetic field (IMF) when it encounters the magnetopause. Today, using real-time data from NASA's Advanced Composition Explorer (ACE) satellite, located at the L1 point in front of Earth about 1% of the way to the sun, we get about 30 minutes to 1 hour lead time on IMF conditions. This has provided a major advance in our ability to forecast geomagnetic activity, but challenges remain. We need to understand better how to interpret the data, especially when ACE is off the sun-Earth line. We also need longer lead times than are possible with an L1 satellite. Opportunities to place a satellite, using solar sails, between sun and Earth at twice the L1 distance, to give twice the lead-time, are being explored. In addition, the NASA TRIANA satellite that will be located at L1 will make full disk images of Earth and will carry instruments to monitor the solar wind. With improved solar observations and improved models of the propagation of disturbances and background solar wind conditions through interplanetary space, there are also expectations for improved forecasts of the IMF and the resultant geomagnetic activity at Earth.

Ultimately, much of the solar wind energy that encounters the magnetosphere goes into energizing particles in the radiation belts, enhancing currents that dissipate energy in the ionosphere, and energizing particles that precipitate into the upper atmosphere. There are numerous outstanding problems in understanding how the energy that enters the magnetosphere is partitioned into the various sinks, as well as the time sequence of these events during storms and substorms. In addition to active research programs, there are many efforts underway to improve our observations and models of the magnetosphere, thermosphere, and ionosphere. There will be new energetic particle and solar EUV measurements from geosynchronous orbit on the next generation of the NOAA operational GOES satellites beginning in about 2002. During the latter part of this decade the NASA LWS program is planning both Ionospheric and Radiation Belt mappers that will contribute measurements that can be used for space weather forecasts,

and NPOESS, the new joint NOAA/DoD low-altitude polar orbiting satellite program will be making many geophysical measurements that can be used to improve space weather forecast and specification. New ground-based observational techniques using magnetometers and radars are also providing new diagnostics of the space environment with spatial and temporal coverage that far exceeds what was possible just a few years ago.

As a result of new research understanding and improved numerical models and observations, there have been outstanding advances in our ability to forecast conditions in the space environment, yet many challenges remain to improve our capability to protect valuable resources and human activities that are affected by the space environment.

Acknowledgments. The authors appreciate the contributions from N. Arge, T. Detman, K. Doggett, T. Fuller-Rowell, J. Joselyn, J. Kunches, T. Onsager, B. Poppe, C. Smithtro, and D. Speich. Special thanks to D. Evans for bringing to my attention the 1959 New Yorker article.

REFERENCES

Brooks, J., The Subtle Storm, *New Yorker Magazine*, 39-77, Feb. 7, 1959.

Baker, D.N., Linear prediction filter analysis of relativistic electron properties at 6.6 Re, *J. Geophys. Res., 95*, 15,133-15,140, 1990.

Garcia, H., S. Greer, and R. Viereck, Predicting solar energetic particle events from flare temperatures, Proc. 9th European Meeting on Solar Physics, 'Magnetic Fields and Solar Processes', Florence Italy, 12-18 September 1999, ESA SP-448, December, 1999.

Joselyn, J.A., Geomagnetic activity forecasting: the state of the art, *Rev. Geophys.,* 33, 383-401, 1995.

Joselyn, J.A., J.B. Anderson, H. Coffey, K. Harvey, D. Hathaway, G. Heckman, E. Hildner, W. Mende, K. Schatten, R. Thompson, A.W.P. Thomson, and O.R. White, Panel achieves consensus prediction of Solar Cycle 23, *EOS, Trans. Amer. Geophys. Union, 78*, 205 and 211-212, 1997.

National Space Weather Program, Strategic Plan, FCM-P30-95, Washington, DC, August, 1995.

Radiation and the International Space Station: Recommendations to Reduce Risk, Committee on Solar and Space Physics and Committee on Solar-Terrestrial Research, National Research Council, National Academy of Sciences, National Academy Press, Washington, DC, 2000.

H.J. Singer, G.R. Heckman, and J.W. Hirman, NOAA Space Environment Center, NOAA R/SEC, 325 Broadway, Boulder, Colorado, 80305, USA. (email: howard.singer@noaa.gov)

Space Weather - Lessons From the Meteorologists

Robert P. McCoy

Space and Remote Sensing Office of Naval Research, Arlington, VA

The ability to measure, model and forecast the global ionosphere lags significantly behind current capabilities to model and forecast global weather. However, there are a number of new techniques for ionospheric remote sensing from space that are currently being tested and will soon be operationally available. A significantly large and continuous real-time ionospheric database could provide the basis for a new type of ionospheric model based on data assimilation. The ionospheric modeling and forecast community stands to benefit substantially from the past 50 years of tropospheric weather modeling and forecasting. By studying the development history and lessons learned along the way, ionospheric modelers should be able to advance more rapidly and avoid problems encountered by the meteorologists. Three important lessons include the need for continuous space-based observations, the use of data assimilation to combine good physics models with multiple data types, and the exploitation of geostationary orbit to provide long-dwell, high temporal and spatial resolution measurements of ionospheric variability.

1. INTRODUCTION

As space environment observational and modeling tools become more sophisticated and numerous, the prospects for significant improvement of "space weather" specification and forecasts are excellent - especially for the Earth's ionosphere. Interest and activity in this area at the national level is increasing as documented in the *National Space Weather Program Strategic Plan* (August 1995) and the subsequent *Space Weather Implementation Plan* (January 1997), [both available from the Office of the Federal Coordinator for Meteorological Services and Supporting Research, 8455 Colesville Road, Suite 1500, Silver Spring MD 20910]. In June 1999, the office of the National Security Space Architect (NSSA), [2461 Eisenhauer Avenue, Alexandria VA, 22331] completed a *Space Weather Ar-*

chitecture Study summarizing current space weather impacts and projected shortfalls. These documents describe the need for improvement in present capability to specify and forecast the global electron density distribution and better capabilities to predict the onset and severity of ionospheric irregularities and scintillation. To achieve these goals there are several important lessons that can be adapted from 50+ years of weather research and development by the meteorological community that comprises the backbone of current weather modeling and prediction.

This year NOAA celebrated the 40th anniversary of the launch of the first Television and Infrared Observation Satellite (TIROS) polar orbiting weather satellite. These satellites, now called Polar-orbiting Operational Environmental Satellites (POES), together with satellites from the Defense Meteorological Satellite Program (DMSP) have provided a continuous long-term observational weather data set to enable improvements in weather modeling and feed operational weather forecasts. In 1975 NASA launched the first Geostationary Observational Environmental Satellite (GOES-1) over the western hemisphere which augmented the TIROS observations providing a

Space Weather
Geophysical Monograph 125

larger view and longer dwell time than what was possible with low Earth orbiting (LEO) satellites. Weather satellites had a fundamental impact on weather prediction by providing both a global view of mesoscale phenomena and a larger and more randomly distributed atmospheric data set than was possible with individual ground-based measurements.

There have been numerous generations of numerical weather forecast models during the past 50 years. As the total volume of surface and space observations reached a critical level, and with the aid of increased computational power, new classes of weather forecast models were developed employing various forms of data assimilation. While these models come in a variety of types and use various approaches, a common characteristic is the ability to ingest large quantities of disparate types of data with variable sampling in space and time. In lieu of simple interpolation or least-square fitting, the assimilation scheme uses large volumes of data to constrain basic physics models. Sufficient quantities and qualities of data allow improved global predictions even in regions with sparse or no data samples.

Three important lessons from the past 50 years of meteorological weather prediction that may greatly benefit the space weather and ionospheric forecast efforts include: 1) continuous satellite observations are necessary to provide sufficient quantities and global sampling of the upper atmosphere and ionosphere; 2) LEO satellite measurements should be augmented with high spatial and temporal resolution measurements from geostationary platforms; and 3) when sufficient long-term global observation data are available, models based on data assimilation can provide an optimal means for ionospheric specification and prediction. Current research initiatives within the Navy addressing these three objectives are discussed in the remainder of this paper. These research initiatives include both in-house programs at the Naval Research Laboratory (NRL) and extramural research funded by the Office of Naval Research (ONR).

The motivation for Naval investment in space weather research is based on the fact that the Navy is permanently forward deployed and depends critically on space-borne platforms for: global communication (EHF- SHF); navigation (GPS and celestial); space tracking radars and over the horizon radars; surveillance and precision geolocation; satellite meteorology and oceanography; and ocean altimetry. Space weather affects the performance of satellite systems and ionospheric variability in particular has a strong influence on the ability to communicate, navigate, etc. The Air Force has a much larger role than the Navy for acquisition and operation of DoD space systems. To ensure that unique Naval needs are met in this area, the Navy has had a long and successful history of maintaining a strong basic space science and technology research program. At key

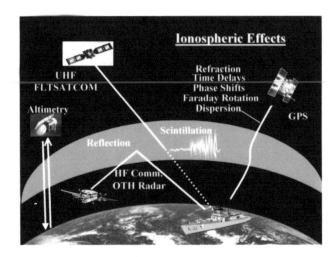

Figure 1. Ionospheric Effects on Naval Systems

times in the past this philosophy has enabled the Navy to have a significant influence on space acquisition, operations and direction of space research.

2. IMPROVED SPECIFICATION AND FORECAST OF THE GLOBAL IONOSPHERE

Any radio frequency (RF) signal at 10 GHz or lower transmitted into or through the ionosphere can be strongly affected by the ionospheric density distribution and irregularity structure. As illustrated in Figure 1, the ionosphere is refractive and reflective for RF frequencies up to and through the HF allowing global communication and over the horizon radar. At higher frequencies, the dominant effects of ionospheric electrons include refraction, time delays, phase shifts, dispersion and Faraday rotation. The presence of ionospheric irregularities can cause a variety of phenomena including scintillation, spread-F, fading and black-outs. The Number 1 Recommendation that resulted from the NSSA Space Weather Architecture Study called for "Improved Ionospheric and Radiation Environment Specifications and Scintillation Forecast" with special emphasis on operational model development. Improved global specification and forecast of the three-dimensional global electron density and variability are required to minimize the impacts of the ionosphere on space systems. Improved specification and forecast of ionospheric irregularities and scintillation are required to provide advanced warning of communication and navigation outages.

Adapting lessons from meteorological weather prediction, an obvious approach to improved specification lies in amassing large amounts of real-time and near real-time ionospheric data to drive an assimilating ionospheric model. One example of current meteorological data assimilation is the Navy Operational Global Atmospheric Prediction System (NOGAPS) [Hogan and Rosmond, 1991] weather prediction model which assimilates more than 5 million observations per day including surface observations, rawinsonde, balloon, aircraft and satellite observations. Until recently there were limited prospects for amassing real-time global ionospheric measurements with volumes approaching this size. Most ionospheric measurement sources were ground-based and local - similar to the situation 40 years ago before the first weather satellites. In the last five years, however, a number of new techniques for satellite remote sensing of the ionosphere have been developed and are under test with the result that it may soon be possible to implement an operational assimilating model.

The Air Force 55th Space Weather Squadron (SWxS) is responsible for space weather and ionospheric products within the Department of Defense (DoD). The 55th SWxS currently uses empirical ionospheric models and electron density measurements from three independent sources including: peak heights and densities from an array of about 17 globally distributed digisondes; total electron content (TEC) measurements from 24 ground-based dual-frequency GPS receivers (part of the International GPS Service [IGS] network); and in-situ measurements of plasma density, temperature and composition from the Special Sensor for thermal Ions, Electrons and their Spatial variations (SSIES) on DMSP weather satellites [Rich, 1994]. While these data are insufficient to support an assimilating model, new space-based remote sensing techniques including ultraviolet remote sensing, GPS occultation [Anthes et al., 1997] and computer ionospheric tomography (CIT) [Bernhardt et al., 1998] will soon allow the production of large databases of electron density measurements. All these data sources (see Figure 2) combined should provide a sufficient sampling of global electron densities to sustain an operational assimilation model.

2. OPERATIONAL ULTRAVIOLET REMOTE SENSING DATA SOURCES

In December 2000, the DMSP is scheduled to launch the first of its next block of weather satellites (Block 5D3). Each of the new satellites will include a pair of ionospheric and thermospheric remote sensing spectrographs called the Special Sensor Ultraviolet Limb Imager (SSULI) [McCoy et al., 1994] and the Special Sensor Ultraviolet Spectrographic Imager (SSUSI) [Paxton et al., 1992]. Both instruments measure airglow radiance in the far-ultraviolet but the SSULI's are primarily limb scanning and include wavelengths into the extreme-ultraviolet while the SSUSI's are primarily nadir cross-track scanning. Each SSULI will measure limb airglow profiles every 90 seconds over the wavelength interval 80 to 170 nm which includes the iono-

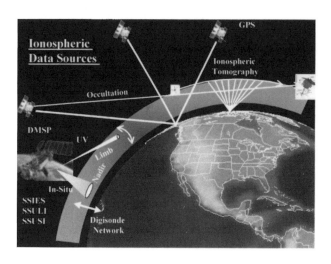

Figure 2. Ionospheric Data Sources for Assimilation. Digisondes, GPS TEC and In-situ are Currently Available. In December 2000, UV Sources Will be Available From DMSP Satellites. GPS Occultation and CIT are Future Candidates.

spheric O^+ resonance multiplet and O recombination emission from atomic oxygen at 91.1, 135.6 and 130.4 nm as well as other dayglow emission features from oxygen and molecular nitrogen. Measured limb radiance profiles from these atoms, ions and molecules can be inverted to determine altitude profiles of the O^+ density (approximately equal to the electron density in the F-region) and neutral density. Each SSUSI will obtain high spatial resolution measurements of ionospheric recombination longward of 120 nm and dayglow oxygen and nitrogen emissions. The SSUSI observations will be particularly useful for mapping the auroral ovals. Five pairs of these instruments on DMSP satellites F16 through F20 will provide a continuous long-term observational database for improved ionospheric specification and forecast. Together these pairs of instruments will provide a three-dimensional map of the ionosphere along the satellite orbit.

4. IONOSPHERIC REMOTE SENSING INSTRUMENTS ON THE ARGOS SATELLITE

A prototype version of SSULI, part of the High Resolution Airglow/Aurora Spectroscopy (HIRAAS) experiment [Dymond et al., 1999; Wolfram et al., 1999] is currently in orbit aboard the Air Force Space Test Program (STP) Advanced Research and Global Observation Satellite (ARGOS). HIRAAS consists of three limb scanning spectrographs with wavelength coverage from 50 to 340 nm. Figure 3 shows an artists sketch of the HIRAAS experiment on the nadir panel of the ARGOS satellite and a photograph of the experiment before integration on the space-

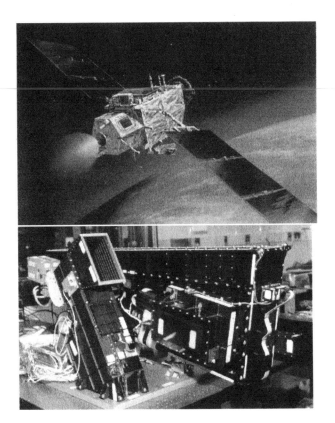

Figure 3. Upper Panel: Artist Sketch of the ARGOS Satellite. Lower Panel: HIRAAS Experiment Including SSULI Prototype (Foreground).

GPS receivers in LEO has already been extensively demonstrated using the GPSMET satellite [Anthes et. al, 1997]. This technique uses a modified GPS receiver on a LEO satellite tuned to an overhead GPS satellite. The changing refraction of the GPS signal as the LEO satellite passes into occultation can be used to determine a profile of electron density on the limb. The CIT technique employs a satellite borne phase-locked dual frequency VHF/UHF transmitter (typically at 150 and 400 MHz) and one or more receivers on the ground. High-resolution two-dimensional images of electron density under the satellite track can be inferred from the differential phase shifts of the two RF signals. The National Space Program Office (NSPO) of Taiwan in conjunction with a consortium of U. S. Agencies are planning to fly a constellation of six microsatellites called the Constellation Observing System for Meteorology and Climate [Rocken et al., 2000]. Each satellite will contain a GPS occultation receiver, CIT beacon and ultraviolet photometer. The COSMIC satellite and other GPS occultation and CIT satellite instruments (e.g. The current DMSP F15 weather satellite contains a CIT beacon) can provide a significant long-term database of ionospheric measurements to support improved specification and forecast models.

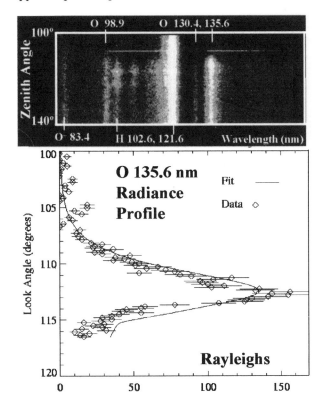

Figure 4. Upper Panel: Sample HIRAAS Dayglow Limb Scan. Lower Panel: Sample Ionospheric Nightglow Radiance Profile. (Both Courtesy Ken Dymond)

craft. A sample dayglow limb scan from HIRAAS is shown in Figure 4 along with a measured nightglow profile of ionospheric radiative recombination from O at 135.6 nm. The lower panel also contains a fit to the data based on a Chapman density profile with a peak height of 326 km and a peak density of 8.2×10^5 cm^{-3} [Ken Dymond, personal communication]. Inversions of the optically thin O 135.6 nm emission are the basis of the SSULI/HIRAAS nighttime ionospheric remote sensing technique. The resulting volume emission rate is proportional to the product of the O$^+$ and e$^-$ density or approximately equal to the square of the electron density throughout most of the F-region. For the dayside, the electron density can be determined from measured profiles of O$^+$ multiple resonant scattering at 83.4 nm.

5. RADIO TECHNIQUES FOR SPACE SENSING OF THE IONOSPHERE

While not currently planned as operational data sources, GPS occultation and CIT may in the future contribute valuable long-term ionospheric data for use in an assimilating model. Determination of electron density profile using

Each of the ionospheric data types shown in Figure 2 have their own inherent strenghts and limitations. Limb techniques like GPS occultation and ultraviolet limb scans provide high vertical resolution at the expense of horizontal resolution. Vertical sounding techniques like CIT, ground-based dual-frequency GPS TEC measurement, ionosondes and nadir ultraviolet provide good horizontal resolution at the expense of vertical resolution. Together, these techniques can provide complementary sampling. With a sufficient data volume the implementation of an assimilating ionospheric model should be feasible.

6. DEVELOPMENT OF AN ASSIMILATING IONOSPHERIC MODEL

Combining the need for improved ionospheric model products with the realization that a significant global database of ionospheric measurements would soon be routinely available, the DoD has sponsored the development of a new generation of ionospheric model based on real-time and near real-time data assimilation. This effort, funded as a Multidisciplinary University Research Initiative (MURI), and managed jointly by ONR and the Air Force Office of Scientific Research (AFOSR), is directed towards the development of a state of the art ionospheric physics model that will employ data assimilation for ionospheric specification and forecast. The five-year development effort began last year and is conducted by two consortia of universities headed by Robert Schunk at Utah State University and Chunming Wang at the University of Southern California. By the time of the completion of the model, most or all of the ionospheric data types discussed above should be tested and be operationally available to the 55th SWxS.

The enormous amount of successful research and development in the past 50 years by the meteorological assimilators has led to numerous assimilation, filtering and variational schemes which should provide valuable insights and lessons for ionospheric assimilation modelers. In general, the rule for weather prediction has been that more data helps, but data quality - or at least precise knowledge of data errors - is very important to improved forecast skill. An additional lesson for ionospheric modelers is that often it has been more advantageous to assimilate data in its raw form rather than to assimilate secondary products derived from data.

Not all aspects of meteorological assimilation are readily adaptable to ionospheric modeling. There are a number of fundamental differences in the physics of the two atmospheric regions. For example the ionosphere is more deterministic, undergoes much larger diurnal variations and is much more strongly coupled to the sun and geomagnetic forcing than the lower atmosphere. Ionospheric data as-

simulation will focus more on determining the forcing parameters that drive the global electron density.

7. REAL-TIME MONITORING OF IONOSPHERIC WEATHER FROM GEOSTATIONARY ORBIT

Another lesson to take from the weather community is the value of long-dwell observations available from a geostationary imager like GOES. The ionospheric analog for GOES would be an ultraviolet imager measuring ionospheric radiative recombination radiation at night and neutral and ion dayglow limb radiances during the day. On the dayside disk of the Earth a geostationary ultraviolet imager could provide images of O/N_2 airglow depletions [Strickland et al. 1999] from thermospheric heating during geomagnetic storms. From geostationary orbit it should be possible to provide real-time maps of nighttime ionospheric TEC with spatial resolutions of about 10 km^2 in a time interval of one to two minutes. By viewing at more than one wavelength (eg. O 135.6 nm, 130.4 nm and 630 nm), it may be possible to make maps of the peak height of the ionosphere [McCoy and Anderson, 1984]. High resolution maps of TEC could be used to improve the performance of a number of satellite systems, but the highest potential benefit would derive from a capability to provide high resolution real-time maps of the location and dynamics of ionospheric irregularities.

Figure 5 is an ultraviolet image of the Earth taken from the Moon on Apollo 16 [Carruthers and Page, 1976]. Obvious features in this image are the ionospheric equatorial tropical arcs in the right side of the image and the northern and southern auroral ovals. While this image has low spatial resolution, it is suggestive of what would be possible closer in at a GEO orbit. Figure 6 is a ground-based all-sky image over Arecibo, Puerto Rico [Kelley et al., 2000] of the atomic oxygen 630.0 nm night airglow produced by charge exchange and radiative recombination. The dark feature extending vertically across the 1000 km x 1000 km image is due to depletion in the electron density column. These features, often measured at low magnetic latitudes, reveal the morphology and dynamics of ionospheric irregularities. The small-scale irregularities within these plumes can produce scintillation and spread-F.

A geostationary ultraviolet imager could produce similar real-time images viewing from above and could track the development and movement of these features in a manner analogous to current geostationary monitoring of the development of large fronts and cyclones. The major differences are that while these irregularity features have cyclone-size spatial scales, they evolve on time scales more typical of tornadoes. With real-time tracking of irregularity regions, advance warning of impending communication or navigation outages could be provided. An additional benefit of a geostationary imager is that simultaneous or near simultaneous (depending on the size of the field of view) view of opposite magnetic conjugate regions could be observed - especially the equatorward edges of opposite auroral ovals.

8. SUMMARY

The prospects for significant improvement in the ability to specify and forecast the state of the global ionosphere are excellent. The ionospheric and space weather community is in many ways at the same point the meteorological community found itself 40 years ago when weather satellites began providing a global picture to augment regional information. By following the past 50 years of meteorological weather forecast development there are important lessons to be learned and a number of pitfalls to be avoided. Within the next five years there should be a significantly new and large volume of real-time and near real-time ionospheric data available. A new generation of ionospheric model, based on data assimilation is under development and where applicable, adaptations from the meteorological assimilation are being included. There are initiatives in the planning stage to place an ultraviolet imager at geostationary orbit to provide real-time, long-dwell observations of electron morphology and track geomagnetic storms and ionospheric irregularities as ionospheric

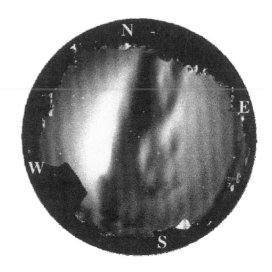

Figure 6. 630.0 nm All-Sky Image Over Puerto Rico Showing Ionospheric TEC Variations and Large Irregularity Region Identifiable as an Airglow and Electron Density Depletion (From Kelley et al., 2000)

weather systems in much the same way weather fronts and cyclones are tracked today. The ionospheric and space weather community may be able to achieve significantly improved specification and forecast skills in next five years by adapting lessons learned by the meteorologists.

REFERENCES

Anthes, R., W. Schreiner, S. M. Exner, D. Hunt, Y. Kuo, S. Sokolovskiy, R. Ware and X. Zou, GPS Sounding of the Atmosphere from Low Earth Orbit: Preliminary Results and Potential Impact on Numerical Weather Prediction, Augmenting the GPS Infrastructure for Earth, *GPS for the Geosciences,* National Academy Press, pp. 114-124, 1997.

Bernhardt, P. A., R. P. McCoy, K. F. Dymond, J. M. Picone, R. R. Meier, F. Kamalabadi, D. M. Cotton, S. Chakrabarti, T. A. Cook, J. S. Vickers, A. W. Stephan, L. Kersley, S. E. Pryse, I. K. Walker, C. N. Mitchell, P. R. Straus, Helen Na, C. Biswas, G. S. Bust, G. R. Kronschnabl, and T. D. Raymund, Two-Dimensional Mapping of the Plasma Density in the Upper Atmosphere With Computerized Ionospheric Tomography (CIT), *Physics of Plasmas, 5,* pp. 2010-2021, 1998.

Carruthers, G. R. and T. Page, Apollo 16 Far Ultraviolet Imagery of the Polar Auroras, Tropical Airglow Belts, and General Airglow, *J. Geophys. Res., 81,* pp. 483-502, 1976.

Dymond, K. F., S. A. Budzien, K. D. Wolfram, C. B. Fortna, and R. P. McCoy, The High Resolution Ionospheric and Thermospheric Spectrograph (HITS) on the Advanced Research and Global Observation Satellite (ARGOS): Quick Look Results, *Ultraviolet Atmospheric and Space Remote Sensing: Methods and Instrumentation II, SPIE, 3818,* pp. 137-148, 1999.

Hogan, T. and T. Rosmond, The Description of the Navy

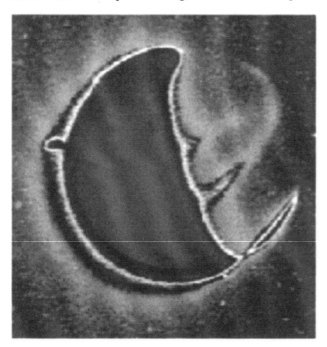

Figure 5. Ultraviolet (130.4, 135.6 nm) Image of the Earth taken from the Moon on the Apollo 16 Mission.

Operational Global Atmospheric Prediction System's Spectral Forecast Model, *Mon. Wea. Rev.*, *119*, pp. 1186-1815, 1991.

Kelley, M. C. F. Garcia. J. Makela, E. Mak, C, Sia and D. Alcocer. Highly structured tropical airglow and TEC signatures during strong geomagnetic activity, *Geophys. Res. Lett*, *27*, pp. 465-468, 2000.

McCoy, R. P., and D. E. Anderson, Jr., Ultraviolet remote sensing of the F_2 ionosphere, *Effect of the Ionosphere on C^3I Systems; Ionospheric Effects Symposium*, pp. 295-302, 1984.

McCoy, R. P., K. F. Dymond, G. G. Fritz, S. E. Thonnard, R. R. Meier, and P. R. Regeon, Special Sensor Ultraviolet Limb Imager: An Ionospheric and Neutral Density Profiles for the Defense Meteorological Satellite Program Satellites, *Optical Eng.*, *33*, pp. 423-425, 1994.

Paxton, L.J., C.-I. Meng, G.H. Fountain, B.S. Ogorzalek, E.H. Darlington, J. Goldsten, K. Peacock, SSUSI: Horizon-to-Horizon and Limb-Viewing Spectrographic Imager For Remote Sensing Of Environmental Parameters, *Ultraviolet Technology IV, SPIE, 1764*, pp. 161-176, 1992.

Rich, F., Users Guide for the Topside Ionospheric Plasma Monitor (SSIES, SSIES-2 and SSIES-3) on Spacecraft of the Defense Meteorological Satellite Program (DMSP), *PL Technical Report PL-TR-94-2187*, Vol. 1: Technical Description, 1994.

Rocken, C., Y.-H. Kuo, W. S. Schreiner, D. Hunt, S. Sokolovskiy, and C. McCormick, COSMIC System Description, Special Issue Of *Terrestrial, Atmospheric and Oceanic Science, 11*(1), pp. 21-52, 2000.

Strickland, D. J., R. J. Cox, R.R. Meier and D.P. Drob, Global O/N_2 Derived From DE 1 FUV Dayglow Data: Technique and Examples From Two Storm Periods, *Journal Geophys. Res.*, *104*, No. A3, pp. 4251-4266, 1999.

Wolfram, K. D., K. F. Dymond, S. A. Budzien, C. B. Fortna, R. P. McCoy, D. D. Cleary, and E. B. Bucsela, The Ionospheric Spectroscopy and Atmospheric Chemistry (ISAAC) Experiment on the Advanced Research and Global Observation Satellite (ARGOS): Quick Look Results, *Ultraviolet Atmospheric and Space Remote Sensing: Methods and Instrumentation II, SPIE, 3818*, pp. 149-159, 1999.

What We Must Know About Solar Particle Events to Reduce the Risk to Astronauts

Ron Turner

ANSER

Solar particle events (SPEs) pose a health risk to astronauts on deep space missions and in high-inclination, low Earth orbit (LEO), as with the International Space Station. Risk mitigation strategies to minimize this threat will require a combination of in situ radiation monitoring with forecasts and observations of coronal mass ejections (CMEs) and the state of the heliosphere. Since only a small fraction of CMEs are sufficiently fast to produce SPEs, a very high confidence in CME prediction will be needed to reduce false alarms to an acceptable level. However, an achievable predictive capability for "no CME" could be very helpful for mission scheduling. Combining published data on dose equivalent per particle at energies from 10 MeV to several GeV with fitted SPE fluence data from cycle 22, one can see that the most important energy range for astronaut safety is from 20 to about 200 MeV. To avoid exposing astronauts to an excessive dose rate, it will be important to accurately characterize the ambient solar wind and to forecast or detect shock-enhanced flux. This is demonstrated by comparing the fluence of the October 1989 event with and without the prominent shock-enhanced peak.

INTRODUCTION

This report examines the threat of SPEs and various ways the space physics community can contribute to efforts to mitigate this risk.

NASA has a moral and legal responsibility to minimize the threat of radiation exposure to its astronauts. NASA meets this responsibility by adhering to a principle called ALARA—"as low as reasonably achievable"—which ensures that radiation limits are treated as upper limits...not to be exceeded or even approached. Radiation limits expressed in terms of dose equivalent are established for blood-forming organs (BFOs), skin, and eye as upper limits to lifetime, annual, and 30-day exposures. The LEO limits have not been adopted for exploration missions, but they are used as a metric for measuring the impact of alter-native architectures. SPEs are important because under nominal shielding or while on extravehicular activity, individual events can exceed the LEO exposure limits, and a significant fraction of the exposure can occur over a period of a few hours.

Limitations on our understanding of SPEs and an inability to predict SPEs or forecast the evolution of ongoing SPEs have an operational impact on human spaceflight. Achievable advances in these areas will support design of future exploration missions and may improve the flexibility of flight rules for the International Space Station.

RADIATION EFFECTS AND MEASURES

High-energy solar and galactic cosmic ray particles affect the body in a qualitatively different way than the lower-energy radiation that most terrestrial radiation-effects studies are based on. The issue is important because radiation effects must be understood to effectively assess the risks of exposure to astronauts.

Ionizing radiation is any form of energy (particles or

Space Weather
Geophysical Monograph 125
Copyright 2001 by the American Geophysical Union

electromagnetic) that can displace target atomic electrons from their orbit, breaking chemical bonds or contributing to changes in chemical properties such as the material's electron affinity. High-energy and high–nuclear charge particle radiation can displace or fragment the nuclei of target atoms, producing recoil or spallation products leading to a cascading effect of lower-level ionization up to several tens of μm around the 1- to 5-nm core of the primary particle's track. A typical cell dimension is approximately 10 μm.

The cumulative effect of exposure to ionizing radiation is a function of the total dose, the location and distribution of the dose, the rate of accumulation of the dose, and the types of ionizing radiation that produce the dose. There are two broad categories of effects: prompt and delayed.

The prompt, or acute, effects range from headaches, dizziness, or nausea to severe illness or death (Wilson et al., 1993). Acute effects can have a serious impact on astronauts' ability to complete the mission. Measures must be provided to minimize or eliminate the possible occurrence of these effects.

Delayed effects may be nonstochastic (severity depends on dose) or stochastic (probability of occurrence depends on dose). Examples of nonstochastic delayed effects include cataracts and nonmalignant skin damage. Examples of stochastic delayed effects include induced cancer and genetic damage. While late effects would not have an impact on the mission, it has long been the responsibility of the U.S. space program to keep the cumulative risk to life as low as reasonably achievable. Precise risk-benefit assessments are made in the context of overall mission risk.

The current measure of merit for radiation impact on astronauts is the dose equivalent, which is related to the total absorbed dose by a quality factor that takes into account the relative cancer risk of primary and secondary particles. It is an attempt to put different biological effects of different types of radiation on a common scale. Dose equivalent for an astronaut exposed to SPEs depends not only on the external environment, but also on the location within the body, due to its self-shielding. It is therefore calculated for several locations, such as the BFOs, skin, eye, breast, and other organs and tissues.

For further discussion of the radiation hazards, see the National Research Council Space Studies Board report *Radiation Hazards to Crews of Interplanetary Missions: Biological Issues and Research Strategies* (1997), the NASA *Strategic Program Plan for Space Radiation Health Research*, 1998, and *NCRP 98: Guidance on Radiation Received in Space Activities*, 1989.

RADIATION EXPOSURE FROM SPES

A significant effort has been expended to determine how severe an SPE can be to astronauts under various shielding configurations (see Wilson et al., 1997a, and Wilson et al., 1991). Most of these studies focus on large events (1956, 1972, 1989), multiples of large events (two

times the flux and fluence of Oct 1989, for example), or worst-case composites of large events (the fluence of the 1956 event with a spectrum like the 1972 event, for example). Dose rates have been substantiated by measurements on the U.S. Space Shuttle and the Russian Mir space station during the limited periods that instruments have been exposed to SPEs in polar regions outside the bulk of the protection offered by the Earth's geomagnetic field (for example, see Badhwar and Atwell, 1999).

Analyses of large events have clearly demonstrated that SPEs have the potential to be hazardous to astronauts both in the short term and in the long term. Dose and dose rates can be severe to a lightly shielded astronaut with no access to shelter. While the dose would not be directly fatal, nausea causing vomiting in a spacesuit could lead to suffocation. It is not possible today to prescreen an astronaut to determine tolerance to radiation, particularly if the radiation effect is compounded by an individual's reaction to weightlessness. As for the long term, it is important to remember that the impact of radiation exposure stays with an astronaut for the rest of his or her life. It is possible to accrue significant exposure under modest shelter. Even if this does not lead to a medically adverse condition, it may be enough to cause an individual to exceed regulated limits, thus reducing or ending his or her career as an astronaut.

While SPEs have all the ions present in the solar corona the balance of this paper will consider the dose contribution from protons only (or a discussion of abundances; see Reames, 1999) Unlike their counterparts in galactic cosmic ray exposure, which are much more penetrating, elements heavier than helium in SPEs are believed to account for less than three percent of the total dose equivalent (see, for example, Kim et al., 1999). Helium can account for a significant fraction of the dose equivalent, typically on the order of ten to fifteen percent but may be of the same order as the proton dose equivalent in skin exposure on EVA (Kim et al., 1999).

Wilson et al., 1997, identified 22 events from solar cycles 19–22 (roughly from 1956 to 1996) that would have exceeded some exposure limits. Of these, they identified the August 1972 event as the most dangerous, in large part because most of the fluence was delivered over a very short period (hours), thus increasing dose rate effects and limiting the time to respond.

Since SPEs can be hazardous, it is important to be able to answer the question "When is an SPE large enough to be significant to astronauts in free space?" The radiation exposure an astronaut experiences during an SPE will depend on both the total size of the event (as measured by the total number of high-energy particles, the event "fluence") as well as the spectrum of the event (how the number of particles varies with energy). A study of 27 events over the most recent period of solar activity (cycle 22, which lasted from 1986 to 1996) (Cleghorn and Badhwar, 1999) suggests that the spectrum of protons can be reasonably represented by a power law in kinetic energy, E,

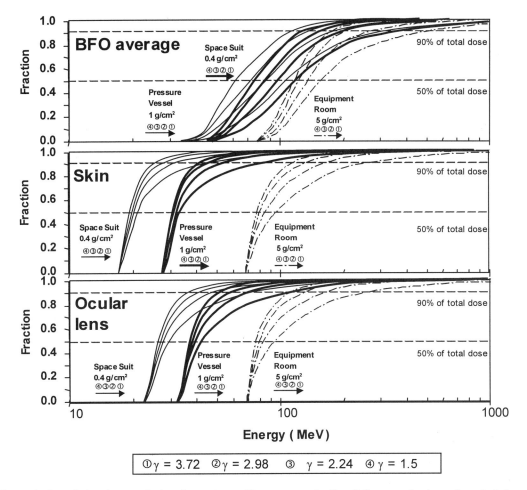

Figure 1. Cumulative dose equivalent for protons with energy greater than E for several values of spectral slope.

$$F(E) = k (E/30)^{-\gamma} \qquad (1)$$

where γ varies from 1 to 4. The corresponding differential fluence model has the form:

$$dF(E)/dE = (\gamma k /30) (E/30)^{-\gamma-1} \qquad (2)$$

Wilson et al., 1997, have used a simulation of proton transport through shielding and through a computerized anatomical man to estimate the contribution to dose equivalent per incident proton per cm^2, h(E), for proton energies from 10 MeV to several GeV:

$$h(E) = (dH/dE) / (dF/dE) \qquad (3)$$

Curves were derived for skin, ocular lens, and several locations representing the BFOs, including one BFO-average response curve. They were generated for three shielding configurations: a space suit (0.4 g/cm^2 alumi-

num), a pressure vessel (1.0 g/cm^2 aluminum), and an equipment room (5.0 g/cm^2 aluminum).

Since dF(E)/dE, equation 2, gives the number of particles with energy near E, it can be combined with h(E), equation 3, to get the differential contribution to the dose equivalent for various values of γ. The differential contribution can then be integrated to get the cumulative contribution to dose equivalent for all protons:

$$H = \int h(E) (dF/dE) dE \qquad (4)$$

Figure 1 shows what proton energies contribute most significantly to the total dose equivalent. The lower cutoff in each of the curves is determined by the minimum energy to penetrate the shielding and reach to the depth at which the dose equivalent is measured. Typical cutoff energies vary from just below 20 MeV for a space suit to more than 60 MeV under 5 g/cm^2 shielding. A peak in the differential dose equivalent occurs due to the increasing response function modulated by the decreasing differential proton fluence. Where the fluence becomes of diminishing concern is subjective. However, it is reasonable to conclude that after 90 percent of the cumulative dose, higher-energy protons become less significant. For ocular lens and skin exposure under light shielding, solar protons with energy significantly less than 100 MeV cause most of the dose equivalent. For modest shielding (5 g/cm^2), most of the dose equivalent is due to incident protons with initial energy between 60 and 200 MeV. Only rarely do protons with energy greater than a few hundred MeV contribute substantially to the total dose equivalent under modest shielding.

SPE OPERATIONAL REQUIREMENTS

There are two classes of SPE models of interest to the operational community: climatological and near-term models. Climatological models provide the long-term view of what to expect over time in terms of number of events of a given size and total mission fluence. Near-term forecast tools provide estimates of the probability of an event within the next few hours to days and the expected flux, fluence, and duration of predicted events. None of the tools available today are adequate to support human spaceflight on missions to Mars or a return to the moon.

The most popular climatological model available, Feynman et al., 1993, produces mission-integrated fluence estimates, but not bounds on the probability of one or more large events or on the probability that one large event will be followed by successive large events. Efforts to adapt the model to produce such products will be inherently limited by the brief period over which observations have been made: four or fewer solar cycles—each with its own character in numbers and intensities of events.

There is no capability to predict SPEs one to a few days in advance. This is a significant deficiency, severely limiting risk management alternatives for human spaceflight. If there is no progress in this area, a crew on a spacecraft or a team of astronauts at a remote site on the Moon will have to react to an event after it begins, limiting the time for evasive actions, if warranted, to a few hours.

Even after an SPE has been observed to be under way, forecasts of peak flux are only good to within an order of magnitude or worse. The best models are phenomenological, based on proxies to the SPE (Balch, 1999, and Smart and Shea, 1989). One of the major weaknesses of each of these models is an inability to predict shock enhancements to the proton flux—potentially the most dangerous period to a human in space.

In 1996 an interagency workshop comprising representatives from the research and operational communities addressed the issue of operational requirements for SPE forecasting (Turner, 1996). The workshop concluded that the operational community would like a 10- to 12-hour forecast prior to a likely event, a 6- to 8-hour forecast of the magnitude and spectral slope of the event after onset, and a 3- to 4-hour rolling forecast of the event as it progressed.

To be credible to the flight director, the information would have to have accuracy better than 90 percent and a false alarm rate less than 10 percent.

ACHIEVABLE ADVANCES

Predicting large SPEs days or even hours in advance will be a formidable challenge. First, the forecaster must predict that a CME is about to lift off from the Sun. Then the properties of that CME (size, speed, location) must be characterized, in advance. Then the forecaster must predict the ability of the shock associated with the CME to accelerate trapped particles to high energy. Finally, the escape and transport of these high-energy particles away from the shock and through the ambient solar wind must be forecast. We cannot do step one yet, and step two must be done to high accuracy because fewer than 5 percent of all CMEs have a speed sufficient to produce significant numbers of high-energy particles. There are too few scientists today seriously addressing the physics of particle acceleration and transport. Notable work is underway as published by Reames, 1999, Lario et al., 1998, and Zank, 2000.

In spite of these challenges, there is reason for hope about near-term opportunities to demonstrate progress if the community focuses on achievable goals. The NASA space science missions SOHO (with the European Space Agency), ACE, TRACE, and others are producing valuable insight into the workings of the Sun. A large scientific community is interested in forecasting and monitoring CMEs. Several significant missions are on the horizon, including especially STEREO, but possibly other relevant solar missions to follow within the "Living With a Star" initiative (see http://lws.gsfc.nasa.gov). A new emphasis on applications should naturally draw more scientists to the SPE problem.

It should be possible, by applying our understanding of the fundamental processes of SPE generation, to better forecast the evolution of an SPE after onset. Also, techniques being tested today to predict the arrival of geoeffective shocks should be applicable to the problem of arrival and magnitude of the shock-enhanced peak during an SPE (see for example, Dryer, 1994). The importance of this problem is illustrated by an SPE that occurred in October 1989. Figure 2 shows the flux and cumulative fluence of that event as measured by the GOES spacecraft. The shock-enhanced peak accounts for over 60 percent of the total fluence. These enhanced forecasts could support better planning for the period of greatest risk, since most of the fluence of an SPE is usually limited to a fairly short time (a few hours).

It may also be possible within a few years to reliably predict periods of no SPE threat. If the challenge of predicting an event in advance proves too formidable, then the community should recognize that mission planners would also benefit from knowing when an SPE is not going to occur. During the active years 1989 to 1991, there were

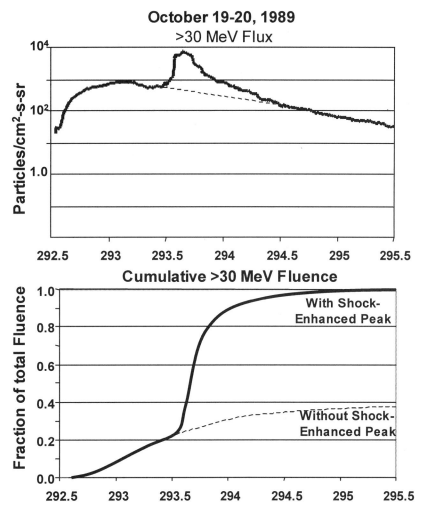

Figure 2. Flux and cumulative fluence of the October 19-20, 1989, particle event, as measured by the GOES spacecraft.

approximately 970 days of no ongoing SPE (as defined by NOAA criteria) and 120 days when there was an SPE in progress. If during this period persistence was used as the forecast, and any day an SPE was not under way the forecast was for three more days of no event, the forecaster would have been right more than 90 percent of the time. However, the forecast would also have, 100 percent of the time, given a false sense of security prior to each of 30 or so events that did occur. So a success criteria that may be achievable within the next few years could be to more than 95 percent of the time, identify conditions that would not lead to an SPE over a period of days; to ensure that fewer than five percent of the SPEs occur while "all clear" is forecast; and all the while to work to reduce the period of uncertainty about the prospects of an event (when one cannot issue a forecast either way).

CONCLUSION

Enhanced confidence in SPE forecast capability could lead to improved risk management strategies. The main benefit would be in planning for mission activities that may place the astronauts at increased risk of exposure, such as EVA. This could have a significant impact on procedures for lunar surface exploration, which may take the astronauts away from base for days at a time. It could also affect scheduling significant events on a Mars mission.

The lack of reliable SPE forecasts has fostered a "cope and avoid" strategy. Clearly today and in the near future the most essential elements of a risk management strategy are shielding and exposure measurements. Additional elements compete on a cost-benefit or value-added basis. Mission designers should be encouraged to explore how they would design systems differently if they had reliable all-clear forecasts; reliable 1-hour, 8-hour, or 24-hour forecasts of flux once an event is under way; or 1, 8, or 24 hours' warning of a pending event.

Better understanding the fundamental physics of the particle acceleration mechanisms will help mission designers by adding confidence to projected worst-case flux, fluence, or particle energy spectrum. This will significantly aid in the design criteria for spacecraft or surface vehicle nominal shielding and provisions for storm shelters. It will help mission planners develop prudent and effective flight rules, contingency plans, and emergency procedures.

Throughout the exploration community, there is a qualitative awareness of the threat posed by SPEs. This awareness ensures that some consideration is given to shielding the astronauts from the threat. There is no quantitative agreement about the risk of SPE radiation. This leads to an approach with the mission planners that may significantly over- or under-design some elements.

We must continue to explore the fundamental physics of SPEs and work to evolve more reliable forecasts. As our understanding increases and our forecasts become mature, the operational community will begin to move beyond a "cope and avoid" philosophy and progress to an "anticipate and react" philosophy of risk management.

Acknowledgments. The work described in this paper was supported by NASA research grant NAG5-3888.

REFERENCES

NCRP Report Number 98, *Guidance on Radiation Received in Space Activities,* National Council on Radiation Protection and Measurements, Washington, D.C., 1989.

Radiation Hazards to Crews of Interplanetary Missions—Biological Issues and Research Strategies, National Academy Press, Washington, D.C., 1996.

Strategic Program Plan for Space Radiation Health Research, NASA, Office of Life and Microgravity Sciences, October 1998.

Badhwar, G., and W. Atwell, Detailed comparison of observed dose-time profile of October 19–20, 1989 SPE on Mir with model calculations, *Radiation Measurements, 30, 3,* pp. 223–230, June 1999.

Balch, C., SEC proton prediction model: verification and analysis, *Radiation Measurements, 30, 3,* pp. 231–250, June 1999.

Cleghorn, T., and G. Badhwar, Comparison of the SPE model with proton and heavy ion data, *Radiation Measurements, 30, 3,* pp. 251–260, June 1999.

Dryer, M., Interplanetary studies: propagation of disturbances between the Sun and the magnetosphere, *Space Science Reviews 67, 30,* pp. 363–419, 1994.

Feynman, J., G. Spitale, and J. Wang, Interplanetary proton fluence model: JPL 1991, *J. Geophys. Res,* 98, A8, pp. 13,281–13,294, August 1993.

Heckman, G., J. Kunches, and J. Allen, Prediction and evaluation of solar particle events based on precursor information, *Adv. Space Res.* 12, 2–3, pp. (2)313–(2)320, 1992.

Kim, M.Y., J. Wilson, F. Cucinotta, L. Simonsen, W. Atwell, F. Badavi, J. Miller, Contribution of High Charge and Energy (HZE) ions during Solar-Particle Event of September 29, 1989.

Lario D., B. Sanahuja, A.M. Heras, Energetic particle events: efficiency of interplanetary shocks as 50 keV < E < 100 MeV proton accelerators, *Astrophysical Journal, 509,* pp. 415–434, 1998.

Reames, D., Particle acceleration at the sun and in the heliosphere, *Space Science Reviews, 90,* pp. 413–491, 1999.

Simonsen, L., J. Nealy, H. Sauer, and L. Townsend, Solar flare protection for manned lunar missions: analysis of the October 1989 proton flare event, *SAE Technical Paper Series, No. 911351,* 1991.

Smart, D., and M. Shea, PPS87—a new event-oriented solar prediction model, *Adv. Space Res., 9, No. 10,* pp. 281–284, 1989.

Turner, R., editor, Foundations of solar particle event risk management strategies: findings of the risk management workshop for solar particle events, *ANSER Technical Report,* Arlington, Virginia, July 1996.

Wilson, J., L. Townsend, W. Schimmerling, G. Khandelwal, F. Khan, J. Nealy, F. Cucinotta, L. Simonsen, J. Shinn, and J. Norbury, Transport methods and interactions for space radiations, *NASA Reference Publication 1257,* 1991.

Wilson, J., W. Schimmerling, F. Cucinotta, and J. Wood, Effects of radiobiological uncertainty on vehicle and habitat design for missions to the Moon and Mars, *NASA Technical Paper 3312,* 1993.

Wilson, J., J. Shinn, L. Simonsen, F. Cucinotta, R. Dubey, W. Jordan, T. Jones, C. Chang, and M. Kim, Exposures to solar particle events in deep space missions, *NASA Technical Paper 3668,* October 1997.

Wilson, J., J. Miller, A. Konradi, and F.A. Cucinotta, editors, Shielding Strategies for Human Space Exploration, *NASA Conference Publication 3360,* pp. 273–281, December 1997.

Zank, G.P., W. Rice, and C. Wu, Particle acceleration and coronal mass ejection-driven shocks: a theoretical model, submitted to *Journal of Geophysical Research (Space).*

Ron Turner, ANSER, Suite 800, 1215 Jefferson Davis Hwy, Arlington, VA 22202

Living With a Star

George L. Withbroe

Office of Space Science, NASA Headquarters, Washington DC

The goal of the Living With a Star (LWS) program is to develop scientific knowledge and understanding of those aspects of the connected Sun-Earth system that directly affect life and society. The NASA LWS initiative includes four major elements: (1) a Space Weather Research Network of solar-terrestrial spacecraft, (2) a theory, modeling, and data analysis program, (3) Space Environment Testbeds for flight testing radiation-hardened and radiation-tolerant systems in the Earth's space environment, and (4) development of partnerships with national and international agencies and industry. This paper provides a brief overview of the LWS program.

1. INTRODUCTION

The Living With a Star (LWS) program seeks to develop the scientific understanding necessary to effectively address those aspects of the connected Sun-Earth system that directly affect life and society. It is the study of the physics of solar variability and its effects.

Why do we care? We have increased dependence on space-based systems, a permanent presence of humans in Earth orbit, and eventually human voyages beyond Earth. Solar variability can affect space systems, human space flight, electric power grids, GPS signals, high frequency radio communications, long range radar, microelectronics and humans in high altitude spacecraft, and terrestrial climate. Prudence demands that we fully understand the space environment affecting these systems. In addition, given the massive economic impact of even small changes in climate, we should fully understand both natural and anthropogenic causes of global climate change. Solar variability is a primary natural driver and may be responsible for as much as 30% of the global warming in the past century. The scientific community has demonstrated its strong interest in the scientific challenges implicit in the LWS goal. The LWS program serves multiple national interests such as those discussed in interagency reports by the National Space Weather Program (1995, 2000) and the National Security Space Architect (1999), and the National Research Council report *Solar Influences on Global Change* (Lean *et al.*, 1995).

The strategy for NASA's program for understanding the connected Sun-Earth system is organized around three fundamental quests or questions:

- Why does the Sun vary?
- How do the planets respond to solar variability?
- How does solar variability affect life and society?

These three interlinked quests can be restated as objectives for LWS: To determine

- How a star works.
- How it affects humanity's home.
- How to live with a star.

2. NASA SOLAR-TERRESTRIAL RESEARCH PROGRAM

Since the early 1960's NASA has supported the scientific study of the phenomena and fundamental physical processes involved in solar-terrestrial physics. In the late 1980's this research program was consolidated under the NASA Space Physics program. In the mid-1990's the name of the program was changed to the Sun-Earth Connection program to convey better the primary objective, the study of the Sun-Earth connected system, a system driven by a variable star, our Sun. During the 1990's NASA and

Space Weather
Geophysical Monograph 125

its international partners launched a powerful set of spacecraft under the International Solar Terrestrial Physics (ISTP) program. These missions, and complementary smaller international and Explorer-class missions, make up the extensive fleet now studying solar-terrestrial events and physics during the current solar maximum. For information on ISTP and other solar-terrestrial missions see the NASA websites given in the references.

Several missions in this research fleet also provide real-time or near real-time data for support of space weather activities. Data from the Advanced Composition Explorer (ACE), Solar and Heliospheric Observatory (SOHO), Yohkoh, and soon Imager for Magnetopause-to-Aurora Global Exploration (IMAGE), are routinely used for space weather purposes by operational agencies (*e.g.,* NOAA, USAF), including the only real-time solar wind monitor providing 30-60 min space weather warnings (ACE). The real-time capability of ACE stems from a NOAA/NASA partnership established when ACE was under development. In order to provide a backup/replacement for ACE, the NASA Sun-Earth Connection program is funding development of a real time solar wind instrument that was added to the Earth Science mission Triana currently scheduled for launch in 2002. The ACE and Triana instruments also will provide two-point measurements of the solar wind at the L1 position approximately 1.5 million km from Earth between the Sun and Earth. EUV and coronagraphic images from SOHO are proving to be very useful tools for detecting coronal mass ejections (CME's) headed toward Earth. Earth-directed CME's appear as "halo" events in the coronagraphic images. The Solar Terrestrial Relations Observatory (STEREO) currently under development, and planned for launch in 2004, will provide the next step via stereo imaging of the Sun and a capability to track solar mass ejections from Sun to Earth using optical and radio instruments.

Many of the new tools for space weather forecasting stem from NASA-sponsored solar-terrestrial research, from imaging coronal mass ejections, discovering x-ray signatures of active regions likely to produce CME's, to detecting solar active regions on the far side of the Sun via helioseismology. These advances and the steady flow of new knowledge from the ISTP and complementary missions, coupled with a renewed interest in the effects and potential importance of solar influences on terrestrial climate, led to the development of the Living With a Star (LWS) Initiative.

3. LIVING WITH A STAR INITIATIVE

3.1 Introduction

LWS is built on the heritage of the past and current international solar-terrestrial physics research effort, the Sun-Earth Connection (SEC) Roadmaps (Burch *et al.* 1997,

Strong *et al.* 2000) developed in conjunction with the NASA Space Science strategic planning process, recent reports of the National Research Council/National Academy of Sciences (Lean *et al.* 1995, Neugebauer *et al.* 1995, Siscoe *et al.* 2000), and reports by the interagency National Space Weather Program (1995, 2000) and the National Security Space Architect (1999). The goal of LWS is to develop the scientific understanding necessary to effectively address those aspects of the connected Sun-Earth system that directly affect life and society. LWS differs from the other programs in the NASA Sun-Earth Connection program in that contribution to the LWS goal is a primary criteria in the selection of the research problems and missions to be undertaken within the program. The LWS program is not limited to space weather. Equally important are solar influences on global change.

Relevant issues involving impacts of solar variability on life and society include:

- Human radiation exposure in space flight and high altitude aircraft,
- Impacts on technology related to space systems, communications, navigation and terrestrial systems, and
- Terrestrial climate change, both short and long term.

LWS, a NASA research program, directly benefits the National Space Weather Program, aerospace industry, satellite operations, air transport industry, communications and navigation. LWS also has links to each of the NASA Enterprises. For Space Science, LWS quantifies the physics, dynamics, and behavior of the Sun-Earth system over the 11-year solar cycle and is the detailed study of the only star and space environment directly accessible by a wide range of remote sensing and *in situ* measurements. For Earth Science, LWS improves the understanding of the effects of solar variability and disturbances on terrestrial climate change. For Human Exploration and Development, LWS provides data and scientific understanding required for advanced warning of energetic particle events that affect the safety of astronauts. Finally, for Aeronautics and Space Transportation, LWS provides detailed characterization of radiation environments useful in the design of more reliable electronic components for air and space transportation systems.

3.2 Science Questions

A flowdown of science questions directed at space weather hazards for human radiation exposure, impacts on technology, and the relationship to terrestrial climate are summarized below. This summary was prepared by the NASA Sun-Earth Connection Advisory Subcommittee chaired by Andrew Christensen.

Human Radiation Exposure: The systems affected are the International Space Station, space exploration, high altitude aircraft, space utilization and colonization. The space weather hazards are:

- Solar energetic particles (SEP) events.
- Relativistic electron events (REE).

The relevant science questions are:

- What determines when SEP or REE will occur?
- What determines their spatial, temporal, and spectral development?
- What are the mitigation strategies?

Mission definition issues include:

- What are the required predictive capabilities?
- What parameters should be monitored?
- Where is the best place to monitor them?
- What models are needed?

Impact on Technology: The systems affected are satellites (Earth orbiting), spacecraft (non-Earth orbiting), human space flight (International Space Station, Space Shuttle), communications and navigation, and some ground systems (e.g. the electric power grid).

The relevant space weather hazards are:

- Variable atmospheric drag.
- Enhanced ionospheric ionization.
- Solar x-ray and solar energetic particle events (SEP).
- Relativistic electron events (REE).
- Magnetospheric particles and fields.

The relevant science questions are:

- What determines the heliospheric, magnetospheric, atmospheric, and ionospheric responses to solar variability?
- What causes onset and development of ionospheric scintillations?
- What determines when solar generated particle and radiation events, magnetospheric storms and substorms occur?
- What determines the spatial, temporal, and spectral development of all these phenomena?
- What drives these phenomena and how much warning can reliably be obtained?
- How do the mean space environmental conditions (relevant to spacecraft design) at a given spatial location vary through the solar cycle and what are the probabilities of different levels of deviations from the mean?

Mission definition issues include:

- What are the required predictive capabilities?
- What variables should be monitored?
- How and where should it be done?
- What models and theory are needed?

Terrestrial Climate: The terrestrial phenomena that are affected are climate change (past and future), surface warming, and ozone depletion and recovery. The relevant solar-terrestrial sources are:

- Solar electromagnetic radiation.
- Solar and galactic cosmic rays.

- Upper atmospheric/ionosphere boundary region.

The related science question is:

- What is the role of the sun and heliosphere in global climate change on multiple time scales (seasonal, decadal, centennial)?

Relating this science question to missions requires addressing the following issues:

- What long-term studies of sources of energy from the Sun should be undertaken to advance understanding of solar effects on climate change?
- What long-term studies are needed to understand the role of the heliosphere, magnetosphere and the upper atmosphere/ionosphere on climate?
- How should development of quantitative models proceed?
- What predictive capabilities will be needed?

These issues have been, and continue to be addressed, through meetings and workshops involving stakeholders who are interested in and/or affected by solar variability and its effects on life and society. These interactions provide the basis for a requirements flowdown for the LWS program.

3.3 LWS Elements

LWS has four elements: (1) a Space Weather Research Network, (2) a Theory, Modeling & Data Analysis Program, (3) Space Environment Testbeds, and (4) Partnerships.

Space Weather Research Network: There are two groups of missions proposed in the LWS Initiative to form a Space Weather Research Network. (a) solar dynamics element (Solar Dynamics Observatory and Solar Sentinels) that observe the Sun and track space disturbances originating there and (b) geospace dynamics elements (Radiation Belt Mappers and Ionospheric Mappers) consisting of constellations of small satellites located in key regions around the Earth to measure effects of solar variability on the geospace environment. Further in the future are missions that depend on development of solar sail propulsion, the Solar Polar Orbiter/Imager and Earth Pole Sitters. The location of the LWS missions in the Sun-Earth environment are depicted schematically in Plate 1.

The Solar Dynamics Observatory (SDO) will observe the Sun's dynamics and help us understand the nature and source of variations, from the solar interior into the solar atmosphere. The Solar Sentinels will provide a global view of the Sun and inner heliosphere and describe the origin and evolution of eruptions and flares from the Sun to Earth. The Radiation Belt Mappers will study the origin and dynamics of the radiation belts and determine the evolution of penetrating radiation during magnetic storms. The Ionospheric Mappers will gather knowledge on how the ionosphere behaves as a system that responds directly to solar photons and energetic particles, as well as solar

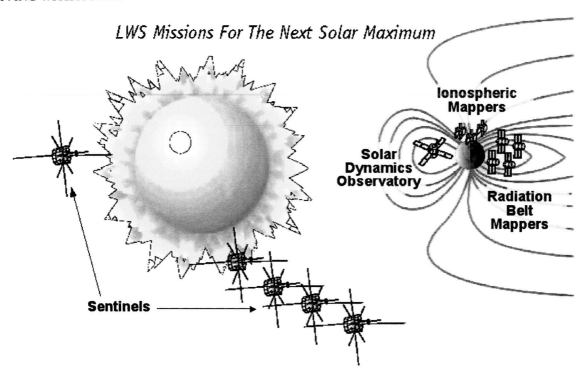

Plate 1. LWS Space Weather Research Network. The four types of missions planned for launch by the next solar maximum are illustrated schematically.

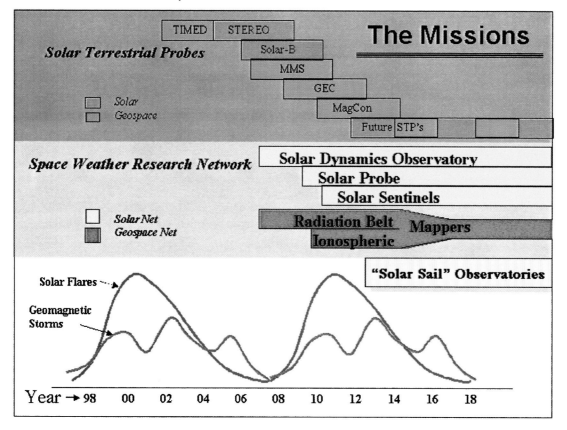

Plate 2. LWS and the Solar Cycle. The approximate times when various Solar Terrestrial Probes and Living With a Star missions are expected to be operating are illustrated schematically.

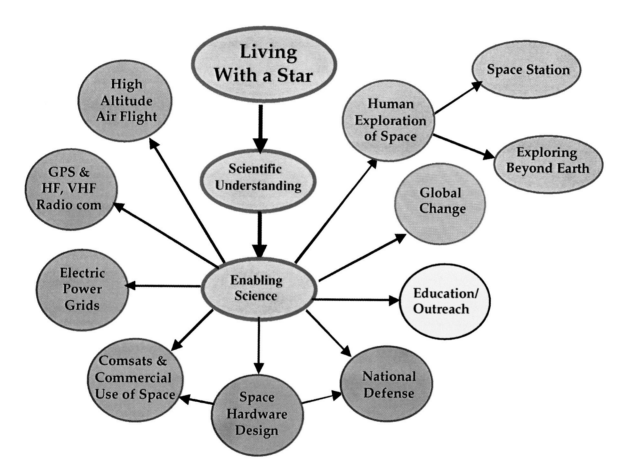

Plate 3. Sun-Earth Connections.

wind energy stored and then suddenly discharged by the magnetosphere during geomagnetic storms and substorms.

The schedule for the launch of these missions is planned so that the basic mission set (SDO, Solar Sentinels, and Magnetospheric and Ionospheric Mappers) will be operational by the next solar maximum, as illustrated in Plate 2. The upper part of the plate shows the Solar Terrestrial Probe (STP) missions currently under development or planned. The LWS and STP missions are intended to operate synergistically. The presently planned STP missions include Thermosphere, Ionosphere, Mesosphere, Energetics and Dynamics (TIMED) mission, Solar-B, STEREO, Magnetospheric Multiscale (MMS), Geospace Electrodynamic Connections (GEC), and Magnetospheric Constellation (MagCon). Additional information about these missions can be found at the NASA Space Science Mission and Solar Terrestrial Probe websites given in the references.

Theory, Modeling and Data Analysis: The LWS research and data analysis program will exploit data from present and past missions to (1) improve the scientific knowledge and understanding of those aspects of the Sun-Earth connected system relevant to space weather and terrestrial climate change, (2) develop new techniques and models for predicting solar/geospace disturbances that affect human technology (e.g. the "S marks the spot" signature of solar regions with enhanced probability of generating coronal mass ejections), and (3) improve knowledge of space environmental conditions and variations for design of cost-effective systems to minimize or eliminate impact of space environmental effects. The first solicitation for proposals for data analysis, theory, and modeling in support of LWS goals was issued by NASA Headquarters for investigations to begin in 2001. Future solicitations will are planned yearly.

An important component of this effort will be the newly initiated Community Coordinated Modeling Center (CCMC). The CCMC's purpose is to take new research-grade space environment simulations codes, modify them to run on Department of Defense supercomputers, and transition the results to operational agencies such as NOAA. Additional information about the CCMC may be found at the CCMC website given in the references.

Space Environment Testbeds: The use of environment-tolerant systems in place of environment-hard systems shifts the focus from risk avoidance to risk management. This requires a better understanding of on-orbit performance for risk assessment analyses. The goals of the Space Environment Testbeds element of LWS are to:

- Understand space environment effects on current and emerging technologies and biosystems.
- Understand application of space weather predictions for risk management of new technologies.
- Establish ground test protocols for new technologies.

- Validate models.
- Demonstrate instrumentation for LWS missions.
- Improve environmental guidance for spacecraft design and operations.

Partnerships: The LWS Initiative is developing partnerships with other federal and international agencies to augment existing capabilities and provide new systems. To this end, LWS will:

- Work with partners to define national needs.
- Prototype new observing capabilities to prove or disprove their usefulness. Useful technologies will be made available to other agencies for monitoring and predictive purposes.
- Make data from research missions available for use by other federal agencies and the scientific and applications communities.
- Work with national and international partners to develop a coherent, synergistic program in studying solar effects on space weather and climate.

3.4 Complementary Missions

LWS will be complemented by data from other missions. These include currently operational or soon to be launched space research missions of NASA and its international partners. The relevant solar missions are Yohkoh (91), SOHO (95), TRACE (98), and HESSI (01) where the number in parentheses is the launch year. The relevant geospace missions are: Geotail (92), SAMPEX (92), WIND (94), POLAR (96), FAST (96), IMAGE (00), Cluster (00), TIMED (01), and TWINS (03,04). In addition, the GOES series of NOAA operational missions will continue to be a source of data at geostationary orbit. The relevant heliospheric missions are IMP-8 (73), Ulysses (90), ACE (97), and TRIANA (02), and Solar Probe (08). As indicated earlier *(cf.* Plate 2), the Solar Terrestrial Probes (STP) are particularly relevant to LWS because they will be operational in the same time frame as the LWS missions, whereas many of the currently operating missions will be well beyond their design lifetimes. See the NASA Space Science Mission website for more information on the above missions.

4. SUMMARY

The NASA LWS Initiative is a scientific research program that will provide the scientific context and understanding necessary to characterize, understand, and predict those aspects of solar variability and its effects that affect life and society. Plate 3 depicts the communities that will benefit from the scientific understanding resulting from the LWS program. The SEC and LWS websites (see references), which are periodically updated, provide additional information about the NASA Sun-Earth Connection program and its LWS component.

Acknowledgements. The author wishes to thank the many colleagues whose dedicated efforts on behalf of science and society led to the development of the Living With a Star initiative. This includes colleagues at NASA Headquarters; the NASA Goddard Space Flight Center; other federal agencies, particularly NOAA, NSF, and USAF; and numerous members of the scientific community, especially the 1997 and 2000 Sun-Earth Connection Roadmap teams and the solar-terrestrial scientists who have contributed time, energy, and creativity to public education and outreach.

REFERENCES

Burch, J.L. et al., *The Sun-Earth Connection Roadmap*, Strategic Planning for the Years 2000-2020, http://umbra.nascom.nasa.gov/spd/secr/, 1997.

CCMC website: http://ccmc.gsfc.nasa.gov.

ISTP website: http://www-istp.gsfc.nasa.gov/.

Lean, J. et al., *Solar Influences on Global Change*, National Research Council, Washington, DC, 1995.

Living With a Star website: http://sec.gsfc.nasa.gov/lws.htm.

NASA Space Science Mission website, http://spacescience.nasa.gov/missions/index.htm.

National Security Space Architect Space Weather Architecture Study, http://www.acq.osd.mil/nssa/majoreff/swx/swx.htm, 1999.

National Space Weather Program Implementation Plan, FCM-P31-2000, Office of the Federal Coordinator for Meteorological Services and Supporting Services, also at http://www.ofcm.gov/NSWP-IP/TableofContents.htm, 2000.

National Space Weather Program Strategic Plan, FCM-P30-1995, Office of the Federal Coordinator for Meteorological Services and Supporting Research, Silver Spring, MD, also at http://www.ofcm.gov/nswp-sp/text/a-cover.htm, 1995.

Neugebauer, M. et al., *A Science Strategy for Space Physics*, National Research Council, Washington, DC, 1995.

Siscoe, G..L. et al., *Radiation and the International Space Station*, National Research Council, Washington, DC, 2000.

Strong *et al*, *The Sun-Earth Connection Roadmap, Strategic Planning for 2000-2025*, http://www.lmsal.com/sec/, 2000.

Sun-Earth Connection website: http://sec.gsfc.nasa.gov/.

Solar Terrestrial Probe website: http://sec.gsfc.nasa.gov/stp_home.htm.

George L. Withbroe, Code SA, NASA Headquarters, Washington DC 20546

Space Weather: European Space Agency Perspectives

E. J. Daly and A. Hilgers

ESA Space Environments and Effects Analysis, ESTEC, Noordwijk, The Netherlands

Spacecraft and payloads have become steadily more sophisticated and therefore more susceptible to space weather effects. ESA has long been active in applying models and tools to the problems associated with such effects on its spacecraft. In parallel, ESA and European agencies have built a highly successful solar-terrestrial physics capability. ESA is now investigating the marriage of these technological and scientific capabilities to address perceived user needs for space weather products and services. Two major ESA-sponsored studies are laying the groundwork for a possible operational European space weather service. The wide-ranging activities of ESA in the Space Weather / Space Environment domain are summarized and recent important examples of space weather concerns given.

1. INTRODUCTION

Space Weather encompasses a broad range of phenomena impacting space and terrestrial technologies. A few high-profile events have drawn attention to the effects of space weather, including hazards to spacecraft from electrostatic charging [*Baker et al*, 1996, *Fredrickson*, 1996, *Wrenn*, 1995] and to ground-based power networks from induced current surges [*Kappenman and Albertson*, 1990, *Lanzerotti*, 1979]. But these are just the tip of an iceberg of more numerous, less well publicized problems. There is a growing appreciation that as society becomes more reliant on space-based systems for services such as communications and navigation, the disruptions to these services from space weather is becoming a serious concern. In addition to these commercial space activities, scientific missions can be seriously affected because of their use of highly advanced technologies. Furthermore, in the next few years manned space flight will undergo a considerable expansion with the exploitation of the international space station. Space weather, in the form of enhancements to energetic particle radiation, is of crucial importance for manned space activities. Space radiation also penetrates the upper atmosphere where crew and electronic systems on aircraft can be affected. Finally, ground-based systems such as power distribution networks, pipelines and ground-to-ground communications can also be seriously affected.

We are near the maximum of the current cycle, cycle 23. The increased chances of solar flares, coronal mass ejections and solar energetic particle events have led to added interest in space weather both from the space community and from the general public. This interest is supported by an array of excellent solar-terrestrial science missions such as SOHO [*Huber and Wilson*, 2000]. It is important to recognize that the effects of Space Weather are present throughout the solar cycle and that some important aspects are more severe away from solar maximum. For example the electron flux levels in the Earth's outer radiation belt are generally higher during the decaying phase of the solar cycle.

European countries host a considerable repository of knowledge and world-class facilities in the fields of space environments and effects analysis and solar-terrestrial physics. Nevertheless, while co-ordinated Space Weather activities are well established in the US, Europe has yet to undertake a co-ordinated programme in this area. An important step towards an autonomous European Space Weather programme has recently been taken with the initiation of broadly based studies in the context of the ESA General Studies Programme. This followed an important ESA Workshop on Space Weather when scientific, technical and strategic issues were debated and co-ordination of national activities discussed [*ESA*, 1998].

Space Weather
Geophysical Monograph 125

Table 1. The most common Space Environmental Effects to consider in space mission development.

Environment	Effects
High-Energy Radiation:	
Cosmic Rays	Upsets in electronics; Long-term hazards to crew; Interference with sensors
Solar Energetic Particle Events	Radiation damage of various sorts; Upsets in electronics; Serious prompt hazards to crew; Massive interference with sensors;
Radiation Belts	Radiation damage of various sorts; Upsets in space electronics; Hazards to astronauts; Considerable interference with sensors; Electrostatic charging/discharge
Near-Earth Plasma Populations:	
Geomagnetic Sub-storms	Electrostatic charging and discharge
Ionospheric Effects	Communications disruption; GPS/GNSS disruption; EO active instrument (SAR, etc.) disruption
Others:	
Atmosphere	Increased drag; Attitude perturbation; GPS/GNSS positioning errors
Meteoroids	Damage, e.g. during Leonids

There is considerable interest in Europe to investigate the marriage of its technological and scientific capabilities to address perceived user needs for space weather products and services.

2. PAST TECHNOLOGY EFFORTS

For many years ESA has been active in applying models and tools to the problems associated with the effects of space environments on its spacecraft. As spacecraft and their payloads have become more sophisticated, their susceptibilities to the effects induced by the environment have increased. Some of these space-system-related effects are summarized in Table 1.

In order to develop spacecraft which operate correctly in the presence of these effects, it is necessary to use models of the environments and effects for analysis, and to undertake appropriate testing. Models are intended to address the needs of the space system developer and for efficiency and usability reasons often simplify the physics involved in the phenomena. Even when physical understanding or information is incomplete, the threat still needs to be countered with some quantitative method, albeit of limited validity. It is nevertheless a long-term objective

for this community to have models available which are both physically accurate and responsive to the users needs. A good example is in the area of radiation environments and effects where for many years developers have used the "standard" AP-8 [*Sawyer and Vette*, 1976] and AE-8 [*Vette*, 1991] models of the radiation belts. These models are known to be weak and do not represent the dynamic ("space weather") behaviour of the electron belt. Nevertheless, in the absence of anything better, they have continued to be used. Developments have recently given hope but there still remains a usability problem.

ESA's Space Environments and Effects Analysis Section has responsibility for supporting the development of ESA missions. The service it provides includes assessments of all elements in Table 1. In parallel with this support function, it is responsible for the initiation and execution of technology R&D as part of a *Space Environments and Effects Major Axis* of ESA's Technology Research Programme [*ESA*, 1999]. This R&D has led to developments of tools and models, as well as R&D for longer-term application. In doing this R&D and support work, the section is closely in touch with the user needs for space weather data for space system applications. Current R&D activities include [*ESA*, 1999]:

- The Space Environment Information System (Spenvis) [*Heynderickx*, 1998]. This is an internet/intranet based system containing a wide range of models, tools and data concerning many aspects of space environments and their effects on space systems. It is targeted at the space systems developer who needs rapid reliable access to authoritative (often standard) methods. The system also contains link to the European ECSS Space Environment Standard [*ECSS*, 2000].

- Modelling of the Earth's Radiation Belts where various high altitude and low-altitude data sets are studied to validate or improve models of the radiation belts. The activity also includes detailed comparison between a physical model, Salammbô, and spacecraft data of energetic electron belt dynamics

- Development of data-based analysis of space environments ("SEDAT") [*CLRC*, 2000] where existing spacecraft data sets are interrogated by standard and user-defined methods to derive custom "models".

- Development of engineering tools for assessment of the hazard from charging of materials inside spacecraft by energetic electrons ("internal charging") [*Sørensen et al.*, 1999]. This research also investigated the way to specify the hazard for design and the associated test methods.

- Participation with the High-Energy Physics community in a world-wide effort to produce a next generation of object-oriented tool-kit for simulations of particle interactions with matter ("Monte-Carlo codes"), Geant4 [*Apostolakis*, 2000]. This effort was initiated by CERN, the European centre for nuclear research. ESA's activity has resulted in space-specific features for Geant4 [*Truscott et al.*, 2000].

- Developments of space environment monitors and the analysis and exploitation of data from them [*Bühler*, 1998, *Desorgher et al.*, 1999, *Daly et al.*, 1999].
- Analysis of electrostatic charging behaviour of spacecraft in polar orbits and analyses of the correlations between anomalies and environmental parameters [*Andersson et al.*, 1998]. These studies also included research on tools for anomaly predictions [*Wu et al.*, 1998].
- Research on AI methods in spacecraft anomaly analysis and prediction – the SAAPS (Spacecraft Anomaly Spacecraft Anomaly Analysis and Prediction System) [*Wintoft*, 1999].

Many other important activities have been undertaken including activities related to Martian environments, micro-particle impacts and contamination [*ESA*, 1999].

3. SUPPORT FUNCTIONS

A good case history of support to ESA projects in this area is the ESA XMM X-ray astronomy mission. X-ray astronomy in space relies on the focussing of X-ray photons by low-angle scattering from shaped "shells". In most cases the "optics" consist of two sets of nested concentric shells with shapes near to sections of cones. Two grazing-incidence scatters result in focussing of the X-rays on the shell axis. ESA's XMM mission has three mirror modules of outer diameter 70 cm, each consisting of 58 nested shells which focus the X-rays onto CCD detectors some 7 m from the mirrors. XMM is in a highly eccentric orbit of apogee 114000km, perigee 7000km and inclination 39°. In this orbit it is subjected to fluxes of electrons and ions of various energies from magnetospheric and heliospheric sources.

Recently, an intensive investigation was undertaken to study potential problems to detector operation from medium-energy (100's of keV's) protons [*Nartallo et al.*, 2000]. CCD detectors are known to be radiation sensitive and much attention is given to shielding them against radiation penetrating spacecraft structural materials. However, it was found that protons of energies in the range of hundreds of keV to a few MeV could scatter at low angles through the mirror shells. These protons, because of their low energy can produce a high non-ionising dose in unshielded CCDs and are therefore a potential threat. Historical data on the interplanetary and magnetospheric low-energy proton environments were interrogated to determine the magnitude of the threat. Complex modelling of particle propagation through spacecraft systems was undertaken with the *Geant4* Monte-Carlo toolkit. The datasets were used to establish details of observing time

expected to be lost in protecting the CCDs from sporadic particle flux enhancements. Several interesting points regarding space weather can be made as a result of this analysis:

- Crucial data sets used for the analysis of "hard" mission-critical engineering problems were produced by science missions (IMP, SOHO, ACE, Equator-S, ISEE) which could never foresee such applications;
- XMM has an on-board radiation monitor, to which there was much resistance early in the project preparation. It is now a vital feature of the spacecraft;
- Spacecraft operators are keenly interested in the state of the space weather and would certainly make use of predictions of sporadic particle enhancements should they be available.

All this effort was in addition to several space environment related analyses carried out in the course of the definition and development of the XMM mission over the preceding 10 years. In such a process, early analyses of the environments of orbit options were undertaken, followed by detailed analyses related to the final orbit and the radiation doses and particle fluxes to be anticipated for electronic components and detectors. Further analyses included assessments of the electrostatic charging hazards, analysis of the potential problems from micro-meteoroids (punctures to telescope tube and hazards to fuel tanks) and detailed analysis of radiation background sources [*Dyer et al.*, 1995, *Hilgers et al.*, 1998]. XMM was launched in December 1999 and is operating well.

4. ESA'S SCIENCE ACTIVITIES

ESA's science programme is related to space weather in two ways. On the one hand its effects are an increasing concern and on the other, science missions contribute crucially to space weather research. As space science missions become more complex and demanding, the need to design tolerance to space weather effects into scientific payloads as well as spacecraft systems is apparent. Examples include sensitivity to radiation, leading to increased backgrounds and even detector damage, as well as the complete failure of key components. As mentioned above, these issues were important to the recently launched XMM-Newton mission [*Nartallo et al.*, 2000].

On the other hand, an important spin-off of scientific missions can be to show what is possible for future service-oriented ventures. The joint ESA-NASA SOHO mission is a key member of the fleet of spacecraft studying the Sun and its effects on the interplanetary environment, and providing Space Weather warnings. ESA's scientific studies related to Space Weather phenomena were further enhanced with the launch of the Cluster II satellites. It is expected that Cluster II will also contribute to space weather monitoring.

As part of the competitive process for selection of future science missions, ESA is studying future medium-sized missions. Among these are the STORMS and Solar Orbiter proposals, which may contribute to the world-wide space weather effort. STORMS is a set of 3 spacecraft in eccentric near-equatorial earth orbits. With apogee at about 8 R_E, the spacecraft pass through the radiation belts and the ring current regions. As the name suggests, the principal motivation for the mission is to study the physics of geomagnetic storms and the inner magnetosphere's responses to them. The spacecraft would carry particle and fields instruments and energetic neutral atom imagers. Solar Orbiter is designed to orbit the sun as close as 40 solar radii (0.19 AU) and to carry out detailed solar remote sensing. Its orbit would also take it to helio-latitudes of about 33°. For part of the time the orbit would be quasi-co-rotational. Spectroscopy and imaging would be performed at high spatial and temporal resolution, along with in-situ sampling of particles and fields.

5. EFFORTS TOWARDS A SPACE WEATHER PROGRAMME

Recognising that there was a growing need for space weather types of data for ESA programmes and also that there were issues related to the impact of space weather on non-space technologies which could be important for Europe, ESA took steps to analyse the subject in detail. While not the first ESA activity, a workshop held in 1998 [*ESA*, 1998] was an important event which brought together the user, science and technology communities to explore the possible ways forward. It was clear that user needs were growing. At the same time, the maturity achieved in solar terrestrial physics, allied to technological advances (in-orbit monitoring, ground-based computing power, etc.) meant that it was certainly feasible to deliver products for users in the short term and contemplate considerable improvements to them over the medium and long terms. These improvements would imply developments in the systems deployed in space and on the ground for space weather monitoring and in the science, simulation, modelling and delivery aspects of the ground-based activities. As a result ESA approved the execution of parallel wide-ranging studies. The two studies [*ESA*, 2000] are led by Rutherford Appleton Laboratory and Alcatel Space. In each consortium there is a strong blend of technology, science and applications. The top-level goals of the studies are to:
- investigate the needs for and the benefits of an ESA or other European Space Weather Programme
- establish the detailed data supply requirements by detailed consideration of the quantification of effects and intermediate tools;
- perform detailed analysis of potential programme contents:
- a detailed definition of the space-segment

- a detailed definition and proto-typing of the service-segment
- perform an analysis of collaborative and organisational structures which need to be implemented by ESA and member states
- provide inputs and advice for preparation of a programme proposal, including project implementation plan, cost estimate and risk analysis.

While there is considerable interest in space weather in Europe, initiating any major new ESA space weather activity requires the agreement of national delegations to ESA's decision-making committees. Such a commitment can only be made after the needs for such expansion and demonstrations of the benefits are clearly established. Purely scientific aspects will be handled in ESA's science programme in the usual way – through competition with quality proposals in all disciplines and thorough peer review. Technological research and developments will probably continue into space environments and effects as described above, at a similar level. But large-scale expansion of these activities in R&D for ground- and space-based space weather infrastructures are conditional upon high-level approval.

In ESA member states many activities related to space weather are being undertaken as part of national programmes [*ESA*, 1998]. These include activities addressing military needs, which ESA has no mandate to address. The interests of ESA's various member-states also differ. For example, Scandinavian and other nations at high latitude are keenly interested in effects on power systems, pipelines and other ground systems from auroral electrojet induced ground-level currents.

It will also be important to consider how any eventual service will be implemented in Europe. ESA's role is as an initiator and developer of technologies. The provision of fully operational end-user services should be provided by other organisation in a way analogous to satellite communications or meteorology.

6. CONCLUSIONS

The wide-ranging activities of ESA in the Space Weather / Space Environment domain have been summarised and recent important examples of space weather concerns given. In particular, the space weather effects on XMM and efforts to analyse these effects and other space environmental hazards illustrated the depth and breadth of work in this domain that is typically necessary in preparing a complex space mission.

We have highlighted the important scientific and technological contributions that ESA in particular and Europe in general have made. We emphasise that while there is considerable interest in Europe in expanding space weather activities toward a fully-fledged programme, this will be as a result of clear demonstration of real needs and benefits. These complex issues are being addressed by on-going studies.

REFERENCES

Andersson L., L. Eliasson, P. Wintoft, Prediction of times with increased risk of internal charging on spacecraft, in *Proceedings of the ESA Workshop on Space Weather*, ESA-WPP-155, ESA, 1998.

Apostolakis J., Geant4 status and results, *CHEP-2000 Conference Proceedings*, in press 2001.

Baker, D. N., J. H. Allen, R. D. Belian, J. B. Blake, S. G. Kanekal, B. Klecker, R. P. Lepping, X. Li, R. A. Mewaldt, K. Ogilvie, T. Onsager, G.D. Reeves, G. Rostoker, H.J. Singer, H. E. Spence, N. Turner, An assessment of space environmental conditions during the recent Anik E1 spacecraft operational failure, *ISTP Newsletter*, Vol. 6, No. 2. June, 1996. Retrievable from http://www-istp.gsfc.nasa.gov/istp/newsletter.html

Bühler P., Desorgher L., Zehnder A. and Daly, E., Observation of the Radiation-Belts with REM, in: *Proceedings of the ESA Workshop on Space Weather*, WPP-155, p.333, ESA, 1998.

CLRC Rutherford Appleton Laboratory, SEDAT Project homepage http://www.wdc.rl.ac.uk/sedat/, 2000

Daly E.J., P. Bühler and M. Kruglanski, Observations of the outer radiation belt with REM and comparisons with models, *IEEE Trans. Nucl. Sci* NS-46, 6, December 1999.

Desorgher L, P Bühler, A Zehnder, E Daly and L Adams, Modeling of the outer electron belt during magnetic storms, *Radiation Measurements* 30, 5 pp 559-567, 1999

Dyer C. and Truscott P., The analysis of XMM instrument background induced by the radiation environment in the XMM orbit, DERA report DRA/CIS(CIS2)/CR95032/1.0, final report on work performed for ESTEC Contract No 10932/94/NL/RE, December 1995; http://www.estec.esa.nl/→ →wmwww/WMA/XMM/XMM.html

ECSS (European Co-operation on Space Standards), *Space Environment*, ed. E. Daly, ECSS-E-10-04, ECSS Secretariat, ESTEC, Noordwijk The Netherlands, 2000.

ESA, *Proceedings of the Workshop On Space Weather*, ESTEC, Noordwijk, The Netherlands, ESA WP-155, ESA, 1998

ESA, Space environments and effects, in: *Preparing for the Future*, Special Issue, 9, 3, ESA, 1999.

ESA *Space Weather Studies* www Site: http://www.estec.esa.nl/→ →wmwww/spweather/spweathstudies.htm , ESA, 2000.

Fredrickson A.R., Upsets related to spacecraft charging, *IEEE Trans. Nucl. Sci*. NS-43, 2, 426, 1996.

Heynderickx, D., Quaghebeur, B., Fontaine, B., Glover, A., Carey, W.C. & Daly, E.J., New features of ESA's space environment information system (Spenvis), in: *Proc. ESA Workshop on Space Weather*, ESTEC, Noordwijk, The Netherlands, ESA WPP-155, pp. 245-248, ESA 1998.

Hilgers A., P. Gondoin, P. Nieminen and H. Evans, Prediction of plasma sheet electron effects on X-ray mirror missions, *Proc. ESA Workshop On Space Weather*, ESTEC, Noordwijk, The Netherlands, ESA WP-155 p.293, 1998.

Huber M.C.E. and Wilson A., eds., *ESA's report to the 33rd COSPAR meeting*, pp.29-33, ESA- SP-1241, ESA, 2000.

Kappenman J.G. and Albertson V.D., Bracing for the geomagnetic storms", *IEEE Spectrum*, p5, March, 1990.

Lanzerotti, L. J., Geomagnetic influences on man-made systems, *J. Atm. Terr. Phys.*, 41, 787-796, 1979.

Nartallo R., E. Daly, H.D. Evans, A. Hilgers, P. Nieminen, J. Sørensen, F. Lei, P.R. Truscott, S. Giani, J. Apostolakis, L. Urban and S. Magni, Modelling the interaction of the radiation environment with the XMM-Newton and Chandra X-ray telescopes and its effects on the on-board CCD detectors, *Proceedings of the Space Radiation Environment Workshop*, DERA Space Department, Farnborough, UK, in Press 2000.

Sawyer D.M. and Vette J.I., AP8 trapped proton environment for solar maximum and solar minimum, NSSDC 76-06, 1976.

Sørensen J., D.J. Rodgers , K.A. Ryden, P.M. Latham, G.L. Wrenn, L.,Levy, G. Panabjere, ESA's tools for internal charging, *IEEE Trans. Nucl. Sci.*, NS-47, 3, 491-497, 2000.

Truscott P., F. Lei, C. Ferguson, R. Gurriaran, P. Nieminen, E. Daly, J. Apostolakis, S. Giani, M.G. Pia, L. Urban, M. Maire, Development of a spacecraft radiation shielding and effects toolkit based on Geant4, *CHEP-2000 Proc.*, 2000.

Vette J.I. The AE-8 trapped electron model, NSSDC/WDC-A-R&S 91-24, 1991.

Wintoft P., Development of AI methods in spacecraft anomaly predictions, IRF Web site http://eos.irfl.lu.se/saaps/ 1998.

Wrenn G.L. Conclusive evidence for internal dielectric charging anomalies on geosynchronous communications spacecraft, *J. Spacecraft and Rockets* 32, pp 514-520, May-June 1995

Wu, J.G, Lundstedt, H., Eliasson, L., Hilgers, A., Analysis and real-time prediction of environmentally induced spacecraft anomalies, in *Proceedings of the ESA Workshop on Space Weather* ESA-WPP-155, ESA, 1999.

E. Daly and A. Hilgers
ESA/ESTEC
Postbus 299
2200 AG Noordwijk,
The Netherlands

Space Weather: Japanese Perspectives

Y. Kamide

Solar-Terrestrial Environment Laboratory, Nagoya University, Toyokawa, Aichi 442-8507, Japan

This paper presents the existing and near-future efforts of space weather-related research and its applications in Japan. The program consists of several core elements, including predictions of major geomagnetic storms based on solar wind information, specifications of the magnetosphere-ionosphere coupling system, and observations of solar activity from the L5 location.

1. INTRODUCTION

The goals of the Japanese space weather program is not, and should not be, any different from the objectives of North American and European space weather programs: see, for example, *Wright et al.* [1995] and *Daly and Hilgers* [2001]. Space weather is becoming even more important in conjunction with the international space station in terms of manned space exploration. In view of the fact that Japan occupies geomagnetically lower latitudes than its counterparts, the needs of our "customers" are somewhat different from those in other high-latitude countries. It is noted that the term "space weather" was first used officially in Japan as early as 1988 when a project called the space weather forecast project began at the Communications Research Laboratory (CRL) and its earlier radiowave alert services were expanded to include space environment forecast services.

Through the joint program on space weather in Japan, we hope to achieve the following objectives:

1. Advance numerical modeling and simulations
2. Construct algorithms for predicting space weather events
3. Improve observational techniques
4. Enhance educational programs on the solar-terrestrial environment and Sun-Earth relationships

Through these operational and practical efforts for the Japanese space weather program, it is, of course, very important to advance fundamental understanding of solar-terrestrial processes

The purpose of this paper is to present Japanese perspectives on the space weather program and define the existing and future efforts of space weather-related research and their applications in Japan.

2. PRESENT PROGRAM

2.1. Structure

The space weather program in Japan, including overall research on the dynamics of the solar-terrestrial environment, is being led by two major institutions: the Communications Research Laboratory (CRL) and the Solar-Terrestrial Environment Laboratory (STEL) of Nagoya University. As shown in Figure 1, other government-supported institutions, such as the National Astronomical Observatory and the Institute of Space and Astronautical Science, as well as major universities, such as Kyoto University, in which the World Data Center for A is housed, are actively involved in this joint endeavor. Since 1996, the CRL and STEL have co-organized symposia/workshops on space weather at least twice a year. The following are the subjects of some of the presentations at the most recent symposium: Solar wind speed and magnetic fields in the corona; Propagation of interplanetary disturbances as observed by radio scintillation measurements; Twenty-two year periodicity of coronal flow; Magnetic flux ropes; Solar cycle dependence of storm-substorm relationships; Magnetospheric configuration for northward

Space Weather Program in Japan
<Operation to Research>

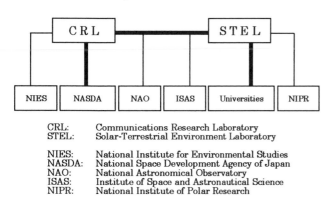

CRL: Communications Research Laboratory
STEL: Solar-Terrestrial Environment Laboratory

NIES: National Institute for Environmental Studies
NASDA: National Space Development Agency of Japan
NAO: National Astronomical Observatory
ISAS: Institute of Space and Astronautical Science
NIPR: National Institute of Polar Research

Figure 1. Institutions in Japan which actively participate in the space weather program.

IMF; Effect of solar wind dynamic pressure on geomagnetic indices; Storm-associated electron flux in the outer radiation belt; AI predictions of the Dst index; A ring current model as a function of solar wind parameters; and Ionosphere-thermosphere dynamics in the auroral region.

A working group on space weather was formed in 1999 within the Society of Geomagnetism and Earth, Planetary and Space Sciences (SGEPSS), which is the Japanese counterpart of the International Association of Geomagnetism and Aeronomy (IAGA). The purpose of the working group is to enhance communications in terms of the common techniques and algorithms for forecasting geomagnetic storms/substorms, as well as to promote collaborative studies in solar-terrestrial physics. The Committee on Solar-Terrestrial Environment Research (CSTER) serves to form a forum in which scientists in the field and engineers engaged in forecasting geoeffective events cooperate with each other toward better serving the scientific community. Several times a year, CSTER conducts workshops to exchange information/ideas on all aspects of solar-terrestrial phenomena whenever major events have ta-

ken place. These activities are closely coordinated with international bodies, such as SCOSTEP (Scientific Committee on Solar-Terrestrial Physics) and COSPAR (Committee on Space Research). For example, in addition to presenting results of Japanese space weather activity at international meetings, the Japanese community regularly sends committee members to these bodies. The first S-RAMP (STEP – Results, Applications, and Modeling Phase) Conference during which one of the main topics is space weather, will be held in Sapporo, Japan in October 2000.

2.2. CRL activity

As indicated in the Introduction, the CRL has been engaged in its own space weather projects since 1988: see *Marubashi* [1989] and *Sagawa* [1997] for details. It is responsible for the operation of space weather forecast in Japan. At present, its Hiraiso Solar Terrestrial Research Center is serving as one of nine Regional Warning Center (RWC) for the International Space Environment Service (ISES). Table 1 shows the data being used for forecasting space weather from the Sun to the Earth's ionosphere. These services are available by telephone, fax, and/or the Internet (URL, http://hirweb.crl.go.jp/index-j.html).

The users of these services are quite a diverse group: see Table 2. This is from the statistics spanning over 10 years. Most of "General Public" are interested in high solar activity and its geoeffective events, such as polar auroras and geomagnetic disturbances. On-line data services are also available to scientists on a real-time basis. The available data include high-resolution solar images, intermagnet magnetometer data, and ionospheric data.

Figure 2 shows how the number of users' inquiries varied as a function of solar activity and geomagnetic activity. Note the peaks, shown by arrows, corresponding to the occurrence of major geomagnetic storms. In January 1999 the CRL began electronic services. The number of subscribers has increased dramatically, following the growth of the current solar cycle.

Table 1. Observation instruments/facilities and data usages at the Communications Research Laboratory.

Instruments/Facilities	Data	Usages of data
H-alpha Solar Telescope	Monochromatic images	Flare monitoring
	Doppler images	Flare forecast
Sunspot Monitoring Telescope	Sunspots structures	Activity forecast
Solar Radio Spectrograph	Dynamic spectrum 25 MHz-2.5 GHz	Coronal disturbances
Solar Flux Meter	Intensity of solar radio at 2.8 GHz	Solar activity
ACE-Receiving Station	Solar wind plasma and magnetic field	Solar wind structure
	Energetic particles at L1	Solar energetic particles
		Dst and ring current particles
Magnetogram Network	Geomagnetic variations	Geomagnetic disturbances
Ionosonde Network	Ionograms	Ionospheric conditions

Table 2. Wide spectrum of the space weather customers in Japan.

Customers	Percent
Communications	27.6 %
General public	25.6
Media	13.2
Amateur radio	12.3
Government	8.7
Universities	5.2
Power companies	2.9
NASDA	2.2
Air transportation	2.2

In addition to forecasting services, the CRL has been engaged in several research projects, covering a wide area between the Sun and the Earth's magnetosphere/ionosphere. Some of the active projects include: Three-dimensional magnetic field modeling of active regions of the Sun; Solar wind turbulence; Magnetic ropes in the solar wind; Electrons in the radiation belt; Ion heating in the ionosphere; and MHD modeling of solar wind/magnetosphere coupling. These are important elements of research to the advancement of space weather forecasting.

2.3. STEL activity

At the STEL, particularly the Integrated Studies Division, space weather related research is being conducted in two ways: understanding basic processes occurring over the boundaries of various plasma regions of the solar-terrestrial system, and the developing algorithms to predict geomagnetic storms/substorms. In solar-terrestrial physics, it is necessary to deal with a huge amount of data assembled in various formats, making interpretation difficult. For instance, the relatively common act of integrating properly ground-based and satellite-based observations creates a number of irreconcilable outcomes. While observations from the Earth's surface are considered "remote" and thus indirect, they nevertheless create high spatial/temporal resolutions. On the other hand, satellite observations, being in-situ and "direct," provide only "point" measurements.

STEL has recently installed a high-technology computer system, called GEDAS, which stands for Geospace Environment Data Analysis System. It represents a new way to promote integrated studies by combining ground-based and satellite-based observations and simulation research. GEDAS connects similar systems throughout the world on a real-time or near real-time basis. More importantly, GEDAS is not only a data exchange or data display system, but a research tool as well. Institutions around the world have their own specialties, and possess their own simula-

tion methods. In addition to common data, each institution has its own up-to-date data. It is hoped that data products generated by combining original data with common data can be instantly provided to the world community via GEDAS. Although GEDAS is not mainly for operational purposes, it will actively be involved in predicting space weather. What GEDAS does can be summarized in the following:

1. Local and global viewpoints: It is important to determine the precise locations of observational staging areas in the entire solar-terrestrial system. By relying on GEDAS, one can identify the location of local explosive phenomena within large-scale energy flows.

2. Connection of data analysis and theoretical approach: Throughout the solar-terrestrial system, energy is accumulated and then abruptly released. To understand interactions among various plasma regions, numerical modeling using basic equations is required. Under the GEDAS system, real-time data can be used as initial and boundary conditions for computer simulations, permitting researchers to forecast important events in the near future. Having access to data in real-time through GEDAS, any researcher can also be an instant leader in any joint research project. Fresh ideas can be tested almost instantly against real data.

Four projects currently underway using GEDAS are described briefly below:

(1) Ionospheric electric fields and currents: Figure 3 outlines data flow of this project. We use real-time ground magnetometer data from some 70 observatories, which will be combined with data from more direct observations by satellites and radars. This project, representing a joint ef-

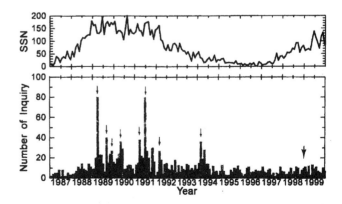

Figure 2. Support number (SSN) and number of inquiries about space weather specifications and predictions at the Hiraiso Solar Terrestrial Research Center of the Communications Research Laboratory. Services have begun electronically in January 1999 (shown by an arrow). Courtesy of T. Nagatsuma.

PROPOSED SCHEME FOR GEDAS

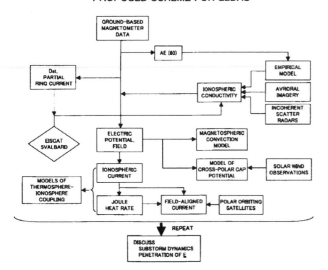

Figure 3. Example of one of the GEDAS projects using the KRM method, in which several ionospheric parameters are to be estimated. Our output will become important input for simulation/modeling studies at other institutions.

fort of STEL, NGDC, SEC, NCAR, and the University of Michigan, will use the KRM and AMIE programs [*Kamide et al.*, 1981; *Richmond and Kamide*, 1988] to compute the instantaneous distribution of ionospheric electrodynamic parameters at high latitudes. The ionospheric conductivities will be normalized using their real-time observations by radars and observations of the global auroral distribution by polar-orbiting satellites. Once the ionospheric parameters are computed, the output will be sent to institutions around the world. It is hoped that our output becomes input for other computer codes. For example, our electric field distribution can be mapped to the magnetosphere and is therefore useful for tracing particles in the magnetosphere: see *Kamide et al.* [2000] for some preliminary studies. Joule heating from our calculations can be used as input for calculating neutral winds in the thermosphere, which will modify the original electric field. Our output is, of course, important to understanding the "present" status of the auroral electrojets, which are critical to forecasting the strength of induced currents.

(2) Solar wind-magnetosphere coupling: An MHD simulation model of solar wind/magnetosphere interactions [*Ogino et al.*, 1994] is run in real-time using data of the solar wind and IMF from ACE. Figure 4 shows one such example in which the cross-polar cap potential difference is nowcasted for a recent magnetic storm. The validity of the potential distribution calculated from this MHD model can be tested immediately by comparing it with the KRM/AMIE calculations. Any inconsistency between the two approaches can be accounted for in terms of the as-

sumptions employed in these modeling techniques. The inconsistency results partially from the "one way" MHD simulation in which no relevant ionospheric boundary condition is included. Relying on the realistic distribution of the ionospheric potential from the KRM/AMIE calculations, the simulations can be upgraded.

(3) Prediction of solar wind speed: To better understand the propagation of heliospheric structures such as CMEs, streamers, and corotating interaction regions, observations of interplanetary scintillation (IPS) of natural radio sources are conducted on a routine basis at STEL. The IPS is capable of measuring solar wind velocity and density fluctuations in three-dimensions at distances of 0.1 - 1 AU. Since IPS measurements are biased by line-of-sight integration, however, a computer assisted tomography technique is employed to obtain longitudinal and latitudinal distributions of speed and electron density fluctuations. Solar wind predictions are being performed using this program, in which IPS data are transferred to the Center for Astrophysics and Space Sciences of the University of California at San Diego automatically in near real-time. Predictions of the solar wind speed near the Earth for the following several days are derived from this tomography technique.

(4) Ionosphere-thermosphere coupling: A high-resolution ionosphere-thermosphere coupling model developed at STEL will be used to study disturbances in the high-latitude upper atmosphere during geomagnetic

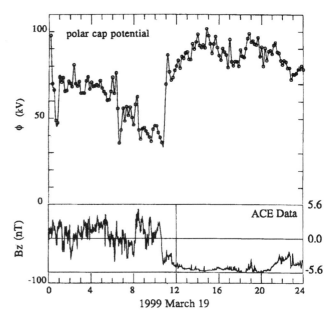

Figure 4. Example of real-time predictions of the cross-polar cap potential, calculated from an MHD simulation model using solar wind observations by the ACE satellite. Courtesy of T. Ogino.

storms. Input such as electric fields, Joule heating, and particle precipitation is obtained from empirical models and data from radars, optical instruments, and satellites. The model is to nowcast and forecast the condition of the thermosphere and the ionosphere. The calculated densities and velocities of ions and neutral particles will be compared with measured data in near real-time. Through this procedure the thermosphere-ionosphere model itself can be improved. When a notable event is observed, the mechanism of the perturbation in the upper atmosphere will immediately be studied through GEDAS.

3. NEAR-FUTURE PLANS

We plan to connect most, if not all, space weather institutions and systems in the Asia and Pacific region. A real-time, high-speed computer network serving the countries in the region, called Asia-Pacific Advanced Network (APAN), is underway. A joint effort toward this goal has already begun with institutions in Taiwan and Korea.

The CRL is proposing even an ambitious new scheme for space weather studies that includes constructing the Space Weather Forecast Center and placing a solar observation spacecraft at the L5 location. This proposal is outlined below:

3.1. Space Weather Forecast Center

The proposed space weather forecast program at the CRL consists of three components: Data collection, Analysis of data in conjunction with MHD simulations and artificial intelligence (AI), and Distribution of data and data products. This center not only offers space weather predictions for operational purposes but also provides the scientific community with a mechanism for research forum. For example, its capabilities include to collect nearly real-time data from the vast space environment from the Sun to the Earth's upper atmosphere and to achieve data for detailed after-the-event analysis. One of the proposed core facilities is the Space Environment Display System, in which teleconferencing and data-sharing functions, in addition to real-time joint studies at multiple institutions, are added. In order for this system not to be too complicated in dealing with different formats of data, it handles only graphic data.

3.2. L5 Mission

A feasible study has begun at the CRL to launch a new satellite at the L5 location to observe the Sun laterally, i.e., at a 60-degree angle. As a next-generation tool for space weather forecast, this mission would not only provide new data from a new angle but also detect geoeffective events several days in advance. A 2005 launch date is planned,

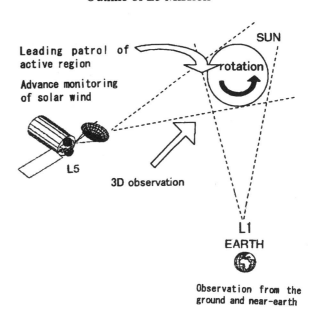

Figure 5. Configuration of L5 observations with respect to the locations of the Sun and the Earth. Courtesy of M. Akioka.

beginning space weather services of monitoring the Sun at the next solar cycle. As shown in Figure 5, observations from the Earth's surface as well as from this spacecraft make it possible to monitor three-dimensional structures of the solar corona and its disturbances.

Candidate mission instruments are estimated to be 450 kg including the spacecraft mass. An on-board data process and selection capability will be employed to minimize the data transmission load. One of the more important tasks the L5 satellite team must accomplish is to develop an advanced telemetry system consisting of automatic event detectors and on-board data processing, at a distance of more than 15 million km away from the Earth.

4. DISCUSSION

It would be useful, in many respects, for us to compare our space weather and surface, or troposphere, weather in terms of their ability as the weather forecaster. It is often pointed out that space weather research is behind surface weather research by some 50 years. It would be the end of research as we know it if all physical/chemical processes from the Sun to the Earth were understood and all of these processes are expressed in equations. When all details are coded in a computer system and when the "present" status of the Sun is given, we could predict accurately space weather events in our (the Earth's) neighborhood.

There can be no doubt that in order to achieve this degree of accuracy, a super computer of infinite speed and capacity would be necessary. In fact, we are far from a complete understanding of the solar-terrestrial environment. Thus, empirical formulas still play an important role in predicting geoeffective events. We have witnessed several instances in which empirical formulas connecting processes between solar wind observations and geomagnetic activity have been used to advance significantly our understanding of the solar wind-magnetosphere coupling. Recent modeling efforts for space weather including global MHD and other modular models have also been improved, beginning to contribute to understanding the effects of solar wind disturbances on the magnetosphere and ionosphere in conjunction with the space weather program [e.g., *Song et al.*, 1999].

Three areas are noted which are of crucial importance toward the success of space weather studies. They are needs for effective contributions of: (1) Observations and Modeling, (2) Nowcasting and Forecasting, and (3) Research and Operations.

(1) Observations and Modeling

Needless to say, spacecraft and radar observations alone are inadequate to cover every plasma region in the solar-terrestrial system. Only under appropriate initial/boundary conditions using real-time observations in the upstream solar wind, a numerical solution can be achieved which would then be used to predict the next magnetospheric and ionospheric event. This represents an example of the "healthy" relationship between observations and modeling.

(2) Nowcasting and Forecasting

Nowcasting refers to specifying the ionosphere and magnetosphere in terms of the distribution of plasmas and of fields. For example, it is now possible to calculate the global distribution of the electric potential in the ionosphere primarily from ground-based magnetometer observations. The time histories of this potential distribution, which is very similar to the weather maps in terms of the air pressure system we see in newspapers everyday, can be extrapolated to the next moment using AI techniques and/or other statistical methods. Efforts have also been made to reproduce or predict every detail of geomagnetic activity using only solar wind variations. We must realize, however, data from the solar wind and the IMF alone cannot determine each and every process in the magnetosphere and ionosphere. The solar wind only gives the boundary condition to the magnetosphere, in which various types of disturbances, such as magnetospheric substorms, are triggered. The history of the inner magnetosphere and the polar ionosphere does determine where and how a substorm of what magnitude will take place. Just as in atmospheric weather forecasts, the number of observatories is very important. "Determining what tomorrow's weather is like" depends on how and who extends today's maps to tomorrow's.

(3) Research and Operations

Simply predicting the next day's *Kp* or *ap* value, which might have been good enough 10 years ago, is not adequate in today's space weather research. A scalar, three-hourly value does not express the dynamic nature of our space environment. Space weather predictions should always update the operational scheme to predict major geomagnetic storms. This is why a joint endeavor between researchers and engineers is acute.

Acknowledgments. I would like to thank K. Marubashi, M. Akioka, T. Nagatsuma, and M. Kunitake, all of CRL, for their useful input in preparing this manuscript. I also appreciate fruitful discussions with T. Ogino, H. Shinagawa, and M. Kojima of STEL and their constructive comments on an earlier version of this paper.

REFERENCES

Daly, E. J., and A. Hilgers, Space weather: European space agency perspectives, this issue, 2001.

Kamide, Y., A. D. Richmond, and S. Matsushita, Estimation of ionospheric electric fields, ionospheric currents, and field-aligned currents from ground magnetic records, *J. Geophys. Res., 86*, 801-813, 1981.

Kamide, Y., J.-H. Shue, B. A. Hausman, and J. W. Freeman, Toward real-time mapping of ionospheric electric fields and currents, *Adv. Space Res., 26*, 213, 2000.

Marubashi, K., The space weather forecast program, *Space Sci. Rev., 51*, 197, 1989.

Ogino, T., R. J. Walker, and M. Ashour-Abdalla, A global magnetohydrodynamic simulation of the response of the magnetosphere to a northward turning of interplanetary magnetic field, *J. Geophys. Res., 99*, 11027, 1994.

Richmond, A. D., and Y. Kamide, Mapping of electrodynamic features of the high-latitude ionosphere from localized observations: Technique, *J. Geophys. Res., 93*, 5741, 1988.

Sagawa, E., Space weather forecast program - Interim report, *Rev. Comm. Res. Lab., 43*, 211, 1997.

Song, P., D. L. De Zeeuw, T. I. Gombosi, C. P. T. Groth, and K. G. Powell, A nemrical study of solar wind-magnetosphere interaction for northward IMF, *J. Geophys. Res., 104*, 28361, 1999.

Wright, J. M., Jr. et al., *National Space Weather Program: Strategic plan*, FCM-P30-1995, Washington, D. C., 1995.

Y. Kamide, Solar-Terrestrial Environment Laboratory, Nagoya University, Toyokawa, Aichi 442-8507, Japan

Space Weather: Russian Perspectives

M. Panasyuk

Skobeltsyn Institute of Nuclear Physics Moscow State University, Moscow, Russia

Different scientific groups in Russia are successfully developing the complex of experiments and models directed to Space Weather problems. At present the complex consists of the following major parts: optics and radio observations of the Sun and a forecasting model of large solar flares; ground based monitoring of the heliospheric (neutron monitors) and geomagnetic (magnetometers) conditions; models of the geomagnetic storm forecasting; a model of galactic cosmic rays and a probabilistic model of solar energetic particles; monitoring of the radiation environment by satellites. The problems of compilation of these parts and perspectives of their future development are discussed.

1. INTRODUCTION

Space weather in Russia has a long history. One of the first mentions of space weather phenomena at the beginning of this century may be found in [*Ryumin*, 1925]. In this book an extremely strong magnetic storm and its destroying consequences, which impacted the ground based electric and communication systems in Russia are described. The physical nature of the phenomena is explained by the influence of a magnetic cloud from the Sun on the Earth. The fundamental studies in Russia devoted to solar-terrestrial linking were performed by *Tchijevsky* [1928]. He has shown clearly that a lot of geophysical, life and social processes on the Earth are to a certain extent connected with solar activity.

The next step in the space weather development was triggered in 60-ies by space flights. Knowledge of the near-Earth radiation environment is extremely important for radiation safety of the satellites. Studies of the high energy particles trapped in the radiation belts and cosmic rays, which can penetrate through the spacecraft shielding, showed that their fluxes have a complicated non-linear dependence on solar and magnetospheric activity. The very

high energy protons originate from galactic cosmic rays (GCR), which are modulated in anti-phase with the solar cycle. The protons can be accelerated also in the solar energetic particle (SEP) events up to very high energies and, therefore, penetrate deeply into the Earth's magnetosphere. The fluxes of high energy trapped radiation vary strongly due to the geomagnetic storms and sub-storms. Shortly, the radiation environment has significant long-term variations, associated with the solar cycle and short time enhancements associated with SEP events and geomagnetic disturbances. To describe this complicated process, responsible for variations of the radiation environment, SINP MSU in 60-s initiated an activity to develop different models that permit to describing any phenomena in solar-terrestrial physics. The results of this activity were published in the form of books *"Models of the Space"* [1976] which combine the quantitative (theoretical and empirical) models of the solar activity, heliosphere, planets and moons, Earth's magnetosphere, ionosphere and atmosphere, GCR, solar energetic particles, trapped radiation and their impact on the satellite's systems and electronic equipment.

During the last decade the space weather research developed intensively due to several important reasons. The first one is the development of the International Space Station (ISS) and as a consequence, a significant increase of the number of manned space flights and duration of the astronauts' work in the space. The second reason is extensive use of high integration space electronic systems. Such micro-chips are very sensitive to both highly ionizing protons and to volume charge induced by high energy penetrating electrons. The space weather events in energetic particles may dramatically effect satellite electronic systems and can lead to faults in their operations and even to the loss of the satellite. The third reason is the discovery that many negative events on the ground based widely used systems (power systems, pipelines, railways and so on) at medium and high latitudes are closely connected with geomagnetic activity. Strong electric currents induced during the geomagnetic disturbances may distort the equipment in electric power systems and lead to faster erosion processes in high conductive materials of pipelines and railways. The fourth reason is expansion of radio communication systems in to space and, as a consequence, the importance of stable radio signal propagation through the ionosphere where conditions significantly depend on the solar and geomagnetic activity. The fifth reason is medical research, showing that space weather events have significant influence on human health. Fluxes of high energy particles in the stratosphere generated in strong SEP events increase the doses acquired by humans during flights onboard high altitude aircraft. Many medical tests confirm that geomagnetic field variations directly impact human health. Therefore, we can conclude that space weather defines the conditions for human activities both in space and on the Earth.

Many countries announced of their own space weather activity programs over the last two years. These activities are based on existing scientific results and accomplished or future experiments (ground-based and space). Russia has many scientific Centers that have been working in the field of solar terrestrial physics for a long time. We also have several experiments in the near-Earth space. The existing potential of the Russian space weather activity permits to successfully perform space weather monitoring and forecasting: from the solar activity via interplanetary and magnetospheric perturbations to the ionosphere-atmosphere coupling. The wide network of ground-based stations created in Russia permits to perform the measurements of the solar activity, Earth's magnetic field variations, solar and ionospheric radio waves and fluxes of cosmic rays (GCR and SEP). The Russian program of near-Earth satellite experiments permits to obtain in-situ information on the magnetosphere and interplanetary medium including radiation environment, plasma characteristics, solar wind and interplanetary magnetic field properties etc. This information is vitally important for diagnostics of the interplanetary medium and Earth's magnetosphere conditions. Many theoretical and empirical models developed by Russian scientists are used successfully for description and prediction of practically any phenomena in solar-terrestrial physics. The current activity of different Russian scientific groups was partially described in [Avdyushin et al., 1999] and presented on the Web-site of Russian Space Weather Initiatives

(RSWI: http://alpha.npi.msu.su/RSWI/rswi.html). This paper describes the Russian space weather data resources and models. The future perspectives of RSWI activity are discussed.

2. SPACE WEATHER DIAGNOSTICS

2.1 Onground Measurements

2.1.1. Observation of the Sun

The monitoring of the solar activity is performed in Russia in the form of radio and optical observations. IZMIRAN Solar Radio Laboratory (LaRS) supports the solar radio patrol by means of 169 MHz, 204 MHz, 3000 MHz radiometers and 45 - 270 MHZ digital spectrograph (http://helios.izmiran.troitsk.ru/lars/LARS.html). Solar radio flux in different frequency bands, spectra and solar radio bursts are measured and presented on-line. Irkutsk Radioastrophysical Observatory presents on-line information about optical and radio observations of the Sun (ftp://ssrt.iszf.irk.ru/pub/data) including bursts of solar emission at 5.7GHz (http://rao.iszf.irk.ru/bursts/stat.html). A complex of radiotelescopes in Laboratory Zimenki performs high resolution (time and angular) monitoring and patrol observations of solar radio emission with frequency from 0.1 MHz to 10 MHz (http://www.nirfi.sci-nnov.ru/english/index3e.html). Complex observations of the Sun in the optical range and in the radio waves carry out information about dynamics of active regions and flare activity including CME and filament eruption. Identification of the solar radio burst type permits to estimate the possible geomagnetic consequences of the observing solar events.

2.1.2. Heliospheric cosmic ray observations

Cosmic rays are sensitive for current global heliospheric conditions and may be used as a probe of the heliospheric disturbances. Knowledge of conditions in the inner heliosphere is extremely important because these conditions have significant influence on the CME and SEP propagation from the Sun to the Earth. High energy solar cosmic rays reach the Earth on several hours before then most intensive SEP with energies less than 100 MeV. This fact permits to estimate possible SEP flux before it arrival to the Earth's magnetosphere.

Cosmic rays are observed by the net of several Russian neutron monitors: in Moscow (IZMIRAN http://helios.izmiran.rssi.ru/cosray/main.htm), highest time resolution (10s) monitor in Apatity (PGI http://pgi.kolasc.net.ru/CosmicRay/), Yakutsk and Tixie (IKFIA http://teor.ysn.ru/rswi/graph-GIF.html). Experimental data about cosmic ray variations are loaded in the data bases presented on-line on Web-pages http://helios.izmiran.rssi.ru/cosray/main.htm (Moscow), http://pgi.kolasc.net.ru/CosmicRay/form.htm (Apatity) and http://teor.ysn.ru/imf/neutron.htm (Yakutsk and Tixie Bay). Monitoring of GCR variations and scintillation is used for short time (days) prediction of IMF disturbance [*Kozlov et al.*, 1999]. For quantitative description of GCR variations hourly index of cosmic ray activity "CR Activity Index" has been introduced [*Belov et al.*, 1999]. This index is developed directly for the purposes of diagnostic and forecasting of interplanetary perturbations and consequent geomagnetic dynamics.

2.1.3. Earth's magnetic field

Geomagnetic field is measured by the magnetometers located at middle and high geomagnetic latitudes. The measured data are presented in real time from Moscow (IZMIIRAN http://helios.izmiran.troitsk.ru/cosray/magnet.htm) and Irkutsk (ISTP http://cgm.iszf.irk.ru/magnet2.htm). The data bases of the geomagnetic variations (time profiles and spectra of ultra low frequency pulsations) are presented on the web-pages http://www.izmiran.rssi.ru/magnetism/mos_data.htm, ftp://vodin.izmiran.rssi.ru/start.htm (IZMIRAN), http://pgi.kolasc.net.ru/Lovozero/ (PGI), http://cgm.iszf.irk.ru/outmag/ (ISTP). The data from the net of Russian geomagnetic observatories are combined on CD-ROM data base on ftp-server ftp://vodin.izmiran.rssi.ru/start.htm. The indexes of the geomagnetic activity are calculated and presented on-line on http://cgm.iszf.irk.ru/magnet2.htm, http://charlamp.izmiran.rssi.ru/ (local 3 hour K-indexes in Irkutsk and Moscow respectively) and on http://www.aari.nw.ru/clgmi/geophys/pc_Data_Intermagnet.html (1 min Ap index). These indexes permit to estimate current geomagnetic activity at the middle and high geomagnetic latitudes.

2.1.4. Ionosphere

A regular control of ionospheric conditions has been carrying out since 1978 at the Radio Astronomical Observatory "Staraya Pustyn" (http://www.nirfi.sci-nnov.ru/english/index3e.html) by the measurements of the ionosphere total electron content (TEC) and its variations using a polarimeter on the basis of 10-m steerable radiotelescope at operating frequency 290 MHz. These observations make it possible to measure the Faraday rotation angle of the polarization plane of the linearly polarized Galactic radio emission in the ionosphere and by its value to determine TEC. The accuracy of TEC determination is about 5% that is not worse than that attained with geosynchronous satellites. The methods have been developed to determine 3D

spectra of ionospheric turbulence based on the analysis of amplitude and phase fluctuations of satellite signals. Temporal variations of electromagnetic emission activity in different frequency ranges as well as variations of near-earth's waveguide and resonator geometrical and electromagnetic parameters exhibit a strong dependence on solar radiations of different nature and solar wind or, in other words, they are high-sensitive indicators of Space Weather. Russia has built a network of oblique sounding stations on the basis of the national low-power chirp ionosonde by cooperative efforts with NIRFI (http://www.nirfi.sci-nnov.ru/english/index3e.html), Mari State Technical University, the Institute of Solar-Terrestrial Physics of the Siberian Branch of the RAS (http://www.iszf.irk.ru/) and IKIR of the Far Eastern Branch of the RAS. The network is used at present to conduct a large-scale investigations of the ionosphere at natural and modified conditions.

2.2. Atmospheric Measurements

2.2.1. Cosmic ray monitoring

At present the data sets on the cosmic ray fluxes in the atmosphere from ground levels up to altitudes of 30-35 km are available from the basic stratosphere stations - Murmansk (1957-1998), Moscow (1957-1998), Alma-Ata (1962-1992) and Mirny, Antarctica (1963-1998). Monthly averaged data sets are used to model the atmosphere with account for cosmic ray induced ionization effects. The correlation between cosmic ray proton fluxes with cloud area and precipitation magnitude, discovered in [*Stozhkov*, 1996], permit to assume, that global climatic changes may be caused by the variations of cosmic rays, entering the solar system from galactic space. The variations of the total cosmic ray flux, entering the terrestrial atmosphere, lead to variations of the cloud-covered, which, in turn, alters the balance between the absorbed and reflected solar emission and infra-red radiation of the Earth. The decrease of the cloud-covered area by 8% is equal to a 2% increase of the solar constant. It is expected that these processes will lead to changes in the temperature of the Earth's surface.

2.2.2. Arctic region observations

A considerable part of the auroral zone and polar cap is located over Russia. We still have several operating observatories in this region. At the western part of the Polar Geophysical Institute and Kola Geophysical Center three observatories in Loparskaya, Lovozero and Apatity provide regular riometer and magnetometer observations and quasi-regular TV and photometer recordings of the aurora borealis.

Closer to the East observatories in Arhangelsk (IZMIRAN), Norilsk, (SibIzmir) and Dixon (Arctic and Antarctic Institute) continue similar regular ground-based observations. In the eastern part of the auroral and sub-auroral zone observations are provided in Yakuts, Tixie and Zhigansk by the Institute of Cosmophysical research. Riometer and magnetometer observations are available from most of the above mentioned observatories. Cosmic ray neutron monitors are in operation in Yakutsk and Tixie.

The polar cap and auroral zone are the scene of direct manifestation for magnetospheric disturbances, storms and sub-storms. On the one hand, this means, that the polar region of the ionosphere is a major scene of this activity and ground-based objects, such as oil pipe lines and electric power lines are mostly affected by strong sub-storms and magnetic storms. On the other hand, geophysical observations in the auroral region are important for magnetic activity monitoring and forecast.

2.3. Space Experiments

In spite of the decrease in the number of satellite launches in Russia over the past several years, there are a number of experiments, the results of which can be used for current space weather diagnostics.

The most significant of these experiments is the 'Interball' project (http://www.iki.rssi.ru/ida.html). The launch of the 'Interball' satellites was made several years ago and the installed instruments still provide high-quality data on the fields and particles both outside and inside the Earth's magnetosphere. This project was developed for studying the different plasma processes in the surrounding space environment and consists of two pairs of space vehicles (satellite - sub-satellite) launched into orbits of up to 200 000 km (the Tail Probe) and up to 20 000 km (Auroral Probe).

The Tail Probe was launched into an ecliptic plane orbit , so that it can reach the high-altitude cusp and sub-solar magnetopause regions on the day side and the neutral sheet in the night side tail. The Auroral Probe is optimized for magnetic conjunctions between the two pairs of satellites around the midnight meridian. This permits to study the cause-and-effect relationships between the plasma processes in the tail and in the auroral particle acceleration region above the auroral oval with a high time-space resolution.

It is important to stress, that the 'Interball' project plays an important role in international cooperation in correlation analysis of plasma phenomena in the interplanetary and near-Earth environment according to data from different space probes: IMP-8, WIND, ACE, Geotail and others.

Another important national space experiment in the filed of space weather is radiation environment monitoring on orbital station (OS) 'Mir' (http://dec1.npi.msu.su/english/data/lasre/mir/mir.html). Here, a set of instruments for radiation studies (measuring both spectral and dose characteristics), which permits to estimate changes in the radiation environment at low altitudes (350-400 km) has been operating for many years. The radiation measurements on OS 'Mir' played a key role in the development of the Solar Cycle Low Altitude Model of Radiation [*Bashkirov et al.*, 1998]. Now we understand, that solar activity variations lead to significant variations in radiation intensity in the South Atlantic Anomaly region: during years of solar activity minimum (in contrast to solar activity maximum) the radiation belt particle flux intensities increase due to decrease of losses in a less denser atmosphere. Currently, in the year 2000, during the year of solar cycle maximum, minimum radiation doses are observed, the doses will increase again towards the solar cycle minimum (2006-2008).

Among the projects to be launched during the next several years, we should mention the following:
-    'CORONAS-F' (http://www.izmiran.rssi.ru/projects/CORONAS/index.html) The scientific goal of this project is to carry out complex research of the powerful dynamic solar activity processes (active regions, flares, mass ejections) in a broad spectral range from radio frequencies to gamma-rays; to study solar cosmic rays accelerated in solar active phenomena besides their release conditions, propagation into the IMF and impact on the Earth's magnetosphere. Helioseismology of the Sun interior is also among the main research targets. This satellite is to be launched at the beginning of 2001 into a circular polar orbit with a ~500 km altitude
-    'Interheliozond' (http://www.izmiran.rssi.ru/projects/INTERHELIOS/) is a mission specifically designed to cover the gaps in our knowledge by exploring the inner solar system with the objectives to understand the processes, which heat the corona and accelerate the solar wind. This solar probe will orbit the Sun at 30-100 solar radii at a speed 3 times as fast as the Earth and will occupy an ideal position near the transient sources in the solar atmosphere. Interheliozond will be an important element in the future program of orbital studies of solar-terrestrial coupling, providing real-time information for the space-weather forecast.

3. NOWCASTING AND FORECASTING MODELS.

The main purpose of the nowcasting and forecasting models is determination of current and possible future conditions that directly influence on the equipment (space and ground based), communication systems and human health. These conditions include radiation fluxes (trapped particles and cosmic rays), geomagnetic field variations (both at high and low latitudes) and ionospheric conductivity for radio signals. Therefore the space weather models have to calculate as output parameters such quantities that reflect above described conditions.

3.1. Nowcasting

Information about geoeffective solar and interplanetary events is extremely important for prediction of geomagnetic disturbances and radiation enhancement. Nowcasting models of solar and heliospheric conditions permit to define such events by means of indirect observations.

To describe the global structure of inner heliosphere the IMF sector structure is continuously restored in the form of the map of the source surface according to original programs developed in IZMIRAN (http://www.geocities.com/romashets/). The model uses current on-line experimental information about solar and interplanetary magnetic field measurements. IMF sector structure (especially sector boundary location) is extremely important for description of a structure of the heliospheric current sheet and coronal holes (associated with formation of the geoeffective corotation interaction region) and for estimation of the SEP and GCR propagation conditions.

Heliospheric disturbances may be observed by indirect method of neutron monitors. This technique is sensitive to variation of GCR that reflect the condition of high energy particle propagation in the heliosphere. Data from the world wide cosmic ray station network (> 40 stations) are processed by the special methods [Belov et al., 1997] to obtain hourly means of density, spectral index of density variations and of the 3D anisotropy parameters of cosmic rays near the Earth (see 2.1.2).

Solar wind - magnetosphere coupling is described by different models. The global structure of the magnetosphere is restored from boundary conditions that associated with shape and location of the magnetopause. Three-dimensional model of the Earth dayside magnetopause is accessible online on SINP web-page http://dec1.npi.msu.su/~alla/. The model permits to calculate 3D shape and size of the dayside magnetosphere boundary in dependence on the solar wind dynamic pressure and IMF B_y and B_z components [Dmitriev and Suvorova, 2000].

Determination of the spatial structure of storm-injected relativistic electrons of outer radiation belt in dependence on the Dst-variation is performed by a model of Tverskaya [2000]. The dependence permits also to predict extreme storm-time location of some very important magnetospheric

plasma domains such as extreme latitude of west electrojet center during a storm, boundary of discrete auroral forms, trapped radiation boundary and intensity maximum of symmetrised storm-time ring current [Tverskaya, 2000].

3.2. Forecasting

RSWI web-site supports references on both long time and short time forecasting of the space weather events. Long term forecasting (months - years) is based on the empirical models of averaged values of different physical parameter depending on the solar activity. Short time forecasting (hours - days) is produced by means of dynamical empirical model applications to the current situation determined from the diagnostic of the space weather conditions.

3.2.1. Long-time forecasting

The long-time forecasting models of the solar activity and heliospheric dynamics have been developed on the base of the artificial neural network (ANN) technique in SINP MSU (http://dec1.npi.msu.su/~dalex/events/iswmc/sept99.htm). The solar activity model permits to predict monthly dynamics of yearly means sunspot number (W) and solar radio flux ($F10.7$) during the current XXIII solar cycle (up to 2005). The ANN model shows that current 23-d solar cycle will be similar to XX solar cycle with maximum in the autumn-winter 1999 and peak mean value W~110. Next solar minimum is supposed between 2006-2008.

The dynamical models of near Earth's radiation permit to describe the dynamics of three main kinds of radiation: galactic cosmic rays (GCR), solar energetic particles (SEP) and trapped radiation. The SINP MSU dynamical model of the GCR is accessible on web-page http://www.npi.msu.su/gcrf/form.html. The GCR model allow calculating differential flux of different nuclear types (from H to U) for definite parameters of the near Earth's space mission orbit (Inclination, Perigee, Apogee, Right Ascension of perigee) as a function of solar activity (W) and geomagnetic conditions (Kp-index). This model is incorporated now in the American model CREME96. Probabilistic model of solar cosmic rays (SINP MSU http://www.npi.msu.su/scrf/form.html) allows prescribe the size of the solar particles fluences and peak fluxes that are expected within a given probability, to be exceeded at a given solar activity level within a given time interval. The improved trapped radiation model based on information system SEREIS (http://dec1.npi.msu.su/~vfb/SEREIS/) is able to describe smoothed variations of energetic particles in the magnetosphere associated with local time, seasonal and solar cycle variations.

3.2.2. Short time forecasting

Short time forecasting of the large solar flare and solar geo-effective events (large flare, filament ejection and coronal holes) impacting on the geomagnetic and radiation condition have been developed in IZMIRAN (http:\\izmiran.rssi.ru\space\solar\forecast.html) and weekly updated. Short time large flare event forecasting is presently based on observation by the process of new magnetic flux emergencies, its evolution: the magnitude and rate of emergence, its localization and interaction with already existing magnetic fields of the active region or outside of it. Taking into account physical and geometrical parameters of the own flare and the flare active region makes possible to predict the space weather: parameters of the solar proton events, the characteristics of geomagnetic activity and other. The method has been put to successful test on Russian scientific satellites such as GRANAT, GAMMA, CORONAS-I [*Ishkov*, 1999].

The other method of short time prediction of the geoeffective solar flares is based on the statistical studies of the special radio observations in the microwave region (wavelength about 3cm) (*Kobrin, et al.*, 1997) An algorithm of the forecast procedure consists of the comparison of the mean long-period ($t>20$min) pulsation amplitude in the current series of observations and that one in a calm (nonflare) period taking into account the specific features of the equipment and observation procedure. The investigations are under way to create procedures for short time forecasting of Coronal Mass Ejection onset on the basis of registration the nonstationary processes in the lower layers of the solar atmosphere at a stage of the coronal mass ejection formation using the patrol observations of solar radio emission.

The neutron monitor technique is also used for short-term predictions of geomagnetic disturbances. This method is based on monitoring GCR variations and scintillations, which can be used for identification and short-term (days) predictions of IMF disturbances. The predictions obtained using this technique can be found at http://teor.ysn.ru/rswi/graph-GIF.html

The empirical dynamic model, developed by *Popov et al.* [1999] can be used for numerical short-term on-line predictions of energetic electron fluxes in the outer RB. The current step of modeling is the evaluation of quantitative relations between the current solar wind parameters (n, V, B_z, B and their combinations) and current smoothed values of electron fluxes in geostationary orbit. The running average procedure is used as a smoothing filter, which resolves the fluxes of trapped and quasi-trapped electrons. It was also obtained, that the impact of *Dst* and V input parameters on the electron fluxes is practically the same. During magnetic storms the smoothed component of electron fluxes in geosynchronous orbit (energies from tens keV to 1.4 MeV) displays a distinct and stable dependence on the current V and *Dst* values. The dynamics of this component (in time and space) is similar to a certain trapped electron diffusion wave, propagating inside the magnetosphere and crossing the geosynchronous orbit.

For short-term predictions of the current state of ionospheric parameters and prompt correction of the ionospheric mode, a special software package was developed in NIRFI (http://www.nirfi.sci-nnov.ru/english/index3e.html). It has been shown on the basis of the experimental data obtained on the chirp sounder traces that the ionospheric model adaptation to current ionospheric conditions permits to forecast MOF on a real-time scale and with an error which is smaller by a factor of 3-5 than that of the long-term forecast method.

4. CONCLUSIONS

As it was shown above, in a number of space environment research fields, which are vitally important for Space Weather progress, Russia plays a key role.

These research fields include:

- ground based and atmospheric measurements of magnetic fields, cosmic rays and optic observations;

-space experiments, including deep space probes and magnetospheric satellites;

- nowcasting and forecasting modeling

We hope that our efforts in these fields will be an important part of the international space weather collaboration.

Acknowledgements. The author thanks all the contributors of the "Russian Space Weather Initiatives" web-site (http://alpha.npi.msu.su/RSWI/rswi.html) and especially the site coordinator Dr. A. Dmitriev for the opportunity to use the materials of the collaboration. The author thanks the organizers of the AGU Chapman Conference on Space Weather "Progress and Applications" for the opportunity to present this report.

REFERENCES

Avdyushin S., A. Belov, A. Dmitriev, et al., Russian Space Weather Initiatives, *Proc. Workshop on Space Weather*, ESA, ESTEC, 185-198, 1999.

Belov A.V., Eroshenko E.A., Yanke V.G. Modulation effects in 1991-1994 years. *Correlated phenomena at the Sun, in the Heliosphere and in Geospace*, SP-415, ESA, ESTEC, 469-475, 1997.

Belov A.V., E.A. Eroshenko, V.G. Yanke, Indices of cosmic ray activity as reflection of situation in interplanetary medium, *Workshop on Space Weather*, ESA, ESTEC, 325-328, 1999.

Bashkirov V., Panasyuk M., Teltsov M., *Proc. of INT. Workshop on Respect to Heavy Particle Radiation*, Chiba, 151-161, 1998.

Dmitriev A.V., and A.V. Suvorova, Artificial neural network model of the dayside magnetopause: physical consequences, *Phys. Chem. Earth*, *25*, 1-2, 169-172, 2000.

Gosstandart RD50-25645.152-90. Methodical Instruction. Solar Cosmic Rays. Calculation Method of temporal variations of proton energy spectra. *M. Gosstandart USSR*, 1991.

Ishkov V.N., The forecast of geoeffective solar flares: resources and restrictions (in Russian), *Izvestija RAN (ser. Phyzicheskaja), 63*, 2148-2151. 1999.

Kobrin,M.M., V.V.Pakhomov, S.D.Snegirev, V.M.Fridman, and O.A.Sheiner, An Investigation of the Relationship between Long-Period Pulsations of cm Radio Emission and Solar Proton Flare Forecasts, *Proc. of Solar-Terrestrial Prediction-V*, Jan 23-27, 1996, Hitachi, Japan, 200-205, 1997.

Kozlov V.I., Starodubtsev S.A., Markov V.V. et al., Forecast of Space Weather on the Ground-Based Radiation Monitoring, *Proceed. 26 ICRC, 7*, 406-109, 1999.

Models of the Space (in Russian), MSU, Ed. S.N. Vernov, 1976.

Popov G.V., V.I. Degtyarev, S.S. Sheshukov, and S.E. Chudnenko, The Solar Wind Control of Electron Fluxes in Geostationary Orbit During Magnetic Storms, *Radiation Measurements, 30,* 5, 679-685, 1999.

Ryumin V.V., Talking About Mignitizm (in Russian), "Petrograd", Leningrad-Moskva, pp. 173, 1925.

Stozhkov Yu., Pokrevsky P., Zullo Zh., Influence of charged particles fluxes on atmospheric precipitation (in Russian), *Geomagn. i Aeron.*, *36*, 4, 211-221, 1996.

Tchijevsky, A.L., Sun-spot and History, *Bulletin of the New York Academy of Sciences*, New York, 1928.

Tverskaya L.V., Diagnosing the magnetospheric plasma structures using relativistic electron data, *Phys. Chem. Earth.*, *25*, 1-2, 39-42, 2000

M.I. Panasyuk, Skobeltsyn Institute of Nuclear Physics, Moscow State University, Moscow, 119899, Russia. (panasyuk@srdlan.npi.msu.su)

Solar Wind and Interplanetary Magnetic Field: A Tutorial

C. T. Russell

Institute of Geophysics and Planetary Physics and
Department of Earth and Space Sciences
University of California Los Angles, California

The convection layer completes the transport of energy from the nuclear furnace at the center of the sun to its radiation into space by the photosphere, but most importantly for the solar wind it sets the temporal and spatial scales for the structure of the coronal magnetic field that in turn controls the properties of the solar wind. In this tutorial review we examine the properties of the fields and particles that constitute the solar wind and ultimately affect space weather and the underlying physical processes. In particular we discuss the role of the coronal magnetic field; the effect of the rotation of the sun; and the properties of the principal solar wind disturbance at 1 Astronomical Unit, the interplanetary coronal mass ejection.

1. INTRODUCTION

We know much about the solar wind at the orbit of the Earth. It flows with a speed of about 400 km s^{-1} and has a density of about 5 electrons and ions cm^{-3}. It carries a magnetic field of about 5 nT, lies on average near the ecliptic plane in an Archimedean spiral pattern, but is highly variable about this direction. We know about anisotropies, heat flow and chemical composition but we know very little about the source of the solar wind. This situation arises both because the origin of the solar wind is complex and because the region of solar wind acceleration has not been probed with in situ investigations.

Addressing the solar wind from first principles is also difficult because the solar wind is a magnetized plasma in which, throughout most of the solar system, the magnetic and plasma pressures are comparable. Often our intuition, trained on the behavior of gases, fluids and magnetic fields in a vacuum, does not strictly apply in a plasma like the solar wind. To treat this regime properly many approaches have been tried, each with a different approximation and with minimal guidance from observables in the critical regions. The fluid approximation assumes that the plasma can be treated by examining the behavior of moments of the distribution such as density and velocity. The behavior of the fluid is governed by a set of equations including Maxwell's laws, for the electromagnetic quantities, and the conservation of mass, momentum and energy, and possibly higher order moments. Frequently these equations are truncated by setting the high order moments to zero and the equation of state of an ideal gas becomes the energy equation. Here $P\rho^{-\gamma}$ is constant where P is the pressure, ρ is the mass density and γ is the polytropic index. This closure in fluid treatments is often arbitrary and based on analogies with ideal gases that may not apply. Further a decision on the number of components of the plasma that need to be separately treated must be made. Should ions and electrons be treated separately? Should different ions be considered separately? Is there more than one population of electrons or ions that requires a separate treatment? Thus even a fluid treatment of the solar wind can be cumbersome. Kinetic treatments avoid some of these difficulties [*Meyer-Vernet*, 2000] but may lead to tractable solutions in only simple geometries. The caveats about the number of components that may be important in a plasma applies equally well to kinetic treatments. As a result unlike many areas of space

Space Weather
Geophysical Monograph 125

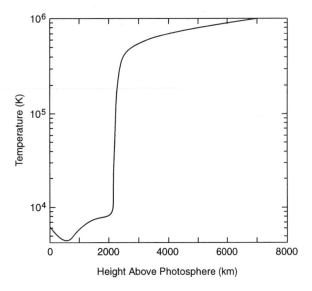

Figure 1. Plasma temperature from the photosphere, through the chromosphere and transition region to the lower corona. Adapted from Noyes [1982].

physics, our understanding of the solar wind is founded on long term observations rather than a solid foundation of theory.

In addition to being unable to decide a priori the number of components of the solar wind that control its behavior, there is controversy as to whether the distribution of solar wind particles can be treated as a Maxwellian or not, and even whether fluid theory can be applied. In a collisional gas the phase space distribution, the number of particles per unit of configuration and velocity space tends toward a Maxwellian distribution. Maxwellians have special properties that make them simple to use, but plasmas are not always Maxwellian. Thus one must be very careful about applying theories or equations that may implicitly assume Maxwellian behavior. To rectify this problem some authors approximate space plasmas with Lorentzian distributions that has a high-energy tail, characterized by a value, kappa. For a kappa of infinity the Lorentzian distribution becomes Maxwellian. The Lorentzian distribution can be used in a fluid approach, and so it is not strictly correct to call plasma studies kinetic just because they use a non-Maxwellian distribution. Nevertheless, this is often done in the literature, adding ambiguity to the use of the term "kinetic" treatment. As we will see below when we examine the detailed properties of the solar wind, the observed behavior of the solar wind may be very strongly affected by its non-Maxwellian distribution.

The energy of the sun derives from the conversion of mass to energy when hydrogen atoms are converted by fusion to helium and helium converted to carbon. This energy is radiated from the core into the layers above. At about 0.75 solar radii (R_s) heat transport by radiation drops in efficiency below that of convection and, the outer layer transports the escaping heat to the surface by convection. This mass motion results in cooler material sinking from the photosphere into the convection layer and hot material rising toward the photosphere. The mass motion drives the magnetic dynamo that we see on the sun and whose time variations are so very important for solar terrestrial relations.

The time scale for the transport of heat from the interior of the Sun is millions of years while the time scales in the photosphere range from hours to years. Once heat reaches the photosphere it radiates to the Earth at the speed of light reaching 1 AU in about 8 minutes or convects in the solar wind at about 400 km s^{-1} reaching the Earth in about 4 days. In this review we examine the behavior of this magnetized solar wind plasma. We discuss its behavior beginning with the simplest possible model: a spherically symmetric, non-magnetized, non-rotating and time-stationary star. Next we examine those properties that are affected by its magnetic field, those associated with rotation, and finally those associated with the time variation of the system. We hope that, by dividing the phenomena this way, a greater understanding of the ultimate causative mechanisms may be more readily obtained.

Finally, we note that the solar wind is important for very practical as well as academic reasons. For those interested in space weather, the interaction of the solar wind with the Earth's magnetic field is very important. The solar wind both controls the size of the magnetic cavity through its momentum flux or dynamic pressure and the energy flow into the magnetosphere coupled from the solar wind mechanical energy flux by the reconnection of the interplanetary magnetic field with the terrestrial field. This coupling is strongly controlled by the direction of the interplanetary magnetic field, being most strong when the interplanetary magnetic field is southward. Thus for space weather applications we are most interested in the solar wind velocity and mass density and the strength and orientation of the interplanetary magnetic field. Nevertheless, the interaction with the Earth is complex and other parameters such as temperature are important as well.

2. THE SOLAR WIND PLASMA

The temperature of the photosphere is less than 6000K and as a result the sun is most luminous in what we call visible light. The temperature at first falls and then rises rapidly as one moves above the photosphere as shown in Figure 1. This behavior is reminiscent of the behavior of the upper atmosphere of the Earth, in which the absorption of solar photons in the stratosphere and the thermosphere

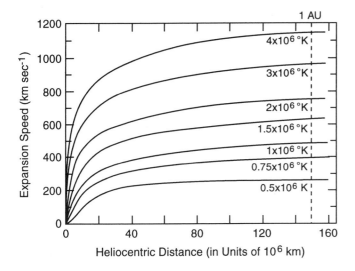

Figure 2. Theoretically derived speed of the solar wind from an isothermal model for varying coronal temperatures. Adapted from Parker [1958].

create warm layers well above the surface of the Earth. Here too the corona appears to be a region of heating from which heat is conducted downward to the chromosphere below and convected outward from the corona by the solar wind. Understanding the source of this coronal heating is one of the outstanding problems of the solar wind.

The high temperature of the corona is important since the solar wind must escape the deep gravitational potential well of the sun. The escape velocity is 625 km s^{-1} from the surface of the sun. Thus a proton requires over 2 keV to escape to infinity from the surface of the sun and about 0.5 keV to escape from 5 R_s. The thermal energy of a 10^6 K proton is only about 100 eV. Understandably there was some doubt expressed before the advent of the first in situ solar wind data about the feasibility of a continual high speed solar wind, but *Parker* [1958] predicted such a wind based upon hydrodynamic theory and the long-term observations of Mariner 2 confirmed its reality [*Snyder et al.*, 1963]. Figure 2 shows the results of an isothermal fluid treatment of the expansion from the corona as first described by *Parker* [1958]. Because the pressure gradient falls off with radial distance more slowly than the gravitational force the solar wind is accelerated even to supersonic velocities. The Earth sits about 230 R_s from the sun. As Figure 2 illustrates, the solar wind has achieved close to its asymptotic velocity by the time it reaches 1 AU. For a coronal temperature of close to 10^6 K the observed median solar wind velocity is obtained in this solution. Nevertheless, modeling the solar wind expansion is still an active area because we have few observational constraints on the nature of the "heating processes" in the acceleration region. In part because of the weakness of our under-

standing of the solar wind in those critical regions of the corona where we have not been able to obtain in situ data we concentrate in this review on solar wind phenomenology at 1 AU.

2.1 Ions

Most solar wind instruments flown to date return counts versus energy per charge which can be interpreted as counts per increment of velocity times square root of mass/charge as shown in Figure 3. In this particular example we see that there are four beams: two proton beams and two alpha beams. In a magnetized plasma the bulk velocity of all particles perpendicular to the field must be equal. The bulk velocity along the field need not be so, and here it is clearly not equal. The reason for such differences must still be speculative, but since interpenetrating ion beams are not rare, it is important to understand their origin. The widths of these peaks are narrow compared to the energy of the peak. Thus the thermal velocity of the particles (the random component perpendicular to and parallel to the magnetic field) is small compared to the bulk velocity. This is an indication that they are supersonic. They move faster than the speed of a sound wave in the plasma. They also move

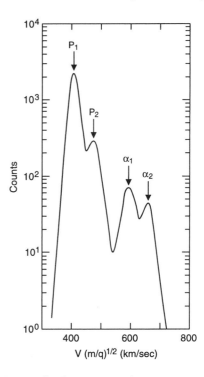

Figure 3. Unnormalized energy per charge spectrum of the solar wind ion population at a time when interpenetrating proton and alpha particles are present. These two alpha particles beams have the same velocities along the magnetic field as the two proton beams. Adapted from Asbridge et al., [1974].

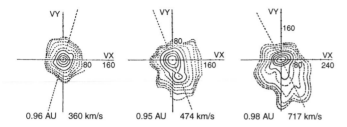

Figure 4. Selected two-dimensional solar-wind ion distributions at varying heliocentric radii as measured by the Helios spacecraft. The dashed line shows the projection of the interplanetary magnetic field in the spin plane of the spacecraft. Adapted from Marsch et al. [1982].

faster than the magnetosonic velocity that is a combination of the sound and Alfven speeds. Not as obvious here is the fact that the alpha particles have temperatures about four times greater than those of the protons. In fact all ions have temperatures that on average are roughly proportional to their atomic mass, i.e. their thermal velocities are roughly equal [*Boschler et al.*, 1985]. This relationship, however, is not strictly obeyed. Coulomb collisions tend to equilibrate these temperatures, especially when the ion temperatures are low [*Klein et al.*, 1985]. Thus deviations from this relation are found in cool dense solar wind flows and He^{++} cools more rapidly with heliocentric radius than protons [*Thieme et al.*, 1989].

The distribution shown in Figure 3 is a one-dimensional cut along the direction toward the sun. A rapidly spinning single detector can resolve a two dimensional or angular distribution, and, if it has multiple sensors such as on Helios, a three-dimensional spectrum can be measured. Since charged particle distributions are usually cylindrically symmetric around the magnetic field (gyrotropic), the angular distribution can be displayed, as in Figure 4, by displaying the flux parallel and perpendicular to the magnetic field. The angular distribution of the solar wind ions is complex, often with quite different shapes in the directions parallel and perpendicular to the magnetic field. Sometimes the contours are elliptical with their greatest dimension along the field, so that $T_{\parallel} > T_{\perp}$. In other cases the contours are elongated perpendicular to the field, so that $T_{\perp} > T_{\parallel}$.

If we assume that the entropy of the solar wind ions does not change as the ions move outward then we can relate pressure and density by the polytropic law $P\rho^{-\gamma}$ = constant. For $P = nkT$, this translates to $n^{(1-\gamma)}T$ being constant. Since γ is greater than unity, T decreases as N decreases. As the solar wind expands radially, and roughly spherically, its density falls off as r^{-2}. We therefore expect the solar wind temperature to fall as the solar wind expands and it does. In a collisionless magnetized plasma, in the absence of fluctuations both near the gyrofrequency and at magnetic

scale sizes near the gyro radius, the first adiabatic invariant, or magnetic moment, of the ions will be conserved. Thus, the perpendicular energy of the ions divided by the local magnetic field strength would be constant, and as the ions propagate outward, their pitch angles would become more field-aligned. This behavior is indeed seen qualitatively, but not quantitatively. There is heat transport along the field and dissipation. Wave particle interactions couple the parallel and perpendicular ion temperatures in the solar wind and limit the ratio of these two temperatures as the solar wind expands. The slow solar wind that is denser and has a longer transit time cannot be assumed to be completely collisionless. The observed distributions are also affected by the electric potential of the solar wind. In the fast solar wind for reason we do not completely understand the temperatures are more isotropic than they are in the slow solar wind.

At the Earth the solar wind is highly variable. In sections 3-5 we discuss the various processes that lead to this variability but here we concern ourselves just with the statistical properties. Figure 5 shows a histogram of the solar wind velocity. It can be as slow as 260 km s^{-1} and faster than 750 km s^{-1}, but typically lies about 400 km s^{-1}. Figure 6 shows the density. It is much more variable than the velocity, ranging from about 0.1 cm^{-3} to 100 cm^{-3}. Thus variations in the dynamic pressure with which the solar wind blows against the magnetosphere and the size of the magnetosphere are controlled principally by the density fluctuations. The factor of three changes in the velocity makes a $\pm$ 20% change in the radius of the magnetosphere, but the range of density makes a $\pm$ 80% change. Over the solar cycle the yearly average dynamic pressure changes about $\pm$ 20% [*Petrinec et al.*, 1991]. This makes a small but perceptible ($\pm$ 3%) change in the size of the magnetosphere. When high densities and high velocities occur together the size of the magnetosphere can be halved, but in general the density of the solar wind and the velocity are inversely correlated as shown in Figure 7. While it is often said that the number flux is roughly constant, the line of constant flux shown in Figure 7 indicates that this constant flux rule is only a rough approximation. In contrast the ion temperature is positively correlated with the solar wind velocity as shown in Figure 8. This correlation is certainly qualitatively consistent with the Parker's model of solar wind expansion illustrated in Figure 2.

2.2 Electrons

Although frequently it is possible to treat ions as a single fluid in the solar wind, this is not true of electrons. Electrons most frequently appear to have at least two components, a core and a halo. As shown in Figure 9, the core is colder than the halo and it is denser. It is important

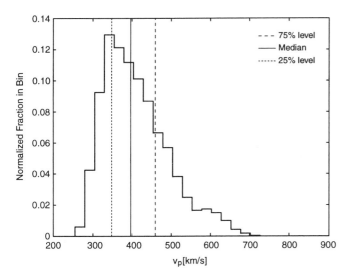

Figure 5. Histogram of the solar wind velocity derived from 18 months of ISEE-3 observations. [Adapted from Newbury, 2000]. Quartiles of the velocity are: 348, 397 and 459 km s⁻¹.

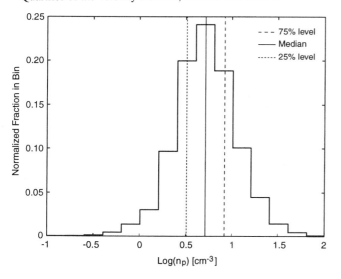

Figure 6. Histogram of the logarithm of the solar wind density derived from 18 months of ISEE-3 observations. Quartiles of the density are: 3.2, 5.2 and 8.4 cm⁻³. Adapted from Newbury [2000].

also to note that the thermal speeds of both populations are greater than the solar wind bulk speed. Thus the electrons can leave the sun much faster than the ions although in the direction perpendicular to the magnetic field they drift together at the same velocity. A paradox is introduced by this difference in velocity that arises because in the corona the ions and electrons have similar temperatures but very

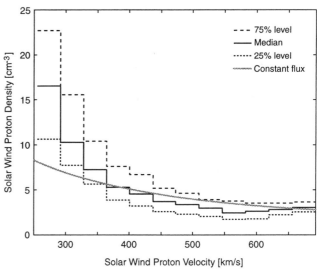

Figure 7. The variation of the solar wind density with solar wind velocity. The median, quartiles and mean values are shown. Adapted from Newbury [2000].

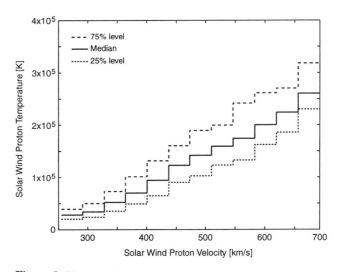

Figure 8. The variation of the solar wind proton temperature with the solar wind velocity. The median and quartiles are shown. Adapted from Newbury [2000].

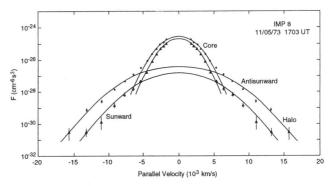

Figure 9. The distribution function of solar wind electrons along the magnetic field direction showing the core and halo populations and their differing bulk velocities and temperatures. Adapted from Feldman et al. [1975].

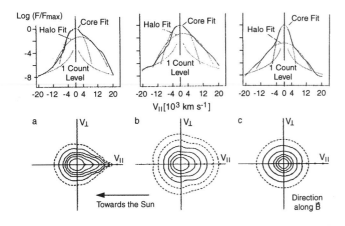

Figure 10. Typical two-dimensional distribution functions of solar wind electrons showing cuts through the distribution function along the magnetic field. On the left is a distribution with a noticeable strahl component along the magnetic field. On the right is a distribution that is almost isotropic and having a very weak halo component. Adapted from Pilipp et al. [1987].

dissimilar masses. In a collisionless gas, like the solar wind over most of its transit from the sun, charged particles do not leave their magnetic field lines. They also maintain quasi charge-neutrality. How can they do this if they are not traveling at the same speed? The answer is trivial if the source regions on the sun are constant in time so that the flux of electrons and ions at the base of a field line is constant. The density is always the same and it does not matter if the protons and ions stream relative to each other along the field. If the solar wind production rate varies in time and the ion density along a flux tube varies with distance, then the electrons have to slow down as they pass through dense regions and then speed up in rarefied regions. This occurs because when there is an over abundance of ions there will be a polarization electric field that attracts the electrons to that region. In regions of under dense ions the electrons will be expelled by the excess negative charge. As a result, charge imbalances are minimal in the solar wind despite the speed differences and time variations. We might expect variations in the electric potential along the magnetic field but no detectable variation in charge neutrality. Because the core electrons travel most slowly they are affected the most by these electric potential variations, and the overall electric potential surrounding the sun.

Figure 10 shows angular distributions of the solar wind electrons and cuts through these distribution functions. Sometimes the electrons can be nearly isotropic. At other times, not only is there the core and halo distribution but also a narrow field-aligned high energy beam that is called the "strahl". Since the collision frequency of charged particles is inversely proportional to square of the particles' energy, the high energy strahl can propagate farthest without scattering. In some sense the strahl provides a look at the

deepest part of the corona. As with the ions we expect gyrotropy about the magnetic field direction, and as with the ions there is great variability in the shapes of the distributions. In stark contrast to the ions the electrons have a nearly constant temperature when averaged over the distribution function which is dominated by the denser core electrons. Figure 11 illustrates this constancy versus the solar wind velocity. This constancy occurs in part because the electrons are strongly coupled to the corona by their high thermal velocities. When they do get heated in the solar wind, say by the passage of a strong shock, they rapidly (in hours) return to their typical temperatures. The main exception to this rule is when the ions are exceptionally hot. This controlling factor is shown by the plot of electron temperature versus ion temperature in Figure 12. If the electron and ion temperatures were completely independent then the distribution of electron temperatures would be the same at each ion temperature. This is largely true except at the lower edge of the electron distribution where a cutoff occurs that moves to higher electron temperatures as the ion temperatures rise. Since we know of no waves that can couple the electrons and ion thermal energies we are left to speculate that the electric potential of the solar wind must in some way be doing this coupling.

We have stressed in this review the non-Maxwellian nature of the solar wind electron and ions. One might think this could have but a minor effect on the behavior of the particles, perhaps because of wave-particle interactions. However, the non-Maxwellian nature of the distributions may have a very important and non-intuitive effect on the plasma. Figure 13 plots Maxwellian and a Lorentzian distribution functions f(E) as a function of energy E

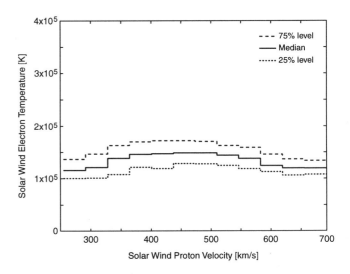

Figure 11. The temperature of solar wind electrons as a function of the solar wind velocity. The median and quartiles are shown. Adapted from Newbury [2000].

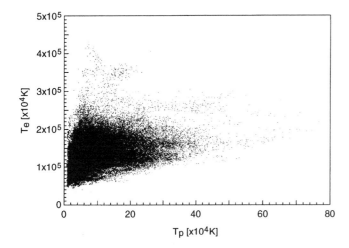

Figure 12. The temperature of solar wind electrons versus solar wind ion temperature observed by ISEE 3. Individual 5-minute observations are shown. Adapted from Newbury [2000].

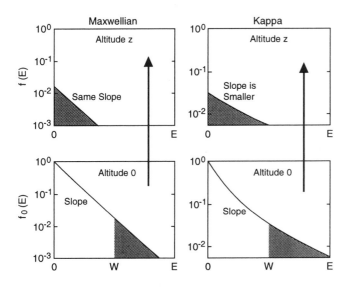

Figure 13. Velocity filtration of non-Maxwellian distributions by an attractive force. When the Maxwellian distribution on the lower left moves upward in Z against the force of the attractive potential it loses potential energy, W, and translates to the left, but maintains its slope. The Lorentzian distribution on the right changes its slope when it moves to higher altitude Z and loses potential energy W. [Meyer-Vernet, 2000].

showing their evolution with height in a potential that is pulling the particles down. When the Maxwellian particles rise to altitude, Z, and thereby lose potential energy, W, their distribution function has the same shape. When the particles with the Lorentzian distribution rise, the distribution function has a new slope near zero energy because the slope of the original Lorentzian varied with energy. Thus the temperature at high altitude appears to be greater.

The gravitational field of the sun is an attractive potential. This potential is modified because the electron thermal speed greatly exceeds the bulk velocity of the ions. In a vacuum the sun would charge up positively to keep the electrons in check and to give the ions a boost. This does happen in a plasma but only over a Debye length. In the solar wind an ambipolar electric field appears that maintains charge neutrality and modifies the gravitational potential. *Scudder* [1992] has used this potential plus Lorentzian particle distribution functions to produce a rise in temperatures in the corona without any additional source of heat. Figure 14 shows the electron and ion temperatures as a function of distance for a particular Lorentzian distribution [*Scudder*, 1992]. A strength of this model is that the temperature of heavier than the average, non-Maxwellian ions should increase as their potential energy, proportional to their mass, and such an increase is seen [*Scudder*, 1992; *Meyer-Vernet*, 2000]. This hypothesis for explaining solar wind properties is the subject of much controversy. If indeed this model were to explain the temperature of the corona, it still has not solved the main solar wind problem, as it does not specify how the non-Maxwellian distributions are produced, only how they evolve. The various postulates about the roles of waves, small scale reconnection and other micro processes in the corona are relevant to either the Maxwellian or Lorentzian approaches.

We cannot presently probe the inner corona directly to measure its electron distributions, but the Earth's magnetosphere surprisingly gives us a way to recreate at least some of the coronal electron distribution function. Figure 15 shows the phase space density of "polar rain" energetic electrons, measured on open tail field lines connected to the polar cap, as a function of electron energy [*Fairfield and Scudder*, 1985]. These originally solar-wind

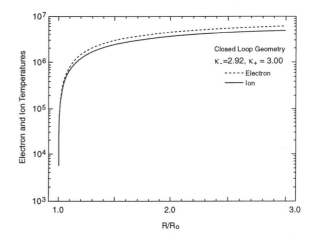

Figure 14. Electron and ion temperatures for solar coronal model using non-Maxwellian distributions and no heat deposition in the coronal [Scudder, 1992].

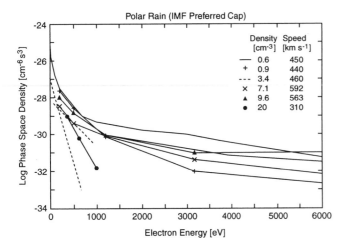

Figure 15. Phase space density of polar rain versus energy for days with differing solar wind conditions. Adapted from Fairfield and Scudder [1985].

electrons have entered the magnetosphere by reconnection of the interplanetary magnetic field with magnetotail field lines. In the tail and especially at low altitudes the magnetic field increases to values approaching those in the corona and the pitch angle distribution present in the corona can be recreated. This mapping is negatively affected by collisions, and these are greatest when the solar wind is densest. Thus we would expect the greatest anisotropies to survive when the solar wind is least dense and because scattering is least at highest energies the anisotropy should be greatest at highest energy. Figure 15 shows that, when the solar wind density is greatest, the polar rain and the solar wind strahl are weakest and, when the density is least, the strahl and polar rain are most intense as evidenced by the high energy tail on the distributions. When the solar wind density gets extremely low as it did for example on May 11, 1999, the strahl and the polar rain can become intense enough to cause x-ray auroras over the polar cap [*D. L. Chenette*, personal communication, 1999].

3. THE SOLAR WIND MAGNETIC FIELD

Processes inside the sun break its temporal and spatial homogeneity, creating an ever-changing photosphere. An important element of this spatial and temporal in-homogeneity is the presence of a magnetic field generated at the base of the convection layer that extends high into the corona and the solar wind. This magnetic field in turn affects the spherical symmetry of the solar wind expansion and suppresses that expansion in some places. Figure 16 shows both an undistorted dipole field and isothermal MHD coronal expansion model of *Pneumann and Kopp* [1971]. The dipolar magnetic field in this model keeps the equatorial corona tied to the sun. At higher latitudes the

solar wind expands along the magnetic field, carrying the solar wind field into interplanetary space from the polar regions of the sun and creating a heliospheric current sheet between the two polar coronal hole outflows.

This magnetic pattern is usually not symmetric about the rotational axis of the sun. Even the tilted stretched dipolar configuration shown in Figure 17 is an oversimplification because the heliomagnetic equator that separates inward and outward interplanetary magnetic field is quite warped. As the sun rotates, the polarity pattern set by the Sun's magnetic field sweeps over the Earth causing a magnetic sector pattern in which the magnetic field is outward from the sun for perhaps 7 to 14 days and then inward as the Earth moves from the magnetic north hemisphere to the south. Furthermore, the Earth's orbit is not in the rotational equator of the sun so that in the course of a year the Earth spends six months above the rotational equator and six months below it. As illustrated in Figure 18 these excursions can result in a spacecraft in the ecliptic plane seeing predominantly inward and then predominantly outward fields six months later, but on average the magnetic field is nearly equally inward and outward.

For the reasons discussed in section 4 the angle that the average magnetic field makes in the solar equatorial plane or the ecliptic plane lies mainly along a spiral defined by the rotation of the sun, the distance from the sun and the solar wind speed. This direction is generally called the Parker spiral. Individual magnetic vectors can lie at any angle. The occurrence rate of the estimated or spiral angle for three sets of solar conditions and at two heliocentric radii are

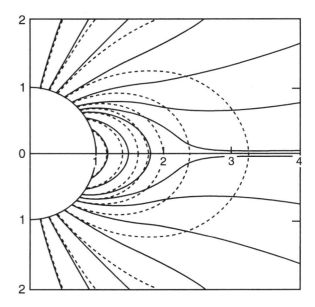

Figure 16. Coronal magnetic field in the Pneumann and Kopp [1971] isothermal solar wind solution. Dashed field lines show the magnetic field lines of a dipole for comparison.

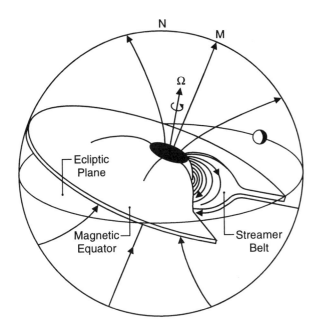

Figure 17. Idealized three dimensional view of the streamer belt and the coronal magnetic field. Orbit of the Earth carries it above and below the sun's equator in the course of the year, modulating but not eliminating the sector structure. Because the streamer belt is tilted with respect to the rotational equator the streamer belt can sweep across the Earth twice every solar rotation.

given in Figure 19 [*Luhmann et al.*, 1993]. We note that for the reasons discussed in section 4, the Parker spiral angle at 0.7 AU is closer to 0 and 180 (radially toward and away from the sun) than it is at 1 AU. This plot also demonstrates the very near equality of the occurrences of magnetic fields toward and away from the sun.

The magnitude of the solar wind magnetic field is also quite variable as illustrated in Figure 20 again for two heliocentric distances and three levels of solar conditions. It is clear that the distribution is not Gaussian but that it has a long high amplitude tail, especially so at active times. This behavior is reflected in the components of the field shown in Figures 21, 22 and 23. Note that here the 0.72 AU measurements for the three components, Bx, By and Bz have been adjusted to 1.0 AU both to be more easily compared with the IMP 8 and to be more relevant to space weather applications. We see again that the strongest fields occur at solar maximum when there is a high field tail on the distribution. These high fields are almost exclusively due to interplanetary coronal mass ejections that are discussed in Section 5. We note that the shape of the Bx and By distributions differ from the shape of the Bz distribution near zero strength components. This occurs because the sector pattern causes the Bx and By components to be distributed around two modal values, one negative and one positive, while the Bz component is distributed around zero. Further there is some evolution in the distributions as

the magnetic field is carried from 0.7 to 1.0 AU. We attribute the appearance of more high fields at 1 AU than in the adjusted 0.72 AU distribution as due to the interaction of fast and slow streams as the solar wind moves from the orbit to Venus to that of Earth.

The closed magnetic regions allow the corona to build up in density and temperature in the magnetic equatorial or streamer belt region, as sketched in Figure 24 and as seen in coronagraphs or with the naked eye at the time of eclipses. The regions on the sun, from which the solar wind flows freely and whose magnetic field lines are open, are called coronal holes. In these regions the density is lower and the temperature cooler. The connectivity of coronal holes and the more distant solar wind has been demonstrated by tracing field lines back to the sun with an MHD model

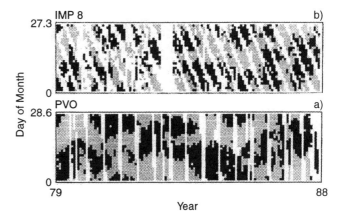

Figure 18. Sector structure plots for 0.7 AU (bottom) and 1.0 AU (top) showing periods when toward (stippled) and away (solid) magnetic fields were seen by Pioneer Venus and IMP-8. Gaps occur when the spacecraft are in the magnetosheath [Luhmann et al., 1993].

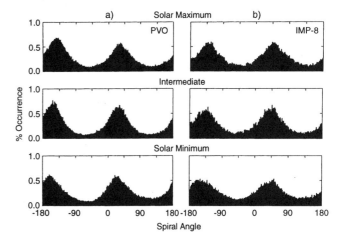

Figure 19. Histograms of the occurrence of various spiral angles in the Pioneer Venus 10-minute averages and the IMP-8 5-minute averages separated according to solar activity level [Luhmann et al., 1993].

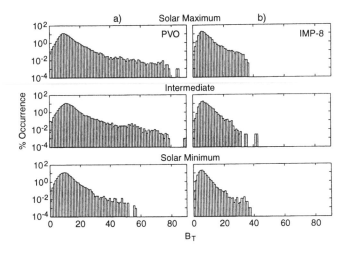

Figure 20. Histograms of the occurrence of field magnitudes in the Pioneer Venus 10-minute data and the IMP-8 5-minute data, separated by solar activity level. [Luhmann et al., 1993].

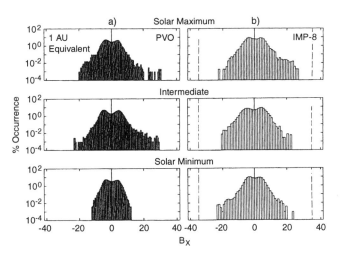

Figure 21. Histograms of the radial or Bx component of the interplanetary magnetic field at IMP-8 and Pioneer Venus for three levels of solar activity. The size of the plot of the IMP-8 data has been scaled to 0.72 AU by a r^{-2} factor to make the plots easily comparable. The dashed vertical lines indicate that the IMP-8 data have a maximum value of 32 nT. [Luhmann et al., 1993].

[*Linker et al.*, 1999]. The model also reproduces the intensity variations seen in the streamer belt of the lower corona and has been used to predict the appearance of the sun during total solar eclipses.

Since the streamer belt is nearly equatorial at least at solar minimum, the slow solar wind is on average found near the ecliptic plane and the fast solar wind at high latitudes. At solar minimum Ulysses measurements out of the ecliptic plane [*McComas et al.*, 1998] shown in Figure 25 emphasize what a biased sample the Earth can receive of the solar wind. The latitudinal profile of the solar wind

velocity measured by Ulysses at solar minimum suggests that at that time there is a dichotomy of solar wind properties organized around the heliospheric current sheet with dense low speed, solar wind near the equatorial regions and rarefied fast flows at high latitudes. However, even though solar wind properties, such as composition, are ordered by the solar wind speed, this control appears as a continuous change and there is no sudden change in properties at some particular speed. Again using MHD models to extrapolate the solar wind back to its solar source

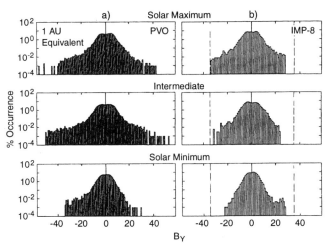

Figure 22. Histograms of the azimuthal or By component of the interplanetary magnetic field. The size of the plots of the IMP-8 values has been scaled to 0.72 AU by an r^{-1} factor. Other comments of the caption of Figure 21 apply. [Luhmann et al., 1993].

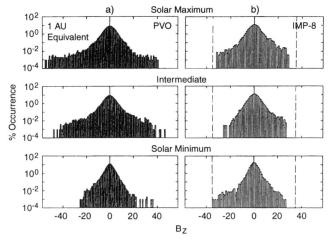

Figure 23. Histograms of the north-south or Bz component of the interplanetary magnetic field. The size of the plots of the IMP-8 values has been scaled to 0.72 AU by an r^{-1} factor. Other comments of the caption of Figure 21 apply. [Luhmann et al., 1993].

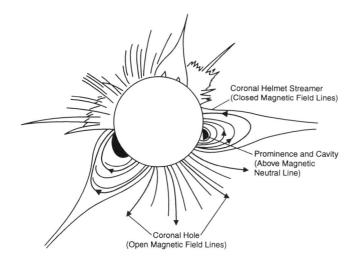

Coronal Helmet Streamer
(Closed Magnetic Field Lines)

Prominence and Cavity
(Above Magnetic
Neutral Line)

Coronal Hole
(Open Magnetic Field Lines)

Figure 24. More realistic sketch of the structure of the corona and its presumed magnetic field. Adapted from Hundhausen [1995].

puts the origin of the fast solar wind in coronal holes and the slow solar wind at the edges of coronal holes [*Neugebauer et al.*, 1998]. A corollary of this magnetic control of the solar wind expansion is that the flux tube divergence of the magnetic field measured at the source surface correlates well but negatively with the solar wind speed [*Wang and Sheeley*, 1990]. Where coronal magnetic flux tubes are straight and radial the solar wind is fast and where they strongly diverge the solar wind is slow.

During solar maximum coronal holes and regions of flux tube divergence are more evenly spread over the Sun's surface than at solar minimum, allowing the Earth to experience a broad range of solar wind velocities, almost independent of its heliographic or heliomagnetic latitude. During the declining phase of solar activity when the current sheet has a large tilt to the solar equator but there is more order to the coronal fields, a more stable fast/slow steam structure arises. It is possible that some slow solar wind arises in the coronal streamer belt [*Gosling*, 1997] but this source is not expected to provide as steady or as wide a region of slow solar wind as that observed.

A solar terrestrial consequence of this control of the solar wind by the magnetic structure of the sun could be that in the course of the year, as the Earth moves from near the Sun's rotational equator to higher (7°) latitudes on roughly March and September 5 each year, it might encounter more frequent high speed solar wind and hence encounter greater geomagnetic forcing. However, there is a second annual variation that affects geomagnetic activity. During the course of the year when the Earth's dipole is tilted closer and further from the ecliptic plane in the direction perpendicular to the Earth-Sun line, it can encounter greater and weaker southward interplanetary magnetic fields even though the field is lying in the ecliptic plane. Both the

former and latter effects could lead to a semiannual variation of geomagnetic activity because geomagnetic activity is controlled both by the velocity and magnetic field of the solar wind.

We can test how the Earth responds to these two possible effects by using a terrestrial geomagnetic index, designed to represent the level of disturbance of the Earth's magnetosphere. One such suitable index is the am index based on worldwide magnetic records at mid-latitudes. It is linearly related to the size of the magnetic disturbance seen in the magnetogram records and has been normalized to be uniform in Universal Time as the Earth rotates. As shown by the annual variation of the Am index in Figure 26, the latter (magnetic) control is stronger than the former (solar wind velocity) effect, since the top two panels that divide the data according to the direction of the east-west component of the IMF would each show evidence for a semiannual variation in geomagnetic activity if the velocity control were dominant. Rather two annual variations are seen, controlled by IMF direction.

Another effect that could be heliolatitude dependent is one due to the variance of the solar wind velocity. X. Li [this volume] has reported that fast solar wind streams have significantly larger variance in the velocity than slow streams possibly explaining enhanced diffusion and

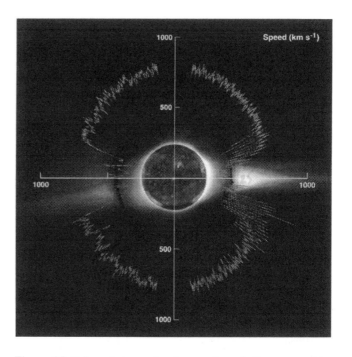

Figure 25. Solar wind speed and magnetic polarity measured by Ulysses as a function of helio-latitude overlaid with three concentric images of the sun and its corona obtained in the extreme ultraviolet by SOHO/EIT, the HAO coronagraph and the SOHO C2 coronagraph [after McComas et al., 1998].

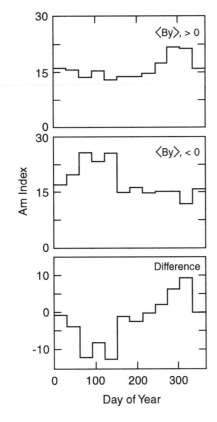

Figure 26. The annual variation of the Am index for positive and negative GSM Y component of the interplanetary magnetic field and the difference between the two variations. This illustrates that the magnetic field and not the solar wind velocity is dominantly responsible for the semi-annual variation of geomagnetic activity. [Russell, 1989].

acceleration of MeV electrons in the terrestrial magnetosphere. The source of this variation may be due to the existence of thin jets of high velocity plasma near the base of coronal holes [*Feldman*, 1997].

4. EFFECTS OF SOLAR ROTATION

Since the sun rotates, material emitted from the same source region lines up along an Archimedean spiral as illustrated in Figure 27. Since the magnetic field lines passing through those parcels also originate in this source region, the magnetic field too forms an Archimedean spiral pattern that at 1 AU makes an angle of about 45° to the solar radial direction for a 400 km s^{-1} solar wind velocity. Variations in solar wind velocity, and processes acting in the solar wind and corona, including reconnection, create a spread in directions about the average spiral angle as shown above in Figure 19.

If the sun's magnetic axis were aligned with the rotation axis, and the solar magnetic field were axially symmetric,

the fast and slow solar wind streams would not interact. However, even at solar minimum when the coronal magnetic structure is simple, the tilt of the streamer belt allows the fast streams to run into the slow streams as shown in Figure 28. This collision compresses the trailing edge of the slow stream and the leading edge of the fast stream. Thus the pattern of solar wind density and speed seen in the solar wind is a combination of a pattern imposed by the magnetic structure of the sun plus the effects of the interaction between streams. Figure 29 shows three solar rotations of data obtained by Mariner 2 in 1962, illustrating the fast-slow stream structure (heavy line) and the density variation (light line) accompanying it. These data, some of the first obtained in the solar wind, illustrate the general anti-correlation of density and velocity of the solar wind illustrated statistically in Figure 7. The structure in these quantities arises principally in their source regions on the sun caused by the magnetic control discussed above. There are kinematic effects as the fronts of the fast streams steepen when the fast core of a stream overtakes the slower leading edge as on October 8, November 15 and November 21. There are also what we term dynamic effects when the fast stream compresses the back end of the slow stream in front of it. Examples of this effect can be found on October 7, November 14 and November 29. Inside 1 AU the "dynamic" interaction between streams is less important in creating the observed density structure than the structure in

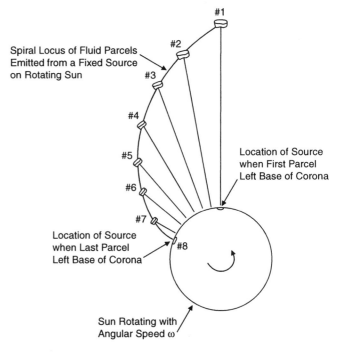

Figure 27. The creation of a spiral magnetic field pattern by the emission of magnetized plasma from a source on the rotating solar surface. Adapted from Hundhausen [1995].

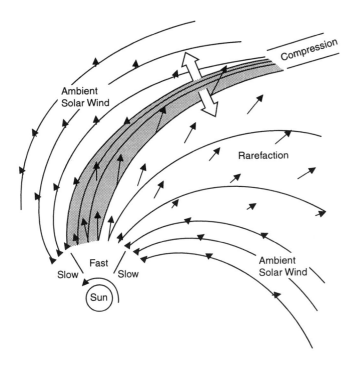

Figure 28. The collision of a fast stream with a slow stream to produce a corotating interaction. Adapted from Pizzo [1985].

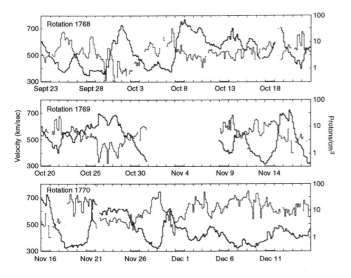

Figure 29. Solar wind velocity (dark line) and density (light line) measured by Mariner 2 on three successive solar rotations. Adapted from Hundhausen [1995].

the coronal source regions and their kinematic evolution within streams but at greater distances dynamic interactions become quite important.

At times during the solar cycle, especially during the declining phase, the stream structure is quite steady from rotation to rotation. This in turn leads to recurrence in terrestrial geomagnetic activity every time the same fast stream sweeps over the Earth. This is illustrated with the geomagnetic C9 index in Figure 30. The C9 index represents the peak of the early twentieth-century geomagnetician's art in attempting to visualize the effect of the Sun on the terrestrial magnetosphere. The font of the numbers printed illustrates on a logarithmic scale the strength of activity. The R9 column gives the average

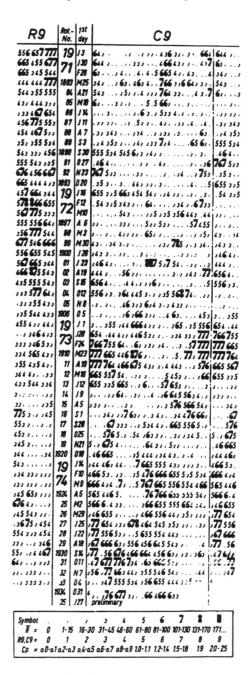

Figure 30. C9 index from 1971 through 1973 showing recurrent geomagnetic activity. Each line corresponds to one solar rotation. [McPherron, 1995].

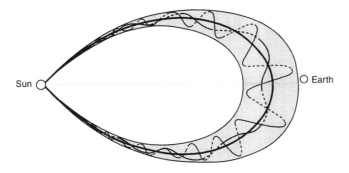

Figure 31. Modern paradigm of a magnetic cloud or magnetic flux rope, a frequent type of ICME. While the cross section is usually drawn to be cylindrically symmetric, it is most likely quite oval at 1 AU.

sunspot number every 3 days for each solar rotation. This is followed by the rotation number and the first day of the solar rotation. Then follows the C9 index for each day of the rotation with a week of overlapping data after the vertical line. The C9 index was only semi-quantitative but still extremely useful as this figure illustrates. The left most column shows the peaking of solar activity in 1971 and 1972 followed by the declining phase in 1973 and 1974. There is little recurrence of geomagnetic activity near solar maximum but at the end of 1973 and throughout 1974 the activity has strong recurrence, certainly associated with two long-lived fast streams.

5. TIME VARIATION

Evidence for century-long changes in the solar wind and interplanetary magnetic field can be obtained from the geomagnetic record [*Russell*, 1975; *Russell and Mulligan*, 1995] and even longer period changes are implied by optical data on sunspots and ^{14}C in tree rings. On shorter scales the sun has a 22-year magnetic cycle and an 11-year sunspot cycle, and on time scales of less than a day the sun manifests disturbances of the corona called coronal mass ejections (CMEs). When seen head on these CME disturbances may envelop the visible sun as they move directly toward the Earth appearing as "halos" around the sun [*Plunkett et al.*, 1998]. When seen above the solar limbs they subtend an angle of about 45° at the sun. Most importantly for Earth these disturbances frequently spawn magnetic clouds travelling through the solar wind, regions of twisted, strong magnetic flux that are very effective at producing geomagnetic activity. The modern paradigm for such a structure is shown in Figure 31 [*Burlaga* 1988; *Lepping et al.*, 1990]. The frequent presence of electrons streaming in both directions along the magnetic field suggests that at least some parts of these structures are connected to the Sun. The gradual rotation in the magnetic

field seen as these structures pass, suggests that they form twisted magnetic tubes.

The way we characterize such a structure is to invert the magnetic time series to obtain a best-fit model. A recent approach of *Mulligan and Russell* [2000] allows both force free and non-force-free ropes to be fit and it allows the rope to expand. The model obtains the total flux, the current flowing and the orientation of each rope. Most ropes are found to be force-free meaning that the outward pressure in the rope due to the increased magnetic field there is balanced entirely by the inward force due to the curvature of the magnetic field. The pressure in the plasma does not play a significant role.

This observation is consistent with the observation that the plasma pressure in these ropes is much less than the magnetic pressure i.e. they are low beta structures. Occasionally we have contemporaneous observations with multiple spacecraft in the same interplanetary coronal mass ejection or ICME [*Mulligan et al.*, 1999; *Mulligan and Russell*, 2000]. An example of the results of the study of such an interval with two relatively close spacecraft is shown in Figure 32. The assumption that the ropes are cylindrically symmetric would have required that two separate but similar ropes erupted adjacent to each other. A more plausible solution, illustrated here, that fits both satellites' observations is that the cross-section of the rope is oval and that there is but a single rope. Thus we must change our paradigm of a magnetic cloud from that shown in Figure 30. The ICME appears to be an azimuthally extended oval rope. The radial thickness of the ICME is

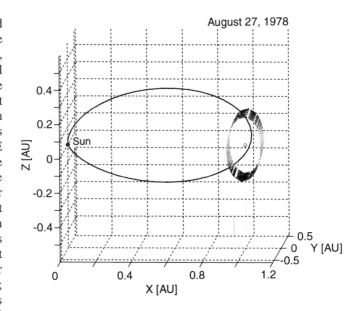

Figure 32. Inferred size, shape and orientation of the magnetic rope seen by PVO and ISEE 3 on August 27 and 28, 1978. [Mulligan et al., 2000].

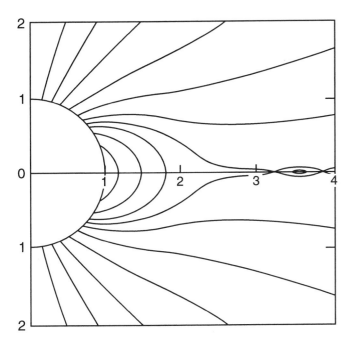

Figure 33. Possible method of forming flux ropes by reconnection across the heliospheric current sheet.

about 0.2 AU and at 1 AU the radius of curvature of the ICME is about 0.3 AU both in the plane of the axial field and perpendicular to it as sketched in Figure 32. In order to predict the terrestrial disturbance from an ICME one needs to ensure that one has measured the flux tube that will impact the Earth, is using an appropriate model with which to invert its properties, and has obtained a good and unique inversion. Multiple spacecraft maybe needed for this. The average orientation of the axes of these flux tubes varies during the solar cycle roughly following the heliomagnetic equator [*Mulligan et al.*, 1998; *Mulligan et al.*, 2000b]. This can affect the nature of the geomagnetic activity associated with the arrival of these ropes. We should emphasize that when the ejecta of a CME travels faster than the ambient solar wind plasma it overtakes the upstream magnetized plasma that then drapes over the structure. If the relative velocity is greater than the magnetosonic Mach number a fast mode shock arises in front of the ejecta.

How does the sun make these ropes? It is possible that the twisting of flux tubes by shearing motions in the photospheric plasma can make ropes. Such velocity shear is thought to be responsible for magnetic flux ropes in the Venus ionosphere [*Russell and Elphic*, 1979; *Elphic and Russell*, 1983 a, b]. The ropes are formed at high altitudes and sink in the flowing ionospheric plasma becoming more twisted. At low altitudes where the twist is the greatest, the ropes appear to become kink unstable [*Elphic and Russell*, 1983c]. Kinking may also occur on the Sun and be responsible for the observed S-shaped arcades. As sketched

in Figure 33 reconnection across the heliospheric current sheet may also be able to make ropes, just like ropes are created in the geomagnetic tail. Such a mechanism was recently proposed by *Moldwin et al.* [2000] to produce small flux ropes seen in the solar wind at 1 AU. It would require reconnection to proceed well into the region of open field lines to produce flux ropes of the size of those observed in ICMEs. In addition it would require the existence of multiple current layers reconnecting simultaneously to produce the multiple flux ropes observed. Finally, it is possible to make flux ropes out of coronal arches by reconnection as shown in Figure 34 [*Gosling*, 1990].

The reason we are interested in interplanetary coronal mass ejections is that they provide the solar wind conditions that lead to the most disturbed geomagnetic conditions. As mentioned in the introduction the energy flow into the magnetosphere is greatest when the interplanetary magnetic field is southward. More precisely it is proportional to the rate at which southward magnetic flux is carried to the magnetosphere by the solar wind [*Burton et al.*, 1975]. The relative occurrence of this quantity, VB_z, that has units of electric field is shown for ICMEs and stream interfaces in Figure 35 [*Lindsay et al.*, 1995]. Thus the occurrence of these structures is controlled by the sunspot cycle and peaks when the sunspots do [*Lindsay et al.*, 1994]. While other types of solar wind disturbances have a role in affecting aspects of geomagnetic interaction such as corotating solar wind stream interaction regions and regions of high Alfven wave content, the magnetosphere responds to these smaller disturbances principally at high geomagnetic latitudes. In fact, the buildup of the ring current, a major signature of geomagnetic storms, seems to require strong, long lasting southward IMF over periods of hours [*Russell et al.*, 2000]. Short period disturbances with durations less than 20 min. appear to be averaged out by the magnetosphere [*Burton et al.*, 1975].

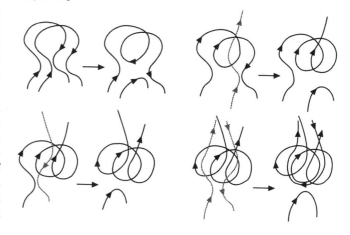

Figure 34. Proposed mechanism for creating interplanetary magnetic ropes from arcades in the lower corona. [*Gosling*, 1990].

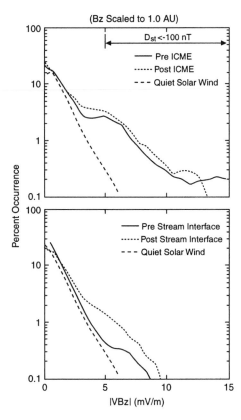

Figure 35. Occurrence rate of different values of the east-west interplanetary electric field in ICMEs (top panel), and corotating interaction regions (bottom panel) contrasted with that in the quiet solar wind [Lindsay et al., 1995].

6. CONCLUDING REMARKS

The key to understanding the heating and acceleration of the solar wind is to determine what mechanisms create the decidedly non-Maxwellian distributions of the electrons and ions. If the high temperatures of the corona can be explained with a theory that uses non-Maxwellian distributions, explicit coronal heating mechanisms such as via waves may be unnecessary, but such a theory relegates the solar wind acceleration problem to what causes these non-Maxwellian distribution. The magnetic field imposes a 3D structure on the solar wind. This allows fast streams to collide with slow streams in corotating interaction regions and helps generate recurrent geomagnetic activity. Interplanetary coronal mass ejections are thin compared to their overall width suggesting they are oval in cross-section. The solar magnetic field imposes magnetic polarity and orientation rules on these flux ropes. We need to understand much better the source regions of the solar wind and CMEs, and the mechanisms that create the observed solar wind disturbances and the distributions of solar wind particles. The STEREO mission will make a major step in the former direction and Solar Probe in the latter.

Acknowledgments. We gratefully acknowledge the assistance of J. G. Luhmann, J. A. Newbury, and T. Mulligan in the preparation of this review. This work was supported by the National Aeronautics and Space Administration under contract NAS5-00133-6.

REFERENCES

Asbridge, J. R., S. J. Bame and W. C. Feldman, Abundance differences in solar wind double streams, *Solar Phys.*, *37*, 451-467, 1974.

Burlaga, L. F., Magnetic clouds and force-free fields with constant alpha, *J. Geophys. Res.*, *93*, 7217, 1988.

Burton, R. K., R. L. McPherron and C. T. Russell, An empirical relationship between interplanetary conditions and Dst, *J. Geophys. Res.*, *80*, 4204-4214, 1975.

Elphic, R. C., and C. T. Russell, Magnetic flux ropes in the Venus ionosphere: Observations and models, *J. Geophys. Res.*, *88*, 58-72, 1983a.

Elphic, R. C., and C. T. Russell, Global characteristics of magnetic flux ropes in the Venus ionosphere, *J. Geophys. Res.*, *88*, 2993-3004, 1983b.

Elphic, R. C., and C. T. Russell, Evidence for helical kink instability in Venus magnetic flux ropes, *Geophys. Res. Lett.*, *10*, 459-462, 1983c.

Fairfield, D. H., and J. D. Scudder, Polar rain: Solar coronal electrons in the Earth's magnetosphere, *J. Geophys. Res.*, *90*, 4055, 1985.

Feldman, W. C., Coronal structure inferred from remote sensing observations, in *Robotic Exploration Close to the Sun*, edited by S. R. Habbal, p.9-24, 1997.

Feldman, W. C., J. R. Asbridge, M. D. Montgomery and S. P. Gary, Solar wind electrons, *J. Geophys. Res.*, *80*, 4181-4196, 1975.

Gosling, J. T., Coronal mass ejections and magnetic flux ropes in interplanetary space, in *Physics of Magnetic Flux Ropes*, *Geophys. Monogr. Ser.*, Vol. *58*, edited by C. T. Russell, E. R. Priest, and L. C. Lee, pp.343-364, AGU, Washington, DC, 1990.

Gosling, J. T., Physical nature of the low-speed solar wind, in *Robotic Exploration Close to the Sun: Scientific Bases*, 17-24, American Institute of Physics, 1997.

Hundhausen, A. J., An interplanetary view of coronal holas, in *Coronal Holas and High Speed Streams*, edited by J. B. Zirker, 225-319, Colorado Associated University Press, Boulder, 1977.

Hundhausen, A. J., The solar wind, in *Introduction to Space Physics*, edited by M. G. Kivelson and C. T. Russell, 91-128, Cambridge University Press, New York, 1995.

Klein, L. W., K. W. Ogilvie, and L. F. Burlaga, Coulomb collisions in the solar wind, *J. Geophys. Res.*, *90*, 7389-7395, 1985.

Lepping, R. P., J. A. Jones, and L. F. Burlaga, Magnetic field structure of interplanetary magnetic clouds at 1 AU, *J. Geophys. Res.*, *95*, 11,957-11,965, 1990.

Li, X. L., this volume, 2000.

Lindsay, G. M., C. T. Russell, J. G. Luhmann, and P. Gazis, On the sources of interplanetary shocks at 0.72 AU, *J. Geophys. Res.*, *99*, 11-17, 1994.

Lindsay, G. M., C. T. Russell, and J. G. Luhmann, Coronal mass ejection and stream interaction region characteristics and their potential geomagnetic effectiveness, *J. Geophys. Res., 100*, 16,999-17,013, 1995.

Linker, J. A., Z. Mikic, D. A. Biesecker, R. J. Forsyth, S. E. Gibson, A. J. Lazarus, A. Lecinski, P. Riley, A. Szabo, and B. J. Thompson, Magnetohydrodynamic modeling of the solar corona during whole sun month, *J. Geophys. Res., 104*, 9809-9830, 1999.

Luhmann, J. G., T-L. Zhang, S. M. Petrinec and C. T. Russell, Solar cycle 21 effects on the interplanetary magnetic field and related parameters at 0.7 and 1.0 AU, *J. Geophys. Res., 98*, 5559-5572, 1993.

Marsch, E., K-H. Muhlhauser, R. Schwann, H. Rosenbauer, W. Pilipp, and F. M. Neubauer, Solar wind protons: Three dimensional velocity distributions and derived plasma parameter measured between 0.3 and 1.0 AU, *J. Geophys. Res., 87*, 52-72, 1982.

McComas, D. J., S. J. Bame, B. L. Barraclough, W. C. Feldman, H. O. Funsten, J. T. Gosling, P. Riley, and R. Skoug, Ulysses' return to the slow solar wind, *Geophys. Res. Lett., 25*, 1-4, 1998.

McPherron, R. L., Magnetospheric dynamics, in *Introduction to Space Physics*, edited by M. G. Kivelson and C. T. Russell, 400-458, Cambridge University Press, New York, 1995.

Meyer-Vernet, N., Large scale structure of planetary environments: The importance of not being Maxwellian, *Planet. Space Sci.*, in press, 2000.

Moldwin, M. B., S. Ford, R. Lepping, J. Slavin, and A. Szabo, Small-scale magnetic flux ropes in the solar wind, *Geophys. Res. Lett., 27*, 57-60, 2000.

Mulligan, T., and C. T. Russell, Multi spacecraft modeling of the flux rope structure of interplanetary coronal mass ejections: cylindrically symmetric versus non-symmetric topologies, *J. Geophys. Res.*, submitted 2000.

Mulligan, T., C. T. Russell and J. G. Luhmann, Solar cycle evolution of the structure of magnetic clouds in the inner heliosphere, *Geophys. Res. Lett., 25*, 2959-2962, 1998.

Mulligan, T., C. T. Russell, B. A. Anderson, D. Lohr, D. Rust, B. A. Toth, L. J. Zanetti, M. H. Acuna, R. P. Lepping, and J. T. Gosling, Intercomparison of NEAR and Wind interplanetary coronal mass ejection observations, *J. Geophys. Res., 104*, 28,217-28,223, 1999.

Mulligan, T., C. T. Russell, and D. Elliott, Modeling of the solar cycle variations in magnetic cloud structure observed at PVO, *Geophys. Res. Lett.*, in press, 2000.

Neugebauer, M., R. J. Forsyth, A. B. Galvin, K. L. Harvey, J. T. Hoeksema, A. J. Lazarus, R. P. Lepping, J. A. Linker, Z. Mikic, J. T. Steinberg, R. von Steiger, Y.-M. Wang, and R. F. Wimmer-Schweingruber, Spatial structure of the solar wind and comparisons with solar data and models, *J. Geophys. Res., 103*, 14,587-14,599, 1998.

Newbury, J. A., Plasma heating and thermal transport in the solar wind near 1 AU, Ph.D. Dissertation, Earth and Space Sciences, UCLA, 2000.

Noyes, R. W., *The Sun our Star*, Harvard University Press, Cambridge, Mass, 1982.

Parker, E. N., Dynamics of the interplanetary gas and magnetic fields, *Astrophys. J., 128*, 664-676, 1958.

Petrinec, S. M., P. Song and C. T. Russell, Solar cycle variations in the size and shape of the magnetopause, *J. Geophys. Res., 96*, 7893-7896, 1991.

Pilipp, W. G., H. Miggenrieder, M. D. Montgomery, K-H. Muhlhausser, H. Rosenbauer, and R. Schwenn, Characteristics of electron velocity distribution functions in the solar wind derived from the Helios plasma experiment, *J. Geophys. Res., 92*, 1103-1118, 1987.

Pizzo, V. J., Interplanetary shocks on the large scale: A retrospective on the last decades' theoretical efforts, in *Collisionless Shocks in the Heliosphere: Reviews of Current Research* (edited by B. T. Tsurutani and R. G. Stone), 51-68, Amer. Geophys. Union, Washington D.C., 1985.

Plunkett, S. P., B. J. Thompson, R. A. Howard, D. J. Michels, O. C. St. Cyr, S. J. Tappin, R. Schwenn and P. L. Lumy, LASCO observations of an Earth-directed coronal mass ejection on May 12, 1997, *Geophys. Res. Lett., 25*, 2477-2480, 1998.

Pneumann, G. W., and R. A. Kopp, Gas-magnetic field interactions in the solar corona, *Solar Phys., 18*, 258, 1971.

Russell, C. T., On the possibility of determining interplanetary and solar parameters from geomagnetic records, *Solar Phys., 42*, 259-269, 1975.

Russell, C. T., and R. C. Elphic, Observation of magnetic flux ropes in the Venus ionosphere, *Nature, 279*, 616-618, 1979.

Russell, C. T., and T. Mulligan, The 22-year variation of geomagnetic activity: Implications for the polar magnetic field of the Sun, *Geophys. Res. Lett., 22*, 3287-3288, 1995.

Russell, C. T., G. Lu, and J. G. Luhmann, Lessons from the ring current injection during the September 24-25, 1998 storm, *Geophys. Res. Lett., 27*, 1371-1374, 2000.

Scudder, J. D., Why all stars should possess circumstellar temperature inversions, *Astrophys. J., 398*, 319-349, 1992.

Snyder, C. W., M. Neugebauer and U. R. Rao, The solar wind velocity and its correlation with cosmic ray variations and with solar and geomagnetic activity, *J. Geophys. Res., 68*, 6361-6370, 1963.

Thieme, K. M., E. Marsch, and H. Rosenbauer, Estimates of alpha particle heating in the solar wind inside 0.3 AU, *J. Geophys. Res., 94*, 1973-1976, 1989.

Wang, Y. M., and N. R. Sheeley, Jr., Solar wind speed and coronal flux-tube expansion, *Astrophys. J., 355*, 726, 1990.

Institute of Geophysics and Planetary Physics, and Department of Earth and Space Sciences, University of California Los Angeles, 405 Hilgard Ave., Los Angeles, CA 90095-1567, USA (email: ctrussell@igpp.ucla.edu).

Space Weather and The Changing Sun

E. N. Parker

Dept. Physics, Enrico Fermi Institute, University of Chicago, Illinois

The free energy that drives space weather is created in the convective zone of the Sun with the generation and convective distortion of magnetic fields. The fields rise to the surface where they provide the vigorous suprathermal activity that is the direct parent of space weather. Some aspects of the hydrodynamics and magnetic field generation are understood, while there remains much that is mysterious. An important part of the mystery centers around the complex hydrodynamics of the convective zone and the dominating micro-scale magnetic fibril structure, motion, and interactions at the surface. The next generation Advanced Solar Telescope-the "solar microscope"- is intended to open up this basic small-scale world to direct observational study. The other major mystery lies in the long- term variations in the general level of solar activity, with the associated variations in space weather and terrestrial climate. Unfortunately long-term variation can be studied only in the long term, although monitoring other solar-type stars has been helpful so far in suggesting the extreme possibilities.

INTRODUCTION

To begin at the beginning, space weather is driven by varying doses of UV, EUV, and X-rays emitted by the Sun, along with fast particles from solar flares, a varying galactic cosmic ray intensity, and a varying solar wind carrying a magnetic field with rapidly fluctuating orientation. The impact of all this on the terrestrial magnetosphere and atmosphere creates a situation of breathtaking, dynamical complexity, with frequent consequences for radio communication and occasional lethal consequences for orbiting spacecraft. The varying total luminosity of the Sun has long-term effects on terrestrial climate, and this Chapman Conference is devoted to the scientific challenge of understanding, and perhaps anticipating, these terrestrial effects.

This review takes a look at the Sun as the origin of the many emanations that bedevil Earth and the other planets,

Space Weather
Geophysical Monograph 125

noting that the Sun plays its cards so close to the vest that there are still many outstanding basic mysteries (Parker, 1985).

To a first approximation the Sun is a placid self-gravitating ball of hydrogen and helium, with a dash of the heavier elements thrown in to provide opacity. The Sun (radius 7×10^5 km) that we see in the sky is essentially an opaque dark shroud enclosing a central thermonuclear core (radius $\sim 1 \times 10^5$ km, $T \sim 10^7$ K, $\rho \sim 10^2$ gm/cm^3) that is more than ten times brighter than a supernova at maximum. The shroud is an insulating blanket that allows only one part in 2×10^{11} of the central brightness to leak out. That number is particularly impressive when contemplated while standing in the noontime sunlight on a hot summer day.

The outward leakage of heat from the core is radiative, but upon reaching a radius of 5×10^5 km the temperature T has fallen to 2×10^6 K and the decline ($\sim T^4$) in the effectiveness of radiative transfer produces an increasing temperature gradient to the point that the gas becomes unstable to convective turnover, i.e. the temperature gradient becomes steeper than the adiabatic gradient. Thus an upward moving volume of gas, expanding and cooling adiabatically, finds itself increasingly warmer, and hence less dense, than its surroundings, thereby accelerating upwards. Similarly the temperature of a downward moving volume of gas heating adiabatically does not keep up with the rising temperature of its surroundings, so it is denser than its surroundings and accelerates downward. So the gas is in a perpetual state of convective turnover above a radius of 5×10^5 km, all the way out to the 5600 K visible surface at 7×10^5 km. The convection takes over the outward heat transfer up to the visible surface where the opacity abruptly becomes negligible and radiation dominates again.

Space Weather

The convection represents a heat engine diverting about 10^{-2} of the total heat flow (4×10^{33} ergs/sec) into convective motion. Then about 10^{-4} of the convective kinetic energy is converted into the magnetic free energy that creates flares, the corona, the solar wind, etc.

The reader is presumably familiar with the standard dynamo concepts for the convective generation of magnetic field. The Sun rotates nonuniformly and the convection is cyclonic as a consequence of the rotation. Together the nonuniform rotation and the cyclonic convection form an $\alpha\omega$-dynamo, generating the migratory toroidal (azimuthal) and poloidal (meridional) magnetic fields on a 22-year cycle. The toroidal fields lie near the base of the convective zone and have a mean intensity of at least a few kilogauss, while the poloidal field is apparently no more than about 10 gauss. The strong toroidal fields are intrinsically buoyant, of course, so they bulge upward here and there to form Ω-loops whose apexes emerge through the surface to provide the observed bipolar magnetic active regions. Above the visible surface the bulging Ω-loops are deformed by the convective motions of their photospheric footpoints, leading to coronal mass ejections and flares, as well as the coronal heating that provides the solar X-ray emission and the solar wind. From there on it is space weather.

Now the inquisitive space weather forecaster might ask for more details in the solar convective dynamo process, and might even go so far as to note that a description is not the same thing as a proper theoretical explanation. At that point the superficial nature of the conventional descriptions is laid bare, because only the sketchiest explanation in terms of the basic laws of physics is available. Enough of the basic physics has been worked out to suggest that the overall description is not in gross error, but there is much about the proposed scheme of things that is not at all understood. Many of the intermediate steps in the process are simply baffling if examined closely and critically–politically incorrect and potentially devastating activity in these days of government financing of research. So from the security of retirement the author proposes to describe some of the more mysterious aspects of the solar activity responsible for space weather. They pose some fascinating scientific challenges.

Solar Convection

Let us start with the convection that is the root of all solar activity and space weather. The fact is that it is not yet possible to simulate the convection either analytically or numerically to a sufficient precision to show how the convection sustains the nonuniform rotation and meridional circulation of the Sun (see discussion in Gilman and Miller, 1986; Durney, 1993, 2000; Gilman and Dikpati, 2000; Miesch et al., 2000). The internal velocity profile of the Sun has been successfully inferred from helioseismology (Thompson et al., 1996), showing that the angular velocity (except for a thin layer immediately below the visible surface) is approximately independent of depth through the convective zone, where it depends mainly on heliographic latitude. The equatorial rotation period is 25 days, increasing with latitude to almost 35 days at both poles. In contrast, the radiative interior below the convective zone rotates nearly rigidly at an intermediate rate, so that there is a strong shear at the base of the convective zone, with the convective zone running ahead of the radiative zone at low latitude and lagging behind at high latitudes. This should all be straightforward classical hydrodynamics, so what is the problem?

First of all, the convection is strongly turbulent (Reynolds number $\sim 10^{12}$), and the standard mixing length representation is used to estimate the effective eddy viscosity and convective heat transport. The idea is that the mean large-scale effects of the small-scale convection at each level r are adequately characterized (described) by a single length $\lambda(r)$ and velocity $v(r)$. Then the large-scale convective transport of a scalar field, e.g., smoke or heat, is assumed to be similar to molecular diffusion in a gas, with a diffusion coefficient of about $0.1\lambda(r)v(r)$ in terms of the mean free path and mean thermal velocity. The next step, then, is to determine $\lambda(r)$ and $v(r)$ throughout the convective zone.

The convective zone is strongly stratified, with a density of 0.2 gm/cm^3 at its base ($r = 5 \times 10^5$ km) and a density of 0.2×10^{-6} gm/cm^3 at the visible surface, above which the atmosphere is stable against convection and the energy is transported by radiation. The density decline by a factor of 10^6 up through the convective zone is equivalent to 14 scale heights (factors of e) and nearly 20 pressure scale heights (the temperature declines by a factor of 300). It is generally assumed that the rapid expansion of a rising plume of hot gas limits its vertical coherence length to something of the order of the pressure scale height $\Lambda = kT/mg$, where m is the mean particle mass and g is the local acceleration of gravity. Thus $\lambda(r)$ is put equal to $\Lambda(r)$, or some fraction thereof, with Λ equal to 4×10^5 km at the base of the convective zone and about 200 km at the top. The effective turbulent velocity $v(r)$ is estimated from the work done by the buoyancy of a rising cell of gas and from the fact that the convection must transport the heat that is not transported by radiation. The buoyancy is calculated by assuming a temperature excess of the order of λ times the slight excess of the mean temperature gradient over the adiabatic temperature gradient (cf. Spruit, 1974). This line of reasoning provides an estimate of the relatively small correction, amounting to a total of about 3×10^3 K, over the purely adiabatic model of the convective zone. The concept appears to be adequate for constructing a model for the temperature and density profile of the convective zone. It should be noted, then, that the theoretical model of the solar interior, from the center to the visible surface, provides a speed of sound (determined by the temperature and mean molecular weight) that nowhere differs by more than one part in 500 from the speed of sound in the model deduced from helioseismology (Bahcall et al., 1996; Bahcall and Pinsonneault, 1995; Bahcall et al., 1997; Bahcall, 1999 and refs. therein). Hence, it appears that the thermal structure of the Sun is pretty well in hand.

But what about the convection and the nonuniform rotation? Numerical modeling is the most promising approach, using the idealizations of the effects of the turbulence already noted. The primary difficulty is the intense vertical stratification, noted above. Each scale height, of which there are nearly 20, is a dynamical problem in itself, and so far it has been possible to treat only about five scale heights with existing supercomputers. The numerical models provide nonuniform rotation, but not in the form inferred from helioseismology.

The work of Cattaneo et al., (1991) shows that, while rising convective cells break up from their rapid expansion in a scale height or so, a downdraft is quite another thing. The downward moving cell becomes increasingly compact, instead of expanding and dispersing, and accelerates downward with surprising vigor and coherence. So it will be some time yet before the full stratification and the intricacies of the internal turbulence can be properly included in the theoretical models of the convective zone. In plainer language, it will be some time yet before we understand the hydrodynamics of the nonuniform rotation and meridional circulation.

Generation of Magnetic Field

Consider the generation of magnetic field, given the helioseismic results for the nonuniform rotation $\Delta\Omega(r,\theta)$ and the expected cyclonic nature of the convective cells, as estimated from mixing length theory. The familiar $\alpha\omega$-dynamo equations can be written down (Parker, 1955, 1970, 1979b; Steenbeck, Krause, and Radler, 1966; Moffatt, 1978; Krause and Radler, 1980; Childress and Gilbert, 1995) with estimates of the α-coeffcient – essentially the mean cyclonic velocity of the gas – and the eddy diffusivity $\eta = 0.1\lambda v$ for the magnetic field. A variety of dynamo models are available in the literature (Parker, 1957, 1971, 1979b and refs. therein, 1993; Steenbeck and Krause, 1969; Roberts, 1970; Roberts, 1972; Stix, 1976; Soward and Roberts, 1977; Krause and Radler, 1980; Priest, 1982; Soward, 1983; Choudhuri, 1984; Glatzmaier,

1985; Ferriz-Mas, Schmitt, and Schussler, 1994; Ferriz-Mas, 1996; Tobias, 1996a,b; Tobias et al., 1998; Markiel and Thomas, 1999), and it is fair to say that the theoretical models can be adjusted to show the 11-year cyclic variation and equatorward migration of the magnetic fields exhibited by the Sun for values of $\Delta\Omega$, α, and η that are entirely within the limits of plausibility. So it is not unreasonable to think that the $\alpha\omega$-dynamo concept may be valid. However, it must be appreciated that the eddy diffusivity η of the vector magnetic field is a problematical conjecture. It is known from extensive laboratory experiments and terrestrial observations that the concept of turbulent diffusion works well for scalar fields, e.g., smoke. But a vector field with internal stresses is a different matter.

First of all, an inventory (Gaizauskas et al., 1983) of the total magnetic flux emerging though the visible surface of a single activity complex of a few degrees on a side shows a total of 3×10^{23} Maxwells over a period of six months. Unless one assumes some effective recycling of magnetic flux, this indicates a reservoir of at least 3×10^{23} Maxwells in the lower convective zone, suggesting an azimuthal magnetic field of at least 3000 gauss through an estimated meridional cross section of the order of 2×10^4 km by 4×10^5 km.

The obvious difficulty is that the magnetic stress and energy density ($B^2 / 8\pi \sim 4 \times 10^5$ ergs/cm^3) of so strong a mean field are comparable to the Reynolds stress and kinetic energy density ($\rho v^2 / 2 \sim 5 \times 10^5$ ergs/cm^3) of the convection (cf. Spruit, 1974). Turbulent diffusion of magnetic fields is based on analogy with the turbulent diffusion of scalar fields, and involves the hydrodynamic deformation of an initial blob of fluid into a long thin ribbon, with ribbon thickness declining over time as $\exp(-vt/\lambda)$ and length increasing as $\exp(+vt/\lambda)$. It is not long before the thickness becomes so small that the ribbon is dominated by molecular diffusion. A magnetic field in the same blob of fluid increases in intensity as $\exp(+vt/\lambda)$, starting from near the equipartition value. One would expect that the field is initially strongly deformed, with $\Delta \mathbf{B} \sim \mathbf{B}$, but the fluid motions simply lack the strength to take the deformation much farther. The magnetic field is too strong to be drawn out into a long thin ribbon, and it is not obvious how the field can be subject to an effective eddy diffusion of the order of $0.1\lambda v$.

Quantitatively, the dynamo requires an effective diffusion coefficient η of such magnitude that the field can diffuse over distances $(4\eta t)^{1/2}$ of the order of a third of a solar radius in a time t of ten years. The requirement is $\eta \sim 3 \times 10^{11}$ cm^2/sec. A mixing length $\lambda \sim 4 \times 10^4$ km and a convective velocity $v \sim 1 \times 10^3$ cm/sec in the deep convective zone (Spruit, 1974) yields $0.1\lambda v \sim 4 \times 10^{11}$ cm^2/sec. The coincidence of the dimensional quantity $0.1\lambda v$ with the diffusion required to make the dynamo models work is suggestive, but the physics underlying such a turbulent diffusion effect is unknown. So there is a fundamental mystery about the solar dynamo.

Fibril Magnetic Fields

Another mysterious point is the intensely fibril structure of the magnetic field extending up through the surface of the Sun. Instead of being a continuum, the field is composed of a collection of slender ($\sim 10^2$ km), intense (B_f ~ 1500 gauss) magnetic flux bundles separated by wide essentially field-free spaces, so that the spacing of the fibrils determines the overall mean magnetic fields, typically 10 – 100 gauss, through the surface of the Sun. Thus, in a quiet region, where the mean magnetic field is perhaps 10 gauss, the individual fibrils are separated by 10^3 km, so that they occupy about 10^{-2} of the total area. In an active region, the fibrils occupy about 10^{-1} of the total area. The individual fibrils evidently have about the same characteristics in both places.

The obvious question is whether the fibril state obtains throughout the convective zone, or is it only a surface effect? The Ω-loops in the azimuthal magnetic field, bulging up through the surface from the deep convective zone, provide an interesting clue. The apex of the Ω-loop forms a bipolar magnetic field at the surface oriented with only a small tilt to the east-west direction. Numerical simulations of the buoyant rise of the Ω-loop find that the magnetic field must be very intense, 0.5-1.0×10^5 gauss, at the base of the convective zone if the Ω-loop is to avoid being overwhelmed by the Coriolis force of the rotating Sun (D'Silva, 1993; D'Silva and Choudhuri, 1993; Fan, Fisher and DeLuca, 1993; Fan, Fisher, and McClymont, 1994). Given so high a field strength the conservative assumption is that the mean azimuthal field is perhaps 3×10^3 gauss while the 0.5-1.0×10^5 gauss is an indication of the intensity of the individual fibrils. The alternative assumption, that the mean field is as strong as 0.5-1.0×10^5 gauss, requires too long a time for the nonuniform rotation to produce the azimuthal field and total flux through shearing the poloidal field.

The next question is why the magnetic field is in such an intense fibril state throughout the convective zone? Nothing in the conventional mean-field theory of dynamos suggests a fibril state. Indeed, the magnetic energy for a given total magnetic flux is increased by a factor $B_f/\langle B \rangle$, where B_f is the characteristic fibril field and $\langle B \rangle$ represents the mean field. On the other hand, the fibril state of the field interferes but little with the convection and the convective heat transport. If the field were spread out into a continuum, threading all of the overturning fluid, it would soon be entangled and intensified so as to impede the continuing convective turnover, damming the heat flow and increasing the temperature in the region below. It is interesting to note that minimizing the sum of the magnetic energy and the thermal energy gives fibril intensities B_f of a couple of kilogauss (Parker, 1984), not unlike the fibrils at the surface of the Sun. This result is suggestive, but it is not an explanation.

It is evident that we really do not understand how the convection generates the magnetic field of the Sun. We wonder if the individual magnetic fibril is the basic magnetic entity rather than the mean field conventionally used in dynamo theory. The effect sometimes called "turbulent diamagnetism" (Blanchflower, Rucklidge, and Weiss, 1998; Tao et al., 1998; Tao, Proctor, and Weiss, 1998) concentrates the magnetic field into regions of weak turbulence and may play a role in forming the fibrils. The continual stretching of a fibril caught up in the convective flow and nonuniform rotation would continually intensify the field of the fibril. The fibril is drawn out and ultimately folded into loops that subsequently are snipped off by rapid reconnection, leaving reduced fibril length and enhanced fibril field (Vishniac, 1995a,b). Nonparallel fibrils quickly reconnect wherever they are pressed together, giving a dissipative effect of the same order as suggested by the mixing length concept of turbulent diffusion and as required to make the conventional dynamo ($\alpha\omega$-dynamo) work for the Sun. But this is all conjectural at the present time, based only on order-of-magnitude estimates.

There are other puzzles, of course. For instance, the familiar sunspot cycle of 11-years indicates the general magnetic cycle in which both the toroidal and poloidal fields reverse from one cycle to the next. This involves the bipolar magnetic regions on the surface of the Sun with lengths in the general range of 3×10^4 to 3×10^5 km on the surface of the Sun. At the same time the Sun exhibits the ephemeral active regions, involving bipolar magnetic fields with lengths of 2×10^4 km or less, showing only a modest variation in number with the 11-year cycle, rather than the extreme "on and off" found with the larger bipolar magnetic regions (Martin and Harvey, 1979). This raises the question of whether there are two distinct dynamo processes at work in the Sun, perhaps loosely coupled. Or whether the ephemeral magnetic regions represent the magnetic snippings from the looping fibrils of the main field? Or do we have the whole idea of field generation wrong?

Finally, note the curious fact that the emergence of Ω-loops through the surface of the Sun to create the bipolar magnetic active regions occurs at special fixed solar longitudes, which persist from one 11-year magnetic cycle into the next (cf. Gaizauskas et al, 1983; Benevolenskaya et al., 1999). This phenomenon is telling us something that we do not yet comprehend.

The reader can see that space weather begins in a "fog" at the Sun.

Magnetic Activity

Suppose that we accept the magnetic fields emerging through the visible surface of the Sun as given and inquire into their behavior at, and above, the surface of the Sun. The observations of the Sun by the SOHO and TRACE spacecraft show that the visible surface of the Sun is paved over with active small-scale magnetic fields (Berger and Title, 1996; Kohl et al., 1997; Schrijver et al., 1997, 1999; Handy et al., 1999). Unfortunately, the individual magnetic fibrils are not resolved in the telescope but appear as fuzzy blobs in the very best CD pictures, such as Alan Title has obtained from the Swedish Vacuum Tower Telescope on Las Palmas, with resolution of about 0.3″ (200 km). A time sequence of pictures shows the individual fibrils kicking around in the dark intergranular lanes (downdrafts) with speeds of the order of 1 km/sec. The flickering between neighboring fibrils suggests both system noise and active flux exchange between fibrils. Two blobs are often seen to approach and merge into one and then separate again, but the observer has no idea what happened in the encounter. Little or nothing is known of the fluid motions and of the atmosphere structure in and around the individual fibril. The degree of twisting of the magnetic field of a fibril is another important quantity that is beyond observation at the present time. The twisting plays a central role in the activity of the fibril a few hundred km above the visible surface, where the fibrils are not confined by the tenuous gas around them, thereby expanding to fill all the available space. The point is that the twisting in the concentrated fibril beneath the visible surface propagates up into the expanded portion of the fibrils, providing vigorous reconnection between contiguous fibrils (Parker, 1974, 1975).

While we are on the subject of fibrils, it is interesting to note that sunspots are formed by the peculiar tendency of magnetic fibrils to cluster together during the times when the fresh magnetic fibrils of an emerging Ω-loop are breaking the surface in the same active region (Zwaan, 1978, 1985). The clustering and the eventual compression to 3000 gauss are in opposition to the magnetic pressure and tension, so, presumably, the clustering is driven by a hydrodynamic flow somewhere unseen below the visible surface (Meyer et al., 1974; Parker, 1979a). Thus we are again puzzling over the hydrodynamics. Recent observations show that the penumbra of a sunspot is a sandwich of vertical slabs of magnetic field alternating with slabs of field inclined about 45°, totally without explanation. Nor is it known whether the magnetic field of a sunspot forms a single column below the surface or is a loose cluster of many separate fibrils with relatively field-free gas between (Parker, 1979a).

Fortunately the mysterious sunspot seems to make no direct contribution to space weather, except for transient decreases in the brightness of the solar disk. Of more concern are the coronal mass ejections, evidently representing a magnetic catapult of coronal gas ($10^{15} - 10^{16}$ gms) into space with speeds upwards of a hundred of km/sec. Coronal mass ejections are often followed by a two-ribbon flare, for reasons that are not clear. The event temporarily disrupts the initial coronal magnetic helmet structure of the active region in which it occurs. The ejection evidently arises when a twisting or shearing of a bipolar field or magnetic arcade is loaded down with gas or tied down with an over arching field, allowing the magnetic energy to build up to where it is sufficient to fling open the upper part of the field structure when the gaseous load or tie

down is removed (Aly, 1991; Low, 1996, 1999; Low and Hundhausen, 1995).

The coronal mass ejection forms a blast wave in the solar wind (Parker, 1963) with a complex internal structure (Rothmer, 1999; Vandas, Fisher, and Geranios, 1999) and vigorous particle acceleration in the shock front ahead (Bieber et al., 1999). The impact of the blast wave front against the terrestrial magnetosphere is the cause of some of the most severe space weather. The complexity of the magnetic structure creating the mass ejection makes it difficult to anticipate the ejection, and, while coronal mass ejections are conspicuous in profile on the limb of the Sun, those directed straight at Earth are detected only with substantial observational effort.

The two-ribbon flare, as well as the more numerous compact flares (Svestka, 1976), provide a variable UV, EUV, and X-ray intensity, with direct consequences for the state of the terrestrial ionosphere. The proportions of radio emission, visible light, UV and X-ray emission, and particle acceleration vary enormously from one flare to the next, while the total energy release may be anything from 10^{33} ergs down at least to the threshold for detection by contemporary instruments, in the neighborhood of 10^{24} ergs. The observational and theoretical literature on this important topic is extensive (cf. Sewwt, 1969; Svestka, 1976; Priest, 1981,1982; Zirin, 1988; Haisch and Rodono, 1989).The diverse forms of flares, microflares and nanoflares in the coronal magnetic fields all appear to involve the rapid dissipation of magnetic free energy where opposite field components are pushed together. However, the theoretical complexity of the dissipation and reconnection of field, together with the variety of the magnetic configurations in which rapid reconnection and dissipation occurs, make the forecasting of flares an empirical art.

The tenuous outer atmosphere of the Sun is heated to temperatures in excess of 10^{6} K, forming the corona with a pressure scale height close to the Sun of 5×10^{4} km for each million degrees of temperature. The density of the X-ray coronal gas, trapped in the 10^{2} gauss bipolar magnetic fields of active regions, can be as high as N = 10^{10} /cm^{3} , with the X-ray emission proportional to N^{2} and showing a mean of 10^{7} ergs/cm^{2} sec for the bright X-ray corona (Withbroe and Noyes, 1977). The temperature of the X-ray coronal gas is strongly inhomogeneous in directions perpendicular to the magnetic field and fluctuating rapidly in time on small scales (Sturrock et al., 1990; Feldman et al., 1992). It has been suggested (Parker, 1988, 1994) that the principal heat input creating the X-ray corona is the nanoflaring in the surfaces of tangential discontinuity (current sheets) created spontaneously by the Maxwell stresses in the small-scale interwoven topology of the bipolar magnetic fields. The concept is simply that the photospheric footpoints at each end of the bipolar magnetic field continually move slowly (1 km/sec) and randomly in the photospheric granules, while the field line connections between the footpoints at opposite ends of the bipolar field are preserved by the near absence of electrical resistivity of the million degree coronal gas. After a time the field lines become interlaced, and static equilibrium in such a bipolar interlaced topology spontaneously creates surfaces of tangential discontinuity throughout the field. Since the electrical resistivity is not identically zero, the surfaces of incipient discontinuity are subject to rapid dissipation and reconnection of the magnetic field. The reconnection presumably occurs in bursts, estimated to lie in the range $10^{22} - 10^{24}$ ergs, i.e. nanoflares. The precise details are not known because the motions of the footpoints and the larger of the nanoflares are only beginning to be subject to observation. The rms angular deflection of the interlaced field lines from the local mean field direction is estimated as $5° - 10°$, based on the rate of 10^7 ergs/cm^2 sec at which the 1 km/sec random motion of the footpoints does work on the field. It is interesting, then, that electric current density in the incipient surface of discontinuity becomes sufficiently intense as to suggest the onset of plasma turbulence and anomalous resistivity, creating a burst of reconnection (Parker, 1994). Needless to say, we cannot be sure of this general scenario until the observational threshold and angular resolution have been substantially improved. Moore et al., (1999) describe a specific scenario for reconnection events involving magnetic arcades, etc.

There is transient local heating of the corona by occasional large flares, when the X-ray brightness of the Sun may increase enormously for brief periods, and by the many microflares in ephemeral active regions, and by the microflares occurring among the small magnetic structures continually herded into the downdrafts in the supergranule boundaries. However, these sources appear to be only a minor contribution to the active X-ray corona (but see Moore et al., 1999). On the other hand, the coronal holes (N ~10^8 ions/cm^3, B ~ 10 gauss), where the heat input is estimated as 0.5×10^6 ergs/cm^2 sec (Withbroe and Noyes, 1977), may be heated primarily by the microflares (Martin, 1984; Porter and Moore, 1988). The question is whether there is a sufficient number of microflares to supply the necessary average of 0.5×10^6 ergs/cm^2 sec. Recent studies by the SOHO and TRACE space missions suggest that the answer is affirmative (cf. Krucker and Benz, 2000).

Most of the heat input to the coronal hole is expended in the gradual outward expansion that becomes the high-speed solar wind, so it appears that the microflaring is the immediate power source of the solar wind. The coronal expansion is possible because the magnetic field is too weak to confine the coronal gas, in contrast with the strong fields that contain the X-ray corona. The precise radial temperature profile and the radial divergence of the expanding coronal hole are essential for developing a quantitative model of the high speed solar wind, and have yet to be determined. The solar wind is the basic constituent of day by day space weather, forming the comet shaped magnetosphere of Earth as well as agitating it to

some degree. The coronal mass ejections, and the UV, X-rays, and solar cosmic rays from the larger flares, embellish and complicate the space weather, creating events that can be exceedingly troublesome. The interplanetary magnetic field carried in the solar wind is the stretched out field of the Sun (Parker, 1958), and, as is now well known, plays a direct role in initiating magnetic substorms (Akasofu, 1966, 1980, 1981; Akasofu and Chapman, 1972; Zhu, 1993, 1995). So the basic concepts are largely in hand (assuming that they all check out with further observational investigation), except perhaps for the slow solar wind component.

The slow component of the solar wind is associated with the magnetic active regions on the Sun. The bipolar magnetic fields of the active regions are too strong to be forced open by the coronal gas, so it is conjectured that the slow wind originates only around the periphery of active regions, where the magnetic field is sufficiently weak to be forced open at some distance out from the Sun. The diffuse X-ray corona enclosing the entire Sun, discovered by Yohkoh, is evidently another phase of the corona whose origins are not understood.

Between the coronal gas at 10^6 K and the photosphere at 5600K, there is an abrupt and spotty transition, containing patches of chromosphere and the so called transition region where the temperature increases through 10^5 K. The layer contains intense activity, with small explosive events and UV and X-ray bright points not immediately identifiable with the conventional flare and microflare (Golub et al., 1974; Brueckner and Bartoe, 1983; Habbal et al., 1986; Innes et al., 1997; Kankelborg, Walker, and Hoover, 1997, Krucker and Benz, 2000) and presumably caused by the carpet of small-scale magnetic fields already mentioned. So there is a whole jungle there to be explored.

The basic need is to see and study the photospheric magnetic fibrils, the exotic structure of the sunspot, the microflares and nanoflares of the quiet regions and the active regions, the microstructure of the H$_\alpha$ prominences and the coronal mass ejections, the microstructure of the flare, and the microstructure of the transition region. This calls for the "solar microscope", capable of observations at 10 second intervals with high dispersion and with angular resolution substantially better that 0.1″ (75 km). The requirement is a 4m telescopic aperture with state-of-the-art adaptive optics located at a site with the best possible natural seeing. Such a telescope is essential for penetrating into the still conjectural scenario outlined above for the dynamics of the Sun and the origins of space weather.

Concluding Thoughts

Looking back over the foregoing discussion, the reader can see the sketchy nature of our understanding of the activity of the Sun and the origins of space weather. The understanding, such as it is, contains many established facts and several islands of established theory, bridged with plausible conjectures and separated by more than one abyss of total ignorance.

Yet there is still another realm that needs to be addressed, and that is the changing personality of the active Sun over decades, centuries, and millennia. The waxing and waning of the general level of solar activity, as shown by the cosmic ray production of ^{14}C in the terrestrial atmosphere (^{14}N + n $\rightarrow$ ^{14}C + p) over the last 7000 years, show 10 centuries when the activity was so low that the intensity of galactic cosmic rays at Earth was of the order of 10 percent higher than nowadays. Then there were 8 centuries when the Sun was so active as to reduce the cosmic ray intensity by about 10 percent below the present level. That is to say, the Sun spends about one out of every four centuries in a conspicuously abnormal state (Eddy, 1973, 1983, 1989), with the Maunder Minimum of the 17th century, the Spörer Minimum of the 15th century, and the Medieval Maximum of the 12th century as the most recent examples. Lesser excursions are evident in the modern sunspot record, spanning nearly four centuries and including the Maunder Minimum. The 20th century has seen a doubling of the sunspot number and a doubling of the solar magnetic field (Lockwood, Stamper, and Wild, 1999).

The mean annual temperature in the northern temperate zone responds (with $\Delta T \sim \pm 1\text{-}2$ K) to the major excursions of solar activity, presumably mainly through the associate systematic change in solar luminosity ΔL. The luminosity varies 0.1-0.2 percent with the normal contemporary 11-year magnetic cycle (Hoyt et al., 1992). The observations of other solar type stars show similar variations in ΔL associated with the variations in activity (Baliunas and Jastrow, 1990; Friis-Christensen and Lassen, 1991; Zhang et al., 1994) and such observational monitoring of other stars greatly accelerates the study of the long term variations of activity and ΔL of the Sun. The study shows that there is an approximate long term proportionality between ΔL and the activity level of stars like the Sun. It is estimated that ΔL varied by about 0.4 percent both up and down from the norm during the extremes of the abnormal centuries, so that $\Delta T/T \sim \Delta L/L$.

There is no clear understanding of why the brightness of the Sun should increase slightly with the level of magnetic activity. We have suggested (Parker, 1995) that the magnetic flux bundles rising from the deep convective zone to the surface stir the fluid in their passing, thereby enhancing the convective heat transport. The order of magnitude of the effect is about right, but the idea is only a conjecture at this stage. The magnetic activity itself involves too little total energy to account for the observed ΔL.

An obvious question is how space weather might have been different during the Maunder Minimum of the 17th century or the Medieval Maximum of the 12th century. Presumably the space weather in abnormal centuries would be made up of the same basic ingredients of solar activity that we see today, but with different intensities and different proportions. Only time can reveal the answer. Unfortunately the time is longer (centuries) than the duration of human projects, scientific or otherwise.

Finally, it is clear from the text that the author feels that it is important to pursue the physics of the mysterious activity of the Sun, both as a purely scientific challenge and as a practical matter in the forecasting of space weather. However, it must be understood that the dynamical activity of the Sun, even if one day we understand it in clear fashion, will not provide the ultimate numerical simulation with predictive power for the wandering of the activity and the associated space weather. We need only think about the prediction of weather here on Earth, where, with detailed input on the 3-D atmospheric conditions over the planet at some given time t, it is possible to see only a few days into the future. For the Sun there is no hope that we should ever be able to state the detailed 3-D conditions throughout the convective zone at any given time, to say nothing of the gigantic computational problem involved in projecting forward in time.

On the other hand, an understanding of the physics can be expected to provide insight into interpretation of the ongoing observations and some basis for guessing what additional observations might provide crucial clues on the state of things now and in the near future. We would hope to be able to judge the relevance, or irrelevance, of each observational aspect of the activity. That is to say, a proper understanding is the best basis for anticipating the twists and turns of the ongoing space weather from the observations of the Sun.

REFERENCES

Akasofu, S. I. Electrodynamics of the magnetosphere: Geomagnetic storms, *Space Sci. Rev.*, 6, 21,1966.

Akasofu, S. I., The solar-magnetosphere energy coupling and magnetospheric disturbances, *Planet. Space Sci.*, 28, 495, 1980.

Akasofu, S. I., Energy coupling between the solar wind and the magnetosphere, *Space Sci. Rev.*, 28, 121, 1981.

Akasofu, S. I. and S. Chapman, *Solar Terrestrial Physics*, Clarendon Press, Oxford, sections 8.72-8.9.

Aly, J. J., How much energy can be stored in a three-dimensional force-free magnetic field?, *Astrophys. J. Lett.*, 375, L61, 1991.

Bahcall, J. N., Solar Neutrinos: An overview, *Current Science*, 77, 1487, 1999.

Bahcall, J. N., F. Calaprice, A. M. McDonald, and Y. Totsuka, Solar neutrino experiments: The next generation, *Physics Today*, 49, July, p. 30, 1996.

Bahcall, J. N. and M. H. Pinsonneault, Solar models with helium and heavy-element diffusion, *Rev, Mod, Phys.*, 67, 781`, 1995.

Bahcall, J. N., M. H. Pinsonneault, S. Basu, and J. Christensen-Dalsgaard, Are standard solar models reliable?, *Phys. Rev. Lett.*, 78, 171, 1997.

Baliunas, S. and R. Jastrow, Evidence on the climate impact of solar variations, *Nature*, 348, 520, 1990.

Benevolenskaya, E. E., J. T. Hoeksema, A. G. Kosovichev, and

P. H. Scherer, The interaction of new and old magnetic fluxes at the beginning of solar cycle 23, *Astrophys. J. Lett.*, *517*, L163, 1999.

Berger, T. E. and A. M. Title, On the dynamics of small-scale solar magnetic elements, *Astrophys. J.*, *463*, 365, 1996

Bieber, J. W., H. Cane, P. Evenson, R. Pyle, and I Richardson, Energetic particle flows near CME shocks and ejecta, in *Solar wind nine*, ed. S. R. Habbal, R. Esser, J. V. Hollweg, and P. H. Isenberg, American Institute of Physics Conference Proceedings 471, Woodbury New York, p. 137, 1999.

Blanchflower, S. M., A. M. Rucklidge, and N. O. Weiss, Modeling photospheric magnetoconvection, *Mon. Not. Roy. Astron. Soc.*, *301*, 593, 1998.

Bothmer, V. Magnetic field structure and topology within CME's in the solar wind, in *Solar wind nine*, ed. S. R. Habbal, R. Esser, J. V. Hollweg, and P, H, Isenberg, American Institute of Physics Conference Proceedings 471, Woodbury, New York, P. 119, 1999.

Brueckner, G. E. and J. D. F. Bartoe, Observations of high energy jets in the corona above the quiet Sun, the heating of the corona, and the acceleration of the solar wind, *Astrophys. J.*, *272*, 329, 1983.

Cattaneo, F., N. H. Brummel, J. Toomre, A. Malagoli, and N. E. Hurlburt, Turbulent compressible convection, *Astrophys. J.*, *370*, 282, 1991.

Childress, S. and A. D. Gilbert, *Stretch, twist fold: The fast dynamo*, Springer, Berlin, 1995.

Choudhuri, R. A., The effect of closed boundary conditions on a stationary dynamo, *Astrophys. J.*, *281*, 846, 1984.

D'Silva, S., Can equipartition fields produce the tilts of bipolar magnetic regions? *Astrophys. J.*, *407*, 385, 1993.

D'Silva, S. and A. R. Choudhuri, A theoretical model for tilts of bipolar magnetic fields, *Astron. Astrophys.*, *272*, 621, 1993.

Durney, B. R., On solar differential rotation:Meridional motions associated with a slowly varying angular velocity, *Astrophys. J.*, *407*, 367, 1993.

Durney, B. R., Meridional motions and the angular momentum balance in the solar convective zone, *Astrophys. J.*, *528*, 486, 2000.

Eddy, J. A., Climate and the changing Sun, *Climate Change*, *1*, 173, 1973.

Eddy, J. A., An historical review of solar variability, weather, and climate, in *Weather and climate responses to solar variations*, ed. B. M. McCormac, Associated University Press, Boulder, Colorado, p. 1.

Eddy, J. A., The Maunder Minimum: a reappraisal, *Solar Phys.*, *89*, 195, 1989.

Fan, Y., G. H. Fisher, and E. E. DeLuca, The origin of morphological asymmetries in bipolar active regions, *Astrophys. J.*, *405*, 390, 1993.

Fan, Y.,G. H. Fisher, and A. N. McClymont, Dynamics of emerging active region flux loops, *Astrophys. J.*, *436*, 907, 1994.

Feldman, U., J. M. Laming, P. Mandelbaum, W. H. Goldstein, and A. Osterheld, A burst model for line emission in the solar atmosphere. II. Coronal extreme ultraviolet lines, *Astrophys. J.*, *398*, 692, 1992.

Ferriz-Mas, A. On the storage of magnetic flux tubes at the base of the solar convective zone, *Astrophys. J.*, *458*, 802, 1996.

Ferriz-Mas, A., D. Schmitt, and M. Schussler, A dynamo effect due to instability of magnetic flux tubes, *Astron. Astrophys.*, *289*, 949, 1994.

Friis-Christensen, E. and K. Lassen, Length of the solar cycles: An indicator of solar activity closely associated with climate, *Science*, *254*, 698, 1991.

Gaizauskas, V., K. L. Harvey, J. W. Harvey, and C. Zwaan, Large-scale patterns formed by solar active regions during the ascending phase of cycle 21, *Astrophys. J.*, *265*, 1056, 1983.

Gilman, P. A. and M. Dikpati, Joint instability of latitudinal differential rotation and concentrated toroidal fields below the convective zone. II. Instability of narrow bands at all latitudes, *Astrophys. J.*, *528*, 552, 2000.

Gilman, P. A. and J. Miller, Nonlinear convection of a compressible fluid in a rotating spherical shell, *Astrophys. J. Suppl.*, *61*, 585, 1986.

Glatzmaier, G. A., Numerical simulations of stellar convective dynamos. II. Field propagation in the convective zone, Astrophys. J., *291*,3000, 1985.

Golub, L., A. S. Krieger, J. K. Silk, A. F. Timothy, and G. S. Vaiana, Solar X-ray bright points, *Astrophys. J. Lett,*, *189*, L93, 1974.

Habbal, S. R., R. S. Ronan, G. L. Withbroe, R. K. Shevgaonkar, and M. L. Kundu, Solar coronal bright points observed with the VLA, *Astrophys. J.*, *306*, 740, 1986.

Haisch, B. M. and M. Rodono, *Solar and stellar flares*, Proc. 104th IAU colloquium, ed. by B. M. Haisch and M. Rodono., Kluwer Academic Publishers, Dordrecht, 1983.

Handy, B. N. et al., The Transition Region and Coronal Explorer (TRACE), *Solar Phys.*, *187*, 229, 1999.

Hoyt, D. V., K. L. Kyle, J. R. Hickey, and R. N. Maschoff, The Nimbus 7 solar total irradiance: A new algorithm for its derivation, *J. Geophys. Res.*, *97*, 51, 1992.

Innes, D, E., B. Inhester, W. I. Axford, and K. Wilhelm, Bi-directional plasma jets produced by magnetic reconnection on the Sun, *Nature*, *386*, 811, 1997.

Kankelborg, C. C., A. B. C. Walker, and R. B. Hoover, Observation and modeling of soft X-ray bright points. II. Determination of the temperature and energy balance., *Astrophys. J.*, *491*, 952, 1997.

Kohl, J. L. et al., First results from the SOHO ultraviolet coronagraph spectrometer, *Solar Phys.*, *175*, 645, 1997.

Krause, F. and K. H. Radler, *Mean-field magnetohydrodynamics and dynamo theory*, Pergamon Press, Oxford, 1980.

Krucker, S. and A. O. Benz, Are heating events in the quiet solar corona small flares?- Multiwave length observations of individual events, *Solar Phys.*, in publication, 2000.

Lockwood, M., R. Stamper, and M. N. Wild, A doubling of the Sun's coronal magnetic field during the past 100 years, *Nature*, *399*, 437, 1999.

Low, B. C., Solar activity and the corona, *Solar Phys.*, *167*, 217, 1996.

Low, B. C., Coronal mass ejections, flares, and prominences, in *Solar wind nine*, Ed. S. R. Habbal, R. Esser, J. V. Hollweg, and P. H. Isenberg, American Institute of Physics Conference Proceeding 471, Woodbury, New York, P. 109, 1999.

Low B. C. and R. J. Hundhausen, Magnetostatic structures of the solar corona.II. The magnetic topology of quiescent prominences, *Astrophys. J.*, *443*, 818, 1995.

Markiel, J. A. and J. H, Thomas, Solar interface dynamo models with realistic rotation profile, *Astrophys. J.*, *523*, 827, 1999.

Martin, S. in *Small-scale dynamical processes in quiet stellar atmospheres*, ed. S. L. Keil, Sacramento Peak, Sunspot, New Mexico; National Solar Observatory, p. 30.

Martin, S. and K. L Harvey, Ephemeral active regions during solar minimum, *Solar Phys.*, *64*, 93, 1979.

Meyer, F., H. U. Schmidt, N. O. Weiss, and P. R. Wilson, The growth and decay of sunspots, *Mon. Not. Roy. Astron., Soc.*, *169*, 35, 1974.

Miesch, M. S., J. R. Elliott, J. Toomre, T. L.Clune, G. A. Glatzmaier, and P. A. Gilman, Three dimensional simulations of solar convection, differential rotation, and pattern evolution achieved with laminar and turbulent states, *Astrophys. J.*, March, 2000.

Moffatt, H. K. *Magnetic field generation in electrically conducting fluids*, Cambridge University Press, Cambridge, 1978.

Moore, R. L., D. A. Falconer, J. G. Porter, and S. T. Suess, On heating the Sun's corona by magnetic explosions: Feasibility in active regions and prospects for quiet regions and coronal holes, *Astrophys. J.*, *526*, 505, 1999.

Parker, E. N., Hydromagnetic dynamo models, *Astrophys. J.*, *122*, 293, 1955.

Parker, E. N., The solar hydromagnetic dynamo, Proc. *Natl. Acad. Sci.*, *43*, 8, 1957.

Parker, E. N. Dynamics fo interplanetary gas and magneic fields, *Astrophys. J.*, *128*, 644, 1958.

Parker, E. N., *Interplanetary dynamical processes*, Interscience Div. John Wiley and Sons, New York, 1963.

Parker, E. N., The generation of magnetic fields in astrophysical bodies.I. The dynamo equations, *Astrophys. J.*, *162*, 665, 1970.

Parker, E. N., The generation of magnetic fields in astrophysical bodies. IV. The solar and terrestrial dynamos, *Astrophys. J.*, *164*, 491, 1971.

Parker, E. N., The dynamical properties of twisted ropes of magnetic field and the vigor of nonactive regions on the Sun, *Astrophys. J.*, *191*, 245, 1974.

Parker, E. N., X-ray bright points on the Sun and the nonequilibrium of a twisted flux tube in a stratified atmosphere, *Astrophys. J.*, *201*, 494, 1975.

Parker, E. N., Sunspots and the physics of magnetic flux tubes. I. The general nature of the sunspot, *Astrophys. J.*, *230*, 905, 1979a.

Parker, E. N. *Cosmical magnetic fields*, Clarendon Press, Oxford, 1979b.

Parker, E. N., Stellar fibril magnetic systems. I. Reduced energy state, *Astrophys. J.*, *283*, 343, 1984.

Parker, E. N., The future of solar physics, *Solar Phys.*, *100*, 599, 1985.

Parker, E. N. Nanoflares and the solar corona, *Astrophys. J.*, *330*, 474, 1988.

Parker, E. N., A solar dynamo surface wave at the interface between convection and nonuniform rotation, *Astrophys. J.*, *408*, 707, 1993.

Parker, E. N., *Spontaneous current sheets in magnetic fields*, Oxford University Press, New York, 1994.

Parker, E. N. Theoretical properties of Ω-loops in the convective zone of the Sun. II. The origin of enhanced solar irradiance, *Astrophys. J.*, *440*, 415, 1995.

Porter, J. G. and R. L. Moore, in Proc. 9th Sacramento Peak Summer Symposium, ed. R. C. Altrock, Sacramento Peak, Sunspot, New Mexico; National Solar Observatory, p.30.

Priest, E. R., *Solar flare magnetohydrodynamics*, ed. E. R. Priest, Gordon and Breach, New York, 1981.

Priest, E. R., *Solar magnetohydrodynamics*, D. Reidel Pub. Co., Dordrecht, 1982.

Roberts, G. O., Spatially periodic dynamos, *Phil. Trans. Roy. Soc. London* A, *266*, 535, 1970.

Roberts, P. H., Kinematic dynamo models, *Phil, Trans, Roy. Soc. London* A, *272*, 663, 1972.

Schrijver, C. J., A. M. Title, A. A.Van Ballegooijen, H. J.Hagenar, and R. A. Shine, Sustaining the quiet photospheric network: The balance of flux emergence, fragmentation, merging, and cancellation, *Astrophys. J.*, *487*, 424, 1997.

Schrijver, C. J. et al., A new view of the solar outer atmosphere by the Transition Region and Coronal Explorer, *Solar Phys.*, *187*, 261, 1999.

Soward, A. M., *Stellar and planetary magnetism*, Gordon and Breach, New York, 1983.

Soward, A. M. and P. H. Roberts, *Magnetohydrodynamics*, *12*, 1, 1977.

Spruit, H. C. A model of the solar convection zone, *Solar Phys.*, *34*, 277, 1974.

Steenbeck, M. and F. Kruase, Zur dynamotheorie stellarer und planetarer Magnetfelder II. Berechnung planeten ahnlicher Gleichfeld generatoren, *Astron nachr.*, *291*, 271, 1969.

Steenbeck, M., F. Krause, and K. H. Radler, Berechnung der mittleren Lorentz-Feldstarke v×B fur ein elektrisch leitendes Medium in turbulenter, durch Coriolis-Krafte beeinflusster bewegung, *Zeit. Naturforsch. A21*, 369, 1966.

Stix, M. Differential rotation and the solar dynamo, *Astron. Astrophys.*, *47*, 243, 1976.

Sturrock, P. A., W. W. Dixon, J. A. Klimchuk, and S. Antiochos, Episodic coronal heating, *Astrophys. J. Lett.*, *356*, L31, 1990.

Svestka, Z., *Solar flares*, D. Reidel Pb. Co., Dordrecht, 1976.

Sweet, P. A., Mechanisms of solar flares, *Ann. Rev. Astron. Astrophys.*, *7*, 149, 1969.

Tao, L., R. E. Proctor, and N. O. Weiss, Flus expulsion from inhomogeneous turbulence, *Mon. Not. Roy. Astron. Soc.*, *300*, 907, 1998.

Tao, L., N. O. Weiss, D. P. Brownjohn, and M. R. E. Proctor, Flux separation in stellar magnetoconvection, *Astrophys. J. Lett.*, *496*, L39, 1998.

Thompson, M. J., Differential rotation and dynamics of the solar interior, *Science*, *272*, 1300, 1996.

Tobias, S. M., The solar cycle: parity interactions and amplitude modulation, *Astron. Astrophys.*, *322*, 1007, 1996a.

Tobias, S. M., Diffusivity quenching as a mechanism for Parker's surface dynamo, Astrophys. J., *467*, 870, 1996b.

Tobias, S. M., N. H. Brummel, T. L. Clune, and J. Toomre, Pumping of magnetic fields by turbulent penetrative convection, *Astrophys. J. Lett.*, *502*, L177, 1998.

Tomczyk, S., J. Snow, and M.J. Thompson, Measurement of the rotation rate in the deep solar interior, *Astrophys. J. Lett.*, *448*, L57, 1995.

Vandas, M., S. Fischer, and A. Geranios, Double flux rope structure of magnetic clouds?, in *Solar wind nine*, ed. S. R. Habbal. R. Esser, J. V. Hollweg, and P. H. Isenberg, American Institute of Physics Conference Proceedings 471, Woodbury, New York, p. 127, 1999.

Vishniac, E. T., The dynamics of flux tubes in a high-β plasma. I. A general description, *Astrophys. J.*, *446*, 724, 1995a.

Vishniac, E. T., The dynamics of flux tubes in a high-β plasma. II. Buoyancy in stars and accretion disks, *Astrophys. J. 451*, 816, 1995b.

Withbroe, G. L. and R. W. Noyes, Mass and energy flow in the solar chromosphere and corona, *Ann. Rev. Astron. Astrophys.*, *15*, 363, 1977.

Zhang, Q., W. H. Soon, S. L. Baliunas, G. W. Lockwood, B. A Skiff, and R. R. Radick, A method for determining possible brightness variations of the Sun in past centuries from observations of solar-type stars, *Astrophys., J. Lett.*, *427*, L111, 1994.

Zhu, X., Magnetospheric convection pattern and its implications, *J. Geophys. Res.*, *98*, 21, 291, 1993.

Zhu, X. How the magnetosphere is driven into substorm, *J. Geophys. Res.*, *100*, 1847, 1995.

Zirin, H., *Astrophysics of the Sun*, Cambridge University Press, Cambridge, Chap. 11, 1998.

Zwaan, C., On the appearance of magnetic flux in the solar photsphere, *Solar Phys.*, *60*, 213, 1978.

Zwaan, C., The emergence of magnetic flux, *Solar Phys.*, *100*, 397, 1985.

SEPs: Space Weather Hazard in Interplanetary Space

Donald V. Reames

NASA Goddard Space Flight Center, Greenbelt MD

In the largest and most hazardous of solar energetic particle (SEP) events, acceleration takes place at shock waves driven out from the Sun by fast CMEs. Multi-spacecraft studies show that the particles from the largest events span more than 180 degrees in solar longitude; the events can last for several days. Protons streaming away from the shock generate waves that trap particles in the acceleration region, limiting outflowing intensities but increasing the efficiency of acceleration to higher energies. Thus, early intensities are bounded, but at the time of shock passage, they can suddenly rise to a peak. These shock peaks extend to >500 MeV in the largest events, creating a serious 'delayed' radiation hazard. At high energies, spectra steepen to form a 'knee.' This spectral knee can vary from ~10 MeV to ~1 GeV depending on shock conditions, greatly affecting the radiation hazard. Elements with different charge-to-mass ratios differentially probe the wave spectra near shocks, producing abundance ratios that vary in space and time. These abundance ratios are a tool that can foretell conditions at an oncoming shock.

1. INTRODUCTION

As we move beyond the protective shield of the Earth's atmosphere and magnetosphere, we are exposed to sources of radiation that can be a serious hazard to humans and machines. High-energy particles in space include the sudden intense bursts of the solar energetic particle (SEP) events that can last several days. Large 'gradual' SEP events occur at a rate of 10-20 yr[-1], but the ones most threatening to human life occur less than once a decade. This makes them especially difficult to study or to predict.

In the preceding article, Kahler described the properties of SEP events [see also *Gosling* 1993; *Kahler* 1994; *Reames* 1997, 1999a, b]. He also discussed the checkered history of the large events that were once mistakenly associated with solar flares rather than CME-driven shock waves. It is impossible to predict SEP events well when you start with the wrong source. Large SEP events with no flares and large flares with no SEP events were among the clues that eventually set us straight. We now know that only the *fastest* CMEs drive the shock waves where acceleration takes place; particle intensity is strongly correlated with CME speed. In fact, CME speed is the best predictor of an intense SEP event.

This article presents our understanding of the underlying physics that controls the energy spectra and element abundances in SEP events and the way that they evolve in space and time. This new understanding has been greatly assisted by the first *dynamic* model of SEP events [*Ng, Reames, and Tylka* 1999].

2. INTENSITY TIME PROFILES

2.1. Streaming Limited Intensities

Observations of 3-6 MeV proton intensities near 1 AU early in large SEP events showed evidence of an intensity limit of ~100 $(cm^2 \ sr \ s \ MeV)^{-1}$, within a factor of ~2 or so [*Reames* 1990]. Later in these same events, near the time of

Space Weather
Geophysical Monograph 125

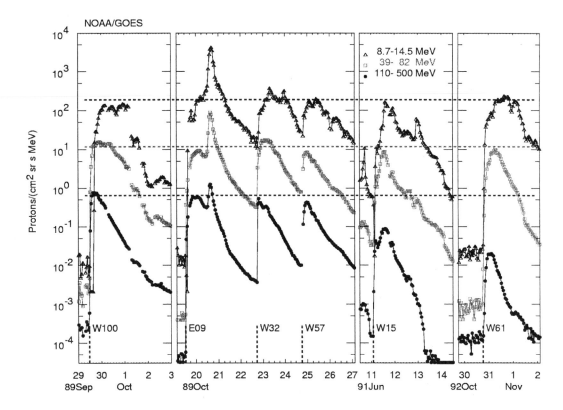

Figure 1. Intensity-time profiles are shown for three energy channels during six large SEP events of the last solar cycle. Streaming-limited intensities for each channel are shown as dashed horizontal lines. CME source longitudes are indicated for each event as dashed vertical lines at the time of onset [*Reames and Ng* 1998].

shock passage, intensities can rise by factors of 100 above the early limit. Large events studied during the next solar cycle [*Reames and Ng* 1998] appeared to have limits that decreased with proton energy, as shown in Figure 1. Dashed lines are drawn at the 'streaming limit' for three energy intervals plotted in the figure; the 100-500 MeV protons do not reach the limiting value in the last two events.

It is well known that distributions of particles streaming along magnetic field lines are unstable to the production of resonant Alfvén waves [*Stix* 1962; *Melrose* 1982]. At high particle intensities, sufficient intensities of resonant waves are produced to scatter the particles that come behind and reduce their streaming. This process serves to trap particles near their source and bound the outward flow at the streaming limit. This limit, of course, depends upon radial distance in the diverging magnetic geometry. *Ng and Reames* [1994] performed a numerical simulation of particle transport out from the Sun through self-generated Alfvén waves. They were able to confirm the value of the low-energy streaming limit. In a subsequent study using Helios data, the radial dependence of the early streaming limit was found to be consistent with the R^{-3} dependence expected theoretically [*Reames and Ng* 1998].

2.2. Longitude Distributions

Kahler [2000] discussed the effect of the solar longitude of the observer relative to the CME on the appearance of the intensity-time profile. Because of the spiral magnetic field, an observer's magnetic connection to the shock swings eastward with time, either approaching or receding from the intense 'nose' of the shock [see Figure 10 of *Kahler* 2000 and *Reames, Barbier, and Ng* 1996].

Source longitudes are shown for the events in Figure 1. Coincidentally, most of these event source longitudes are west of the observer so the shock nose does not reach 1 AU. No strong peaks are seen at times of shock passage for these events, although, the events from W15 and W32 do show shock increases at low energies. However, the event on 1989 October 19 from E09 shows a strong peak in all energy channels when the shock reaches Earth a day later. This shock has an *average* speed of ~1500 km s⁻¹. Shock intensity peaks are produced by CMEs from central meridian that produce sufficiently fast shocks to continue acceleration out to 1 AU. Shock peaks in >100 MeV protons are rare, but can be a serious hazard when they occur.

At times, longitude distributions of a single SEP event can be measured using multiple spacecraft [*Reames, Bar-*

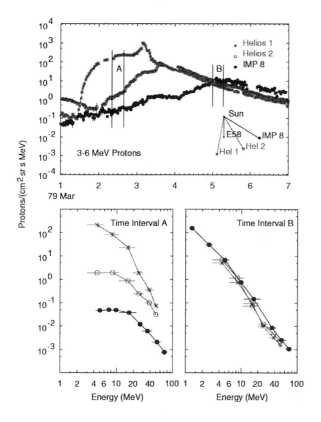

Figure 2. Intensity-time profiles for 3-6 MeV protons for three spacecraft at different longitudes (see inset) are shown in the upper panel. Proton energy spectra at time A are contrasted with invariant spectra at time B in the lower panels.

bier, and Ng, 1996; *Reames, Kahler, and Ng*, 1997]. Figure 2 shows intensity-time plots for three spacecraft distributed in longitude about a CME as shown in the inset. Helios 1 passes near the nose of the shock and sees a flat, streaming-limited profile followed by a peak at the shock. Helios 2 and IMP 8 farther around the west flank see increasingly slower rises. This small event has a relatively narrow longitude span; other events have high intensities over a span of more than 90° to the east and west of the central longitude of the CME.

Figure 2 also shows that the intensities at the three spacecraft merge, within a factor of ~2, late in the event, long after shock passage in this case. This is a region of spatially and temporally invariant spectra [*Reames, Kahler, and Ng*, 1997]. Energy spectra observed early and late in the event are contrasted in the lower panels of the figure. These invariant spectra are produced in regions where particles are trapped or quasi-trapped in magnetic bottles. Adiabatic deceleration of the particles preserves the spectral shape as the volume of the bottle expands. At times, preexisting leaky bottles formed by old CMEs can fill with particles from a new event at the Sun, causing invariant

spectra to be seen ahead of the shock [*Reames* 1999a]. Bottles can also be formed when particles are quasi-trapped behind the wave turbulence near a shock, either the shock that accelerated them or one from an earlier event.

3. SHOCK ACCELERATION

The same proton-generated waves, that limit streaming and trap particles near the source, greatly increase the acceleration efficiency of that source. The importance of self-generated waves has been recognized for shock acceleration in many astrophysical contexts; self-generated waves were first applied to shock acceleration in SEP events by Lee [1983]. Although this is a static, equilibrium model with a planar shock, the Lee model has been seminal in promoting our understanding of the physics of SEP acceleration. SEP acceleration at shocks can only be sustained by proton-generated waves; ambient turbulence is completely inadequate to support acceleration above ~1 MeV. Only a small fraction of the ambient turbulence resonates with energetic ions.

A simple cartoon illustrating shock acceleration is shown in Figure 3. Acceleration actually occurs as particles are scattered back and forth across the shock by waves carried at the different velocities of the upstream and downstream plasma. Injected at low energy, probably from the tail of the thermal plasma distribution function, particles begin to scatter, first on ambient turbulence then on reso-

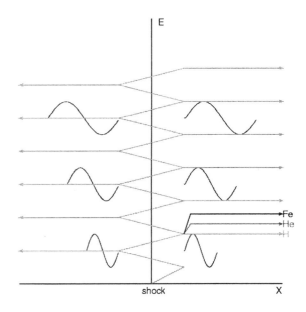

Figure 3. The cartoon illustrates efficient energy gain of particles at a shock because of trapping by self-generated Alfvén waves. Particles of increasing energy resonate with waves of lower wave number, *k*, and greater wavelength.

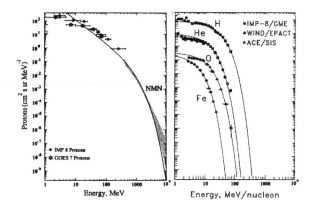

Figure 4. The proton energy spectrum in the 1989 September 29 SEP event (left panel) is measured by spacecraft and the neutron monitor network (NMN) [*Lovell, Duldig, and Humble* 1998]. The right panel shows spectra of ions in the 1998 April 20 event [*Tylka et al.* 2000]. Proton spectral knees occur at much different energies in the two events.

nant waves they generate as they stream away from the shock. Particles of magnetic rigidity (momentum per unit charge) P, resonate with waves of wave number k, when $k=B/\mu P$, where B is the magnetic field strength and μ is the cosine of the particle pitch angle with respect to the field. Most of the waves are produced by the dominant species, protons.

As trapping increases for particles of one rigidity, they are more likely to be accelerated to a higher rigidity, where they again stream out and produce resonant waves, *etc*. As this process continues, a 'wall' of resonant waves grows up the energy axis. Since the streaming of particles away from the shock is limited, the height of the 'wall' depends upon the number of particles injected at the bottom. Thus, increasing the injection increases the maximum energy to which particles can be accelerated.

Given infinite time, acceleration would produce a power-law energy spectrum at the shock. Ions of different elements are accelerated to the same velocity or energy/nucleon at the shock; these spectra differ in proportion to the coronal abundances of the elements [*Reames* 1999a]. However, escape from the shock depends upon the rigidity of an ion and hence upon it's ionization state or charge-to-mass ratio, Q/A. Hence, different species, of the same velocity, probe different parts of the wave spectrum as they escape. Element abundance ratios, such as Fe/O, viewed away from the shock, can be enhanced if Fe escapes more easily than O; near the shock Fe/O would be suppressed. The degree of enhancement or suppression depends upon the slope and intensity of the wave spectrum.

The CME-driven shocks that accelerate SEPs are uniquely dynamic when compared with other shock waves in the heliosphere. These shocks are born anew in each

event in the plasma of the high corona and they expand in a roughly spherical shell across the magnetic environment. By the time they reach 1 AU, days later, the plasma densities and magnetic fields have decreased by orders of magnitude, and even the fastest shocks have slowed considerably. Simulation of the shock and the particles in this environment requires a numerical model. It is only recently that the first dynamic models of shock acceleration [*Zank, Rice, and Wu* 2000] and of particle transport, solving coupled equations for the transport of both particles and waves [*Ng, Reames, and Tylka* 1999], have become available.

4. SPECTRAL KNEES

At high particle energies, intensities may become too low to sustain wave growth so that scattering is reduced and the particles begin to leak away from the shock. This causes the energy spectrum to depart from its nominal power-law form and to steepen exponentially, forming a spectral 'knee.' Figure 4 shows examples of spectral knees in SEP events. The left panel shows proton data taken on spacecraft and measured by the ground-level neutron monitor network (NMN, shaded region in figure) in the 1989 September 29 SEP event. The fit to the data is the power-law times exponential form used by *Ellison and Ramaty* [1985]. The e-folding energy E_{knee} = 1 GeV for this event [*Lovell, Duldig, and Humble* 1998]. The right-hand panel in Figure 4 shows spectra of H, He, O, and Fe in the 1998 April 20 event. In this event, E_{knee} =15 MeV for protons and scales as Q/A for the other species. In those few other events that can be measured, the e-folding energy/nucleon does not always scale linearly with Q/A [*Tylka et al.* 2000].

It is difficult to understand the origin of the large difference in E_{knee} for protons in the two events in Figure 4. Both events occur near the west limb of the Sun and the CME speeds are ~1800 and 1600 km s^{-1} in the 1989 and 1998 events, respectively. The detailed physics of shock acceleration that determines the knee energies is not well known. Worse, many of the largest events have knee energies above the region we can observe; the required measurements are simply not available.

However, the practical effect of differing knee energies becomes clear when we directly compare, in Figure 5, fits to the two proton spectra from Figure 4. Also shown in Figure 5 are levels of radiation hazard to astronauts outside the magnetosphere. 'Soft' radiation occurs where protons begin to penetrate space suits or spacecraft walls. 'Hard' radiation begins at a proton energy that will penetrate ~5 cm of Al; here, shielding becomes impractical. Even though the two events have similar proton intensities from 10 to 100 MeV, the different values of E_{knee} cause vastly different levels of hazard. The spectrum shown for the

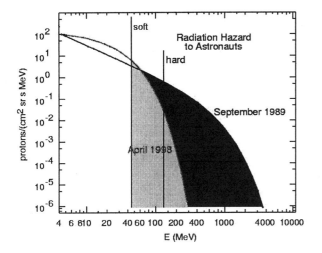

Figure 5. The proton spectra from the two events in Figure 4 are compared to show substantial differences in their radiation hazard inside a spacecraft wall (soft) and inside ~5 cm of Al (hard).

1989 September event would produce ~4 rem hr^{-1} behind 10 g cm^{-2} of shielding. The annual allowed dose for astronauts, 50 rem, would be accumulated with exposures of half a day to such intensities. Radiation workers at ground level are only allowed 15 rem yr^{-1}, and reduction of the level for astronauts has been suggested.

A truly serious situation would result if a high-energy knee persisted until the large peak at the time of shock passage. In this case, streaming limits would not apply as they do in the 1989 September event. The event of 1972 August 4 is an example of high intensities of high-energy protons occurring at a shock peak; unfortunately, instrument saturation prevented definitive spectral measurements in that event.

It is generally accepted that radiation levels in the 1972 August 4 event would have been fatal to inadequately shielded astronauts. The issue is the thickness of shielding required for protection. The thickness required to stop protons of given energy goes as the 1.6 power of the energy. Increasing E_{knee} from 50 to 500 MeV would increase the thickness and weight of the required shielding by a factor of 40. Mission costs increase at least linearly with payload weight, and manned missions to Mars, for example, are already expensive. Our present knowledge does not allow us to define a meaningful value of E_{knee} that is appropriate for shielding design.

5. ABUNDANCE VARIATIONS

The discovery of regular methodical time variations in element abundance ratios in SEP events [*Tylka, Reames, and Ng* 1999] has been a key to our new understanding of the dynamics of acceleration at CME-driven shocks. The

theory that evolved to explain these observations [*Ng, Reames, and Tylka* 1999] has also explained other abundance anomalies in He/H that have puzzled observers for 20 years [*Reames, Ng, and Tylka* 2000].

Figure 6 shows dramatic time variations of abundance ratios, normalized to the corresponding coronal abundances, in a large SEP event. The right-hand panel in the figure shows a theoretical simulation of the event. The detailed time behavior of the abundances depends upon the time behavior of shock parameters such as the shock compression ratio, which has been assumed to decrease linearly in the simulation. Variations of this kind were essentially unknown a few years ago; abundances were studied only by averaging over entire events.

The variation of abundances with energy and with time results from changes in the spectrum of resonant waves at the shock. Since proton intensities are often streaming-limited early in large events, these abundance variations may be our only probe of conditions at an oncoming shock. For example, He/H ratios that are strongly suppressed and rising early in an event are already a powerful indication of high proton intensities at the oncoming shock [*Reames, Ng, and Tylka* 2000]. Before we can 'calibrate' abundances as a remote-sensing device of proton intensities and spectra at a shock, we must refine our models to include all processes that influence the relationship between protons, resonant waves, and abundance variations. Often we must work with particles from distant shocks whose properties are not independently known. Thus, models of SEP acceleration and

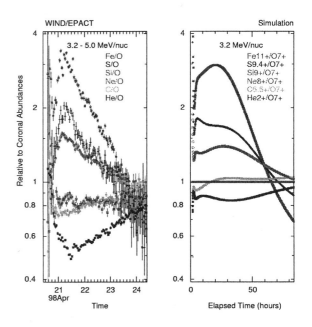

Figure 6. Measured and simulated element abundance ratios vary with time during an SEP event. These variations result from changes in the proton-generated wave field near the shock.

transport must eventually be combined with models of the evolution of the shock itself.

6. SUMMARY AND PROSPECTS

Large SEP events can be a significant hazard to humans and equipment outside the Earth's magnetosphere. In the largest SEP events, particles are accelerated at CME-driven shock waves. Events where particles are stochastically accelerated in solar flares also occur, but they are small and are not considered in this article [see *Reames* 1999a].

Particle intensities from large SEP events may be streaming-limited, early in the events, by proton-generated waves that throttle the outflow from the shock. This trapping increases the efficiency for acceleration to higher energy. Self-generated waves dominate all aspects of the events, including intensities, element abundances, spectra and angular distributions.

The spatial distribution of accelerated particles around a shock in solar longitude controls the time profile seen by an observer whose magnetic connection swings eastward across the face of the shock with time. Intensities increase (decrease) as the connection point moves toward (away from) the nose of the shock. The shock strength may also decrease with time as the shock expands radially. While flow of particles from the distant shock is streaming-limited, very large intensity peaks can occur when a strong shock from central meridian passes over the spacecraft.

At some high energy, resonant wave generation at the shock can no longer be sustained. At this energy, particles begin to leak from the shock and a spectral knee forms. The position of this knee can vary greatly from event to event and during an event. Events with a high-energy knee are the most threatening to astronauts, especially if that knee occurs during a shock peak where streaming limits do not apply.

Events with high intensities of >100 MeV protons at shock peaks are rare, 1972, 1989, …. However, their occurrence becomes more likely during long-duration missions to the moon or Mars. Furthermore, the rarity of these events itself makes them difficult to study and their probability of occurrence difficult to assess.

Prospects for improving SEP models in the future include the following:

1) Improved theoretical models of shocks and SEP acceleration and transport are likely. Several research groups now work in this new field of dynamic SEP acceleration at shocks [*e.g. Zank, Rice and Wu* 2000].

2) The STEREO mission, scheduled for launch in 2004, will provide stereoscopic images of CMEs permitting three-dimensional modeling of the acceleration region along with multi-spacecraft observation of the SEP events.

3) We have proposed a large-geometry, high-energy experiment, SPARKLE, for the International Space Station that will measure spectra and abundances of particles out to 2 GeV/amu to pursue the study of spectral knees into this new and important energy region that is currently inaccessible. SPARKLE would measure spectral knees at the same time that STEREO is mapping the source region.

A serious problem during the last 10 to 20 years has been the dearth of scientists actively studying the properties of protons in SEP events. Former workers in the field have moved their interests to the outer heliosphere, to abundances and isotopes, or to energies below 1 MeV. Few papers in the refereed literature contribute information of use to space weather. In this context, the renewed theoretical interests in SEP events are especially gratifying.

New models have guided our understanding of SEP acceleration, but we cannot yet make detailed forecasts of SEP properties for a given shock. Nevertheless, recent progress suggests that such forecasts are within our grasp.

Acknowledgments. I am pleased to acknowledge many helpful discussions with Steve Kahler, Chee Ng, and Allan Tylka during the preparation of this paper. The content of this article has been coordinated with that of the preceding article by Kahler.

REFERENCES

Ellison, D., and R. Ramaty, Shock acceleration of electrons and ions in solar flares, *Astrophys. J.* 298, 400, 1995.
Gosling, J. T., The solar flare myth, *J. Geophys. Res.* 98, 18949, 1993.
Kahler, S. W., Injection profiles for solar energetic particles as functions of coronal mass ejection heights, *Astrophys. J.* 428, 837, 1994.
Kahler, S. W., this volume, 2000.
Lee, M. A., Coupled hydromagnetic wave excitation and ion acceleration at interplanetary traveling shocks, *J. Geophys. Res.*, 88, 6109, 1983
Lovell, J. L., M. L. Duldig, and J. E Humble, An extended analysis of the September 1989 cosmic ray ground level enhancement, J. Geophys. Res. 103, 23,733, 1998.
Melrose, D. B., *Plasma Astrophysics*, Vol. 1, Gordon and Breach, New York, 1980.
Ng, C. K., D. V. Reames, and A. J. Tylka, Effects of proton-generated waves on the evolution of solar particle composition in gradual events, *Geophys. Res. Lett.* 26, 2145, 1999.
Ng, C. K., and D. V. Reames, Focused interplanetary transport of ~1 MeV solar energetic protons through self-generated Alfven waves, *Astrophys. J.*, 424, 1032, 1994.
Reames, D. V. 1990, Acceleration of energetic particles by shock waves from large solar flares, *Astrophys. J.* (Letters), 358, L63.
Reames, D. V., Energetic particles and the structure of coronal mass ejections, in: *Coronal Mass Ejections*, edited by N. Crooker, J. A. Jocelyn, J. Feynman, Geophys. Monograph 99, (AGU press) p. 217, 1997.

Reames, D. V., Particle acceleration at the sun and in the heliosphere, *Space Sci. Revs.*, 90, 413, 1999a.

Reames, D. V., Solar energetic particles: is there time to hide? *Radiation Measurements* 30/3, 297, 1999b.

Reames, D. V., L. M. Barbier., and C. K. Ng, The Spatial Distribution of Particles Accelerated by CME-Driven Shocks, *Astrophys. J.* 466, 473, 1996.

Reames, D. V., S. W. Kahler and C. K. Ng, Spatial and temporal invariance in the spectra of energetic particles in gradual solar events, *Astrophys. J.* 491, 414, 1997.

Reames, D. V., and C. K. Ng, Streaming-limited intensities of solar energetic particles, *Astrophys. J.* 504, 1002, 1998.

Reames, D. V., C. K. Ng, and A. J. Tylka, Initial time dependence of abundances in solar particle events," *Astrophys. J. Letters* 531, L83, 2000.

Stix, T. H., *The Theory of Plasma Waves*, McGraw-Hill, New York, 1962.

Tylka, A. J., D. V. Reames, and C. K. Ng, Observations of systematic temporal evolution in elemental composition during gradual solar energetic particle events, *Geophys. Res. Letters*, 26, 2141, 1999.

Tylka, A. J., P. R. Boberg, R. E. McGuire, C. K. Ng, and D. V. Reames, Temporal evolution in the spectra of gradual solar energetic particle events, *Acceleration and Transport of Energetic Particles Observed in the Heliosphere*, eds. R.A. Mewaldt, J.R. Jokipii, M.A. Lee, E. Moebius, and T.H. Zurbuchen, AIP Conference Proceedings (2000), in press.

Zank, G. P., W. K. M. Rice, and C. C. Wu, Particle Acceleration and Coronal Mass Ejection-Driven Shocks: A Theoretical Model, *J. Geophys. Res.,* submitted, 2000

Donald V. Reames, Code 661, NASA Goddard Space Flight Center, Greenbelt, MD 20771.

Origin and Properties of Solar Energetic Particles in Space

S. W. Kahler

Space Vehicles Directorate, Air Force Research Laboratory, Hanscom AFB, Massachusetts

Transient energetic (E > 10 MeV) particle events from the Sun have been observed at the Earth for half a century using several detection techniques. We review these observations and the change in focus from solar flares to fast coronal mass ejections (CMEs) and shocks as the causal agents of solar energetic particle (SEP) events. We review the properties of SEP events that are important for space weather: 1) the number distributions of peak SEP intensities and of SEP fluences; 2) the relationship of SEP intensities to CME speeds; 3) when SEP events occur during the solar cycle; and 4) where SEP events occur relative to the CME-driven shocks. We conclude with a discussion of two cases of direct negative effects of SEPs on space experiments.

1. THE DISCOVERY AND EARLY OBSERVATIONS OF SOLAR ENERGETIC PARTICLE EVENTS

The first indication that the Sun produces energetic particles came with three cases of abrupt increases of cosmic ray intensity measured with ionization chambers in the 1940s. Each of these increases closely followed a solar flare, and it was suggested that the energetic particles had been emitted from the Sun (*Forbush* 1946). A system of neutron monitors at different geomagnetic latitudes enabled Meyer et al. (1956) to observe the rigidity spectrum and time variations of an intense P > 2 GV SEP event on February 23, 1956. They modeled the event as particle emission from a solar flare into a magnetic-field free cavity extending to 1.4 AU and surrounded by disordered magnetic fields which scattered the particles.

The number of detected SEP events increased significantly in the late 1950s when it was understood that energetic particles of 5 < E < 50 MeV precipitating onto the ionosphere increase the electron density, causing enhanced absorption of cosmic radio noise (Figure 1). Since charged particles of low energy preferentially precipitate over the geomagnetic polar caps, the events were known as polar cap absorption (PCA) events (*Reid and Leinbach* 1959) and constitute a space weather threat to communications at high latitudes. PCA events were associated with various types of solar flare radio emission (e.g., *Warwick and Haurwitz* 1962) to establish more firmly a link between the SEPs observed at 1 AU and the generation of nonthermal electrons at the associated flare site.

The study of SEPs became really serious when space-based particle experiments provided measurements over energy ranges down to several MeV and from solar-orbiting as well as Earth-orbiting satellites. The good association of SEP events with flare metric slow-drift type II bursts suggested that protons were accelerated to low energies in flare impulsive phases and then to higher energies in the shocks producing type II bursts (*Svestka and Fritzova-Svestkova* 1974). A particularly influential statistical work by Reinhard and Wibberenz (1974) concluded that particle transport consisted of an energy-independent propagation through the solar corona followed by energy-dependent travel through interplanetary space. Rapid (t < 1 hr) particle transport filled a "fast propagation region" identified with a unipolar magnetic cell and extending up to 60° from the flare site. This region occupied the solar longitudes of ~ 0° to W100°. All investigators of the time assumed that SEP production and coronal injection oc-

Space Weather
Geophysical Monograph 125

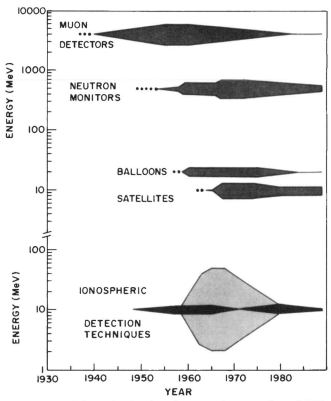

Figure 1. Schematic showing the general progression of SEP measurements toward lower energies with time. The thickness of the each line indicates the relative number of type of detector in use. Satellite measurements now extend the energy range down to solar wind energies. Two ionospheric techniques were used in the polar caps, the second of which observed attenuation of astronomical radio sources to deduce the 5 to 50 MeV SEP fluxes. From Shea and Smart (1990).

curred only during the flare impulsive phase, which was short (t < 1 hr) compared to the time scales of tens of hours to days for the SEP event duration. While understanding the SEP acceleration processes was a primary goal, the more practical studies of the time included finding the magnetic context for SEP coronal propagation (e.g., *Roelof and Krimigis,* 1973) and relating the flare microwave burst parameters to the associated SEP event intensities (e.g., *Croom* 1971).

2. CORONAL MASS EJECTIONS AND SHOCKS

The first observations of coronal mass ejections (CMEs) may well have occurred in photographs taken during solar eclipses. Cliver (1989) has argued that the presumed comets observed near the Sun in the July 1860 and April 1893 eclipses were in fact CMEs. The first clear observation of a CME, seen as several bright clouds followed by a diffuse cloud rapidly ejected from the Sun, was on December 14,

1971 with the NRL coronagraph on the OSO-7 satellite. That event, observed to have a speed of ~ 1000 km/s, occurred over the southeast limb of the Sun at a position angle ~ 100° (Figure 2) and was preceded by intense metric type II and type IV radio bursts beginning about 0240 UT. The event was quickly appreciated in the popular press (*Lyons* 1972; *Maran and Thomas* 1972) as a very energetic solar eruption with space weather implications. Brueckner (1974) estimated the CME mass to be 4×10^{16} g with an energy of 1×10^{32} ergs.

The lack of an observed Hα or soft X-ray flare compelled observers to interpret this event as the result of a flare about 25° behind the east limb of the Sun (*Brueckner* 1974; *Kosugi* 1976; *Hudson* 1978). A reexamination of the data suggests that the actual CME source region may well have been on the visible disk, perhaps associated with the disappearance of a quiescent filament at S21 E22 within 12 hrs of the CME, as noted in the catalog of Wright (1991). A large eruptive prominence or spray was observed at 0252 UT at a limb position angle 130° (*Kosugi* 1976), appropriate for that disappearing filament location. A long-lived noise storm was also observed with the Nancay interferometer at 169 MHz at about E40° beginning December 14 (*Solar-Geophysical Data,* 1971), suggesting an energetic eruptive event in association with the filament disappearance. In addition, a pair of SSCs were recorded at the Earth at 1904 UT on December 16 and at 1418 UT on December 17, each of which was followed by a Forbush decrease (*Cane et al.* 1996). A shock detection at Earth is possible but unlikely for a source region 25° behind the limb.

The SEP observations of this event are more definitive for establishing the source region. A rapid rise of 3 to 30 MeV SEPs with velocity dispersion was seen in the IMP 6 observations beginning about 04 UT on December 14 (Figure 2). If the source region lay 25° behind the east limb, it would have been about 180° away from the field lines connecting the Earth to the Sun, and therefore the SEP event should have been observed with only a very gradual onset, if at all. SEP protons of energies E > 175 MeV were also observed from this event at Pioneer 6 (*Solar-Geophysical Data* 1972). Interplanetary SEPs were thought to be produced exclusively in solar flares, so this first CME observation and its associated SEP event was the first hint of a serious challenge to the basic paradigm of SEP production in active region flares. If the CME source was the filament at S21 E22, then it stands as one of the very few energetic SEP events without an associated flare, as we discuss later.

The discovery that CMEs were a spectacular but common solar phenomenon was established with Skylab coronagraph observations (*Gosling et al.* 1974). CME properties such as speeds (*Gosling et al.* 1976) and solar event

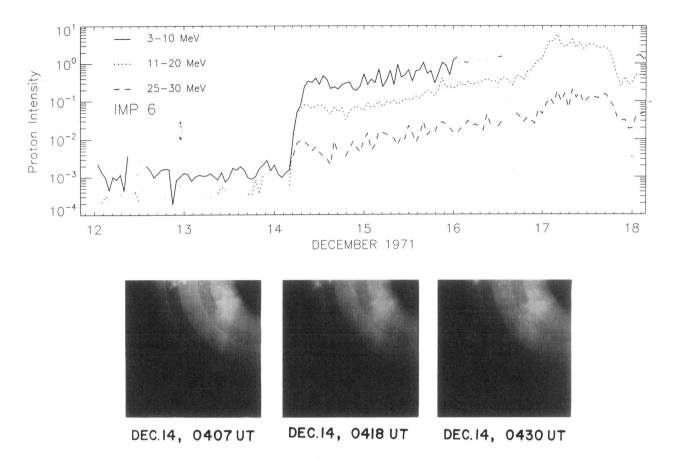

CORONAL TRANSIENT EVENT, RECORDED
BETWEEN 3 AND 10R☉, DEC. 13-14,1971.
NAVAL RESEARCH LAB. EXPT. ON NASA OSO-7

Figure 2. The first observed CME, on December 14, 1971, observed with the OSO-7 NRL coronagraph (bottom). Although the source region was presumed to lie 25° behind the east limb, the associated SEP event (top) has an onset profile characteristic of a front-side, near central-meridian source region.

associations (*Munro et al.* 1979) were measured. A correlation was then found between peak intensities of 4 to 23 MeV SEP events and associated CME speeds (*Kahler et al.* 1978). It was proposed that SEPs were accelerated in broad shocks driven by the fast (v > 400 km/s) CMEs, as first suggested by Wild et al. (1963), and that the angular extents (~ 60°) of the CMEs could explain the fast propagation region of Reinhard and Wibberenz (1974).

With the launch of the P78-1 spacecraft a new set of CME observations was obtained with the NRL Solwind coronagraph. Several comparative studies of Solwind CMEs with IMP-8 and ISEE-3 SEP events (*Kahler et al.* 1987; *Kahler* 1996) confirmed the earlier Skylab results. Similar comparisons combining IMP-8 SEP events with SMM

CMEs (*Kahler et al.* 1999), Helios SEP events with Solwind CMEs (*Reames et al.* 1997), and Wind EPACT SEP events with SOHO Lasco CMEs (Figure 3) have further confirmed the basic result of a correlation between peak SEP intensities and CME speeds. Other studies have compared SEP events with halo CMEs observed with the SOHO Lasco coronagraph (*Cane et al.* 1998). It is important to note that to date no case of a gradual SEP event without an associated CME has been found.

The refutation of the old paradigm that solar flares are the sources of SEP events is more convincing if one can find cases in which the most likely solar associations are filament eruptions rather than flares. In the low energy (E < 5 MeV) regime at least 14 such cases have now been

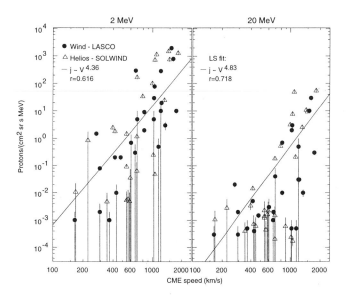

Figure 3. Proton peak intensities at two different energies versus the speeds of the associated CMEs for Wind/Lasco events (●) and Helios/Solwind events (△). Both event sets were selected for reasons other than their SEP-CME associations. Diagonal lines are least-squares best fits to the data.

identified (*Sanahuja et al.* 1991). In the E > 50 MeV range two cases have been reported, one with a clear quiescent filament disappearance and observed CME (*Kahler et al.* 1986), the other with a huge southern polar crown X-ray arcade observed with the Yohkoh SXT (*Kahler et al.* 1998), a clear proxy for a CME. We can now count the December 14, 1971 CME without an observed flare, described above, as a candidate third example.

Since SEP injections occur at shocks driven by CMEs, the heights of the CMEs during SEP injections are important to determine. In general this is difficult to do, especially for low energy (E < 100 MeV) protons and ions, because scattering of the particles in the interplanetary medium destroys information about injection profiles. Kahler (1994) examined onsets of several ground level events (GLEs) of E > 500 MeV, where scattering effects are minimal, to deduce the times of injections at the Sun. Since the associated CME heights were also known as a function of time, he plotted the SEP injection profiles against the simultaneous CME heights, as shown in Figure 4. That result showed that for all energies above 470 MeV the injection begins when the CME height is about 3 to 5 Rs and continues to increase as the CMEs reach heights of 10 to 12 Rs.

As a CME-driven shock moves out from the Sun, the shock magnetic fields are weaker and the injection energy spectrum of the SEPs grows softer. In addition, the geometry of the field lines connecting the observer to the

shock is continually changing, with the connection point always moving to a more eastward part of the shock. This results in systematic changes of SEP profiles for different observer longitudes. Cane et al. (1988) used 235 events to deduce average time scales and intensity-time profiles of SEP events in the 1 to 300 MeV range as a function of the solar source longitude.

The changing views of our understanding of SEP events can be seen from a glance at the basic solar-heliospheric (SH) topic groups of the biennial International Cosmic Ray Conferences. The more basic topics such as particle accel-

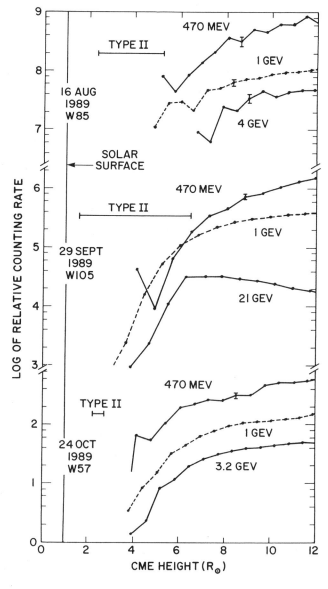

Figure 4. Coronal injection profiles of high-energy SEPs as functions of associated CME heights for three SEP events in 1989. From Kahler (1994).

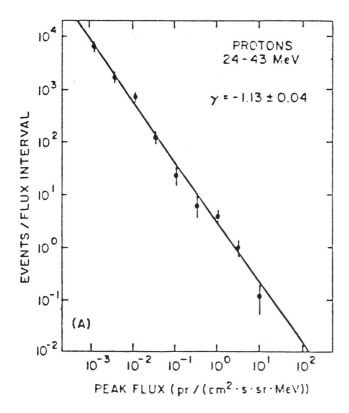

Figure 5. Distributions of peak intensities of SEP events on a log-log plot. The least-squares best-fit slope is $\gamma = 1.13$. From Cliver et al. (1991).

eration and composition, solar neutrons, solar photon radiation, and interplanetary propagation, have been on the agenda for several decades. However, beginning with the 1997 meeting in Durban, South Africa, the topic of coronal propagation has been dropped, and the topics of coronal mass ejections, propagating and corotating shocks, and SEP charge states have been added. These changes reflect not only an acceptance of the importance of CMEs and shocks as the sources of gradual SEP events, but also the emergence of detailed measurements of charge to mass ratios for a variety of ions which can serve as diagnostics of the acceleration and propagation processes.

Gradual SEP events produced in shocks constitute the largest and most important energetic particle events for space weather, but two other classes of relatively small energetic particle events are often observed at 1 AU (*Reames* 1999b). The first is the impulsive SEP events, with abundances and charge states of elements with $Z \geq 6$ enhanced over those of the gradual SEP events. These events are associated with solar flares and type III radio bursts, and their elemental abundances are similar to those of ions producing γ-ray lines in solar flares. The second class of energetic particle events is produced at shocks originating

in corotating interaction regions (CIRs). These particle populations have positive radial spatial gradients of intensities, since their shock acceleration occurs beyond 1 AU.

3. SIZES OF SEP EVENTS

3a. Size Distributions

With a good statistical sample of SEP events it was possible to construct plots of event size distributions for peak intensities, I, in a given energy range. This was first done by van Hollebeke et al. (1975) for 20 < E < 80 MeV proton events. For 125 events in the peak intensity range of 10^{-4} to 10^1 p/cm^{2}s-sr-MeV they obtained a differential power law of the form $dN/dI = CI^{-\gamma}$, where dN/dI is the number of events per logarithmic intensity interval I, and C and γ are parameters of the fit, with $\gamma = 1.15 \pm 0.1$. Cliver et al. (1991) fitted a later data set of 24 < E < 43 MeV proton events over the range 10^{-3} to 10^1 p/cm^{2}s-sr-MeV and found a best fit of $\gamma = 1.13 \pm 0.04$ for 92 events (Figure 5). These values of γ are significantly smaller than those for the power-law size distributions of flare parameters, such as hard and soft X-ray and microwave peak fluxes, which lie in the range $1.5 < \gamma < 2$ (*Crosby et al.* 1993).

This distinction between the exponents of SEP event intensities and those of flare intensities may reflect the physical differences between flares and SEP production. Since SEP events are produced by shocks driven by fast CMEs, we might expect that CME statistics, such as their speeds, would be more relevant to SEP event statistics than would flare statistics. CME speed distributions are usually plotted linearly with speed, but I have done least-squares fits of published speed v distributions of Solwind (*Howard et al.* 1985), SMM (*Hundhausen et al.* 1994), and Lasco (*St. Cyr* private communication) CMEs to integral N versus v plots in log-log format (Figure 6). Over the range 500 < v < 1500 km/s these fits give differential power-law exponents of $\beta = 4.59$, 4.20, and 4.18 for Solwind, SMM outer loops, and Lasco, respectively. These values far exceed the range of 1.5 - 2.0 characterizing the flare parameters as well as the ~1.1 value for the SEP event peak intensities.

The occurrence probability of a SEP event resulting from a CME-driven shock depends on both the CME speed distribution and the probability that a CME of a given speed will drive a shock. The CME speed distribution is quite steep, but the shock probability should be increasing sharply in the relevant (v > 400 km/s) speed range. Comparing the speed distribution of Solwind CMEs known to drive strong interplanetary shocks (*Cane et al.* 1987) with the steeper distribution of all Solwind CMEs in Figure 6, we see that the probability of a strong shock rises from ~ 0.1 in the range 400 < v < 800 km/s to perhaps unity at ~

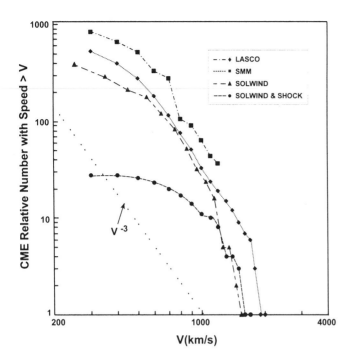

Figure 6. Integral log-log distribution of CME leading-edge speeds v from three different coronagraph data sets. The SMM set was restricted to the 331 Outer Loop measurements (*Hundhausen* 1994). The least-squares best fits for each set gave slopes < −3 (shown by the dotted line), corresponding to exponents β < −4. The + symbols are the Solwind CMEs associated with fast interplanetary shocks (*Cane et al.* 1987).

1200 km/s. This result compares well with the conclusions that SEP events will nearly always occur when v > 750 km/s (*Reames et al.* 1997) or v > 800 km/s (*Kahler et al.* 1987).

We have seen that the CME speed distribution for fast CMEs can be fitted approximately by $dN/dv = Bv^{-\beta}$ (Figure 6) and that the SEP intensity as a function of v is given by $dI/dv = Av^{\alpha}$ (Figure 3). We integrate the latter expression as $I = A(\alpha + 1)^{-1} v^{(\alpha + 1)}$. We can then calculate the distributions of the peak SEP intensities shown in Figure 5 to be given by

$$dN/dI = (dN/dv)(dv/dI) = Bv^{-\beta} \times A^{-1}v^{-\alpha}$$
$$= B A^{-1}v^{-(\alpha + \beta)} = DI^{[-(\alpha + \beta)/(\alpha + 1)]},$$

where D is a constant of the fit. For the approximate values of α = 4.5 and β = 4.2 derived from the plots of Figures 3 and 6, we get a power law fit of γ = 1.58 for the dN/dI versus I fit. This is steeper than the fit of Figure 5, but we have seen that the distribution of CME speeds which give rise to shocks is flatter than the distribution of all CME speeds (Figure 6), particularly in the range 400 km/s to 1200 km/s, so the appropriate value of β for the

above expression has been overestimated. A value of β = 1.71 would yield the result of Figure 5 that γ = 1.13. The important point here is that the distribution of SEP event intensities of Figure 5 can be understood as a convolution of the steep decline of the speed distribution of CMEs driving shocks (Figure 6) with the strong increase of SEP intensity with CME speed (Figure 3).

For space-weather purposes of determining total radiation exposures, the SEP event fluences F, defined as the intensities I integrated over time and solid angle, are more useful quantities. The determination of F is obviously more involved than that of I, which is measured at a single time. We generally find that higher intensity SEP events are also longer in duration, sometimes because multiple events close in time are treated as a single event. Therefore F should increase faster than linearly with I such that $dF/dI \sim I^{\delta}$, where $0 < \delta < 1$. This is the case with solar hard X-ray bursts, where $\delta \sim 0.2$ (*Crosby et al.* 1993). Then the differential fluence distribution

$$dN/dF = dN/dI \times dI/dF = AI^{-\gamma} \times I^{-\delta} \sim I^{-\varepsilon},$$

where $\varepsilon = \gamma + \delta$, which means that the slope of dN/dF versus F is steeper than that of dN/dI versus I.

Goswami et al. (1988) fitted E > 10 MeV SEP event fluences from cycles 19 to 21 to a differential power law distribution and claimed a fit of ε = 1.30 in the F range 10^7 to 10^{10} p/cm^2. That contradicted a value of ε = 1.15 from a similar earlier fit by Lingenfelter & Hudson (1980) that was flawed by the explicit assumption that the differential intensity and differential fluence slopes should be equal, contrary to our discussion above. However, using the data in the plot of Figure 4 of Goswami et al. (1988), I find that ε = 0.30, suggesting that they misstated their value, perhaps by confusing integral and differential distributions, for which the power law exponents differ by unity. Their differential F distribution for the period 1972-1987 also decreased in the range below 10^8 p/cm^2, leading them to do log-normal fits to their distributions. They may have been biased against small F events, since we expect the number of events to increase as F decreases. Feynman et al. (1990) noted this disagreement with Goswami et al. when they produced a differential fluence plot of events over 1964-1985 and found no decrease at small F events. Although Feynman et al. did not do a power-law fit to their data, I calculate a value of ε = 0.47 for their differential F distribution, similar to that of Goswami et al. (1988). The differential F distributions of both Goswami et al. (1988) and Feynman et al. (1990), with many events common to both, are flatter than we expect from the discussion above. An integral distribution of E > 10 MeV fluences over the range 10^7 to 10^{10} p/cm^2 from the period 1954-1991 published by Reedy (1996) is fitted by a power-law exponent of 0.4, im-

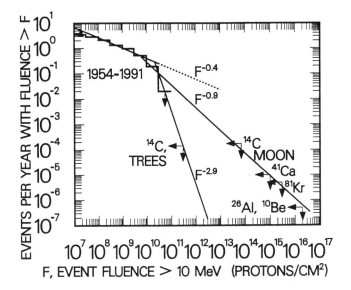

Figure 7. The integral distribution of SEP fluences from Reedy (1996). The distribution above 10^{10} p/cm^2 falls sharply and is limited by lunar radionuclide and terrestrial C^{14} measurements.

plying an F differential power-law exponent $\varepsilon = 1.4$ for their data. Gabriel and Feynman (1996) compiled a distribution of fluence events from 1965 to 1990 over the broader range 10^5 to 10^{10} p/cm^2, getting a range of $1.2 < \varepsilon < 1.4$, depending on the particle energy range. Their exponent fits are consistent with that of Reedy (1996), with $\varepsilon > \gamma$, as we expect.

3b. The Largest Events

An important goal for space weather is to determine the largest SEP events that we can expect. Lingenfelter & Hudson (1980) considered upper limits to fluctuations in the terrestrial ^{14}C activity to argue that they precluded the occurrence of any E > 10 MeV SEP event with F > 3 × 10^{11} p/cm^2 over the past 7000 years. They extended this limit to 10^7 years on the basis of lunar radionuclide measurements. In addition, they deduced that the differential power-law distribution of E > 10 MeV fluence events must have a sharp break or cutoff at about 10^{10} p/cm^2. Reedy (1996) came to the same result, based on more complete lunar radionuclide data (Figure 7).

More recently, Smart & Shea (1997) examined the event distribution of E > 10 MeV integral peak intensities in the range 10^1 to 10^4 p/cm^{2}s-sr. Up to $I = 10^3$ p/cm^{2}s-sr they found a differential power-law fit of $\gamma = 1.47$, somewhat steeper than those of van Hollebeke et al. (1975) and Cliver et al. (1991) discussed above, followed by a break in the distribution to a steeper slope above 10^3 p/cm^{2}s-sr. Their table lists only one event (August 4, 1972) exceeding

10^4 p/cm^{2}s-sr, and that event may have been the result of an unusual reflection from two shocks (*Pomerantz & Dugal* 1974). Assuming that peak value of 8.6 × 10^4 p/cm^{2}s-sr as a high intensity cutoff and an event duration of 5 hours around the time of the shock passage, this cutoff corresponds to F = 10^{10} p/cm^2, the value of Lingenfelter & Hudson (1980) and Reedy (1996). Shea et al. (1993) listed the largest E > 10 MeV fluence events over the period 1955 to 1992, with only four events exceeding a fluence of 10^{10} p/cm^2, the largest being 3 × 10^{10} p/cm^2 for 12-15 November 1960.

There are at least two additional reasons for accepting the E > 10 MeV fluence of 10^{10} p/cm^2 as a near upper limit for future SEP events. The first is that Ng & Reames (1994) modeled particle acceleration at an interplanetary shock and performed numerical calculations that followed the spatial and temporal evolution of the spectra of both Alfven waves and particles. Increases in streaming particle intensities produced wave growth that increased the particle scattering and produced streaming limits to the particle intensities as a function of distance from the shock. Subsequent comparison with observed GOES and Helios 1 and 2 SEP events (*Reames & Ng* 1998) showed good agreement with their calculations. At 1 AU the E ~ 10 MeV differential intensity is limited to ~ 10^2 p/cm^{2}s-sr-MeV. If we assume an average E^{-3} differential energy spectrum and an event lasting 3 days, we get an event fluence of ~ 10^9 p/cm^2. An additional factor of 10 can arise from harder spectra, longer event durations, and significant SEP intensity enhancements over the streaming limits expected at the shock passage itself, to bring the limit to 10^{10} p/cm^2.

The second, more speculative reason for the fluence upper limit is based on a comparison of the range of the solar Ca H and K line index with that of solar-type stars. Over cycle 22 the variation of the solar H and K index (0.17 to 0.20) (*Hoyt & Schatten* 1997), an indicator of magnetic activity, occurred in the top range of the observed variation (0.13 to 0.21) for a sample of 74 solar-type stars (*Balliunas & Jastrow* 1990). The implication is that the recent solar magnetic activity, and therefore the accompanying CME and SEP events, have been at the high end of the expected intensity range.

4. WHEN DO SEP EVENTS OCCUR?

The annual numbers of E > 10 MeV SEP events at the NOAA threshold of 10 p/cm^{2}s-sr have been compiled for solar cycles 19 through 22 by Shea & Smart (1999). Events from cycle 19 (1954-1964) were inferred from riometer observations (*Shea & Smart* 1990). Each separate SEP injection was counted as an individual event, even if the preceding background intensity exceeded the NOAA threshold of 10 p/cm^{2}s-sr, and intensity increases at shock

Table I. The E > 10 MeV Proton Solar Cycle Fluences

Cycle	Start Date	Max SS Nbr	Number of SEP Events	Total Fluence (p/cm^2)
19	May 1954	190	65	7.2×10^{10}
20	Nov. 1964	106	72	2.2×10^{10}
21	Jul. 1976	155	81	1.8×10^{10}
22	Oct. 1986	158	84	5.8×10^{10}

(adapted from Shea & Smart (1999))

passages were not counted as separate events (*Shea & Smart* 1992). When averaged over the four solar cycles for each year following the point of solar minimum, they found a close correlation between (a) the total 12-month sums of the number of NOAA SEP events and (b) the sunspot number per 12-month period. I calculate a correlation coefficient of r = 0.74 between their 12-month numbers of SEP events and their 12-month average sunspot numbers, a total of 43 data points, with a least-squares best fit of

$$N = 0.079(SS\#) + 1.02,$$

where N is the number of SEP events per 12-month period, and SS# is the matching 12-month average sunspot number. Using this equation for the first three 12-month periods of cycle 23 (October 1996 through September 1999) and rounding to the nearest integer, we expect 2, 5, and 8 SEP events. The observed numbers of SEP events were 0, 8, and 7, suggesting that SEP events are on course for the rise of cycle 23. Thus, on an annual basis the sunspot numbers give us a rough guide to the numbers of SEP events we can expect. Since SEP events result from fast CMEs, it is not surprising that the annual rate of CMEs, determined for the epoch 1975-1989, correlated very well (r = 0.94) with the annual sunspot numbers (*Webb & Howard* 1994).

The solar-cycle dependence of SEP fluences, i.e., the event-integrated intensities, may be of more interest than the numbers of SEP events. Feynman et al. (1990) compared annual and cycle-integrated fluences of E > 10 MeV and E > 30 MeV proton events with sunspot numbers for cycles 19 through 21. They found high fluence levels from 2 years before maximum to 4 years after maximum. No correlation between annual fluence and sunspot number was obvious during the 7 years of high fluences when the average annual sunspot number was greater than 50. A comprehensive proton fluence model, JPL 1991, published by Feynman et al. (1993), included cycle 22 and confirmed the bimodal result. Shea & Smart (1999) have updated their solar-cycle fluence studies through the end of cycle 22. Cycle 19 had the largest annual sunspot number and E > 10 MeV solar-cycle fluence, but otherwise there was no obvious correlation, as shown in Table I. Note that the cycle-integrated fluences varied by only a factor of 4.

There is a further temporal organization of SEP events within solar cycles. The discovery of a 154-day periodicity in solar hard X-ray events by Rieger et al. (1984) set off a widely successful effort to search for that periodicity in other solar observations (*Lean & Brueckner* 1989). Gabriel et al. (1990) found the 154-day periodicity in 200 E > 30 MeV proton events over solar cycles 19 to 21, although it was weak in cycle 20. Bai & Cliver (1990) examined 385 E > 10 MeV proton events over the same cycles with the same general conclusion. Their more detailed analysis allowed them to separate two epochs (1958 to 1971 and 1978 to 1983) during which the periodicity was operative. They also found a phase shift of 0.5 cycles between the two epochs. The extent to which the 154-day modulation is apparent is shown in Figure 8, which covers their second epoch of periodicity. The work of Gabriel et al. (1990) and Bai & Cliver (1990) predated proton events of the recent 22nd cycle. Although the cause of the 154-day periodicity remains a mystery, it would now be worthwhile to see whether the periodicity has continued through the recent 22nd cycle, and if so, the strength and phase of the periodicity. Recent observations of ~ 1 MeV proton intensities during 1998-99 on Ulysses suggested recent evidence of a 154-day recurrence pattern (*Dalla & Balogh* 2000).

We would like to extend our specification of the numbers and sizes of SEP events as far back in time as possible. SEPs interact with constituents of the terrestrial polar atmosphere, so it is possible in principle to find signatures of past large SEP events in various natural records. The nitrate ion NO_3^- is the terminal product of the dissociation

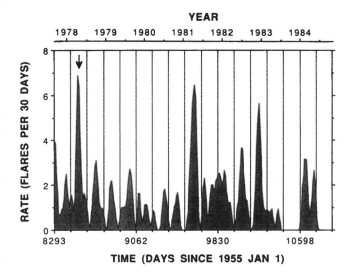

Figure 8. Smoothed occurrence rate of SEP events for the semi-periodic interval from February 1978 through August 1983. The vertical grid lines are positioned at the minimum phases of the 154-day periodicity. From Bai & Cliver (1990).

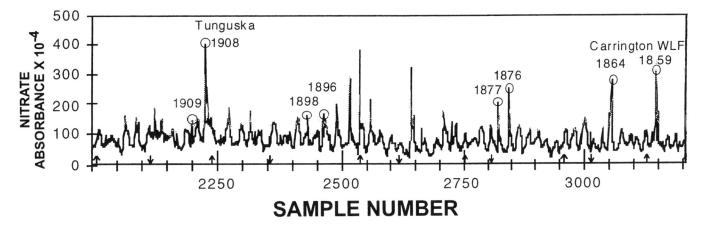

Figure 9. Part of the nitrate absorbance record from the central Greenland ice sheet. Some prominent dates are indicated, including the white light flare of 1859. From Zeller & Dreschhoff (1995).

and ionization of atmospheric oxygen and nitrogen by SEPs. These ions are precipitated out in the snow layers of polar ice sheets (*Zeller & Dreschhoff* 1995) which can be sampled and studied in sections as fine as one-month intervals. A 20th-century snow core sample from Antarctica yielded a number of nitrate peaks, with some of those exceeding 4 standard deviations corresponding to the major SEP events of August 1972, July 1959, July 1946, and July 1928 (*Dreschhoff & Zeller* 1990; *Shea et al.* 1993). An ice core sample from the Greenland ice sheet revealed the nitrate record back to the late 16th century (*Zeller & Dreschhoff* 1995). A number of nitrate peaks are seen in that record, including one in late 1859, possibly associated with the white-light flare observed by Lord Carrington, and other peaks during periods of major SEP events in 1989, 1990, and 1991 (*Dreschhoff et al.* 1997). Periods of low, but significant activity were seen during the Dalton (1792-1828) and Maunder (1645-1715) minima (*Dreschhoff & Zeller* 1998).

Unfortunately, there are sources of NO_3^- other than ionization by SEPs, and atmospheric transport of the NH_3^- is locally and seasonally dependent, with polar winter favored for nitrate precipitation by snow (*Dreschhoff & Zeller* 1990; *Zeller & Dreschhoff* 1995). In the Greenland sample the largest known wintertime GLE events of February 1956, November 1949, and February 1942 and the largest recent E > 10 MeV fluence event of November 1960 (*Shea et al.* 1993) appear as only minor peaks (*Zeller & Dreschhoff* 1995). The nitrate record may yet prove useful in the study of past SEP events, but the tantalizing correspondence between the nitrate peaks (Figure 9) and known SEP events has yet to be critically examined. A recent interpretation (*Kocharov et al.* 1999) of a 5.3-year periodicity in the Greenland ice cores as evidence that SEP events preferentially occur during periods of increasing or

decreasing activity should therefore be considered with caution.

SEPs can also produce ^{14}C in the terrestrial atmosphere through the $^{14}N(n, p)^{14}C$ reaction of secondary neutrons. However, the annual production of ^{14}C is only about 1 percent of the total atmospheric ^{14}C content (*Peristykh & Damon* 1999), and the residence time of ^{14}C in the atmosphere is 10 to 20 years, so that significant variations of ^{14}C in tree rings are measured only over the Schwabe 11-year cycle, precluding the detection of individual SEP events.

5. WHERE DO SEP EVENTS OCCUR?

We have seen that the sources of gradual SEP events are shocks driven through the corona and interplanetary space by CMEs. The relationship between the shock location and the SEP intensity profile was described by Cane et al. (1988) using a large sample of 235 SEP events observed at Earth over a 20-year period. A later study by Reames et al. (1996) used multispacecraft observations to study the SEP spatial distributions. Their schematic (Figure 10) shows that solar events from the western hemisphere have rapid rises to maxima followed by weak or decreasing intensities during shock passages. This is because the observer is initially connected to the nose of the shock near the Sun and then to a more easterly part of the shock flank where the shock is weaker. Near central meridian the observer is first connected to the western flank, then progressively toward the nose of the shock so that at low energies the SEP intensity peaks near shock passage. For events with solar origins at eastern longitudes, connection to the shock nose is achieved only after the shock passage.

This pattern is the result of three basic rules about SEP acceleration at shocks. The first is that the acceleration to high energies is most effective early when the shock is be-

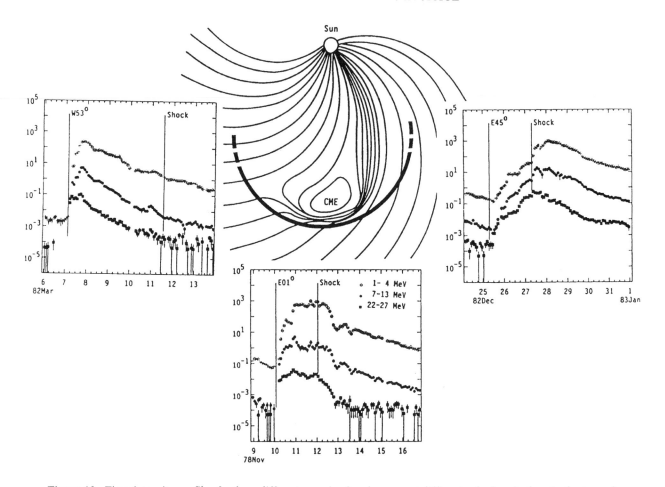

Figure 10. Time-intensity profiles for three different energies for observers at different solar longitudes. In the tens of MeV energy range, times of connection to the nose of the shock dictate the times of peak intensities. From Reames et al. (1996).

ing rapidly driven in the high magnetic fields and particle densities of the outer corona. The second is that the peak spatial intensity of SEPs lies in the Alfven wave fields (*Ng & Reames* 1994; *Reames et al.* 1996) at the shock where the SEPs are accelerated. The third is that the shock acceleration is most effective at the nose of the shock and least effective at the flanks. The first of these three rules is dominant for high energy (E > 100 MeV) SEPs and the second and third for low energy (E < 30 MeV) SEPs (*Cane et al* 1988; *Smart & Shea* 1995), as indicated in Figure 11.

We would now like to extend this basic scheme to predict SEP profiles at other regions of space which might be encountered by future space travelers. The high energy SEPs at a point a distance r AU from the Sun will be most intense when the source region is located at a west longitude of

$$\Phi \text{ (deg)} = \Omega \times r/v = 51.4 \times r$$

where the average solar wind speed v is assumed to be 450 km/s (*Richardson et al.* 1986), and the solar rotation rate Ω

is taken as 360/27 deg/day. The SEPs must travel a distance L along the spiral field line to reach the observer where

$$L = \int \sqrt{(1+r^2\Omega^2/v^2)}\,dr$$

For our assumed values of v and Ω, the distance L = 1.32 AU at 1 AU. Travelers venturing to other solar system regions would find the connection longitudes (measured west of central meridian) and path lengths indicated in Table II. We also show T(min), the minimum time required for a 100 MeV proton to reach that distance from the Sun, assuming a pitch angle cosine of unity.

As astronauts venture further out into the solar system from 1 AU, the magnetic connection longitude moves westward, but even for Mars-bound observers the connection region is still on the visible side of the west limb, allowing them to monitor solar activity that could result in a high-energy SEP event. However, for travelers beyond the asteroids the only signature of a solar eruptive event producing energetic particles would be a fast halo or large-

GLE

P >1000 MeV

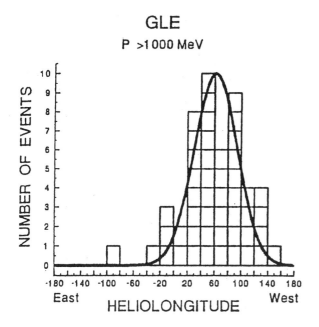

PROTONS >10 MeV

Flux >10 (Cm² S Ster)⁻¹

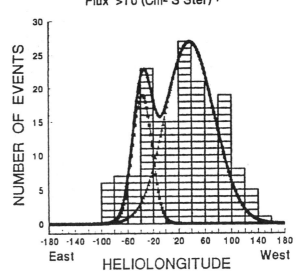

Figure 11. Associated flare longitudes for GLEs (left), reflecting the importance of an early connection to the nose of the shock. For large E > 10 MeV SEP events (right), in which connection to the shock nose occurs early for the western events and late for the eastern events, the time of connection is less important, and the longitude distribution is broader. Gaussian fits to profiles by the authors (*Smart & Shea* 1995).

angle west limb CME showing no associated front-side solar activity. Beyond Mars the path lengths and transit times of SEPs are considerably longer than at 1 AU due to the winding of the magnetic fields into a predominantly azimuthal direction. The scattering of SEPs by magnetic irregularities on their outbound trajectories results in significant decreases of peak SEP intensities with distance. An additional decrease of SEP intensity results from the diverging magnetic field. For a population of high energy SEPs confined to a magnetic flux tube of flux B × A, where B is the field intensity and A is the cross-sectional area, the decrease of the SEP intensity with distance should match the decrease of B. Using equation (2.6) of Burlaga (1995) for B, which varies roughly as r^{-2} near the Sun, but more nearly as r^{-1} beyond several AU, we show in the last column of Table II the decrease of B/B_0 with distance r. At the distance of Jupiter that decrease is less than a factor of 8 from the value at 1 AU.

Observational studies have determined how SEP intensities decrease with radial distance in the solar system. Hamilton et al. (1990) examined multiple spacecraft observations of five well connected 10-20 MeV SEP events and derived power-law decreases for peak intensities as $r^{-3.3\pm0.4}$ and for fluences as $r^{-2.1\pm0.3}$. Lario et al. (2000) recently compared SEP events at the WIND spacecraft with those at Ulysses during 1997-98 when Ulysses was near the ecliptic plane and at a distance of 5.2 to 5.4 AU. Figure 1 of their work shows a rough correspondence between the major E

~ 10 MeV SEP events at the two spacecraft, despite the fact that the connection longitudes of each spacecraft to the source shocks varied significantly throughout the study period. Comparing the fourth largest event at each spacecraft, we get a ratio of 2.7 for the ratio of the logs of the peak intensities, suggesting an $r^{-3.75}$ decrease of the peak intensity for those events. The event time scales at Ulysses clearly increase, however, so the decrease in the fluence will be less. These results appear consistent with the earlier work of Hamilton et al. (1990).

Unfortunately for space weather purposes, the 10-30 MeV SEP temporal profiles are often complicated by several factors. One is that the particle trapping region around the shock, which is not subject to the streaming limit, may or may not be the region with the highest SEP intensity, depending on whether shock acceleration continues or diminishes with distance (*Reames* 1999a). At distances beyond 1 AU there can also be interaction or merging of different transient shocks, and the corotating shocks begin to play a role, possibly by reaccelerating some of the SEPs from transient shocks (*Lario et al.* 2000).

6. SOME SEP EFFECTS ON SPACE EXPERIMENTS

The negative effects of SEPs on high-latitude communications, satellite operations, and men in space is well known. However, SEPs also present problems for space observations. Here I discuss two such cases.

6.1. The effects of SEPs on data compression

The use of CCDs in space imaging instruments has produced high data rates, which in turn have required the use of various data compression schemes to maximize the rate at which images are transmitted from space. Trapped energetic particles are a problem for CCD experiments on low-Earth orbiting spacecraft, such as the NASA TRACE explorer, because they distort the images and automatic exposure control systems. However, the particles also produce in the images tracks of high contrast and high spatial frequency which do not compress effectively (*Handy et al.* 1999). This is also the case when SEPs access CCD experiments outside the magnetosphere. A remarkable example is that of the effect of SEPs on the SOHO Lasco coronagraph images during the very fast west limb CME of November 6, 1997. A C3 coronagraph image of that CME with extensive SEP tracks was published by *Maia et al.* (1999). Table III shows the comparison of the compression ratios achieved before and during the SEP event, when a 20 to 30% decrease occurred in the compression ratio. The scale factor, which dictates the maximum allowable difference in counts between any pixel and the average of the transmitted subblock of the image containing that pixel, can be changed by ground command. When it was realized that the rate of image transmission had decreased, operators increased the scale factors, as shown in the last column of Table III, with the result that the compression ratios were then much higher, but only at the cost of a serious degradation of the image quality (*D. Wang and R. Howard*, private comm.)

6.2. The reflections of SEPS from grazing-incidence X-ray telescope mirrors

Grazing incidence optics is the technique used by the Chandra Observatory to image X-rays through small-angle reflections of the X-rays off surfaces of a set of nested mirrors. Soon after the launch of the Chandra Observatory in July 1999 a problem was found with the performance of the Advanced CCD Imaging Spectrometer (ACIS) which was traced to the effects of low-energy ($E \leq 1$ MeV) ions incident on the ACIS. The ions, in this case trapped magnetospheric particles which entered the telescope aperture, were reflected by the mirrors and focussed on the ACIS in

Table III. Effects of SEP Event on Coronagraph Images

Coronagraph	Pre-event	SEPs	New scale factor
C1 compression ratio	6.98	5.45	9.36
scale factor	32	32	192
C2 compression ratio	7.38	5.23	6.17
scale factor	64	64	192
C3 compression ratio	6.05	4.51	9.30
scale factor	32	32	192

the optical focal plane. The front illuminated CCDs were vulnerable to damage from the ions. The solution was to stow the ACIS out of the focal plane during passage through the regions of trapped particles.

Although the theory of scattering off surfaces by charged particles at grazing incidence angles had been published decades ago (*Firsov* 1967), the application to grazing incidence telescopes was only recently appreciated. There are two somewhat surprising features of this scattering. First, the reflection probability approaches unity with decreasing angle of incidence for all particle energies. Second, the reflected flux density is maximum when the angle of incidence equals the angle of reflection, i.e., when the reflection is specular. Both these factors act to focus the incident ions along with the X-rays in the focal plane. The interaction of the radiation environment with grazing incidence optics is also being modeled for the ESA XMM-Newton telescope, launched in December 1999 (*Nartallo et al.* 2000).

Acknowledgements. My thanks to D. Baker, E. Cliver, B. Dichter, J. Feynman, R. Howard, C. Lopate, D. Reames, M. Shea, C. St. Cyr, D. Smart, and D. Wang for their contributions.

REFERENCES

Bai, T., and E.W. Cliver, A 154 day periodicity in the occurrence rate of proton flares, *Ap. J.*, 363, 299, 1990.

Baliunas, S., and R. Jastrow, Evidence for long-term brightness changes of solar-type stars, *Nature,* 348, 520, 1990.

Brueckner, G.E., The behaviour of the outer solar corona (3 Ro to 10 Ro) during a large solar flare observed from OSO-7 in white light, in *Coronal Disturbances*, edited by G. Newkirk, Jr., IAU Symp. 57, 334, 1974.

Burlaga, L.F., *Interplanetary Magnetohydrodynamics*, Oxford University Press, New York, 1995.

Cane, H.V., D.V. Reames, and T.T. von Rosenvinge, The role of interplanetary shocks in the longitude distribution of solar energetic particles, *J. Geophys. Res.*, 93, 9555, 1988.

Cane, H.V., I.G. Richardson, and T.T. von Rosenvinge, Cosmic ray decreases: 1964-1994, *J. Geophys. Res.,* 101, 21561, 1996.

Cane, H.V., N.R. Sheeley, Jr., and R.A. Howard, Energetic interplanetary shocks, radio emission, and coronal mass ejections, *J. Geophys. Res.*, 92, 9869, 1987.

Cane, H.V., I.G. Richardson, and O.C. St. Cyr, The interplanetary events of January-May, 1997 as inferred from energetic parti-

Table II. Solar Connection Parameters in the Solar System

Distance (AU)		Φ (deg)	L (AU)	T(min)	B/B_0
Venus	(0.72 AU)	37	0.77	15	1.69
Earth	(1.0 AU)	51	1.32	26	1.00
Mars	(1.52 AU)	78	1.91	37	0.55
Asteroids	(2.77 AU)	142	4.63	90	0.264
Jupiter	(5.20 AU)	268	13.67	265	0.134

cle data, and their relationship with solar events, *Geophys. Res. Let.*, 25, 2517, 1998.

Cliver, E.W., Was the eclipse comet of 1893 a disconnected coronal mass ejection?, *Solar Phys.*, 122, 319, 1989.

Cliver, E., et al., Size distributions of solar energetic particle events, *22nd ICRC*, 3, 25, 1991.

Croom, D.L., Forecasting the intensity of solar proton events from the time characteristics of solar microwave bursts, *Solar Phys.*, 19, 171, 1971.

Crosby, N.B., M.J. Aschwanden, and B.R. Dennis, Frequency distributions and correlations of solar x-ray flare parameters, *Solar Phys.*, 143, 275, 1993.

Dalla, S., and A. Balogh, Recurrence in MeV proton fluxes and anisotropies at 5 AU from the Sun, *Geophys. Res. Let.*, 27, 153, 2000.

Dreschhoff, G.A.M., and E.J. Zeller, Evidence of individual solar proton events in antarctic snow, *Solar Phys.*, 127, 333, 1990.

Dreschhoff, G., and E.J. Zeller, Ultra-high resolution nitrate in polar ice as indicator of past solar activity, *Solar Phys.*, 177, 365, 1998.

Dreschhoff, G.A.M., et al., Evidence for historical solar proton events from no(x) precipitation in polar ice cores, *25th ICRC*, 1, 89, 1997.

Feynman, J., Solar proton events during solar cycles 19, 20, and 21, *Solar Phys.*, 126, 385, 1990.

Feynman, J., et al., Interplanetary proton fluence model: JPL 1991, *J. Geophys. Res.*, 98, 13281, 1993.

Firsov, O.B., Reflection of fast ions from a dense medium at glancing angles, *Soviet Physics – Doklady*, 11, 732, 1967.

Forbush, S.E., Three unusual cosmic-ray increases possibly due to charged particles from the sun, *Phys. Rev.*, 70, 771, 1946.

Gabriel, S.B., and J. Feynman, Power-law distribution for solar energetic proton events, *Solar Phys.*, 165, 337, 1996.

Gabriel, S., R. Evans, and J. Feynman, Periodicities in the occurrence rate of solar proton events, *Solar Phys.*, 128, 415, 1990.

Gosling, J.T., et al., Mass ejections from the sun: a view from Skylab, *J. Geophys. Res.*, 79, 4581, 1974.

Gosling, J.T., et al., The speeds of coronal mass ejection events, *Solar Phys.*, 48, 389, 1976.

Goswami, J.N., et al., Solar flare protons and alpha particles during the last three solar cycles, *J. Geophys. Res.*, 93, 7195, 1988.

Hamilton, D.C., G.M. Mason, and F.B. McDonald, The radial dependence of the peak flux and fluence in solar energetic particle events, *21st ICRC*, 5, 237, 1990.

Handy, B.N., et al., The transition region and coronal explorer, *Solar Phys.*, 187, 229, 1999.

Howard, R.A., et al., Coronal mass ejections: 1979-1981, *J. Geophys. Res.*, 90, 8173, 1985.

Hoyt, D.V., and K.H. Schatten, *The Role of the Sun in Climate Change*, Oxford University Press, New York, 1997.

Hudson, H.S., A purely coronal hard X-ray event, *Ap. J.*, 224, 235, 1978.

Hundhausen, A.J., J.T. Burkepile, and O.C. St. Cyr, Speeds of coronal mass ejections: SMM observations from 1980 and 1984-1989, *J. Geophys. Res.*, 99, 6543, 1994.

Kahler, S., Injection profiles of solar energetic particles as functions of coronal mass ejection heights, *Ap. J.*, 428, 837, 1994.

Kahler, S.W., Coronal mass ejections and solar energetic particle events, in *High Energy Solar Physics*, edited by R. Ramaty et al., AIP Conf. Proc. 374, 61, 1996.

Kahler, S.W., E. Hildner, and M.A.I. van Hollebeke, Prompt solar proton events and coronal mass ejections, *Solar Phys.*, 57, 429, 1978.

Kahler, S.W., et al., Solar filament eruptions and energetic particle events, *Ap. J.*, 302, 504, 1986.

Kahler, S.W., et al., Solar energetic proton events and coronal mass ejections near solar minimum, *20th ICRC*, 3, 121, 1987.

Kahler, S.W., et al., The solar energetic particle event of April 14, 1994, as a probe of shock formation and particle acceleration, *J. Geophys. Res.*, 103, 12069, 1998.

Kahler, S.W., J.T. Burkepile & D.V. Reames, Coronal/interplanetary factors contributing to the intensities of E > 20 MeV gradual solar energetic particle events, *26th ICRC*, 6, 248, 1999.

Kocharov, G.E., M.G. Ogurtsov, and G.A.M. Dreschhoff, On the quasi-five-year variation of nitrate abundance in polar ice and solar flare activity in the past, *Solar Phys.*, 188, 187, 1999.

Kosugi, T., Type II-IV radio bursts and compact and diffuse white-light clouds in the outer corona of December 14, 1971, *Solar Phys.*, 48, 339, 1976.

Lario, D., et al., Energetic proton observations at 1 and 5 AU. I: January-September 1997, *J. Geophys. Res.*, in press, 2000.

Lean, J.L., and G.E. Brueckner, Intermediate-term solar periodicities: 100-500 days, *Ap. J.*, 337, 568, 1989.

Lingenfelter, R.E., and H.S. Hudson, Solar particle fluxes and the ancient sun, in *Proc. Conf. Ancient Sun*, edited by R.O. Pepin et al., 69, 1980.

Lyons, R.D., Navy releases first pictures of a solar flare from sun's far side, *N.Y. Times*, 11 Jan., p.26, 1972.

Maia, D., et al., Radio signatures of a fast coronal mass ejection development on November 6, 1997, *J. Geophys. Res.*, 104, 12507, 1999.

Meyer, P., E.N. Parker, and J.A. Simpson, Solar cosmic rays of February, 1956 and their propagation through interplanetary space, *Phys. Rev.*, 104, 768, 1956.

Munro, R.H., et al., The association of coronal mass ejection transients with other forms of solar activity, *Solar Phys.*, 61, 201, 1979.

Nartallo, R., et al., Modelling the interaction of the radiation environment with the XMM-Newton and Chandra x-ray telescopes and its effects on the on-board CCD detectors, *Proc. Space Rad. Envir. Wkshp*, in press, Defence Evaluation and Research Agency, 2000.

Ng, C.K., and D.V. Reames, Focused interplanetary transport of ~ 1 MeV solar energetic protons through self-generated waves, *Ap. J.*, 424, 1032, 1994.

Peristykh, A.N., and P.E. Damon, Multiple evidence of intense solar proton events during solar cycle 13, *26th ICRC*, 6, 264, 1999.

Pomerantz, M.A., and S.P. Dugal, Interplanetary acceleration of solar cosmic rays to relativistic energy, *J. Geophys Res.*, 79, 913, 1974.

Reames, D.V., Solar energetic particles: is there time to hide?, *Rad. Measurements*, 30, 297, 1999a.

Reames, D.V., Particle acceleration at the sun and in the heliosphere, *Space Sci. Rev.*, 90, 413, 1999b.

Reames, D.V., L.M. Barbier, and C.K. Ng, The spatial distribution of particles accelerated by coronal mass ejection-driven shocks, *Ap. J.*, 466, 473, 1996.

Reames, D.V., S.W. Kahler, and C.K. Ng, Spatial and temporal invariance in the spectra of energetic particles in gradual events, *Ap. J.*, 491, 414, 1997.

Reames, D.V., and C.K. Ng, Streaming-limited intensities of solar energetic particles, *Ap. J.*, 504, 1002, 1998.

Reedy, R.C., Constraints on solar particle events from comparisons of recent events and million-year averages, in *Solar Drivers of Interplanetary Disturbances*, edited by K.S. Balasubramaniam, et al., ASP Conf. Series, 95, 429, 1996.

Reid, G.C., and H. Leinbach, Low-energy cosmic-ray events associated with solar flares, *J. Geophys. Res.*, 64, 1801, 1959.

Reinhard, R., and G. Wibberenz, Propagation of flare protons in the solar atmosphere, *Solar Phys.*, 36, 473, 1974.

Richardson, J.D., K.I. Paularena, A.J. Lazarus and J.W. Belcher, Radial evolution of the solar wind from IMP 8 to Voyager 2, *Geophys. Res. Let.*, 22, 325, 1995.

Rieger, E., et al., A 154-day periodicity in the occurrence of hard solar flares? *Nature*, 312, 623, 1984.

Roelof, E.C., and S.M. Krimigis, Analysis and synthesis of coronal and interplanetary energetic particle, plasma, and magnetic field observations over three solar rotations, *J. Geophys. Res.*, 78, 5375, 1973.

Sanahuja, B., et al., Three solar filament disappearances associated with interplanetary low-energy particle events, *Solar Phys.*, 134, 379, 1991.

Shea, M.A., and D.F. Smart, A summary of major solar proton events, *Solar Phys.*, 127, 297, 1990.

Shea, M.A., D.F. Smart, Solar proton events: history, statistics and predictions, *STP Workshop IV*, 48, 1992.

Shea, M.A., and D.F. Smart, Patterns of solar proton events over four solar cycles, *26th ICRC*, 6, 374, 1999.

Shea, M.A., et al., The flux and fluence of major solar proton events and their record in Antartic snow, *23rd ICRC*, 3, 846, 1993.

Sky and Telescope, Eruption in solar corona observed by OSO 7, 43, 158, 1972.

Smart, D.F., and M.A. Shea, The heliolongitude distribution of solar flares associated with solar proton events, *24th ICRC*, 4, 313, 1995.

Smart, D.F., and M.A. Shea, The > 10 MeV solar proton event peak flux distribution, in *ST Predictions* V, Hitachi, 449, 1997.

Solar-Geophysical Data, Nat. Oceanic and Atmos. Admin., No. 329, p.23, 1972.

Svestka, Z., and L. Fritzova-Svestkova, Type II radio bursts and particle acceleration, *Solar Phys.*, 36, 417, 1974.

Van Hollebeke, M.A.I., L.S. Ma Sung, and F.B. McDonald, The variation of solar proton energy spectra and size distribution with heliolongitude, *Solar Phys.*, 41, 189, 1975.

Warwick, C.S., and M.W. Haurwitz, A study of solar activity associated with polar-cap absorption, *J. Geophys. Res.*, 67, 1317, 1962.

Webb, D.F., and R.A. Howard, The solar cycle variation of coronal mass ejections and the solar wind mass flux, *J. Geophys. Res.*, 99, 4201, 1994.

Wild, J.P., S.F. Smerd, and A.A. Weiss, Solar Bursts, *Ann. Rev. Astron. Astrophys.*, 1, 291, 1963.

Wright, C.S., Catalogue of solar filament disappearances 1964-1980, *Report UAG-100*, Nat. Geophys. Dat. Ctr, 1991.

Zeller, E.J., and G.A.M. Dreschhoff, Anomalous nitrate concentrations in polar ice cores – do they result from solar particle injections into the polar atmosphere? *Geophys. Res. Let.*, 22, 2521, 1995.

Stephen Kahler, AFRL/VSBS, 29 Randolph Rd., Hanscom AFB, MA 01731-3010.

The Solar Sources of Geoeffective Structures

D. F. Webb,[1,2] N. U. Crooker,[3] S. P. Plunkett,[4] and O. C. St. Cyr[5]

We review our current understanding of the solar sources of interplanetary structures that result in geomagnetic disturbances, especially storms. These sources have been broadly classified as: 1) transient or sporadic, or 2) recurrent, although we now know that these categories are not exclusive. Transients usually refer to coronal mass ejections (CMEs). Recurrence refers to disturbances that repeat with the 27-day synodic rotation period of the Sun. CMEs are expulsions of large quantities of plasma and magnetic field from the Sun's corona. The occurrence rate of CMEs approximately follows that of the solar (sunspot) activity cycle. Recurrent sources are usually attributed to high-speed solar wind streams emanating from coronal holes. However, compression at the leading edge of the stream as it runs into the higher-density slow flow surrounding the heliospheric current sheet leads to a corotating interaction region (CIR), and it is passage of the CIR, not the high speed flow, that results in enhanced recurrent geomagnetic disturbances. Moreover, the largest recurrent geomagnetic storms are actually caused by CMEs caught up in the CIRs.

1. OVERVIEW OF SOLAR AND GEOMAGNETIC ACTIVITY OVER THE SOLAR CYCLE

In this paper we review our present understanding of the solar sources of those interplanetary structures that create geomagnetic (GM) disturbances, especially storms. Since coronal mass ejections (CMEs) are an im-

[1]Institute for Scientific Research, Boston College, Chestnut Hill, Massachusetts.

[2]Also at Air Force Research Laboratory, Space Vehicles Directorate, Hanscom Air Force Base, Massachusetts.

[3]Center for Space Physics, Boston University, Boston, Massachusetts.

[4]USRA, NASA Goddard Space Flight Center, Greenbelt, Maryland.

[5]CPI/Naval Research Laboratory, NASA Goddard Space Flight Center, Greenbelt, Maryland.

Space Weather
Geophysical Monograph 125
Copyright 2001 by the American Geophysical Union

portant source of major transient activity both at the Sun and in the heliosphere at all phases of the solar activity cycle, our review emphasizes what we know about CMEs and their influence on space weather.

It has long been known that the level of geomagnetic activity at Earth tends to follow the sunspot cycle. Sunspot counts are considered a useful measure of the general level of magnetic activity emerging through the surface, or photosphere of the Sun. The smoothed sunspot number rises and falls relatively uniformly over the cycle, as shown in Figure 1a. Other major classes of solar activity also tend to track the sunspot number during the cycle. These include active regions, flares, filaments and their eruptions, and CMEs [e.g., *Webb and Howard*, 1994]. This activity is transmitted to Earth through the solar corona and its expansion into the heliosphere called the solar wind.

Figure 1, from *Luhmann et al.* [1998], illustrates three fundamental aspects of the solar-cycle variation of the global solar magnetic field. The plots include data from 1977–1991 covering all of cycle 21 and the first half of cycle 22. The top plot of monthly sunspot number shows that the maxima of the two cycles were in 1980 and 1989-90 with a broad minimum in 1986.

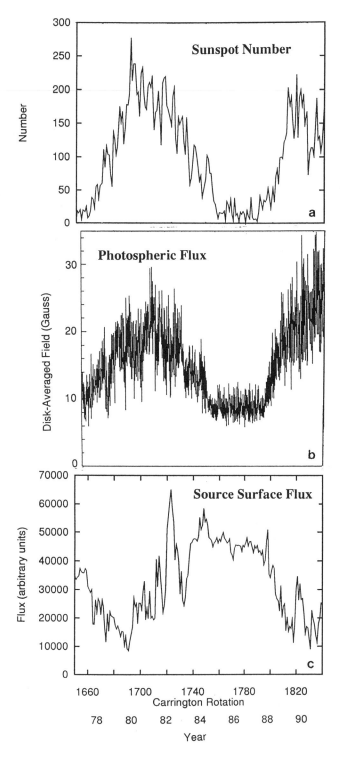

Figure 1. Signatures of solar global magnetic flux over about $1\frac{1}{2}$ solar cycles: (a) Monthly sunspot number; (b) Disk-integrated daily photospheric flux with rms deviations; (c) Carrington rotation averaged values of the total flux through the source surface at $2.5R_S$ computed from a Wilcox Solar Observatory potential field model. From *Luhmann et al.* [1998].

The middle plot shows the average daily photospheric magnetic flux, and the bottom a potential field model of the surface field extrapolated to the source surface at $2.5\ R_S$.

In essence, these plots address the following characteristics of the solar magnetic field: (1) The sunspot number indicates the time scale of the emergence through the solar surface of the areas of strongest flux. (2) The photospheric flux is indicative of the strength of the toroidal flux which, assuming a Babcock-type dynamo model of solar activity, is built up from the basic solar dipole field over a solar cycle. The total photospheric flux varies by about a factor of three over the cycle. (3) The source surface flux is constructed from synoptic, 27-day maps of the observed photospheric flux as the inner boundary of the calculation and extrapolated to an outer boundary, the source surface, above which the field is assumed to be radial [*Luhmann et al.*, 1998 and *Hoeksema and Scherrer*, 1986]. The source surface flux tracks the lower order, dipole contribution of the solar field over the cycle and is indicative of the variation of the total open flux from the Sun, which emanates mainly from coronal holes. Although the total interplanetary magnetic flux (IMF) flux at 1 AU is stronger at maximum and weaker near minimum than the total open flux [*Slavin et al.*, 1986; *Wang et al.*, 2000], the interplanetary flux remains strong during the declining phase when it is dominated by the open flux from coronal holes.

Since solar activity is transmitted to Earth via the solar wind, we can ask how closely the cycle of GM activity follows that of solar activity. The overall correspondence between the sunspot and GM activity cycles can be seen in long-term plots comparing various indices of the two kinds of activity [Figure 2]. For example, the so-called centennial series of the *aa* GM index has been used to show its long-term trend compared with sunspot numbers. The *aa* index measures GM activity in 3-hour intervals at two antipodal stations and was compared with sunspot number for the years 1868–1967 by *Mayaud* [1975], and *Legrand and Simon* [1993] updated these data through cycle 21.

On annual time scales the GM activity cycle clearly has more structure than the solar cycle. This is illustrated in Figure 2, which shows the annual number of disturbed days with $Ap > 50$ and sunspot number over the last 6 solar cycles, 17–22. The Ap index is a linearized version of the quasi-logarithmic, 3-hour Kp index, which averages the range of the magnetic H and D values measured at a number of worldwide observatories. The Ap activity shown in the figure is more variable than sunspot number, but tends to track the sunspot cycles in amplitude. For example, the least "active" solar cycles, 17 and 20, also had the lowest levels of GM storms. The plot extends through the end of 1999. GM activity was very low in 1999 which may presage another low activity cycle (23).

Figure 2 also demonstrates the double-peaked nature of the GM activity cycle. On average, GM activity exhibits a peak near sunspot maximum and another during the declining phase of the cycle. These peaks can vary in amplitude and timing, and the peak around maximum may itself at times consist of two peaks [*Richardson et al.*, 2000]. It is commonly thought that the two main peaks represent the maximum phases of two components of GM activity that have very different solar and heliospheric sources. The first (or double) peak is considered associated with transient solar activity that tracks the solar cycle in amplitude and phase. The later peak is usually attributed to recurrent high speed streams from coronal holes. The late peak is often higher than the early peak. *Legrand and Simon* [1993] studied the distribution of "severe" storms ($aa \geq$ 100 nT) from 1868–1980 and concluded that there are, on average, two intervals of activity during the sunspot cycle: a 6–year interval of severe storms peaking one year before sunspot maximum, and a 3–4 year period of storms with a peak ~3 years after maximum. *Gonzales et al.* [1990] performed a similar study of the distribution of aa and *Dst* storms over cycles 13–21, finding that both intense and moderate storms had the first peak occurring 8 months before sunspot maximum and the later peak 2 years after maximum.

However, *Richardson et al.* [2000] recently studied the relative contributions of different types of solar wind structures to the aa index over solar-cycle time scales. They identified CME-related flows, corotating high-speed streams, and slow flows near the Earth from 1972–1986, and found that each of these types contributed significantly to aa at all phases of the cycle. For example, CMEs contribute ~50% of aa at solar maximum and ~10% outside of maximum, and high speed streams contribute ~70% of aa outside of maximum and ~30% at maximum. Slow solar wind contributes ~20% throughout the cycle. Thus, although it is convenient and approximately correct to ascribe the two average peaks in each GM cycle to two different solar sources, both types of sources, CMEs and coronal holes/high speed streams, contribute to all phases of the cycle.

Using ISEE-3 counterstreaming electron events as proxies for CMEs, *Gosling et al.* [1991] found that nearly all of the largest storms ($Kp > 5$) from 1978–1982 were associated with the passage of CME material in the interplanetary medium, here called ICMEs, and/or shocks. These events were most geoeffective when both a CME and its shock passed over Earth; in this case the disturbance is encountered more head-on where the flow parameters are maximized. However, the association was much reduced for the more frequent

smaller storms. Slower CMEs not driving shocks are probably associated with many smaller storms, but this has yet to be demonstrated. Using ISEE data from this same period, *Zhao et al.* [1993] found that 78% of all periods with southward IMF $\leq$ –10 nT for durations $\geq$3 hr were associated with one or more CME signatures. *Gosling and McComas* [1987] argued that magnetic field compression in the leading edge of a CME and compression and draping in the preceding ambient solar wind are prime causes of strong out-of-the-ecliptic fields, especially when the CME speed is higher than the ambient wind speed. Field line draping alone can enhance southward IMF but if, in addition, there are strong southward fields in the leading edge of the CME, a long period of strong southward field may ensue causing a severe storm at Earth. Thus, CMEs are responsible for the most geoeffective solar wind disturbances, particularly the largest storms, so the ultimate goal of solar-terrestrial physics should be to predict when a CME will arrive at Earth and how geoeffective it will be.

In this review we discuss the solar activity and interplanetary signatures associated with geoeffective CMEs (next section), and the sources and characteristics of geoeffective recurrent or periodic disturbances (section 3). Recent advances in the remote sensing of CMEs by spaceborne coronagraphs, especially observations of "halo" CMEs, combined with near-surface observations of CMEs and coronal holes at EUV and soft X-ray wavelengths, have resulted in significant improvements in forecasting these structures for space weather purposes.

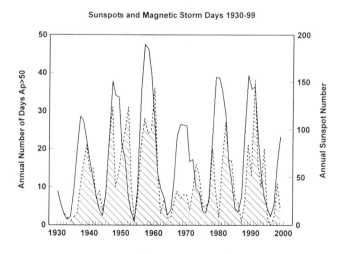

Figure 2. Annual number of geomagnetic disturbed days with Ap index > 50, (dashed line and hatched area) vs. annual sunspot number (solid line) for solar cycles 17–23. Courtesy NOAA National Geophysical Data Center, Boulder, CO.

2. CMEs: SOURCES OF TRANSIENT GEOMAGNETIC STORMS

In this section we discuss the transient solar and interplanetary activity related to the first peak of GM activity and the newly-realized importance of CMEs and interplanetary shocks to GM storms. CMEs, not solar flares, are now considered the key causal link between solar activity and major transient interplanetary and geomagnetic disturbances [*Kahler*, 1992; *Gosling*, 1993]. Coronagraphs image CMEs as bright features moving outward from the Sun with speeds ranging from tens to about 2000 km s^{-1}. The associated material can subsequently be detected by spacecraft in the solar wind by its anomalous plasma and magnetic field characteristics. Fast CMEs propel shock waves ahead of them which trigger sudden storm commencements when they reach the earth. These transient shocks can also accelerate energetic particles which can penetrate the magnetosphere and occasionally create hazardous conditions to spacecraft and astronauts in Earth orbit. The CME itself constitutes the "ejecta" following the shock and can contain high speed plasma and strong southward magnetic fields which, when coupled to the magnetosphere, lead to intensification of the ring current, effectively causing the main phase of a storm.

2.1. The Source Regions of CMEs

CMEs are large-scale transient events in which plasma and magnetic field are expelled from the Sun into interplanetary space, on timescales of minutes to hours. For a review of the general properties of CMEs and a summary of CME theories, see the papers by *Burkepile et al.* [2000] and *Klimchuk* [2000], respectively.

Many CMEs observed by the Solar Maximum Mission (SMM) coronagraph appeared to arise from large-scale, closed structures, most of which were preexisting coronal streamers [e.g., *Hundhausen*, 1993]. Although the solar origins of CMEs remain obscure, they are often associated with the disruption of large-scale structures such as prominences and loop arcades. Many energetic CMEs actually involve the disruption ("blowout") of a preexisting streamer, which can increase in brightness and size for days before erupting as a CME [*Howard et al.*, 1985; *Hundhausen*, 1993; *Subramanian et al.*, 1999].

Statistical association studies indicate that erupting filaments and X-ray events, especially of long duration, are the most common near-surface activity associated with CMEs [e.g., *Webb*, 1992]. Most optical flares occur independently of CMEs and even those accompanying CMEs may be a consequence rather than a cause of CMEs [*Gosling*, 1993; but see *Hudson et al.*, 1995]. The fastest, most energetic CMEs, however, are usually also associated with surface flares.

Comparisons of soft X-ray data with the white light observations have provided many insights into the source regions of CMEs. *Sheeley et al.* [1983] showed that the probability of associating a CME with a soft X-ray flare increased linearly with the flare duration, reaching 100% for flare events of duration ≥6 hours. The SMM CME observations indicated that the estimated departure time of flare-associated CMEs typically preceded the flare onsets. *Harrison* [1986] found that such CMEs were initiated during weaker soft X-ray bursts that preceded any subsequent main flare by tens of minutes, and that the main flares were often offset to one side of the CME [c.f., *Webb*, 1992]. This offset is also shown by *Plunkett et al.* [2000a] for CMEs observed by the Solar and Heliospheric Observatory (SOHO) Large Angle Spectrometric Coronagraph (LASCO). The latitude distribution of the CMEs peaks at the equator, but the distribution of EUV activity observed by the SOHO Extreme ultraviolet Imaging Telescope (EIT) telescope associated with these CMEs is bimodal with peaks about 30° north and south of the equator.

This pattern supports the view that many CMEs can involve more complex, multiple polarity systems [*Webb et al.*, 1997]) as recently modeled by *Antiochos et al.* [1999]. This multipolar pattern was evident in the LASCO observations during the recent solar minimum in 1996 to early 1997, in which the simple, bipolar streamer belt observed in the outer coronagraphs overlay twin arcades flanking the equator as observed by the inner coronagraph [*Schwenn et al.*, 1997]. The eruption of such a streamer as a CME would imply that it had a quadrupolar magnetic source region. In general, CMEs seem to involve the eruption of large-scale coronal structures which are associated at the surface with the stronger but small-scale magnetic fields typically found at low heliolatitudes, and with the weaker, large-scale fields at higher heliolatitudes.

The most obvious X-ray signatures of CMEs in the low corona are the arcades of bright loops which develop after the CME material has apparently left the surface [*Kahler*, 1977; *Hudson and Webb*, 1997]. Prior to the eruption, an S-shaped structure, called a sigmoid by *Rust and Kumar* [1996], often develops in association with a filament activation. The sigmoid structure, one of the most important discoveries in the Yohkoh spacecraft observations, denotes the dominant helicity in a given hemisphere, being S-shaped or right-handed in the south and reverse-S or left-handed in the north. This type of structure is now considered to be an important precursor of a CME [*Canfield et al.* 1999], and is indicative of a highly sheared, non-potential coronal magnetic field. Eventually an eruptive (rather than a confined) flare can occur, resulting in the bright, long-duration arcade of loops. *Sterling et al.* [2000] call this

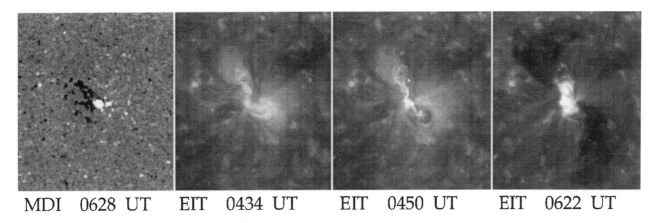

MDI 0628 UT   EIT 0434 UT   EIT 0450 UT   EIT 0622 UT

Figure 3. Example of surface activity associated with the May 12, 1997, halo CME. The first image is a SOHO/Michelson-Doppler Imager (MDI) photospheric magnetogram. The other three images are differences between consecutive pairs of full-disk EIT 195Å images; the later image time is listed. The EIT images show at 0434 UT the preexisting sigmoidal active region, at 0450 UT the event onset, and at 0622 UT the fully developed arcade and dual dimmings. During the onset a filament was beginning to erupt from the southern part of the region. Sigmoid structure is often better delineated in soft X-rays rather than EUV images [e.g., *Sterling et al.,* 2000]. Adapted from *Thompson et al.* [1998].

process "sigmoid-to-arcade" evolution. Figure 3 shows this development for the source regions of the May 12, 1997 halo CME. These arcades suggest the eruption and subsequent reconnection of the strongest magnetic field lines associated with the CME. In models describing this process, field lines stretched open during the eruption reconnect near the surface to form a magnetic loop system [c.f., *Svestka and Cliver,* 1992]. Such arcades appear to involve a simplification of the pre-event magnetic structure.

Clearly visible in Figure 3 is the formation of two regions of depleted emission measure on either side of the bright arcade. Such "dimming regions", best observed in X-ray and EUV images, imply that material is evacuated from the low corona. Transient dimmings in soft X-rays observed during Skylab were called "transient coronal holes" by *Rust* [1983]. The images obtained by the Yohkoh Soft X-ray Telescope (SXT) now reveal areas of subtle dimming, or depletion above or near brightening X-ray arcades. One interpretation of the dimming signature is that the initially closed field lines are opening during the early phase of a CME, in analogy to transient coronal holes observed against the disk. The dimming events appear to be one of the earliest and best-defined signatures in soft X-ray and EUV emission of the mass ejection in the low corona.

2.2. Halo CMEs

The classic form of a CME is a bright, loop-like leading edge, followed by a less dense cavity and a bright, dense core. This type of CME typically lies close to the limb of the Sun, and, thus, moves outward at a large angle to the Sun-Earth line. Since CMEs can be approximated as spherically symmetric structures, CMEs directed toward or away from the earth should appear as expanding halo-like brightenings surrounding the occulting disk in a coronagraph [*Howard et al.* 1982]. Since halo CMEs are visible only when far from the Sun, they are expected to be much fainter than CMEs observed near the limb. For this reason, the identification of halo CMEs with older coronagraphs had been difficult and only infrequently reported before the SOHO mission. Since the SOHO LASCO coronagraphs have improved sensitivity and dynamic range, and a larger field of view, halo events are now routinely observed by LASCO. Coronagraph observations alone cannot distinguish whether a structure is moving toward or away from Earth. Complementary observations of possibly associated activity on the solar disk are necessary to distinguish whether a halo CME was launched from the frontside or backside of the Sun.

Halo CMEs are important to study for three reasons. First, they are known to be the key link between solar eruptions and many space weather phenomena such as major GM storms and solar energetic particle events. Thus, there is a practical impetus to understand and predict the occurrence and evolution of halo CMEs in order to improve forecasts of space weather events. Second, the source regions of frontside halo CMEs are usually located within a few tens of degrees of Sun center, as viewed from Earth [*Webb et al.,* 2000a]. Thus, the source regions of halo events can be studied in greater detail than for most CMEs which are observed near the limb. The knowledge gained by studying the sources of

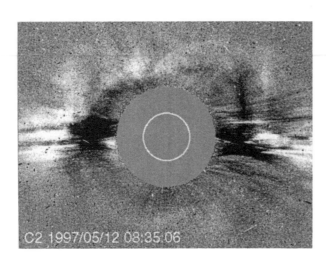

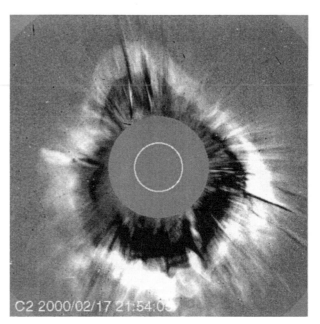

Figure 4. Examples of halo CMEs observed by LASCO. These are differences between direct images, to enhance the visibility of the CME features. Solar north is up and east is to the left. (Left) The event of May 12, 1997 and (right) the event of February 17, 2000.

halo events can be applied to understanding the physical processes involved in CMEs in general. Third, frontside halo CMEs must travel approximately along the Sun-Earth line, so their internal material can be sampled in situ by spacecraft near the Earth. Three spacecraft, SOHO, Wind and ACE, now provide solar wind measurements upstream of Earth and, since early 1996, have obtained measurements on the internal structure of 50–100 halo CMEs.

Many halo CMEs appear as expanding circular fronts that completely surround the occulter. Other CMEs clearly have a larger apparent angular size than typical limb CMEs, but do not appear as complete halos. Some of these unusually large events are called 'partial halo' CMEs, with a significant component of motion along the Sun-Earth line. The definition of a partial halo CME is somewhat subjective, but values of angular spans between 100° and 140° have been used as a lower limit.

An example of a complete halo CME on May 12, 1997 [*Plunkett et al.*, 1998] is shown in Figure 4a. In this case, the CME front is faint, and has a diffuse, featureless appearance. Figure 4b shows a halo on February 17, 2000 that has a markedly different appearance; it is much brighter and has a more ragged, sharply defined, front. *Sheeley et al.* [1999] concluded that events with a smooth, featureless appearance usually had fairly low speeds and often accelerated as they moved outwards. The bright, structured events were usually faster, with

apparent speeds sometimes in excess of 1,000 km s^{-1}, and often decelerated at large distances from the Sun.

Solar activity associated with frontside halo CMEs can be identified in extreme ultraviolet (EUV) and X-ray images of the low corona, especially the SOHO EIT and the Yohkoh SXT. *Hudson et al.* [1998] and *Webb et al.* [2000a] examined halo CMEs observed during the first half of 1997, finding that about half of the halo events were associated with frontside activity as expected. Studies of CME source regions using EIT and LASCO data have been reported by *Thompson et al.* [1999] and *Plunkett et al.* [2000b]. The coronal activity associated with CMEs includes the formation of dimming regions, formation of long-enduring (≥30 min.) loop arcades, associated with flaring active regions and sometimes extending over a substantial fraction of the visible solar disk, large-scale coronal waves that propagate outward from the CME source region, and filament eruptions. Optical flares and filament eruptions can also be identified in Hα images of the chromosphere. Metric radio bursts are also often associated with the CMEs. These association results are typical of those in previous studies comparing X-ray and white light data on CMEs which were usually only observed near the limb (see Section 2.1).

Figures 3 and 5 show activity in EIT associated with the May 12, 1997 halo shown in Figure 4a [*Thompson et al.*, 1998; *Webb et al.*, 2000b]. This event occurred in an active region close to the central merid-

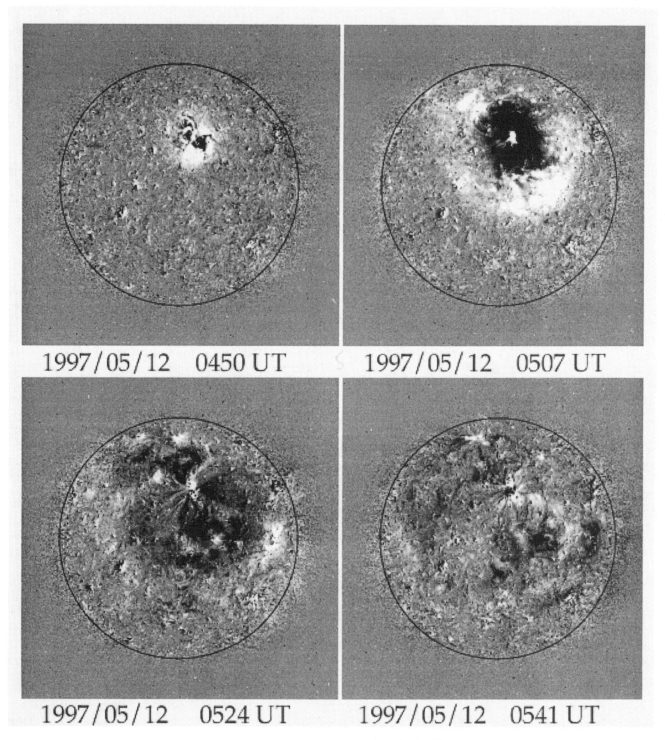

1997/05/12 0450 UT 1997/05/12 0507 UT

1997/05/12 0524 UT 1997/05/12 0541 UT

Figure 5. Example of a large-scale coronal wave associated with the May 12, 1997, halo CME. These are differences between consecutive pairs of full-disk 195Å images; the later image time is listed.

ian at N23°W07°. The coronal structure at this time near solar minimum was particularly simple, and this was the only active region on the visible disk. Prior to the eruption, the active region had the appearance of a sigmoid-shaped arcade of loops in the 195Å line. A long

duration X-ray flare began at 04:42 UT and was associated with an Hα filament eruption. The active region structure evolved from its initial sigmoid shape into a much less sheared arcade orthogonal to the underlying photospheric magnetic neutral line. The evolution of

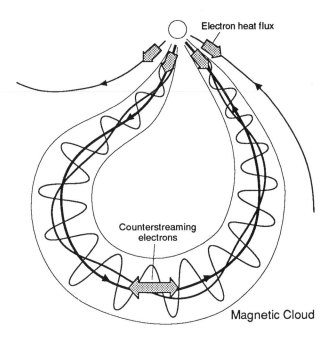

Figure 6. Schematic diagram of the projection of a magnetic cloud onto the ecliptic plane. The cloud has the form of a flux rope attached to the Sun at both ends. Heat flux electrons streaming away from the Sun along the magnetic field lines travel in both directions (counterstream) on the closed field lines of the flux rope. Reprinted from JASTP, Vol. 62, Crooker, Solar and heliospheric geoeffective disturbances, Pp. 1071–1085, Copyright (2000), with permission from Elsevier Science.

the active region structure in EUV is shown in Figure 3. The dimming regions noted above in Figure 3 are almost symmetric about the magnetic neutral line, and are located near the footpoints of the pre-eruption sigmoid structure. *Webb et al.* [2000b] interpreted these dimming regions as the footpoints of a magnetic flux rope that expanded outward from the Sun as part of the halo CME. Most dimmings are not as symmetrical as those shown here, tend to have a ragged appearance, and often involve only one dimming region.

The images in Figure 5 are differences of consecutive pairs of full-disk EIT Fe XII 195Å images. These 'running differences' clearly show an expanding wave front that moves outward from the active region. The wave consists of a rim of enhanced coronal emission travelling quasi-radially across the disk from the flare site almost symmetrically with a speed of ~250 km s⁻¹, covering almost the entire visible disk in about an hour. The frequent detection of these wave events is probably the most exciting new aspect of the EIT observations. EIT has now observed many of these waves in association with CMEs. *Thompson et al.* [1999] interpreted them as fast mode MHD waves, similar to the Moreton waves observed in chromospheric lines. Again, most waves are

not as symmetrical as the May 1997 wave, which was almost circularly symmetric because of the simple nature of the global magnetic field at this time.

2.3. Interplanetary Signatures of CMEs and Geomagnetic Activity

CMEs carry into the heliosphere increased amounts of coronal magnetic fields and plasma, which can be detected by remote sensing and in-situ spacecraft observations. The passage of this material past a single spacecraft is usually marked by certain distinctive signatures, but the degree of variation makes identification by a single parameter less reliable than identification by a group of parameters [e.g., *Gosling,* 1993; *Richardson and Cane,* 1993; *Neugebauer and Goldstein,* 1997]. These signatures include transient interplanetary shocks, depressed proton temperatures, cosmic ray depressions, flows with enhanced helium abundances, unusual compositions of ions and elements, and magnetic field structures consistent with looplike topologies [c.f., *Gosling,* 1993]. Recent efforts have focused on signatures indicative of the topology of the ejected magnetic fields, such as magnetic clouds and bidirectional particle flows [see *Russell,* 2000, and *Gosling,* 2000, for more details].

Perhaps the most widely used single-parameter signature of ejecta is the occurrence of counterstreaming suprathermal electrons. Since suprathermal electrons carry electron heat flux away from the Sun along magnetic field lines, when found streaming in both directions along the field they are interpreted as signatures of closed field lines and, thus, as a good proxy for CMEs in the solar wind [e.g., *Gosling,* 1993]. Figure 6 shows the most likely closed configuration, with the field lines imbedded in a flux rope with both ends connected to the Sun.

An important multiple signature of an ICME, is a magnetic cloud [e.g., *Burlaga,* 1991]. Magnetic clouds are defined as large-scale rotations in the magnetic field accompanied by low ion temperatures and strong magnetic fields. The magnetic field data from clouds often provide good fits to flux rope models, such as those of *Marubashi* [1986; 1997] or *Burlaga* [1988] (also see *Russell,* 2000). A flux rope consists of helical fields with increasing pitch angle from core to boundary and is assumed to form large loops with ends extending back toward the Sun (Figure 6).

Another class of ejecta plasma signatures are the abundances and charge state compositions of elements and ions which are systematically different in ICMEs compared with other kinds of solar wind [*Galvin,* 1997]. As the corona expands outward, the electron density decreases so rapidly that the plasma becomes collisionless and the relative ionization states become constant, thus reflecting the conditions in the corona where this occurs. The charge states of minor ions (Z>2) in CME flows usually suggest slightly hotter than normal coronal conditions (i.e., > 2 MK) at this "freezing in" location. In addition, transient flows often exhibit element and ion abundances that are enhanced relative to the typical solar wind. Unusually low ionization states of He and minor ions have also been detected in CME flows. Although rarely observed before, enhanced He^+ flows have been detected in the flows from several recent halo CMEs with more sensitive instruments on the SOHO, Wind and Advanced Composition Explorer (ACE) spacecraft. In each of these events an erupting filament–halo CME could be associated with either a dense and compact 'plug' or an extended flow of cool plasma in the trailing edge of a magnetic cloud. This is likely material from the filament itself, consistent with near-Sun observations showing that erupting filaments lag well behind the leading edge of their associated CMEs.

A recent example of ICME signatures is shown in Figure 7. The 6 January 1997 halo CME was associated with a clear magnetic cloud/flux rope on 10 January near Earth [Burlaga et al., 1998]. The characteristic field rotation is evident in the polar angle, θ, where the north-south component of the magnetic field changes systematically from strongly southward to northward in the cloud interval. Other typical characteristics of clouds are also evident. Counterstreaming electrons were present intermittently within the cloud, which also contained material at unusually low temperatures [Larson et al., 2000]. The cloud's average speed (bottom) of 450 km s^{-1} was moderate but faster than the ambient medium ahead of it and, as a result, the cloud drove a shock wave ahead. An unusual feature of the ejecta was a narrow, cool, dense 'plug' of material at the cloud's trailing edge. This was followed by an interface region and a high speed stream.

With the advent of the SOHO observations of halo CMEs, we can now more directly study the influence of Earth-directed CMEs on geoactivity. Recent analyses of the relation between halo CMEs and GM storms have been conducted by Brueckner et al. [1998], Webb et al. [2000a, c] and St. Cyr et al. [2000] and indicate a high degree of correlation. Figure 8, adapted from Webb et al. [2000a], shows the onset times of halo CMEs marked as vertical lines on a 27-day recurrence plot of the Dst index of GM activity. The vertical lines are solid or dashed according to whether the halo CME was classified as frontside or backside, respectively, and the filled triangles mark moderate or greater level ($Dst \leq -50$nT) storms. Webb et al. [2000a] found that all six halo CMEs accompanied by near-Sun center activity were associated with magnetic clouds, or cloud-like structures,

and moderate storms at 1 AU 3–5 days later. Cane et al. [1998; 1999] studied this same period and also found a strong association between frontside halo CMEs and ejecta signatures at 1 AU. St. Cyr et al. [2000] found a similar, but not as strong association as Webb et al. between halo CMEs and storms. They identified 40 frontside halo CMEs using only LASCO and EIT observations. There were 20 major GM storms during the same period (defined as periods when the Kp index was $\geq$ 6), 14 (70%) of which were preceded by frontside halo CMEs. Accounting for data gaps suggests that the association was as high as 17 out of 20, or 85%.

All these studies included partial halo CMEs. LASCO has now observed a sufficient number of CMEs to permit statistical studies of only full (360°) halo events, those most likely to be along the Sun-Earth line. In a preliminary study of all full halos observed in the 3 years 1996–1998, Webb et al., [2000c] found that about 70% of all the frontside halos were associated with all of the following signatures: moderate or greater level storms, shocks, magnetic clouds or cloud-like structures, and counterstreaming electrons at 1 AU.

Webb et al., [2000c] found that the full halo events had an average travel time of 3.6 days from their onsets to the onsets of the ejecta at 1 AU. This is consistent with the average solar wind speeds of ejecta associated with halo CMEs determined by Gopalswamy et al., [2000]. However, the initial speeds of the leading edges of these halos near the Sun varied over an order of magnitude, and were not proportional to their in situ speeds at 1 AU. This is perhaps not surprising because coronagraphs measure speeds projected on the skyplane; since halo CMEs are significantly out of this plane, their speeds are comprised of both outward motion, usually radial, and expansion velocities. In addition, we expect significant acceleration of initially slow CMEs and deceleration of fast events, such as measured for selected halos by Sheeley et al. [1999], as the CMEs expand into the solar wind. This complicates our ability to use halo events to forecast the arrival times of ICMEs at Earth.

For an intense magnetic storm to occur, there must be a sustained period of strong southward IMF to provide an efficient transfer of energy and momentum from the solar wind to the magnetosphere [e.g., Tsurutani et al.,1988]. For example, in the January 1997 event, the magnetosphere's response to the magnetic cloud (Dst) closely followed the southward field. Dst rose in response to the shock compression, then decreased in proportional response to the southward IMF in the cloud. Thus, a key reason why the above studies found that the observation of a frontside halo CME is not a sufficient condition for a large storm may be that not all CMEs contain or create sufficiently strong, sustained

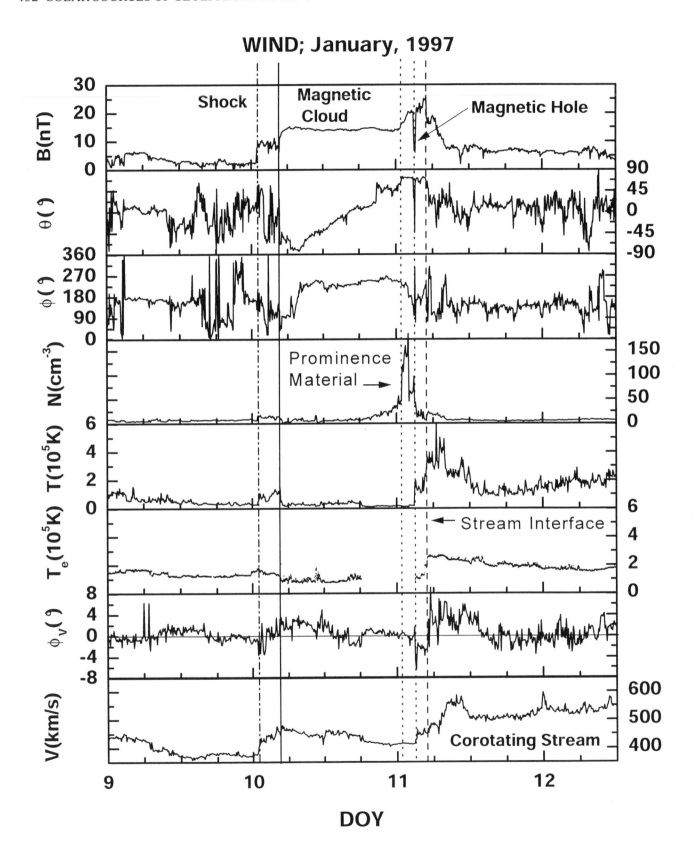

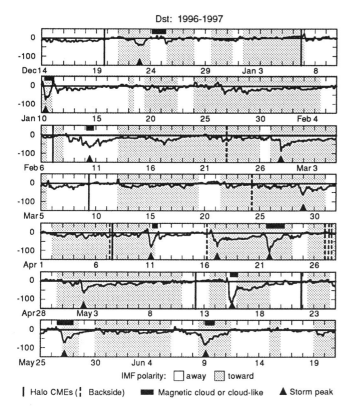

Dst: 1996-1997

IMF polarity: ☐ away ▨ toward

❘ Halo CMEs (⦙ Backside)　■ Magnetic cloud or cloud-like　▲ Storm peak

Figure 8. Stackplot of seven Bartels rotations (Numbers 2231–2237) of the *Dst* level (nT, left scale) showing geomagnetic storms at Earth. The horizontal line marks the zero nT level. The occurrences at Wind of magnetic clouds or cloud-like structures are denoted by the horizontal bars above the zero line on each rotation. Vertical lines mark the onsets at the Sun of LASCO halo CMEs: Solid lines are probable frontside events, and dashed lines are probable backside events. Solid triangles mark the peak times of moderate storms, i.e., *Dst* < –50 nT.

southward field. On the other hand, some interval of southward magnetic field will probably occur within or ahead of most CMEs as they travel through interplanetary space, and this may explain why some level of enhanced GM activity is usually observed following any frontside halo CME.

From these results we can conclude that halo CMEs with associated surface features, especially if near Sun center, usually presage at least moderate geomagnetic

storms, but that at least half as many such storms occur without halo-CME forewarning. Therefore, the simple observation of the occurrence of a probable frontside halolike CME has already significantly increased our ability to forecast the occurrence of storms of moderate or greater levels at Earth, though with a day or so uncertainty in onset time. Moderate storms not associated with CMEs are usually caused by Earth passage through the heliospheric current sheet and related corotating interaction regions (CIRs) [see Section 3].

2.4. CMEs, Streamers and the Heliospheric Current Sheet

Recently, a new paradigm has formed regarding our understanding of the degree to which the magnetic topology of the Sun, CMEs, and the heliosphere itself are interrelated. On a global scale the slow wind is confined to the vicinity of the heliospheric extension of the streamer belt, which is the locus of the heliospheric current sheet (HCS) at the Sun. It now appears that this belt is the source of most CMEs [e.g., *McAllister and Hundhausen*, 1996]. Thus, CMEs tend to carry with them the imprint of the lower harmonics of the Sun's magnetic field and, as a result, their magnetic structure tends to merge with that of the heliosphere. As we now show, these findings have important implications for predicting the geoeffectiveness of CMEs.

Figure 9 is a simplified schematic view of solar topology which we use to place the observations in the larger heliospheric context. This view is for the simplest magnetic configuration of the Sun around activity minimum, when the Sun's magnetic field can be approximated as a dipole whose axis is tilted slightly with respect to the axis of rotation. For other phases of the solar cycle, the dipole tilt is larger and higher harmonics distort the field, but observations suggest that the arguments below still apply except, perhaps, close to solar maximum.

The dark shading in the polar regions in Figure 9 represent coronal holes, regions of mostly open fields which are of opposite polarity in each hemisphere. Separating them is a wide band forming the base of the streamer belt, within which appear the small-scale, closed structures near the surface, i.e., active regions, sunspots and

Figure 7. Stackplot of solar wind plasma and IMF data for January 9–12, 1997. From top to bottom are plotted: magnetic field strength *B*, the elevation θ and azimuthal φ angles of the field direction, proton density *N*, the moment proton and electron temperatures, *T* and T_e, azimuthal flow angle φV, and bulk flow velocity *V*. Vertical lines denote, from left to right, the times of the shock, the leading edge of the magnetic cloud, the leading edge of the high density filament at the rear of the cloud, a magnetic hole, and the stream interface. The following CIR was overtaking the cloud. Courtesy of L. Burlaga; from *Burlaga et al.* [1998].

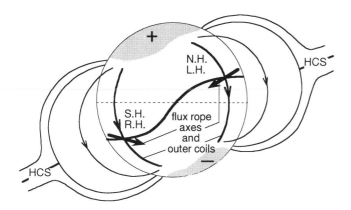

Figure 9. Schematic diagram illustrating the imprint of the solar magnetic field on two CME flux ropes forming under the umbrella of the helmet streamer belt, one in the northern hemisphere (N.H.) with left-handed (L.H.) chirality, and one in the southern hemisphere (S.H.), with right-handed (R.H.) chirality. Reprinted from JASTP, Vol. 62, Crooker, Solar and heliospheric geoeffective disturbances, Pp. 1071–1085, Copyright (2000), with permission from Elsevier Science.

filaments. These structures are often associated with CME eruptions. In addition, since CMEs can disrupt or destroy streamers, it is clear that CMEs as a class are intimately tied to the streamer belt. The heavy curve is the projection of the HCS onto the solar surface as the heliomagnetic equator. Dipolar field lines form an arcade from whose apex arises the heliospheric current sheet. The high speed wind flowing from the polar holes constricts the oppositely-directed fields over the streamers to a narrow current sheet which, in 3-dimensions, appears like a ballerina's skirt. The rotating, warped HCS appears as a sector boundary crossing at Earth, such that during one solar rotation Earth will be immersed in two large sectors of opposite polarity, each with the relatively high speed wind of its parent hole.

The short heavy arrows in Figure 9 mark the locations of two CMEs forming under the helmet streamer belt. Each is assumed to take the form of a flux rope in which arcade field lines become the outermost coils through reconnection behind the CME as it lifts off the Sun [e.g., *Gosling*, 1993; *Crooker et al.*, 1998a]. The direction and tilt of each arrow represents the orientation of the projected axis of that flux rope, roughly parallel to the helimagnetic equator. Given the overall polarity of the dipole field, the directions of the two arrows determine the chirality (sense of twist or helicity) of the resulting flux rope. The axis fields drawn here produce a left-handed (right-handed) rope in the northern (southern) hemisphere to match a hemispherically-dependent magnetic field pattern associated with filaments lower in the solar atmosphere [e.g., *Martin et al.*, 1994]. *Martin and McAllister* [1997] have shown that

arcade fields and the much smaller-scale filament fields are intimately related. Thus, Figure 9 indicates how a CME flux rope might blend into the Sun's magnetic topology, with its axis aligned with the heliomagnetic equator and its overlying leading fields in the direction of the Sun's dipolar fields (solid north-south loops).

Marubashi [1986, 1997]) was one of the first to connect the interplanetary and solar aspects of CMEs. He associated a set of erupting filaments with magnetic clouds and, using a cylindrical flux rope model, found that the orientations of the rope axes correlated with those of the filaments, consistent with *Martin and McAllister* [1997]. Thus, the arrows in Figure 6 could represent filament axes as well as flux rope axes. In addition, *Zhao and Hoeksema* [1997, 1998] showed a correlation between the orientations of cloud axes fit with flux ropes and the duration and maximum intensity of southward IMF in those clouds, consistent with expectations based on Figure 9. For example, both of the flux ropes created there would have leading southward and trailing northward fields, whereas a flux rope with a northward axis parallel to a highly inclined heliomagnetic equator would have little southward field. These relationships provide a predictive capability for the strength and duration of southward IMF in a magnetic cloud based on the orientation of the associated filament. In a related study, *Webb et al.* [2000a] assumed that the symmetric dimmings in the 12 May 1997 halo CME [Figure 3] marked the feet of the flux rope that was detected at Earth on 15 May. They found that the orientation of the associated filament and calculated magnetic flux in the dimmings agreed with those parameters of the flux rope.

As mentioned above, the direction of the leading field in a magnetic cloud, composed of the overlying, large-scale arcade fields, should point in the same direction as the solar dipole field. Since the dipolar field changes polarity at the maximum of every solar cycle, it follows that the direction of the leading field in clouds should also change at maximum, and this was confirmed by *Bothmer and Rust* [1997] and *Bothmer and Schwenn* [1998]. The pattern was also confirmed by *Mulligan et al.* [1998] who, in addition, found a solar cycle variation in the orientation of cloud axes, with high inclinations, relative to the ecliptic, dominating during solar maximum and the declining phase and low inclinations during minimum and the ascending phase [also see *Russell*, 2000]. This approximately matches the well-known solar cycle variation of HCS inclination [e.g., *Hoeksema*, 1991], supporting the picture in Figure 9.

Rust [1994], *Bothmer and Schwenn* [1998] and others have associated cloud flux ropes with erupting filaments, and have tested cloud chirality against predictions based on the hemispherically sorted filament

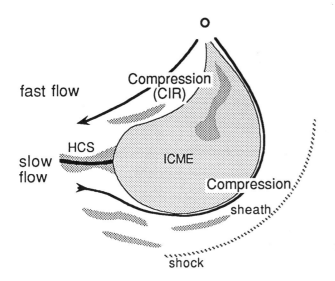

O

fast flow

Compression
(CIR)

slow
flow

HCS

ICME

Compression

sheath

shock

Figure 10. Schematic diagram illustrating compression effects at a corotating interaction region (CIR) in ecliptic cross-section. An ICME, a low-density feature, travels through the slow flow of the streamer belt and compresses plasma in a preceding sheath behind a shock, while an oncoming high-speed stream compresses plasma on its trailing edge, in the CIR. These interplanetary compression processes create large-scale, high-density regions. The highest density features are the smaller-scale, dark-shaded pressure balance structures of solar origin. Reprinted from JASTP, Vol. 62, Crooker, Solar and heliospheric disturbances, Pp. 1071–1085, Copyright (2000), with permission from Elsevier Science.

skew patterns mentioned above. Despite uncertainties about the relationship of the filament and the magnetic cloud fields, they found that the cloud helicity is well-correlated with the hemisphere of origin of the associated erupting filament and can be predicted from observations of the filament or arcade skew pattern.

Finally, enhanced density in the solar wind provides a modulation effect on geoeffectiveness. High density can have an important impact on storm strength. Recently, *Fenrich and Luhmann* [1998] confirmed a weak correlation between negative *Dst*, which measures storm strength, and dynamic pressure, which includes density. Density itself, however, may be a more important factor when appropriate time delays are accounted for. *Smith et al.* [1999] realized that, for southward IMF, the magnetosphere's response to a change in density is a change in plasma sheet density, with a time scale much longer (~5 hours) than its response to southward IMF (<1 hour). *Thomsen et al.* [1998] point out that, because of this effect, the January 1997 cloud [Figure 7] would have been much more geoeffective had the extremely high density at its trailing edge occurred when the IMF was southward. Figure 10 illustrates where high-density regions can be found in the solar wind. An ICME can be

associated both with high-density features of interplanetary origin, at the two compression regions ahead of and trailing the CME at a corotating interaction region (CIR), and of solar origin in the slow flow itself (such as filament material) and in the smaller-scale pressure balance structures embedded in the slow flow [*Crooker, 2000*].

3. SOLAR AND HELIOSPHERIC SOURCES OF GEOMAGNETIC ACTIVITY DURING THE DECLINING PHASE

3.1. CIRs and High Speed Streams

In this section we discuss how our new understanding of the streamer belt and the solar imprint on CMEs affects geoeffectiveness, especially during the declining phase of the solar cycle. The most important solar-related periodicity that is important for GM activity, other than the basic long-term solar 11-year and 88-year cycles, is, of course, the 27-day recurrence pattern imposed by the corotation of high-speed solar wind streams. This recurrent stream pattern dominates the solar wind during the declining phase of the solar cycle.

An important question then is what are the solar sources of high speed solar wind and of high IMF strengths, the parameters which best correlate statistically with geoactivity, during the declining phase? Both of these parameters can be enhanced for extended periods during this phase [e.g., Figure 1c]. Since longer periods of GM activity correlate better with solar wind speed and shorter periods with enhanced southward field [e.g., *Crooker et al.*, 1977; *Crooker and Gringauz*, 1993], it is the wind speed that has usually been considered the most important storm driver during this phase. With Skylab observations, coronal holes were determined to be the primary sources of open field lines and high speed, low density flows in the solar wind [e.g., *Hundhausen*, 1977]. They were long-lived and corotated with the Sun. High speed wind streams were largest and strongest during the declining phase of the cycle when the holes were largest and extended toward the helioequator. The association between the number of near-equatorial coronal holes and storms is greatest during the declining phase. We emphasize three aspects of high-speed streams here: Their ability to compress the solar wind, their impact on CMEs, and the dependence of their geoeffectiveness on the polarity of their magnetic fields.

A common misunderstanding about high speed streams is that the high-speed flow itself causes GM storms. Studies before Skylab had shown that peak storm strength coincides with passage of the leading edge of a high-speed stream, not with the subsequent high-speed flows, and ascribed this pattern to compression of pre-existing

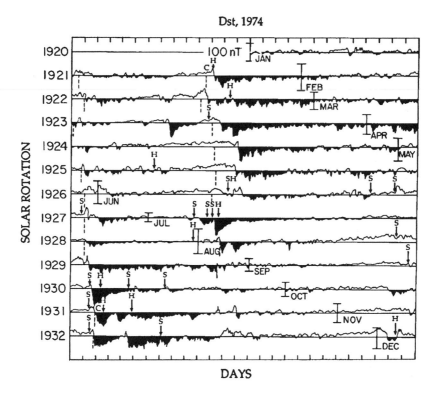

Figure 11. 27-day recurrence plot of the *Dst* index in 1974. Vertical dashes mark sector boundary crossings at 1 AU and arrows mark passage times of transient shocks (S), helium abundance enhancements (H), and magnetic clouds (C), all signatures of ICMEs. From *Crooker and Cliver* [1994].

southward IMF at the leading edge [e.g., *Dessler and Fejer*, 1963; *Hirshberg and Colburn*, 1973]. *Crooker and Cliver* [1994] and *Tsurutani et al.* [1995] reviewed the earlier studies and emphasized that the peak recurrent GM activity coincides with the passage of CIRs and their associated sector boundaries.

In the solar wind CIRs separate regions of high and low speed flows, which are identified with coronal holes and streamers, respectively, at the Sun. Thus, the solar sources of the recurrent activity must include the boundaries between coronal holes and streamers. Figure 11 illustrates an example of this effect during a period of particularly prominent recurrent activity in the declining phase of cycle 20 in 1974. This is a plot of the 27-day recurrence pattern of the *Dst* index. The solar wind data show two high speed streams during the year with opposite magnetic polarities; the boundaries between these two sectors are marked by vertical dashes. The peak GM activity follows the boundaries by 1-2 days, near the leading edges of the high-speed streams.

A less appreciated aspect is that the largest, usually recurrent GM storms associated with high-speed streams are often caused by CMEs at the leading edges of high-speed streams [e.g., *Crooker and Cliver*, 1994].

The reason for this is primarily geometrical. Since CMEs arise from the closed field line regions of the Sun, as pictured in Figure 9, they expand at least partially within the slow flows associated with the streamer belt, which border fast flows from coronal holes. For a tilted streamer belt which gives high-speed flows access to the ecliptic plane, a spacecraft located there will observe any CMEs in the slow flow immediately preceding the leading edge of a high-speed stream. Since the HCS acts as a channel for CMEs, the probability of encountering a CME in the ecliptic plane increases near sector boundaries [*Crooker et al.*, 1993], a prediction well supported during the recurrent phase of the cycle [e.g., *Crooker and Cliver*, 1994]. Although CMEs might often expand the confines of the original corridor of slow flow, average CMEs with widths of 45–50° should still pass within about a 3-day interval centered on the sector boundary. Figure 11 shows how the interplanetary signatures of CMEs at Earth tended to cluster around the sector boundary crossings during the decline of cycle 20.

An ecliptic cross-section of this stream pattern is illustrated in Figure 10. The ICME is expanding and traveling in the corridor of slow flow surrounding the HCS. Behind the slow flow is fast flow from a coronal hole. In this case the IMF in the fast flow has away polarity. The fast flow pushes into the slow flow ahead and compresses it, creating the CIR. In principle, this situation arises only for ICMEs which are slower than the fast stream flow. However, this seems typical of ICMEs, which can rapidly decelerate even only out to 1 AU [e.g., *Gopalswamy et al.*, 2000]. ICMEs also can contain southward fields which can be compressed in the CIR. The CIR immediately follows the ICME so that southward IMF both in the ICME and in the CIR contribute to storm strength. Figure 10 also shows how the sector structure and ICMEs are related. Most ICMEs do not disrupt sector structure [*Zhao and Hoeksema*, 1996], but blend into the heliosphere, consistent with the solar imprint on their topology [Figure 9]. For example, the flux rope loop in Figure 6 representing the ICME/magnetic cloud effectively carries the sector boundary, as drawn in Figure 10 [cf. *Crooker et al.*, 1998a]. The leading leg has fields pointing toward the Sun, and the trailing leg has fields pointing away from the Sun. A spacecraft being engulfed by the flux rope loop would first see the toward polarity of the "old" western sector, then the rotating fields of the cloud, and finally the away polarity of the "new" eastern sector.

As mentioned above, occasionally the *trailing* section of the ICME might be driven into by the CIR so that any southward fields intensify through compression [*Crooker and Cliver*, 1994; *Odstrcil and Pizzo*, 1999]. *Fenrich and Luhmann* [1998] also pointed out that the effect will have a solar cycle variation: as discussed above, the north to south turning magnetic clouds that are favorable for this effect occur between those solar maxima, such as cycle 21 to 22, when the Sun's dipolar field is opposite that shown in Figure 9.

The geoactivity associated with the January 1997 event shows profiles characteristic of high speed streams. Comparing Figure 7 with Figure 8, we see that there are two periods of negative Dst, the first corresponding to the southward fields in the leading part of the magnetic cloud on January 10 and the second to the fluctuating IMF in the following high speed stream from January 11–14. This pattern repeats during many of the storms in Figure 8. However, the peak wind speeds reach only moderate levels ($\sim$550 km s^{-1}), and the streams tend not to be recurrent, consistent with their occurrence in the *ascending* phase of the solar cycle, when the solar magnetic field varies considerably from rotation to rotation, compared to the declining phase, when recurrent storms reflect the stability of the solar field.

Finally, *Crooker and Cliver* [1994] noted that there is a period of "sustained" GM activity which follows and is weaker than the CIR-associated activity, but can last 10 days or more [see Figure 11]. This weaker activity is what is associated with the high speed stream itself and,

therefore, with the coronal hole. This activity is related to the continuous auroral activity caused by southward components of Alfven waves in the streams [*Tsurutani et al.*, 1995]. Because of the interrelationship of CMEs with the streamer belt and the HCS, it is clear that forecasting the time of arrival of sector boundary crossings at Earth should also improve forecasts of GM stormy periods at nearly all phases of the solar cycle.

3.2. The Seasonal Variation

There is another periodic variation in GM activity which, although not directly related to solar activity, has important consequences for geoeffectiveness. This is the semiannual or seasonal variation characterized by a higher average level of geoactivity at the equinoxes than at the solstices. This effect is quite apparent in statistical analyses using common indices of geoactivity, such as *am*, *AE* and *Dst* [*Cliver and Crooker*, 1993; *Cliver et al.*, 2000]. Surprisingly, a pronounced seasonal effect is also seen in the occurrences of the largest GM storms [*Crooker et al.*, 1992].

Historically, three hypotheses have been developed to explain the seasonal activity variation. These are the *Russell-McPherron* [1973] effect, the equinoctial hypothesis and the axial hypothesis [see *Cliver et al.*, 2000 for a review]. The Russell-McPherron (RM) mechanism is a projection effect which depends upon season. The IMF, ordered in the Sun's heliographic coordinate system, projects a southward component in Earth's dipole-ordered coordinate system when it points toward the Sun during spring (March peak) and away from the Sun during fall (September peak). In the equinoctial hypothesis GM activity is maximum when the angle between the Sun-Earth line (i.e., the solar wind flow direction) and Earth's dipole axis is 90°, and less at other times. This effect works by reducing the coupling efficiency of the magnetosphere at the solstices. Finally, in the axial hypothesis, GM activity peaks when Earth is at its highest heliographic latitudes, where it is best aligned with any radially propagating transient ICME flows and recurrent high-speed streams from mid-latitude coronal holes.

Figure 11 demonstrates how effective the seasonal variation can be during the declining phase of the cycle. As noted above, sustained, low-level activity associated with high-speed streams often follows the peak activity at the CIR. But not all high-speed flows generate sustained activity. The southward IMF responsible for the sustained activity can be ineffective if the polarity of the magnetic field in the stream is unfavorable for the RM effect. Although there were two strong wind streams of opposite polarity in 1974, GM activity was clearly stronger in the second stream of toward polarity during the first half of the year and in the first stream of away polarity during the second half of the year.

Although all of these mechanisms may play some role in the seasonal variation, the RM effect has generally been the accepted mechanism because of its apparent success in explaining geoactivity data sets such as in Figure 11 [*Crooker and Siscoe*, 1986; *Crooker and Cliver*, 1994]. However, recently *Cliver et al.* [2000] suggested that the equinoctial hypothesis better accounts for most of the semiannual modulation than either the RM or axial mechanisms. They estimate that the equinoctial effect is responsible for ~65% of the seasonal modulation whereas the other two mechanisms provide only 15–20% each.

4. SUMMARY AND CONCLUSIONS

The distribution of geomagnetic disturbances over the solar activity cycle statistically exhibits two peaks, a single or double peak near solar maximum and a later peak during the declining phase. These represent two major components of GM activity that have different but overlapping solar and heliospheric sources: one associated with transient solar activity that peaks with the sunspot cycle, and the other associated with recurrent CIRs, CMEs and high speed streams from coronal holes during the declining phase. The first single or double peak is dominated by transient solar activity, which generally follows the solar cycle, in the form of coronal mass ejections and their associated shocks, the key link between solar activity and major GM storms at maximum. However, both CMEs and CIRs-high speed stream ensembles can be geoeffective at all phases of the cycle. As we have shown, the coronal streamer belt and its extension as the heliospheric current sheet appears to play a major role in both kinds of geoactivity. For example, streamers are the source of many, if not most CMEs. During the declining phase, the passage of CIRs by Earth is associated with the strongest recurrent GM activity, often because CMEs tend to cluster there. CIRs arise from interactions at the boundaries between streamers (slow wind) and coronal holes (fast wind).

What makes solar and heliospheric disturbances geoeffective, in the sense that they cause GM storms, can be summarized in two terms: southward IMF and compression. Southward IMF is important because it allows merging of the IMF and Earth's magnetic field, which transfers solar wind energy and mass into the magnetosphere. Compression is important primarily because it strengthens existing southward IMF and, to a lesser extent, increases density. CMEs, the most geoeffective structures, usually contain long duration flows of southward IMF and fast CMEs compress any southward IMF at their leading edges and behind shocks created by the speed difference. In addition, CMEs themselves can carry high-density structures, such as solar filaments. High-speed streams are geoeffective primarily because they compress any southward IMF in CIRs. This compression can be enhanced when CIRs interact with CMEs erupting through the HCS, as is often the case during periods outside of solar maximum. The subsequent, sustained GM activity caused by southward IMF in Alfvenic fluctuations in the high-speed flow itself reaches significant levels only when supplemented by southward IMF from the RM polarity effect in the spring and fall. The high solar wind density structures which can appear in the leading and trailing edges of ICMEs are expected to be geoeffective only if associated with southward IMF. Speed, along with the strength of the southward magnetic field, is a factor in the solar wind electric field, which controls the merging rate at the boundary of the magnetosphere. But its overall contribution to the driving electric field is not large because speed varies much less than the IMF. Speed differences in the solar wind, of course, contribute to compression, but it is the compression itself that is the important geoeffective parameter.

Finally, recent studies have demonstrated that CMEs carry the imprint of the solar magnetic field out into the heliosphere. These findings imply that many characteristics of ejecta from CMEs can be predicted, at least statistically, from characteristics of the solar field and of associated solar source features. Potentially predictable characteristics include the orientation of magnetic cloud axes and their leading fields, the chirality of their helical fields, and the maximum strength and duration of their southward IMF.

Acknowledgments. The authors thank the convenors of the Chapman Conference on Space Weather: Progress and Challenges in Research and Applications for the invitation to present this material at the conference and in this paper. We benefited from data from the SOHO mission which is an international collaboration between NASA and ESA and is part of the ISTP. We appreciate the support of the P.I. of the SOHO LASCO experiment, Russell Howard. We are grateful to Stephen Kahler and Edward Cliver of the Air Force Research Lab for helpful comments on the manuscript. DFW was supported by AFOSR grant AF49620–98–1–0062 and by NSF grant ATM-9819483; NUC was supported by NSF grant ATM–9805064; SPP was supported by NASA grants NAG5–8116 and S–86760–E; and OCS was supported by NASA contract S–86760–E and by the National Space Weather Program under NSF grant ATM–9819668.

The Editor thanks the referees for their assistance in evaluating this paper.

REFERENCES

Antiochos, S.K., C.R. DeVore, and J.A. Klimchuk, A model for solar coronal mass ejections, *Astrophys. J., 510,* 485, 1999.

Bothmer, V. and D.M. Rust, The field configuration of magnetic clouds and the solar cycle, in *Coronal Mass Ejections, Geophys. Monogr. Ser.,* vol. 99, edited by N. Crooker et al., p. 139, AGU, Washington, D.C., 1997.

Bothmer, V., and R. Schwenn, The structure and origin of magnetic clouds in the solar wind, *Ann. Geophys., 16,* 1, 1998.

Brueckner, G.E., et al., Geomagnetic storms caused by coronal mass ejections (CMEs): March 1996 through June 1997, *Geophys. Res. Lett., 25,* 3019, 1998.

Burkepile, J., H. Gilbert, and T. Holzer, this volume, 2000.

Burlaga, L.F., Magnetic clouds and force-free fields with constant alpha, *J. Geophys. Res., 93,* 7217, 1988.

Burlaga, L.F.E., Magnetic clouds, in *Physics of the Inner Heliosphere,* Vol. 2, edited by R. Schwenn and E. Marsch, p. 1, Springer-Verlag, New York, 1991.

Burlaga, L., et al., A magnetic cloud containing prominence material: January 1997, *J. Geophys. Res., 103,* 277, 1998.

Cane, H.V., I.G. Richardson, and O.C. St. Cyr, The interplanetary events of January May, 1997 as inferred from energetic particle data, and their relationship with solar events, *Geophys. Res. Lett., 25,* 2517, 1998.

Cane, H.V., I.G. Richardson, and O.C. St. Cyr, Correction to "The interplanetary events of January May, 1997 as inferred from energetic particle data, and their relationship with solar events", *Geophys. Res. Lett., 26,* 2149, 1999.

Canfield, R.F., H.S. Hudson, and D.E. McKenzie, Sigmoidal morphology and eruptive solar activity, *Geophys. Res. Lett., 26,* 627, 1999.

Cliver, E.W., and N.U. Crooker, A seasonal dependence for the geoeffectiveness of eruptive solar events, *Sol. Phys., 145,* 347, 1993.

Cliver, E.W., Y. Kamide, and A.G. Ling, Mountains versus valleys: Semiannual variation of geomagnetic activity, *J. Geophys. Res., 105,* 2413, 2000.

Crooker, N. U., Solar and heliospheric geoeffective disturbances, *J. Atmos. Sol. Terr. Phys., 62,* 1071, 2000.

Crooker, N.U., and E.W. Cliver, Postmodern view of M-regions, *J. Geophys. Res., 99,* 23,383, 1994.

Crooker, N.U., and K. Gringauz, On the low correlation between long-term averages of solar wind speed and geomagnetic activity after 1976, *J. Geophys. Res., 98,* 59, 1993.

Crooker, N.U., and G.L. Siscoe, The effect of the solar wind on the terrestrial environment. In *Physics of the Sun,* edited by P.A.. Sturrock, D. Reidel, Dordrecht, Holland, 1986.

Crooker, N.U., J. Feynman, and J.T. Gosling, On the high correlation between long-term averages of solar wind speed and geomagnetic activity, *J. Geophys. Res., 82,* 1933, 1977.

Crooker, N.U., E.W. Cliver, and B.T. Tsurutani, The semiannual variation of great geomagnetic storms and the postshock Russell-McPherron effect preceding coronal mass ejecta, *Geophys. Res. Lett., 19,* 429, 1992.

Crooker, N.U., G.L. Siscoe, S. Shodan, D.F. Webb, J.T. Gosling, and E.J. Smith, Multiple heliospheric current sheets and coronal streamer belt dynamics, *J. Geophys. Res., 98,* 9371 1993.

Crooker, N.U., J.T. Gosling, and S.W. Kahler, Magnetic clouds at sector boundaries, *J. Geophys. Res., 103,* 301, 1998a.

Crooker, N.U., et al., Sector boundary transformation by an open magnetic cloud, *J. Geophys. Res., 103,* 26,859, 1998b.

Dessler, A.J., and J.A. Feyer, Interpretation of Kp index and M-region geomagnetic storms, *Planet. Space Sci., 11,* 505, 1963.

Fenrich, F.R., and J.G. Luhmann, Geomagnetic response to magnetic clouds of different polarity, *Geophys, Res. Lett., 25,* 2999, 1998.

Galvin, A.B., Minor ion composition in CME-related solar wind, in *Coronal Mass Ejections,* Geophys. Monogr. Ser., vol. 99, edited by N. Crooker et al., p. 253–260, Washington, D.C., AGU, 1997.

Gonzalez, W.D., A.L.C. Gonzalez, and B.T. Tsurutani, Dual-peak solar cycle distribution of intense geomagnetic storms, *Planet. Space Sci., 38,* 181, 1990.

Gopalswamy, N., A. Lara, R.P. Lepping, M.L. Kaiser, D. Berdichevsky, and O.C. St. Cyr, Interplanetary acceleration of coronal mass ejections, *Geophys. Res. Lett., 27,* 145, 2000.

Gosling, J.T., The solar flare myth, *J. Geophys. Res., 98,* 18,937, 1993.

Gosling, J.T., this volume, 2000.

Gosling, J.T. and D.J. McComas, Field line draping about fast coronal mass ejecta: a source of strong out-of-the-ecliptic magnetic fields, *Geophys. Res. Lett., 14,* 355, 1987.

Gosling, J.T., D.J. McComas, J.L. Phillips, and S.J. Bame, Geomagnetic activity associated with earth passage of interplanetary shock disturbances and coronal mass ejections, *J. Geophys. Res., 96,* 7831-7839, 1991.

Harrison, R.A., Solar coronal mass ejections and flares, *Astron. Astrophys., 162,* 283–291, 1986.

Hirshberg, J., and D. S. Colburn, Geomagnetic activity at sector boundaries, *J. Geophys. Res., 78,* 3952-3957, 1973.

Hoeksema, J. T., Large-scale solar and heliospheric magnetic fields, *Adv. Space Res., 11,* (1)15-(1)24, 1991.

Hoeksema, J.T. and P.H. Scherrer, The solar magnetic field 1976 through 1985, *Rep. UAG- 94,* Natl. Oceanic and Atmos. Admin., Boulder, Colo., 1986.

Howard, R.A., D.J. Michels, N.R. Sheeley, Jr., and M.J. Koomen, The observation of a coronal transient directed at Earth, *Astrophys. J.,263,* L101–L104, 1982.

Howard, R.A., Sheeley, N.R. Jr., Koomen, M.J., and Michels, D.J., Coronal mass ejections: 1979-1981, *J. Geophys. Res., 90,* 8173–8191, 1985.

Hudson, H.S. and Webb, D.F., Soft X-ray signatures of coronal mass ejections, in *Coronal Mass Ejections,* Geophysical Monograph Vol. 99, edited by N. Crooker et al., p. 27-38, AGU, Washington, DC., 1997.

Hudson, H., Haisch, B. and Strong, K.T., Comments on 'The solar flare myth', *J. Geophys. Res., 100,* 3473–3477, 1995.

Hudson, H.S., J.R. Lemen, O.C. St. Cyr, A.C. Sterling, and D. F. Webb, X-ray coronal changes during halo CMEs, *Geophys. Res. Lett., 25,* 2481-2484, 1998.

Hundhausen, A.J., An interplanetary view of coronal holes,

in *Coronal Holes and High Speed Wind Streams*, edited by J. Zirker, p.225, Colorado Assoc.Univ. Press, Boulder, 1977.

Hundhausen, A.J., Sizes and locations of coronal mass ejections: SMM observations from 1980 and 1984–1989, *J. Geophys. Res., 98,* 13,177 1993.

Kahler, S.W., The morphological and statistical properties of solar X-ray events with long decay times, *Astrophys. J., 214,* 891–897, 1977.

Kahler, S.W., Solar flares and coronal mass ejections, *Ann. Rev. Astron. Astrophys., 30,* 113-141, 1992.

Klimchuk, J.A., this volume, 2000.

Larson, D.E., R.P. Lin, and J. Steinberg, Extremely cold electrons in the January 1997 magnetic cloud, *Geophys. Res. Lett., 27,* 157, 2000.

Legrand, J.P. and P.A. Simon, Geomagnetic storms and their associated forecasts, in *Solar- Terrestrial Predictions - IV*, Vol. 3, edited by J. Hruska et al., p. 191, Natl. Oceanic Atmos. Admin., Boulder, Colo., 1993.

Luhmann, J.G., J.T. Gosling, J.T. Hoeksema, and X. Zhao, The relationship between large-scale solar magnetic field evolution and corona mass ejections, *J. Geophys. Res., 103,* 6585, 1998.

Martin, S.F., and A.H. McAllister, Predicting the sign of helicity in erupting filaments and coronal mass ejections, in *Coronal MassEjections*, Geophys. Monogr. Ser., vol. 99, edited by N. U. Crooker et al., p. 127, AGU, Washington, D. C., 1997.

Martin, S.F., R. Bilimoria, and P.W. Tracadas, Magnetic field configurations basic to filament channels and filaments, in*Solar Surface Magnetism*, edited by Rutten, R. J., and C. J. Schrijver, Kluwer Academic, Holland, 1994.

Marubashi, K., Structure of the interplanetary magnetic clouds and their solar origins, *Adv. Space Res. 6,* 335-338, 1986.

Marubashi, K., Interplanetary magnetic flux ropes and solar filaments, in *Coronal Mass Ejections*, Geophys. Monogr. Ser., vol. 99, edited by N. Crooker, et al., p. 147, AGU, 1997.

Mayaud, P.N., Analysis of storm sudden commencements for the years 1868-1967, *J. Geophys. Res., 80,* 111, 1975.

McAllister, A. H., and A. J. Hundhausen, The relation of Yohkoh coronal arcade events to coronal streamers and CMEs, in *Solar Drivers of Interplanetary and Terrestrial Disturbances*, edited by K. S. Balasubramaniam, S. L. Keil, and R. N. Smartt, p. 171, Astron. Soc. of the Pacific, San Francisco, Calif., 1996.

Mulligan, T., C.T. Russell, and J.G. Luhmann, Solar cycle evolution of the structure of magnetic clouds in the inner heliosphere, *Geophys. Res. Lett., 25,* 2959-2963, 1998.

Neugebauer, M., and R. Goldstein, Particle and field signatures of coronal mass ejections in the solar wind, in *Coronal Mass Ejections*, Geophys. Monogr. Ser., vol. 99, edited by N. Crooker et al., p. 245, AGU, 1997.

Odstrcil, D., and V. J. Pizzo, Three-dimensional propagation of coronal mass ejections (CMEs) in a structured solar wind flow, 1. CME launched within the streamer belt, *J. Geophys. Res., 104,* 483-492, 1999.

Plunkett, S.P., Thompson, B.J., Howard, R.A., Michels, D.J., St. Cyr, O.C., Tappin, S.J., Schwenn, R. and Lamy, P.L., LASCO observations of an Earth-directed coronal mass ejection on May 12, 1997, *Geophys. Res. Lett., 25,* 2,477, 1998.

Plunkett, S.P. et al., New insights on coronal mass ejections from SOHO, in Proceedings of First ISCS Workshop, *Adv. Space Res.*, in press, 2000a.

Plunkett, S.P., Thompson, B.J., St. Cyr, O.C. and Howard, R.A., Solar source regions of coronal mass ejections and their geomagnetic effects, *J. Atmos. Sol. Terr. Phys.*, in press, 2000b.

Richardson, I. G., and H. V. Cane, Signatures of shock drivers in the solar wind and their dependence on the solar source location, *J. Geophys. Res., 98,* 15295, 1993.

Richardson, I.G., E.W. Cliver, and H.V. Cane, Sources of geomagnetic activity over the solar cycle: Relative importance of CMEs, high speed streams, and slow solar wind, *J. Geophys. Res., 105,* 18,203, 2000.

Russell, C.T., this volume, 2000.

Russell, C.T. and R.L.McPherron, Semiannual variation of geomagnetic activity, *J. Geophys. Res., 78,* 92, 1973.

Rust, D.M., Coronal disturbances and their terrestrial effects, *Space Sci. Rev., 34,* 21-36, 1983.

Rust, D.M., Spawning and shedding helical magnetic fields in the solar atmosphere, *Geophys. Res. Lett., 21,* 241–244, 1994.

Rust, D.M. and Kumar, A., Evidence for helically linked magnetic flux ropes in solar eruptions, *Astrophys. J., 464,* L199–L202, 1996.

Schwenn, R. et al., First view of the extended green-line emission corona at solar activity mini mum using the LASCO-C1 coronagraph on SOHO, *Solar Phys., 175,* 667–684, 1997.

Sheeley, N.R., Jr., Howard, R.A., Koomen, M.J. and Michels, D.J., Associations between coronal mass ejections and soft X-ray events, *Astrophys. J., 272,* 349-354, 1983.

Sheeley, N.R., Jr., Walters, J.H., Wang, Y.-M. and Howard, R.A., Continuous tracking of coronal outflows: Two kinds of coronal mass ejections, *J. Geophys. Res., 104,* 24,739-24,767, 1999.

Slavin, J.A., G. Jungman, and E.J. Smith, The interplanetary magnetic field during solar cycle 21: ISEE-3/ICE observations, *Geophys. Res. Lett., 13,* 513, 1986.

Smith, J.P., M.F. Thomsen, J.E. Borovsky, and M. Collier, Solar wind density as a driver for the ring current in mild storms, *Geophys. Res. Lett., 26,* 1797-1800, 1999.

Sterling, A.C., Hudson, H.S., Thompson, B.J. and Zarro, D.M., Yohkoh SXT and SOHO EIT observations of 'sigmoid-to-arcade' evolution of structures associated with halo CMEs, *Astrophys. J., 532,* 628–647, 2000.

St. Cyr, O.C., et al., Properties of coronal mass ejections: SOHO LASCO observations from January 1996 to June 1998, *J. Geophys. Res., 105,* 18,169–18,185, 2000.

Subramanian, P., Dere, K.P., Rich, N.B., and Howard, R.A., The relationship of coronal mass ejections to streamers, *J. Geophys. Res., 104,* 22,321–22,330, 1999.

Svestka, Z. and E.W. Cliver, History and basic characteristics of eruptive flares, in *Eruptive Solar Flares*, edited by Z. Svestka et al., p. 1, New York, Springer-Verlag, 1992.

Thompson, B.J., Plunkett, S.P., Gurman, J.B., Newmark, J.S., St. Cyr, O.C. and Michels, D.J., 1998, SOHO/EIT observations of an Earth-directed coronal mass ejection on May 12, 1997, *Geophys. Res. Lett., 25,* 2,465, 1998.

Thompson, B.J., et al., The correspondence of EUV and white light observations of coronal mass ejections with SOHO EIT and LASCO, in *Sun-Earth Plasma Connections*, Geophys. Monogr. Vol. 109, edited by J.L. Burch, p. 31, AGU, Washington, DC., 1999.

Thomsen, M.F., J.E. Borovsky, D.J. McComas, R.C. Elphic, and S. Maurice, The magnetospheric response to the CME passage of January 10-11, 1997, as seen at geosynchronous orbit, *Geophys. Res. Lett., 25*, 2545-2548, 1998.

Tsurutani, B.T., Gonzalez, W.D., Tang, F., Akasofy, S.-I. and Smith, E.J., Origin of interplanetary southward magnetic fields responsible for major magnetic storms near solar minimum (1978-1979), *J. Geophys. Res., 93*, 8,519-8531, 1988.

Tsurutani, B.T., W.D. Gonzalez, A.L.C. Gonzalez, F. Tang, J.K. Arballo, and M. Okada, Interplanetary origin of geomagnetic activity in the declining phase of the solar cycle, *J. Geophys.Res., 100*, 21,717-21,733, 1995.

Wang, Y.-M., J. Lean, and N.R. Sheeley, Jr., The long-term variation of the Sun's open magnetic flux, *Geophys. Res. Lett., 27*, 505, 2000.

Webb, D.F., The solar sources of coronal mass ejections, in *Eruptive Solar Flares*, edited by Z.Svestka et al., p. 234, Springer-Verlag, New York, 1992.

Webb, D.F., and R.A. Howard, The solar cycle variation of coronal mass ejections and the solar wind mass flux, *J. Geophys. Res., 99*, 4201, 1994.

Webb, D.F., S.W. Kahler, P.S. McIntosh, and J.A. Klimchuk, Large-scale structures and multiple neutral lines associated with coronal mass ejections, *J. Geophys. Res., 102*, 24,161-24,174, 1997.

Webb, D.F., Cliver, E.W., Crooker, N.U., St. Cyr, O.C. and Thompson, B.J., Relationship of halo coronal mass ejections, magnetic clouds, and magnetic storms, *J. Geophys. Res., 105*, 7,491-7,508, 2000a.

Webb, D.F., Lepping, R.P., Burlaga, L., DeForest, C.E., Larson, D., Martin, S., Plunkett, S.P. and Rust, D.M., The origin and development of the May 1997 magnetic cloud, *J. Geophys. Res.,* in press (December 2000), 2000b.

Webb, D.F., R.P. Lepping, N.U. Crooker, S.P. Plunkett, and O.C. St. Cyr, Studying CMEs using LASCO and in-situ observations of halo events, *Bull. Am. Astron. Soc., 32*, 825, 2000c.

Zhao, X.-P., and J.T. Hoeksema, Effect of coronal mass ejections on the structure of the heliospheric current sheet, *J. Geophys. Res., 101*, 4825-4834, 1996.

Zhao, X.-P., and J.T. Hoeksema, Is the geoeffectiveness of the 6 January 1997 CME predictable from solar observations?, *Geophys. Res. Lett., 24*, 2965-2968, 1997.

Zhao, X.-P., and J.T. Hoeksema, Central axial field direction in magnetic clouds and its relation to southward interplanetary magnetic field events and dependence on disappearing solar filaments, *J. Geophys. Res., 103*, 2077-2083, 1998.

Zhao, X.P., J.T. Hoeksema, J.T. Gosling, and J.L. Phillips, Statistics of IMF Bz events, in *Solar- Terrestrial Predictions - IV*, Vol. 2, edited. by J. Hruska et al., p. 712, Natl. Oceanic and Atmos. Admin., Boulder, Colo., 1993.

N. U. Crooker, Center for Space Physics, 725 Commonwealth Ave., Boston University, Boston, MA 02215.

S. P. Plunkett, USRA, Code 682.3, NASA Goddard Space Flight Center, Greenbelt, MD 20771.

O. C. St. Cyr, CPI/Naval Research Laboratory, Code 682, NASA Goddard Space Flight Center, Greenbelt, MD 20771.

D. F. Webb, AFRL/VSBS, 29 Randolph Road, Hanscom AFB, MA 01731-3010
(e-mail: david.webb@hanscom.af.mil).

Theory of Coronal Mass Ejections

James A. Klimchuk

Space Science Division, Naval Research Laboratory, Washington, DC

Coronal mass ejections (CMEs) are extremely important phenomena, both for understanding the evolution of the global corona and for understanding and predicting space weather. Despite this importance, the physical explanation of CMEs remains largely confused. A variety of theoretical models have been proposed, and we here attempt to organize them according to a classification scheme that identifies and differentiates their essential physical attributes. We propose five distinct classes of models, which we present with the aid of simple analogues involving springs, ropes, and weights. We also indicate how some of the models appear to be inconsistent with certain observations.

1. INTRODUCTION

Coronal mass ejections (CMEs) are the most energetic events in the solar system. They are spectacular displays representing the sudden eruption of up to 10^{16} gm of coronal material at speeds of typically several hundred kilometers per second [e.g., *Hundhausen,* 1999; *St. Cyr et al.,* 2000]. In more colloquial terms, they are a billion tons of super heated matter flying away from the Sun at a million miles per hour!

We now know that CMEs are more than just interesting natural phenomena. They are also the primary cause of the largest and most damaging space weather disturbances [e.g., *Gosling,* 1993; *Webb et al.,* 2000]. Effects include the temporary and sometimes permanent failure of satellites, the degradation or disruption of communication, navigation, and commercial power systems, and the exposure of astronauts and polar-route airline crews to harmful doses of radiation. Some of these effects are delayed approximately 3 days, the time it takes a CME to propagate to Earth and interact with the magnetosphere. Others begin almost immediately after the CME lifts off from the Sun due to the production of solar energetic particles (SEPs) that travel at relativistic speeds. Clearly, there is a need to understand CMEs and ultimately to predict them before they occur. This need will grow even greater as society's dependence on space and space-based technologies steadily increases over time.

A number of interesting theoretical models have been proposed to explain the nature and origin of CMEs. This article attempts to review these models in something of a tutorial fashion. The goal is not to produce an exhaustive compilation of all the work that has been done on the subject, but rather to identify the essential physics involved in the CME problem, and to distinguish the various types of models in terms of their most basic physical differences.

We find that existing CME models can be logically organized into five distinct classes. These are in turn grouped into two major divisions called "storage and release" models and "directly driven" models. We present the five classes with the aid of analogues involving everyday items such as springs, ropes, and weights. Our hope is that these simpler systems will help reveal the fundamental ways in which the more complex CME models differ. Note that this is not the first CME classification scheme to be proposed. *Forbes* [2000], for example, has recently advocated an approach that em-

Space Weather
Geophysical Monograph 125

phasizes resistive versus ideal instabilities. This is a complementary and equally useful way of looking at the problem. Additional physical insights of a general nature can be found in *Low* [1994, 1996, 1999].

Ultimately, we must rely on observations to determine which of the proposed models is correct. Perhaps none is, or perhaps several are. There are hints that CMEs originating in active regions are different from those originating in quiet areas [e.g., *MacQueen and Fisher*, 1983; *St. Cyr et al.*, 1999; *Delannée, Delaboudinière, and Lamy*, 2000]. Additional studies are necessary to quantify the differences. Whether they are due to fundamental physical differences in the CME mechanism or are simply a reflection of the different environments in which the CMEs occur is a completely open question.

We will show that most of the proposed models have difficulty explaining one or more aspects of the observations. In some cases the discrepancies appear to be fatal, while in other cases their significance needs further evaluation. For a more complete discussion of the observational properties of CMEs, see one of the many excellent reviews [e.g., *Wagner*, 1984; *Hundhausen*, 1988, 1999; *Kahler*, 1992; *St. Cyr et al.*, 2000; *Webb et al.*, 2000].

As a background for our discussion of the different classes of models, we begin with a generic treatment of the structure, dynamics, and energetics of CMEs. These three aspects are of course closely related, but it is instructive to consider each of them separately.

2. STRUCTURE

The structure of CMEs and the pre-eruption configurations which spawn them are not well determined at this time. This is due largely to the optically thin nature of coronal observations. What we see in images are two-dimensional projections of three-dimensional structures, and it is often difficult to disentangle the overlapping features. The situation should improve dramatically with the STEREO mission, since its two spacecraft will provide views from different angles, allowing us to resolve many of the line-of-sight ambiguities.

In most models, the pre-eruption magnetic field has one of the two basic topologies illustrated in Figure 1. Arcade field lines arch directly over the magnetic neutral line to connect the opposite polarity parts of the photosphere. The configuration may be sheared, in which case the positive and negative footpoints are displaced in opposite directions parallel to the neutral line. Sheared fields are stressed and contain magnetic free en-

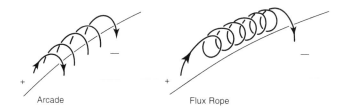

Figure 1. Arcade (left) and flux rope (right) magnetic topologies adopted by most CME models. Representative field lines are shown.

ergy that can be released during an eruption to power the mass motions. The single arcade in Figure 1 corresponds to a dipolar configuration. We can also imagine a quadrupolar configuration consisting of three arcades lined up side-by-side. Multipolar fields of this type are a fundamental property of the "magnetic breakout" model discussed later.

The flux rope topology is very different. Field lines form a helical structure that lies above the neutral line, disconnected from the photosphere except at the ends (in 2D models it is completely disconnected). Our cartoon drawing shows much more twist than we would expect in a real structure. There may be situations of one turn or less, in which case the distinction between flux rope and sheared arcade becomes blurred [*Titov and Démoulin*, 1999]. Many CMEs have what is commonly known as a "classic three-part structure" consisting of a bright frontal loop, a dark cavity underneath, and an embedded bright core [e.g., *Hundhausen*, 1988]. Plate 1 shows a typical example observed by the LASCO C3 coronagraph on the *Solar and Heliospheric Observatory (SOHO)*. It has been suggested that cavities correspond to flux ropes seen edge-on [e.g., *Chen et al.*, 1997]. Support for this interpretation comes from the fact that the tops of most cavities are well rounded and the recent result that upwardly curved striations are sometimes detected at the trailing edge of the cavity [*Dere et al.*, 1999]. It is further believed that the embedded core is prominence material that is trapped at the bottoms of the helical field lines and dragged upward during the eruption. We must remember, however, that not all CMEs have a clear three-part structure. In fact, only a minority do [*Burkepile and St. Cyr*, 1993; see also *Dere et al.*, 1999]. Most CMEs are much more complex in appearance, as in the example of Plate 2. It is not obvious that these are flux ropes.

Additional evidence of a flux rope topology, at least in some evolved events, is the *in situ* observation of a rotating magnetic field pattern within interplanetary magnetic clouds. About one-third of interplanetary structures identified with CMEs have this signature [*Gosling, 1990*]. We note that magnetic reconnection may cause flux ropes to form naturally from erupting sheared arcades [*Gosling, 1993*], so the observation of a flux rope CME high in the corona or in the heliosphere does not necessarily imply that a flux rope was part of the initial configuration. Recent observations support the view of flux rope formation *after* the eruption begins [*Dere et al., 1999*].

3. DYNAMICS

Presumably a CME occurs when the balance of forces that maintains an equilibrium is upset. Something causes the upward forces to become dominant over the downward ones. What are these forces? Gravity and gas pressure play important roles in some models. However, the magnetic field dominates the plasma throughout much of the corona ($\beta \equiv 8\pi P/B^2 \ll 1$), especially within active regions and at lower altitudes, and many models ignore the plasma altogether. In this case the only forces are magnetic.

The two competing magnetic forces are magnetic pressure and tension. Regions of strong magnetic field have enhanced pressure and naturally tend to expand into regions of weak field. An isolated flux rope like that shown in Figure 1 would grow in diameter and its axis would rise upward away from the surface were it not held in place by an overlying field. On the real Sun, magnetic field permeates the entire corona, and arcade-like field lines arch over any flux ropes that may exist. These arcade field lines are rooted in the surface and their tension acts to hold the flux rope in place.

This same interplay between magnetic pressure and tension exists even in simple arcades where no flux rope is present. The field strength is greatest in the center of an equilibrium arcade, and the outward force produced by the gradient in the magnetic pressure is exactly balanced by the inward tension force. An eruption begins when something tips the balance in favor of the outward pressure gradient. If this occurs unstably, as is usually the case, the force imbalance grows with time, and the eruption may become violent.

4. ENERGETICS

Perhaps as much as 10^{32} ergs of energy is required to lift the mass of a CME against solar gravity and accelerate it to the observed high velocities [e.g., *Vourlidas et al., 2000*]. What is the source of this energy? There is little indication that energy passes through the solar surface from below the corona during the time of

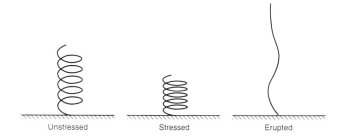

Figure 2. Simple spring analogue to the solar corona. The three states represent the magnetic field when it is unstressed (potential), stressed (current-carrying), and erupted (also current-carrying).

an eruption. Consequently, the energy is likely to be stored in the corona before the eruption begins. Since the plasma β is small throughout much of the corona, we can safely conclude that the energy is probably magnetic. Recall that only that part of the magnetic energy that is associated with electric currents, the so-called "free magnetic energy," is available to be converted to other forms. Energy cannot be extracted from a current-free potential field. The field must be stressed. It has been amply demonstrated that the free energy stored in pre-eruption coronal fields is at least as great as the gravitational and kinetic energies of a typical CME [e.g., *Klimchuk and Sturrock, 1992; Wolfson, 1993*].

The energy problem is not so easily solved, however. A CME opens up a large portion of the corona, meaning that the field lines are stretched outward into the distant heliosphere. Such open magnetic configurations are themselves highly stressed, containing a current sheet that extends vertically above the neutral line. In the limiting case of a fully open field in which all of the field lines extend to infinity (idealized to be sure), the magnetic energy would actually increase during the eruption [*Aly, 1984, 1991; Sturrock, 1991*]. So, the question becomes one of how to open the field to the extent required by observations and at the same time decrease its energy by a sufficient amount to power the mass motions. Models which answer this question are classified as "storage and release" models. Storage refers to the slow buildup of magnetic free energy from the gradual stressing of the field by footpoint motions or mass accumulation. It is a phase of quasistatic evolution. Release refers to the highly dynamic phase when rapid energy conversion and eruption take place.

As a simple analogue, we can represent the corona by a spring attached to a rigid base, as sketched in Figure 2. On the left, the spring is in its natural, or unstressed,

Plate 1. Three-part CME observed near the solar north pole by the LASCO C3 coronagraph on 27 February 2000. The white circle indicates the limb of the solar disk.

Plate 2. CME observed by LASCO on 28 November 1996. The image is a composite of C2 and C3 coronagraph images. The white circle indicates the limb of the solar disk.

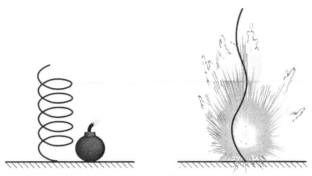

Thermal Blast

Figure 3. Thermal blast model analogue. The spring is blown open from its initial unstressed state by a sudden pressure pulse corresponding to, e.g., a solar flare.

state; in the middle, it is compressed and contains free energy associated with the stresses; on the right, it is stretched out beyond its natural length and is therefore also stressed. We refer to this last state as the erupted state for obvious reasons. For the sake of discussion, we assume that the spring has more free energy in the compressed (stressed) state than in the erupted state, so that the ordering of the energies is:

$$E_U < E_E < E_S, \qquad (1)$$

where the subscripts refer to unstressed, erupted, and stressed.

In storage and release models, the system evolves slowly from the unstressed to stressed states, and then rapidly from the stressed to erupted states. Because $E_E < E_S$, it is energetically favorable for the latter transition to occur, so it can do so in an unstable fashion without any driving by external forces.

The second major category of CME models, called "directly driven" models, behave entirely differently. They bypass the intermediate stressed state and go directly from the unstressed state to the erupted state. There is no slow energy buildup phase. Because $E_E > E_U$, external drivers must rapidly pump energy into the system in real time as the eruption occurs. As alluded to above, and as we now discuss, there is little observational evidence to support this idea.

5. DIRECTLY DRIVEN MODELS

5.1. Thermal Blast

Let us now consider the five different classes of CME models. We begin with "thermal blast" models, largely for historical reasons. As represented in Figure 3, these

models are characterized by a sudden release of thermal energy in the low corona. The greatly enhanced gas pressures associated with this release cannot be contained by the magnetic field, and the corona is literally blown open. This was the first explanation given for CMEs and was inspired by the fact that many CMEs occur in conjunction with solar flares. It seemed quite logical at the time to suppose that CMEs are simply a response to the impulsive energy release of the flare [e.g., *Dryer*, 1982; *Wu*, 1982]. We now know that this interpretation is not correct, at least not in a majority of cases. Improved observations have demonstrated that many CMEs are not associated with flares, and that when a flare does occur, it often begins well after the CME is underway [e.g., *Harrison*, 1986]. In many events the timing is very close, however [e.g., *Dryer*, 1996; *Delannée, Delaboudinière, and Lamy*, 2000; *Zhang et al.*, 2000].

At the risk of confusing the reader, we note that the thermal blast model is in some ways better suited to the storage and release category. It is widely accepted that flares derive their energy from the stressed coronal magnetic field, yet early simulations, and even some recent ones, did not take this into account. A thermal pulse (temperature increase) was simply input to the corona from an unspecified source. Very recently, *Krall, Chen, and Santoro* [2000] have considered a more self-consistent scenario in which hot plasma is injected into the corona from below, and they find that the resulting evolved structures do not resemble interplanetary magnetic clouds.

5.2. Dynamo

The second class of directly driven models are called "dynamo" models, because the real-time stressing of the field involves the rapid generation of coronal magnetic flux. We can represent this in our analogue system by the sudden stretching of the spring from its initial unstressed state, as indicated in Figure 4. The external driving is in this case provided by a crank and pulley system. On the Sun, the driving can take the form of rapid displacements of magnetic footpoints in the photosphere, due, for example, to the upward propagation of magnetic stresses generated deeper in the Sun. The footpoint displacements cause an increase in magnetic flux in the direction of motion (i.e., they amplify the shear component of the field). There is an associated increase in the magnetic pressure, and the system responds by inflating [e.g., *Klimchuk*, 1990]. If the driving is rapid enough, the inflation may resemble a CME.

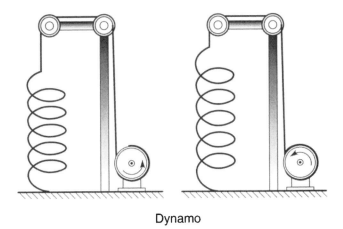

Figure 4. Dynamo model analogue. The spring is rapidly stretched open from its initially unstressed state by a crank and pulley system. On the Sun, this corresponds to a driver external to the corona (i.e., below the solar surface).

Chen [1989, 1997, 2000] and colleagues have proposed a model involving a large coronal flux rope. They prescribe a time-dependent increase in the poloidal (azimuthal) magnetic flux of the rope through a generic process they call "flux injection." Physically, this must correspond to one of three possible scenarios: 1. pre-existing coronal field lines become twisted, as described above; 2. new ring-shaped field lines rise upward into the corona from below, becoming completely detached from the photosphere once they have fully risen; or 3. new arch-shaped field lines emerge into the corona, but retain their footpoint connections to the photosphere. In a highly-conducting plasma like the corona, it is *not* physically possible for new field lines to suddenly appear; the frozen field condition requires that they come from somewhere (i.e., below the photosphere).

It is important to consider which, if any, of these three scenarios is plausible. It has been pointed out by many people [e.g., *Krall, Chen, and Santoro,* 2000] that the twisting of pre-exising field lines can produce CME-like eruptions only if the footpoint motions are at least two orders of magnitude faster than those observed. The first possibility can be safely ruled out. Ring-shaped field lines that rise into the corona from below the photosphere would be entrained by an enormous amount of mass. Such material is never seen in the corona, and furthermore, there are no obvious forces of sufficient magnitude to lift it. Possibility two is therefore also unreasonable. This leaves the third possibility of emerging loops. The subphotospheric material contained in such loops is free to slide down the legs as the loops emerge, so there is no problem with entrained mass.

Perhaps the biggest issue is the requirement of increasing flux through the photosphere. Because the loops retain their photospheric connections, the total unsigned vertical flux through the photosphere must increase in order for the poloidal flux in the corona to increase. Simplistic considerations suggest that the increase in vertical flux is at least twice that of the poloidal flux. Whether this is compatible with photospheric magnetogram observations has yet to be investigated.

6. STORAGE AND RELEASE MODELS

6.1. *Mass Loading*

We now move on to storage and release models, which represent the bulk of the more recent theoretical work on CMEs. Recall that these models are characterized by a slow buildup of magnetic stress before the eruption begins. As we have indicated, shearing the magnetic footpoints is one way to do this. Another is to load the field with mass. This is represented in Figure 5, where our spring is being compressed by a heavy weight. If the weight is shifted to the side, the spring will suddenly uncoil, releasing much of the stored energy. Models which do this on the Sun are called "mass loading" models.

The existing work on mass loading has mostly involved comparisons of pre- and post-eruption equilibrium configurations in an effort to show that the mechanism can produce an energetically favorable situation for eruption to occur [e.g., *Low and Smith,* 1993; *Chou and Charbonneau,* 1996; *Wolfson and Dlamini,* 1997, 1999; *Wolfson and Saran,* 1998; *Guo and Wu,* 1998; *Low,* 1999]. It is presumed that the pre-eruption field is metastable, so that mass loading can build up sufficient free energy before the system destresses. A CME is then possible with a large perturbation. If the system were to destabilize prematurely, something less energetic than a CME would occur.

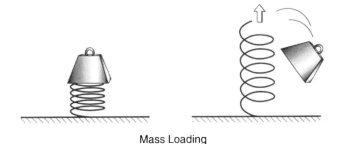

Figure 5. Mass loading model analogue. The spring is compressed by a heavy weight and explosively uncoils when the weight is shifted to the side.

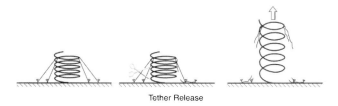

Figure 6. Tether release model analogue. The spring is held in a compressed state by rope tethers. The tethers are slowly released, one by one, until the remaining tethers break from the additional strain. The spring explosively uncoils.

The substantial mass necessary for adequate compression can come in two possible forms: higher-than-normal density coronal material distributed over a large volume, or extremely dense prominence material confined to a small volume. Prominences are structures of chromospheric temperature and density that are suspended in the low corona and that frequently erupt at the same time as CMEs. Their mass can sometimes rival that of the CME itself. It has been suggested by *Low* [1996, 1999] and others that prominences play a fundamental role in the CME process. For observational reasons, we must question whether this can be universally true: first, many CMEs occur without a prominence being present; and second, in at least some of the cases where a prominence is involved, most of the prominence mass appears to rise rather than fall, further increasing the energy requirements on the pre-eruption magnetic field [*Gilbert et al.,* 2000]. We conclude that mass loading by prominences can explain at most a subset of CMEs.

"Ordinary" coronal material may also provide the mass necessary to explain CMEs. Unfortunately, there are observational difficulties here, too. The plasma β must be larger than what we normally attribute to the corona (β approaches unity in portions of the low corona in the models of *Wolfson and Dlamini* [1997]). Furthermore, the material must be distributed in a rather specific way, with higher density material overlying lower density cavities [*Wolfson and Saran,* 1998]. Many CMEs, especially those having the classic three-part structure, originate from coronal helmet streamers that contain such cavities [e.g., *Hundhausen,* 1988, 1999], but many other CMEs come from configurations that show no obvious signs of a low-density internal region. The existence of such regions cannot be ruled out at this time because of the possibility of bright foreground or background structures obscuring the signal. It is hoped that STEREO will resolve this ambiguity.

We end our discussion of mass loading by noting that some models of this variety have been referred to as "buoyancy driven" models. Buoyancy is the tendency for low-density material to rise upward by trading places with higher density material, thus decreasing the total gravitational energy. Under these circumstances, an eruption could be completely gravitationally powered, with no need for the magnetic energy to decrease. As we have indicated, there is little observational evidence that buoyancy plays a significant role in real CMEs. Prominences, when they are present, often rise instead of fall, and cavities, when they exist, are not usually replaced by higher density coronal material until long after the eruption has occurred.

6.2. Tether Release

Mass plays no significant role in the next class of models, which we refer to as "tether release" models. As discussed in Section 3, magnetically dominated configurations generally involve a balance between the upward force of magnetic pressure and the downward force of magnetic tension. The field lines that provide the tension are sometimes called tethers. In our analogue system, they are represented by ropes that hold the compressed spring in place (Figure 6). Imagine that the tethers are slowly and systematically released, as indicated in the middle sketch. Each time a new tether is released, the strain on the remaining tethers increases. Eventually the strain becomes so great that the tethers start to break. This proceeds catastrophically, and the spring uncoils in an explosive fashion.

We note that the slightly different term "tether cutting" has been used in the past, usually in reference to the final explosive phase. We distinguish this from tether release, which is the gradual phase leading up to the explosion.

We also caution that our rope analogy for magnetic field lines must not be taken too far. When a rope is cut, two free ends are created. This can never happen to a magnetic field line, since it would imply the existence of magnetic monopoles. Instead, a pair of magnetic field lines (or two sections of a single field line that doubles back on itself) reconnect at a point of contact to produce two new field lines with different connectivity (i.e., different topology) from the original pair.

The translationally-symmetric model of *Forbes and Isenberg* [1991] is a good example of how tether release might work on the Sun [see also *Isenberg, Forbes, and Démoulin,* 1993; *Lin et al.,* 1998; *Van Tend and Kuperus,* 1978; *VanBallegooijen and Martens,* 1989]. It consists of an infinitely long flux rope and overlying arcade. Figure 7 shows a projection of the field onto a vertical plane orthogonal to the main axis. As described previously, the arcade field lines act as the tethers that prevent the flux rope from rising. Their footpoints are slowly brought together by a converging flow imposed at the photosphere. When the footpoints meet at the neutral line, they reconnect to form a new circular field line at the perimeter of the flux rope, disconnected from the photosphere (panels *a* through *c*). After enough of the arcade field lines have been converted to flux rope field lines, force balance is no longer possible, equilibrium is lost, and the flux rope abruptly rises (panel *d*).

The model of Forbes and colleagues is actually an equilibrium sequence. There is a continuous change in the field in going from *a* to *b* to *c* and then a discontinuous jump from *c* to *d*. The field in panel *d* is nonetheless a valid equilibrium, suggesting that the eruption may be aborted shortly after it begins. This is not likely to be the case in a resistive system, however. The partial eruption produces a vertical current sheet extending downward from the flux rope. Reconnection at this current sheet should allow the eruption to proceed to completion [*Forbes*, 1991; *Lin and Forbes*, 2000]. This has recently been verified in fully time-dependent 3D versions of the model that include a more realistic flux rope of finite length [*Mikić and Linker*, 1999; *Amari et al.*, 2000]. Thus, an ideal loss of equilibrium leads to fully dynamic resistive evolution.

The reconnection necessary for full eruption has an important observational consequence. It produces closed loops *underneath* the erupting flux rope. They form a new arcade that steadily grows with time as more and more reconnected flux accumulates (Figure 8). Many such arcades are observed in association with CMEs, but whether their timing, size, and location are consistent with the model predictions is not so clear [e.g., *Hundhausen*, 1988, 1999]. In many instances, the first indication of the arcade does not appear until well after the eruption has begun. Hence their common description as "post-eruption arcades" [e.g., *Klimchuk*, 1996]. Furthermore, coronagraph observations suggest that the horizontal scale of the opened field can be many times greater than that of the reconnection arcade, and this may be difficult to reconcile with the geometry of the model (although see *Delannée and Aulanier* [1999]).

6.3. Tether Straining

Our last class of model is called "tether straining." It is similar to tether release in that a slow evolution leads to a situation where the tethers can no longer withstand the upward forces and catastrophically break. In tether release, the total stress is approximately constant in time but is distributed over fewer and fewer tethers. In tether straining, the number of tethers is constant but the total stress increases. In both cases, the stress per tether steadily builds to the breaking point. Our analogue representation of tether straining is shown in Figure 9. The spring now sits on a platform that is slowly lifted. Ropes attached to the ground hold the top of the spring at a fixed height. As the spring becomes compressed, the strain on the ropes increases until they finally break, releasing the spring.

One example of tether straining is the "breakout" model shown in Plate 3 [*Antiochos*, 1998; *Antiochos, DeVore, and Klimchuk*, 1999]. It is fundamentally quadrupolar in nature, with four distinct flux systems that are color coded blue (central arcade), green (two side arcades), and red (overlying field). Shearing motions imposed near the equator stretch the inner field lines of the central arcade in an east-west direction. They are shown as thicker blue lines. This type of situation is well suited to prominence support [*Antiochos, Dahlburg, and Klimchuk*, 1994], so one might associate these thick lines with a prominence, although this has no direct bearing on the model.

The enhanced magnetic pressure associated with the shear causes the core of the central arcade to inflate. Red field lines and unsheared (thin) blue field lines are the tethers which counter this tendency. As the system becomes more and more stressed, the magnetic X-point above the central arcade distorts, closing like a pair of scissors to form a horizontal region of enhanced electric current (the red and blue field lines are oppositely directed). Once the stress is sufficiently great and the current layer sufficiently thin, the adjacent red and blue field lines reconnect to become green field lines that pull away from the X-point. With fewer tethers to resist, the central arcade bulges more, and a runaway eruption ensues.

The system shown in Plate 3 is global and perfectly symmetric. This need not be the case for the breakout mechanism to work. In fact, *Aulanier et al.* [2000] discuss a well observed active region eruption that appears to be due to breakout.

Another example of tether straining is the model of *Forbes and Priest* [1995]. The basic configuration resembles Figure 7, except that all of the photospheric flux is concentrated in two point sources (line sources in 3D), as shown in Figure 8. A converging flow is imposed, as in the earlier model, but because the footpoints never meet, no reconnection occurs and none of

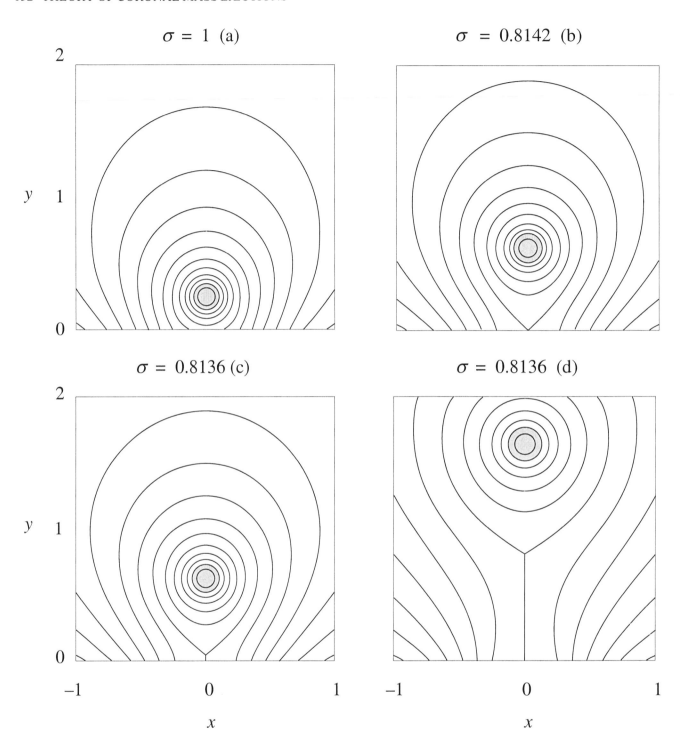

$\sigma = 1$ (a)

$\sigma = 0.8142$ (b)

$\sigma = 0.8136$ (c)

$\sigma = 0.8136$ (d)

Figure 7. Equilibrium sequence showing the evolution of a flux rope and overlying arcade when the footpoints are subjected to a converging flow. The flux rope rises slowly from a to b to c and then jumps abruptly from c to d because of an ideal loss of equilibrium. It is presumed that magnetic reconnection at the vertical current sheet in d would allow the eruption to proceed to completion. From *Isenberg, Forbes, and Démoulin* [1993].

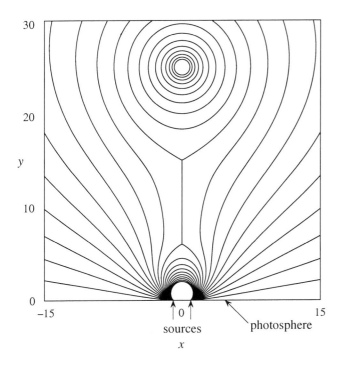

Figure 8. Magnetic reconnection at a vertical current sheet in configurations like Figure 7d produces a growing arcade of closed loops underneath the erupting flux rope. After *Lin and Forbes* [2000].

the tethers are released. Instead, there is a buildup of magnetic pressure underneath the flux rope as the point sources are brought closer together. This increases the strain on the tethers, and equilibrium is eventually lost. Once again, the flux tube abruptly rises, a vertical current sheet is formed, and reconnection allows the eruption to proceed.

One of the important differences between this and the breakout model is that this model requires reconnection underneath the erupting structure, whereas breakout requires reconnection above it. Breakout thus avoids the observational difficulties discussed earlier. It may face other observational challenges, but detailed comparison with data has only just begun. We note that breakout should be accompanied by post-eruption arcades, since it too produces a vertical current sheet, but this is only a secondary process not integral to the eruption process, and the arcades might not appear until long after the eruption is underway.

Other examples of tether straining include, but are not limited to, the sheared arcade models of *Mikić and Linker* [1994], *Linker and Mikić* [1995], *Choe and Lee* [1996], and *Amari et al.* [1996], and the flux rope mod-

els of *Wu, Guo, and Wang* [1995] and *Wu et al.* [2000]. Like *Forbes and Priest* [1995], these models ultimately require reconnection at the base of the arcade, but unlike *Forbes and Priest*, the eruption tends not to happen until the arcade has become highly distended during the quasistatic buildup phase. It is well known observationally that streamers often swell for 2-3 days leading up to a CME, but generally not to the degree of the simulations. How serious this discrepancy is must be looked at more carefully.

7. SUMMARY AND CONCLUSION

We have proposed a classification scheme for CME models that we believe identifies and differentiates the essential physical attributes of the models. Storage and release models are powered by magnetic free energy that is slowly built up in stressed configurations by footpoint shearing motions and/or the accumulation of mass. The fields could also be pre-stressed before they emerge (slowly) from below the solar surface. In marked contrast are the directly driven models, which acquire the energy for eruption in real time.

We have cited many specific models that are published in the literature, although we have not attempted a complete compilation. We have excluded some models from discussion because they are not so easy to classify within our adopted scheme. For example, *Sturrock et al.* [2000] describe a magnetic configuration that is metastable and thus susceptible to eruption with a large perturbation, but they do not address whether the configuration is achieved through a tether straining or tether release type of process. Similarly, *Moore*

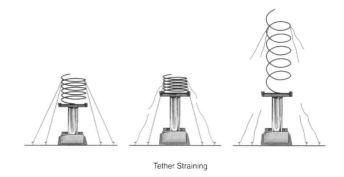

Tether Straining

Figure 9. Tether straining model analogue. The bottom of the spring is slowly raised on a moveable platform while its top is held fixed by rope tethers attached to the ground. The strain on the tethers builds to the breaking point, and the spring explosively uncoils.

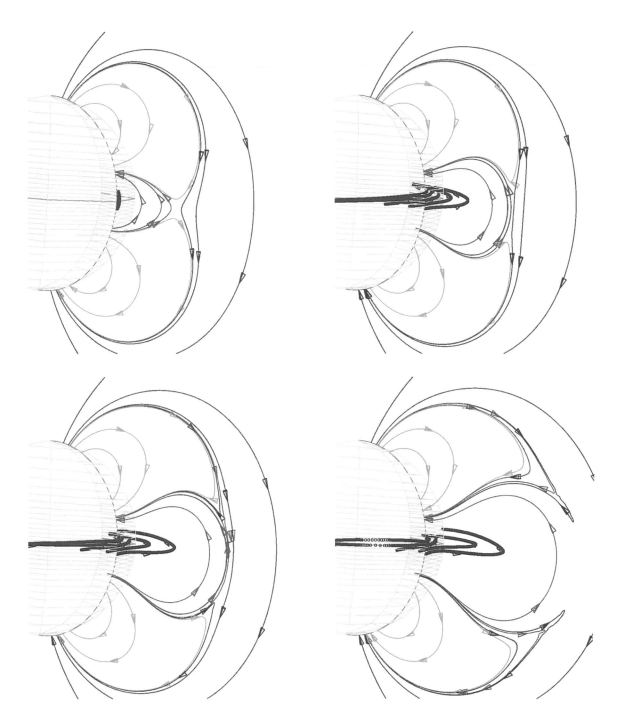

Plate 3. Breakout CME model showing the evolution of a quadrupolar system in which the inner part of the central arcade are sheared by antiparallel footpoint motions near the neutral line (equator). The field bulges slowly until the red and blue field lines begin to reconnect, and a runaway eruption ensues. From *Antiochos, DeVore, and Klimchuk* [1999].

[2000] describes a possible eruption scenario, but does not consider the slow evolution leading up to the eruption.

We have demonstrated that many models have difficulty explaining one or more aspects of the observations. In some cases the inconsistency is quite severe, whereas in others it is relatively modest. The real Sun is very complex, and all of the models are highly idealized by comparison. Whether the addition of more complexity to the models will improve their agreement with the observations remains to be seen.

It would be nice if there were a universal explanation of CMEs, but we must leave open the possibility that more than one model might be correct. The fact that prominence mass loading cannot be important in all events does not rule it out as the primary mechanism in some cases. As we have already indicated, observational evidence is accumulating for at least two different types of CMEs on the Sun.

It is clear that much more work needs to be done, both theoretically and observationally. On the theoretical front, models must be made more realistic and constructed on the basis of actual observations, such as photospheric magnetograms and velocity maps. Although challenging, this should now be possible with the recent advances in computer hardware and software. Adaptive mesh refinement (AMR) schemes, which provide high numerical resolution only where it is needed, are especially promising in this regard. Not only can AMR simulations handle more complex boundary conditions, but they can also treat magnetic reconnection in a much more realistic fashion.

On the observational front, specific and meaningful tests of the models must be devised. These should include existing data, which is far from fully exploited, as well as new data to come from missions like STEREO, *Solar-B*, *Solar Probe*, and the *Solar Dynamics Observatory*. Among the key questions that must be addressed are the following. Does the pre-eruption structure necessarily have a flux rope topology? Does it necessarily have a low-density cavity? Does eruption begin as an ideal process, in which case there is no reconnection heating in the early stages? Does reconnection occur above the main erupting structure, as in the breakout model, or below it, as in the models represented in Figures 7 and 8? Are all erupting systems multipolar, as required for breakout, or can bipolar systems also erupt? Is there a particular photospheric flow pattern (e.g., converging flow) leading up to eruption?

The realization of the practical importance of space weather has ushered in a marvelous new era for space physics. With carefully planned missions and a strong theory, modeling, and data analysis program, we can hope to make great progress in the exciting years ahead!

Acknowledgments. I gratefully acknowledge numerous discussions with Spiro Antiochos on many aspects of the CME problem. I also thank Guillaume Aulanier, James Chen, Jonathan Krall, Peter Sturrock, Richard Wolfson, and two anonymous referees for commenting on the manuscript. This work was supported by grants from NASA's Sun-Earth Connections Theory Program and Supporting Research and Technology Program.

REFERENCES

Aly, J. J., On some properties of force-free magnetic fields in infinite regions of space, *Astrophys. J., 283,* 349-362, 1984.

Aly, J. J., How much energy can be stored in a three-dimensional force-free magnetic field?, *ApJ (Lett), 375,* L61-64, 1991.

Amari, T., J. F. Luciani, J. J. Aly, and M. Tagger, Plasmoid formation in a single sheared arcade and application to coronal mass ejections, *Astron. Astrophys., 306,* 913-923, 1996.

Amari, T., J. F. Luciani, Z. Mikić, and J. Linker, A twisted flux rope model for coronal mass ejections and two-ribbon flares, Astrophys. J. (Lett.), 529, L49-L52, 2000.

Antiochos, S. K., The magnetic topology of solar eruptions, *Astrophys. J. (Lett.), 502,* L181-L184, 1998.

Antiochos, S. K., R. B. Dahlburg, and J. A. Klimchuk, The magnetic field of solar prominences, *Astrophys. J. (Lett.), 420,* L41-L44, 1994.

Antiochos, S. K., C. R. DeVore, and J. A. Klimchuk, A model for solar coronal mass ejections, *Astrophys. J., 510,* 485-493, 1999.

Aulanier, G., E. E. DeLuca, S. K. Antiochos, R. A. Mc-Mullen, and L. Golub, The topology and evolution of the Bastille day flare, *Astrophys. J.,* in press, 2000.

Burkepile, J. T., and O. C. St. Cyr, A revised and expanded catalogue of mass ejections observed by the Solar Maximum Mission coronagraph, *Tech. Note TN-369+STR,* 233 pp., National Center for Atmospheric Research, Boulder, Colorado, 1993.

Chen, J., Physics of coronal mass ejections: a new paradigm for solar eruptions, *Space Sci. Rev.,* in press, 2000.

Chen, J., Effects of toroidal forces in current loops embedded in a background plasma, *Astrophys. J., 338,* 453-470, 1989.

Chen, J., et al., Evidence of an erupting magnetic flux rope: LASCO coronal mass ejection of 1997 april 13, *ApJ (Lett), 490,* L191-194, 1997.

Chen, J., Coronal mass ejections: causes and consequences: a theoretical view, in *Coronal Mass Ejections, Geophys. Monogr. Ser.* vol. 99, edited by N. Crooker, J. Joselyn, and J. Feynman, pp. 65-81, AGU, Washington, DC, 1997.

Choe, G. S., and L. C. Lee, Evolution of solar magnetic arcades. II. Effect of resistivity and solar eruptive processes, *Astrophys. J., 472,* 372-388, 1996.

Chou, Y.-P., and P. Charbonneau, A numerical study of the pre-ejection, magnetically sheared corona as a free boundary problem, *Solar Phys., 166,* 333-369, 1996.

Delannée, C. and G. Aulanier, CME associated with transequatorial loops and a bald patch flare, *Solar Phys., 190,* 107-129, 1999.

Delannée, C., J.-P. Delaboudinière, and P. Lamy, Observation of the origin of CMEs in the low corona, *Astron. Astrophys., 355,* 725-742, 2000.

Dere, K. P, G. E. Brueckner, R. A. Howard, D. J. Michaels, and J. P. Delaboudiniere, LASCO and EIT observations of helical structure in coronal mass ejections, *Astrophys. J., 516,* 465-474, 1999.

Dryer, M., Coronal transient phenomena, *Space Sci. Rev., 33,* 233-275, 1982.

Dryer, M., Comments on the origins of coronal mass ejections, *Solar Phys., 169,* 421-429, 1996.

Forbes, T. G., Magnetic reconnection in solar flares, *Geophys. Astrophys. Fluid Dynamics, 62,* 15-36, 1991.

Forbes, T. G., A review on the genesis of coronal mass ejections, *J. Geophys. Res., 105,* 23153-23165, 2000.

Forbes, T. G., and P. A. Isenberg, A catastrophe mechanism for coronal mass ejections, *Astrophys. J., 373,* 294-307, 1991.

Forbes, T. G., and E. R. Priest, Photospheric magnetic field evolution and eruptive flares, *Astrophys. J., 446,* 377-389, 1995.

Gilbert, H. R., T. E. Holzer, J. T. Burkepile, and A. J. Hundhausen, Active and eruptive prominences and their relationship to coronal mass ejections, *Astrophys. J., 537,* 503-515, 2000.

Gosling, J. T., Coronal mass ejections and magnetic flux ropes in interplanetary space, in *Physics of Magnetic Flux Ropes, Geophys. Monogr. Ser.* vol. 58, edited by C. T. Russell, E. R. Priest, and L. C. Lee, pp. 343-364, AGU, Washington, DC, 1990.

Gosling, J. T., The solar flare myth, *J. Geophys. Res., it 98,* 18937-18949, 1993.

Guo, W. P., and S. T. Wu, A magnetohydrodynamic description of coronal helmet streamers containing a cavity, *Astrophys. J., 494,* 419-429, 1998.

Harrison, R. A., Solar coronal mass ejections and flares, *Astron. Astrophys., 162,* 283-291, 1986.

Hundhausen, A. J., The origin and propagation of coronal mass ejections, in *Proceedings of the Sixth International Solar Wind Conference, TN-306,* edited by V. J. Pizzo, T. E. Holzer, and D. G. Sime, pp. 131-214, National Center for Atmospheric Research, Boulder, Colorado, 1988.

Hundhausen, A. J., Coronal mass ejections: a summary of SMM obserations from 1980 and 1984-1989, in *The Many Faces of the Sun,* edited by K. T. Strong, J. L. R. Saba, B. M. Haisch, and J. T. Schmelz, pp. 143-200, Springer-Verlag, New York, NY, 1999.

Isenberg, P. A., T. G. Forbes, and P. Démoulin, Catastrophic evolution of a force-free flux rope: a model for eruptive flares, *Astrophys. J., 417,* 368-386, 1993.

Kahler, S. W., Solar flares and coronal mass ejections, *Ann. Rev. Astron. Astrophys., 30,* 113-141, 1992.

Klimchuk, J. A., Shear-induced inflation of coronal magnetic fields, *Astrophys. J., 354,* 745-754, 1990.

Klimchuk, J. A., and P. A. Sturrock, Three-dimensional force-free magnetic fields and flare energy buildup, *Astrophys. J., 385,* 344-353, 1992.

Klimchuk, J. A., Post-eruption arcades and 3-D magnetic reconnection, in *Magnetic Reconnection in the Solar Atmosphere, ASP Conf. Ser.* vol. 111, edited by R. D. Bentley and J. T. Mariska, pp. 319-330, Astron. Soc. of Pacific, San Francisco, California, 1996.

Krall, J., J. Chen, and R. Santoro, Drive mechanisms of erupting solar magnetic flux tubes, *Astrophys. J., 539,* 964-982, 2000.

Lin, J., and T. G. Forbes, Effects of reconnection on the coronal mass ejection process, *J. Geophys. Res., 105,* 2375-2392, 2000.

Lin, J., T. G. Forbes, P. A. Isenberg, and P. Démoulin, The effect of curvature on flux-rope models of coronal mass ejections, *Astrophys. J., 504,* 1006-1019, 1998.

Linker, J. A., and Z. Mikić, Disruption of a helmet streamer by photospheric shear, *Astrophys. J. (Lett.), 438,* L45-L48, 1995.

Low, B. C., Magnetohydrodynamic processes in the solar corona: flares, coronal mass ejections, and magnetic helicity, *Phys. Plasmas, 1,* 1684-1690, 1994.

Low, B. C., Solar activity and the corona, *Solar Phys., 167,* 217-265, 1996.

Low, B. C., Coronal mass ejections, flares, and prominences, in *Solar Wind Nine, CP471,* edited by S. R. Habbal, R. Esser, J. V. Hollweg, and P. A. Isenberg, pp. 109-114, Am. Inst. Physics, Woodbury, NY, 1999.

Low, B. C., and D. F. Smith, The free energies of partially open coronal magnetic fields, *Astrophys. J., 410,* 412-425, 1993.

MacQueen, R. M., and R. R. Fisher, The kinematics of solar inner coronal transients, *Solar Phys., 89,* 89-102, 1983.

Mikić, Z., and J. A. Linker, Disruption of coronal magnetic field arcades, *Astrophys. J., 430,* 898-912, 1994.

Mikić, Z., and J. A. Linker, Initiation of coronal mass ejections by changes in photospheric flux, *Bull. Am. Astron. Soc., 31,* 918, 1999.

Moore, R., Solar Prominence Eruption, in *Encyclopedia of Astronomy and Astrophysics,* Institute of Physics Publishing, Bristol, in press, 2000.

St. Cyr, O. C., J. T. Burkepile, A. J. Hundhausen, and A. R. Lecinski, A comparison of ground-based and spacecraft observations of coronal mass ejections from 1980-1989, *J. Geophys. Res., 104,* 12493-12506, 1999.

St. Cyr, O. C., et al., Properties of coronal mass ejections: SOHO LASCO observations from January 1996 to June 1998, *J. Geophys. Res., 105,* 18169-18185, 2000.

Sturrock, P. A., Maximum energy of semi-infinite magnetic field configurations, *Astrophys. J., 380,* 655-659, 1991.

Sturrock, P. A., M. Weber, M. S. Wheatland, and R. Wolfson, Metastable magnetic configurations and their significance for solar eruptive events, *Astrophys. J.,* in press, 2000.

Titov, V. S., and P. Démoulin, Basic topology of twisted magnetic configurations in solar flares, *Astron. Astrophys., 351,* 707-720, 1999.

VanBallegooijen, and P. C. H. Martens, Formation and eruption of solar prominences, *Astrophys. J., 343,* 971-984, 1989.

Van Tend, W., and M. Kuperus, The development of coronal electric current systems in active regions and their relation to filaments and flares, *Solar Phys., 59,* 115-127, 1978.

Vourlidas, A., P. Subramanian, K. P. Dere, and R. A. Howard, Large-angle spectrometric coronagraph measurements of the energetics of coronal mass ejections, *Astrophys. J., 534,* 456-467, 2000.

Wagner, W. J., Coronal mass ejections, *Ann. Rev. Astron. Astrophys., 22,* 267-289, 1984.

Webb, D. R., N. U. Crooker, S. P. Plunkett, and O. C. St. Cyr, The solar sources of geoeffective structures, this volume, 2000.

Wolfson, R., Energy requirements for opening the solar corona, *Astrophys. J., 419,* 382-387, 1993.

Wolfson, R., and B. Dlamini, Cross-field currents: an energy source for coronal mass ejections?, *Astrophys. J., 483,* 961-971, 1997.

Wolfson, R., and B. Dlamini, Magnetic shear and cross-field currents: roles in the evolution of the pre-coronal mass ejection corona, *Astrophys. J., 526,* 1046-1051, 1999.

Wolfson, R., and S. Saran, Energetics of coronal mass ejections: role of the streamer cavity, *Astrophys. J., 499,* 496-503, 1998.

Wu, S. T., Numerical simulation of magnetohydrodynamic shock propagation in the corona, *Space Sci. Rev., 32,* 115-129, 1982.

Wu, S. T., W. P. Guo, S. P. Plunkett, B. Schmieder, and G. M. Simnett, Coronal mass ejections (CMEs) initiation: models and observations, *J. Atm. and Solar Terrestrial Phys.,* in press, 2000.

Wu, S. T., W. P. Guo, and J. F. Wang, Dynamical evolution of a coronal streamer-bubble system, *Solar Phys., 157,* 325-348, 1995.

Zhang, J., K. P. Dere, R. A. Howard, M. R. Kundu, and S. M. White, LASCO and EIT observations of CMEs associated with flares, *Bull. AAS, 32,* 841, 2000.

J. A. Klimchuk, Code 7675JAK, Naval Research Laboratory, Washington, DC 20375-5352. (e-mail: klimchuk@bandit.nrl.navy.mil)

MHD Modeling of the Solar Corona and Inner Heliosphere: Comparison with Observations

Pete Riley, Jon Linker, Zoran Mikić, and Roberto Lionello

Science Applications International Corporation, San Diego, CA

In this paper, we describe a 3-dimensional MHD algorithm that provides a global picture of the solar corona and inner heliosphere. For computational efficiency, we split the modeling region into two distinct parts: the solar corona (1 solar radius, R_s, to 30 R_s) and the inner heliosphere (30 R_s to 1-5 AU). This combined model is driven solely by the observed line-of-sight photospheric magnetic field and thus can provide a realistic global picture of the corona and heliosphere for specific time periods of interest. We validate the model results by comparing them with a variety of observations, including white-light images and in situ measurements. We also point out the main limitations of the model and our approaches toward resolving them. Ultimately, we plan to migrate this research model into an operational tool, with the goal of predicting geo-effective solar wind conditions with up to 4 days advance warning.

1. INTRODUCTION

Space weather attempts to describe the conditions in space that affect both the Earth and its technological systems. It is a consequence of the behavior of the Sun and the interaction of the Earth's geomagnetic environment with the solar wind. Thus an integral component of ultimately predicting space weather effects is an accurate model of the Sun and its extended environment.

Global models of the solar corona [e.g., *Schatten et al.*, 1971; *Mikić and Linker*, 1996; *Zhao and Hoeksema*, 1995] and heliosphere [e.g., *Pizzo*, 1994a,b] can provide a necessary contextual basis with which to interpret and connect widely disparate data sets. This is particularly true for models driven by observed parameters [e.g., *Wang and Sheeley*, 1990; *Pizzo et al.*, 1995; *Riley et al.*, 1999; *Riley et al.*, 2000; *Arge and Pizzo*, 2000], which also have the potential for predictive capabilities.

Space Weather
Geophysical Monograph 125
Copyright 2001 by the American Geophysical Union

In this paper, we provide a brief description of our efforts to model the solar corona and inner heliosphere using a 3-D resistive MHD model. After introducing the main features and approximations of the model, we illustrate its abilities and limitations in predicting coronal structure near solar maximum by comparing simulation results with white-light coronal images. We then show how this coronal solution is used to drive the heliospheric portion of the model, and compare this solution with Ulysses and WIND in situ plasma measurements. Finally, we use the global nature of the heliospheric current sheet to illustrate how the large-scale structure of the inner heliosphere changes during the solar cycle.

2. MODELING THE SOLAR CORONA

In recent years, we have developed a 3-D, time-dependent resistive MHD model to investigate the structure of the solar corona [e.g., *Mikić et al.*, 1999; *Linker et al.*, 1999]. In its simplest form, we approximate the energy equation with a simple adiabatic energy equation and choose the polytropic index (γ) to be 1.05. This value reflects the fact that temperature does not vary significantly in the corona. While this approximation

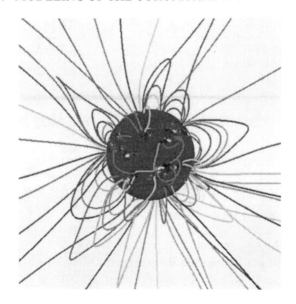

Figure 1. MHD computation of the solar corona for August, 1999. Line-of-sight photospheric magnetic field is mapped onto the solar surface. Field lines are shaded for clarity.

significantly simplifies the calculation and reduces the time necessary to complete a simulation, the resultant plasma parameters predicted by the model do not show the same degree of variability as is inferred from in situ and white light measurements. On the other hand, the structure of the magnetic field appears to be well determined [e.g., *Mikić et al.,* 1999].

Calculations are typically performed between 1 solar radius (R_s, i.e., the photosphere) and 20-30 R_s, although we have positioned the outer boundary as far a way as 1-2 AU [e.g., *Linker and Mikić,* 1997]. The spherical (r,θ,ϕ) grid permits non-uniform spacing of mesh points in both r and θ, providing better resolution of narrow structures, such as the current sheet. At the lower boundary, we specify the radial component of the magnetic field, B_r, based on the observed line-of-sight measurements of the photospheric magnetic field, and uniform, characteristic values for the plasma density and temperature. An initial estimate of the field and plasma parameters are found from a potential field model and a Parker transonic solar wind solution [*Parker,* 1963], respectively. This initial solution is advanced in time until a steady state is achieved.

The details of the algorithm used to advance the MHD equations are provided elsewhere [*Mikić and Linker,* 1994; *Lionello et al,* 1998; and *Mikić et al.,* 1999]. Here we briefly make a few remarks. In the radial (r) and meridional (θ) directions we use a finite-difference approach. In azimuth (ϕ), the derivatives are calculated pseudo-spectrally. We impose staggered meshes in r and θ

which has the effect of preserving $\nabla\cdot\mathbf{B} = 0$ to within round-off errors for the duration of the simulation. The simulations discussed here were performed on (r,θ,ϕ) grids ranging from 81×81×64 to 121×121×128.

Previously, we have predicted the structure of several total eclipses at various phases of the solar cycle: October 1995 (1 year prior to solar minimum); March 1997 and February 1998 (ascending phase); and August 1999 (1 year before the predicted maximum of solar cycle 23). These predictions are compared with actual eclipse photographs at http://haven.saic.com/corona/coronal_modeling.html. The comparisons suggest that near solar minimum the model reproduces the observed coronal structure remarkably well. Moving toward solar maximum, however, the comparisons worsen. This deterioration can probably be attributed to three main effects. First, temporal variability of the photospheric magnetic field increases toward solar maximum. The photospheric magnetic field used to drive the simulation, however, is observed at central meridian. Thus an implicit assumption within the model is that this field does not vary significantly during a solar rotation. At the very least, the field should not change appreciably over a quarter of a rotation, by which time the observed photospheric field has rotated underneath the region of the corona primarily responsible for producing the structures observed in white light images of the west limb. Second, approaching solar maximum, fine-scale structure plays an increasingly important role in the make-up of the overall features of the corona. Near solar minimum, on the other hand, the corona is dominated by large-scale structures that are well resolved by the limited spatial resolution of the simulation. Third, it may be that the twist in the magnetic field, which is currently not measured routinely (since it requires the use of a vector magnetogram) has not been incorporated into the solution.

To illustrate some of the features of the model, in Figure 1 we show a selection of magnetic field lines based on our prediction of the August 1999 total eclipse [*Mikić et al.,* 2000]. The input photospheric magnetic field (derived from line-of-sight magnetic field at central meridian from the National Solar Observatory at Kitt Peak) has been projected onto the photosphere and the field lines are shaded solely for clarity. Plasma flow occurs along the open field lines: where the field lines are closed, the flow is stagnant. The coronal structure inferred from this picture is substantially different from solutions obtained near the minimum of the solar activity cycle [e.g., *Linker et al.,* 1999]. The organized pattern of a low-latitude streamer belt encircling the Sun is replaced by a complex mixture of field topologies. There appears to be little, if any semblance of an axis of symmetry.

June 25, 1999   June 28, 1999   July 2, 1999   July 6, 1999

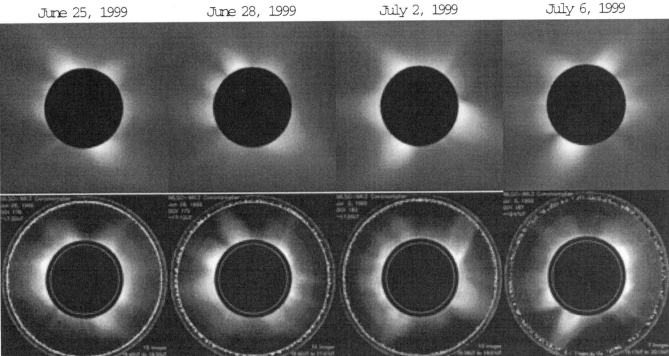

Figure 2. Comparison of the predicted polarization brightness (top) from our MHD simulation with daily white-light coronal images from the Mauna Loa Solar Observatory's Mark III K-coronameter (bottom) during June/July, 1999.

By integrating the model densities with a suitable scattering function, it is possible to obtain simulated polarized brightness images [e.g., *Linker et al.,* 1999]. In Figure 2, we compare a series of such images with daily white-light coronal images from the Mauna Loa Solar Observatory's Mark III K-coronameter. The model reproduces many aspects of the observations, however, there is not a one-to-one correspondence between the predicted and observed streamers. The coronal hole (dark region) off the south-east limb is the most prominent feature that appears in both model and simulation. The limited spatial resolution of the simulation precludes us from obtaining the fine scale structure that is apparent in the observations.

A potentially significant oversimplification of the model is the use of the polytropic approximation in the energy equation. This leads to velocities that are lower than observed in interplanetary space, and underestimates density and temperature variations. To resolve this, we have incorporated the effects of parallel thermal conduction, radiation loss, parameterized-coronal heating, and a self-consistent model for Alfven wave acceleration into the model [*Mikić et al.* 1999]. Figure 3 illustrates the improvement in flow speed that is achieved with these modifications. The left panel shows a meridional view from the solar surface to 30 R_s. This profile matches – in

an average sense – speeds observed by Ulysses [e.g., *Riley et al.,* 1997]: the flow about the equator is slow (~350 km/s) and (not shown) dense. Above the poles, the flow is fast (~750 km/s) and tenuous. The right panel is an expanded view near the Sun showing field lines superimposed on the flow and highlights the super-radial expansion of the flow close to the Sun.

3. MODELING THE INNER HELIOSPHERE

It is convenient to separate the region between the solar photosphere and the Earth into two parts. We distinguish between the more complex "coronal" region, which includes the region from the photosphere up to 20-30 R_s, and the "heliospheric" region, which covers the region of space between 30 R_s and 1-5 AU. In the latter, the flow is everywhere supersonic, gravity can be neglected (although it is included for completeness), and the energy equation can be reasonably approximated by a polytropic relationship. Thus the time step required to advance the solution in the heliospheric model is considerably larger than the time step required for the coronal solution. Computationally then, it would be inefficient to advance the heliospheric portion of the simulation at the coronal time step.

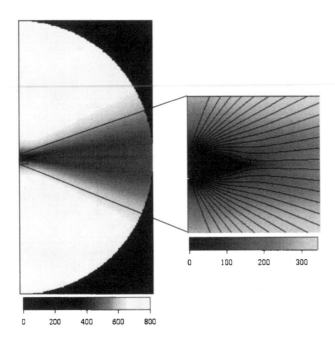

Figure 3. (left) Meridional cross section (r,θ) of a two-dimensional MHD calculation incorporating improved thermodynamic treatment of the energy equation. The outer boundary of the simulation is located at $r=30\ R_s$. (right) Expanded view of region near Sun illustrating the super-radial expansion of the field lines.

Ultimately, our goal is to use the output of the coronal solution to provide the inner boundary condition of the heliospheric model. At present, however, as we have noted, the plasma speed predicted by the polytropic coronal model is not sufficiently high to drive the heliospheric model. The recent addition of a more complex thermodynamic treatment of the energy equation promises to yield realistic plasma parameters; however, this refinement is still under development and requires further validation. Thus both as an interim solution, and as a practical and computationally inexpensive method of defining boundary conditions for the heliospheric solution, we have developed a technique for deducing speed, density, and temperature, based on the magnetic topology of the coronal solution (which is probably the most accurately determined property of the coronal model).

We make a number of additional assumptions. First, we neglect the effect of pickup ions, which are thought to dominate the internal energy of the solar wind beyond 6-10 AU [*Axford,* 1972]. Thus we limit our modeling region to ≤ 5 AU. Second, we neglect the effects of differential rotation, which may play a role in connecting high latitude field lines near the Sun with lower-latitude interaction

regions much further away [*Fisk,* 1996]. We assume that the inner boundary rotates rigidly with a period of 25.38 days. Third, although the MHD model is time-dependent, we assume that the flow at the inner boundary is time-stationary. Thus the flow is "corotating", so that rotating spatial variations are responsible for the generation of dynamic phenomena in the solution. Fourth, we assume a polytropic approximation to the energy equation. This is a much more reasonable assumption for the solar wind, where a polytropic relationship between pressure and density is observed to hold, with $\gamma = 1.5$ for protons (e.g., *Totten et al.* [1996]; *Feldman et al.* [1998]).

To determine the inner (or lower) boundary conditions for the heliospheric solution, we utilize the magnetic field topology of the coronal solution. We assume that within coronal holes (i.e., away from the boundary between open and closed magnetic field lines) the flow is fast. At the boundary between open and closed field lines, the flow is slow. Over a relatively short distance, we smoothly raise the flow speed to match the fast coronal hole flow. Although this approach is somewhat ad hoc, it is based on the commonly held views that slow flow originates from the boundary between open and closed field lines and fast flow originates in coronal holes [*Wang & Sheeley,* 1990]. Once this speed map is determined in the photosphere, it is mapped outward along field lines to generate the inner boundary of the heliospheric model at 30 R_s.

We illustrate this technique in Figure 4 for the time period August 10 to September 8, 1996. This interval coincided with the campaign known as "Whole Sun Month," (WSM) occurring just four months after the termination (minimum) of solar cycle 22, and has been the focus of considerable research (see papers in the special issue of *Journal of Geophysical Research,* May, 1999). In panel (a) we have traced coronal field lines to deduce whether they are open or closed. Regions in which the field lines are closed are shaded gray, and open field line regions are shaded black. In panel (b), we have shaded the photosphere according to the algorithm described above. Note that the central band, while colored with the lowest speed, does not contribute to the inner boundary conditions of the heliospheric model since the field lines here are all closed.

Using this recipe, we derive the flow speed at 30 R_s. By imposing momentum flux balance at the inner boundary, we derive the plasma density. Theoretically, there is little justification for this assumption; however, Ulysses observations suggest that momentum flux is roughly conserved [*Riley et al.,* 1997]. To derive the temperature we impose pressure balance. This is again, an ad hoc assumption, however, any significant pressure gradients would be quickly minimized by the flow. To

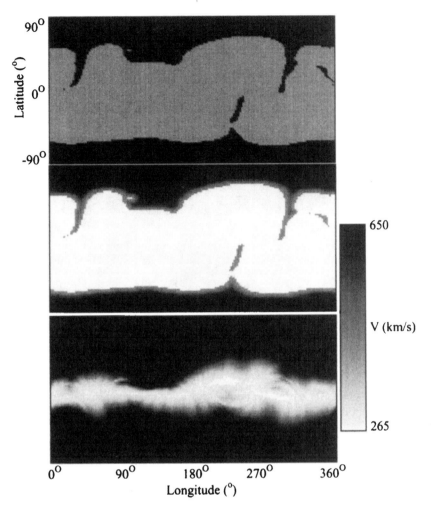

Figure 4. (a) Topology of the magnetic field in the photosphere: Black indicates open field lines and gray indicates closed field lines. (b) Imposed speed profile in the photosphere which is mapped along field lines to the inner boundary of the heliospheric solution at 30 R_s. (c) Speed profile at the inner boundary of the heliospheric calculation after mapping (b) along field lines.

complete the necessary inputs, we use the radial component of the magnetic field, B_r, directly from the coronal solution.

In Figure 5, we illustrate the structure of the heliosphere using the flow speed shown in Figure 4c. The heliospheric current sheet (inferred from the iso-surface $B_r=0$) is displayed out to 5 AU. A meridional slice of the radial velocity is shown at an arbitrary longitude. White corresponds to slowest speeds (~350 km/s) and black corresponds to fastest speeds (~750 km/s). Superimposed is a selection of interplanetary magnetic field lines, as well as the trajectories of the WIND and Ulysses spacecraft. The structure portrayed by Figure 5 fits well with the general picture deduced from solar and interplanetary observations during this time period [e.g., *Riley et al,*. 1999; *Linker et al,* 1999]. In particular, the streamer belt showed little (<10°) inclination relative to the

heliographic equator, but was deformed northward at longitudes of ~250° - 300° (at the Sun) due to the presence of an active region. Note that in spite of the low inclination of the current sheet, by 4 AU it has developed considerable structure, including a fold back on itself.

At the time of WSM, Ulysses was returning to lower latitudes and was located at ~28° North heliographic latitude, at a distance of ~4.3 AU, and on the opposite side of the Sun from Earth. Thus the measurements of the line-of-sight photospheric magnetic field used to drive the solution were always separated by ~180° from the actual solar wind that Ulysses observed. In Figure 6 we compare plasma observations by Ulysses of solar wind speed, scaled number density, and temperature with simulation. The profile at Ulysses consisted of a simple, single stream pattern. The model appears to reproduce the essential features of the large-scale variations, i.e., the interaction

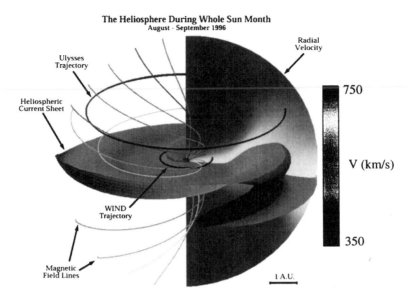

The Heliosphere During Whole Sun Month
August - September 1996

Figure 5. Heliospheric solution for August 22 – September 18, 1996. The heliospheric current sheet (inferred from the iso-surface B_r=0) is displayed out to 5 AU. The central sphere marks the inner boundary at 30 R_s. A meridional slice of the radial velocity is shown at an arbitrary longitude. Superimposed is a selection of interplanetary magnetic field lines. The trajectories of the WIND and Ulysses spacecraft are also shown (in a frame co-rotating with the Sun).

region and the expansion wave; however, the interaction region is not as steep as the observations indicate, and the model does not reproduce fluctuations on scales less than a few days.

In Figure 7, we make an analogous comparison with WIND plasma data. WIND was located in the ecliptic plane at 1 AU and observed a more complex pattern of variations that are only partially reproduced by the MHD solution. The observations indicate the presence of two fast (>600 km/s) streams, of which only one is obviously found in the simulations (on day 255). The remaining fast stream (on day 243) may be related to the broader speed increase in the simulations on day 240; however, such an association is, at best, tentative. There are other minor perturbations in the model; however, none of them can be reliably matched with WIND observations.

Using these simulations, we can investigate the evolution of the large-scale structure of the inner heliosphere during the course of a solar cycle. The heliospheric current sheet (HCS) represents a convenient feature to follow: being passive, it acts as a tracer for the macroscopic structure of the heliosphere [e.g., *Riley et al.,* 1996]. In Figure 8, we use the simulation results described here to illustrate the shape of the HCS at 3 epochs of the solar cycle. The first panel (a) shows the HCS near solar minimum. This is the same surface as shown in Figure 5 but viewed from a different position. Note that the single fold is not radially symmetric, with its outer edge being sharper than its inner edge: faster solar wind overtaking

slower wind ahead accelerates it and steepens the HCS profile at the outer edge, whereas at the inner edge of the deformation, the HCS is "stretched out" as faster wind outruns slower wind, creating a rarefaction region. This picture differs from those produced by kinematic models of the HCS, which assume that the speed of the plasma remains constant as plasma moves out from the Sun [e.g., *Sanderson et al.,* 1999]. The second panel (b) shows the HCS during the declining phase of the solar cycle (January 27 – February 23, 1995, corresponding to Carrington Rotation 1892). Two folds dominate the HCS profile and they are in qualitative agreement with the picture inferred by *Smith et al.* [1995] based on in situ crossings of the HCS by Ulysses. In the final panel (c), we present the HCS approaching solar maximum (March 7 – April 4, 1999, corresponding to Carrington Rotation 1947). The HCS is again dominated by two folds that extend up to ~55°. In contrast to solar minimum, however, the inner and outer edges of the folds are more symmetric, since the surrounding wind speed, while still variable, does not consist of the organized flow patterns characteristic of solar minimum that is capable of producing strong interaction regions.

4. CLOSING REMARKS

Global MHD models provide a useful contextual basis with which to interpret coronal and solar wind observations. The coronal model reproduces many of the

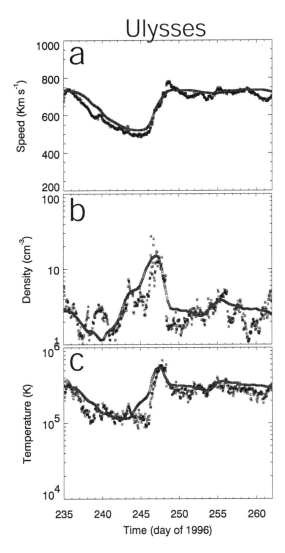

Figure 6. Comparison of plasma speed, density, and temperature observations at Ulysses with simulation results (smooth profiles) for the same time period as Figure 5.

the magnetic field topology of the coronal solution appears to yield useful results.

Currently, the model produces a "rotationally averaged" picture of the corona and heliosphere, in the sense that the observations of the photospheric magnetic field that drive the model are taken at central meridian. To improve our capabilities would require observations centered at other longitudes. A mission like STEREO, but which includes magnetographs, could significantly improve our abilities to predict time-dependent global structure. Nevertheless, by virtue of the fact that the observations are made at central meridian, our results are most accurate along the Sun-Earth line. Thus in the near

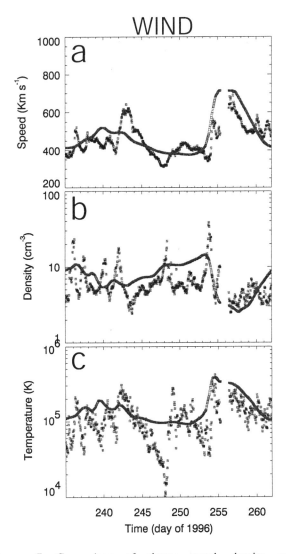

Figure 7. Comparison of plasma speed, density, and temperature observations at WIND with simulation results (smooth profiles) for the same time period as Figure 5.

features seen in white-light observations, but performs best when photospheric structure does not change appreciably during the course of a solar rotation. The heliospheric model also reproduces the large-scale features of the inner heliosphere, as inferred from spacecraft data. Again, however, the model becomes less accurate as temporal variations begin to dominate.

Ultimately, we plan to drive the inner boundary of the heliospheric solution directly with output from the time-dependent coronal solution. However, this requires further validation of the model that incorporates improved thermodynamics. As an interim solution, our ad hoc specification of the inner boundary conditions based on

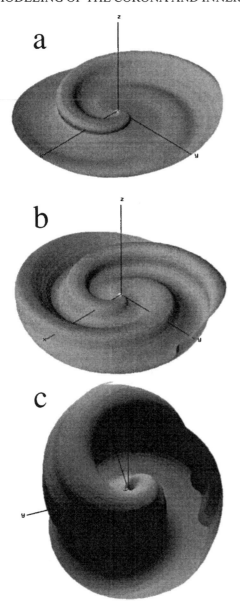

Figure 8. The heliospheric current sheet at three epochs of the solar cycle: (a) solar minimum; (b) declining phase; and (c) approaching solar maximum.

future, with further development, we believe the model has significant potential to become an operational tool, providing a reliable prediction of solar wind conditions in the vicinity of Earth up to 4 days in advance.

Acknowledgments. The authors gratefully acknowledge the support of the National Aeronautics and Space Administration (Sun-Earth Connections Theory Program, Supporting Research and Technology Programs, and SOHO Guest Investigator Program) and the National Science Foundation (Space Weather Program ATM9613834) in undertaking this study. We also thank National Science Foundation at the San Diego Supercomputer Center and the National Energy Research Supercomputer Center for providing computational support.

REFERENCES

Arge, C. N., and V. J. Pizzo, Improvement in the prediction of solar wind conditions using near-real time solar magnetic field updates, 105, 10,465, 2000.

Axford, W. I., Interaction of the solar wind with the interstellar medium, in *Solar Wind,* edited by C. P. Sonett, P. J. Coleman, and J. M. Wilcox, *NASA Spec. Publ., NASA SP-308, 609,* 1972.

Feldman, W. C., B. L. Barraclough, J. T. Gosling, D. J. McComas, P. Riley, B. E. Goldstein, and A. Balogh, Ion energy equation for the high--speed solar wind: Ulysses observations, *J. Geophys. Res., 103,* 14547, 1998.

Fisk, L. A., Motion of the footpoints of heliospheric magnetic field lines at the Sun: Implications for recurrent energetic particle events at high heliographic latitudes, *J. Geophys. Res., 101,* 15547, 1996.

Linker, J. A., and Z. Mikić, Extending coronal models to Earth orbit, in *Coronal mass ejections, Geophys. Monograph 99,* edited by N. Crooker, J. Joselyn, and J. Feynmann, American Geophysical Union, Washington, DC, 269, 1997.

Linker, J. A., Z. Mikić, D. A. Bisecker, R. J. Forsyth, S. E. Gibson, A. J. Lazarus, A. Lecinski, P. Riley, A. Szabo, and B. J. Thompson, Magnetohydrodynamic modeling of the solar corona during whole sun month, *J. Geophys. Res. 104,* 9809, 1999.

Lionello, R., Z. Miki\'c, and J. A. Linker, Magnetohydrodynamics of solar coronal plasmas in cylindrical geometry, *J. Comput. Phys.,* 140, 172, 1998.

Mikić, Z., and J. A. Linker, Disruption of coronal magnetic field arcades, *Astrophys. J., 430,* 898, 1994.

Mikić, Z., and J. A. Linker, The large-scale structure of the solar corona and inner heliosphere, in *Solar Wind Eight,* edited by D. Winderhalter et al., *AIP Conf. Proc., 382,* 104, 1996.

Mikić, Z., J. A. Linker, D. D. Schnack, R. Lionello, and A. Tarditi, Magnetohydrodynamic modeling of the global solar corona, *Phys. Plasmas, 6,* 2217, 1999.

Mikić, Z., J. A. Linker, P. Riley, and R. Lionello, Predicting the structure of the solar corona during the 11 August 1999 total solar eclipse, in *The last Solar Eclipse of the Millenium, ASP Conference Series, 205,* 162, 2000.

Parker, E. N., *Interplanetary Dynamical Processes,* Wiley-Interscience, New York, 1963.

Pizzo, V. J., Global, quasi-steady dynamics of the distant solar wind, 1, Origin of north-south flows in the outer heliosphere, *J. Geophys.Res., 99,* 4173, 1994a.

Pizzo, V. J., Global, quasi-steady dynamics of the distant solar wind, 2, Deformation of the heliospheric current sheet, *J. Geophys. Res., 99,* 4185, 1994b.

Pizzo, V. J., D. S. Intriligator, and G. L. Siscoe, Radial alignment simulation of solar wind streams observed by Pioneers 10 and 11 in 1974, *J. Geophys. Res., 100,* 12,251, 1995.

Riley, P., J. T. Gosling, L. A. Weiss, and V. J. Pizzo, The tilts of corotating interaction regions at mid-heliographic latitudes, *J. Geophys. Res., 101,* 24349, 1996.

Riley, P., S. J. Bame, B. L. Barraclough, W. C. Feldman, J. T. Gosling, G. W. Hoogeveen, D. J. McComas, J. L. Phillips, B. E. Goldstein, and M. Neugebauer, Ulysses solar wind plasma observations at high latitudes, *Adv. Space Res., 20,* 15, 1997.

Riley, P., J. T. Gosling, D. J. McComas, V. J. Pizzo, J. G. Luhmann, D. Biesecker, R. J. Forsyth, J. T. Hoeksema, A. Lecinski, and B. J. Thompson, Relationship between Ulysses plasma observations and solar observations during the Whole Sun Month campaign, *J. Geophys. Res.,* 104, 9871, 1999.

Riley, P., J. A. Linker, and Z. Mikić, Solar cycle variations and the structure of the heliosphere: MHD simulations, submitted to *J. Geophys. Res.,* 2000.

Sanderson, T. R., D. Lario, M. Maksimovic, R. G. Marsden, C. Tranquille, A. Balogh, R. J. Forsyth, and B. E. Goldstein, Current sheet control of recurrent particle increase at 4-5 AU, *Geophys. Res. Lett., 26,* 1785, 1999.

Schatten, K. H., Current sheet model of the solar corona, *Cosmic Electrodyn., 2,* 232, 1971.

Smith, E. J., A. Balogh, M. E. Burton, G. Erdös, and R. J. Forsyth, Results of the Ulysses fast latitude scan: Magnetic field observations, *Geophys. Res. Lett., 22,* 3325, 1995.

Totten, T. L., J. W. Freeman, and S. Arya, Application of the empirically derived polytropic index for the solar--wind to models of solar-wind propagation, *J. Geophys. Res., 101,* 15,629, 1996.

Wang, Y. -M., and N. R. Sheeley, Jr., Solar wind speed and coronal flux-tube expansion, *Astrophys. J., 355,* 726, 1990.

Zhao, X. P., and J. T. Hoeksema, Prediction of the interplanetary magnetic field strength, *J. Geophys. Res.,* 100, 19, 1995.

R. Lionello, J. A. Linker, Z. Mikić, and P. Riley, Science Applications international Corporation, 10260 Campus Point Dr., MS W2M, San Diego, CA 92121. (e-mail: lionel@iris023.saic.com; linker@iris023.saic.com; Mikić@iris023.saic.com; uk2@haven.saic.com)

From Sun to Earth:
Multiscale MHD Simulations of Space Weather

Tamas I. Gombosi[1], Darren L. DeZeeuw[1], Clinton P. T. Groth[2],
Kenneth G. Powell[1], C. Robert Clauer[1], and Paul Song[1]

There is an increasing need to develop physics-based, high performance models of the Sun-Earth system — from the solar surface to the Earth's upper atmosphere — which can operate faster than real time and which can provide reliable predictions of the near Earth space environment based upon solar observations and upstream solar wind measurements. Taking advantage of the advent of massively parallel computers, sophisticated solution-adaptive techniques, and recent fundamental advances in basic numerical methods we have developed a high performance, multiscale MHD code capable of resolving many of the critical processes in the Sun-Earth system which range over more than 9 orders of magnitude. We report on the first comprehensive numerical simulation of a "synthetic" space weather event, starting with the generation of a CME and subsequently following this transient solar wind disturbance as it evolves into a magnetic cloud and travels through interplanetary space towards Earth where its interaction with the terrestrial magnetosphere–ionosphere system is also predicted as part of the simulation.

1. INTRODUCTION

"Space Weather" has been used to refer to the conditions on the Sun and in the solar wind, magnetosphere, ionosphere, and thermosphere that can influence the performance and reliability of space-borne and ground based technological systems or can endanger human life or health. A major goal of space weather research is to unify our physical understanding of the Sun-Earth system into a more comprehensive mathematical framework that can predict the deterministic properties of space weather.

Global computational models based on first principles represent a very important component of efforts to under-

---

[1] The University of Michigan, Ann Arbor, Michigan, USA

[2] University of Toronto, Toronto, Ontario, Canada

stand the intricate processes coupling the Sun to the geospace environment. The hope for such models is that they will eventually fill the gaps left by measurements, extending the spatially and temporally limited observational database into a self-consistent global understanding of our space environment. Presently, and in the foreseeable future, magnetohydrodynamic (MHD) models are the only models that can span the enormous distances present in the magnetosphere. However, it should not be forgotten that even generalized MHD equations are only a relatively low-order approximation to more complete physics; they provide only a simplified description of natural phenomena in space plasmas.

This paper describes our multiscale MHD model and its application to the Sun-Earth system. The code itself is described in Section 2; the application of the code to simulate a "synthetic" space weather event is described in section 3.

2. SIMULATION CODE

2.1. Description of BATS-R-US

The BATS-R-US code solves the governing equations of magnetohydrodynamics. All terms describing deviations

Space Weather
Geophysical Monograph 125

from ideal MHD are included through appropriate source terms. A detailed description of the code can be found in *Powell et al.*, [1999], *Groth et al.*, [1999a] and *De Zeeuw et al.*, [2000]. The code uses a limited reconstruction that ensures second-order accuracy away from discontinuities, while simultaneously providing the stability that ensures nonoscillatory solutions. In addition, the code employs two accurate approximate Riemann solvers: *Roe*'s [1981] scheme [*Powell et al.*, 1999] and *Linde*'s [1998] solver. The resulting scheme solves for the hydrodynamic and electromagnetic effects in a tightly coupled manner, yielding a scheme that works equally well across a range of several orders of magnitude in plasma β.

The basic data structure used in the BATS-R-US approach is that of adaptive blocks [*Stout et al.*, 1997, *Powell et al.*, 1999]. Adaptive blocks partition space into regions, each of which is a regular Cartesian grid of cells, called a block. If the region needs to be refined, then the block is replaced by 8 child subblocks (one for each octant of the parent block), each of which is a Cartesian grid of cells containing the same number of cells as the parent block. If coarsening is needed, then the 8 children are replaced by their parent. The blocks in the grid, at their various levels of refinement, are stored in a tree-like data structure.

2.2. Non-Ideal MHD Terms

Ideal MHD is based on the asssumption that the plasma is in local thermodynamic equilibrium. An immediate consequence of this assumption is that no dissipative process can be described by ideal MHD. Even non-ideal MHD equations are only a relatively low order approximation to more complete physics. These equations replace a detailed description of the micro-physics with transport coefficients such as diffusion coefficients, resistivity, and viscosity. In high Reynolds number flows (such as the magnetosphere) viscous effects are usually negligible. However, resistivity and diffusion play important roles in some critical regions of the magnetosphere, and therefore these effects must be included in realistic simulations.

Spatial discretization (an essential element of all numerical solutions) results in numerical dissipation which can mimic the appropriate physical processes. However, numerical resistivity and diffusion in high-resolution modern codes are quite low, and the inclusion of physics-based empirical dissipation coefficients is important. Resistivity is of particular importance, because it controls the rate of magnetic reconnection which is a very important element of the solar wind interaction and global magnetospheric configuration.

Non-ideal MHD processes are included in our MHD model through appropriate terms in the source vector.

2.3. Simulation Domain

The computational domain used in the space weather calculation is a heliocentric rectangular box defined by $-32\,R_\odot \leq x \leq 224\,R_\odot$, $-192\,R_\odot \leq y \leq 192\,R_\odot$, $-192\,R_\odot \leq z \leq 192\,R_\odot$. The adapted computational grid for the initial solution of the solar wind (which is a steady-state solution in the corotating frame) consists of 15,768 self-similar $4 \times 4 \times 4$ blocks and 1,009,152 cells with 8 refinement levels and a minimum cell size at the solar surface of 1/16 $R_\odot$. During the time-dependent calculation of the CME, the grid dynamically adapted to the varying solution according to the refinement criteria and the size of the computational mesh varied from under 800,000 cells to in excess of 2,000,000 cells. The length of the entire simulation was 120 hours starting from the initiation of the CME ($t = 0$).

The procedure for prescribing boundary conditions at the solar surface is dependent on local flow conditions. Plasma can freely leave the reservoir, but no "backflow" is allowed. In addition, no deviation is allowed form the intrinsic magnetic field at the interface. At the outer boundaries of the rectangular solution domain, the solar wind flow is essentially super-fast (and hence super-Alfvénic). Simple zero-gradient or constant extrapolation boundary conditions are therefore appropriate and are used to specify the plasma properties at the outer boundary. A detailed description of the boundary conditions can be found in *Groth et al.*, [2000].

The resolution of the initial computational grid at Earth is approximately $10\,R_\odot$, but during the passage of the interplanetary transient this resolution increased by about a factor of four. The embedded magnetosphere is simulated in a moving computational box defined by $-384\,R_E \leq x \leq 128\,R_E$, $-128\,R_E \leq y \leq 128\,R_E$, $-128\,R_E \leq z \leq 128\,R_E$ with 8 levels of refinement, 2004 self-similar blocks of $4 \times 4 \times 4$ cells and 128,256 computational cells. The magnetosphere is "turned on" for a total of 27 hours starting at $t = 70$ h.

The inner boundary of the magnetosphere simulation is a sphere at 3 R_E. This sphere is connected to the height-integrated electrostatic ionosphere.

2.4. Magnetosphere-Ionosphere Coupling

The coupling between the magnetosphere and the ionosphere is taken into account through an electrostatic ionosphere model. It is assumed that magnetospheric field-aligned currents can penetrate into the height-integrated electrostatic ionosphere, where the current closure is calculated with the help of Ohm's law using a height-integrated conductivity (conductance) tensor. The model directly relates the normal component of the magnetospheric currents entering the ionosphere through the magnetosphere-ionosphere interface to the electrostatic potential distribution in the ionosphere. Once the ionospheric electrostatic potential is derived, one can calculate the ionospheric convection velocity, which is used as boundary condition for the magnetosphere simulation at the magnetosphere-ionosphere interface.

2.5. Applications of BATS-R-US

BATS-R-US has been extensively applied to global numerical simulations of the Sun-Earth system *Groth et al.*, [1999a], *Gombosi et al.*, [2000c], *Groth et al.* [2000], the coupled terrestrial magnetosphere-ionosphere *Gombosi et al.*, [1998], *Gombosi et al.* [2000a], *Gombosi et al.*, [2000c], *Song et al.*, [1999], *Song et al.*, [2000], and the interaction of the heliosphere with the interstellar medium *Linde et al.*, [1998].

3. A SYNTHETIC SPACE WEATHER EVENT

3.1. The Inner Heliosphere Near Solar Minimum

An initial solution representative of the state of the solar wind before the initiation of the CME is required as a starting point for this simulation of a space weather event. A "pre-event" solution has been developed for these purposes using the BATS-R-US simulation code which attempts to reproduce many of the observed global features of solar corona and the inner heliosphere for conditions near solar minimum.

Global computational models based on first principles mathematical descriptions of the physics represent a very important component of efforts to understand the initiation, structure, and evolution of CMEs. Recent examples of the application of MHD models to the study of coronal and solar wind plasma flows include the studies by *Steinolfson* [1994], *Mikić and Linker* [1994], *Suess et al.*, [1996], *Wang et al.*, [1998], *Guo and Wu* [1998], *Lionello et al.*, [1998], *Dryer* [1998], and *Odstrčil and Pizzo* [1999].

In the first part of our simulation we attempted to reproduce the three-dimensional bulk features of the solar wind for solar minimum conditions. We applied the BATS-R-US simulation code to solve the three-dimensionsional ideal MHD equations from the solar surface to beyond Earth's orbit. BATS-R-US has the capability to include the effects of solar rotation, solar gravity, a multipole intrinsic magnetic field (up to octupole), as well as including user-defined volumetric sources of mass, momentum and energy. This feature is very important in the present simulation, since BATS-R-US solves the full MHD energy equation and uses $\gamma = 5/3$ for the ratio of specific heats (since ideal MHD assumes perfect gases).

For the purposes of the present simulation, it is assumed that at the top of transition region, at a radial distance of 1

$R_\odot$, the base of the solar corona is a large rigidly rotating reservoir of hot, stationary plasma with an embedded magnetic multipole field. The plasma temperature (the sum of the ion and electron temperatures) in the reservoir is taken to be $T_\odot = 2.85 \times 10^6$ K and the plasma density is assumed to be $n_\odot = 1.5 \times 10^8$ cm^{-3}.

The intrinsic solar magnetic field at the solar surface is defined in terms of a multipole expansion that includes terms up to the octupole moment. In the present simulation there was no quadrupole moment, while the octupole and dipole were tilted in the (x, z) plane of the corotating coordinate system. The magnetic axis tilt angle was $-15°$ and the solar magnetic field was azimuthally symmetric about the magnetic axis. The surface field strength was 8.4 G at the magnetic poles and 2 G at the solar magnetic equator.

In our simulations we used a volumetric heating function to mimic the combined effects of energy absorption above the transition region, heat conduction (which is not included in ideal MHD) and radiative losses. Presently, the physical understanding of these processes is quite limited, therefore one has considerable freedom in choosing the volumetric heating function. Our approach is to reverse the problem and see whether one can find a heating function which results in reasonable plasma parameters near the Sun and around 1 AU. We adopted a heating function that includes both local energy depositions and losses, thus mimicking the real situation in the lower corona. Specifically, the heating function was assumed to be proportional to $(T_0 - T)$, where T_0 is a prespecified "target" temperature. We have chosen $T_0 = 1.75T_\odot$ inside the coronal hole and $T_0 = T_\odot$ outside (the coronal hole boundary was at $72.5°$ latitude). We note that our heating function has a sharp gradient at the edge of the coronal hole, however, the resolution of the computational grid (about $5°$ near the Sun) limits the sharpness of the transition. Finally, the heating scale-height slowly varies from about 4.5 $R_\odot$ near the equator to about 9 $R_\odot$ at the poles. Overall, the motivation for this choice of heating function was to reproduce many of the observed global plasma properties of the solar wind. A detailed description of the heating function is given in *Groth et al.*, [2000].

Plate 1 shows a 3D representation of the initial solution for the pre-event solar wind near the Sun. The solution is shown in the meridional and equatorial cuts (the (x, z) and (x, y) planes, where the z axis is along the rotation axis, while the magnetic axis is in the (x, z) plane). The color code represents the logarithm of the magnitude of the magnetic field (in units of G) in the two planes. Solid lines are magnetic field lines: magenta denotes the last closed field lines, red is open field lines expanding to the interplanetary medium just above the heliospheric current sheet, and finally, white lines show open magnetic field lines in the (y, z) plane.

The narrow dark blue region in Plate 1 is the beginning of the heliospheric current sheet. It originates around 4 $R_\odot$ where the equatorial portion of the closed magnetic field lines become highly stretched, and extends throughout the entire solution. The current sheet is warped and tilted due to the combined effect of magnetic tilt and rigid solar rotation.

The solution, which is dictated by the complex balance of pressure, magnetic, gravitational, and inertial forces, has regions of open and closed magnetic field lines and leads to the formation of a "helmet" streamer magnetic streamer configuration with associated neutral point and equatorial current sheet. The solution correctly mimics some of the dual state features of the solar wind. It produces fast solar wind (~ 800 km/s) above $\sim 30°$ heliolatitude, slow (~ 400 km/s) solar wind near the solar equator, and provides reasonable values for the solar wind temperature and density and interplanetary magnetic field at 1 AU.

Overall, our initial condition represents a very reasonable description of three-dimensional inner heliosphere for solar minimum conditions. We used physically reasonable input parameters near the Sun and were able to obtain a solution out to 1 AU which is in good overall agreement with average observed solar wind conditions.

3.2. Simulation of a CME

Coronal mass ejections (CMEs) are highly transient solar events involving the expulsion of mass and magnetic field from the solar surface. On the order of 10^{12} kg of plasma may be expelled from the solar surface during a typical event. These dynamic events originate in closed magnetic field regions of the corona. They produce large-scale reconfiguration of the coronal magnetic field and generate large solar wind disturbances that, as mentioned above, appear to be the primary cause of major geomagnetic storms at Earth.

The physical mechanisms involved in the initiation of CMEs are not well understood. After release, CMEs accelerate and become part of the outward flow of the solar wind. They are either accelerated by the solar wind so as to come into equilibrium with the ambient wind or act as drivers moving faster than the background solar wind. Close to the Sun, the typical dimension of a CME is less than a solar radius. As the CMEs propagate outward from the corona, they expand dramatically and may extend over tenths of an AU by the time Earth's orbit is reached at 1 AU. Moreover, many, if not all, CMEs are associated with magnetic clouds and the plasma properties within these clouds can differ substantially from those of the ambient solar wind.

Global computational models based on first principles mathematical descriptions of the physics represent a very important component of efforts to understand the initiation, structure, and evolution of CMEs. Here we show our numerical results for a CME driven by local plasma density enhancement. This initial condition represents a simplified model of the situation just before eruption, when significant amount of mass is elevated into the lower corona. Later, we will apply more sophisticated initial conditions. At this time we just wanted to initiate a mass ejection and see how it moves through interplanetary space. In this calculation, the background solar wind solution described above was used as an initial solution and then a localized isothermal density enhancement was introduced at the solar surface just above the equatorial plane (as a consequence of the isothermal assumption the plasma pressure was also increased by the same factor as the density). This localized isothermal density enhancement initiated the CME. In this enhancement the density and pressure are locally increased by a factor of 135 in a small region just above the solar equator for a duration of about 12 hours.

Plate 2 shows a three-dimensional representation of the magnetic field configuration 9 hours after the initiation of the CME. The color code represents $\log(B)$, white lines are open magnetic field lines, magenta lines represent magnetic field lines with both ends connected to the Sun. At this time the density enhancement is in the declining phase, but it is still more than 100 times higher than the background density. The density enhancement first leads to the "filling" of the closed magnetic field lines with additional plasma and subsequent expansion of the closed field line region. One can see that the closed field lines became greatly stretched by the outward moving plasma. This is due to the fact that the plasma β (the ratio of the kinetic and magnetic pressures) is quite large and the magnetic field is "carried" by the outward moving plasma. We also note the decrease of magnetic field strength behind the leading edge of the outward moving disturbance.

A very interesting aspect of simulation results is the anisotropic expansion of the CME. The CME is fairly concentrated near the disrupted heliospheric current sheets, while it is broadly spread in the equatorial plane.

3.3. Interaction with the Magnetosphere

The magnetosphere is a complex nonlinear system. The direction of the interplanetary magnetic field (IMF) fundamentally controls the large-scale topology of the magnetospheric configuration. The magnetospheric topology in turn controls the entry of mass, momentum, energy, and magnetic flux into the magnetosphere. The entry of these physical quantities from the solar wind into the magnetosphere produces various transition layers, the extended geomagnetic tail and the plasma sheet, current systems and auroral phenomena.

Three-dimensional global MHD simulations have been used for a long time to simulate the global magnetospheric configuration and to investigate the response of the magnetosphere-ionosphere system to changing solar wind conditions. The first global-scale 3D MHD simulations of the solar wind-magnetosphere system were published in the early 1980s. Since then, global 3D MHD simulations have been used to study a range of processes [*Ogino and Walker*1984, *Lyon et al.*, 1986, *Ogino* 1986, *Ogino et al.*, 1986, *Fedder and Lyon* 1987, *Watanabe and Sato* 1990, *Usadi et al.*, 1993, *Walker et al.*, 1993, *Fedder et al*, 1995a,b, *Berchem et al.*, 1995, *Rader et al.*, 1995, *Tanaka* 1995, *Janhunen and Koskinen* 1997, *Raeder et al.*, 1997, *Winglee et al.*, 1997, *Gombosi et al.*, 1998, 2000a, *White et al.*, 1998].

Here, for the first time, we simulate the dynamic response of the global magnetospheric configuration to changing solar wind conditions self-consistently simulated all they way from the Sun to 1 AU. The solar wind changes at the location of the Earth due to the rotation of the tilted solar magnetic field and due to the passage of the coronal ejection described in the previous section.

In the simulation Earth is assumed to move on a perfectly circular orbit about the Sun with an orbital radius of 215.5 R_s and an orbital period of 365.25 days. The orbital plane is assumed to be inclined at an angle of 7.25° to the solar equator with a node line aligned with the x-axis and the maximum and minimum excursions of the planet in z-direction occurring in the $y = 0$ plane. Earth's initial position at $t = 0$ in the numerical simulation is heliolatitude 7.24° and longitude 11.9° (the CME takes place at 0° longitude and 11.5° latitude).

At this position Earth is located near the heliospheric current sheet when the CME is initiated. At Earth's location the first signatures of the CME can be seen in the magnetic field components at around 55 hours after initiation. The driving mass (the piston) arrives around 72 hours after the beginning of the event. The B_y component of the magnetic field remains pretty steady during the entire event, its value only changes by ~ 0.5 nT. The B_x and B_z components, however, exhibit a significant rotation around the y axis. This, in effect, is the signature of the passage of a CME-related flux rope. During the event the magnitude of the IMF increases from about 2 nT to approximately 4 nT. A more detailed description of the solar wind parameters at 1 AU is given in *Groth et al.*, [2000].

The plasma temperature $(T_e + T_p)$ in the background solar wind is about 2×10^5K. During the CME (beginning at around 55 hours) the plasma temperature significantly decreases dipping below 1.5×10^5K. The solar wind velocity remains nearly radial during the entire event with the speed gradually decreasing from about 550 km/s to about 450 km/s. The undisturbed solar wind density is fairly high before the CME event (~ 38 cm^{-3}), but it decreases to a more typical value of ~ 18 cm^{-3} just at the arrival of the piston. At the peak of the event the density increases to about 45 cm^{-3}. The solar wind dynamic pressure increases from its pre-CME value of 2.25 nP (at 72 hours) to 4.6 nP at the peak of the event.

Plate 3 shows the change of the global magnetospheric configuration during the passage of the CME event. The Plate shows two 3D snapshots at $t = 70.5$h and 94.5h. The color code represents the electric current density in the plane of the terrestrial equator, solid lines show last closed magnetic field lines. One can see the magnetopause current (or Chapman-Ferraro current) near the subsolar magnetopause and the tail current on the nightside.

The global magnetospheric configuration is primarily controlled by the B_z component of the interplanetary magnetic field. For $B_z < 0$ the magnetosphere exhibits an open configuration with significant dayside reconnection and open magnetic field lines connected to large regions near the magnetic poles. For strong northward IMF conditions ($B_z > 5$ nT) the magnetosphere becomes practically closed with magnetic reconnection limited to small regions near the cusps. For "intermediate" values of B_z (between about 0 and 5 nT) the global magnetospheric configuration is "partially closed" (or "partially open"). This configuration is characterized by significant dayside reconnection and large open cusps, by a narrow near-Earth reconnection line around the center of the magnetotail (the length of this reconnection line decreases with increasing B_z), and by long, stretched magnetospheric wings connected to the dawn and dusk sides of the ionosphere. These magnetic wings are formed by highly stretched closed magnetic field lines. They represent the transition from magnetopause reconnection to tail reconnection. In Plate 3 the last closed filed lines at the magnetosphere are shown by green, while the last closed field lines in the tail are shown by red.

During the simulated CME event B_z varies between about 0 nT and 3 nT, therefore the global magnetospheric configuration never switches to pure "South" or "North" configuration. However, the magnetosphere significantly changes during the event. As the solar wind dynamic pressure increases the dayside magnetopause moves inward and the current densities significantly increase. In addition, the magnetosphere becomes narrower and the length of the magnetic wings increases from about 60 R_E to 90 R_E. Overall, the energy and magnetic flux stored in the magnetosphere increased substantially.

A very interesting feature seen in Plate 3 is the clockwise "twist" of the magnetosphere perpendicular to the solar wind direction. This twist, which is particularly visible in the wings, is due to the presence of a non-zero B_y component.

Plate 4 shows the change of ionospheric potential and convection pattern during the CME event. The color code repre-

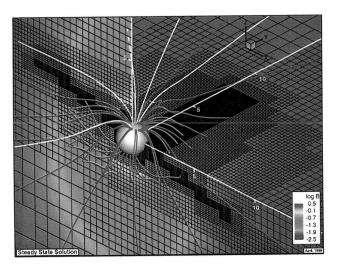

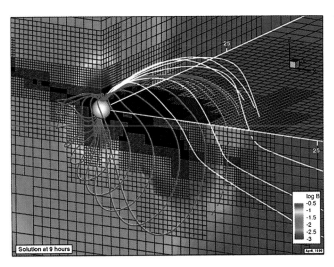

Plate 1. 3D representation of the near solar region before the initiation of the CME. The color code represents $\log(B)$ in the (x, z) and (x, y) planes. Solid lines are magnetic field lines: magenta denotes the last closed field lines, red is open field lines expanding to the interplanetary medium just above the heliospheric current sheet, and finally, white lines show open magnetic field lines in the (y, z) plane.

Plate 2. 3D representation magnetic field lines 9 hours after the initiation of a CME. Color code represents $\log(B)$, white lines are open magnetic field lines, magenta lines represent magnetic field lines with both ends connected to the Sun.

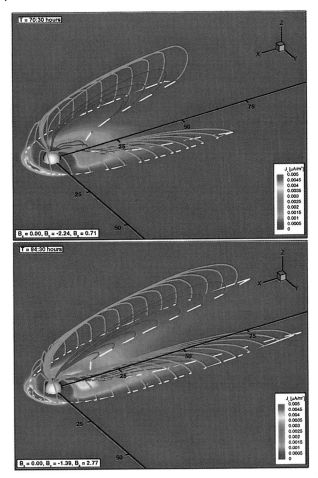

Plate 3. The response of the magnetosphere to the CME.

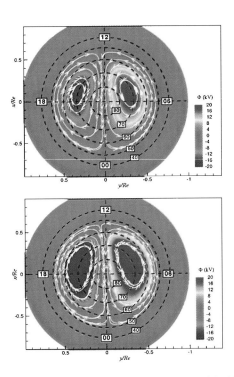

Plate 4. Polar plot of the ionospheric convection (white lines) and potential distribution (color code) in the northern polar ionopshere at $t = 70.5$ h (left panel) and $t = 97.3$ h (right panel).

sents the calculated electric potential in the height-integrated ionosphere, white lines show ionospheric convection patterns. One can see the two-cell pattern of ionospheric convection typical for southward-type IMF conditions. The convection pattern is also "twisted" due to the presence of a non-zero IMF B_y component. The most important change in the ionosphere is the doubling of the cross-cap potential drop from 30 kV at 70.5h to 60 kV some 27 hours later.

Overall, the terrestrial consequences of the simulated space weather event were not "dramatic" due to our choice of initial conditions to drive it. During the event the B_z component of the interplanetary magnetic field never exceeded $+1$ nT, therefore the geoeffectiveness of the CME was quite small. However, the capabilities demonstrated by doing the calculation are themselves dramatic, and nothing precludes us from using the code to simulate more geoeffective events, and ultimately even for forecasting real storms in space. This will be the focus of our upcoming investigations.

Acknowledgments.   This work was supported by the NSF KDI grant NSF ATM-9980078, by NSF CISE grant ACI-9876943 and by NASA AISRP grant NAG5-9406.

REFERENCES

Berchem, J., J. Raeder, and M. Ashour-Abdalla, Reconnection at the magnetospheric boundary: Results from global magnetohydrodynamic simulations, in *Physics of the Magnetopause*, edited by P. Song, B. Sonnerup, and M. Thomsen, p. 205, AGU, Washington, D.C., 1995.

De Zeeuw, D. L., T. I. Gombosi, C. P. T. Groth, K. G. Powell, and Q. F. Stout, An adaptive MHD method for global space weather simulations, *IEEE Trans. Plasma Sci.*, in press, 2000.

Dryer, M., Multidimensional, magnetohydrodynamic simulation of solar-generated disturbances: Space weather forecasting of geomagnetic storms, *AIAA Journal*, 3, 365–370, 1998.

Fedder, J. A., and J. G. Lyon, The solar wind-magnetosphere-ionosphere current-voltage, *Geophys. Res. Lett.*, 14, 880–883, 1987.

Fedder, J. A., J. G. Lyon, S. P. Slinker, and C. M. Mobarry, Topological structure of the magnetotail as a function of interplanetary magnetic field direction, *J. Geophys. Res.*, 100, 3613–3621, 1995a.

Fedder, J. A., S. P. Slinker, J. G. Lyon, and R. D. Elphinstone, Global numerical simulation of the growth phase and the expansion onset for a substorm observed by Viking, *J. Geophys. Res.*, 100, 19,083–19,093, 1995b.

Gombosi, T. I., D. L. De Zeeuw, C. P. T. Groth, K. G. Powell, and P. Song, The length of the magnetotail for northward IMF: Results of 3D MHD simulations, in *Physics of Space Plasmas*, edited by T. Chang, and J. R. Jasperse, vol. 15, pp. 121–128, MIT Press, Cambridge, Mass., 1998.

Gombosi, T. I., D. L. De Zeeuw, C. P. T. Groth, and K. G. Powell, Magnetospheric configuration for Parker-spiral IMF conditions: Results of a 3D AMR MHD simulation, *Adv. Space Res.*, 26(1), 139–149, 2000a.

Gombosi, T. I., D. L. De Zeeuw, C. P. T. Groth, K. G. Powell, and Q. F. Stout, Multiscale MHD simulation of a coronal mass ejection and its interaction with the magnetosphere-ionosphere system, *J. Atmos. Solar-Terr. Phys.*, 62, 1515–1525, 2000b.

Gombosi, T. I., K. G. Powell, and B. van Leer, Comment on "Modeling the magnetosphere for northward interplanetary magnetic field: Effects of electrical resistivity" by Joachim Raeder, *J. Geophys. Res.*, 105, 13,141–13,147, 2000c.

Groth, C. P. T., D. L. De Zeeuw, T. I. Gombosi, and K. G. Powell, A parallel adaptive 3D MHD scheme for modeling coronal and solar wind plasma flows, *Space Sci. Rev.*, 87, 193–198, 1999a.

Groth, C. P. T., D. L. De Zeeuw, K. G. Powell, T. I. Gombosi, and Q. F. Stout, A parallel solution-adaptive scheme for ideal magnetohydrodynamics, in *Proc. 14th AIAA Computational Fluid Dynamics Conference*, Norfolk, Virginia, AIAA Paper No. 99-3273, 1999b.

Groth, C. P. T., D. L. De Zeeuw, T. I. Gombosi, and K. G. Powell, Global 3D MHD simulation of a space weather event: CME formation, interplanetary propagation, and interaction with the magnetosphere, *J. Geophys. Res.*, 105(A11), 25,053 – 25,078, 2000.

Guo, W. P., and S. T. Wu, A magnetohydrodynamic description of coronal helmet streamers containing a cavity, *Astrophys. J.*, 494, 419–429, 1998.

Janhunen, P., and H. E. J. Koskinen, The closure of Region-1 field-aligned current in MHD simulation, *Geophys. Res. Lett.*, 24(11), 1419–1422, 1997.

Linde, T. J., A three-dimensional adaptive multifluid MHD model of the heliosphere, Ph.D. thesis, Univ. of Mich., Ann Arbor, 1998.

Linde, T. J., T. I. Gombosi, P. L. Roe, K. G. Powell, and D. L. De Zeeuw, The heliosphere in the magnetized local interstellar medium: Results of a 3D MHD simulation, *J. Geophys. Res.*, 103(A2), 1889–1904, 1998.

Lionello, R., Z. Mikić, and D. D. Schnack, Magnetohydrodynamics of solar coronal plasmas in cylindrical geometry, *J. Comput. Phys.*, 140, 172–201, 1998.

Lyon, J. G., J. Fedder, and J. Huba, The effect of different resistivity models on magnetotail dynamics, *J. Geophys. Res.*, 91, 8057–8064, 1986.

Mikić, Z., and J. A. Linker, Disruption of coronal magnetic field arcades, *Astrophys. J.*, 430, 898–912, 1994.

Odstrčil, D., and V. J. Pizzo, Distortion of the interplanetary magnetic field by three-dimensional propagation of coronal mass ejections in a structured solar wind, *J. Geophys. Res.*, 104, 28,225–28,239, 1999.

Ogino, T., A three-dimensional MHD simulation of the interaction of the solar wind with the Earth's magnetosphere: The generation of field-aligned currents, *J. Geophys. Res.*, 91, 6791–6806, 1986.

Ogino, T., and R. J. Walker, A magnetohydrodynamic simulation of the bifurcation of tail lobes during intervals with a northward interplanetary magnetic field, *Geophys. Res. Lett.*, 11, 1018–1021, 1984.

Ogino, T., R. Walker, M. Ashour-Abdalla, and J. Dawson, An MHD simulation of the effects of the interplanetary magnetic field B_y component on the interaction of the solar wind with the Earth's magnetosphere during southward interplanetary magnetic field, *J. Geophys. Res.*, 91, 10029, 1986.

Powell, K. G., An approximate Riemann solver for magnetohydrodynamics (that works in more than one dimension), Tech. Rep. 94-24, Inst. for Comput. Appl. in Sci. and Eng., NASA Langley Space Flight Center, Hampton, Va., 1994.

Powell, K. G., P. L. Roe, T. J. Linde, T. I. Gombosi, and D. L. D.

Zeeuw, A solution-adaptive upwind scheme for ideal magneto-hydrodynamics, *J. Comput. Phys.*, *154*(2), 284–309, 1999.

Raeder, J., R. J. Walker, and M. Ashour-Abdalla, The structure of the distant geomagnetic tail during long periods of northward IMF, *Geophys. Res. Lett.*, *22*, 349–352, 1995.

Raeder, J., J. Berchem, M. Ashour-Abdalla, L. A. Frank, W. R. Paterson, K. L. Ackerson, S. Kokubun, T. Yamamoto, and J. A. Slavin, Boundary layer formation in the magnetotail: Geotail observations and comparisons with a global MHD simulation, *Geophys. Res. Lett.*, *24*, 951–954, 1997.

Roe, P. L., Approximate Riemann solvers, parameter vectors, and difference schemes, *J. Comput. Phys.*, *43*, 357–372, 1981.

Song, P., D. L. De Zeeuw, T. I. Gombosi, C. P. T. Groth, and K. G. Powell, A numerical study of solar wind–magnetosphere interaction for northward IMF, *J. Geophys. Res.*, *104*(A12), 28,361–28,378, 1999.

Song, P., T. Gombosi, D. De Zeeuw, K. Powell, and C. P. T. Groth, A model of solar wind - magnetosphere - ionosphere coupling for due northward IMF, *Planet. Space Sci.*, *48*, 29–39, 2000.

Steinolfson, R. S., Modeling coronal streamers and their eruption, *Space Sci. Rev.*, *70*, 289–294, 1994.

Stout, Q. F., D. L. De Zeeuw, T. I. Gombosi, C. P. T. Groth, H. G. Marshall, and K. G. Powell, Adaptive blocks: A high-performance data structure, in *Proc. Supercomputing '97*, 1997.

Suess, S. T., A.-H. Wang, and S. T. Wu, Volumetric heating in coronal streamers, *J. Geophys. Res.*, *101*(A9), 19,957–19,966, 1996.

Tanaka, T., Generation mechanisms for magnetosphere-ionosphere current systems deduced from a three-dimensional MHD simulation of the solar wind-magnetosphere-ionosphere coupling process, *J. Geophys. Res.*, *100*(A7), 12,057–12,074, 1995.

Usadi, A., A. Kageyama, K. Watanabe, and T. Sato, A global simulation of the magnetosphere with a long tail: Southward and northward interplanetary magnetic field, *J. Geophys. Res.*, *98*, 7503–7517, 1993.

Walker, R. J., T. Ogino, J. Raeder, and M. Ashour-Abdalla, A global magnetohydrodynamic simulation of the magnetosphere when the interpalnetary magnetic field if southward: The onset of magnetotail reconnection, *J. Geophys. Res.*, *98*, 17235, 1993.

Wang, A.-H., S. T. Wu, S. T. Suess, and G. Poletto, Global model of the corona with heat and momentum addition, *J. Geophys. Res.*, *103*, 1913–1922, 1998.

Watanabe, K., and T. Sato, Global simulation of the solar wind-magnetosphere interaction: The importance of its numerical validity, *J. Geophys. Res.*, *95*, 75–88, 1990.

White, W. W., G. L. Siscoe, G. M. Erickson, Z. Kaymaz, N. C. Maynard, K. D. Siebert, B. U. Ö. Sonnerup, and D. R. Weimer, The magnetospheric sash and the cross-tail S, *Geophys. Res. Lett.*, *25*(10), 1605–1608, 1998.

Winglee, R. M., V. O. Papitashvili, and D. R. Weimer, Comparison of the high-latitude ionospheric electrodynamics inferred from global simulations and semiempirical models for the January 1992 GEM campaign, *J. Geophys. Res.*, *102*, 26,961–26,977, 1997.

T. I. Gombosi, D. L. DeZeeuw, C. R. Clauer, and P. Song, Department of Atmospheric and Oceanic Sciences, The University of Michigan, Ann Arbor, MI 48109. (e-mail: tamas@umich.edu, darrens@umich.edu, rclauer@umich.edu, psong@umich.edu)

K. G. Powell, Department of Aerospace Engineering, The University of Michigan, Ann Arbor, MI 48109 (e-mail: powell@umich.edu)

C. P. T. Groth, University of Toronto Institute for Aerospace Studies, 4925 Dufferin St., Toronto, Ontario, Canada, M3H 5T6 (e-mail: groth@utias.utoronto.ca)

Visualizing CMEs and Predicting Geomagnetic Storms from Solar Magnetic Fields

Yan Li and Janet G. Luhmann

Space Sciences Laboratory, University of California, Berkeley, USA.

J. Todd. Hoeksema and Xuepu Zhao

Center for Space Science and Astrophysics, Stanford University, Stanford, USA.

C. Nick Arge

Cooperative Institute for Research in Environmental Sciences, University of Colorado and Space Environment Center, National Oceanic and Atmospheric Administration, Boulder

Because solar photospheric magnetic fields are the main source of the magnetic field in the corona and interplanetary space, changes in the photospheric field may be expected to drive transients in both regions. However, because the solar field is both complex and influenced by the solar wind, it is difficult to determine whether specific photospheric field features underlie the important transients called Coronal Mass Ejections (CMEs). Models enable us to link the photospheric field to coronal field changes and visualize the response. The potential field source surface (PFSS) model is known for its combination of relative simplicity and ability to approximate coronal hole geometry and eclipse images. We utilize PFSS models and specialized synoptic maps of the solar magnetic field to study the relationship between coronal field changes and CMEs, and to attempt to visualize CMEs and predict geomagnetic storms. For prediction purposes, updated synoptic maps in real time are needed. Several solar observatories and SOHO/MDI are currently providing such maps on a roughly daily basis, although the potential exists for hourly updates. Our results to date suggest that the combination of photospheric field observations and coronal models can provide a useful addition to the collection of space weather forecast tools.

1. INTRODUCTION

The existence of the solar magnetic field makes the sun more fascinating. The well known solar activity cycle, sunspots and many other interesting features are all at least in part due to magnetic fields. Essentially

Space Weather
Geophysical Monograph 125
Copyright 2001 by the American Geophysical Union

all solar activity, including the spectacular solar flares and CMEs, is due to magnetic energy release. Solar magnetic fields are carried out by solar wind and transient activities such as CMEs into interplanetary space. When the solar wind with the magnetic fields reaches the Earth, interaction with the geomagnetosphere occurs. Transients in the solar wind often create disturbances in the geomagnetosphere including geomagnetic storms. The solar magnetic fields are both complex and influenced by the solar wind. It is difficult to determine whether specific photospheric field features underlie the important transients, CMEs. Coronal models enable us to link the photospheric magnetic fields to coronal field changes and visualize the responses. Existing coronal models include potential field source surface (PFSS), force free, and MHD models [Altschuler et al., 1977; Sakurai, 1981; Wu et al., 1990; Linker et al., 1999]. Some of them have been applied in CME studies [Luhmann et al., 1998; Mikic and Linker, 1994; Linker and Mikic, 1995]. Each model has its particular advantages and problems. The PFSS model is known for its combination of relative simplicity and ability to approximate coronal hole geometry and eclipse images [Altschuler et al., 1977; Levine, 1982; Gibson et al., 1999; Zhao et al., 1999]. We utilize PFSS models and specialized synoptic maps of the solar magnetic field to study the relationship between coronal field changes and CMEs, and attempt to visualize CMEs and predict geomagnetic storms. This approach was first suggested in Luhmann et al. [1998].

2. SOLAR MAGNETOGRAPH AND SYNOPTIC MAPS

Full disk magnetograms are regularly available from several providers. Among these, Wilcox Solar Observatory (WSO) takes one or two magnetogram(s) per day, National Solar Observatory (NSO) at Kitt Peak takes one magnetogram per day, and Mt. Wilson Observatory (MWO) makes 10 observations per day. Global Oscillation Network Group (GONG) takes one magnetogram per hour at each site, and every 20 min within the network of six sites. SOHO/MDI takes one magnetogram per 96 min at the L1 point. All the observatories listed above make standard Carrington Rotation synoptic maps. Although the formats are basically the same, the resolutions and details of constructing the maps can be different from observatory to observatory. The Carrington Rotation synoptic maps cannot meet the need of having new information on the magnetic fields in real time. "Daily updated" synoptic maps [Hoeksema and Zhao, private communication, 1997; Arge and Pizzo,

2000] are now routinely constructed at WSO, MWO, NSO and from SOHO/MDI data. Unlike the Carrington Rotation synoptic map, a daily updated synoptic map is constructed by merging the latest observation into the left side of the previous synoptic map and remove a corresponding portion of the old map from the right side. This provides the latest synoptic magnetic field information. This does not have to be daily, as updated maps can be made whenever there is a new full disk magnetogram. Therefore any time cadence of magnetograms can be used in the attempt to keep up with near realtime. Higher cadence updates are now available for MDI and GONG magnetograms, for which 24-hour observation also exist. Daily updated synoptic maps use magnetic field measurements from the last observation to about 27 days ago. These measurements are merged to approximate the instantaneous global solar magnetic field. To better approximate the global field, a special map referred to as a synoptic frame was proposed by Zhao et al. [1999]. To approximate the global magnetic field at a certain time, these authors make a global map centered at the time of interest using measurements 13.5 days before and 13.5 days after, then replace the center portion using the full disk magnetogram at the time of interest. This gives the best approximation until we have measurements from the back of the sun. But synoptic frames obviously cannot be used in real time, because they depend on 13.5 days of later data. In this paper we show results only from the previously described daily updated maps.

3. OUR APPROACH

3.1. The Method

The solar coronal magnetic field must respond to photospheric field changes. We assume that some of the responses give rise to CMEs. Our approach is based on the idea that we may be able to visualize the responses using photospheric field measurements and coronal models. Our first attempt, described here, uses PFSS models. PFSS models use photospheric magnetic synoptic maps as the inner boundary, and a "source surface" (usually at 2.5 Rs) as an outer boundary, where the assumption of radial magnetic field is applied. La Place's equation for the magnetic potential is solved in the space between the photosphere and the source surface [Altschuler et al., 1977; Hoeksema, 1995; Wang, 1995]. The solution of the PFSS model is represented by spherical harmonics. Using these solutions, we trace the coronal field lines from a fixed grid of starting points on the photosphere. We find that the first 9 harmonics are adequate for the WSO magnetograph resolution.

1998/08/24 00h:33m:57s

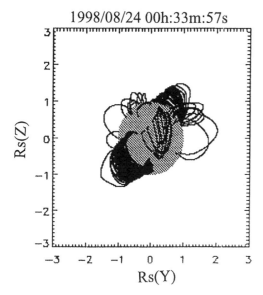

1998/08/24 21h:32m:45s

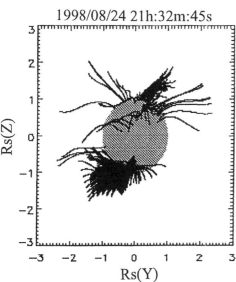

Figure 1. Newly opening field lines between August 24 00:34 UT and August 24 21:33 UT, 1998. These field lines, starting from the same grid points on the photosphere, were all closed at 00:34 UT, and all open at 21:33 UT. The group of field lines at the disk center are thought to indicate the occurrence of an Earthward-directed CME between the observation times. See text for details of the method to isolate newly opening field lines.

But note that WSO resolution may become inadequate around solar maximum due to the many active regions present on the disk. Field lines are traced for two consecutive updated synoptic maps starting from the same fixed grid, and are compared. Those field lines that change from closed at the earlier time to open at the next time are isolated and plotted. We refer to these field lines as newly opening field lines. For further de-

scription of the method, see Luhmann et al. [1998]. If there are some newly opening field lines between the times of two maps, we assume there was a possibility for CME occurrence(s) between these times. In particular, if there are newly opening field lines near the solar disk center, an Earthward-directed CME may have been generated.

3.2. Examples

Coronal field lines were traced for the two maps updated on August 24 00:34 UT and August 24 21:33 UT, 1998, resulting in the newly opening field lines shown in Figure 1. A group of field lines at the southeast limb, another at the northwest limb, and a group of field lines at the disk center were closed at 00:34 UT and open at 21:33 UT, at least as seen in the PFSS models. According to our assumption, the group of field lines near the disk center may indicate the occurrence of an Earthward-directed CME sometime between the observation times of these two magnetograms. Newly opening field lines between September 2, 20:02 UT and September 3, 22:44 UT, 1998 are shown in Figure 2. There are many newly opening field lines around the limb and the poles, but no field lines near the disk center. According to our assumption, this picture indicates that there should be no Earthward-directed CME between these two observations. Using the above results of the modeling, on August 24 we predict that there will be a geomagnetic storm during the following few days; and on September 3 we predict that there will be no geomagnetic storm in the following few days. Dst measurements from August 21 to September 10, 1998 by WDC at Kyoto are shown in Figure 3. A magnetic storm started on August 26 and reached the minimum Dst = -155 on August 27. We see quiet geomagnetic condition several days following September 3.

3.3. Routine Prediction Web Site and Summary of the Results

Our procedure for visualizing newly opening field lines has been automated to update every day using WSO daily synoptic maps, and posted at the WWW (http://sprg.ssl.berkeley.edu/mf_evol/) since mid 1998. Newly opening field line calculations and our estimated Dst index using real time ACE solar wind data [Fenrich and Luhmann, 1998] are running in parallel on the WWW. We monitor the newly opening field lines and cross check with the Dst predicitons routinely, although we chose to use Kyoto observed Dst index in this paper. We present results for two months with the best coverage of the magnetograph observations at WSO. The results of the visualized newly opening field lines

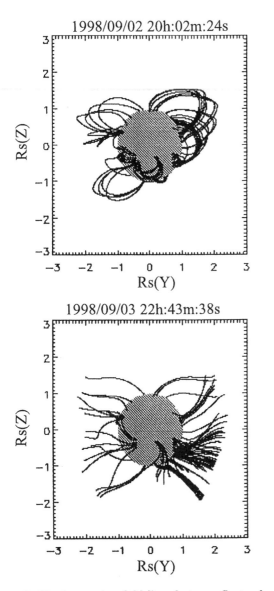

Figure 2. Newly opening field lines between September 2, 20:02 UT and September 3, 22:44 UT, 1998. There is no newly opening field lines near the disk center in this case. Therefore no Earthward-directed CME is expected between the observation times.

are in grey scale bars and displayed against Dst index in Figure 4, where black bars indicate that there are substantial numbers of newly opening field lines near the disk center on that day, light grey bars indicate no newly opening field lines near the disk center, and dark grey bars indicate a few newly opening field lines near the disk center, but we are uncertain about their significance. July 1999 in the top panel was a geomagnetically quiet month. Our newly opening field line calculation showed mostly no field lines near the disk center (light grey), from which we predict no geomagnetic storms in the following few days. During this quiet month there were three cases with uncertain situations (dark grey). Only on one day were there newly opening field lines near the disk center (black bar) followed by a small storm about two days later. September 1999, the bottom panel in Figure 4, has more geomagnetic activity or disturbances. We have many uncertain days in this month. The "Alarm" case (black bar) on the 4th was followed by only minor disturbances, which should be counted as a false alarm. As to the two "Alarm" cases (black bar) on the 11th and 12th, a storm of minimum Dst = -71 began on September 12, which was followed by a long disturbed period (about 10 days). The two "Alarm" cases (black bar) on September 20 and September 21 followed by a large storm of minimum Dst = -164 on September 23. In general, from the relative frequency of the "Uncertain" (dark grey bar) and "Alarm" (black bar) conditions, we would have predicted September to have been a more disturbed month overall. Newly opening field line results from July 1998 to December 1999 have all been summarized and plotted against the Kyoto Dst index as in Figure 4. Since there have been many solar magnetogram data gaps mainly due to weather conditions, it is difficult to give statistics of successful and unsuccessful predictions with reasonable confidence. One interesting and important point is that almost all large storms were preceded, by a few days, by substantial newly opening field structure near the central meridian, subject to the availability of the

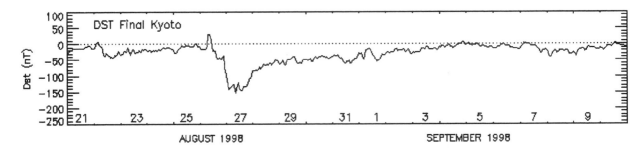

Figure 3. Dst measurements from August 21 to September 10, 1998 by WDC at Kyoto.

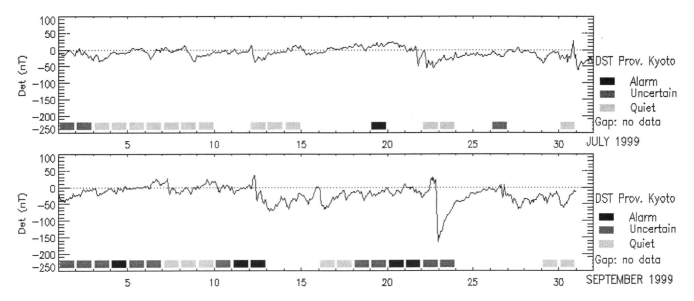

Figure 4. The results of the visualized newly opening field lines from two months (July and September, 1999) with the best coverage of magnetograph observations are in grey scale bars and displayed against Kyoto Dst index. Black bars: storm expected; Dark grey bars: uncertain; Light grey bars: no storm expected. See text for detailed descriptions.

solar magnetograms. But there are some "false alarms". As to "missed" cases, we cannot judge because of data gaps. Magnetograms and updated synoptic maps from other observatories can be used to confirm (or not) a particular prediction from WSO. This is expected to increase confidence in any given case, an approach we are currently testing.

4. REMARKS

Solar magnetic fields and their variations are complex. We have the following questions and concerns regarding our approach. (1). Can we distinguish CME related field changes from other changes? (2). What if the solar magnetic field changes more rapidly than our updates (especially near solar maximum)? (3). How well can the PFSS model represent large scale coronal fields? (4). To what degree do magnetic changes on the invisible side of the sun affect the Earthward facing solutions? (5) Can our method tell us much about the potential geoeffectiveness of a particular CME (e.g., its strength of southward interplanetary magnetic field and/or speed)? To address the above questions and concerns we have the following plans for future studies. (1). To search for additional information and criteria for selecting changing field lines that are related with CMEs or geoeffective CMEs. (2). To use SOHO/MDI data and GONG network data to look at higher cadence changes. (3). To look for more realistic model(s) (e.g., an affordable MHD model). (4). To use vector full disk

magnetograms for boundary conditions (e.g., from SO-LIS). (5). To use data from future missions that obtain global solar magnetic fields.

Acknowledgments.   The work at UCB was supported by NASA SEC-GI Program NAG5-7951 and NSF Space Weather Program 98-19776.

REFERENCES

Altschuler, M. D.; R. H. Levine; M. Stix and J. Harvey, High resolution mapping of the magnetic field of the solar corona, *Solar Phys.*, *51*, 345-375, 1977.

Arge, C. N. and V. J. Pizzo, Improvement in the Prediction of Solar Wind Conditions using near-real time solar magnetic field updates, *J. Geophys. Res.*, , 2000 (in press).

Fenrich, F. and J. G. Luhmann, Geomagnetic response to magnetic clouds of different polarity, *Geophys. Res. Lett.*, , *25*, 2999, 1998.

Hoeksema, J. T., The Large-Scale Structure of the Heliospheric Current Sheet During the ULYSSES Epoch, *Space Science Reviews*, *72*, 137, 1995.

Levine, R. H., Open magnetic fields and the solar cycle I, *Solar Phys.*, *79*, 203, 1982.

Linker, J. A.and Z. Mikic, Disruption of a helmet streamer by photospheric shear, *Astrophys. Journal*, Part 2 - Letters, *438*, 1, L45-L48, 1995

Linker, J. A.; Mikic, Z.; Biesecker, D. A.; Forsyth, R. J.; Gibson, S. E.; Lazarus, A. J.;Lecinski, A.; Riley, P.; Szabo, A.; Thompson, B. J., Magnetohydrodynamic modeling of the solar corona during Whole Sun Month, Journal of Geophysical Research, *104*, A5, 9809-9830, 1999.

Luhmann, J. G., J. T. Gosling, J. T. Hoeksema, X. Zhao, The relationship between large-scale solar magnetic field evolution and coronal mass ejections, *J. Geophys. Res.*, *103*, 6585, 1998.

Mikic, Z. and J. A. Linker, Disruption of coronal magnetic field arcades, *The Astrophys. Journal*, *430*, 2, pt. 1, 898-912, 1994.

Sakurai, T., Calculation of Force-Free Magnetic Field with Non Constant Alpha, *Solar Phys.*, *69*, 343, 1981.

Wang, Y.-M., Latitude and Solar-Cycle Dependence of Radial IMF Intensity, *Space Science Reviews*, *72*, 193, 1995.

Wu, S. T.; Sun, M. T.; Chang, H. M.; Hagyard, M. J.; Gary, G. A. On the numerical computation of nonlinear force-free magnetic fields, *Astrophys. Journal*, Part 1, *362*, 698-708, 1990.

Zhao, X. P., J. T. Hoeksema, P. H. Scherrer, Changes of the boot shaped coronal hole boundary during Whole Sun Month near sunspot minimum, *J. Geophys. Res.*, *104*, 9735, 1999.

C. N. Arge, NOAA, Space Environment Center, 325 Broadway, MS: R/E/SE, Boulder, Colorado, USA 80303.

J. T. Hoeksema and X. P. Zhao, Stanford University, H.E.P.L. Annex B213 Stanford, CA 94305-4085, USA.

Y. Li and J. G. Luhmann, Space Sciences Laboratory, University of California, Berkeley, CA 94720, USA.

Prediction of Southward IMF Bz

J. K. Chao and H. H. Chen

Institute of Space Sciences, National Central University, Chung-li, Taiwan

Intense southward interplanetary magnetic field is believed to be one of the most important causes for a major magnetic storm, which can produce severe space weather. We first review some current understanding of the origin of this IMF Bz. Then, some results for the prediction of IMF Bz will be reviewed. Our studies for the prediction of Bz will be demonstrated using a kinematic model first designed by Hakamada and Akasofu (1982).

1. INTRODUCTION

Intense geomagnetic storms and magnetospheric substorms are one of the very important causes for adverse space weather, which can cause disruption of satellite operations, communications, navigation, and electric power distribution grids, leading to a variety of socioeconomic losses. And the primary cause of these storms and substorms are the long-duration southward interplanetary magnetic fields (IMF) in solar magnetospheric coordinate system (GSM), usually called -Bz or Bs events, which play a crucial role in determining the amount of solar wind energy to be transferred to the magnetosphere [Arnoldy, 1971; Tsurutani and Meng, 1972; Russell and McPherron, 1981; Akasofu, 1981; Gonzalez and Tsurutani, 1987]. In order to predict the occurrence of intense geomagnetic storms, it is necessary to know the origin of the large southward IMF Bz. Thus understanding the causes of and predicting the length and strength of IMF Bz are key goal of space weather research.

The intense southward IMF Bz is associated with two types of origins [Gonzalez and Tsurutani, 1987]. The first type has an intrinsic solar origin. The southward IMF Bz events are just a part of the internal magnetic field in the

ejected plasmas from the Sun. These ejected plasmas may be identified as one of the solar eruptive phenomena: CME, flare ejecta, eruptive filament or prominence, etc., which all may be called "ejecta" [Chao, 1973; Bravo, 1997]. The ejecta do not imply a particular shape or magnetic topology nor refer to a solar region. The term "ejecta" applies to any parcel of solar wind with "unusual" characteristics that results from a transient solar event. The other type of source has an interplanetary origin. Several physical processes in the solar wind have been suggested to generate the southward IMF Bz events. The draped interplanetary magnetic field [Gosling and McComas, 1987], kinky heliospheric current sheets [Akasofu, 1981; Tsurutani et al. 1984, 1985] and compressions of plasma and magnetic field fluctuations in the corotating interaction regions [Tsurutani et al., 1995] are examples of this type. The temporal evolution of Alfven intermittent turbulence in the corotating interaction region may become unstable and generate large IMF Bz [Chian et al., 1998 and also in this conference].

Although the source or the origins of southward IMF Bz still need further investigation, quantitative prediction of southward IMF Bz has been attempted [Wu and Dryer, 1996, 1997; Hoeksema and Zhao, 1992]. The purpose of this paper is to give a brief review of previous attempts for predicting the large southward IMF Bz's. For this purpose we will review the studies on the origin of southward IMF Bz's. Finally, we will present our preliminary predictions using the Hakamada-Akasofu scheme [Hakamada and Akasofu, 1982] for 17 events, which have been analyzed previously [McComas et al., 1989; Wu and Dryer, 1996].

Space Weather
Geophysical Monograph 125

2.   SOLAR SOURCE OF LARGE SOUTHWARD IMF Bz

In a steady state solar wind model, solar wind velocity and magnetic field in the coronal region are all in the radial direction. The off-radial component of magnetic field most likely originates from the large-scale coronal structures, such as the magnetic loops, CME's and filaments etc., which propagate through the ambient solar wind. These ejected structures are called "solar transients". Thus, in the past, most investigators assume that the transient Bz component is associated with the so-called "magnetic tougue"[Gold, 1962], or "magnetic cloud" [Klein and Burlaga, 1982].

The magnetic tongue model has been partially supported by Pudovkin and Chertkov [1976], who claimed that the strength of geomagnetic disturbances could be predicted by the north-south disturbances of the photospheric magnetic field at the site of a solar flare. However, Tsurutani et al.[1997] were unable to find any out of 57 events occurring in 1978-1979 looked like a "magnetic tougue".

In order to study the relationship between solar magnetic fields and the transient variations of IMF Bz, at 1AU, Tang et al. (1985, 1989) examine the relationship between the polarity of the transient variation of IMF Bz and the associated flare field and vice versa. Their studies show that a simple relationship between the orientation of the IMF Bz and the magnetic orientation of the associated flare region is not apparent. On the other hand, Hoeksema and Zhao (1992) studied three solar flare associated CME events and their transient variations of IMF Bz at 1AU identified by Tsurutani et al [1988, 1992] using solar wind plasma and magnetic field data from ISEE 3. Based on an exploratory assumption that the free magnetic energy of a CME when it is initiated is transferred almost totally to other forms of energy by the time the CME stops acceleration, the potential field model can be used to calculate the magnetic orientation in the ejected plasma. They found the calculated magnetic orientations at the "release height" (the height where the front of a CME ceases to accelerate) for these three flare associated CMEs agree with their IMF directions at 1AU. On the contrary, events associated with eruptive filament do not have such a relationship. Zhao et al (1993) have used a potential field model of the corona to estimate the direction of the coronal field at 0.03 Ro above the solar active region. They predicted the correct orientation of the interplanetary magnetic field driver gas. On seven out of eight times for active regions but zero out of one time for prominence. This study agrees with the conclusion by Hoeksema and Zhao (1992). However, without applying the potential field model to the field in the flare regions, Tang et al (1989) studied these three events in which they found no correspondence between the magnetic orientations in flare regions and the IMF directions in the driver gas.

The interplanetary counterparts of CMEs that exhibit the topology of helical magnetic flux ropes are commonly called magnetic clouds (MCs) [Burlaga et al., 1981; Bothmer and Schwenn, 1996]. Bothmer and Schwenn [1997] found that there is a solar cycle variation appearing in magnetic orientation in the clouds. From an investigation of the structures of MCs identified in near-Earth solar wind data for the year 1965-1993, they find evidence for a dependence of the magnetic structure of MCs on the phase of solar cycles. From about the time of sunspot maximum in 1981, most MCs had northward-directed fields leading southward-directed ones (NS type) while for the following sunspot maximum in 1991, southward-directed fields have led northward fields (SN type). Over the years studied, the total number of SN-type MCs equal to the total number of NS-type MCs and the number with right-hand magnetic fields is roughly equal to the number of left-hand fields. They also found associations of MCs with quiescent filament disappearances. The magnetic structures of these filaments are interpreted based on results from Bothmer and Schwenn [1994] and the helicity rule of Rust [1994]. From a comparison of the magnetic structure of the MCs with that inferred for the associated filaments, Bothmer and Rust [1997] found agreement of chirality (handedness) in 24 of 27. This association implies the existance of a solar cycle variation of magnetic field SN-NS of MCs. The results are important for space weather forecast. Based on the findings, they predict that the majority of MCs, up to the peak of the next cycle (number 23), will be SN type. However, most filaments associated with MCs in their study were at heliographic latitudes 30 to 40 (i.e. out-side active regions). This result is different from those of a magnetic tongue model by Hoeksema and Zhao [1992] and

Zhao et al. [1993] in which the magnetic topology of those CMEs stems from active solar region accompanied by flares. Recently, Marubashi [1997] inferred the axial field directions of 12 magnetic clouds and compared them with the orientations of nine-associated disappearing filament (DSFs). He found that the orientations of the MS's axis generally are coincided with the orientations of the associated DSFs. Further more, Zhao and Hoeksema [1998] analyzed 26 well-characterized MCs, which are associated with extended periods of strong southward IMF Bz called " Bs events". They find that (1) the central axial field directions of the MCs are almost evenly distributed between −90 and 90 in ecliptic latitude and the longitudinal distribution is slightly peaked around the east and west, (2) the duration and intensity of Bs events correlate linearly with the direction of the clouds central axial field, and (3) cloud central axial field directions are correlated with the

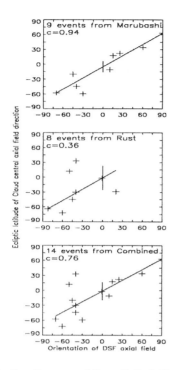

Figure 1. Scatter diagrams of the ecliptic latitude of magnetic cloud central axial field direction versus orientation of disappearing filament central axial field where "c" denotes the correlation coefficient. The line and accompanying vertical bar show the least squares fit to the scatter diagram and its standard deviation, respectively (from Zhao and Hoekesema, 1998).

central axial field directions of the associated disappearing filament on the Sun. In that study, the central axial field orientations are quantitatively determined using Martin's rule of handedness of filament field rotation [Martin et al., 1994]. In Figure 1, the ecliptic latitudes of MC's central axial field directions versus the orientations of the central axial fields of DSFs are shown for the database of Marubashi [1997], Lepping et al., [1990] and combined above data sets, respectively. Also shown in the diagram are their least square fits and standard deviations. Their results suggest that MC's axial field direction undergo only slight changes while propagating through interplanetary space. This supports the possibility of predicting the axial field directions of MCs from the solar observations of DSFs and the photospheric magnetic field, which is essential for prediction of the southward IMF Bz.

As shown in the above studies, there is still disagreement for predicting the IMF inside MCs (driving gases). More statistics are needed for such studies. On the theoretical side, using a 3-dimensional MHD model, Wu et al. [1999] study the dynamical evolution of interactions between a coronal streamer and flux rope from the Sun to 1 AU and applying this model for the physical processes of the

January 6-12, 1997 event. From their Figure 9, the predicted IMF B and solar wind parameters inside the MC are in qualitative agreement with the data observed by WIND at 1 AU. Chen and Garren [1993] and Chen [1996] gave physical description of the evolution of a CME related to a flux rope in a streamer from near the Sun to a MC in IP space.

3. INTERPLANETARY SOURCE OF LARGE SOUTHWARD IMF Bz

Because of the rotation of the Sun, the solar wind and the disturbance entering the interplanetary space will interact with the ambient solar wind originating from different longitudes on the solar surface. This interaction can create additional source for geomagnetic storms. Since the direct cause for storms is a large southward IMF Bz , we look for processes that can generate such a component.

A CME in general is composed of a bright loop, a dark region and a filament or prominence close to the Sun. [Hundhausen, 1993]. When entering IP space, the material of the CME is called a driven gas. A ratio of 1/3 [Gosling, 1990] or 1/6 [Tsurutani et al. 1997] of CMEs has the form of a magnetic cloud or giant flux rope. This flux rope may have a Bz component. When the material carrying the magnetic cloud has a speed greater than the ambient solar wind by more than the ambient fast wave speed, a fast shock will form. This MHD fast shock can compress the upstream magnetic field substantially. If a moderate southward Bz already exists upstream, a large southward Bz will be generated. When it reaches the magnetosphere, a large storm will be initiated. Fast MHD shock can generate the storm efficiently. Recently, Tsurutani et al. [1998] and Burlaga et al. [1998] identify the CME loop and filament of the January 10-11, 1997 event, respectively, from the interplanetary observations. It is interesting to note that the peak magnetic field strength and the peak velocity values of the magnetic clouds are related [Gonzalez et al., 1998].

Interplanetary shock waves can be grouped into two types. The first type consists of corotating shocks, which are generated by interactions of solar wind streams usually formed outside 1AU. The lifetime of these streams may be longer or shorter than one solar rotation period. Hence, the corotating shocks do not necessarily have a recurrence tendency of 27 days (a solar rotational period). The second type consists of transient shocks generated by IMCs. Non-linear large amplitude waves can steepen into fast shocks (Chao, 1973). Both these two types of shocks can amplify the ambient southward Bz to produce the interplanetary cause for geomagnetic storms. Numerical and empirical models have been proposed for this generation mechanism.

Draping of the interplanetary magnetic field can occur when a highly conducting magnetized plasma flow relative to a magnetized obstacle imbedded in it. Gosling and McComas [1987] proposed that large southward IMF Bz can be generated by fast CMEs interacting strongly with the ambient interplanetary plasma and magnetic field. Figure 2 illustrates this drapping in the solar meridinal plane. A north-south draping of a purely radial ambient IMF about a fast-moving plasmoid may generate strong southward IMF Bz in the shocked plasma in the ecliptic plane by the combined effects of draping and compression. The compressed region between the driver gas and the shock wave can be called the sheath region, which is generated in interplanetary space. [Illing and Hundhausen, 1986]. In principle, the strength and the direction of this Bz can be predicted when the undisturbed source surface magnetic fields and solar wind speeds are known (Wu and Dryer, 1996). Large amplitude Alfven waves and turbulence when compressed by the shocks may also be the source for storms when large Bz's are present. Tsurutani and Gonzalez (1997) have listed six types of possibilities of how large southward Bz are created: (1) shocked southward fields (Tsurutani et al., 1988), (2) bending of the heliospheric current sheets (HCS) (Tsurutani et al., 1984), (3) amplification of Alfven waves and turbulence (Tsurutani et al., 1995), (4) draped magnetic fields in the sheath region (Zwan and Wolf, 1976; McComas et al., 1989), (5) equinoctial By effect (Russell and McPherron, 1973) and (6) fast stream-HCS interactions (Odstrcil and Pizzo, 1999).

4. PREDICTIONS OF IMF Bz FROM IMF DRAPING

McComas et al. [1989] first use the draping scenario to look for evidence of magnetic field draping ahead of fast MCs near 1 AU and try to predict the IMF Bz perturbations from their draping model. They argue that IMF Bz is, in principle, predictive with the knowledge of the direction of the CME's motion and the orientation of the ambient IMF into which the MC is propagating. As an illustration in Figure 2, the MC (i.e. Plasmoid) is propagating north of the ecliptic plane, and the radial field in the ecliptic ahead of the disturbance created by the MC is sunward. The draped Bz perturbation in the portion of the sheath, which is directly upstream from the Earth, is clearly southward. If the radial component in the ecliptic were outward, the Bz perturbation would be northward. For an MC directed southward of the ecliptic, the perturbations are reversed. Using this scheme, they predict Bz perturbation correctly for only 13 of the total 26 cases (50%) and suggested the usefulness of their idea for real-time predictions may be limited. Wu and Dryer [1997] adopted the same concept of IMF draping by Gosling and McComas [1987] to compute

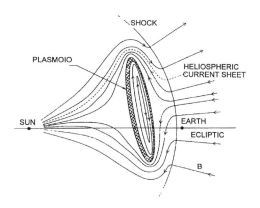

Figure 2. Central meridional cut through an idealized draped field configuration about a fast MC for an initially purely radial IMF(adapted from Gosling and McComas [1987].). For an MC propagating north of the ecliptic plane and an inward (sunward) IMF in the ecliptic ahead of the MC, a southward field perturbation in the sheath field would be expected.

the Bz using a 3-D MHD numerical simulation. Their results confirmed the model proposed by McComas et al. [1989]. In both of these two models the predictions of IMF Bz depend on the initial IMF configuration and the location of the solar disturbance with an initially-flat heliospheric current sheet. These initial conditions may not be accurate enough to render the actual Bz being observed. Using their recipe derived from a 3-D MHD simulation, five independent data sets were used to study the initial turning of the Bz during a solar maximum period (1978-1982). They found the prediction accuracy to be 62/73 (~83%). However, a critical problem resides in the assumption of their model about the flat heliospheric current sheet.

In this study, we use a kinematic code first designed by Hakamada and Akasofu (1982). This method combines the magnetic field frozen-in property and some observational property of the solar wind to construct a 3-D solar wind model. The solar source magnetic field at 2.5 R is used as the initial condition for the code. Therefore, the solar wind variations on source surface are following the variations of the magnetic field. Solar disturbances caused by a CME or a filament eruption event are assumed to be spherical symmetric to the radial direction on the source surface and their intensities decrease from the center of the source following a Gaussian distribution. This assumption is consistent with the concept of magnetic field draping. This model cannot simulate the internal fields inside the driver gases or CMEs. Details for using this code are given by Akasofu (2000). In order to test the ability of the kinematic code for predicting IMF Bz, we use the same events studied by McComas et al. (1989). Figure 3 shows the Feb. 17-19, 1979 event. The solar wind speed, number density of proton, and IMF are shown in dotted lines. The sheath is marked in shaded region. Using the solar source surface

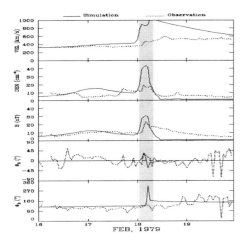

Figure 3. The interplanetary magnetic field and plasma data for the Feb.17-18, 1979 magnetic cloud (dotted). The sheath is in shaded region. The same parameters are simulated from the kinematic code of Hakamada and Akasofu [1982](solid).

magnetic field for Carrington Rotation 1678 and a filament disappear event at 0150 UT, Feb. 16, 1979 located at N16E59, the kinematic code predicts the same interplanetary parameters as shown in solid lines. Comparing the observations with the prediction, one finds that the disturbance is well predicted. Not only the magnitude but also the changes of the IMF are in good agreement. Particularly, the IMF Bz (or θ_B) is predicted. We simulate all seventeen events and have correctly predicted Bz

perturbations for 14 of the total 17 events (~83%) as listed in Table 1. The identification letters, the times and locations of the solar sources and shock arriving time at 1AU by ISEE 3, the observed and model predicted Bx, the predicted directions of IMF Bz and their comparisons with observations are given from left to the right, respectively.

5.   CONCLUDING REMARKS

Intense IMF Bz associates with two types of origins: (1) solar source and (2) interplanetary source. The intense IMF Bz inside a MC (or driver gas) are solar source originated and that outside a MC is mainly caused by IMF draping and shock compression. The solar source connection is difficult to predict. However, recent study [Zhao and Hoeksema, 1998] indicates a linear correlation between the direction of a cloud's central axial field and the associated filament's central axial field. This result is encouraging for the possibility to predict intense IMF Bz. Clearly, further study is needed to work out a scheme for prediction. Draping of IMF by MCs is most likely to generate intense IMF Bz which can also be reinforced by shock compression. Our kinematic simulation shows that intense IMF Bz is possible to predict from the knowledge of solar source surface magnetic field and eruptive-event observations.

Acknowledgments. This work is supported by grants from the National Science Council in Taiwan, ROC to National Central University (NSC89-2111-M-008-003). J.K.C. would like to thank

Table 1. Prediction of IMF Bz from solar source.

Solar Event	Time(UT)	Source Location (deg)	Shock Data Time (UT)	Upstream Bx	Model Bx	Predicted Bz (S/N)	Real Bz (nT)	Score
			1978					
A	Sept. 22; 2129	S23W13	Sept. 25; 0705	in	in	S	-1.6	1
B	Nov. 10; 0110	N17E01	Nov. 12; 0028	in	in	N	5.1	1
			1979					
C	Feb. 16; 0150	N16E59	Feb. 18; 0220	out	out	N	3.3	1
D	March 11; 1054	S24W76	March 15; 0454	in	in	S	3.4	0
E	April 3; 0111	S25W14	April 5; 0121	in	in	S	2.3	0
F	July 2; 1921	N11E36	July 6; 1853	out	out	N	2.0	1
G	Aug. 17; 2221	S33W35	Aug. 20; 0552	in	in	N	11.9	1
			1980					
I	April 4; 1503	N27W35	April 6; 1019	out	out	S	-1.5	1
J	July 14; 0818	S17E43	July 18; 1845	out	out	S	-13.2	1
			1981					
K	April 1; 0138	S43W52	April 3; 0308	in	out	S	3.7	0
L	April 24; 1355	N18W50	April 26; 0750	in	in	S	-2.1	1
N	May 13; 0357	N29W10	May 14; 1827	out	out	N	1.5	1
O	May 14; 0844	N20E35	May 16; 0513	in	in	S	-1.4	1
Q	Aug. 21; 0832	S16E02	Aug. 13; 1217	out	out	S	-2.6	1
R	Oct. 7; 2259	S17E83	Oct. 10; 1354	out	out	S	-0.8	1
S	Oct. 12; 0627	S18E31	Oct.13; 2217	in	in	N	1.0	1
U	Dec. 27; 0251	S13E16	Dec. 29; 0423	out	in	S	-4.6	1

Correct Prediction / Total Events

14/17

Drs. X. P. Zhao and B. T. Tsurutani for useful comments and Dr. Ching-Huei Lin for preparing figures.

REFERENCES

Akasofu, S. -I., 'Energy Coupling between the Solar Wind and the Magnetosphere', *Space Sci. Rev. 28*, 111, 1981.

Akasofu, S. I., Predicting geomagnetic storms as a space weather project, This issue of *AGU monography*, 2000.

Arnoldy, R. L., Signature in the interplanetary medium for substorms, *J. Geophys. Res., 76*, 5189, 1971.

Bothmer, V., and D.M. Rust, The field configuration of magnetic clouds and solar cycle, in *Coronal Mass Ejections* edited by N. Crooker, J.A. Joselyn and T. Feynman Amer. Geophys. Union Press, Washington D.C.,Geophys. Mon. Ser, 99, 139-146, 1997.

Bothmer, V., and R. Schwenn, Eruptive Proceedings of the Soho Workshop at Elba, Italy, 1993, *Space Sci. Rev., 70*, 215-220, 1994.

Bothmer, V., and R. Schwenn, Signatures of fast CMEs in interplanetary space, *Adv. Space Res., 17*, 319-322, 1996.

Bravo, S., The forecasting of intense geomagnetic storms, *Geofisica Internacional, 36*, 127-135, 1999.

Burlaga, L. F., et al., A magnetic cloud containing prominence material: January 1887, J. Geophys. Res., 103, 277,1998.

Chian, A. C. L., F. A. Borotto, and W. D. Gonzalez, Alfven intermittent turbnlence driven by temporal chaos, *Ap. J., 505*, 993-998, 1998.

Gonzalez, W. D. and B. T. Tsurutani, Criteria of interplanetary parameters causing intense geomagnetic storms (Dst< -100nT). *Planet. Space Sci., 35*,1101-1109, 1987.

Burlaga, L. F., E. Sittler, F. Mariani, and R. Schwenn, Magnetic loop behind an interplanetary shock : Voyager, Helios and IMP-8 observations, *J. Geophys. Res., 86*, 6673, 1981.

Chao, J. K, Steepening of Nonlinear Waves in the Solar Wind, *J. of Geophys. Res.,78*, 5411-5424, 1973.

Chen, J., Theory of prominence eruption and propagation: *Interplanetary consequences, 101*, 27499-27519, 1996.

Chen, J. and D. A. Garren, Interplanetary magnetic clouds: Topology and driving mechanism, *Geophys. Res. Letts., 20*, 2319-2322, 1993.

Gold, T., Magnetic storms., *Space Sci. Rev., 1*, 100, 1962.

Gonzalez, A. L. et al., Magnetic cloud field intensities and solar wind velocities, Geophys. Res. Lett., 25, 963-966, 1998.

Gosling, J. T., Coronal mass ejections and magnetic flux ropes in interplanetary space, in *Physics of Magnetic Flux Ropes*, Geophys. Mongr. Ser., vol. 58, edited by C. T. Russell, E. R. Priest, and L. C. Lee, p.343, AGU, Washington, D. C., 1990.

Gosling, J. T., and D. J. McComas, Field line draping about fast coronal mass ejecta: A source of strong out-of-the-ecliptic interplanetary magnetic fields, *Geophys. Res. Lett., 14*, 355-358, 1987.

Hakamada, K. and S. I. Akasafu, Simulation of three- dimensional solar wind disturbances and resulting geomagnetic storms, *Space Science Rev., 31*, 3-70, 1982.

Hoeksema, J. T. and X. Zhao, Prediction of magnetic orientation in driver gas associated -Bz events. *J. Geophys. Res., 97*, 3151-3156, 1992.

Hundhansen, A. J., Sizes and locations of coronal mass ejections: SMM observations from 1980 and 1984-1989, *Jouranl of Geophysical Research, 98*, 13, 177-13200, 1993.

Illing, R. M. E., and A. J. Hundhausen, Disruption of a coronal streamer by an eruptive prominence and coronal mass ejection, *J. Geophys. Res., 91*, 10951, 1986

Klein, L. W. and Burlaga, L. F.: 'Interplanetary Magnetic Clouds at 1 AU', *J. Geophys. Res., 87*, 613-624, 1982.

Lepping, R. P., J. A. Jones, and L. F. Burlaga, Magnetic field structure of interplanetary magnetic clouds at 1 AU, *J. Geophys. Res., 95*, 11957, 1990.

Martin, S. F., R. Bilimoria, and P. W. Tracadas, Magnetic field configurations basic to filament channels and filaments, in *Solar Surface Magnetism,* edited by R. J. Rutten and C. J. Schrijver, p. 308, Kluwer, Norwell, Mass., 1994.

Marubashi, K., Interplanetary flux ropes and solar filaments, in *Coronal Mass Ejections* edited by N. Crooker, J.A. Joselyn and T. Feynman Amer. Geophys. Union Press, Washington D. C., Geophys. Mon. Ser, 99, p. 147, 1997.

McComas D. J.,J. T. Gosling, and S. J. Bame, A test of magnetic field draping induced B_Z perturbations ahead of fast coronal mass ejecta, *J. Geophys. Res., 94*, 1465-1471, 1989.

Odstrcil, D.and V. J. Pizzo, Three-dimensional propagation of coronal mass ejections (CMEs) in a structured solar wind flow 1, CME launched within the streamer belt, *J. Geophys. Res., 104*,483-492, 1999.

Pudovkin, M. I., and A. D. Chertkov, Magnetic field of the solar wind, *Sol. Phys., 50*, 213, 1976.

Russell, C. T. and R. J. McPherron, Semiannual variation of geomagnetic activity. *J. Geophys. Res., 78*, 92-95, 1973.

Rust, D. M., Spawning and shedding helical magnetic fields in the solar atmosphere, *Geophys. Res. Lett., 21*, 241, 1994.

Tang, F., S. –I. Akasofu, E. Smith, and B. Tsurutani, Magnetic fields on the sun and the north-south component of transient variations of the interplanetary magnetic field at 1 AU, *J. Geophys. Res., 90*, 2703-2712. 1985.

Tang, F., B. T. Tsurutani, W. D. Gonzalez, S. –I. Akasofu, and E. J Smith, Solar sources of interplanetary southward Bz events responsible for major magnetic storms, *J. Geophys. Res., 94*, 3535-3541, 1989.

Tsurutani, B. T., and C. –I. Meng, Interplanetary magnetic-field variations and substorm activity, *J. Geophys. Res, 77*, 2964-2970, 1972.

Tsurutani, B. T., Ho, C. M., Arballo, J. K., Goldstein, B. E. and Balogh, A.: 'Large Amplitude IMF Fluctuations in Corotating Interaction Regions: Ulysses at Midlatitudes', *Geophys. Res. Lett. 22*, 3397, 1995.

Tsurutani, B. T., Russell, C. T., King, J. H., Zwickl, R. J. and Lin, R. P.: 'A Kinky Heliospheric Current Sheath: Causes of the CDAW6 Substorms', *Geophys. Res. Lett. 11*, 339, 1984.

Tsurutani, B. T., W. D Gonzalez, Tang, F., S. –I. Akasofu, and Smith E. J, Origin of interplanetary southward magnetic field responsible for major magnetic storms near solar maximum 1978-1979, *J. Geophys. Res., 93*, 8519, 1988.

Tsurutani, B. T., W. D. Gonzalez, F. Tang, and Y. T .Lee, Great magnetic storms, *Geophys. Res. Lett., 19*, 73, 1992.

Tsurutani, B. T., W. D. Gonzalez and Y. Kamide, Magnetic Storms, Surveys in Geophysics, 18, 363-383, 1997.

Tsurutani, B. T. et al., The January 10, 1997 auroral ho spot, horseshoe aurora and first substorm: A CME loop?, Geophys. Res. Lett., 25, 3047-3050, 1998.

Wu Chin-Chun, and Dryer M. Predicting the initial IMF Bz polarity's Change at 1 AU caused by shocks that precede coronal mass ejections, *Geophys. Res. Lett., 23,* 1709-1712, 1996.

Wu, Chin-Chun, and M. Dryer, Three-dimensional MHD simulation of interplanetary magnetic field changes at 1AU caused by a simulated solar disturbance and a tilted helio-spheric current/plasma sheet, *Sol. Phys., 173,* 391-408, 1997.

Wu, S. T., W. P. Guo, D. J. Michels, and L. F. Burlaga, MHD description of the dynamical relationships between a flux-rope, streamer, coronal mass ejection (CME) and magnetic cloud: an analysis of the 1997 January sun-earth connection event, *J. Geophys. Res. 104(7)*, 14789-14801, 1999.

Zhao X. P., Interaction of fast steady flow with slow transient flow: A new cause of shock pair and interplanetary Bz event, *J. Geophys. Res ., 97,* 5051, 1992.

Zhao, X. P., J. T. Hoeksema, J. T. Gosling, and J. L. Phillips, Statistics of IMF Bz events, Solar-Terrestrial Predictions Workshop-IV, 2, 712, 1993.

Zhao, X. P., J. T. Hoeksema, Central axial field direction in magnetic clouds and its relation to southward and dependence on disappearing solar filaments, *J. Geophys. Res., 103,* 2077-2083, 1998.

Zwan, B. J. and Wolf, R. A.: 'Depletion of the Solar Wind Plasma Near a Planetary Boundary', *J.Geophys.Res., 81,* 1636-1648, 1976.

J. K. Chao and H. H. Chen, Institute of Space Sciences, National Central University, Chung-li, Taiwan 32001, R.O.C.

Specifying Geomagnetic Cutoffs for Solar Energetic Particles

John W. Freeman

Department of Physics and Astronomy, Rice University, Houston, TX

Seth Orloff

MIT Lincoln Laboratory, Lexington, MA

Solar Energetic Particles (SEPs) produce detrimental effects on spacecraft and possibly high altitude aircraft. They represent a potential hazard for astronauts and cause the disruption of polar cap communications. They are an ionization source for the polar cap and as such should be included in ionospheric models. Finally, SEPs are an important source for the radiation belts and should be included in dynamic radiation belt models. All of these reasons imply the need for a Solar Energetic Particle model for real-time operational use. Given the flux of Solar Energetic Particles in space beyond the magnetosphere, the main parameter that describes the SEP population near the Earth is what is called the "geomagnetic cutoff". For a given latitude and local time or longitude at the Earth, the cutoff is the lowest energy (or rigidity) particle that can reach the Earth. Because magnetospheric currents have a significant effect on the cutoffs and because these "external" currents change significantly during a magnetospheric storm, it is necessary to calculate cutoffs for each storm and throughout the storm at a multiplicity of latitudes and local times. Geomagnetic cutoffs are traditionally calculated by tracing test particle trajectories in a model magnetic field. Cutoffs are not sharp but exhibit alternate forbidden and allowed regions. As a result, the global calculation of storm-time cutoffs is computationally intensive and if performed in traditional ways may not meet time constraints associated with real-time operations. This challenge can be met by specialized cutoff search strategies. This paper will review one effort that is underway to construct a fast, useful, solar energetic particle model.

I. INTRODUCTION

Solar energetic particles (SEPs) are hazardous in several ways. First, they are known to cause irreversible degrada-tion to spacecraft solar arrays as well as component damage. For example, the GOES 6 and 7 solar panel output dropped several tenths of an amp as a result of the March 1991 solar proton storm (*Allen and Wilkinson*, 1993). Second, SEPs are suspected to result in single-event upsets in high-altitude aircraft avionics (*Johansson and Dyreklev*, 1998). Third, they present a hazard to exposed astronauts during extravehicular operations (Report of the Committee on Solar and Space Physics, 2000). Fourth, they produce ionization in the polar ionosphere that results in radio

Space Weather
Geophysical Monograph 125

blackouts in the HF and VHF range over the polar cap (*Allen and Wilkinson*, 1993).

This latter item implies the need for information on SEPs as input data for ionospheric models. Equally important is the fact that SEPs find their way in to the magnetosphere and co-mingle with the radiation belt population. For this reason their flux is needed as input to radiation belt models. What is needed to address the space weather requirements is the global distribution of solar energetic particle fluxes, particularly protons, at the top of the ionosphere and on the equatorial plane throughout the magnetosphere.

Geomagnetic field lines that extend out from the polar regions connect with the interplanetary magnetic field and present little or no barrier to incoming energetic charged particles. At lower latitudes the Earth's magnetic field acts as a giant filter that removes lower energy (or rigidity) particles from the interplanetary solar energetic particle or cosmic ray spectrum by bending these particles in tight arcs that prevent them from reaching Earth. The energies excluded increase toward the equator.

Particle trajectories that allow particles to reach Earth from outside the magnetosphere are called allowed trajectories and those for which particles cannot reach Earth are called forbidden trajectories. The energy (or rigidity) boundary between allowed and forbidden trajectories is called the cutoff. The exact trajectory depends on the direction of arrival of the incoming particle in space and hence the cutoffs are a function of direction at a specified point near the Earth.

Trajectories exhibit chaotic behavior and hence cutoffs are not sharp but consist of bands of allowed and forbidden regions. The usual method of determining cutoffs is to compute trajectories of particles from a given point near the Earth outwards at successively lower energies until the forbidden trajectories are found. The upper cutoff energy is the highest detected allowed/forbidden transition and the lower cutoff energy is the lowest allowed/forbidden transition. The region in between is called the penumbra (*Cooke et al.*, 1991).

The determination of the spectrum of SEPs at a point near the Earth in a given direction involves applying the cutoff for that point and direction to the spectrum in interplanetary space. The problem of building an SEP specification model is reduced to 1.) Determination of the SEP spectrum just outside the magnetosphere, and 2.) Determination of the cutoff energies at all relevant points. It is known that the cutoffs change during a large solar energetic particle event (*Mason et al.*, 1995 and *Leske et al.*, 1997) and so these spectra and cutoffs must be specified dynamically throughout a SEP event.

The problem of the specification of solar energetic protons has been addressed in the past by several approaches:

• Smart and Shea [1979, 1989, and 1992] developed a phenomenological model called the Proton Prediction System. The PPS uses empirical information about solar flares and the production of solar energetic protons to specify SEP fluxes when fed a variety of information by an operator.

• A classical cutoff model using a reference magnetic field, IGRF 1980, was built by Smart and Shea [1985]. Smart and Shea updated this model with a 1990 epoch reference field for two altitudes, 20 km (*Smart and Shea*, 1997a) and 450 km (*Smart and Shea*, 1997b).

• Smart *et al.* [1999a, 1999b and 1999c] also developed a dynamic model by combining a reference field with a Tsyganenko model (*Tsyganenko*, 1990).

More recently we have developed a real-time dynamic cutoff particle trace model that uses the magnetic field from a MHD model to compute cutoffs coupled with real-time SEP flux data from spacecraft, when available, or statistical spectra when satellite data are not available. This model is called the Solar Energetic Particle Tracer (SEPTR).

II. THE SOLAR ENERGETIC PARTICLE TRACER

A. General Methodology

The Solar Energetic Particle Tracer (SEPTR) is a computer model that calculates the trajectories of high-energy (> 1 MeV) charged particles in magnetic fields (*Orloff*, 1998) and combines these cutoffs with SEP spectra from satellites or a default statistical model to produce differential spectra in the ionosphere and equatorial plane. SEPTR uses the Mission Research Corporation Integrated Space Model [*White*, 2000] MHD magnetic field and integrates the relativistic Lorentz force equations with a 5th order Runge-Kutta integration. Particles are launched vertically from a latitude and local-time grid near the surface of the Earth and also from a grid in the magnetic equatorial plane. Cutoffs are determined by varying the energy of the particles until the minimum energy at which the particle escapes from the magnetosphere is determined.

B. Cutoffs and Search Strategy

The methodology for finding the cutoffs is called the search strategy. The goal of an intelligent search strategy is to balance the computer time required against accuracy for the cutoffs. A *regular search* would start at a very high energy and reduce the energy in fixed steps until the first escaping trajectory was found. This approach can be accurate but is very time consuming if the steps are small enough to be accurate. A *bounded search* starts high and then takes a big step down in energy to get below the cutoff. Once below the cutoff it steps back up and gradually reduces the upper and lower bounds to approach the upper cutoff. SEPTR employs a *hybrid search*, which starts with a bounded search and shifts to a regular search after the range has been defined. In this way the compute time and accuracy are optimized.

C. Validation

Our first order validation of the integration of the equations of motion consisted of verifying that the particle energy and adiabatic invariants were adequately conserved along the trajectory. A second approach used was to compute the time-reverse trajectory and verify that the time-reverse particle returns to the origin. Having met accuracy criteria on these points, we next compared the cutoffs generated by SEPTR with cutoffs computed independently but for the same magnetic field model. We were generously provided with effective vertical cutoffs generated from the IGRF 1980 magnetic field model by D. Smart. These differed slightly from the cutoffs published by Smart and Shea, [1985] (*D. Smart*, private communication). For computational expediency we concerned ourselves only with the vertical cutoffs. For each grid point we computed the absolute difference between the SEPTR and Smart and Shea unpublished cutoffs from the same field model. We computed the standard deviation of these differences. The result, with one anomalous value excluded, was $\sigma = 0.03$ GV. We find the highest discrepancies at mid-latitudes where the spread in the allowed and forbidden bands is the largest.

It would be desirable to obtain a more complete validation of SEPTR using actual polar orbit satellite data; however, because of funding limitations this has not been done. Mason *et al.*, [1995] showed that cutoffs seen by SAMPEX can be substantially different from computed cutoffs. Moreover, it is known that solar energetic protons are promptly incorporated into the radiation belts and new belts are formed [*Hudson et al.*, 1997; *Obara*, 2000]. This greatly complicates the observational determination of cutoffs.

D. Obtaining SEP Spectra

The SEPTR strategy for obtaining real-time spectra involves the following steps: First, data are obtained from any one of several spacecraft in as near real-time as possible. These data must then be corrected for cosmic ray or radiation belt background. A spectrum can then be computed and the appropriate cutoff applied for each of the model grid points. A default spectrum based on data averaged over nine historical solar energetic particle storms has been prepared for use in the event of absence of spacecraft data. The default spectrum can also be triggered when the x-ray flux from the Sun exceeds a threshold or it can be triggered independently by the operator.

The ACE spacecraft is the prime SEP real-time data source. Data on energetic protons and electrons may also be available in near real-time from the geostationary orbit GOES and Los Alamos National Laboratory satellites. For the energies involved, spectra from these satellites are considered a good proxy for interplanetary spectra at 1 AU.

III. SUMMARY

We have developed a code (SEPTR) which calculates upper cutoffs for solar energetic protons and electrons when given a model of the Earth's magnetic field. The code also provides solar energetic proton spectra. The cutoffs and spectra are computed on two grids; one in the ionosphere and the other in the equatorial plane and as such can provide inputs to ionospheric or radiation belt models respectively. SEPTR is designed to work with the Integrated Space Model developed by Mission Research Corporation.

Acknowledgments. Seth Orloff performed most of the work on SEPTR described here. The development of SEPTR was sponsored by the U. S. Defense Threat Reduction Agency through a subcontract with Mission Research Corporation as a prime contractor for the Integrated Space Weather Prediction Model (ISM). We appreciate the unpublished cutoffs made available by Smart and Shea.

REFERENCES

Allen, J.H., and D.C. Wilkinson, Solar-Terrestrial Activity Affecting Systems in Space and On Earth, Solar-Terrestrial Predictions-IV, Proceedings of a Workshop at Ottawa, Canada, May 18-22, 1992, National Oceanic and Atmosphere Administration Environmental Research Laboratories, Boulder, CO., 1993.

Committee on Solar and Space Physics, Space Studies Board, National Academy of Science, Radiation and the International Space Station, National Academy Press, 2000.

Cooke, D.J., J.E. Humble, M.A. Shea, D.F. Smart, N. Lund, I.L. Rasmussen, B. Byrnak, P. Goret, and N. Petrou, On Cosmic Ray Cutoff Terminology, Nuovo Cimento C, 14, 213, 1991.

Hudson, M. K., S. R. Elkington, J. G. Lyon, V. A. Marchenko, I. Roth, M. Temerin, J. B. Blake, M. S. Gussenhoven, and J. R. Wygant, Simulation of Radiation Belt Formation During Storm Sudden Commencement, *J. Geophys. Res.*, 102, 14,087, 1997.

Johansson, Karin and Peter Dyreklev, Space Weather Effects on Aircraft Electronics, AI Applications in Solar-Terrestrial Physics, ESA WPP-148, 1998.

Leske, R.A., R.A. Mewalt, E.C. Stone, and T.T. von Rosenvinge, Geomagnetic Cutoff Variations During Solar Energetic Particle Events – Implications for the Space Station, *Proc. of the 25th International Cosmic Ray Conf.*, 2, 381, 1997.

Mason, G.M., J.E. Mazur, M.D. Lopper, and R.A. Mewalt, Charge State Measurements of Solar Energetic Particles Observed with SAMPEX, *Astrophys. J., 452,* 901, 1995.

Obara, Takahiro, New Proton Radiation Belt Formation Due to Injection of Tens of MeV Protons During Magnetospheric Compression Event, submitted to *22nd International Symposium on Space Technology and Science,* May 28-June 4, Moriaka, Japan, 2000.

Orloff, Seth, A Computational Investigation of Solar Energetic Particle Trajectories in Model Magnetospheres, Rice University Ph.D. Thesis, 1998.

Smart, D. F., and M. A. Shea, PPS76 - A Computerized "Event Mode" Solar Proton Forecasting Technique, in *Solar Terrestrial Predictions Proceedings,* edited by R. F. Donnelly, U. S. Department of Commerce, NOAA/ERL, *1,* 406, 1979.

Smart, D. F., and M. F. Shea, Galactic Cosmic Radiation and Solar Energetic Particles, in *Handbook of Geophysics and the Space Environment,* Air Force Geophysics Laboratory, 1985.

Smart, D. F., and M. A. Shea, PPS-87: A New Event Oriented Solar Proton Prediction Model, *Adv. Space Res., 9,* No. 10, 281, 1989.

Smart, D. F., and M. A. Shea, Modeling the Time-Intensity Profile of Solar Flare Generated Particle Fluxes in the Inner Heliosphere, *Adv. Space Res., 12,* No. 2-3, 303, 1992.

Smart, D. F., M. A. Shea, World Grid of Calculated Cosmic Ray Vertical Cutoff Rigidities for Epoch 1990.0, *Proc. of the 25th International Cosmic Ray Conf.,* Vol. 2, 401-404, 1997a.

Smart, D. F., M. A. Shea, Calculated Cosmic Ray Cutoff Rigidities at 450 km for Epoch 1990.0, *Proc. of the 25th International Cosmic Ray Conf.,* Vol. 2, 397-399, 1997b.

Smart, D. F., M. A. Shea, E. O. Flückiger, A. J. Tylka, and P. R. Boberg, Calculated Vertical Cutoff Rigidities for the International Space Station During Magnetically Active Times, *26th International Cosmic Ray Conference, 7,* 398-401, 1999a.

Smart, D. F., M. A. Shea and E. O. Flückiger, Calculated Vertical Cutoff Rigidities for the International Space Station During Magnetically Quiet Times, *26th International Cosmic Ray Conference, 7,* 394-398, 1999b.

Smart, D. F., M. A. Shea, E. O. Flückiger, A. J. Tylka, and P. R. Boberg, Changes in Calculated Vertical Cutoff Rigidities at the Altitude of the International Space Station as a Function of Geomagnetic Activity, *26th International Cosmic Ray Conference, 7,* 337-340, 1999c.

Tsyganenko, N.A., Quantitative models of the magnetospheric magnetic field: methods and results, Space Sci. Rev., 54, 75, 1990.

White, W. W., MHD Simulation of Magnetospheric Transport at the Mesoscale, This Volume, 2000.

Dr. John W. Freeman, Rice University, Physics and Astronomy Department, MS 108, P. O. Box 1892, Houston, Texas 77251-1892.

Status of Cycle 23 Forecasts

David H. Hathaway, Robert M. Wilson, Edwin J. Reichmann

NASA/Marshall Space Flight Center, Huntsville, Alabama

Forecasts for the amplitude of cycle 23 that were reported prior to the start of the cycle covered a full range of values from very small to very large. A forecast reached by the consensus of a panel of forecasters convened at the time of minimum in 1996 [*Joselyn et al.*, 1997] suggested that this cycle would be much larger than average with the smoothed International Sunspot Number reaching a maximum of 160 ± 30 in the middle of the year 2000. A recent survey of solar cycle prediction techniques [*Hathaway et al.*, 1999] found that the two most reliable techniques for forecasting the cycle prior to its start give similar predictions for this cycle's maximum — 154 ± 26 and 153 ± 33. Curve-fitting and regression techniques can be used with some confidence now that cycle 23 is well underway. These techniques indicate a more modest sunspot cycle with a maximum of 114 ± 10 — only slightly larger than average. The current (July 2000) prediction using the combined predictions from both precursors and curve-fitting gives a cycle amplitude of about 136 ± 20. This is within the errors given by the consensus and the precursor technique predictions but very close to their lower bounds.

1. INTRODUCTION

Accurate predictions of the levels of solar activity are increasingly important as we become more reliant upon technology that is sensitive to this activity. Ideally, we would like to predict solar activity using a model of the Sun's magnetic dynamo along with observations of current and past conditions to initialize that model. Unfortunately, both the model and many of the important observations do not exist at present. Given this state of the art we, as a community, have been forced to predict solar cycle behavior by using statistical methods that rely on relationships between past and future behavior. These relationships are not only useful for predicting solar cycle activity but they also reveal several interesting characteristics of the solar activity cycle.

Space Weather
Geophysical Monograph 125

Dozens of predictions for the amplitude of cycle 23 were offered prior to its start in 1996. Most of these predictions were for a larger than average cycle as measured by the smoothed (13-month running mean) International Sunspot Number. A panel of solar cycle forecasters was convened in September of 1996 to sort through these predictions and arrive at some consensus concerning the amplitude of cycle 23, [*Joselyn et al.*, 1997]. Several classes of prediction techniques were identified and their most likely amplitudes for cycle 23 were noted. The even/odd cycle behavior [*Kopecký*, 1991; *Letfus*, 1994] indicated an amplitude near 200. Geomagnetic precursors gave an amplitude of 160 while spectral methods and the recent climatological mean gave 155. Neural networks suggested an amplitude near 140. The consensus opinion of this panel was that cycle 23 would have an amplitude of 160±30 for the smoothed sunspot number with the time of maximum in mid-year 2000.

In a recent paper [*Hathaway et al.*, 1999], we undertook a systematic survey and testing of a number of

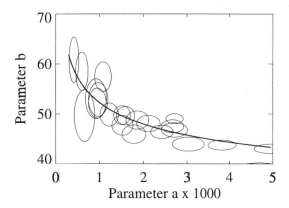

Figure 1. Rise-time parameter b (in months) as a function of cycle amplitude parameter a. Error ovals are plotted for each of the last 22 cycles. The solid line through these ovals is given by equation (2).

solar cycle prediction techniques and derived a similar prediction (154 ± 21) based on the two most reliable geomagnetic precursor techniques. Now that the cycle is near its maximum we should be in a good position to assess these predictions and to examine the characteristics of the current cycle.

2. EARLY PREDICTIONS FOR CYCLE 23

A number of forecasting techniques have been devised to predict the amplitude of a cycle using observations from times near or before the time of sunspot number minimum. Although the term "precursor technique" is usually reserved for those techniques that use geomagnetic activity, here we consider any technique that provides a prediction before the start of the cycle as a precursor technique. Many of these techniques are describe in detail in *Hathaway et al.* [1999] and we refer the reader to that paper for those details. Several of these methods were found to be unreliable in the sense that the relationships derived between the given quantities and the amplitude of the next maximum changed significantly from cycle-to-cycle. Other methods were unreliable in the sense that the errors in their predictions were not significantly smaller than those associated with using the climatological mean (the average cycle).

Among the methods found to be most reliable, the top three relied on geomagnetic activity to predict solar activity. The method derived from the work of *Feynman* [1982] uses the geomagnetic aa index and separates it into a component that varies directly with sunspot number and a second, remaining, component that reaches a peak at about the time of sunspot minimum. This

method gives an amplitude for cycle 23 of 154 ± 26. The method due to *Thompson* [1993] uses the number of geomagnetically disturbed days (days with geomagnetic index $Ap \geq 25$) during the previous cycle to predict the amplitude of the next cycle. This method gives an amplitude for cycle 23 of 153 ± 34. The third of these methods is due to *Ohl* [1966]. Ohl's method uses the minimum in the aa index near the time of sunspot minimum to predict the next sunspot maximum. This method gives an amplitude for cycle 23 of 135 ± 35. Since the average amplitude of cycles 1 through 22 is about 100, all of these geomagnetic precursor techniques predict a larger than average cycle for cycle 23.

Two other techniques were found by *Hathaway et al.* [1999] to be reliable but less accurate than the previous three. The Maximum-Minimum technique uses the size of the sunspot minimum to predict the next maximum. This method gives an amplitude for cycle 23 of 115 ± 52. The Amplitude-Period technique uses the length, or period, of the previous cycle to predict the amplitude of the next cycle. This method gives an amplitude for cycle 23 of 121 ± 54. Here again, both methods predict a larger than average cycle but closer to the average than those from geomagnetic precursors.

Wilson et al. [1998] examined a number of techniques for estimating the size of cycle 23 based on near minimum conditions. All of the techniques examined gave a larger than average amplitude for cycle 23. The region of overlap between the various predictions suggested that cycle 23 would be very large — comparable to or even larger than the last two cycles.

Many other techniques have been used to predict the amplitude of cycle 23 but were of a nature that made it difficult or impossible to test using the procedures described in *Hathaway et al.* [1999]. Noteworthy among these techniques is one described by *Schatten et al.* [1978] which uses the strength of the sun's polar magnetic field near minimum to predict the size of the next sunspot maximum. Unfortunately, the data required for this method (polar magnetic fields) have only been available for two sunspot cycles and both cycles had similar amplitudes. While *Schatten et al.* [1978] suggested a number of proxies for the polar magnetic field that might extend the method to earlier times, *Layden et al.* [1991] found that these proxies were not reliable. Thus, rigorous testing of this promising technique has not been possible.

Schatten and Pesnell [1993] did employ this method to provide an early estimate for the size of cycle 23 and found an amplitude of 170 ± 25. As sunspot minimum approached *Schatten et al.* [1996] provided a later estimate of 138 ± 30 for the amplitude of cycle 23.

3. CURVE FITTING CYCLE 23

Hathaway et al. [1994] constructed a mathematical function that mimics the behavior of the sunspot cycle. The function is asymmetric, with a rapid rise and slow decay. This function gives the sunspot number R as a function of time t with two free parameters — the amplitude a and the starting time t_0. The function is given by

$$R(t) = a(t - t_0)^3 / [e^{(t-t_0)^2/b^2} - 0.71] \qquad (1)$$

where the dependent parameter b gives the number of months for the rise to maximum and the fixed parameter, 0.71, gives the sunspot cycle asymmetry. This function was tailored to provide a good fit to each of the previous cycles and requires only two parameters per cycle to produce that fit.

Figure 1 shows the relationship between the rise-time parameter b and the amplitude parameter a. Error ovals are plotted for each of the last 22 cycles (long rise times are more uncertain than short rise times and large amplitudes are more uncertain than small amplitudes). The thick line passing through the data is given by

$$b(a) = 27.12 + 25.15/(a \times 1000)^{1/4}. \qquad (2)$$

This relationship between rise-time and amplitude was first noticed by *Waldmeier* [1935, 1939] who found that large amplitude cycles rise to maximum in a shorter time than small amplitude cycles.

Predictions for the smoothed sunspot numbers are obtained with this function by finding the values for

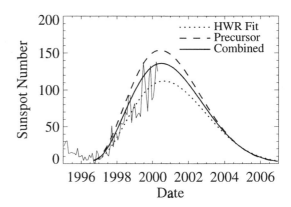

Figure 3. Cycle 23 predictions and current status. The thin ragged line shows the monthly averages of the International Sunspot Number through June 2000. The thick dotted line is the current fit to those averages using the function given by equation (1). The thick dashed line shows the cycle behavior using the geomagnetic presursor prediction for the amplitude. The thick solid line shows the combined forecast using a weighted average of the geomagnetic precursor and curve fitting techniques.

the amplitude and starting time that give the best fit. These two parameters are always well established before the time of sunspot maximum. The starting time is usually well determined within a year after minimum.

The behavior of solar activity near the minimum between cycles 22 and 23 provoked some confusion concerning the actual month of the minimum for this cycle. The minimum in the 13-month running mean of the International Sunspot Number occurred in May 1996 but other indicators suggest September as a more appropriate month for minimum [*Altrock et al.* 1999, *Harvey and White,* 1999]. Fitting the function given by equation (1) to cycle 23 indicates a start in May 1996 but with minimum (due to overlap with cycle 22 activity) in September 1996. This curve-fitting has consistently given this result since the Fall of 1997.

The amplitude of a cycle is not well determined by the curve-fitting technique until much closer to the time of maximum. Figure 2 shows the standard deviations between the predicted and observed cycle behavior as functions of time since minimum averaged over the last 11 cycles. Using the precursor technique for the amplitude of the function along with the early determination of the starting time gives a good fit throughout each of the cycles. The curve-fitting technique itself doesn't surpass this precursor prediction until about 48 months after minimum on average. The weighted combination of the two [*Hathaway et al.,* 1999] provides some overall improvement.

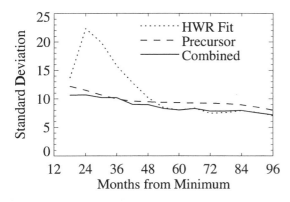

Figure 2. Standard deviations between predicted and observed cycle behavior as functions of time since minimum. The curve-fitting predictions (dotted line) become more reliable than the precursor predictions (dashed line) at about 48 months after minimum. The weighted combination of the two (solid line) gives a slight improvement over using the presursor derived amplitude alone.

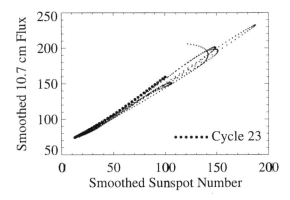

Figure 4. Smoothed 10.7 cm radio flux (in solar flux units) vs. smoothed sunspot number. Cycle 23 behavior, as represented by the large dots, lies at the upper limit for the 10.7 cm flux observed at the corresponding sunspot numbers.

4. CHARACTERISTICS OF CYCLE 23

Cycle 23 is currently near its maximum. The preliminary smoothed (13-month running mean) International Sunspot Number for December 1999 is 111. Although the size and timing of the actual maximum will require another year or more of activity to be set, many of the characteristics of cycle 23 are already evident.

Figure 3 shows the monthly averaged International Sunspot Numbers through June 2000. The current fit to this activity using equation (1) is given by the dotted line. This fit gives a minimum in September 1996 and a maximum of 114 ± 10 in August 2000. The prediction using the geomagnetic precursors is shown by the dashed line with a maximum of 154 ± 21 in June 2000. The combined technique described by *Hathaway et al.* [1999] takes a weighted average of these two predictions with the current weighting being nearly 50/50. This prediction is shown by the thick, solid line in Figure 1. It gives a maximum of 136 ± 20 in June or July 2000. This current predicted value lies within the range of values given by both the precursor prediction and the concensus prediction (Panel), but near the lower bound and above the range of values given by the current best fit. These values for maximum are all above average but are more moderate than most of the early predictions.

The International Sunspot Number is only one of many indicators of solar activity. The radio flux at 10.7 cm is often taken as the indicator of choice for ionospheric and thermospheric modeling. These two indicators of solar activity are extremely well correlated as shown in Figure 4 by plotting the smoothed radio flux against the smoothed sunspot number. (This smoothing is done with the 24-month Gaussian weighted average described by *Hathaway et al.* [1999].) The be-

havior of cycle 23 is shown by the large dots in Figure 4. Although this behavior is very close to that of the previous four cycles it does lie at the upper extreme for corresponding sunspot numbers. Cycle 23 has had the *largest* radio flux as a function of sunspot number for any of the observed cycles (cycles 19-22 overlap each other while cycle 18 is significantly smaller).

Sunspot area is another indicator of solar activity that is well correlated with the International Sunspot Number. Figure 5 shows the smoothed sunspot area plotted against the smoothed sunspot number for the last 11 cycles. Cycle 23, again represented by the large dots, exhibits behavior close to the other cycles but at the lower extreme for corresponding sunspot numbers. Cycle 23 has had the *smallest* sunspot area as a function of sunspot number for any of the observed cycles.

Geomagnetic fluctuations are also associated with the sunspot cycle as suggested by their predictive ability in the previous sections. These fluctuations are usually well correlated with sunspot number during the rising phase of the cycle but then have another component that peaks much later in the cycle due to recurrent solar wind streams that are associated with coronal holes (rather than sunspots). Figure 6 shows the smoothed geomagnetic *aa* index plotted against smoothed sunspot number for the last 12 cycles. The *aa* index continues to rise after sunspot maximum. This behavior gives the steady or even increasing geomagnetic activity during the declining phase of the sunspot cycle. The rising phase of cycle 23 is shown with the large dots. Cycle 23 appears quite normal compared to the previous cycles for the relative levels of the sunspot number and the *aa* index.

The number of geomagnetically disturbed days (as

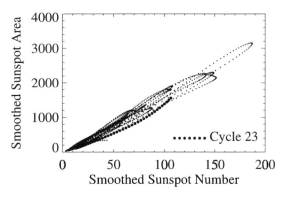

Figure 5. Smoothed sunspot area (in millionths of a hemisphere) vs. smoothed sunspot number. Cycle 23 behavior, as represented by the large dots, lies at the lower limit for the sunspot area observed at the corresponding sunspot numbers.

given by $Ap \geq 25$) has a similar behavior with sunspot number. The average number of disturbed days per month increases as the sunspot number increases during the rising phase of the sunspot cycle but then continues to rise to a peak much later. Figure 7 shows the smoothed number of disturbed days per month plotted against smoothed sunspot number for the last 6 cycles. Cycle 23 appears normal but much closer in behavior to the two smallest cycles of the group (cycles 17 and 20).

5. CONCLUSIONS

All of the reliable precursor techniques predicted a larger than average amplitude for cycle 23. The behavior thus far supports this assessment but also indicates that the actual size of cycle 23 may be at the lower limit of the predicted range. The relative success of these predictions suggests a physical basis behind the techniques. The geomagnetic precursor techniques imply a significant overlap between cycles so that geomagnetic disturbances late in a cycle are connected to the activity of the next cycle. The Maximum-Minimum and Period-Amplitude techniques suggest that large cycles tend to start early — cutting off the previous cycle and producing more activity at minimum.

Some of the characteristics of cycle 23 suggest that it is behaving oddly. For the observed sunspot numbers the 10.7 cm radio flux is higher than that of any other cycle while the sunspot area is lower than that of any other cycle. The geomagnetic indices appear quite normal for the sunspot numbers that have been seen so far. But, the number of disturbed days is more in-line with the average sized cycles (cycles 17 and 20).

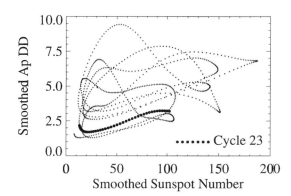

Figure 7. Smoothed geomagnetically disturbed days per month (days with $Ap \geq 25$) vs. smoothed sunspot number. Cycle 23 behavior, as represented by the large dots, lies in the lower regions of the values observed since 1932 for the rising phase of each solar cycle.

This sunspot cycle has not yet reached maximum so some surprises may still be in store. Nonetheless, the current predicted maximum of 136 ± 20 in June or July 2000 still seems reasonable.

Acknowledgments. This study was funded by the U. S. National Aeronautics and Space Administration through the Office of Space Science. The International Sunspot Number is compiled by the Sunspot Index Data Center in Brussels, Belgium. The 10.7 cm radio flux is acquired by the Canadian National Research Council at the Dominion Radio Astrophysical Observatory. The geomagnetic indices are acquired by a number of observatories and reported by the National Oceanic and Atmospheric Administration. Sunspot area information was produced by the Royal Greenwich Observatory until 1976 and has continued to be produced by the USAF/NOAA Solar Optical Observing network. Our community is greatly indebted to these institutions and their observers for these valuable datasets.

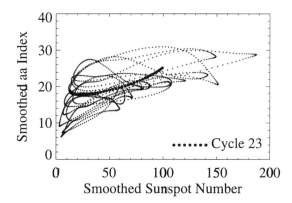

Figure 6. Smoothed geomagnetic *aa* index (in nT) vs. smoothed sunspot number. Cycle 23 behavior, as represented by the large dots, lies in the midst of the values observed since 1868 for the rising phase of each solar cycle.

REFERENCES

Altrock, R. C., M. Rybanský, V. Rušin, and M. Minarovjech, Determination of the solar minimum period between cycles 22 and 23 from the coronal index of solar activity, *Sol. Phys. 184*, 317-322, 1999.
Feynman, J., Geomagnetic and solar wind cycles, 1900-1975, *J. Geophys. Res., 87*, 6153-6162, 1982.
Harvey, K. L., and O. R. White, What is solar minimum?, *J. Geophys. Res., 104*, 19,759-19,764, 1999.
Hathaway, D. H., R. M. Wilson, and E. J. Reichmann, The shape of the solar cycle, *Sol. Phys. 151*, 177-190, 1994.
Hathaway, D. H., R. M. Wilson, and E. J. Reichmann, A synthesis of solar cycle prediction techniques, *J. Geophys. Res., 104*, 22,375-22,388, 1999.
Joselyn, J., et al., Panel achieves consensus prediction of solar cycle 23, *Eos Trans. AGU 78*, 205, 211-212, 1997.
Kopecký, M., Forecast of the maximum of the next 11-year

cycle of sunspots No. 23, *Bull. Astron. Inst. Czechlo. 42,* 157-158, 1991.

Layden, A. C., P. A. Fox, J. M. Howard, A. Sarajedini, K. H. Schatten, and S. Sofia, Dynamo-based scheme for forecasting the magnitude of solar activity cycles, *Sol. Phys., 132,* 1-40, 1991.

Letfus, V., Prediction of the height of solar cycle 23, *Sol. Phys. 149,* 405-411, 1994.

Ohl, A. I., Forecast of sunspot maximum number of cycle 20, *Solice Danie 9,* 84, 1966.

Schatten, K., D. J. Myers, and S. Sofia, Solar activity forecast for solar cycle 23, *Geophys. Res. Lett., 23,* 605-608, 1996.

Schatten, K. H. and W. D. Pesnell, An early solar dynamo prediction: Cycle 23   Cycle 22, *Geophys. Res. Lett., 20,* 2275-2278, 1993.

Schatten, K. H., P. H. Scherrer, L. Svalgaard, and J. M. Wilcox, Using dynamo theory to predict the sunspot num-

ber during solar cycle 21, *Geophys. Res. Lett., 5,* 411-414, 1978.

Thompson, R. J., A technique for predicting the amplitude of the solar cycle, *Sol. Phys. 148,* 383-388, 1993.

Waldmeier, M., Neue eigneschaften der sonnenfleckenkurve, *Astron. Mitt. Zürich 14(133),* 105-130, 1935.

Waldmeier, M., Sunspot Activity, *Astron. Mitt. Zürich 14(138),* 439 and 470, 1939.

Wilson, R. M., D. H. Hathaway, and E. J. Reichmann, An estimate for the size of cycle 23 based on near minimum conditions, *J. Geophys. Res., 103,* 6595-6603, 1998.

D. H. Hathaway, R. M. Wilson, and E. J. Reichmann, Mail Code SD50, NASA/Marshall Space Flight Center, Huntsville, AL 35812. (e-mail: david.hathaway@msfc.nasa.gov; robert.wilson@msfc.nasa.gov; ed.reichmann@msfc.nasa.gov)

Solar Activity Predicted with Artificial Intelligence

Henrik Lundstedt

Swedish Institute of Space Physics, Solar-Terrestrial Physics Division, Scheelev. 17, SE-223 70 Lund, Sweden

The variability of solar activity has been described as a non-linear chaotic dynamic system. AI methods are therefore especially suitable for modelling and predicting solar activity. Many indicators of the solar activity have been used, such as sunspot numbers, F10.7 cm solar radio flux, X-ray flux, and magnetic field data. Artificial neural networks have also been used by many authors to predict solar cycle activity. Such predictions will be discussed. A new attempt to predict the solar activity using SOHO/MDI high-time resolution solar magnetic field data is discussed. The purpose of this new attempt is to be able to predict episodic events and to predict occurrence of coronal mass ejections. These predictions will be a part of the Lund Space Weather Model.

1. INTRODUCTION

One of the biggest scientific challenges in solar-terrestrial physics today is to develop improved predictions of the solar activity and to create a better understanding of the solar activity. Many have taken the challenge to predict the current solar cycle 23 (Figure 1), the amplitude and time of the maximum. The Sun varies on many time scales, on around 5 minutes associated with oscillations, on hours associated with the emerging of magnetic flux, on 11-22 years associated with the solar cycle, on 100-500 years associated with modulation of cycles and finally on billions of years associated with evolution of stars. Many indicators of solar activity have been used, among them the sunspot number, the integrated 10.7-cm (2.8 GHz) radio flux, the total solar magnetic flux, the solar mean field and so on.

The solar activity has been claimed to exhibit a chaotic behaviour [Weiss, 1998; Mundt et al., 1991] on time scales, associated with the eruption of active regions, the 22-year magnetic cycle and modulation at grand minimum (e.g. the Maunder minimum). If the solar activity can be described as a low dimensional chaotic dynamic system, then the consequences are profound: it provides knowledge of the complexity of the system, i.e. the number of equations needed to describe the system's time evolution. The underlying deterministic nature of the system allows for accurate short-term predictions. Once the Lyapunov exponent has been calculated then the demarcation between short-term and long-term can be defined. Mundt et al., [1991] reconstructed an attractor from sunspot numbers data using the methods of time delay [Takens, 1981]. They found a time delay of 10 months, a dimension of D=2.3 and Lyapunov exponent = .02 per month. Predictions longer than about a year are therefore not possible according to them if no further external information is added.

Today we have so much more, and new information about the solar activity besides the information about the variability of the sunspot numbers. Observations made with the ESA/NASA spacecraft SOHO [Fleck et al., 1995] have opened up a new window to solar activity. The instrument MDI [Scherrer et al., 1995] on board SOHO has, by using the helioseismological methods, given us a possibility to observe the plasma conditions below the solar surface. The global solar dynamo [Parker, 1979], is thought to be generated in a thin region, called tachocline, at the base of the convection zone. Dynamic variations at this base of the convective zone, and therefore variations of the solar dynamo have recently been detected [Howe et al., 2000]. The

Space Weather
Geophysical Monograph 125
Copyright 2001 by the American Geophysical Union

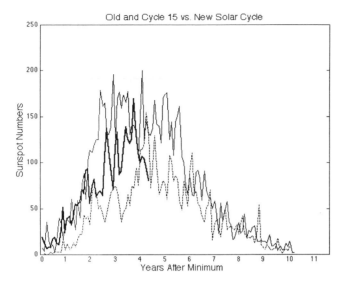

Figure 1. The diagram shows the sunspot number for cycle 15, 22 and the current 23. The last monthly value plotted is for February 2001. Cycle 23 has been suggested to resemble both cycle 22 and 15. Cycle 23 (thick solid line) had a minimum in October 1996. Cycle 22 (solid line) had a minimum in August 1986 and a maximum in June 1989. Cycle 15 (dashed line) had a minimum in June 1913 and a maximum in June 1917.

rotation was found to vary with a period of 1.3 years at low latitude and possibly with 1.0 year at high latitude. These results might have a crucial importance to our understanding of the solar cycle variation.

The evolution of solar activity, hours before it manifestates itself on the solar surface, is now also possible to study using inversion methods in helioseismology and solar tomography [Kosovichev, 1999].

SOHO observations have also shown that the small-scale magnetic field of the solar photosphere is replaced within 48 hours. That corresponds to enough energy to heat the corona. The long lasting problem of the heating of the corona has therefore come much closer to a solution.

The most geoeffective coronal phenomena, coronal mass ejections and coronal holes are related to changes of the solar magnetic field. If the pattern for these changes are found, then also predictions could be made of the coronal phenomena and effects in the interplanetary space and interaction with earth magnetosphere.

2. PREDICTIONS OF SUNSPOT CYCLE ACTIVITY

To study the solar activity and dynamics we need nonlinear methods. Neural networks are such methods. Let me therefore briefly describe somewhat neural networks.

An artificial neural network (ANN) [Haykin, 1994 and Lundstedt, 1997] is essentially a group of interconnected

computing elements (neurons). Typically a neuron computes the sum of its inputs and passes this sum through a non-linear function (an activation function). Each neuron has only one output, but this output is multiplied by a weighting factor if it is used as the input to another neuron. The neural networks typically exhibit two types of behaviour. If no feedback (recurrent) loops connect neurons, the signal produced by an external input moves only in one direction and the output of the network is just the output of the last group of neurons in the network. In that case the network behaves mathematically like a non-linear function of the inputs. The second type of network behaviour is observed when there are feedback loops in the neuron connections. In this case the network behaves like a dynamical system, so the output of the neurons varies with time. The neuron output can oscillate, or settle down into steady state values, or since the threshold function introduces nonlinearity into the system, become chaotic. The modelling capability of the ANN can be ascribed to its ability to learn the mathematical function underlying the system operation. If the network is designed and trained properly, it can perform generalisation rather than simple curve fitting. Any continuous function can be implemented by a three-layer feedforward neural network [Lorentz, 1976]. A combination of neural networks can describe a discontinuous function.

In summary the benefits of neural networks are their nonlinearity, that they are dynamical, they can create from examples an input-output mapping, and that they can describe the interaction between microscopic and macroscopic phenomena.

3. PREDICTIONS OF SUNSPOT CYCLE ACTIVITY

Many [Koons et al., 1990; Weigend et al., 1992; Calvo et al., 1995; Macpherson et al., 1995; Brown et al., 1994; Liszka 1993, 1998; Conway et al., 1998; Ashmall and Moore, 1998; and Tian and Fan, 1998] have trained neural networks to predict the sunspot number as an indicator of the solar activity. The neural networks have been trained to predict the sunspot number on different time scales from days, months to years ahead. Most often a multilayer perception neural network has been used. However, also other neural networks such as a recurrent neural network Ashmall and Moore [1998] and also combinations of networks such as MBP and SOM have been used by Liszka [1993].

In their study Calvo [1995] started by constructing an attractor. In this way they obtained the embedded dimension and therefore how many variables they need to describe the dynamic system. From that they also learned how many input nodes they need for the neural network. They found they needed twelve input nodes i.e. 12 yearly values for a prediction of next year. Ashmall and Moore, on the other hand found they needed monthly values (one monthly value each year) to predict next year. Ashmall and Moore, [1998] and Calvo et al., they both predicted a maximum of 160 for cycle 23 year 2000. Conway et al., [1998] on the other hand predicted 130±30, 2001.

4.PREDICTIONS OF SOLAR ACTIVITY AND CMES

During the progress of current solar cycle 23, other cycles have been suggested as similar. In the beginning of solar cycle 23 the cycle followed closely cycle 22. However, the activity then drastically decreased, and the small solar cycle 15 was suggested to resemble cycle 23. Then the activity again increased and a monthly value of 169.1 for July 2000 was reached. This value is above what most often is suggested as maximum for solar cycle 23. The difficulties with the predictions of current monthly solar activity cycle (Figure 1) and solar maximum Lantos and Richard [1999] shows the urgent need of new data and new methods.

We need a much deeper knowledge of the solar activity and the causes of solar activity in order to improve the predictions. Observations with SOHO spacecraft are now giving us that new and deeper knowledge.

Solar magnetograms obtained by the MDI instrument on board SOHO represent new and very interesting data for studies of the solar activity. The MDI instrument produces 15 magnetograms each 96 minutes or a magnetogram each one or five minute. That makes it possible to study rapid variations of the large-scale magnetic field variations. Zhao et al., 1997 showed, for the famous January 6-11 1997 halo-CME event, that the global magnetic field changed dramatically at time of the CME event. We extended that study by analyzing a list of 15 CMEs [St. Cyr et al., 1999]. In all cases did the global magnetic field change. However, we also had times when the global magnetic field changed but no CME was reported.

Another method to find times of CMEs, has been by using solar mean field data observed by the MDI instrument on-board SOHO. It's based on the idea, that a CME is a global phenomenon and a way for the Sun to get rid of magnetic flux [Luhmann et al., 1998]. We therefore expect to see a change of the mean field.

We have studied time series of mean field data, observed with 96 minute's and one minute's resolution. Sequences of minute's resolution mean field data were analyzed using wavelet transforms [Torrence and Compo, 1998] and compared with a data set of times of CMEs.

A significant fraction of the resulting wavelet power spectra showed a characteristic peak at the time of the coronal mass ejection [Boberg and Lundstedt, 2000]. The

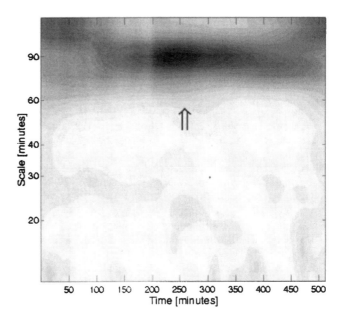

Figure 2. A sum of 53 wavelet power spectra during CME activity. The arrow indicate the CME onsets according to SOHO/LASCO observations.

wavelet power spectrum given in Figure 2 is a sum of 53 individual power spectra, each obtained from a 512 minute long solar magnetic field mean field period containing a reported CME at minute 256 (indicated by the arrow).

We have trained recurrent neural networks to predict the mean field on different time scales. The results from these runs and the results from the wavelet studies are now being combined into a method to predict CMEs. However, these attempts are also part of a bigger plan, namely to develop a model of the solar activity based on intelligent systems. That model will also include helioseismological results and results from studies of active regions and the total magnetic flux.

5. SUMMARY

We are entering a very exciting time with the new SOHO observations and results. The results about conditions below the solar surface, in convective zone and just below convective zone, are very interesting. The results will lead to a much deeper understanding of the solar activity and the solar dynamo. Developing an intelligent hybrid system, including both these scientific results and the data based predictions mentioned above, is now possible. That's the goal of the Lund Space Weather Model [Lundstedt, 1999].

Acknowledgments. I am very grateful for the SOHO MDI data made available by Stanford team and for very interesting discussions with Dr. Todd Hoeksema and Prof. Phil Scherrer at Stanford University in California.

REFERENCES

Ashmall, J. and V. Moore, Long term prediction of solar activity using neural networks, in the proceedings of "*AI Applications in Solar-Terrestrial Physics*", July 29-31, 1997 in Lund, Sweden, edited by I. Sandahl and E. Jonsson, ESA WPP-148, April 1998.

Boberg, F. and H. Lundstedt, Coronal mass ejections detected in solar mean magnetic field, *Geophys. Res. Lett.* Vol. 27, No. 19, pp 3141-3143, Oct. 1, 2000.

Brown, J.C., K.P. Macphersson, A.J. Conway, C.R. Mcinnes, and G. Janin, Neural network approach to solar activity prediction, EU. Space Oper. Cent. contract 9810/92/D/IM, *final report, Univ. of Glasgow, Scotland*, 1994.

Calvo, R.A., H.A. Ceccatto, and R.D. Piacentini, Neural network prediction of solar activity, *Astrophys. J.,* 444, 916, 1995.

Conway, A.J., K.P. Macpherson, G. Blacklaw, and J.C. Brown, A neural network prediction of solar cycle 23, *J. Geophys. Res.*, 103, 29,733-29,742, 1998.

Fleck, B., V. Domingo, and A.I. Poland, The SOHO Mission: An Overview, *Solar Physics*, 162, 1-37, 1995.

Haykin, S. Neural Networks – A Comprehensive Foundation, *Macmillian College Publishing Company, Inc.* 1994.

Howe, R., J. Christensen-Dalsgaard, F. Hill, R.W. Komm, R.M. Larsen, J. Schou, M.J. Thompson and J. Toomre, Dynamic variations at the base of the solar convection zone, *Science*, Vol. 287, 244-248, 31 March 2000.

Koons, H.C. and D.J. Gorney, A sunspot maximum prediction using neural network, *Eos* Trans, AGU, 71, No. 18, 677, 1990.

Kosovichev, A.G., Inversion methods in helioseismology and solar tomography, *J. of Comp. and Appl. Math.*, 109, 1-39, 1999.

Lantos, P. and O. Richard, On the prediction of them maximum amplitude for solar cycles using geomagnetic precursors, *Solar Physics*, 1999.

Liszka, L., Modelling of pseudo-indeterministic processes using neural networks, in proceedings of "*AI Applications in Solar-Terrestrial Physics*", Lund 22-24 September 1993, 1993.

Liszka, L., Decomposition of time series, in the proceedings of "*AI Applications in Solar-Terrestrial Physics*", July 29-31, 1997 in Lund, Sweden, edited by I. Sandahl and E. Jonsson, ESA WPP-148, April 1998.

Lorentz, G.G., The 13th problem of Hilbert, in *Mathematical developments arisning from Hilberts problem*, American mathematical Society, Province, RL, 1976.

Luhmann, J.G., J.T. Gosling, J.T. Hoeksema, and X.P. Zhao, The relationship between large-scale solar magnetic field evolution and coronal mass ejections, *J. Geophys., Res.,* Vol. 103, 6585-6593, April 1, 1998.

Lundstedt, H.. AI Techniques in Geomagnetic Storm Forecasting, in AGU Geophysical Monograph 98 on *Magnetic Storm*s, edited by B.T. Tsurutani, W.D. Gonzalez, Y. Kamide and J.K. Arballo, 1997.

Lundstedt, H. The Swedish Space Weather Initiatives, in proceedings of *ESA Space Weather Workshop* 11-13 November 1998, ESTEC, Noordwijk, The Netherlands, WPP-155, 1999.

Macpherson, K.P., A.J. Conway and J.C. Brown, Prediction of solar and geomagnetic activity data using neural networks, *J. Geophys. Res.,* Vol. 100, No. All, pp 21,735-21,744, 1995.

Mundt, M.D., W.B. Maguire II, and R.P. Chase, Chaos in sunspot cycle: Analysis and prediction, *J. Geophys., Res.*, Vol. 96, 1705-1716, February 1, 1991.

Parker, E., Cosmical magnetic fields, *Oxford, Clarendon Press*, 1979.

Scherrer, P.H.., R.S. Bogart, R.I. Bush, J.T. Hoeksema, A.G. Kosovichev, J. Schou, W. Rosenberg, L. Springer, T.D. Tarbell, A. Title, C.J. Wolfson, I. Zayer and the MDI Engineering Team, The Solar Oscillations Investigation-Michelson Doppler Imager, *Solar Physics* 162, 129-188, 1995.

St. Cyr, C., R. Howard, R. Schwenn, G. Brueckner, J.B. Gurman, B.J. Thompson, D.J Michels, S.P. Plunkett, and S.E. Paswaters, SOHO LASCO and EIT- Space Weather Forecasting Observatory, in proceedings of *ESA Space Weather Workshop* 11-13 November 1998, ESTEC, Noordwijk, The Netherland, WPP-155, 1999.

Takens, F., Detecting strange attractors in turbulence, in Dynamical Systems and Turbulence, Vol. 898 of *Lecture Notes in Mathematics*, edited by D.A. rand and L. S. Young, pp. 361-381, Springer, New York, 1981.

Tian, J. and Q. Fan, Application of a neural network in forecasting the characteristic values for solar cycle 23, in proceedings of "*AI Applications in Solar-Terrestrial Physics*", July 29-31, 1997 in Lund, Sweden, edited by I..Sandahl and E. Jonsson, ESA WPP-148, April 1998, 1998.

Torrence, C. and G.P. Compo, A practical guide to a wavelet analysis, *Bulletin of the American Meteorological Society*, 79, No. 1, 61-78, 1998.

Weigend, A.S., B.A. Huberman, and D.E. Rumelhart, Predicting sunspots and exchange rates with connectionist networks, in proceedings of the *Workshop on Nonlinear Modelling and Forecasting,* pp 395-432, SFI Studies in the Sciences of Complexity, proc. Vol. XII, Eds. Casdagli, M. and Eubank, S. Addison-Wesley, 1992.

Weiss, N.O., Is the solar cycle an example of deterministic chaos?, in *Secular Solar and Geomagnetic Variations in the Last 10,000 Years,* p69-78, F.R. Stephenson and A.W. Wolfendale eds., Kluwer Academic Publishers, 1998.

Zhao, X.P., J.T. Hoeksema and P.H. Scherrer, Application of SOI-MDI images: 1. Changes of large-scale photospheric magnetic field and the 6 january 1997 CME, April 8-9, *1997 ISTP Workshop*, 1997.

Henrik Lundstedt, Swedish Institute of Space Physics, Solar-Terrestrial Physics Division, Scheelev. 17, SE-223 70 Lund, Sweden.

The STEREO Space Weather Broadcast

O.C. St.Cyr[1] and J.M. Davila

Laboratory for Astronomy and Solar Physics, NASA Goddard Space Flight Center Greenbelt, Maryland

The NASA STEREO mission offers exciting possibilities for near-real-time transmission of important measurements for space weather. The STEREO payload will provide solar wind plasma, magnetic field, and energetic particle parameters, as well as optical and radio views of the Sun that cannot be obtained from groundbased observers or spacecraft near Earth. This space weather data will be transmitted continuously from each spacecraft over the X-band frequency (8.4 GHz) at a data rate of about 500 bps. Processing of the space weather broadcast data into useful online displays will be performed at the STEREO Science Center located at Goddard Space Flight Center. NASA will provide for partial coverage from each spacecraft through the Deep Space Network, and we are looking for partners who have ground stations to provide complementary coverage. We anticipate that these data will be very useful to forecasters of space environment conditions.

INTRODUCTION

The primary scientific objectives of NASA's STEREO (Solar TErrestrial RElations Observatory) are to advance the understanding of the origins of coronal mass ejections (CMEs) at the Sun; to track the evolution of CMEs through the interplanetary medium; and to study the dynamic coupling between CMEs and Earth's environment. As of this writing the STEREO mission is completing Phase A development and is scheduled for a mid-2004 launch. The mission includes two essentially identical three-axis stabilized spacecraft, each equipped with a payload of both remote sensing and *in situ* instrumentation.

The current baseline calls for both spacecraft to be launched on a single Delta II expendable launch vehicle from Kennedy Space Center and, following a series of Earth-Moon phasing orbits, to be injected into heliocentric orbits using gravitational assists from lunar fly-bys. The STEREO spacecraft will then drift away from the Sun-Earth line symmetrically, with one spacecraft "leading" Earth and the other "trailing." The planned drift rate is 22° per year for each spacecraft, so at the end of the nominal two year mission the spacecraft will be ~90° apart (Figure 1). The engineering design goal for the mission is a five-year lifetime.

The idea to include a "space weather beacon" capability on each STEREO spacecraft was noted in the NASA Science Definition Team Report [*Rust et al.*, 1997]. That document described a plan to alert Earth via a radio signal if pre-defined thresholds of solar wind parameters were exceeded. However, the current concept is that the broadcast will operate continuously, sending to Earth a highly compressed stream of solar and heliospheric images and *in situ* measurements of the solar wind and energetic particle environment at each spacecraft.

This manuscript describes the current conceptual design for the STEREO space weather broadcast. The second section provides a brief discussion of the scientific and applied (e.g., space environment forecasting) rationale for the beacon; and the third section describes the current imple-

---

[1] *Also at Computational Physics, Inc., Fairfax, Virginia, and The Catholic University of America, Washington, D.C.*

Space Weather
Geophysical Monograph 125
Copyright 2001 by the American Geophysical Union

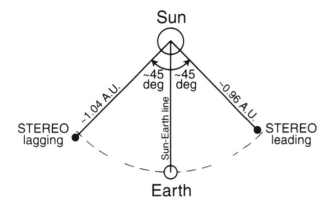

Figure 1. STEREO mission configuration after two years.

mentation concepts. Table 1 shows the payload, instrument teams, measurements, and proposed data content for the space weather broadcast.

RATIONALE

A limitation of space weather forecasting using ground-based and Earth-orbiting (or even L-1) platforms has been the difficulty in remotely sensing coronal mass ejections that might be headed toward Earth [e.g., *Gosling*, 1993]. Recent observations by the *SOHO* LASCO coronagraphs [*Brueckner et al.*, 1995] have been routinely detecting halo CMEs, and in many cases these have been Earth-directed [e.g., *Webb et al.*, 2000]. While these observations represent a significant step toward producing more accurate 2-5 day forecasts of geomagnetic activity, the timing of the arrival of any particular CME at Earth remains ambiguous [e.g., *St.Cyr et al.*, 2000].

One goal in obtaining the STEREO space weather data in near-real-time is to assist NOAA's Space Environment Center and other international space environment forecasters. The baseline concept of the STEREO space weather broadcast has been to build on the highly successful *ACE* real-time solar wind monitoring capability [*Zwickl et al.*, 1998] and to expand those *in situ* measurements to include remotely sensed images. In fact, assuming *ACE* or a follow-on mission is operating near L-1, then the addition of two STEREO spacecraft means there will be three-point *in situ* measurements available. This should provide significantly better understanding of the extent and uniformity of large scale structures at 1 A.U.

At different times during the STEREO mission the space environmental forecasting utility may emphasize different portions of the payload. Early in the mission the *in situ* instruments (IMPACT and PLASTIC) will likely be considered more important as they detect large scale structures in the solar wind while the spacecraft are still relatively close

to Earth. An example of this is that the spacecraft trailing Earth may encounter interplanetary CME (ICME) shocks prior to their arrival at Earth. Also, solar energetic particles from activity east of about 45° west heliographic longitude will travel along interplanetary magnetic field lines and be detected by the trailing spacecraft minutes to hours before their arrival at Earth [e.g., *Cane et al.*, 1988]. Further, corotating interaction regions can be geoeffective [e.g., *McAllister and Crooker*, 1997], and these will be detected by the trailing spacecraft several days prior to their passage by Earth.

Identification of potentially geoeffective features in the plasma electron, magnetic field, and energetic particle data from IMPACT should be straightforward [e.g., *Neugebauer and Goldstein*, 1997]; and the ion compositional content provided by PLASTIC will be useful for identifying ICMEs and other structures in the solar wind [e.g., *Galvin*, 1997].

SWAVES will be used as a remote sensing instrument, producing low frequency radio dynamic spectra for the space weather broadcast. These spectra will be used to track the heliospheric propagation of shocks associated with Type II interplanetary radio bursts [e.g., *Kaiser et al.*, 1998]. The comparison of dynamic spectra from the two spacecraft will give an estimate of the true location of the emission, and hence of the shock, by calculating the time delay between the observations. SWAVES will also measure *in situ* plasma waves, but these data are not part of the baseline space weather broadcast.

As the STEREO spacecraft drift farther away from Earth, the SECCHI remote sensing instrument suite will become more important for space weather forecasting when they provide observations that are not available from the Sun-Earth line. Using images from the EUVI telescope, one will be able to locate CMEs low in the corona [e.g., *Thompson et al.*, 1999] to identify the timing and direction of the launch more precisely than from coronagraphic data alone. Also, images from EUVI on the trailing spacecraft will show newly formed active regions prior to their appearance at the Sun's east limb (as seen from Earth). EUVI will also image other potentially geoeffective structures such as coronal holes and filament channels.

The near-Sun fields of view from the SECCHI coronagraphs (COR1 and COR2) and the wide field views from the heliospheric imagers (HI1 and HI2) will provide detection of and directional information about CMEs. This, along with more precise speed determination as the ICME propagates through the interplanetary medium, will provide significantly better predictions for Earth-arrival times than has been possible in the past, at least for those ICMEs detected by HI. Several studies combining coronagraphic and *in situ* data have demonstrated the difficulty in com-

Table 1. The STEREO mission instrument complement and proposed space weather broadcast content.

Instrument Name and Collaborating Institutions	Measurement and Proposed Space Weather Broadcast Content
IMPACT (*In situ* Measurement of Particles and CME Transients) *Principal Investigator: Dr. J. G. Luhmann, University of California, Berkeley,* NASA-GSFC, Caltech, U. Md, U. Kiel, CESR, MPAe, JPL, ESTEC, UCLA, NOAA, LANL, KFKI, *et al.*	Solar wind plasma characteristics; magnetic field parameters; solar energetic particles One minute average solar wind electron fluxes (6 energy bands); magnetic field strength and direction; energetic electron, proton, ion (He,CNO,Fe) fluxes (multiple bands)
PLASTIC (PLAsma and SupraThermal Ion and Composition) *Principal Investigator: Dr. A. B. Galvin University of New Hampshire* University of Bern, MPE-Garching, *et al.*	Ions in the energy-per-charge range of 0.2 to 100 keV/e One minute average solar wind proton density, bulk speed, thermal speed, and direction; alpha density; representative charge (or abundance) state distributions; suprathermal rates
SECCHI (Sun-Earth Connection Coronal and Heliospheric Investigation) *Principal Investigator: Dr. R. A. Howard Naval Research Laboratory, Washington, D.C.* Lockheed-Martin Solar and Astrophysics Lab, NASA-GSFC, University of Birmingham (U.K.), IAS, RAL, MPAe, U. Kiel, CSL, *et al.*	EUV imager, two coronagraphs with overlapping fields of view; two heliospheric imagers with overlapping fields of view 256x256 pixel highly compressed images from EUVI, COR1, COR2, HI1, HI2
SWAVES (STEREO/WAVES) *Principal Investigator: Dr. J.-L. Bougeret CNRS, Observatoire de Paris,* University of Minnesota, UC Berkeley, NASA-GSFC, U. Colorado	Interplanetary radio bursts from 40 kHz to 16 MHz One minute average radio dynamic spectrum (Intensity, frequency, time)

paring CME speed measurements made near the Sun with ICME speed measurements in the solar wind [e.g., *Lindsay et al.*, 1999; *Gopalswamy et al.*, 2000]. Of course, models will still be necessary to predict ICME properties such as magnetic field strength and direction, and density enhancements due to compression.

Another beneficial aspect to space weather forecasting and to public outreach will be the combination of STEREO *in situ* measurements and images with other available groundbased and spacebased observations. Based on these, one can imagine producing near-real-time visualizations of inner heliospheric conditions [e.g., *Jackson et al.*, 1998].

The STEREO space weather broadcast can also be considered a "test bed" for future monitors of conditions affecting humans in both near-Earth and interplanetary space. As humans venture into space to establish a more permanent presence with the International Space Station,

knowledge of environmental conditions will be crucial to their protection. Future human voyages to Mars or other destinations will require communications from remote monitoring platforms because warnings from Earth of environmental hazards would be too late.

IMPLEMENTATION

Onboard each STEREO spacecraft the space weather broadcast data will be encoded into one of four types of telemetry packets -- one each for the instruments described in Table 1. The IMPACT, PLASTIC, and SWAVES packets will contain the most recent one-minute average for the parameters listed in Table 1. These packets will be "stand alone" in that receipt of any single packet will provide a snapshot of the previous minute's values for the solar wind, energetic particle, and low frequency radio environment at that spacecraft. In contrast, the individual SECCHI packets will contain only a part of a single highly compressed image. Assuming a packet size of 500 bits and excluding overhead, then a compressed 256x256 pixel image will require about 200 packets (and an equivalent number of seconds) to acquire an entire frame. Since transmission of a SECCHI image would delay receipt of data from the other instruments, the *in situ* packets will be placed in the telemetry downlink at least once every minute.

Communications with the STEREO spacecraft will be through the Deep Space Network (DSN) 34 meter antennas. The RF system on the spacecraft will transmit in the X-band frequency range (8.4 GHz), and the space weather broadcast packets will be downlinked at a rate of about 500 bits per second.

The baseline operations concept is that the flight operations control center will be in contact with each spacecraft for four hours each day. It is likely that the DSN contacts will be at different times during the day for each spacecraft. During those DSN contacts commands will be uplinked to each spacecraft, and the solid state recorders containing the stored data will be transmitted to the ground. Some data will be available in real-time such as housekeeping parameters, the space weather packets, and some science data. This real-time telemetry stream will be transmitted to the STEREO Science Center (SSC), which will be located at NASA's Goddard Space Flight Center. At some later time, the recorder data (containing the full resolution data) will be transmitted to the SSC, reformatted at instrument team workstations, and put online for public access. The SSC, modeled after successful facilities for *SOHO* [St.Cyr *et al.*, 1995] and *ACE* [Garrard *et al.*, 1998], will also maintain an archive of the space weather broadcast data.

Upon receipt in the SSC, the space weather broadcast packets will be reformatted with software provided by the instrument teams and made available for public access on an Internet server. Users of the STEREO broadcast data can expect to see plots of the most recent *in situ* data, as well as recent movies of direct and differenced images from SECCHI. An archive of space weather broadcast data from the previous solar rotation should be available online. We encourage modelers to incorporate predictions and novel visualization displays of these data, either on the STEREO site or via links to their own Internet sites.

NASA plans to provide four hours of space weather broadcast coverage for each of the two STEREO spacecraft. Ground stations for the remaining 20 hours per day are not presently accounted for, but we have initiated discussions with potential partners to increase the space weather broadcast coverage. Ideally, complete coverage for each spacecraft will be attained through multiple private,

university, or national facilities. Since the downlink rate is rather small (~500 bps), we believe that even a modest Internet connection from a remote antenna will be sufficient to transmit the raw packets to the SSC for reformatting and display. Of course, there will be light-travel time latency in the data, and that will increase as the spacecraft separate from Earth. But ground processing should be rapid after receipt of the data stream at the SSC.

Although detailed engineering design work is in progress, our present estimate is that a groundbased antenna and receiver with a gain/temperature (G/T) ratio of about 26.6 dB/K would be sufficient to acquire the space weather broadcast packets through the entire two-year nominal mission. At that time, the range to each spacecraft will be more than 0.7 A.U. A three year extension to the nominal mission is the engineering design goal for major spacecraft subsystems. A groundbased system with G/T~33.2 dB/K would be sufficient to obtain the space weather broadcast packets five years after launch. At that time the distance to each spacecraft will be slightly greater than 1.5 A.U. Given these values and an assumed ground system noise value of 440° K, antenna dish diameters of 7.2 meters (for the two year nominal mission) and 15.3 meters (for the extended mission to five years) would be sufficient.

SUMMARY

The real-time broadcast of space weather data from the two STEREO spacecraft offers new possibilities for environmental forecasters. The STEREO payload will provide solar wind plasma, magnetic field, and energetic particle parameters, as well as optical and radio views of the Sun that cannot be obtained from groundbased observers or spacecraft near Earth. This space weather data will be transmitted continuously from each spacecraft over the X-band frequency range at a data rate of about 500 bps. The combination of STEREO *in situ* measurements with those obtained from a solar wind monitor at L-1 offers three-point "ground truth" for modeling geomagnetically-effective heliospheric structures. This capability provides an opportunity to test one concept of an environmental monitoring system that will be necessary for long duration human space flight in the future.

Acknowledgments. We wish to acknowledge the continuing contributions by the STEREO instrument teams and by the Johns Hopkins University Applied Physics Laboratory (JHU/APL) engineers who are developing the spacecraft. We are particularly appreciative of the ongoing efforts of NASA Project Manager A. Harper and systems engineer H. Maldonado, JHU/APL Project Manager J.T. Mueller, systems engineer A. Driesman, and RF engineer J. von Mehlem. We thank J.G. Luhmann, M.L. Kaiser, K. Goetz, D. Curtis, C.T. Russell, R. Mewald, and J. Wolfson for insightful comments about the manuscript. We appreciate the efforts of S. St.Cyr in formatting the manuscript. One of us (OCS) received partial support from the National Space Weather Program under NSF grant ATM-9819668

REFERENCES

Brueckner, G.E., and 14 co-authors, The large angle spectroscopic coronagraph (LASCO), *Solar Physics*, *162*, 357-402, 1995.

Cane, H.V., D.V. Reames, and T.T. von Rosenvinge, The role of interplanetary shocks in the longitude distribution of solar energetic particles, *Journal of Geophysical Research*, *93*, 9,555-9,567, 1988.

Galvin, A.B., Minor ion composition in CME-related solar wind, in *Coronal Mass Ejections*, editors N. Crooker, J.A. Joselyn, and J. Feynman, AGU Monograph 99, American Geophysical Union, Washington, D.C., 253-260, 1997.

Garrard, T.L., A.J. Davis, J.S. Hammond, and S.R. Sears, The *ACE* Science Center, *Space Science Reviews*, *86*, 649-663, 1998.

Gopalswamy, N., A. Lara, R.P. Lepping, M.L. Kaiser, D. Berdichevsky, and O.C. St.Cyr, Interplanetary acceleration of coronal mass ejections, *Geophysical Research Letters*, *27*, 145-148, 2000.

Gosling, J.T., The solar flare myth, *Journal of Geophysical Research*, *98*, 18,937-18,949, 1993.

Jackson, B.V., P.L. Hick, M.Kojima, and A. Yokobe, Heliospheric tomography using interplanetary scintillation observations: 1. Combined Nagoya and Cambridge data, *Journal of Geophysical Research*, *103*, 12,049-12,067, 1998.

Kaiser, M.L., M.J. Reiner, N. Gopalswamy, R.A. Howard, O.C. St.Cyr, B.J. Thompson, and J.-L. Bougeret, Type II radio emission in the frequency range from 1-14 MHz associated with the April 7, 1997 solar event, *Geophysical Research Letters*, *25*, 2501-2504, 1998.

Lindsay, G.M., J.G. Luhmann, C.T. Russell, and J.T. Gosling, Relationships between coronal mass ejection speeds from coronagraph images and interplanetary characteristics of associated interplanetary coronal mass ejections, *Journal of Geophysical Research*, *104*, 12,515-12,523, 1999.

McAllister, A.H. and N.U. Crooker, Coronal mass ejections, corotating interaction regions, and geomagnetic storms, in *Coronal Mass Ejections*, editors N. Crooker, J.A. Joselyn, and J. Feynman, Geophysical Monograph 99, 279-289, American Geophysical Union, Washington, D.C., 1997.

Neugebauer, M. and R. Goldstein, Particle and field signatures of coronal mass ejections in the solar wind, in Coronal Mass Ejections, editors N. Crooker, J.A. Joselyn, and J. Feynman, Geophysical Monograph 99, American Geophysical Union, Washington, D.C., 245-251, 1997.

Rust, D., and 17 co-authors, The Sun and Heliosphere in Three Dimensions: Report of the NASA Science Definition Team for the STEREO Mission, The Johns Hopkins University Applied Physics Laboratory, Laurel, Maryland, December 1997.

St.Cyr, O.C., L. Sanchez-Duarte, P.C.H. Martens, J.B. Gurman, and E. Larduinat, *SOHO* ground segment, science operations, and data products, *Solar Physics*, *162*, 39-59, 1995.

St.Cyr, O.C., and 13 co-authors, Properties of coronal mass ejections: SOHO LASCO observations from January 1996 to June 1998, *Journal of Geophysical Research*, *105*, 18,169-18,185, 2000.

Thompson, B.J., O.C. St.Cyr, S.P. Plunkett, J.B. Gurman, N. Gopalswamy, H.S. Hudson, R.A. Howard, D.J. Michels, and J.-P. Delaboudiniere, The correspondence of EUV and white light observations of coronal mass ejections with *SOHO* EIT and LASCO, in Sun-Earth Plasma Connections, Geophysical Monograph 109, American Geophysical Union, Washington, D.C., 31-46, 1999.

Webb, D.F., E.W. Cliver, N.U. Crooker, O.C. St.Cyr, and B.J. Thompson, Relationship of halo coronal mass ejections, magnetic clouds, and magnetic storms, *Journal of Geophysical Research*, *105*,.7,491-7,508, 2000.

Zwickl, R.D., and 11 co-authors, The NOAA Real-Time Solar-Wind (RTSW) system using ACE data, *Space Science Reviews*, *86*, 633-648, 1998.

O. C. St.Cyr, Computational Physics, Inc., Code 682, NASA-Goddard, Greenbelt, Maryland 20771

J. M. Davila, Code 682, NASA-Goddard, Greenbelt, Maryland 20771

70 Years of Magnetospheric Modeling

Center for Space Physics, Boston University, Boston, Massachusetts

Magnetospheric modeling began in the 1930s with Chapman and Ferraro. They made the first "modular" magnetospheric model. This overview covers the subsequent 70-year progression of modular models. The progression has reached a point where modular models, to be competitive, are becoming modules within global numerical codes.

1. TYPES OF MODELS AND CHOICES

Magnetospheric modeling could fill a book, and in 1979 it did (*Quantitative Modeling of Magnetospheric Processes*, 655 pages [*Olson*, 1979]). The field has mushroomed since then, which means that ruthless selection is vital. A tutorial on models should move through time, begin at the roots and work up the trunk, pursue a main branch or two and reserve for following papers the harvest of space-weather fruits that betip the branches.

Models evolve by ramification. There is a moment of creation when the first model emerges. It engenders a follow-on line of generically related models. There comes a time when a new branch shoots off, and still later another branch, and so on. Magnetospherists recognize four great branches of models: empirical, modular, single-particle (or kinetic), and MHD. Originally "magnetospheric model" meant a model that gave quantitative information about the magnetic field within the magnetosphere [*Roederer*, 1969]. Since then magnetospheric models have progressed so that now they give quantitative information on nearly every facet of magnetospheric structure and dynamics.

Regarding uses, models sort into five kinds. First (in a non-chronological, non-prioritized sense suited for exposition) are models that represent. These are products of data fitting and AI techniques. They are empirical proxies of the surveyed magnetosphere. They tell a user what to expect at a certain place and time by encoding what was there before at a similar time. Next are models that explain, which include modular and single-particle models. These are analytical constructs based on physical idealizations. They check basic concepts and let a user draw solid ancillary consequences from a physical idea. Then there are models that reveal. These are mainly kinetic and MHD codes. They are powerful enough to reveal unforeseen phenomena. Forth are models that interpret, enabling data-theory closure. They render observations meaningful in light of their constituting assumptions. Any type of model can, in principle, serve this function. Indeed, models are tested by how well they perform it. Finally, there are models that predict. Theses are the Holy Grail of those who trade in operational space weather services. They take data from places where instruments happen to be to predict conditions at places and times where information is wanted. Again, any type of model can, in principle, serve this function, but criteria for achieving operational status make these the hardest models to develop. A model might, of course, serve more than one purpose. In keeping with the stated plan of limiting the scope of this tutorial by starting at the roots and working up the trunk and a main branch, I will cover modular models that explain, and illustrate them with results from global MHD simulations. The emphasis is on models that test our understanding of the physics behind magnetospheric structures and processes and that enlarge our awareness of such structures and processes. (For a comprehensive review of empirical models, see *Jordan* [1994]. Unfortunately, there is no corresponding review of MHD models, and I must apologize

Space Weather
Geophysical Monograph 125
Copyright 2001 by the American Geophysical Union

for leaving the reader with no guidance respecting this most important branch of the tree.)

2. THE ROOTS OF MAGNETOSPHERIC MODELING

The tree of magnetospheric modeling is rooted in the 1930s, in the Chapman-Ferraro model that explained the sudden commencement of magnetic storms. This first modular model introduced the concept of a magnetic cavity around the earth, which we now call the magnetosphere. Beginning already in 1913, Chapman pursued the then prevailing notion that streams of "corpuscular radiation" from the sun cause magnetic storms on earth. Presuming the particles to be electrically charged, he tried first to find a way to get them through earth's magnetic field into the upper atmosphere where they could modify and amplify the dynamo currents, then thought to generate the daily variation of the magnetic field. These efforts failed for various good reasons. Then after nearly 20 years, he with his protégé Ferraro hit on the idea that the particles should act collectively, not individually, with the result that when a particle stream approaches earth's magnetic field, the whole stream behaves like a deformable, electrical conductor. Magnetic fields repel conductors; so instead of reaching the atmosphere, as Chapman originally sought, the approaching stream flows around the geomagnetic field like a brook around a trout. Instead of an atmospheric interaction, there was a magnetic interaction, which introduced a set of questions wholly new to the field. What was the size and shape of the magnetic cavity thus formed? How were the currents that repelled the stream distributed? What magnetic field did these currents generate? These are questions that would occupy magnetospheric modelers through the first decade of the space age.

The Chapman-Ferraro model marks a discontinuity in the history of magnetospheric physics. It had no precedents. Nothing led to it logically. Before it there was no conception of a magnetosphere. After it, there was, and the big magnetospheric questions that had quantitative answers were then posed. The original model was quantitative enough to infer from the size of a storm sudden commencement information about the physical properties of solar corpuscular stream that caused it and the size of the magnetic cavity. For example, in a later treatment yet 5 years before Sputnik, Ferraro estimated that the streams carried between about 1 and 25 protons per cubic centimeter and that the streams stopped about 5 earth radii before hitting the earth. Though only about half the now-known distance, it was nonetheless a remarkable estimate for 1952 [*Ferraro*, 1952]. A year after Sputnik but still before the observation of the solar wind or the magne-

topause, Dungey refined the calculation and, as something new, calculated the latitude at which polar cusps predicted by the model touch the ionosphere, finding $72°$ [*Dungey*, 1958]. The point is not that the model made wonderfully accurate estimates (they weren't bad) but that it made estimates at all (see Figure 1). For the first time, a model was generating information about the magnetosphere, giving the founders of magnetospheric physics their first mental pictures of the size, shape and structure of the invisible object that, with the advent of the space age, was increasingly attracting their interest.

3. CULMINATION OF THE CHAPMAN-FERRARO PROJECT

Chapman and Ferraro gave the first magnetospheric modelers a clear project: find the size and shape of the boundary between the solar corpuscular radiation and the earth-enclosing cavity it forms, and determine the magnetic field everywhere within the cavity. The project, as it was first formulated, fell under the heading of a free boundary problem. The boundary takes its size and shape to satisfy two conditions: no magnetic field outside the cavity and balance between magnetic pressure inside the cavity and stream pressure outside. Stream pressure was related to boundary shape by the supposition that the particles composing the stream specularly reflect off the boundary. Though mathematically well posed, the problem was difficult for the time, which was the first half of the 1960s. Nonetheless, spurred by the excitement of a new and glamorous field, a small army of modelers emerged to tackle it. They applied techniques that evolved from 1D solutions for the stream standoff distance to 2D solutions for the shape to full 3D solutions for everything. The 3D solutions, in turn, evolved from analytic solutions based on approximations to series solutions of the exact problem, which could give results to arbitrary accuracy. The history of the rapid convergence on a solution to the Chapman-Ferraro problem was reviewed virtually in real time by *Blum* [1963] and many others including Chapman himself [*Chapman*, 1963].

Midgley and Davis [1963] gave the first series solution to the exact problem. Figure 2a from their paper answers the "shape" part of the Chapman-Ferraro project. It shows what a Chapman-Ferraro magnetopause looks like in the noon-midnight meridian plane. Regarding "size," Midgley and Davis obtained $8.9 \, R_e$ as the distance between earth's dipole and the solar wind stagnation point where the solar wind comes closest to the earth. This is up from Ferraro's $5 \, R_e$ but still less than 10 to 11 R_e as commonly observed [*Fairfield*, 1971]. Where Ferraro had only the size of a storm sudden commencement to go on, Midgley and

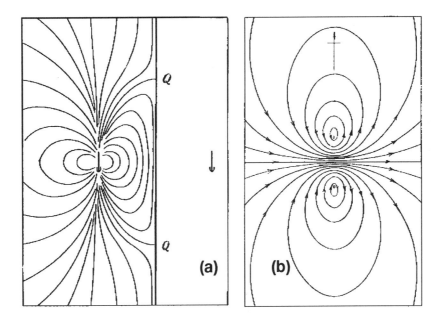

Figure 1. This simplest of all modular models of the magnetosphere appears in *Chapman and Ferraro* [1931, p.179]. Figure 1a shows the 'magnetospheric' field, and Figure 1b shows the current on the 'magnetopause.' These are prototypes of models shown in Figure 2.

Davis, 11 years later, had solar wind measurements. But like Ferraro, they used the expression for specular reflection to calculate the solar wind ram pressure. Gas dynamics, now known to work better, gives a value nearly 2.5 times less [*Spreiter et al.*, 1966], which would bring Midgley and Davis's size up to a more factual value around $10.4 R_e$.

The set of nested, curvy lines in Figure 2a that decorate the now-familiar shape, with its dimple marking the cusp, trace the course of electrical current on the Chapman-Ferraro surface. This "Chapman-Ferraro current" generates a magnetic field that cancels earth's dipole field outside the surface and adds to it inside, which brings us to the "field" part of the Chapman-Ferraro project. The calculated strength of the magnetospheric field (Figure 2b) was an important product of the model for it gave information that modelers of trapped particles needed to calculate particle drifts. As Figure 2b shows, Chapman-Ferraro currents shift lines of constant field strength—drift paths of equatorially mirroring particles—sunward.

Mead [1964], with a forth-order spherical harmonic fit to a series solution to the exact problem [*Mead and Beard*, 1964], addressed the trapped-particle market directly. He gave plots of field strength and field lines that fairly covered the 3D magnetosphere. Figure 2c from his paper answered the question, compared to a dipole field, what do field lines inside a Chapman-Ferraro cavity look like? The qualitative answer "flatter in front and rounder in back"

was made quantitative. Mead showed that at the stagnation point, the magnetospheric field is about 2.3 times stronger than the dipole field. The field is more compressed in front than behind by an amount that can be expressed as a gradient of 1.5 nT per R_e (approximately). Interacting with the geomagnetic dipole, this gradient creates a force of about 2×10^7 N acting to shove the earth away from the sun. This is how nature transfers the push that the solar wind gives to the magnetopause to the earth—the only mass around to absorb the force [*Siscoe*, 1966]. (Incidentally, solar wind pressure exerts less force on Earth than solar light pressure. It is so slight that it amounts to less than a one minute lag of the Earth in its orbit since the solar system was born.)

The Chapman-Ferraro project and its successful conclusion set the style for magnetospheric models that followed, when the magnetotail and internal plasma pressure complicated things. At the magnetosphere's birth as a concept, people perceived it to be a globally integrated unit. Local constraints (zero normal field and pressure balance at every point on the magnetopause) combine to create a global structure with a field distributed just right to guarantee global force balance. One could view global integration to be a necessary consequence of global currents or, equivalently, of global stresses imparted by the cavity-confined magnetospheric field [*Parker*, 1962].

In Figures 2d, e, and f, an MHD code of today imitates Figures 2a, b, and c from Chapman-Ferraro models of 40

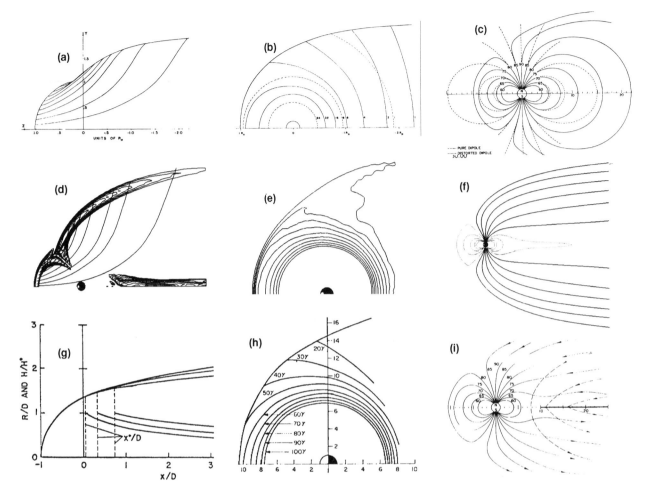

Figure 2. (a) Current streamlines on the magnetopause as calculated by *Midgley and Davis* [1963]. (b) Contours of constant field strength in the equatorial plane as determined with the Midgley and Davis model. (c) Field lines in a vacuum magnetosphere calculated by *Mead* [1964]. (d)-(f) Corresponding figures generated by a global MHD code (the Integrated Space weather Model, ISM) under the condition of zero IMF. (g) Shape of the magnetotail boundary and field strength in the tail determined with the tail model of *Spreiter and Alksne* [1969]. (h) Field strength contours in the equatorial plane as determined with the *Williams and Mead* [1965] model. (i) Field lines in the Williams and Mead model.

years ago. Figure 2d shows contours of current intensity computed for the case in which the interplanetary magnetic field (IMF) is effectively zero. In an MHD code, the zero-IMF situation reproduces an essentially Chapman-Ferraro condition—a magnetosphere that is completely closed. The magnetopause is dimpled at the cusp around which Chapman-Ferraro current streamlines circle, as in the Midgley and Davis picture. Unlike the Chapman-Ferraro model, however, this model contains a magnetotail current, which 'raises' the magnetosphere's 'back' relative to the distance to its nose, shifts the cusp sunward and equatorward, and closes the Chapman-Ferraro currents sooner. In the figure, one sees that the current streamlines close over the magnetosphere's back on the outside of the

current contours. Tail-current streamlines (not shown) close under them [*Tanaka*, 2000]. Figure 2e shows that the magnetotail markedly steepens the nightside gradient of field strength, a condition critical for evaluating particle drift paths. Figure 2f shows that the magnetotail also greatly stretches Mead's rounded, nightside field lines.

In summary, by 1965 the Chapman-Ferraro epoch of magnetosphere modeling reached its culmination. (For a post-epoch review by a protagonist, see *Beard* [1973].) The Chapman-Ferraro problem was mathematically well posed. The first magnetospheric modelers solved it analytically. They calculated the size and shape of the magnetopause and found its dependencies on solar wind ram pressure and (though not discussed here) on dipole tilt.

(The mentioned reviews discuss tilt dependence.). They determined the magnetospheric magnetic field everywhere. They showed how local constraints give rise to global properties, for example, how local pressure balance results in global force balance. In a sense, it marked the acme of magnetospheric modeling. All model parameters could be fully predicted in terms of specified boundary conditions. Its model parameters were abridged; however. They omitted the auroral oval and the ring current—features fully known to model developers starting with Chapman and Ferraro. When in 1965 Ness announced the discovery of the magnetotail, the model's shortcomings became unignorable. A new project, the magnetotail project—find the size and shape of the magnetotail and determine the magnetic field it generates—then replaced the Chapman-Ferraro project. Thirty-five years later, we are still working on this project. We still cannot predict from first principles the magnetic flux in the tail, or equivalently, the size of the auroral oval, from boundary conditions alone. (We can simulate it with global MHD codes, but that is a different thing.) To a large extent the magnetotail project has driven subsequent evolution of magnetospheric models. Naturally, those who first undertook the new project started by applying approaches that worked for the old Chapman-Ferraro project.

4. MODELING THE CLOSED MAGNETOTAIL

In the spirit of a Chapman-Ferraro problem in which all parameters are determined from boundary conditions, theorists have never solved the problem of finding the size and shape of the magnetopause for the realistic case of a magnetosphere with a magnetotail. In the simplest case of a closed magnetosphere, boundary conditions leave unspecified the magnetic flux in the tail. It must be given, directly or indirectly, as an undetermined model parameter. Boundary conditions also leave unspecified the plasma population needed to support the internal magnetic stresses that a magnetotail necessarily implies. In the realistic case of an open magnetosphere, it is possible in principle to determine the tail's magnetic flux as part of a transport problem that cycles solar wind plasma through the magnetosphere, but no one has managed to do this. The same as-yet-unsolved problem should, when solved, yield the internal, stress-bearing plasma population. Independent lines of contemporary thinking seem to be converging on the idea that there is no equilibrium solution to the open magnetosphere problem. Some who work on this problem think the dynamical mode that the transport exhibits resembles a system displaying self-organized criticality. If so then only in a climatological sense might one hope to represent the magnetosphere-with-tail as a set of

Chapman-Ferraro-type still pictures, one for each significantly different IMF orientation and solar wind condition. Such pictures necessarily omit the dynamics that characterize magnetospheric weather. Nonetheless, they have their uses, both practical and pedagogical. Moreover, they are steppingstones leading from the Chapman-Ferraro era to the present. The idea that no equilibrium models exist emerged from trying to evolve toward such models, a project which has been carried farthest in a long-continuing research program at Los Alamos National Laboratory [Birn, 1995; Birn et al., 1996a,b]. So we return to 1965, the year of the magnetotail.

As already noted, interest in specifying drift paths of particles in the outer radiation belt, a space-weather concern, spurred rapid development of quantitative magnetotail models. The first of these [*Williams and Mead*, 1965] was analogous to the 1931 Chapman-Ferraro plane current sheet representation of the magnetopause. It represented the magnetotail by an equatorial current sheet, infinite in the dawn-dusk direction (y direction) and truncated in the noon-midnight direction (x direction). It started $10 R_e$ behind the earth and extended $30 R_e$ downtail. The current sheet's magnetic field was simply superimposed on the Mead-Beard magnetospheric field. As Figures 2h and 2i show, this simple model did a remarkably good job of reproducing the steep nightside field gradient and field-line stretching. For particle drift studies, it was highly serviceable. For studies of the solar wind-magnetosphere interaction, however, it lacked a magnetopause.

Since it was unfeasible given only solar wind boundary conditions to determine in a fully self-consistent way the size and shape of the magnetopause of a combined magnetosphere-plus-tail, early modelers used different approximations to approach the problem; each sought verisimilitude in one facet while sacrificing it elsewhere. *Spreiter and Alksne* [1969] were the first (and for 20 years the last) to focus on determining the shape of the 3D magnetotail boundary and, what was for them the point of the problem, the strength of the magnetic field as a function of distance from earth, empirical information on which had by then been published. They envisioned the tail to be a cylinder partitioned into north and south halves, each with constant field strength over a cross section, and each holding its magnetic flux constant with distance from earth. In their model, the cylinder grows (and so field strength drops) with distance to keep magnetic pressure inside matched to gas pressure outside as the latter declines downwind along the boundary until it returns to the static pressure of the ambient (pre-shocked) wind. As an innovation, the model represents the gas pressure outside by the sum of a gas-dynamic ram pressure, sensitive to the boundary's slope, and a constant static pressure. In hypersonic gas dynamics, this is known as the Newtonian approximation. The static pressure stops the tail from expanding indefinitely. They attach the front end of their cylindrical tail to a computed Chapman-Ferraro boundary truncated behind the earth. At the juncture, they put the field strength on the tail's side equal to the Chapman-Ferraro value on the magnetosphere's side. This guarantees that the juncture is smooth and indirectly fixes the magnetic flux in the tail. Figure 2g shows the resulting profile in the equatorial plane (hence, the absence of a cusp dimple) for three choices of truncation distance. It also shows how field strength decreases with distance from earth. The authors note that the model underestimates field strength in the tail, especially near the front.

The model's field strength comes up short because by initializing the tail's field to the Chapman-Ferraro field at the juncture, it ignores the tail field itself, which should increase the field at the juncture compared to the field that would be there in the absence of a tail. This basic point had been demonstrated a year earlier. *Unti and Atkinson* [1968] used the technique of conformal transformation to find an exact solution to the 2D Chapman-Ferraro problem with a mathematical current sheet (no plasma sheet) representing the tail and no static pressure outside. The magnetic flux in the tail (per unit out-of-plane distance) was a model parameter. They showed that, as tail flux increases, the current sheet moves earthward and the boundary expands substantially. The contrast between Figures 2d (a 3D MHD solution) and 2g (the Spreiter-Alksne model) illustrates by how much the tail's magnetic field increases the magnetosphere's cross-sectional breadth. The earthward movement of the current sheet as tail flux increases and the boundary expands allows the tail to increase the earth-grabbing magnetic-field gradient by which the magnetosphere's currents systems hold on to the earth while the solar wind tries to blow them away. The wind's push grows as the boundary expands necessitating a bigger field gradient, and the current sheet moves earthward to provide it.

After the Spreiter-Alksne model, emphasis among those pursuing the magnetotail project shifted from determining the shape of the boundary to determining the magnetic field everywhere within a magnetically closed magnetosphere-plus tail volume. The requirement of force balance at the magnetopause (solar wind pressure outside matching magnetic pressure inside) was sacrificed for the sake of choosing a boundary with a geometry that facilitated the solution of the interior-field problem. "Prescribed magnetopause" is the name given to such models. In this genre two innovative approaches emerged: *Voigt* [1981, 1984] and *Schulz and McNab* [1987, 1996].

Voigt chose the simplest of all boundary shapes: a hemispheric head attached to a cylindrical tail. In three steps he solved the Laplace equation for a dipole field in a vacuum with the condition that no magnetic flux cross his chosen boundary. Step 1: solve the stated problem for a sphere using spherical harmonics. Step 2: solve it for an infinite cylinder using cylindrical harmonics. Step 3: match the two solutions at the "seam" by an iterative procedure that insures smoothness. To add a current sheet to the tail, he introduced a "stretch" transformation that replaced the field at every (x, y, z) point in the tail by the vacuum field at a closer x distance while reducing the y and z components to keep the field divergence free. The transformation automatically maintains continuity of B_x at the seam, but not B_y and B_z. Thus, the cross section at the seam becomes a current sheet, which generates a field that can be represented by spherical harmonics fore and cylindrical harmonics aft, each representation chosen to satisfy the no-flux-through condition at the boundary. Subtracting this solution then gives the desired field of a magnetosphere-plus-tail that satisfies the boundary condition and is continuous at the seam. The degree of stretching determines the strength of the tail current sheet, the range going from the original pure vacuum (no current sheet) to a pure B_x field, which has the form of a Harris sheet. (For a review and generalization of this model, see *Stern* [1987].)

Schulz and McNab chose for their boundary a figure of revolution generated by an analytic fit to the equatorial profile of the Mead-Beard solution to the Chapman-Ferraro problem. Self-similar forms of the same analytic expression that generates the boundary give the shape of field lines in the tail, thus guaranteeing that the tail boundary is a field-line surface (i.e., magnetically closed). Within the tail, field lines extend tailward from a surface representing the transition between the magnetosphere and the magnetotail. They called this surface the "source surface" and the entire model the "source surface model" (SSM). The source surface, whose distance from earth is a model parameter, is slightly concave earthward to be orthogonal to the field lines extending tailward from it. The magnetic field in the magnetospheric volume sunward of the source surface is determined by a solution to the Laplace equation for a dipole field in a vacuum with a boundary condition that minimizes a weighted sum of two terms, one expressing how much magnetic flux crosses the boundary and the other expressing how much the field lines kink as they cross the source surface. The first aims at satisfying the closed magnetosphere condition, the second at achieving a smooth transition between the magnetosphere and the tail. Once the field in the magnetosphere is thereby determined, the strength of the field in the tail is

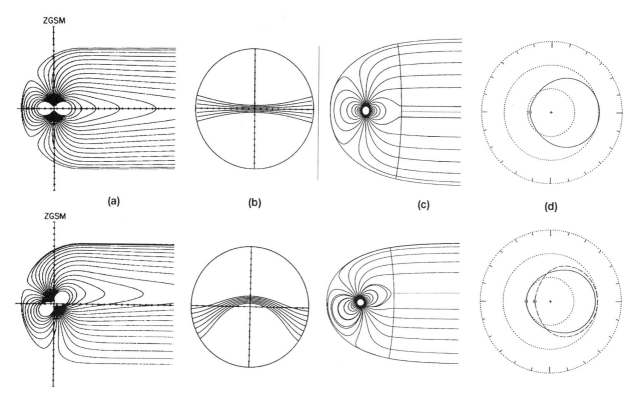

Figure 3. (a) Field lines as determined with the *Voigt* [1984] prescribed magnetopause model for the cases of no dipole tilt and maximum dipole tilt. (b) The cross-sectional shape of the current sheet in the tail for these two cases. (c) Same as (a) for the *Schulz and McNab* [1987] source surface model. (d) Polar cap boundaries determined with the SSM for the two dipole tilt cases. The two polar caps for the titled case show the winter and summer hemispheres.

fixed by continuity of the normal component of the field through the source surface. (For a review of this model and its evolution from the Mead-Beard model, see *Schulz* [1991].)

Figure 3 gives representative results from the Voigt and Schulz-McNab prescribed-magnetopause models. Figure 3a shows field lines in the noon-midnight meridian plane for Voigt's hemisphere-plus-cylinder model. The hemisphere and cylinder have 20 R_e radii, and their juncture lies 9 R_e tailward of earth. The degree of tail-field stretching in this case was chosen to give the observed rate at which field strength in the tail decreases with distance from earth. Voigt's approach clearly succeeds in confining the field to the magnetosphere-plus-tail volume, and it succeeds in smoothly transitioning the field from the magnetosphere to the tail. The bottom panel shows that the approach can also treat a tilted dipole, in this case $35°$, which is the maximum tilt for Earth. Figure 3b gives streamlines of electrical current in a cross-sectional plane 15 R_e behind the earth. The model displays a widening of the current sheet near the flanks and a tilt-induced curvature of the

current sheet, both of which are in reasonable accord with observations. Figure 3c shows field lines in the noon-midnight meridian plane for the Schulz-McNab (Mead-Beard profile) SSM model. The figure also shows the source surface that distinguishes the model. Field confinement and smooth tail transition are reasonably achieved here as well as for the Voigt model. Also like the Voigt model, SSM can treat a tilted dipole (lower panel) producing a warped current sheet (not shown here, but see *Schulz and McNab* [1996]). The main difference in the two models is that whereas the Voigt model has a finite-thickness current sheet that field lines cross from south to north the SSM has a mathematically thin current sheet that separates field lines of the south from those of the north. In effect, in the SSM, all tail field lines are "open" in the sense that they go to infinity, whereas in the Voigt model, all field lines are closed, except in the Harris limit, where they are all open. The Voigt model can therefore describe how B in the tail varies with distance from earth, but except in the Harris limit it cannot be used to map the size and position of the polar cap that connects magnetically to

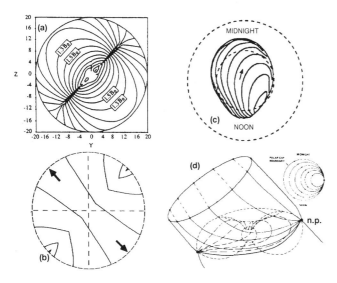

Figure 4. (a) Contours of constant B-normal on the magnetopause in the *Toffoletto and Hill* [1989] model for the case of component merging geometry. (b) Flow streamlines for this case. (c) Polar cap potential pattern for this case. (d) Magnetic field topology of the magnetosphere illustrates the origin of the Stern singularity. Inset shows polar cap potential pattern. (Modified from *Siscoe* [1988].)

the tail. As Figure 3d shows, because its tail field lines are open, the SSM does give a definite size and position for the polar cap (~ 15° radius displaced ~ 5° tailward), which match observations reasonable well.

We see that the Voigt model and the SSM capture some of the major structures that the magnetotail adds to the magnetosphere. But they also miss a few. First, as already noted, by prescribing the shape of the magnetopause, they cannot adjust its shape to achieve pressure balance with the solar wind. Second, since their boundaries are magnetically closed, they cannot address phenomena associated with electrodynamic coupling to the solar wind. Third, the models' defining innovations (the stretching transform for Voigt and the source surface for Schulz-McNab) are not well suited for achieving force equilibrium between the stressed magnetic field and the plasma within the magnetosphere needed to balance the stress. They are able to achieve stress balance perpendicular to the tail axis (at least partially), but not along it.

5. MODELING THE OPEN MAGNETOTAIL: THE STERN-SINGULARITY PROBLEM

In a significant evolutionary step, the Voigt and Schulz-McNab prescribed-magnetopause models were adapted to incorporate open magnetic flux. The Voigt model evolved

into the Toffoletto-Hill (T-H) model and the Schulz-McNab model evolved into the Peroomian-Lyons-Schulz (PLS) model. The T-H model [*Toffoletto and Hill*, 1989, 1993] has overcome a serious difficulty, known as the Stern-singularity problem, that is inherent to open modular models. A short account of the development of the T-H model will expose the Stern-singularity problem and give the T-H solution to it, the only one so far found.

The T-H model incorporates open magnetic flux into the Voigt model by specifying, everywhere on the boundary, an explicitly given normal component of the magnetic field (B-normal). B-normal then becomes a source for a Neumann boundary-value problem, the solution of which, both inside and outside the magnetosphere-plus-tail volume, they call an "interconnection field." Superposing this interconnection field onto the Voigt field gives a modular model of an open magnetosphere magnetic field, which was the first of its kind. The T-H model also augments, as a not insignificant item of model engineering, the Voigt tail field to remove unphysical properties that otherwise arise when the interconnection field is superimposed. By blending aerodynamic flow around a blunt body with deflections caused by magnetic forces that merging brings into play, the modelers have contrived to tailor patterns of B-normal that represent competing hypothesis about where magnetic merging occurs (specifically, the so-called component merging and anti-parallel merging hypotheses).

Figure 4 gives results from the 1989 T-H model for the case in which the IMF is straight duskward (pure IMF B_y). Figure 4a depicts lines of constant B-normal. The conspicuous, diagonal feature at a clock angle of 45° is the merging line that represents the component-merging hypothesis. In concept, the merging line continuously generates normal magnetic flux, which the magnetosheath flow carries away as a steady stream that 'paints' the entire magnetopause in B-normal. Figure 4b shows representative streamlines of the flux-distributing magnetosheath flow. One sees that the flux has a marked tendency to flow towards the upper-left and lower-right quadrants. This tendency reflects the effect of magnetic forces, which pull the flow along the magnetopause in a direction generally perpendicular to the merging line.

From B-normal thus specified on the boundary, the model generates, as already described, B everywhere inside and outside the boundary. The entire magnetic flux that penetrates the boundary can then be followed from the solar wind to the ionosphere to map out the open-field-line polar cap. From this mapping, the model lets one determine, under the standard (ideal MHD) assumption that magnetic field lines are electrical equipotentials, the solar wind's motional (V×B) electric field as it is translated to the polar cap. Figure 4c shows the equipotentials that result in the polar cap for this pure IMF B_y case. Instead of a nice day-to-night pattern of lines that, as might be expected, mimics at the ionospheric level the equipotentials in the magnetosheath, we find that the equipotentials form a nested set of nearly closed lines all tending to come together at a point on the boundary of the polar cap around 15 hours local time. As this is not the way real equipotentials behave, there is a serious problem generic to this modular approach, which is called the Stern-singularity problem. Since, as Figure 4c shows, electric potentials in the ionosphere tend to touch at a point, the electric field at that point tends to become infinite. *Stern* [1973] was the first to recognize this generic property of superposition models. For all but perfect alignments between the IMF and Earth's dipole, a uniform, motional electric field in the solar wind maps into an infinite electric field at a point (the Stern singularity) on the boundary of the polar cap [see also *Lyons*, 1985].

Figure 4d reveals the cause of the singularity. Here, for the case in which a uniform field is superposed on a dipole field, are surfaces that separate volumes in which field lines are disconnected, open, and closed. These are topological separatrix surfaces. (Ignore for a moment the nested circles inserted to the upper right.) Separatrix surfaces are field line surfaces, and only field lines lying on these surfaces are shown. The outer, cylindrically shaped surface separates disconnected (IMF) field lines from open (tail lobe) field lines. The inner, torus-shaped surface separates open field lines from closed field lines. The solar wind flowing across open field lines provides the motional electric field that maps to the polar cap. Dashed lines within the open field line cylinder indicate equipotentials of the motional electric field. As inferred from data taken near or on the earth in the polar cap, a potential typically of the order of 60 kV drops across the open field line cylinder. The origin of the Stern singularity can now be seen. The field lines that generate the open field line cylinder all map to a single point (labeled n.p. for null point), which implies that the full (typically) 60 kV is impressed across a point. In one hemisphere this null point communicates with the ionosphere along a single field line that intersects the polar cap at its boundary. This point becomes a topological 'node' where all equipotentials in the polar cap touch tangentially. The insert shows the resulting topology of ionospheric equipotentials. The Stern singularity acts like a 'black hole' for equipotentials, preventing their escape from the polar cap. The resemblance between these nested ionospheric potentials associated with the Stern singularity and those of Figure 4c is obvious. One difference is also obvious: the magnetopause current in the T-H model shifts the locus of the singularity

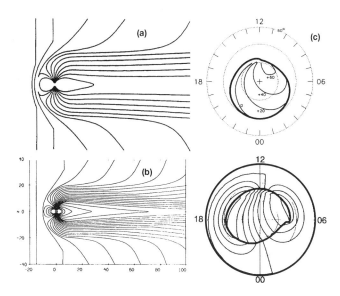

Figure 5. (a) Field lines in the *Toffoletto and Hill* [1989] model and (b) in the [1993] model sowing the presence of an 'expansion fan.' (c) Polar cap potential pattern for the [1989] model showing the effect of the Stern singularity. (d) Same for the [1993] model showing the absence of this effect.

sunward from the dusk terminator, where the dipole-plus-uniform-field superposition model of Figure 4d puts it [*Siscoe*, 1988]. Unless defeated, the Stern-singularity precludes modular models treating realistically the electrodynamic coupling between the solar wind and the magnetosphere.

The 1993 version of the T-H model overcame the problem of the Stern singularity by absorbing the topological null point into a current sheet. Then separatrix field lines, instead of converging on the null point (which no longer exists), penetrate the current sheet and remain separated. To achieve this innovation in modular engineering, the modelers subjected the interconnection field to a tailored stretch transformation expressly to produce a tail field that incorporates the geometry of a slow-mode expansion fan, which theory says is the MHD structure that ushers magnetosheath plasma into the tail. In the 1993 T-H model, the earthward edge of this ersatz expansion fan is a current sheet that automatically absorbs the topological null point that otherwise must exist. Figures 5a and 5b give 'side' views of the T-H magnetic field before and after the stretch transformation was applied to the interconnection field. The current sheet at the earthward edge of the 'expansion fan' is marked by a clear discontinuity, which runs from the 'shoulders' of the magnetosphere obliquely tailward toward the tail axis. Figures 5c and 5d demonstrate how much one gains in verisimilitude by eliminating the Stern singularity before computing the polar cap potential. The two figures show for the same external conditions the polar cap potential computed by the SSM, which is still afflicted with the Stern singularity (Figure 5c, [Peroomian et al, 1998]), and the 1993 T-H model, which has been cured of it (Figure 5d [Hill and Toffoletto, 1998]). Whereas in Figure 5c the equipotentials, having been captured by the Stern singularity, are confined to a single convection cell in the polar cap, in Figure 5d, they cross the polar cap boundary as two convection cells into the closed field line domain, in accordance with observations.

Next in the evolution of modular models comes the inclusion of Birkeland currents (a.k.a. field-aligned currents) conventionally designated Region 1 and Region 2 currents. These carry typically about as much total current (of the order of 1 to 2 million amps) as the Chapman-Ferraro current and as much as a fair stretch ($\sim 10\ R_e$) of the tail current. Observationally, the Region 1 and Region 2 currents are known from data taken by magnetometers on low-altitude, high-inclination satellites and by high-latitude networks of surface magnetometers. The distribution and intensity of field-aligned currents at low altitudes can be inferred from the magnetic perturbations that they generate as measured by these magnetometers. Observations do not, however, uniquely determine where in the magnetosphere these currents go or from where they come. Here theoretical ideas have provided suggestions, and global MHD simulations have even mapped out entire current systems attached to the Regions 1 and 2 currents [e.g., Tanaka, 2000].

To summarize the magnetotail project to this point, with the development of two versions of vacuum open-tail modular models--the Toffoletto-Hill model and the Peroomian-Lyons-Schulz model--modelers have gone about as far as they can by splicing together vacuum modules. Like the closed-tail models from which they branch, these models are mathematically ill posed in the sense that all magnetospheric parameters cannot be determined from boundary conditions alone; some assumption must be made to determine the amount of open magnetic flux and its distribution along the magnetopause. Results depend on the assumptions made to achieve this end. Both the T-H and P-L-S models give the magnetic field everywhere in the magnetosphere and tail. Both also give polar cap potentials, but only the T-H model has overcome the problem associated with the Stern singularity, which confines equipotentials to the polar cap. The P-L-S model has incorporated Regions 1 and 2 current systems, but referring to our opening classification of model types, these incorporations, though ingenious, fall under the head of representations rather than under models that explain. A fundamental limitation pertaining to vacuum models is that they do not embody plasma physics, which models need in order to compute correctly currents interior to the magnetosphere. So at this point we jump to another branch of modular models where we find research projects that treat interior plasma physics, the price for which they have had to pay is loss of global completeness, especially in regard to solar wind-magnetosphere coupling.

6. MAGNETOSPHERE-IONOSPHERE COUPLING

If the Chapman-Ferraro epoch of magnetospheric modeling is distinguished by determinability and universality of results, the post-C-F epoch is distinguished by the elegance of the theory that has been developed to address magnetosphere-ionosphere coupling and by the power of models based on this theory. The subject itself—magnetosphere-ionosphere coupling--is uniquely magnetospheric. Other subjects that any treatment of the magnetosphere as a whole must cover—the magnetopause, the tail, substorms, energetic particles—are magnetospheric examples of generic structures and processes that magnetized plasmas exhibit in various cosmic settings. In contrast to this, magnetosphere-ionosphere coupling is magnetospheric solely, a product of two media--the ionosphere and the magnetosphere--electrically coupled and governed separately by different physics that use different equations to compute the same quantities--electric fields and currents—at the same place and time. The requirement that the answers they return must match turns out to modify strongly the answers that either would return on its own. Region 2 currents and shielding are the signature products that emerge from the requirement of mutual consistency in M-I coupling.

Figure 6's five parts illustrate aspects of M-I coupling relevant to modeling. In Figure 6a, vectors and shading reveal, in this dawn-dusk meridian-plane view, current directions and intensities as derived from a global MHD model for the case in which the IMF strength is effectively zero. The earth is the central, shaded circle. The currents flowing above and below along the magnetopause from dusk to dawn are part of the Chapman-Ferraro current system and also, as discussed in the previous section, are part of the Region 1 current system, the earth-connecting legs of which are prominent. Less prominent, lower in latitude, and reversed in direction compared to Region 1 currents, Region 2 currents are seen occupying an area wholly interior to the magnetosphere. This figure substantiates that, unlike Region 1 currents which are govern by the physics of processes that couple the outer magnetosphere to the solar wind—viscosity in this case--Region 2 currents are governed by the physics of processes that couple the inner magnetosphere to the ionosphere.

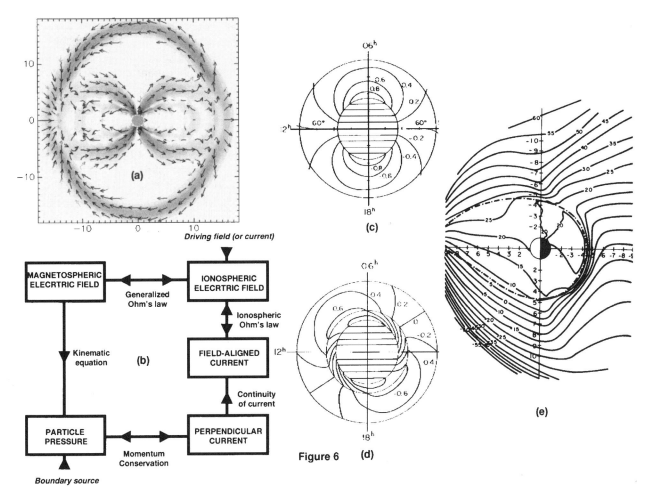

Figure 6. (a) Shading and arrows showing electrical current in the dawn-dusk meridian plane (x=0) as determined with the ISM MHD code for the case of zero IMF. Region 1 and 2 currents are evident. (b) The Vasyliunas loop for magnetosphere0ionosphere coupling [*Vasyliunas*, 1970]. (c) Ionospheric equipotentials showing the distortion that results from M-I coupling [*Vasyliunas*, 1970]. (d) Equipotentials in the equatorial plane showing the effect of M-I coupling in the magnetosphere as determined by *Jaggi and Wolf* [1973].

Figure 6b shows the famous *Vasyliunas* [1970] schematic of the feedback loop between the magnetosphere and ionosphere. Starting in the lower-left corner, we are given a magnetospheric particle population from some boundary source (usually the plasma sheet). This population, under the physics of "momentum conservation," or force balance, in order to create pressure-balancing magnetic stresses, generates an electrical current perpendicular to the magnetic field. This step takes us to the lower-right box in the loop. In general, the perpendicular current thus generated will not be divergenceless, which necessitates the existence of a parallel component of current (a.k.a. a field-aligned current) to make the current as a whole divergenceless. Now we are at the middle box on the right side of the loop. The field-aligned current thus necessitated follows the magnetic field to the ionosphere, where it

becomes a current source for the 2D ionospheric Ohm's law. This magnetospheric current source—plus any source of current or voltage imposed at the boundary of the ionospheric area into which the inner magnetosphere maps—determines the solution of the ionospheric Ohm's law within this area. The solution so obtained gives the ionospheric electric field, which provides the contents of the upper-right box in the loop. But the electric field whose strength and distribution over the mentioned area of the ionosphere has been thus determined maps back along magnetic field lines to the magnetosphere. In the figure, the mapping step is labeled "generalized Ohm's law," but in practice the mapping is done assuming either no parallel electric field or a parallel electric field governed by a semi-empirical formula. Now we see that the initially given magnetospheric particle population, in actuating the looped chain of causal linkages, has created in its own spatial domain an electric field, which was not there before—the last box in the loop. The arrow labeled "kinetic equation" connecting the last box to the first tells us that the electric field now acts to modify the initial population, so that we must go around the loop again, and, in principle, ad infinitum. Fortunately in nature the process converges fairly rapidly—of the order of tens of minutes, the so-called shielding time scale—to a steady condition which leaves all quantities sensibly unchanged with each subsequent iteration of the loop. Mathematically, the final, steady state can be found in one step as the solution of a Poisson equation.

Vasyliunas' quinpartite loop (and other formulations before his) prescribes a formalism that solves a difficult problem: How to determine the distribution within the magnetosphere of hot plasma (that is, plasma governed by full drift transport in distinction to simply ExB transport) that results from sunward convection of plasma sheet plasma? Solutions for the convective transport of hot ions and electrons separately had been well studied. But these suffer from a fatal deficiency: under full drift transport, ions and electrons soon separate, creating an unsupportable charge imbalance. Discharge of the imbalance through the ionosphere offers a solution readily available to the magnetosphere. This is the origin of M-I coupling.

Figure 6d displays in an idealized model a marked counterclockwise twist that M-I coupling, as formulated according to the Vasyliunas loop, forces on an initially symmetrical ionospheric potential pattern (Figure 6c). The twist occurs in an annulus that in this model is the ionospheric area onto which the pressure-bearing plasma of the inner magnetosphere magnetically maps. The twist strengthens the electric field within the annulus and weakens it equatorward of the annulus. The equatorward weakening is called shielding. The discontinuities at the poleward and equatorward borders of the annulus imply the existence of field-aligned current sheets, Region 1 at the poleward border and Region 2 at the equatorward border. An interesting property of the twisted potential pattern and Region 2 currents, which can be derived from the Vasyliunas formulation of the M-I coupling problem, is that after the system achieves steady state, the streamlines of Region 2 current enter and leave the ionosphere at the same potential. The physical significance of this property is that, in steady state, Region 2 currents carry no energy between the magnetosphere and the ionosphere [Siscoe, 1982]. In the language of electrical engineering, because in steady state the ionosphere is an equipotential as seen by the Region 2 circuit, it acts as neither a generator nor a load to the circuit.

The time-dependent situation is different, however. If the potential applied to the Region 1 border of the annulus increases, the increase is 'felt' at the Region 2 border. No longer an equipotential to the Region 2 current, the ionosphere plays the part of a generator. It is, of course, not a real generator but merely a transferor of energy from the Region 1 'generator' (the solar wind) to the Region 2 'load' (the pressure-bearing magnetospheric particle population). The pressure-bearing particle population consumes the energy as it rearranges itself in response to the voltage transient by compressing into regions of higher magnetic field strength to reestablish, via the Vasyliunas loop, an equipotential relation with the ionosphere. If, on the other hand, the potential applied to the Region 1 border of the annulus decreases, everything reverses, and energy flows the opposite way. (The solar wind, which is perpetually working to drag field lines across the polar cap, receives for once a little relief; while energy flows from the Region 2 circuit to the Region 1 circuit, the drag lessens.)

But real life is more complicated than Figure 6d depicts. Applied potentials are probably never symmetric; the magnetosphere's pressure-bearing particle population is certainly never map-able onto a uniform annulus; field-aligned currents do not flow in 2D sheets; and time dependence is the rule—steady state being a perhaps never realized exception. To go from an idealized model that one can treat analytically to models that can simulate observed magnetospheric behavior requires the development of numerical codes. Figure 6e, taken from *Jaggi and Wolf* [1973]—the first paper to report results from a numerical code that incorporates time-dependent M-I coupling formalism to simulate a non-ideal situation—shows equipotentials in the equatorial plane at a late stage in the solution to an initial-value problem that is approaching steady state. A plasma sheet population, which is initially confined to the perimeter of a model magnetosphere with a modified Mead-Williams magnetic field geometry, moves under the combined action of gradient and electric field drifts, the latter from a specified dawn-to-dusk convection potential. At the start of the computation, equipotentials fill the area with reasonably uniform spacing. By the stage illustrated here they have been contorted to create an area around the earth about 5 R_e in radius with wide spacing, and so, of weak electric field. Here is shielding as manifested in the output of this code.

Counterclockwise twisting of the equipotentials with an attendant local increase in electric field strength is evident just tailward of the shielded area. This twisting distorts the equipotentials such that they nearly enclose the shielding area, creating an equipotential arch in the region from which Region 2 currents arise. Thus, the main elements of M-I coupling seen in the idealized model (Figure 6d)—shielding, twisting with attendant electric field amplification, and "shorting out" of the Region 2 circuit—are seen here, too. But now the figure has a realistic, magnetospheric aspect. Moreover, as a product of a time-dependent calculation, Figure 6e is preceded by a series of intermediate figures showing the evolution from the initial uniform state with no shielding and no Region 2 currents to this mature, non-uniform state in which the features of M-I coupling are well expressed.

7. THE RICE CONVECTION MODEL

Because it illustrates M-I coupling operating within a model that attempts to represent magnetospheric geometry and particle distributions with some realism, the Jaggi and Wolf paper is a landmark. It marked the visible beginning of an important modeling program centered at Rice University the goal of which is to develop a global modular model that incorporates advanced M-I coupling physics within a time-dependent magnetosphere. Time dependence here means capable of changing the magnetic field model as the M-I computation proceeds in ways that capture the essence of substorms and magnetospheric compressions and expansions. A further goal of the program is to achieve plasma-field self consistency, that is, to assimilate into the magnetic field model the self field that the pressure-bearing particles generate. This is difficult because the self field varies as the particles encroach and retreat in response to fluctuating solar wind conditions.

The model that is evolving under this program is the Rice Convection Model (RCM). It is built around the Vasyliunas quinpartite loop with boxes feeding into it to provide ionosphere-thermosphere quantities and to specify parameters that apply at the boundaries of the model's computational domain, that is, its outer boundary in the magnetosphere and its poleward boundary in the ionosphere. The modeling achievable with the RCM can be highly dynamic. The algorithms that specify the magnetospheric magnetic field can simulate substorms through imposing dipolarization (this is called "declaring a substorm"), and they can simulate compressions and expansions caused by increases and decreases in solar wind ram pressure. (For a review of the RCM and the formalism behind it, see *Wolf*, [1983].)

Perhaps the most far-reaching contribution to emerge from the RCM is the concept of "pressure inconsistency" or "pressure catastrophe" that results from pressure build up near the transition between the magnetosphere and the magnetotail [*Erickson and Wolf*, 1980]. The build up, which in RCM runs is unavoidable without intervention, results in weakening the magnetic field. Erickson and Wolf and others after them credit this effect with leading to substorm onset.

It deserves to be mentioned, particularly in a book on space weather, that the RCM is the father of the first, and

so far only, numerical magnetospheric code that has been put into service by an operational space weather service provider, NOAA's Space Environment Center. The Magnetospheric Specification Model (MSM) is a highly parameterized, and, so, faster and more robust but less flexible, version of the RCM. It takes real-time solar wind and magnetospheric data as input and returns as output real-time values of fluxes of electrons and ions in the 1 to 100 kilovolt range everywhere in the inner magnetosphere where satellites orbit. Of the types of modeling mentioned at the outset of this tutorial, the modular type in the form of the MSM is the first to have undergone the rigorous ordeal of iterative revision, testing, and validation to rise to the status of a numerical space weather prediction code.

As a more tractable version of the RCM, the MSM is also a powerful tool for research. It retains the RCM's capability to model time-dependent convection potential, solar wind pressure, and substorm dipolarization. Figure 7a from *Wolf et al.* [1997] provides an example of its application to research. It shows a gray-scale representation of MSM-computed contours in the equatorial plane of the abundance of H^+ ions that have energy invariance equivalent to 8.6 keV at geosynchronous orbit. In the top two segments of this figure we see a before-and-after comparison: first a picture of the contours at the end of a quiet period with no magnetospheric compression and weak convection, then a picture of the contours after an hour of change in response to magnetospheric compression and strong convection. The bottom two segments continue the sequence showing first the change resulting from two hours of continued strong convection then the change resulting from three hours that follow of weak convection, all under a condition of solar wind compression. Besides demonstrating the ability of the code to follow details, this sequence of pictures demonstrates the inherently non-MHD character of plasma transport in the inner magnetosphere. The contours, especially in the last segment, are obviously governed by gradient drifts. On the other hand, much of what the MSM and the RCM require as initial and boundary conditions global MHD codes simulate well.

8. COMPLEMENTARITY OF RCM AND MHD MODELS

Figure 7b, which is a 3D extension of Figure 6a, illustrates the complementary capabilities to which we just alluded of the global MHD approach to magnetospheric modeling. In global MHD models, the entire magnetosphere lies in the computational domain. Quantities that ordinarily must be input to the RCM, such as the transpolar potential (or the Region 1 currents) and the magnetospheric magnetic field, in global MHD models are model outputs. Figure 7b demonstrates that the two approaches

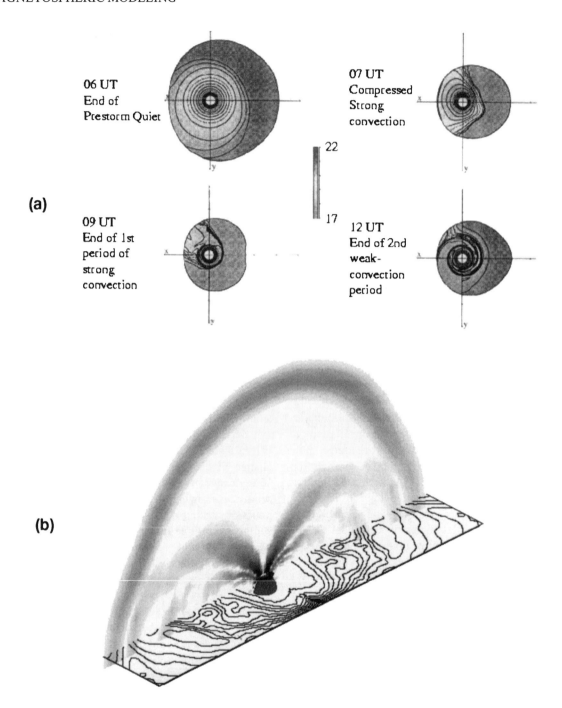

Figure 7. (a) Contours of energetic ion abundance as determined with the Magnetospheric Specification Model (MSM) in a storm simulation [*Wolf et al.*, 1997]. (b) Shading shows current and contours show particle pressure as determined with the ISM MHD code.

use the same physics where they should do so. Notice that the Region 2 currents seen in the dawn-dusk meridian plane (refer also to Figure 6a) arise where strong azimuthal pressure gradients exist as seen in this plane 1 R_e above the equatorial plane. An azimuthal pressure gradient, according to the Vasyliunas equation, should produce a field-

aligned current in the inner magnetosphere, for the gradient in flux tube volume is approximately radial, and field-aligned currents arise wherever these two gradients are other than parallel or anti-parallel.

But pressure gradients that existing (ideal) MHD codes compute for the inner magnetosphere can be assumed to be inaccurate. Velocity dispersion resulting from gradient and curvature drifts distorts 6D phase space volume elements. In the inner magnetosphere this distortion makes the single velocity obtained by the moment technique behind MHD formalism (which gives the velocity of the center of momentum reference frame) a poor means of representing actual mass transport. The RCM, on the other hand, which breaks up phase space along the energy axis so as to reduce distortions that gradient and curvature drifts impose, and thus computes many velocities, is more accurate. Where MHD has 1 velocity, RCM has 16, and so does a better job in situations where these 16 velocities differ significantly.

As a discipline, magnetospheric modeling faces a fortunate situation in having two model types that nearly ideally complement each other. Global MHD codes take solar wind conditions as input and provide electric and magnetic fields everywhere within the magnetosphere as output. The RCM, by contrast, treats these quantities as external inputs or as in house, provide-on-demand products served up by algorithmically constrained modular or empirical sub-models. Global MHD codes also balance plasma and magnetic stresses everywhere and so automatically create a reactive magnetosphere. The RCM program has as a goal yet to be achieved the construction of a reactive magnetosphere. The RCM, for its part, computes pressure within the inner magnetosphere accurately, a capability that global MHD codes, as just noted, famously lack. It is obvious, therefore, that an urgent next step in magnetospheric modeling is the development of a hybrid code that marries the positive features of global MHD and RCM formalisms.

9. CURRENT ACTIVITIES

At the time of writing, nearly all, if not all, groups that are developing global MHD codes in the United States are trying to adapt their codes to be able to simulate the effects of curvature and gradient transport in the inner magnetosphere. Some groups are working to embed RCM-type formalism integrally into their codes. The idea is to let the MHD part of the code compute global electric and magnetic fields for each time step, the latter in stress balance with the plasma. The RCM part of the code then uses these to advance the pressure to the next time step and feeds the answer back to the MHD part of the code. In a demonstration of concept, Frank Toffoletto has used the UCLA MHD code to drive the RCM, but he did not attempt to feed the resulting pressure back to the MHD code. This one-way combination nevertheless improves both codes. Viewed as an improved MHD code, it produces realistic Region 2 currents. Viewed the other way, it is RCM

with an improved magnetic field model. The results have not been published but instead exist only in the form of figures with little text at a web site, http://rigel.rice.edu/~toffo/research.html.

The Integrated Space Weather Model (ISM) being developed by a group at Mission Research Corporation exists in a version that contains both MHD and RCM components. It has been run with coupling between the two components operating in both directions. Preliminary results, presented at the 1998 GEM Summer Workshop, show that plasma sheet plasma encroaches closer to earth with the RCM activated than if it is not activated. These results are preliminary and have not been published, but at least they demonstrate that the two codes run together in mutual feedback mode without "crashing" or without instantly going unstable. The problem that this exercised revealed is that the run time per computational step greatly increases because of the requirement on the MHD code to compute flux tube volumes for the RCM.

At the 2000 GEM Summer Workshop, developers of the Michigan MHD code revealed innovations aimed at reducing to an acceptable level the run-time penalty incurred in embedding RCM numerics into their code. So imminent the development of this innovation seems to be and so clear its need, that before the book you are holding is printed results from the hybridized Michigan—or other—code already might have appeared.

Other MHD modelers, notably at Dartmouth and MRC, are exploring as an alternative to RCM formalism the possibility of modifying the MHD equations to achieve the same result. *Heinemann* [1999], for example, has shown that aspects of gradient and curvature transport can be simulated in a fluid approach by including a collisionless heat flux in the energy equation. Should an alternative approach like this succeed, it could simplify magnetospheric modeling, and perhaps shorten run times, by uniting in one formalism the benefits of the separate RCM and MHD formalisms.

10. SUMMARY

We have followed the history of a main branch of magnetospheric modeling that goes under the heading 'modular.' We chose this branch instead of another in part because Chapman and Ferraro, who started the field, did so with a modular model. But also in part because modular modeling is an attempt to construct the magnetosphere piece by piece. Each piece is a module whose physics, as the modeler conceives it, must be explicitly encoded. Then the pieces must be connected, which is a job that tests the modeler's physical insight into the magnetosphere as a system. Developing a modular model is, in other words, the severest test of our understanding of the magnetosphere as a complex physical system. By comparing a modular model's output against data, the modeler gets guidance that bears pointedly on the applicability of a physical assumption. This is physics in the laboratory mode. Those who pursue it can usually trace their intellectual heritage to a physics department.

Modular modeling has moved from its initial success at solving definitively the Chapman-Ferraro problem—determine the size and shape of a closed, vacuum magnetosphere and the magnetic field within it—into a protracted struggle, still not won, with trying to solve the corresponding problem for an open magnetosphere that is filled with pressure-bearing plasma and that is in strong interaction with the variable solar wind above and the ionosphere below. The effort has evolved along two complementary branches, one that exploits the relative simplicity of vacuum fields by representing the role of plasmas with current sheets (the T-H and SSM models) and one that exploits the relative simplicity of particle drift physics by prescribing the electric and magnetic fields (the RCM and its progeny). Models on the first of these two branches are best suited to treat the interaction with the solar wind, and they have achieved some successes in this arena. Models on the second of the branches are tailored to treat the interaction with the ionosphere, and they have made fundamental contributions to this area. Curiously, despite being complementary and being housed at the same institution, the T-H model and the RCM have not been coupled together. The reason is that, as they stand, their magnetic field models are incommensurable [*Toffoletto*, private communication, 2000].

Instead the RCM seeks to supply parts of global modeling capability that it lacks by becoming an inner-magnetosphere module in a global MHD code. Development of this type of hybrid code appears to be the most likely next step in the evolution of physics-based magnetospheric models.

Acknowledgments. The author is supported by grants from NSF's Upper Atmospheric Section of the Atmospheric Sciences Division (AM98-12678) and NASA's Sun-Earth Connections Theory Program (NAG5-8135). The global MHD code used generate the MHD-comparison figures is the Integrated Space Weather Model developed by Mission Research Corporation under contract from the Defense Threat Reduction Agency, Willard White, principal investigator, and Keith Siebert principal code developer. Daniel Weimer developed the graphics package (ISM-View) used to generate the MHD-comparison pictures. Richard Wolf prepared the gray-scale version of Figure 7 for this paper. Frank Toffoletto helped considerably by supplying information concerning modeling projects at Rice.

REFERENCES

Beard, D. B., The interaction of the solar wind with planetary magnetic fields: Basic principles and observations, *Planet. Space Sci., 21*, 1475-1496, 1973.

Birn, J., Magnetotail stability and dynamics: Progress 1991-1993, *Surveys in Geophysics, 16*, 299-330, 1995.

Birn, J., F. Iinoya, J. U. Brackbill, and M. Hesse, A Comparison of MHD simulations of magnetotail dynamics, *Geophys. Res. Lett., 23*, 323, 1996a.

Birn, J., M. Hesse, and K. Schindler, MHD simulations of magnetotail dynamics, *J. Geophys. Res., 101*, 12,939, 1996b.

Blum, R. The interaction between the geomagnetic field and the solar corpuscular radiation, *Icarus, 1*, 459-488, 1963.

Chapman, S., Solar Plasma, Geomagnetism and Aurora, in *Geophysics: The Earth's Environment*, edited by C. DeWitt, J. Hieblot, and A. Lebeau, Gordon and Breach New York, 1963.

Chapman, S., and V. C. A. Ferraro, A new theory of magnetic storms, *Terr. Mag. Atmos. Elec., 36*, 77-97 and 171-186, 1931; *Terr. Mag. Atmos. Elec., 37*, 147-156, 1932; and *Terr. Mag. Atmos. Elec., 38*, 79-96, 1933.

Dungey, J. W., *Cosmic Electrodynamics*, Cambridge University Press, Chapter 8, 1958.

Erickson, G. M., and R. A. Wolf, Is steady convection possible in the Earth's magnetotail?, *Geophys. Res. Lett., 7*, 897-900, 1980.

Fairfield, D. H., Average and unusual locations of the earth's magnetopause and bow shock, *J. Geophys. Res., 76*, 6700-6716, 1971.

Ferraro, V. C. A., On the theory of the first phase of a geomagnetic storm, *J. Geophys. Res, 57*, 15-49, 1952.

Harel, M., R. A. Wolf, P. H. Reiff, and M. Smiddy, Computer modeling of events in the inner magnetosphere, in *Quantitative Magnetospheric Magnetic Modeling*, edited by W. P. Olsen, Geophysical Monograph 21, American Geophysical Union, pp. 499-512, 1979.

Harel, M., R. A. Wolf, R. W. Spiro, P. H. Reiff, C.-K. Chen, W. J. Burke, F. J. Rich, and M. Smiddy, Quantitative simulation of the magnetospheric substorm: 2. Comparison with observations, *J. Geophys. Res., 86*, 2242-2260, 1981.

Heinemann, M, Role of collisionless heat flux in magnetospheric convection, *J. Geophys. Res., 104*, 28.397-28, 410, 1999.

Hill, T. W., and F. R. Toffoletto, Comparison of empirical and theoretical polar cap convection patterns for the January 1992 GEM interval, *J. Geophys. Res., 103*, 14, 811-14, 817, 1998.

Jaggi, R. K., and R. A. Wolf, Self-consistent calculation of the motion of a sheet of ions in the magnetosphere, *J. Geophys. Res., 78*, 2852-2866, 1973.

Jordan, C. E., Empirical models of the magnetospheric magnetic field, *Rev. Geophys., 32*, 139-157, 1994.

Lyons, L. R., A simple model for polar cap convection patterns and generation of Θ-auroras, *J. Geophys. Res., 90*, 1561-1567, 1985.

Mead, G. D., Deformation of the geomagnetic field by the solar wind, *J. Geophys. Res., 69*, 1181-1195, 1964.

Mead, G. D., and D. B. Beard, Shape of the geomagnetic field solar wind boundary, *J. Geophys. Res., 69*, 1169-1179, 1964.

Midgley, J. E., and L. Davis, Calculation by a moment technique of the perturbation of the geomagnetic field by the solar wind, *J. Geophys. Res., 68*, 5111-5123, 1963.

Olsen, W. P., *Quantitative Magnetospheric Magnetic Modeling*, Geophysical Monograph 21, American Geophysical Union, 1979.

Parker, E. N., Dynamics of the geomagnetic storm, *Space Sci. Rev., 1*, 62-99, 1962.

Peroomian, V., L. R. Lyons, and M. Schulz, Inclusion of shielded Birkeland currents in a model magnetosphere, *J. Geophys. Res., 103*, 151-163, 1998.

Peroomian, V., L. R. Lyons, M. Schulz, and D. C. Pridmore-Brown, Comparison of assimilative mapping and source surface model results for the magnetospheric events of January 27 to 28, 1992, *J. Geophys. Res., 103*, 14,819-14,827, 1998.

Roederer, J. G., Quantitative models of the magnetosphere, *Rev. Geophys., 7*, 77-96, 1969.

Schulz, M. and M. C. McNab, Source-surface model of the magnetosphere, *Geophys. Res. Lett., 14*, 182-185, 1987.

Schulz, M., The Magnetosphere, in *Geomagnetism Vol.4*. edited by J. A. Jacobs, Academic Press, pp. 87-293, 1991.

Schulz, M. and M. C. McNab, Source-surface model of planetary magnetospheres, *J. Geophys. Res., 101*, 5095-5118, 1996.

Siscoe, G. L.: A unified treatment of magnetospheric dynamics, *Planet. Space Sci., 14*, 947-967, 1966

Siscoe, G. L., Energy coupling between Regions 1 and 2 Birkeland current systems, *J. Geophys. Res., 87*, 5124-5130, 1982.

Siscoe, G. L., The magnetospheric boundary, in *Physics of Space Plasmas (1987)*, edited by T. Chang, G. B. Crew, and J. R. Jasperse, pp. 3-78, Scientific Publishers, Inc., 1988.

Spreiter, J. R., A. L. Summers, and A. Y. Alksne, Hydromagnetic flow around the magnetosphere, *Planet. Space Sci., 14*, 223-253, 1966.

Spreiter, J. R., and A. Y. Alksne, Effect of neutral sheet currents on the shape and the magnetic field of the magnetotail, *Planet. Space Sci., 17*, 233-246, 1969.

Stern, D. P., Tail modeling in a stretched magnetosphere: 1. Methods and transformations, *J. Geophys. Res., 92*, 4437-4448, 1987.

Tanaka, T., Field-aligned current systems in the numerically simulated magnetosphere, In *Magnetospheric Currents II*, edited by S.-I. Ohtani et al., Geophysical Monograph 118, American Geophysical Union, 53-59, 2000.

Toffoletto, F. R., and T. W. Hill, Mapping of the solar wind electric field to the earth's polar caps, *J. Geophys. Res., 94*, 329-347, 1989.

Toffoletto, F. R., and T. W. Hill, A nonsingular model of the open magnetosphere, *J. Geophys. Res., 98*, 1339-1344, 1993.

Unti, T., and G. Atkinson, Two-dimensional Chapman-Ferraro problem with neutral sheet: 1. The boundary, *J. Geophys. Res., 73*, 7319-7327, 1968.

Vasyliunas, V. M., The interrelationship of magnetospheric processes, in *Earth's Magnetospheric Processes*, edited by B. M. McCormac, pp. 29-38, D. Reidel, Dordrecht, 1970.

Voigt, G.-H., A mathematical magnetospheric model with independent physical parameters, *Planet. Space Sci., 29* 1-20, 1981.

Voigt, G.-H., The shape and position of the plasma sheet in Earth's magnetotail, *J. Geophys. Res., 89*, 2169-2179, 1984.

Williams, D. J., and D. G. Mead, Night-side magnetospheric configuration as obtained from trapped electrons at 1100 kilometers, *J. Geophys. Res., 70*, 3017-3029, 1965.

Wolf, R. A. The quasi-static (slow-flow) region of the magnetosphere, in *Solar-Terrestrial Physics*, edited by R. L. Carovillano and J. M. Forbes, D. Reidel Publishing Co., Dordrecht, Holland, pp. 303-368, 1983.

Wolf, R. A., Modeling convection effects in magnetic storms, in *Magnetic Storms*, edited by B. T. Tsurutani et al., Geophysical Monograph 98, American Geophysical Union, pp. 161-172, 1997.

George Siscoe, Center for Space Physics, Boston University, 725 Commonwealth Avenue, Boston, MA 02215

MHD Simulation of Magnetospheric Transport at the Mesoscale

W. W. White, J. A. Schoendorf, K. D. Siebert, N. C. Maynard, D. R. Weimer, and G. L. Wilson

Mission Research Corporation, Nashua, New Hampshire

B. U. Ö. Sonnerup

Thayer School of Engineering, Dartmouth College, Hanover, New Hampshire

G. L. Siscoe and G. M. Erickson

Boston University, Center for Space Physics, Boston, Massachusetts

Global MHD simulations of the magnetosphere reveal unsteady vortical plasma flows at the mesoscale, similar in appearance to fluid turbulence at moderate Reynolds numbers, in the magnetotail under a variety of solar-wind conditions and interplanetary magnetic field (IMF) orientations. Such results are not unexpected in view of satellite data for magnetotail plasma-flow properties and the chaotic nature of the model MHD equations. Simulations generated with the Integrated Space Weather Prediction Model that are driven by either steady-state or highly variable solar-wind and IMF conditions suggest mesoscale vortical flows to be pervasive and robust features of magnetotail dynamics. Intermittent surges in simulated plasma flows are suggestive of bursty bulk flows. Autocorrelation functions for fluctuating properties in the simulated plasma sheet compare favorably to corresponding statistical reductions of satellite data.

1. Introduction

The environment of near-Earth space, where satellites orbit and manned space operations are conducted, is highly dynamic due in large measure to the constantly changing state of the ionized solar wind and interplanetary magnetic field (IMF) that emanate from the Sun. Like many man-made contrivances, systems operating in space or through the ionosphere (e.g., satellites, com-

munications links, radars) have limited capacity to function within the extremes of their environment, so operators and users of such systems stand to benefit from a forecast capability for dynamic properties of near-Earth space, much as terrestrial activities benefit from routine meteorological forecasts.

Present models of near-Earth space are for the most part "climatological." They are based on data taken by one or a few satellites, rocket probes, or ground-based instruments at different times and places. Data thus accumulated are patched together into mosaics (e.g., [*Weimer*, 1996; *Kivelson and Russell*, 1993]) that illustrate, for example, how the magnetosphere looks during equinox or solstice, or how its degree of storminess depends on the condition of the solar wind and

Space Weather
Geophysical Monograph 125

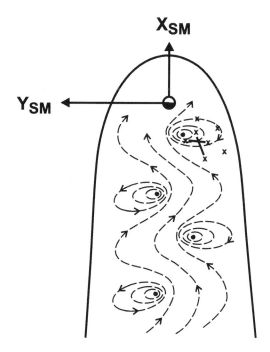

Figure 1. Equatorial plane of the magnetosphere with plasma vortex street (dashed curves) convecting earthward. The locations where plasma vortices were observed with the ISEE satellites are indicated by crosses. Lines joining the crosses depict satellite paths during the prolonged encounter on Jan. 28-29 and March 2 (1978). Orientations of the plane of rotation other than shown here do also occur. (Figure and caption from Figure 6 in *Hones et al.* [1978].)

its magnetic field, which activate the magnetosphere. By analogy with meteorology, this kind of information is like knowing that winds and temperatures are organized into zones and that low pressure is associated with storms, high pressure with calms. In other words, it is magnetospheric climatology. Yet, every satellite experiences continuous departures from the climatological mean. It seems that, like the atmosphere, the magnetosphere is a weather machine, the modes of which are presently lost in climatology.

Although the magnetospheric equivalents of extratropical cyclones, hurricanes, tornadoes, microbursts, mesoscale convective complexes, etc. are not fully resolved by data-gathering capabilities of present satellites, it has been possible to piece together a limited picture of mesoscale phenomena. Waves on the magnetopause surface have been associated with the Kelvin-Helmholtz instability [*Kivelson and Chen*, 1995]. Data for plasma flows in the central plasma sheet indicate vortical convection cells [*Hones et al.*, 1978, 1981, 1983] and transient bursts of high-speed flow [*Angelopolous et al.*, 1992a,b; *Borovsky et al.*, 1997a] in regions where slow flows tend to dominate [*Baumjohann et al.*, 1988,

1989]. Evidence for bursty flows has been interpreted as flow "turbulence" with characteristic mixing length of roughly 2 Earth radii (R_E)[*Borovsky et al.*, 1997a]. More highly resolved data, however, await NASA launches of multi-satellite constellation missions capable of simultaneous multi-point sampling of the magnetospheric volume.

Meanwhile, progress in the area of modeling "weather" in near-Earth space is being made with global magnetohydrodynamic (MHD) simulations. Recent efforts have been largely confined to simulating specific solar-wind/IMF conditions and comparing simulated results with sparse data [*Fedder et al.*, 1995; *Raeder et al.*, 1995, 1996, 1997; *White et al.*, 1998; *Slinker et al.*, 1999; *Song et al.*, 1999; *Siscoe et al.*, 2000]. Such analyses serve two purposes. First, they provide global context within which to interpret available data and the connections between them. Second, they validate the state-of-the-art of global MHD simulations of magnetospheric processes. In this way, global MHD simulation has become a powerful tool for conducting magnetospheric research.

The study reported here extends global MHD simulations to characterize magnetospheric weather systems. Instead of simulating events to provide global context for data analyses, we ask what are the generic modes of mesoscale (a few to 10's of R_E) magnetospheric dynamics. That is, what forms do mesoscale systems take, how are they arranged in space and time, and how do they depend on solar-wind driving conditions? Based on ISEE satellite data, Hones et al. [1978] suggested mesoscale plasma-sheet flows might at times appear as shown in Figure 1. Presumably the spectrum of magnetospheric responses to specific space weather "events" includes instances of one or more such generic modes, just as the recent Oklahoma tornado outbreak was an instance of severe mesoscale storms. We are after the categories of mesoscale magnetospheric dynamical modes of which magnetospheric events are specific instances.

To this end we use the Integrated Space Weather Prediction Model (ISM), a model being developed to simulate properties of near-Earth space—the magnetosphere, ionosphere, and thermosphere as driven by the solar wind—to forecast environmental conditions encountered by operational spacecraft. ISM implements coupled plasma fluid-neutral fluid MHD methods (i.e., two-material MHD equations) optionally augmented with particle-drift kinetics, macro-scale approximations for micro-physics of magnetic reconnection, and related physical processes to simulate near-Earth space from the bottom of the ionosphere (taken to be ~ 100 km altitude), upward through the magnetosphere, and into the solar wind. Recent ISM simulations based on a reduced set of the full two-material MHD equations (reduced to simplify treatment of the ionosphere) reveal complex mesoscale dynamics within the macroscale envelope of the magnetosphere, even when solar-wind conditions are steady. (To switch from a meteorological analog to a fluid dynamical one, mesoscale dynamics in the magnetosphere are like the vortices and turbulence that occur in the wake of a blunt body immersed in fluid flow, except that for the magnetosphere the magnetic field introduces new modes of coupling.) In a space weather context, data suggest dynamics at such intermediate scales mediate transport and energization processes within the magnetosphere [*Kivelson*, 1976; *Angelopolous et al.*, 1992, 1994; *Borovsky et al.*, 1997a] and contribute to injection of hot magnetospheric plasma encountered by geosynchronous satellites [*Mauk and Meng*, 1987].

Such results are not unexpected given the body of knowledge about fluid turbulence (e.g., [*Van Dyke*, 1982; *Meneguzzi et al.*, 1995]). The suggestion that the plasma sheet is a manifestation of the wake in the solar wind behind the Earth [*Montgomery*, 1987] in analogy to high Reynolds-number fluid flow behind an object is supported by the strong correlation of the plasma-sheet number density with the solar-wind number density [*Borovsky*, 1997b], by the correlation of the plasma-sheet ion temperature with the solar-wind velocity [*Borovsky et al.*, 1998], and by the mean flow direction Earthward or tailward also being correlated with the solar-wind velocity. Inside about 45 R_E the mean convection is Earthward [*Hayakawa, et al.*, 1982], while tailward of that, the mean flow is tailward [*Zwickl et al.*, 1984]. However, non-MHD processes make the analogy to fluid wakes complex (see *Borovsky et al* [1997a] and references therein). Other investigators have successfully used MHD models to examine "localized" phenomena such as magnetic flux ropes [*Ogino, et al.*, 1990; Berchem et al., 1995; Walker and Ogino, 1996] and tail lobe reconnection [Crooker et al., 1998], but we are not aware of prior MHD simulations of mesoscale flow structure of the type reported here.

2. ISM SIMULATION METHODS

The Integrated Space Weather Prediction Model operates within a three-dimensional cylindrical computational domain with its origin centered on the Earth and extending 40 R_E sunward, 300 R_E anti-sunward, and 60 R_E radially from the Earth-Sun line. The domain has an interior spherical boundary at the approximate bottom of the E-layer (at 100 km in simulations described here). Figure 2 provides a schematic illustration of the ISM finite-difference grid within this domain.

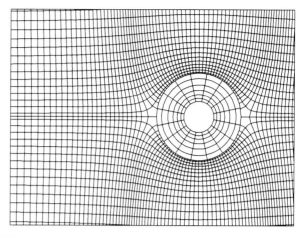

Figure 2. Schematic diagram of ISM computational grid in noon-midnight meridian plane. Interior boundary of grid is at 100 km altitude.

ISM is based on standard MHD equations for ionospheric and magnetospheric plasma augmented with hydrodynamic equations for a collisionally coupled neutral (un-ionized) thermosphere. These "two-material" MHD equations, first put on a firm theoretical basis in the early 1970s and later published by *Kilb* [1977], are statements of mass, momentum, and energy conservation for both plasma and neutral fluids with collision terms that account for momentum and thermal energy exchanges between fluids and frictional heating from differential motions of the fluids. They offer the advantage of transitioning continuously from pure MHD for plasma in the solar wind and magnetosphere to proper ionospheric/thermospheric equations at low altitudes. ISM implements four conservation equations for internal energies to support distinct temperatures for ions, electrons, neutrals, and vibrational modes of N_2 molecules. The last of these temperatures is necessary for computing reactions between oxygen ions and nitrogen molecules in the ionosphere/thermosphere. Chemical reactions between species in the ionosphere and thermosphere require that the full set of two-material MHD equations also include terms to account for mass, momentum, and energy exchanges between fluids resulting from ionization and recombination processes. In the ISM formalism, Ohm's Law is derived from electron momentum conservation with electron inertia neglected.

The focus of the simulations reported here is magnetospheric plasma dynamics, with particular interest in the plasma sheet. Consequently, for purposes of computational economy a static thermosphere was assumed for these simulations. Thermospheric neutral fluid transport, associated chemical and thermal coupling terms between ionosphere and thermosphere were

Table 1. Reduced ISM Two-Material MHD Equations

Mass Conservation

$$\frac{\partial \rho_i}{\partial t} + \nabla \cdot (\rho_i \vec{V}_i) = 0 \tag{1}$$

Momentum Conservation

$$\frac{\partial (\rho_i \vec{V}_i)}{\partial t} + \nabla \cdot (\rho_i \vec{V}_i \vec{V}_i) = -\nabla (P_i + P_e) + \vec{J} \times \vec{B} + \rho_i \nu_{in} (\vec{V}_n - \vec{V}_i) + \rho_i \mu_R \nabla^2 \vec{V}_i \tag{2}$$

Energy Conservation

$$\frac{\partial \epsilon_i}{\partial t} + \nabla \cdot (\epsilon_i \vec{V}_i) = -P_i \nabla \cdot \vec{V}_i + \frac{1}{2} \eta_R J^2 \tag{3}$$

$$\frac{\partial \epsilon_e}{\partial t} + \nabla \cdot (\epsilon_e \vec{V}_i) = -P_e \nabla \cdot \vec{V}_i + \frac{1}{2} \eta_R J^2 \tag{4}$$

$$\frac{\partial}{\partial t} \left(\frac{1}{2} \rho_i V_i^2 + \epsilon_i + \epsilon_e \right) + \nabla \cdot \left[\frac{1}{2} \rho_i V_i^2 + \gamma (\epsilon_i + \epsilon_e) \right] \vec{V}_i = \vec{J} \cdot \vec{E} + \rho_i \nu_{in} (\vec{V}_n - \vec{V}_i) \cdot \vec{V}_i \tag{5}$$

Ohm's Law

$$\vec{E} = -\vec{V}_i \times \vec{B} + \eta_R \vec{J} \tag{6}$$

with

$$\frac{\partial \vec{B}}{\partial t} = -\nabla \times \vec{E} \tag{7}$$

and

$$\vec{J} = \frac{1}{\mu_o} \nabla \times \vec{B} - \epsilon \frac{\partial \vec{E}}{\partial t} \tag{8}$$

where subscripts i, e, and n indicate, respectively, ions, electrons, and neutrals; m's are masses; ρ's are mass densities; N's are number densities; $\vec{V}$'s are velocities; P's are pressures; $\vec{E}$ and $\vec{B}$ are electric and magnetic fields; $\vec{J}$ is current density; ϵ_i and ϵ_e are internal energies; ϵ is a dielectric constant used to control the effective speed of light for numerical purposes; γ is the ratio of specific heats; ν_{in} is ion-neutral collision frequency; and e is the electron charge. Explicit resistivity and viscosity coefficients are designated by η_R and μ_R.

---

disabled. However, essential ion-neutral collision aspects of the full two-material equations were retained for these simulations because these terms provide the basis for ionospheric conductivity and the ion-neutral drag force. The reduced set of two-material equations used for these simulations is summarized in Table 1. The ion-neutral "frictional" drag term in Eq. 2, and the corresponding frictional work term in Eq. 5 are both proportional to the ion-neutral collision frequency ν_{in}. A combination of semi-implicit finite-difference implementation and time-step control are used in ISM to deal with stiffness engendered by the collisional terms at ionospheric altitudes. Above $1000\ km$, the collision terms (and all neutral fluid equations when the full two-material equations are used) are disabled owing to the low neutral density there.

For the present simulations finite-difference grid resolution varied from a few hundred kilometers in the

coarsely resolved ionosphere to several R_E at the outer boundary of the computational domain. At the magnetopause, resolution ranged between 0.2 to 0.8 R_E. Explicit viscosity in the plasma momentum equation was set to zero ($\mu_R = 0$). To approximate non-linear aspects of magnetic reconnection within the context of a finite-difference grid, the coefficient for explicit resistivity η_R in the Ohm's Law equation was zero when current density perpendicular to B is less than $3.16 \times 10^{-3}\ \mu a/m^2$ and was $2 \times 10^{10}\ m^2/s$ in regions with perpendicular current density above this threshold. In practice, this choice of η_R produced non-zero explicit resistivity primarily in the subsolar region of the dayside magnetopause, and then typically only for zero, strongly southward, or dawn/dusk-directed IMFs. Dissipation for numerical stability was based on a variant of the partial donor-cell method (PDM) developed by Hain [1987]. Except as noted, solar-wind inflow boundary conditions for the simulations used typical values: speed = 400 km/s, density = 5 $protons/cc$; ion temperature = electron temperature = 20 eV; IMF strength = 5 nT.

Ionospheric electron density was initialized with a profile uniform in latitude and longitude, giving a Pedersen conductance above the magnetic pole of 6 $Siemens$. Given the magnetospheric focus of these simulations, we elected to eliminate temporal variations in ionospheric conductance from the simulations, so ionospheric electron density was frozen below 1000 km altitude. No Hall conductance was used. On the ionospheric bottom grid boundary at 100-km altitude, imposed boundary conditions require tangential components of the magnetic field to remain fixed at dipole field values; normal derivatives of tangential electric field are set to zero. Simulations reported here did not use a rotating Earth. The assumed ionospheric profile and static thermosphere effectively suppressed ionospheric inflow/outflow. Analysis of boundary layer transport properties in ISM simulations indicates the effective kinematic viscosity due to artificial numerical dissipation in ISM's finite-difference approximations to be of the order of 4×10^8 m^2/s. This level of dissipation is consistent with estimates of kinematic viscosity and mass-diffusion coefficient in the magnetospheric low-latitude boundary layer derived from spacecraft observations of boundary-layer width and mass flow [*Sonnerup*, 1980; *Sckopke et al.*, 1981].

Eq. 8 of Table 1 includes a displacement current with dielectric constant, ϵ, used for numerical purposes to limit the effective speed of light. The value of ϵ is selected to provide a limiting speed of light large enough for a reasonable representation of the operative physics yet small enough to prevent unnecessarily small time steps. For the simulations reported here, the effective speed of light was set at 600 km/s. Test calculations with effective limiting speeds of 600, 1000, and 3000 km/s demonstrated mesoscale results insensitive to the particular limit used.

3. MESOSCALE MODE SIMULATIONS

ISM simulations using steady solar-wind and IMF conditions as described above have been completed for IMF orientations in purely northward and purely southward directions with the geomagnetic dipole normal to the ecliptic plane. An additional simulation with zero IMF (actually 5×10^{-3} nT) is used for reference as a magnetospheric "ground state". The assumed geomagnetic-dipole orientation and IMF geometry gave each of these simulations reflection symmetry about the noon-midnight meridian and equatorial planes, so for purposes of computational economy only one quadrant was simulated. Each of these simulations was run for 2

to 4 hours with fixed parameters to reach a quasi-steady state. Growth and dissipation of mesoscale flow features were examined with the aid of computer-generated movies. Selected frames from the movies are provided in Plate 1, which shows snapshots of plasma densities and flow fields in the central plasma sheet (equatorial plane of the grid). Plasma velocity vectors are color coded by speed—black: $V \geq 400$ km/s, yellow: $V < 100$ km/s, with red, green, blue indicating intermediate speeds and where vectors corresponding to speeds less than 10 km/s are not plotted. The frames of Plate 1 include continuous color contour shading indicative of the density of plasma *of solar-wind origin*. It is important to appreciate that these density contours do not represent *total* plasma density: magnetospheric plasma from grid initialization that remains in the grid is not included in these color contours.

The magnetic cloud event of October 1995 was also simulated for purposes of examining mesoscale structure generated under varying solar-wind and IMF conditions. Geomagnetic dipole orientation and variable IMF in this event simulation required a full four-quadrant grid without assumptions of symmetry.

3.1. Zero IMF

The top panel of Plate 1 illustrates properties in the equatorial (X_{SM}-Y_{SM}) plane through the central plasma sheet from the simulation with zero IMF. Although zero IMF is largely an artificial situation, it is instructive for purposes of sorting out categories of magnetospheric behavior as a function of IMF properties. After being run for about 3 hours, the zero-IMF case serves as a model for a magnetospheric "ground state" because dayside magnetic reconnection between the IMF and the Earth's magnetic field is eliminated as a contributor to interior magnetospheric plasma dynamics. Color contours of solar-wind plasma density indicate uniform conditions in the solar wind, an abrupt density jump at the bow shock, and compressed, expanding flow behind the shock in the magnetosheath. Flows in these regions appear to be laminar. Within the cavity bounded by the magnetopause numerous vortical flow structures evolve continuously in an apparently random manner. The maximum duskward (dawnward) extent of the turbulent cavity is slightly greater than 20 R_E. Vortical flows in the tail sunward of $X \approx -10$ R_E tend to evolve slowly, and the circulation centered at $X \approx -5$ R_E just inside the magnetopause tends to stand in place. Vortices tailward of $X \approx -10$ R_E are more active in mean convective terms, moving sunward and driving smaller-scale vortices near the inner edge of the plasma sheet or moving tailward and eventually decaying to waves. A thin, highly elon-

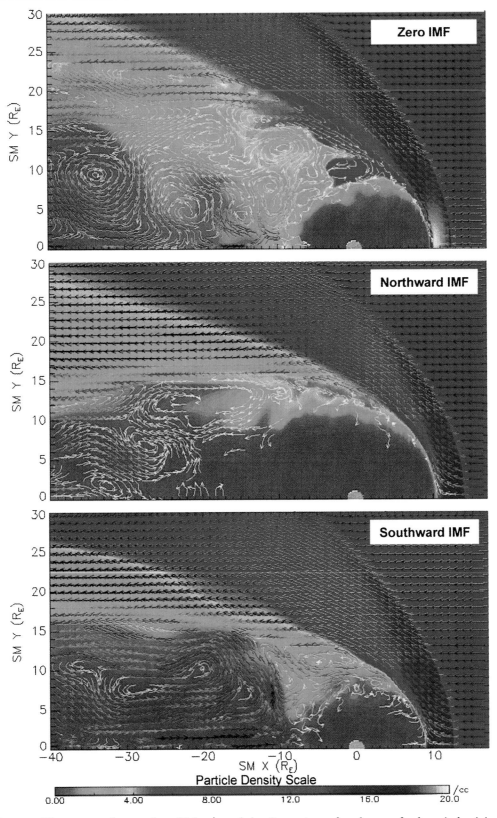

Plate 1. Flow vectors for speeds $> 10 \ km/s$ and density contours for plasma of solar wind origin are shown in the equatorial plane for ISM simulations with zero, northward, and southward IMF.

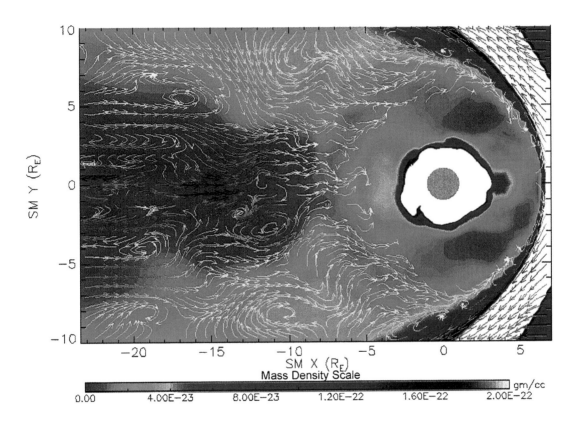

Plate 2. Flow vectors for speeds $> 10\ km/s$ and contours of total plasma mass density are shown in the $(X_{SM}\text{-}Y_{SM})$ equatorial plane for an ISM simulation of the October 1995 magnetic cloud event at 1815 UT.

gated circulation cell extends along the magnetopause at the low latitude boundary layer.

Color contours of solar-wind plasma density indicate entry of plasma through the magnetopause, as denoted by irregular protrusions of red and green contours into the night sector of the interior magnetosphere. Plasma entry of this type occurs in a thin (in Z) layer owing to combined effects of (i) the weak magnetic field of the neutral sheet that extends to and partially through the magnetopause and (ii) the condition of zero IMF that provides a reservoir of unmagnetized solar-wind plasma. Even a small amount of dissipation (real or numerical) at the subsolar magnetopause permits solar-wind plasma to cross the compressed dayside geomagnetic field, be convected around the flanks, then seemingly explode into the weak-field region of the neutral sheet. Solar-wind plasma is excluded from the inner magnetosphere by the strong magnetic field there.

3.2. Northward IMF

The middle panel of Plate 1 illustrates central plasma-sheet properties from a simulation with constant solar-wind parameters and fixed northward IMF. The cavity filled with "turbulent" plasma is bounded by a roughly tear-drop shaped magnetopause and extends beyond the left edge of the plot to $\sim 60\ R_E$ down the magnetotail. Overall flow speeds are smaller than those found in the zero-IMF case, as evidenced by the predominance of yellow flow vectors, and by the substantially larger inner-magnetosphere zone wherein flow speeds are below the plotting threshold. Three prominent vortices with dimensions of several R_E are accompanied by several small vortical flow structures in the area of $-12\ R_E \leq X \leq 6\ R_E$, $9\ R_E \leq Y \leq 15\ R_E$ at the interface with the slow-flow inner magnetosphere, where speeds are below the plotting threshold. Color contours of solar-wind plasma density indicate markedly less intrusion into the interior of the magnetosphere relative to the zero-IMF case.

3.3. Southward IMF

The bottom panel of Plate 1 depicts the plasma-sheet half-plane for conditions of steady solar wind and fixed southward IMF. As in the preceding cases, solar-wind and magnetosheath flows appear laminar. However, the character of flow inside the cavity defined by the magnetopause is distinctly different from corresponding flows in the preceding zero- and northward-IMF cases. Ripples that appear to be of Kelvin-Helmholtz origin are visible in the color contours along the magnetopause in the interval $-8\ R_E \leq X \leq 7\ R_E$, but detailed analysis has not yet been made to confirm the Kelvin-Helmholtz mechanism. Examination of a time-

sequence movie of this simulation shows that as interior vortices convect along the magnetopause flanks the magnetopause undergoes oscillations in the dawn-dusk direction with periods of 10's of minutes and amplitudes on the order of 5 R_E. Compared to the zero-IMF case, the nearly standing vortex just inside the magnetopause at about $X \approx -5\ R_E$ is substantially smaller in size. Several small vortices are also noted at the interface with the inner magnetosphere. Color contours indicate entry of solar-wind plasma (note green "finger" protruding downward at $X \approx -8\ R_E$). Patches of black arrows in the magnetotail indicate transient flow bursts Earthward and duskward associated with intermittent patchy reconnection.

3.4. Magnetic Cloud Event of October 1995

Each preceding case illustrates magnetospheric responses to static solar-wind and IMF parameters. Although this type of simulation is useful for diagnosing fundamental magnetospheric processes, steady conditions over long periods rarely occur in nature. The solar-wind magnetic-cloud event of October 1995 serves as a counterpoint to the preceding simulations. Solar-wind and IMF values as measured by the WIND satellite (see Figure 3) were used to drive an ISM simulation. WIND data were lagged by 35 minutes to account for propagation from the satellite to ISM's inflow boundary. The vertical solid line at 16 UT in Figure 3 indicates initial use of the WIND data stream to drive the ISM simulation. The vertical dashed line indicates the time lag for solar-wind propagation. Solid horizontal lines indicate values used in lieu of WIND data as ISM inflow conditions. Thus, the magnetic cloud simulation was initialized with a period of zero-IMF, followed by WIND data, excepting that relatively small transverse solar-wind inflow velocity components V_y and V_z were set to zero, as was the B_x component of the IMF. From Figure 3 one notes that prior to arrival of the magnetic cloud (southward turning of IMF to $B_Z = -22\ nT$) at 1900 UT solar-wind density was quite variable and roughly an order of magnitude larger than a nominal 5 protons/cc. It is this high-density, high-ram pressure phase of the magnetic cloud event that is reported here. The geomagnetic dipole in this simulation was oriented as appropriate for the event—tipped 0.4° sunward of the dawn-dusk terminator and 17.1° dawnward of the X-Z GSE meridian plane. All four quadrants of the ISM grid were used for this simulation because the dipole orientation and variable IMF break the symmetries noted for the previous simulations.

Plate 2 provides simulated conditions in the central plasma sheet at 1815 UT, prior to the strong southward turning of the IMF. At this time the magnetosphere is strongly driven by high but fluctuating solar-wind ram pressure, so the bow shock and magnetopause are relatively close to the Earth. Examination of a sequence of plots during the period prior to magnetic cloud arrival shows this figure to be representative of the general magnetospheric response to high ram pressure. Magnetosheath flows appear to remain laminar, but flows inside the magnetopause exhibit the same general mesoscale flows as noted in prior simulation results. Mean flow in the magnetotail is Earthward, as expected, but it is punctuated by a rich set of vortical flow features. Sporadic high-speed sunward and anti-sunward flow bursts (black vectors) are noted. Examination of an extended sequence of plots shows such transient patches of high-speed flow to occur over a range of X but primarily between $Y \approx \pm 5~R_E$, and to last for a few minutes each.

4. DISCUSSION

For both constant and variable IMF and solar wind, MHD simulations with ISM show regions of mesoscale dynamical flow to be confined within a magnetospheric cavity bounded by the magnetopause, with cavity shape and dimensions dependent on the orientation of the IMF. The primary motive processes for these mesoscale flows are viscous boundary-layer coupling between the solar wind and the interior of the magnetosphere combined with magnetic reconnection in the magnetotail. The relative importance of each of these processes appears not surprisingly to be a function of IMF orientation. For conditions with southward-directed IMF, or for zero IMF, the cavity extends hundreds of R_E down the magnetotail. In this cavity mesoscale dynamics within approximately 20 R_E of the Earth, but outside the inner magnetosphere region of slow flow, are dominated by quasi-standing vortices. The cavity region tailward of this point exhibits prominent traveling waves and vortices. For conditions of northward-directed IMF, the tail portion of the cavity contracts Earthward on a time scale of less than two hours to a tear-drop shape extending approximately 60 R_E tailward, with quasi-standing vortex motion throughout. In the simulation driven with WIND data for the magnetic cloud event, turbulent cavity dimensions comparable to zero- and southward-IMF cases were found. Transient bursts of sunward and anti-sunward flows were noted within the $Y \approx \pm 5~R_E$ range with speeds on the order of 400 km/s, a result consistent with spatial distributions inferred from statistical analysis of bursty bulk flows by *Angelopolous et al.* [1994]. Characteristic scales for simulated vortical flows appear to be no smaller than $\sim 2~R_E$, consistent with the size suggested

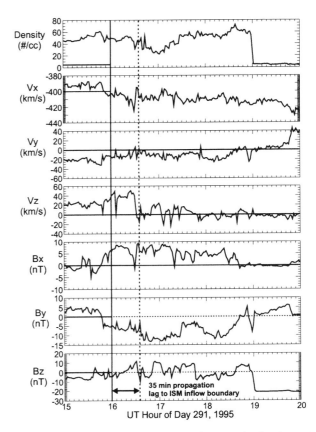

Figure 3. WIND data for period of magnetic cloud event of October 1995 used to drive ISM simulation.

by statistical analysis of ISEE-2 data by *Borovsky et al.* [1997a].

Based on the estimated level of kinematic viscosity of numerical origin in the ISM code and taking the dawn-dusk dimension of the magnetopause as a characteristic scale size, we estimate the code to be operating with Reynolds numbers in excess of 10,000. Consequently, it is no surprise that highly structured flows are generated. There remains the question of the extent to which simulated mesoscale flows represent, in statistical terms, mesoscale flows generated by nature. A preliminary answer is suggested in Figure 4 by comparisons of autocorrelation functions for fluctuating components of a variety of plasma-sheet variables as determined from ISEE-2 data [*Borovsky et al.*, 1997a] and from the interval from 3 to 4 hours in the ISM simulation for zero IMF. The autocorrelation function for fluctuations in quantity g is defined as

$$A_g(\tau) = \frac{\int [g(t) - <g>][g(t + \tau) - <g>]\,dt}{\int [g(t) - <g>][g(t) - <g>]\,dt}$$

where $<g>$ is the mean value of g. Autocorrelation functions based on ISEE-2 are averages derived from

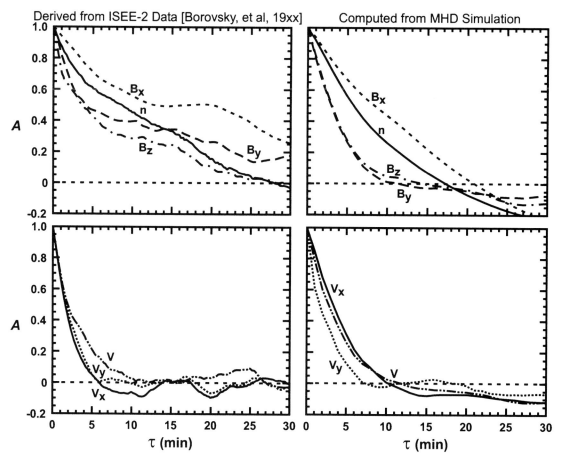

Figure 4. Average autocorrelation functions computed from 10 multi-hour ISEE-2 data sets (6 data sets for B field) at $X \approx -20\ R_E$ for mix of solar wind and IMF conditions [*Borovsky et al.*, 1997] are compared to average autocorrelation functions computed from the ISM zero-IMF (magnetospheric "ground state") simulation. In lieu of several multi-hour data sets, the ISM functions were obtained by averaging autocorrelation functions at all grid points in a rectangular volume at $-30\ R_E < X < -20\ R_E$; $-10\ R_E < Y < 10\ R_E$; $-3\ R_E < Z < 3\ R_E$, thus substituting a spatial average for the ISEE-2 temporal data average.

data sets of 6 and 10 hours duration at $X \approx -20R_E$ corresponding to a mix of solar-wind and IMF conditions. Autocorrelation functions derived from the ISM zero-IMF case substitute spatial averages over the region $-30R_E \leq X \leq -20R_E$, $-10R_E \leq Y \leq 10R_E$, $-3R_E \leq Z \leq 3R_E$ for the ISEE-2 mix of solar-wind and IMF conditions. The resemblance between autocorrelation functions based on data and those derived from simulation is gratifying; orderings of curves for B_x, density n, B_y, and B_z match, and velocity components (where $V = (V_x^2 + V_y^2)^{1/2}$) decorrelate more rapidly than magnetic field components and density. Although the simulation does not reproduce the long tails found on the *Borovsky et al.* [1997a] curves, the general character of the simulated functions from our "ground state" magnetosphere appear to be reasonable approximations to observations.

With insights into a number of distinct modes of mesoscale dynamics exhibited by the simulated magnetosphere, a follow-up task is to investigate in detail the life cycle of these modes, their nature as MHD entities, and their role in global magnetospheric dynamics. Complexities of mesoscale space weather systems revealed by our MHD simulations indicate the need for a large number of satellites—a magnetospheric constellation mission—making simultaneous measurements throughout a substantial volume of the magnetosphere. When such a mission is flown, perhaps a decade in the future, then advanced MHD models such as described here can assume yet another role, that of an integrator of large numbers of discrete measurements into a composite descriptive fabric for the near-Earth space environment. In this manner, predictions of space weather will have evolved to the current state of meteorological forecasting, wherein data from distributed collection points are integrated by sophisticated forecasting models into a unified description covering macroscales and mesoscales.

Acknowledgments. This work was sponsored by the National Aeronautics and Space Administration under grants NAG5-8135 and NASW-99014. The Integrated Space Weather Prediction Model (ISM) was developed under sponsorship of the Defense Threat Reduction Agency, 45045 Aviation Drive, Dulles, VA 20166-7517. WIND proton data courtesy of K. W. Ogilvie, J. T. Steinberg, A. J. Lazarus. WIND electron data provided by K. W. Ogilvie and D. Fitzenreiter in response to a special request. These data were compiled and organized for us by R. Hilmer.
The Editor would like to thank the reviewers of this manuscript.

REFERENCES

Angelopolous, V., W. Baumjohann, C. F. Kennel, F. V. Coroniti, M. G. Kivelson, R. Pellat, R. J. Walker, H. Lühr, and G. Paschmann, Bursty bulk flows in the inner central plasma sheet, *J. Geophys. Res., 97*, 4027, 1992a.

Angelopolous, V., C. F. Kennel, F. V. Coroniti, R. Pellat, M. G. Kivelson, R. J. Walker, W. Baumjohann, G. Paschmann, and H. Lhr, Bursty bulk flows in the inner plasma sheet; An effective means of earthward transport in the magnetotail, in Proceedings of the First International Conference on Substorms, Eur. Space Agency Spec. Pub., ESA SP-335, 303, 1992b.

Angelopolous, V., C. F. Kennel, F. V. Coroniti, R. Pellat, M. G. Kivelson, R. J. Walker, C. T. Russell, W. Baumjohann, W. C. Feldman, and J. T. Gosling, Statistical characteristics of bursty bulk flow events, *J. Geophys. Res., 99*, 21257, 1994.

Baumjohann, W., G. Paschmann, N. Sckopke, C. A. Cattell, and C. W. Carlson, Average ion moments in the plasma sheet boundary layer, *J. Geophys. Res., 93*, 11507, 1988.

Baumjohann, W., G. Paschmann, and C. A. Cattell, Average plasma properties in the central plasma sheet, *J. Geophys. Res., 94*, 6597, 1989.

Berchem, J., J. Raeder, and M. Ashour-Abdalla, Magnetic flux ropes at the high-altitude magnetopause, *Geophys. Res. Lett., 22*, 1189, 1995.

Borovsky, J. E., R. C. Elphic, H. O. Funsten, and M. F. Thomsen, The Earth's plasma sheet as a laboratory for flow turbulence in high-β MHD, J. Plasma Phys., *57*, 1, 1997a.

Borovsky, J. E., M. F. Thomsen, and D. J. McComas, The superdense plasma sheet; Plasmaspheric origin, solar wind origin, or ionospheric origin, *J. Geophys. Res., 102*, 22089, 1997b.

Crooker, N. U., J. G. Lyon, and J. A. Fedder, MHD model merging with IMF B_y: Lobe cells, sunward polar cap convection, and overdrapped lobes, *J. Geophys. Res., 103*, 9143, 1998.

Fedder, J. A., J. G. Lyon, S. P. Slinker, and C. M. Mobarry, Topological structure of the magnetotail as a function of interplanetary magnetic field direction, *J. Geophys. Res., 100*, 3616, 1995.

Hain, K., The partial donor cell method, J. Comp. Physics, *73*, 131, 1987.

Hayakawa, H., A. Nishida, E. W. Hones, and S. J. Bame, Statistical characteristics of plasma flow in the magnetotail, *J. Geophys. Res., 87*, 277, 1982.

Hones, E. W., Jr., G. Paschmann, S. J. Bame, J. R. Asbridge, N. Sckopke, and K. Schindler, Vortices in Magnetospheric Plasma Flow, *Geophys. Res. Lett., 5*, 1059, 1978.

Hones, E. W. Jr., J. Birn, S. J. Bame, J. R. Asbridge, G. Paschmann, N. Sckopke, and G. Haerendel, Further Determination of the Characteristics of Magnetospheric Plasma Vortices with Isee 1 and 2, *J. Geophys. Res., 86*, 814, 1981.

Hones, E. W., Jr., J. Birn, S. J. Bame, and C. T. Russell, New Observations of Plasma Vortices and Insights into Their Interpretation, *Geophys. Res. Lett., 10*, 674, 1983.

Kilb, R. W., Chapter 13, Striation Formation, in Physics of High-Altitude Nuclear Burst Effects, *Tech. Rpt. DNA 4501F*, p. 625, Defense Nuclear Agency, Washington, D.C., 1977.

Kivelson, M. G., Magnetospheric electric fields and their variation with geomagnetic activity, Rev. Geophys., *14*, 189, 1976.

Kivelson, M. G., and S.-H. Chen, The Magnetopause: Surface Waves and Instabilities and their Possible Dynamical Consequences, in *Physics of the Magnetopause*, edited by P. Song, B. U. Ö. Sonnerup, and M. F. Thomsen, p. 257, Geophysical Monograph 90, American Geophysical Union, Washington, DC, 2000.

Kivelson, M. G., and C. T. Russell, *Introduction to Space Physics*, pp. 260, 293, Cambridge Univ. Press, Melbourne, 1995.

Mauk, B. H. and C.-I. Meng, Plasma injection during substorms, Phys. Scr., *18*, 128, 1987.

M. Meneguzzi, A. Pouquet, and P.-L. Sulem (Eds.), *Small-Scale Turbulence in Three-Dimensional Hydrodynamic and Magnetohydrodynamic Turbulence*, Springer-Verlag, Berlin, 1995.

Montgomery, D., Remarks on the MHD problem of generic magnetospheres and magnetotails, in *Magnetotail Physics*, edited by A. T. Y. Lui, p. 203, Johns Hopkins Univ. Press., 1987.

Ogino, T., R. J. Walker, and M. Ashour-Abdalla, Magnetic flux ropes in 3-dimensional MHD simulations, in *Physics of Magnetic Flux Ropes*, edited by C. T. Russell, E. R. Priest, L. C. Lee, p.669, Geophysical Monograph 58, American Geophysical Union, Washington, DC, 1990.

Raeder, J., R. J. Walker, and M. Ashour-Abdalla, The structure of the distant geomagnetic tail during long periods of northward IMF, *Geophys. Res. Lett., 22*, 349, 1995.

Raeder, J., J. Berchem, and M. Ashour-Abdalla, The importance of small scale processes in global MHD simulations: Some numerical experiments, in *The Physics of Space Plasmas*, edited by T. Chang and J. R. Jasperse, vol. 14, p. 403, MIT Cent. for Theoret. Geo/Cosmo Plasma Phys., Cambridge, Mass., 1996.

Raeder, J., J. Berchem, M. Ashour-Abdalla, L. A. Frank, W. R. Paterson, K. L. Ackerson, S. Kokubun, T. Yamamoto, and J. A. Slavin, Boundary layer formation in the magnetotail: Geotail observations and comparisons with a global MHD model, *Geophys. Res. Lett., 24*, 951, 1997.

Sckopke, N., G. Paschmann, G. Haerendel, B. U. Ö. Sonnerup, S. J. Bame, T. G. Forbes, E. W. Hones, Jr., and C. T. Russell, Structure of the low latitude boundary layer, *J. Geophys. Res., 86*, 2099, 1981.

Siscoe, G. L., N. U. Crooker, G. M. Erickson, B. U. Ö. Sonnerup, K. D. Siebert, D. R. Weimer, W. W. White, N. C. Maynard, Global geometry of magnetospheric currents, in *Magnetospheric Current Systems*, edited by S.-I. Ohtani, p. 41, Geophysical Monograph 118, American Geophysical Union, Washington, DC, 2000.

Slinker, S. P., J. A. Fedder, B. A. Emory, K. B. Baker, D. Lummerzheim, J. G. Lyon, and F. J. Rich, Comparison of Global MHD simulations with AMIE simulations for events of May 19-20, 1996, *J. Geophys. Res., 104*, 28379, December, 1999.

Song, P., D. L. DeZeeuw, T. I. Gombosi, C. P. T. Groth, and K. G. Powell, A numerical study of solar wind-magnetosphere interaction for northward interplanetary magnetic field, *J. Geophys. Res., 104*, 28361, 1999.

Sonnerup, B. U. Ö., Theory of the low-latitude boundary layer, *J. Geophys. Res., 85*, 2017, 1980.

Van Dyke, M., *An Album of Fluid Motion*, 176 pp., Parabolic Press, Stanford, Calif., 1982.

Walker, R. J., and T. Ogino, A global magnetohydrodynamic simulation of the origin and evolution of magnetic flux ropes in the magnetotail, *J. Geomag. Geoelectr., 48*, 765, 1996.

Weimer, D. R., A flexible, IMF dependent model of high-latitude electric potentials having "space weather" applications, *J. Geophys. Res., 23*, 2549, 1996.

White, W. W., G. L. Siscoe, G. M. Erickson, Z. Kaymaz, N. C. Maynard, K. D. Siebert, B. U. Ö. Sonnerup, and D. R. Weimer, The Magnetospheric Sash and Cross-Tail S, *Geophys. Res. Lett., 25*, 1605, 1998.

Zwickl, R. D., D. N. Baker, S. J. Bame, W. C. Feldman, J. T. Gosling, E. W. Hones, Jr., D. J. McComas, B. T. Tsurutani, and J. A. Slavin, Evolution of the Earth's distant magnetotail: ISEE-3 electron plasma results, *J. Geophys. Res., 89*, 11007, 1984.

———————

N. C. Maynard, J. A. Schoendorf, K. D. Siebert, D. R. Weimer, W. W. White, and G. L. Wilson, Mission Research Corporation, One Tara Blvd, Suite 302, Nashua, NH 03062. (e-mail: nmaynard@mrcnh.com, jschoendorf@mrcnh.com, ksiebert@mrcnh.com, dweimer@mrcnh.com, bwhite@mrcnh.com, gwilson@mrcnh.com)

B. U. Ö. Sonnerup, Thayer School of Engineering, Dartmouth College, Hanover, NH 03755. (e-mail: Bengt.U.O.Sonnerup@dartmouth.edu)

G. M. Erickson, G. L. Siscoe, Boston University, Center for Space Physics, 725 Commonwealth Avenue, Boston, MA 02543, (e-mail: Gary.Erickson@hanscom.af.mil, siscoe@bu.edu,)

Modeling Extreme Compression of the Magnetosphere:
Results From a Global MHD Simulation of the May 4, 1998 Event

J. Berchem, M. El-Alaoui, and M. Ashour-Abdalla

Institute of Geophysics and Planetary Physics
University of California, Los Angeles, California 90095-1567

This paper reports results from a global magnetohydrodynamic (MHD) simulation of the extreme compression of the magnetosphere that occurred on May 4, 1998. Simultaneous Wind spacecraft measurements of solar wind ions and the interplanetary magnetic field (IMF) upstream of the Earth were used as driving input for the MHD simulation. After assessing the validity of the model by comparing time series from the simulation with Geotail and Polar measurements, we examined the interaction of the leading edge of the solar disturbance with the magnetosphere. The simulation shows that the initial interaction of the interplanetary shock with the Earth bow shock produced a sunward motion of the resultant bow shock and that the transmitted shock wave launched a fast compressional wave at the magnetospheric boundary. We conclude the paper by examining the magnetotail's reaction to a sustained period of strong dynamic pressure. The simulation results suggest that the magnetotail can attain highly organized states during which large-scale flows are more laminar than expected.

1. INTRODUCTION

Although global magnetohydrodynamic (MHD) simulations have been used for more than 20 years to model the time-dependent interaction of the solar wind with the Earth's magnetosphere [see review by *Walker and Ashour-Abdalla*, 1994], only recently have model refinements and computational advances allowed the studies envisioned at the inception of the International Solar Terrestrial Physics (ISTP) program. The principle of these studies is to use solar wind plasma and magnetic field measurements as input parameters to drive three-dimensional global MHD simulations and then to compare the results from these simulations with observations. To date most of these comparisons have been carried out for periods of quiet or moderate solar activity and used to model the growth phase and

expansion onset of substorms [*Fedder et al.*, 1995], the large-scale flows of cold and dense ions in the distant tail [*Frank et al.*, 1995], the boundary layer formation in the magnetotail [*Raeder et al.*, 1997], the response of the distant magnetotail to the east-west component of the interplanetary magnetic field (IMF) [*Berchem et al.*, 1998], the auroral brightening and the onset of lobe reconnection during an isolated substorm [*Lyon et al.*, 1998], and the response of the polar cap and high-latitude convection to a sudden southward turning of the IMF [*Lopez et al.*, 1998]. The very good agreement between simulation results and observations found in those studies established the validity of the approach. With the arrival of the solar maximum (Solar Cycle 23), the use of global models to study actual events presents new challenges. Recently the spectacular interaction of the Earth's magnetosphere with a magnetic cloud driven by a coronal mass ejection (CME) on January 10-11, 1997 [e.g., *Fox et al.*, 1998] was modeled by a global MHD simulation [*Goodrich et al.*, 1998].

In this paper, we present results from a global MHD simulation of the extreme compression of the magneto-

Space Weather
Geophysical Monograph 125

sphere observed on May 4, 1998. During that day, the Polar spacecraft repeatedly crossed the dayside magnetopause well inside the geosynchronous orbit and experienced two brief excursions into the solar wind [e.g., *Russell et al.*, 2000]. Section 1 presents the magnetic field and plasma parameters from May 4, 1998, measured by experiments onboard the Wind spacecraft. Section 2 briefly describes the model and the simulation setup. In section 3, we compare the time series from the Geotail and Polar spacecraft with those obtained from the simulation. Next we show cuts along the noon-meridian and equatorial planes of contours of fields and plasma parameters from the simulation to support a discussion of the interaction of the leading edge of the solar disturbance with the magnetosphere and the behavior of large-scale flows in the magnetotail during a period of a sustained high-density plasma stream.

2. SOLAR WIND OBSERVATIONS

On May 2, 1998, SOHO observed a coronal disturbance spreading outward into the interplanetary medium [e.g., *Baker and Carlowicz*, 1999]. The disturbance was so powerful that two days later, solar wind speeds reaching about 900 km/s were recorded by the Wind spacecraft 215 R_E upstream of the Earth. This is one of the highest solar wind speeds measured near Earth during the last several years. Wind magnetic field and plasma measurements from 0000 UT to 1200 UT on May 4, 1998, are shown in Figure 1. The data are one-minute averaged and displayed in Earth-centered solar ecliptic (GSE) coordinates. The remainder of this paper uses the GSE coordinate system to indicate spacecraft locations and to display the observations and simulation results. At 0600 UT the Wind spacecraft was located at $\mathbf{R}_W = (214.2, 6.8, 27.1)$ R_E. Figure 1 displays from top to bottom, the thermal velocity in km/s, the ion density in particles per cm^3, the three components of the ion bulk velocity in km/s [*Ogilvie et al.*, 1995], and the three components of the magnetic field in nT from the GSFC magnetometer [*Lepping et al.*, 1995]. The bottom panel shows the dynamic pressure calculated from the ion density and the bulk speed of the solar wind.

The progression of the solar disturbance toward Earth is indicated by multiple enhancements of the dynamic pressure between about 0200 UT and 1130 UT. These increases are highly variable, but three main segments can readily be discerned in the data. The first time period, from about 0230 to 0450 UT, starts with a sharp jump in the solar wind velocity from about 550 km/s to 700 km/s. Peaks in both the thermal velocity and the density and a strong increase of the IMF magnitude (≈ 30 nT) occur with the velocity jump; taken together they indicate the formation of an interplane-

tary shock at the leading edge of the solar disturbance. A slow ramping of the thermal energy, bulk velocity, and dynamic pressure follow the shock during the following 2 hours during which the IMF direction is largely southward and has a large B_Y component. The second period, from about 0430 to 0830 UT, is marked by four strong pulses in the solar wind dynamic pressure of various durations with peaks from 40 to about 60 nPa. The strong magnetic fluctuations observed during the period indicate that the solar disturbance comprises filamentary structures consistent with large-scale magnetic flux ropes [e.g., *Burlaga et al.*, 1990]. The last period, from 0830 UT to 1130 UT, is marked by a long period of very high (≈ 30 nPa) sustained dynamic pressure, after which the dynamic pressure decreases abruptly to values in the range observed before 0100 UT. It is interesting that, after its initial jump around 0230 UT, the solar wind flow reaches its maximum velocity (≈ 900 km/s) around 0407 UT and slowly decreases after that time. This indicates that most of the extreme variations

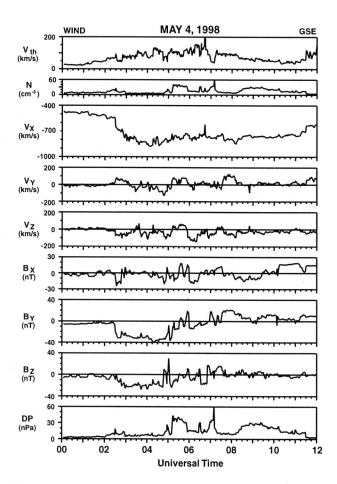

Figure 1. Plasma and field parameters measured by Wind on May 4, 1998 and plotted using the GSE system of coordinates.

observed in the dynamical pressure during the second and third periods result from strong fluctuations in the plasma density rather than from large velocity fluctuations.

3. SIMULATION MODEL

The magnetospheric part of our simulation model is based on a single fluid MHD description. The code solves the normalized resistive MHD equations as an initial value problem, using an explicit conservative predictor-corrector scheme for time stepping and hybridized numerical fluxes for spatial finite differencing [*Berchem et al.*, 1995a, b; *Raeder et al.*, 1995]. A nonlinear function of the local current density is used to model the resistivity term in Ohm's law. This model includes a threshold that is a function of the local normalized current density; this threshold is calibrated to avoid spurious dissipation [*Raeder et al.*, 1996]. Similar phenomenological resistivity models have been used in local MHD simulation models [e.g., *Sato and Hayashi*, 1979; *Ugai*, 1985] and are based on the assumption that current driven instabilities are responsible for the anomalous resistivity that produces reconnection.

A spherical shell with a radius of 3 R_E is placed around the Earth to exclude the region where the Alfvén velocity becomes too large to be computed in the simulation. Inside the shell, the MHD equations are not solved, and a static dipole magnetic field is assumed. We assume a two-dimensional ionosphere to close the field aligned currents and solve the ionospheric potential equation to determine the electrostatic potential self-consistently. A proxy of three ionization sources (solar EUV ionization, precipitating electrons, diffuse electron precipitation) is used to compute the ionospheric Hall and Pedersen conductances that are needed to solve the potential equation [*Raeder et al.*, 1996]. The ionospheric potential is then mapped to the shell where it is used as a boundary condition for the magnetospheric flow velocity.

The simulation was started by switching on an input solar wind flow with the magnetic field and plasma density, temperature, and velocity values measured by the Wind spacecraft at 2300 UT on May 3, 1998. This set of input parameters was maintained for 90 minutes to give the simulation system time to evolve towards a physical state independent of the initial conditions. This period is longer than the transit time of the solar wind through the entire system, which is, assuming a constant speed of 500 km/s, about an hour for the 300 R_E long system used. Furthermore, the initial southward IMF conditions are very efficient in convecting the initial unphysical state out of the simulation domain. At the end of this phase we began using actual one-minute averaged measurements from the Wind

spacecraft. The solar wind magnetic field and plasma parameters measured by Wind are advected from the spacecraft location to the inflow boundary of the simulation box located at 20 R_E in front of the Earth. This process assumes that velocities of the plasma measured remain constant during its convection toward Earth. However, because solar wind measurements are available at only a single location, and Faraday's law prevents the advection of the magnetic field component in the direction parallel to the flow at the inflow boundary, additional hypotheses are required. We used the simplest model, which assumes that the incoming IMF discontinuities are planar and that the B_X component of the IMF is constant. Since the average value of the B_X component remains close to zero until 1000 UT, we assume that the IMF B_X component vanishes at the simulation boundary. As we show below, although these simple assumptions ignore the fast flips observed in the Wind magnetic field data during the periods marked by strong density pulses, they lead to results that compare very well with those observed just downstream of the bow shock.

4. TIME SERIES COMPARISONS

Figure 2 shows a time series of the magnetic field extracted from the simulation that was superimposed over Geotail measurements from 0200 to 1200 UT. Solid traces display the magnetic field measured by Geotail, whereas the dotted traces show the computed values along the spacecraft trajectory inside the simulation domain. As is clear in Figure 2, the simulation tracks the data extremely well. This result is not surprising since Geotail was exploring the dusk magnetosheath during the time period considered [the spacecraft location was G_I= (-11.4, 24.1, -0.7) R_E at 0200 UT, and G_O=(-14.7, 18.7, 0.1) R_E at 1100 UT]. Indeed, the evolution of plasma and field parameters in that region is mostly, at least on the MHD scale length, governed by jump conditions at the bow shock and thus is relatively easy to predict with good accuracy. However, although trivial, the good agreement between the simulation results and magnetosheath data is essential for validating the data stream used as input to the simulation. Firstly, it allows us to verify that the available solar wind measurements from the Wind spacecraft are well suited for the simulation study. Indeed, because solar disturbances have complex geometries and interact with the interplanetary medium during their expansions, the Wind spacecraft frequently observes discontinuities traveling away from the Sun that do not reach the magnetosphere or are significantly altered during their convection from the sunward Lagrangian point to the Earth [*Collier et al.*, 1998]. Since the simulation uses a simple model to propagate the Wind measurements to the inflow bound-

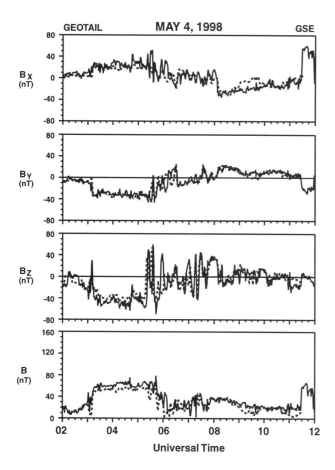

Figure 2. Comparison of time series from the global MHD simulation (dotted traces) with the magnetic field measured by Geotail (solid traces) on May 4, 1998.

ary of the simulation box, Wind measurements are sometimes inappropriate for use as input. The good agreement found here is thus reassuring. Secondly, comparing the simulation results and magnetosheath observations allows us to verify the accuracy of the time delay between the spacecraft measurement and the simulation inflow boundary. Finally, comparing time series from the simulation with magnetosheath data allows us to assess the validity of our solar wind input assumption, which in the present case is the removal of the IMF B_X component. A detailed inspection of Figure 2 reveals small discrepancies that occur predominantly in the Z component rather than in the X component of the field. The largest variations are observed during fast oscillations of the field because the simulation has a tendency to diffuse sharp discontinuities. The overall fit is nevertheless remarkable, indicating that our assumption for the B_X component is reasonable.

Figure 3 shows our simulation's time series superimposed over Polar magnetic field measurements for the 7-

hour period from 0500 to 1200 UT. During that period the Polar spacecraft crosses the magnetopause and the bow shock several times. A detailed description of these unusual events can be found in *Russell et al.* [2000]. Solid traces indicate the magnetic field measured by Polar whereas dotted traces show the computed values along the spacecraft trajectory inside the simulation domain. Simulation results show good agreement with observations of the first crossing of the magnetopause by Polar at about 0544 UT when the spacecraft was well inside the geosynchronous orbit at 5.4 R_E. The time delay seen in the simulation indicates that this exceptional crossing is due to the strong enhancement of the dynamic pressure observed around 0510 UT by the Wind spacecraft. However, the simulation is less successful at accurately reproducing Polar's successive reentry into the magnetosphere between 0600 and 0645 UT; the simulation shows only partial reentries. This is of course the most difficult period to track because of the large magnetic field fluctuations, observed by the Wind spacecraft between 0515 and 0600 UT, which result from the complex structure of the stream. The simulation time series and the spacecraft data agree better after 0700 UT, when the IMF

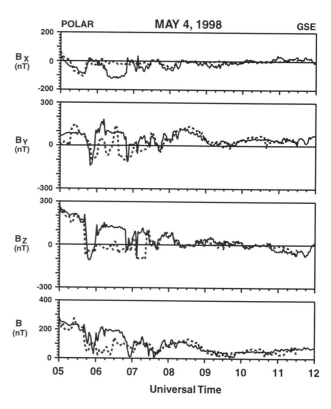

Figure 3. Comparison of times series from the global MHD simulation (dotted traces) with the magnetic field measured by Polar (solid traces) on May 4, 1998.

magnetic field is less perturbed. The simulation does a particularly good job of reproducing the two periods when Polar crosses the bow shock, the first period between 0735 and 0738 UT and the second between 0944 and 0950 UT [e.g., *Russell et al.*, 2000].

5. MAGNETOSPHERIC DYNAMICS

The solar disturbance observed on May 4, 1998, was marked by a series of high-density streams with speeds between 700 to 900 km/s that repeatedly pounded the Earth magnetosphere. In addition to their unusual strength, these streams are notable for their diversity in both duration and magnitude. From the leading edge of the disturbance around 0230 UT, marked by a fast forward shock, to the long period of sustained high-density plasma observed between 0830 UT and 1130 UT, these high-speed streams offer a unique opportunity to study the response of the magnetosphere to solar disturbances. Although the rapid succession of the streams prevents analysis of the interactions of each with the magnetosphere as isolated events, the sequence includes a fairly large variety of parameter regimes. Simulation of the period from 0000 to 1130 UT thus produced a very rich set of results to study the effects of high-speed streams on the global dynamics of the magnetosphere. Below, we briefly examine two aspects of the interaction: the initial impact of the leading edge of the solar disturbance on the magnetosphere and the tail behavior during the period of sustained high-density plasma stream. During these periods Polar and Geotail were not favorably located to observe the details of the interaction, and thus the global simulation is particularly important, as it provides invaluable information about some of the dynamical features that were not measured directly by the spacecraft.

About 40 minutes after being observed by Wind, the leading edge of the solar disturbance interacted with the Earth's magnetosphere. Plate 1 shows the results from the simulation of the interaction. Organized as a matrix of cuts along the noon-midnight direction, the plate shows color contours of plasma and field parameters at three different times: 0306, 0309, and 0312 UT. Displayed from top to bottom are: plasma density, thermal pressure, plasma beta, Alfvén Mach number, current density, and Poynting flux. The left column shows the magnetosphere at 0306 UT, just before the stream collides with the Earth's bow shock. The front of the stream is clearly identifiable at the left boundary of the simulation box; in particular, a thin line resulting from the enhancement of the current density at the leading edge of the stream indicates the interplanetary shock. Because the solar wind's dynamic pressure is enhanced by this time ($\approx$ 8 nPa) the magnetosphere is already significantly

compressed. This is indicated by the large density observed in the magnetosheath and the closer than average proximity of the bow shock and the magnetopause to the Earth, about 10 R_E and 8.7 R_E, respectively, at their subsolar points. At 0309 UT (middle column) the interplanetary part of the front of the stream has reached X= -3 R_E, while in the subsolar region, the front shock has just collided with the Earth's bow shock. The effect of the interplanetary shock interaction with the bow shock is obvious in the contour plots, especially in the plot showing the current density. The resultant bow shock from the reflection of the incident shock wave on the original bow shock has moved slightly sunward, while the transmitted shock wave through the bow shock is seen in front of the magnetopause. A fast compressional wave launched by the penetration of the shock through the magnetopause can also be seen to travel tailward at the stream speed, as indicated by the bump in the high-latitude region of the magnetospheric boundary. The transmitted shock wave and the magnetopause bracket a wedge-shaped region of the magnetosheath, where the plasma density and thermal pressure are significantly enhanced; this is also indicated by the strong Alfvén Mach number and the high plasma beta. Three minutes later, at 0312 UT (right column), the leading edge of the disturbance has progressed tailward another 20 R_E. By this time the bow shock has moved sunward by about 2 R_E from its 0306 UT location. The transmitted shock wave delimiting the region of high Alfvén Mach number is between X= -10 and -15 R_E, and the bump marking the progression of the compression of the magnetospheric field has moved further tailward, to around X= -20 R_E. The dynamics of the shock interaction with the bow shock-magnetopause system seen in the simulation are in agreement with theoretical models [e.g., *Grib et al.*, 1979; *Zhuang et al.*, 1981] and observations [*Winterhalter et al.*, 1981]. A detailed analysis comparing our simulation results with those from analytical MHD models will be reported elsewhere. The plasma density and thermal pressure plots from the simulation also show that hot magnetosheath plasma accumulates in the cusp, suggesting that some of the plasma was trapped as the transmitted shock propagated tailward. The Poynting flux plots displayed in the bottom panel of Plate 1 indicate that the electromagnetic energy spreads very quickly from the dayside magnetopause to the nightside magnetosphere, though most of the momentum is still carried by the magnetosheath flow.

Another interesting feature revealed by the simulation is the structure and dynamics of the magnetotail resulting from the period of sustained strong dynamic pressure observed after 0830 UT. Figure 4 shows contours of the bulk flow velocity calculated from the simulation at time 0936

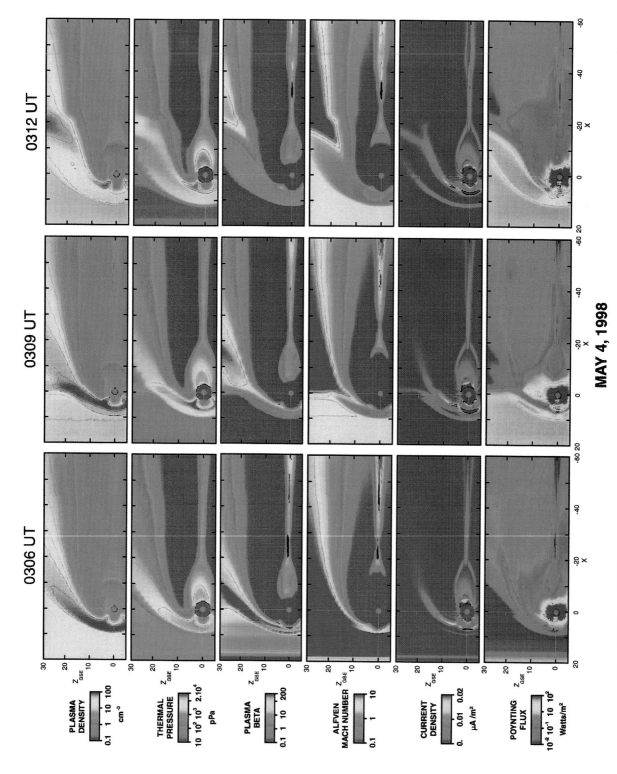

Plate 1. Noon-midnight cuts of color coded contours of plasma and field parameters calculated from the simulation at time (from left to right column) 0306, 0309 and 0312 UT on May 4, 1998.

UT. The upper panel is a noon-midnight cut, whereas a cut through the equatorial plane is displayed in the lower panel. Strong earthward (light contours) and tailward (dark contours) flows in the tail indicate that magnetic reconnection took place at around X= -20 R_E. Cursory inspection of the solar wind data, assuming the approximately 40-minute time lag deduced earlier, indicates that the merging observed in the tail is consistent with the several IMF southward turnings occurring about 0900 UT. The interesting feature of the reconnection process that is predicted by the simulation is not its occurrence, but rather the topology of the flow observed during that time. Indeed, the flows observed in Figure 4 are fairly well organized compared to the turbulent flows observed prior to that time (not shown here). In particular, the equatorial trace of the merging region, which is associated with the light grey region between the light and dark contours, has a regular U shape that is very similar to that deduced from ISEE 3 observations further down the tail [*Slavin et al.,* 1985]. These results suggest that, during a sustained period of strong dynamic pressure, the magnetotail could reach a highly organized state with more laminar flows than would otherwise be expected.

6. SUMMARY

We used a three-dimensional global MHD simulation to model observations of the interaction between a strong solar disturbance and the magnetosphere. Overall, the results from the simulation compare surprisingly well with the observations, if the unusual strength and complex topology of the solar wind disturbance observed on that day are taken into account. In particular, results from the simulation agree remarkably well with spacecraft measurements taken during the interaction from Geotail's position in the magnetosheath. The simulation does a respectable job of reproducing the main features in the Polar data, including the spacecraft crossing the magnetopause well inside the geosynchronous orbit and its several brief excursions into the solar wind. However, the simulation is less successful at reproducing in detail the Polar measurements during the one-hour period marked by large fluctuations of the magnetic field. One reason for the discrepancies between the observed and simulated data might lie in the model's inherent inability to take into account the effects of the ring current. Another cause could be that the model over estimated the erosion of the dayside magnetosphere. More work is required to determine the causes of this discrepancy. Nevertheless, our simulation results bode well for using global MHD models in predicting effects of major storms. In par-

ticular, our success in modeling the interplanetary shock interaction with the bow shock-magnetopause system demonstrates that our current model is fairly adept at simulating the primary effects of large and violent solar disturbances on the Earth environment. Another encouraging result from the simulation is that, during a sustained period of strong dynamic pressure, the magnetotail is capable of attaining a relatively well-organized state, in marked contrast to the very turbulent behavior observed during substorms. As a result, modeling the magnetosphere under such conditions may be more straightforward than expected.

Acknowledgments. The Wind plasma and magnetic field measurements in the solar wind were kindly supplied by K. Ogilvie and R. Lepping (Goddard Space Flight Center, NASA). We thank S. Kokubun (STELAB, Nagoya University) and C. T. Russell (IGPP, UCLA) for providing magnetic field measurements from the Geotail and Polar spacecraft respectively. Computations were performed at the San Diego Supercomputer Center. This research was supported NASA grants NAG5-56689 and NAG5-8074. IGPP UCLA publication # 5475.

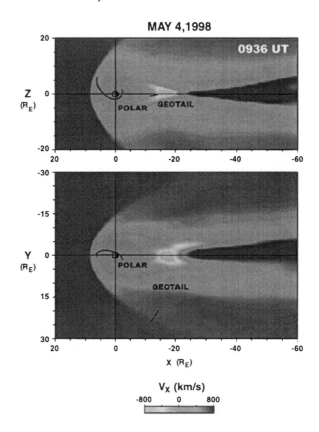

Figure 4. Contours of X-component of the bulk velocity (V_X) calculated from the simulation at time 0936 UT on May 4, 1998, and displayed in the noon-meridian (top panel) and equatorial (bottom panel) planes.

REFERENCES

Baker, D. N, and M. J. Carlowicz, ISTP and beyond: A solar-system telescope and a cosmic microscope, in *Sun-Earth Plasma Connections*, edited by J. L. Burch, R. L. Carovillano, and S. Antiochos, Geophys. Monograph Series 109, pp. 1-9, American Geophysical Union, Washington D. C., 1999.

Berchem, J., J. Raeder, and M. Ashour-Abdalla, Reconnection at the magnetospheric boundary: Results from global magnetohydrodynamic simulation, in *Physics of the Magnetopause*, edited by P. Song, B. U. Ö. Sonnerup, and M. F. Thomsen, Geophys. Monograph 90, pp. 205-213, American Geophysical Union, Washington, D. C., 1995a.

Berchem, J., J. Raeder, and M. Ashour-Abdalla, Magnetic flux ropes at the high-latitude magnetopause, *Geophys. Res. Lett., 22*, 1189, 1995b.

Berchem, J., J. Raeder, M. Ashour-Abdalla, L. A. Frank, W. R. Paterson, L. Ackerson, S. Kokubun, T. Yamamoto, and R. P. Lepping, The distant tail at 200 R_E: Comparison between Geotail observations and the results of a global simulation, *J. Geophys. Res., 103*, 9121, 1998.

Burlaga, L. F., R. P. Lepping, and J. Jones, Global configuration of a magnetic cloud, in *Physics of Flux Ropes*, edited by C. T. Russell, E. R. Priest, and L. C. Lee, Geophys. Monograph 58, pp. 373-377, American Geophysical Union, Washington, D. C., 1990.

Collier, M. R., J. A. Slavin, R. P. Lepping, A. Szabo, and K. Ogilvie, Timing accuracy for the simple planar propagation of magnetic field structure in the solar wind, *Geophys. Res. Letter, 25*, 2509, 1998.

Fedder, J. A., S. P. Slinker, J. G. Lyon, and R. D. Elphinstone, Global numerical simulation of the growth phase and the expansion onset for a substorm observed by Viking, *J. Geophys. Res., 100*, 19083, 1995.

Fox, N. J., M. Peredo, and B. J. Thompson, Cradle to grave tracking of the January 6-11, 1997 Sun-Earth connection event, *Geophys. Res. Letter, 25*, 2461, 1998.

Frank, L. A., M. Ashour-Abdalla, J. Berchem, J. Raeder, W. R. Paterson, S. Kokubun, T. Yamamoto, R. P. Lepping, F. V. Coroniti, D. H. Fairfield, and K. L. Ackerson, Observations of plasmas and magnetic fields in Earth's distant magnetotail: Comparison with a global MHD model, *J. Geophys. Res., 100*, 19177, 1995.

Goodrich, G. C., J.G. Lyon, M. Wiltberger, R. Lopez, and K. Papadopoulos, An overview of the impact of the January 10-11, 1997 magnetic cloud on the magnetosphere via global MHD, *Geophys. Res. Letter, 25*, 2537, 1998.

Grib, S. A., B. E. Brunelli, M. Dryer, and W. W. Shen, Interaction of interplanetary shock wave with the bow shock-magnetopause system, *J. Geophys. Res., 84*, 5907, 1979.

Lepping, et al., The Wind magnetic field investigation, *Space Sci. Rev., 71*, 207, 1995.

Lopez, R. E., C. C. Goodrich, M. Wiltberger, K. Papadapoulos, and J. G. Lyon, Simulation of the March 9, 1995 substorm and initial comparison to data, in *Geospace mass and energy flow:*

Results from the International Solar-Terrestrial Physics Program, edited by J. Horwitz, D.L. Gallager, and W.K. Peterson, Geophys. Monograph Series 104, pp. 237-245, American Geophysical Union, Washington D. C., 1998.

Lyon, J. G., R. E. Lopez, C. C. Goodrich, M. Wiltberger, and K. Papadapoulos, Simulation of the March 9, 1995, substorm: Auroral brightening and the onset of lobe reconnection, *Geophys. Res. Lett., 25*, 3039, 1998.

Ogilvie, K. W., et al., SWE, a comprehensive instrument for the Wind spacecraft, *Space Sci. Rev., 71*, 55, 1995.

Raeder, J., R. J. Walker, and M. Ashour-Abdalla, The structure of the distant geomagnetic tail during a long period of northward IMF, *Geophys. Res. Lett., 22*, 349, 1995.

Raeder, J., J. Berchem, and M. Ashour-Abdalla, The importance of small scale processes in global MHD simulations: Some numerical experiments, in *Physics of Space Plasma*, edited by T. Chang, and J. R. Jasperse, vol. 14, p. 403, Cambridge, MA, MIT Center for Theoretical Geo/Cosmo Plasma Physics, 1996.

Raeder, J., J. Berchem, M. Ashour-Abdalla, L. A. Frank, W. R. Paterson, K. L., Ackerson, S. Kokubun, T. Yamamoto, and J. A. Slavin, Boundary layer formation in the magnetotail: Geotail observations and comparisons with a global MHD simulation, *Geophys. Res. Lett., 24*, 951, 1997.

Russell, C. T., G. Le, P. Chi, X.-W. Zhou, J.-H. Shue, S. M. Petrinec, P. Song, F. R. Fenrich, and J. G. Luhmann, The extreme compression of the magnetosphere on May 4, 1998 as observed by the Polar spacecraft, in *Coordinated Measurements of Magnetospheric Processes*, *Adv. Space Res.*, 25, 1369-1375, 2000.

Sato, T., and T. Hayashi, Externally driven magnetic reconnection and a powerful magnetic energy converter, *Phys. Fluids, 22*, 1189, 1979.

Slavin, J. A., E. J. Smith, D. G. Sibeck, D. N. Baker, R. D. Zwickl, and S.-I. Akasofu, An ISEE 3 study of average and substorm conditions in the distant tail, *J. Geophys. Res., 90*, 10,875, 1985.

Ugai, M., Temporal evolution and propagation of a plasmoid associated with asymmetric fast reconnection, *J. Geophys. Res., 90*, 9576, 1985.

Walker, R. J., and M. Ashour-Abdalla, The magnetosphere in the machine, large scale theoretical mode of the magnetosphere, IUGG Quadrennial Report, *Rev. of Geophys.*, supplement, 639, 1994.

Winterhalter, D., M.G. Kivelson, C.T. Russell, and E.J. Smith, ISEE-1 and −3 observation of the interaction between an interplanetary shock and the Earth's magnetosphere: A rapid traversal of the magnetopause, *Geophys. Res. Lett., 8*, 911, 1981.

Zhuang, H. C., C.T. Russell, E. J. Smith, and J. T. Gosling, Three-dimensional interaction of interplanetary shock waves with the bow shock and the magnetopause: A comparison of theory with ISEE observations, *J. Geophys. Res., 86*, 5590, 1981.

J. Berchem, M. El-Alaoui, M. Ashour-Abdalla, IGPP-UCLA, Slichter Hall, Los Angeles, CA 90095-1567.

Model Predictions of Magnetosheath Conditions

P. Song

*Department of Environmental, Earth and Atmospheric Sciences
and Center for Atmospheric Research, University of Massachusetts Lowell, USA*

The magnetosheath and its two boundaries - the bow shock and the magnetopause - interface the magnetosphere with the solar wind. This interface specifies the upstream conditions for magnetospheric models. Major progress has been made in the last few years in specifying the magnetosheath conditions. The classical Spreiter magnetosheath model is modified to include more physical processes. The modified model has been tested under various upstream conditions. Magnetohydrodynamic models have been developed. The spatial and temporal resolutions of these models are limited by the computer capability and costs for running. The performance of these models remains to be tested. The modified Spreiter model provides fast and inexpensive specification of the magnetosheath conditions. The direction of the magnetic field in the magnetosheath, which is most important for space weather magnetospheric models, is the best-predicted quantity. This paper summarizes the model approximations, procedures and validations.

1. INTRODUCTION

The existence of the magnetosheath makes space weather predictions in the magnetosphere more complicated. The solar wind does not directly interact with the magnetosphere, but through an intermediate medium. The dayside magnetosheath is formed because of two facts. First, the solar wind is supersonic. Second, the dayside magnetosphere is a blunt-nosed obstacle to the solar wind flow. Because the solar wind is supersonic, a shock is formed marking the outermost location of the region where the information about the existence of the magnetosphere can reach. This shock separates the region of supersonic solar wind from the subsonic shocked solar wind flow. Because the magnetosphere is blunt-nosed, the bow shock is detached from the nose of the magnetosphere.

Downstream of the bow shock, the solar wind plasma is heated, decelerated, compressed, and diverted around the magnetosphere.

A predictive magnetosheath model needs to specify the locations of the magnetopause and the bow shock, as well as the physical quantities in the magnetosheath. Since the solar wind constantly varies, the locations of the two boundaries change. As a result, the location of a satellite in the magnetosheath often changes rapidly with respect to the two boundaries. The physical quantities in the magnetosheath also vary with the upstream conditions.

Modeling the magnetosheath conditions was started in early 60s. As we know, earlier supersonic jet airplanes all had a sharp nose. When the Mach number, the speed of the airplane divided by the sound speed, increases to above 3, the nose of the airplane begins to burn. This is caused by the large surface area over which the nose interacts with the ambient air while the small volume stores and transports the heat generated by the friction between the air and the nose. A blunt-nosed object can solve the problem. However, at that time, without the help of computers, the

Space Weather
Geophysical Monograph 125

solution of the flow around a blunt-nosed body was unknown. In particular, the stability of such an airplane was not a trivial problem to solve. After the problem was solved in aerodynamics, application to the magnetosheath faces two additional problems that need to be addressed. First the solar wind is collisionless and electrically conductive, and is not an ordinary gas. Second, there is a magnetic field, which is referred to as the interplanetary magnetic field (IMF), carried with the solar wind from the sun. The first problem has been proven important to the heating processes at the bow shock. It does not have much affect on the description of the magnetosheath itself. The second problem appears more critical. The electromagnetic force affects the thickness of the magnetosheath and the physical quantities in the magnetosheath especially near the magnetopause. In addition, the magnetic field in the magnetosheath needs to be provided in a magnetosheath model because the field orientation near the magnetopause is essential to magnetospheric models. The challenge to predicting the magnetosheath conditions is essentially how to address the second problem in an economical manner.

To model the magnetosheath quantitatively, there are three approaches for consideration. The first is the gasdynamic convected field (GDCF) model which is often referred to as the Spreiter model [Spreiter et al., 1966; Spreiter and Stahara, 1980]. With some simplifications, this model provides a complete description of the locations of the bow shock and magnetopause and physical quantities in the magnetosheath with a given solar wind condition. It runs fast and inexpensively. For example, it takes less than few seconds to make a prediction using an old workstation. It has a very high resolution in both space and time. It resolves down to 5 % spatially in the magnetosheath and the temporal resolution below 1 min has been extensively tested. The weakness of the model is that the solution is not fully self-consistent. As will be discussed in detail, a modified version of the model describes most of the important physical processes. This model is now available for space weather prediction. The second approach is the so-called magnetohydrodynamics (MHD) simulations. It solves a complete set of fluid equations that describe both the fluid and electromagnetic properties of the magnetosheath. It provides a self-consistent description of all major low-frequency processes occurring in the magnetosheath. To solve this strongly coupled equation set is difficult and expensive, and requires large computer power. One has to choose between a better resolution for a longer computing time and a quicker result with a low resolution. For space weather prediction, time is important. Using the available computing resources for real-time predictions, MHD

models provide only very low spatial and temporal resolutions. The accuracy of the predictions has not been tested systematically. The third approach is the so-called hybrid particle simulation models. These models treat electrons as fluid and ions as individual particles. This is a more complete description of the magnetosheath processes than MHD because it includes some of the ion gyromotion effects. At the present time, no 3-D hybrid model with sufficient simulation domain and resolutions exists. For the purposes of space weather, the improvement by including the additional processes may not be crucial.

In summary, the GDCF model is now available for space weather forecasts, MHD models may be the next generation models for space weather forecasts when computers become much faster and less expensive. The next section provides detailed discussion about the present version of the GDCF model, our understanding of the physical processes included and not included in the model. These explain why sometimes the predictions are good and sometimes not. In section 3, a few cases are shown.

2. GASDYNAMIC CONVECTED FIELD MODEL

2.1. Basic Equations.

Before we describe how the GDCF model treats the magnetosheath, it is useful to look at the MHD treatment so that we know what we need. Standard isotropic ideal MHD equations are
Mass conservation

$$\frac{\partial \rho}{\partial t} + \nabla \bullet \rho \vec{V} = 0 \qquad (1a)$$

Momentum equation

$$\rho \frac{\partial \vec{V}}{\partial t} + \rho (\vec{V} \bullet \nabla) \vec{V} = -\nabla p - \nabla \frac{B^2}{2\mu_0} + (\vec{B} \bullet \nabla) \frac{\vec{B}}{\mu_0} \qquad (1b)$$

Energy equation

$$\frac{\partial}{\partial t} \left(\frac{\rho V^2}{2} + \frac{p}{\gamma - 1} + \frac{B^2}{2\mu_0} \right) + \nabla \bullet \vec{V} \left(\frac{\rho V^2}{2} + \frac{\gamma p}{\gamma - 1} + \frac{B^2}{\mu_0} \right) = 0 \qquad (1c)$$

Maxwell equations

$$\nabla \bullet \vec{B} = 0 \qquad (1d)$$

$$\nabla \times (\vec{V} \times \vec{B}) = \frac{\partial B}{\partial t} \qquad (1e)$$

where ρ, **V**, p, γ, **B**, and μ_o, are the mass density, velocity, thermal pressure, ratio of specific heats, magnetic field, permeability in vacuum, respectively. The equation set is closed.

One of the most important effects that slow down the MHD calculation is the time dependence of the equations. With a steady state approximation, variations on time-scales much shorter than the intermediate of L/C_{MS}, L/V, and L/C_A are neglected, where L, $C_A = B/(\mu_o \rho)^{1/2}$, $C_{MS} = (C_S^2$

$+ C_A{}^2)^{1/2}$, and $C_S = (\gamma k T)^{1/2}$ are the spatial scale of interest, the Alfvén speed, the magnetosonic speed, and sound speed, respectively. Noting that the three characteristic velocities change throughout the magnetosheath, the affected temporal scale varies. Typically the three velocities in the dayside magnetosheath are about 50 to 200 km/s. Given the thickness of the magnetosheath about $3R_E$, the flow time and Alfvénic transient time are about 1 min, and the magnetosonic transient time is significantly shorter. Therefore temporal variations longer than 5 min should be adequately represented by the GDCF model.

The other effect that slows down the MHD calculation is caused by the magnetic field terms in the momentum and energy equations. Without these terms, the subset of equations 1a to 1c could be solved without incorporation of equations 1d and 1e, and equations 1d and 1e could then be solved using the velocity solved from the first subset.

Spreiter and Stahara [1980] proposed neglecting the magnetic field in the momentum and energy equations and all time varying terms. This allows very fast calculations. The predictions sometimes are very impressive. In an extensive test of the model performance, it was realized that the performance of the model could be improved if the Mach number in the model is defined differently from Spreiter and Stahara's original definition. The sonic Mach number in the original GDCF model can be replaced with the fast mode Mach number $M_f = V/C_f$, where C_f is the magnetosonic fast mode velocity, $2C_f{}^2 = (C_S{}^2 + C_A{}^2) + [(C_S{}^2 + C_A{}^2)^2 - 4C_S{}^2 C_A{}^2 \cos^2\vartheta_{bn}]^{1/2}$, and $\cos\vartheta_{bn} = B_x/B$ is the cone angle of the field. In 1-D and in high beta plasmas or for nearly perpendicular propagation, the equivalent pressure is $p_{GDCF} = p + (B^2\sin^2\vartheta_{bn})/2\mu_0$. In mathematics, this replacement is equivalent to solving an equation set of

Mass conservation

$$\nabla \bullet \rho \vec{V} = 0 \qquad (2a)$$

Momentum equation

$$\rho(\vec{V} \bullet \nabla)\vec{V} = -\nabla p - \nabla\frac{B^2}{2\mu_0} + \frac{\nabla B_x{}^2}{2\mu_0} \qquad (2b)$$

Energy equation

$$\nabla \bullet \vec{V}\left(\frac{\rho V^2}{2} + \frac{\gamma p}{(\gamma - 1)} + \frac{5}{4}\frac{B^2 - B_x{}^2}{\mu_0}\right) = 0 \qquad (2c)$$

Maxwell equations

$$\nabla \bullet \vec{B} = 0 \qquad (2d)$$

$$\nabla \times (\vec{V} \times \vec{B}) = 0 \qquad (2e)$$

In steady state, the x component of equation 2b is the same as equation 1b in one-dimension. Equation 2c is very similar to equation 1c. However, one of the important differences between the GDCF model and MHD is that there is a phase lock in the terms on the right-hand-side terms in equation (2b) but not in equation (1b). In other words, the variation in the (tangential) magnetic pressure in equation (2b) has to be of the same sign as of the thermal pressure. In equation (1b), on the other hand, the variations in the thermal pressure and in the magnetic pressure can be either in phase or in antiphase. This difference leads to some important physical consequences to be discussed in the next section.

2.2. Upstream Boundary Conditions and Solar Wind Propagation Time.

The upstream boundary is taken as the plane normal to the solar wind flow and containing a solar wind monitor based on which a prediction is made. The solar wind flow direction is aberrated from the Sun-Earth line by adding the Earth's orbital velocity V_E (=30 km/s) to the y component of the solar wind velocity. The upstream boundary conditions are uniform and specified according to the measurements from the solar wind monitor. As the solar wind changes, the boundary conditions change. The temporal variations at the upstream boundary seemingly introduce a time dependence of the model output. However, the fundamental difference between this time dependence and that solved from temporally varying basic equations is that the output given by the GDCF model has no history relating to the previous solution and has no effects on the following solution. In the standard model, the solar wind convection time is assumed to be $\Delta t = l/V_{SW}$, where l is the distance from the solar wind monitor to the earth along the sun-earth line. Since solar wind discontinuities, for example, are sometimes oriented at an angle to this plane, when the solar wind monitor is not on the stagnation streamline, the time shift may not be accurate. A constant additional time shift is included for each run to account for this effect. If several solar wind discontinuities having completely different orientations occurred on the same pass, the method for the time shift may have large errors.

2.3. Magnetopause Boundary.

The magnetopause is considered to be an impermeable solid body. Therefore, the flow velocity is tangential to the surface at the magnetopause. In the original GDCF model, the location of the subsolar magnetopause D_{MP} is determined by

$$K\rho_{SW}V_{SW}^2 = \frac{B_{MP}^2}{2\mu_0} \qquad (5)$$

where $B_{MP} = 2B_E/D_{MP}{}^3$ and B_E is the Earth's magnetic field at the equatorial surface. The constant K reflects the pressure reduction at the stagnation point from the solar wind value due to the deceleration and diversion process and is 0.88 for $\gamma = 5/3$ and for moderate to high Mach numbers [*Spreiter et al.*, 1966]. The magnetopause standoff distance in this vacuum dipole field magnetopause model depends only on the solar wind dynamic pressure.

Observations have shown that the standoff distance also depends on the IMF Bz [Aubry et al., 1970; Fairfield, 1971]. Now an empirical magnetopause model [Shue et al., 1998] has been included in the GDCF model. This magnetopause model specifies the standoff distance as a function of the solar wind dynamic pressure as well as the IMF Bz.

In the GDCF model the shape of the magnetopause is taken to balance the pressure of the form of $K\rho_{sw}v_{sw}^2\cos^2\psi$, where $\psi < \pi/2$ is the angle between the solar wind direction and the normal to the magnetopause surface. This form assumes that the magnetopause or the pressure of the magnetosphere is axisymmetric with respect to the stagnation streamline. The form of the $\cos^2\psi$ dependence can be derived from the momentum balance of a cold beam incident upon a reflecting surface and is consistent with a gasdynamic semi-empirical expression for a supersonic flow [e.g., *Spreiter and Stahara*, 1985]. The empirical magnetopause model [Shue et al., 1998] uses a different functional form of the shape. However, the empirical shape is very similar to the gasdynamic shape in a very large spatial region and a large range of upstream conditions. Therefore, using the empirical magnetopause standoff-distance model to replace the vacuum dipole model will not produce problems in terms of the shape of the magnetopause.

2.4. Bow Shock Location and Jump Conditions.

The subsolar distance of the bow shock is determined according to the gasdynamic result, or

$$\frac{D_{BS}}{D_{MP}} = 1 + 1.1\frac{(\gamma - 1)M_{SW}^2 + 2}{(\gamma + 1)M_{SW}^2} \qquad (5)$$

where D_{BS} is the standoff distances of the bow shock, and M_{SW} should be understood as the solar wind magnetosonic Mach number as defined in section 2.1. The GDCF model uses an implicit Euler equation solver to solve the asymptotic solutions of the equations with a time-marching procedure. The remainder of the flow field is determined by a shock-capturing marching procedure that spatially advances the solution downstream as far as is required by solving the steady Euler equations [*Spreiter and Stahara*, 1980].

2.5. Calibration Factors.

In the present form of the GDCF model, three constants are allowed to calibrate with the in situ observations. They are a constant time shift to align the predicted and observed major field rotations in a pass, a factor to scale the magnetopause location with the observed magnetopause crossing(s), and a factor to scale the thickness of the magnetosheath in order to align the predicted bow shock location with the bow shock crossing(s). The practical meaning of these three parameters will become clear in the next section. This procedure is equivalent to a three-point calibration for each case. The integrity of the model ensures the remaining predictions to be systematic. Statistical values of the three parameters under various solar wind conditions and at different spatial locations will point to the direction for further improvement of the model predictions. We are in the process to obtain these values.

In summary, the modified GDCF model describes many of the most important low-frequency physical processes from the magnetopause to the solar wind. More detailed discussion can be found in Song et al. [1999a]. When combined with the Shue et al. [1998] magnetopause standoff-distance model, the uncertainty in the magnetopause location is reduced to a few percent that is the generic uncertainty of the magnetopause location associated with the magnetopause oscillations.

3. MODEL TESTS

Figure 1 shows an example using the original GDCF model. The solid lines are in situ observations and the dashed lines are model predictions. The magnetopause and bow shock crossings are marked by arrowheads with labels MP and BS, respectively. The two vertical dashed lines show the predicted magnetopause and bow shock locations. It is clear that there are significant errors in the predictions using the original version of the GDCF model. The predicted magnetopause and bow shock are about 20 min too late; indicating that the predicted magnetosphere is slightly too large. The predicted magnetosheath is about 10% thinner than observed. The timings of the field variations are reasonably good in the outer and middle magnetosheath. Figure 2 shows the results using the three parameters discussed in section 2.5 to calibrate the model predictions for this case. The predictions are significantly improved. Most of the major field variations are reasonably correctly predicted. Because in the model the locations of the magnetopause and the bow shock at each time are also predicted, the location with respect to the two boundaries can be determined at each time. The third panel of Figure 3 shows the distance of the satellite from the magnetopause normalized by the thickness of the magnetosheath. This normalized distance equals zero at the magnetopause and one at the bow shock. During this

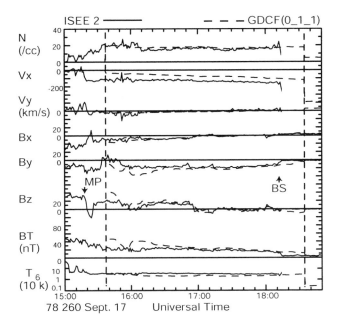

Figure 1. Comparison of the ISEE 2 in situ measurements (solid lines) with the original (without adjustments) GDCF model prediction (dashed lines) during a magnetosheath pass. From top are the density, x and y components of the velocity in GSE, three components (in GSE) and strength of the field, and the plasma temperature. The magnetopause and bow shock crossings of the satellite are indicated by the arrows labeled with MP and BS, respectively.

event, the satellite went out relatively gradually. The bottom panel of Figure 3 shows the ratio of the observed density to the predicted one. The fluctuations of this quantity provide a measure of the deficiency of the model. As we will show later in this section, these seemingly noisy fluctuations have led to an important scientific discovery. Figure 4 shows an event when the solar wind was varying. The solar wind changes lead to large variations in the predicted density and the field direction (dashed lines in the upper two panels). The prediction of the field direction made by the GDCF model correlates very well with the magnetosheath observations. More documented cases and the comparison between the vacuum dipole magnetopause model and the Shue et al. [1998] model can be found in Song et al. [1999b, 2000].

In addition to making space weather forecasts, the GDCF model has been used as a research tool. It has been used in estimating the quantities near the magnetopause. As shown in Figures 3 and 4, the density predicted by the GDCF model is not very accurate, although the magnetic field predictions are very good. An important piece of information that the GDCF model provides is the location of the sheath satellite at each time, as is shown in Figures 3

and 4. In the GDCF model, the density along the stagnation streamline is close to that just downstream of the bow shock. Therefore, the normalized density shown in Figures 3 and 4 removes the effects of the variations in the solar wind density and the bow shock compression. If

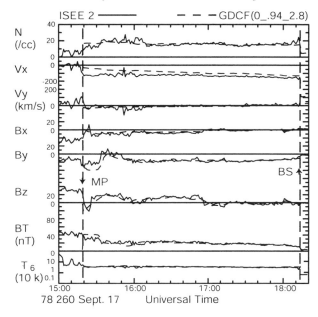

Figure 2. Comparison of the ISEE 2 observations with adjusted GDCF model predictions for the pass shown in Figure 1. The additional time shift is zero, the magnetopause scale factor is 0.94, and the solar wind temperature factor is 2.8. The solar wind temperature factor is not linearly proportional to the sheath thickness but is used to scale the thickness of the magnetosheath.

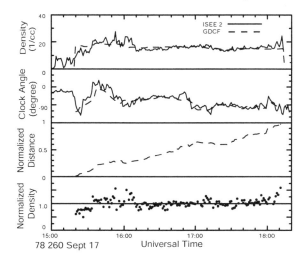

Figure 3. The normalized quantities for the pass shown in Figure 2. The top two panels show the ISEE 2 measurements (solid lines) and GDCF model prediction (dashed lines) of the density and the clock angle of the magnetic field for reference. The next two panels show the normalized distance and density.

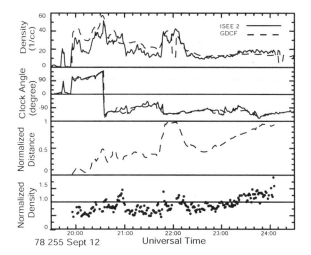

Figure 4. Another example of the magnetosheath passes by the ISEE satellite when the solar wind and IMF varied strongly, in the same format as Figure 3.

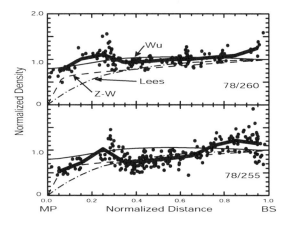

Figure 5. The density profile for the two cases shown in Figures 3 and 4. The horizontal axis is the normalized distance from the magnetopause to the bow shock (from 0 to 1). The vertical axis is the normalized density with the predicted density as 1. The dots are observation, and the thick solid lines are bin averages. The density profiles along the stagnation streamline from MHD models are also shown. The thin solid lines, dashed lines and dash-dotted lines are for *Wu* [1992], *Zwan and Wolf* [1976] and *Lees* [1964], respectively.

plotting the normalized density against the normalized location, the effects of the solar wind temporal variations can be removed, and the density profile from the magnetopause to the bow shock can be derived. This profile is extremely important for theoretical modeling of the magnetosheath. Figure 5 shows such profiles for the two cases shown in Figures 3 and 4. In time series, the density profiles for the two cases appear very different, as shown in the solid lines in the upper panel of each figure.

After the effects of temporal variations are removed, the two profiles shown as the dots in Figure 5 are similar: A gradual density decrease from the bow shock to about one-third of the distance from the magnetopause to the bow shock. Then, a compressional front that doubles the density stands in the sheath. A rapid drop in the density near the magnetopause is consistent with what has been referred to as the plasma depletion layer [Zwan and Wolf, 1976]. The addition of this new compressional front, which has been referred to as the slow mode front [Song et al., 1992; Southwood and Kivelson, 1992, 1994], alters our understanding of the magnetosheath and leaves many questions to be answered. Also shown in Figure 5 are the results from several theoretical models. None of them seems to describe adequately the magnetosheath density profile. In this regard, especially because of the large inconsistency of Wu [1992] calculation with observations, a simple use of global MHD simulations may not predict correctly the magnetosheath processes. Some processes may not be modeled adequately in some numerical methods and others may be sensitive to numerical dissipations. In the latter case, spatial resolution may play essential roles.

4. CONCLUSIONS

The GDCF model is now available for space weather predictions of magnetosheath conditions and upstream boundary conditions for magnetospheric models. It provides high-resolution predictions and is inexpensive to operate. The physical processes included in the model and the limitation of the model are relatively well understood. The magnetic field prediction is quite reliable.

Acknowledgments. The author is grateful to C. T. Russell, J. R. Spreiter, S. S. Stahara, and X. X. Zhang with whom some of the works discussed here were done. This work was supported by the National Science Foundation-Office of Naval Research under award NSF-ATM 9713492, and by NSF under award NSF-ATM 9729775.

REFERENCES

Aubry, M. P., C. T. Russell, and M. G. Kivelson, Inward motion of the magnetopause before a substorm, *J. Geophys. Res.*, *75*, 7018, 1970.

Fairfield, D. H., Average and unusual locations of the Earth's magnetopause and bow shock, J. *Geophys. Res.*, *76*, 6700, 1971.

Lees, L., Interaction between the solar plasma wind and the geomagnetic cavity, *AIAA J.*, *2*, 1576, 1964.

Petrinec, S. M., P. Song, and C. T. Russell, Solar cycle variations in the size and shape of the magnetopause, *J. Geophys. Res.*, *96*, 7893, 1991.

Shue, J.-H., P. Song, C. T. Russell, J. T. Steinberg, J. K. Chao, G. Zastenker, O. L. Vaisberg, S. Kokubun, H. J. Singer, T. R. Detman, and H. Kawano, Magnetopause location under extreme solar wind conditions, *J. Geophys. Res., 103*, 17,691, 1998.

Song, P., C. T. Russell, and M. F. Thomsen, Slow mode transition in the frontside magnetosheath, *J. Geophys. Res., 97*, 8295, 1992.

Song, P., C. T. Russell, X. X. Zhang, S. S. Stahara, J. R. Spreiter, and T. I. Gombosi, On the processes in the terrestrial magnetosheath 1. Scheme development, *J. Geophys. Res., 104*, 22345, 1999a.

Song, P., C. T. Russell, X. X. Zhang, S. S. Stahara, J. R. Spreiter, and T. I. Gombosi, On the processes in the terrestrial magnetosheath, 2. Case study, *J. Geophys. Res., 104*, 22357, 1999b.

Song, P., et al., POLAR observations and model predictions during May 4, 1998, event, *J. Geophys. Res., 105, in press,* 2000.

Southwood, D. J., and M. G. Kivelson, On the form of the flow in the magnetosheath, *J. Geophys. Res., 97,* 2873, 1992.

Southwood, D. J., and M. G. Kivelson, Magnetosheath flow near the subsolar magnetopause: Zwan-Wolf and Southwood-Kivelson theories reconciled, *Geophys. Res. Lett., 22,* 3275, 1995.

Spreiter, J. R., and S. S. Stahara, A new predictive model for determining solar wind-terrestrial planet interaction, *J. Geophys. Res., 85,* 6769, 1980.

Spreiter, J. R., and S. S. Stahara, Planetary bow shocks, in *Collisionless Shocks in the Heliosphere: Reviews of Current Research,* edited by B. T. Tsurutani and R. G. Stone, p. 85, AGU, Washington, D.C., 1985.

Spreiter, J. R., A. L. Summers, and A. Y. Alksne, Hydromagnetic flow around the magnetosphere, *Planet. Space Sci., 14,* 223, 1966.

Wu, C. C., MHD flow past an obstacle: Large scale flow in the magnetosheath, *Geophys. Res. Lett., 19,* 87, 1992.

Zwan, B. J., and R. A. Wolf, Depletion of solar wind plasma near a planetary boundary, *J. Geophys. Res., 81,* 1634, 1976.

P. Song, Center for Atmospheric Research, University of Massachusetts Lowell, 600 Suffolk Street, Lowell, MA 01854-3629. (paul_song@uml.edu)

Nowcasting and Forecasting the Magnetopause and Bow Shock Locations Based on Empirical Models and Real-Time Solar Wind Data

S. M. Petrinec

Lockheed Martin Advanced Technology Center, Palo Alto, CA 94304-1181

Many empirical models have been developed over the years to describe the average location and shape of the Earth's geophysical boundaries; i.e., the bow shock and magnetopause. Several of these models are parameterized according to the external conditions of the solar wind. Because of the comparatively simple nature of these models, they are able to be run quickly and provide reasonable though crude estimates of the shapes of the geophysical boundaries into the near future based upon real-time solar wind observations taken far upstream of the Earth. This work describes the strengths and weaknesses of using such empirical models in a dynamic manner to nowcast and forecast the locations of the geophysical boundaries, and how they are being used for the purposes of space weather.

1. INTRODUCTION

As civilization becomes more technologically advanced, it has also become more sensitive and vulnerable to its surroundings. It is now no longer enough simply to be able to understand and predict the weather conditions of the atmosphere. There is increasing demand to understand and predict the ionospheric environment, near-Earth space, and the solar wind all the way back to the Sun. One important element of this system is the location of the geomagnetic boundaries, or interaction regions, between the solar wind and the magnetosphere. These boundaries (the bow shock and the magnetopause) have been probed by spacecraft since the 1960's, and numerous empirical models have been developed from the spacecraft observations. Initial empirical models simply fit spacecraft observations of the bow shock and magnetopause to conic sections or other functional forms. Such models of the bow shock were developed by

Fairfield, 1971, *Formisano*, 1979, *Slavin and Holzer*, 1981 while early magnetopause models were developed by *Fairfield*, 1971; *Howe and Binsack*, 1972; *Holzer and Slavin*, 1978; and *Formisano et al.*, 1979. As the number of spacecraft crossings increased and with the availability of a large solar wind data set, the empirical models were then parameterized according to the conditions of the solar wind. In the case of the magnetopause, the parameterization was at first univariate (by solar wind pressure or as a function of the interplanetary magnetic field (IMF), as in *Sibeck et al.*, 1991, *Petrinec and Russell*, 1993) and was later followed by bivariate empirical models (*Roelof and Sibeck*, 1993; *Petrinec and Russell*, 1996; *Shue et al.*, 1997, 1998; *Kuznetsov and Suvorova*, 1998; *Kawano et al.*, 1999). For the bow shock, several recent empirical studies have examined the role of solar wind on the shape of the Earth's bow shock (*Nemecek and Safrankova*, 1991; *Farris et al.*, 1991; *Cairns et al.*, 1995; *Peredo et al.*, 1995; *Bennett et al.*, 1997).

In this paper we discuss how the parameterized steady-state empirical models may be used with real-time solar wind observations to nowcast and forecast the locations of the bow shock and magnetopause, including the strengths and weaknesses of the methodologies and assumptions. Assessment criteria for forecasting accuracy are also discussed.

Space Weather
Geophysical Monograph 125

2. INSTRUMENTATION AND DATA

In order to produce a forecast of the locations of the bow shock and magnetopause, it is critically important to have access to real-time solar wind observations sufficiently far upstream of the Earth that they can be used to make predictions. At the time of this writing, the Advanced Composition Explorer (ACE) spacecraft is in a halo orbit about the L1 Lagrangian point, approximately 200 R_E sunward of the Earth. Solar wind plasma moments (ion density, speed, and temperature) are determined from distribution functions measured by ACE, and are telemetered to Earth in real-time along with the interplanetary magnetic field (GSM coordinates) The data are typically available to users via ascii files on the internet (at ftp://solar.sec.noaa.gov/pub/lists/ace/) within one to three minutes after being measured on the spacecraft. The cadence is one minute for solar wind and IMF data. Orbital elements of the ACE spacecraft are also available at much lower time resolution, at ftp://solar.sec.noaa.gov/pub/lists/ace2/. We are therefore able for the first time to obtain solar wind parameters in real-time, enabling us to predict the locations and shapes of the geomagnetic boundaries before the solar wind travels from the ACE spacecraft to the Earth. Using the data contained within these files, one is able to immediately forecast the conditions of the near-Earth environment about 30-60 minutes into the future, depending upon the exact location of the ACE spacecraft and the speed of the solar wind.

While much of the information needed to model the boundaries is available in real-time, there are still some quantities which are not available. One of these parameters is the abundance of minor ion species in the solar wind. While much less abundant than the H^+ which dominates the solar wind, the heavy ions nevertheless influence the momentum flux of the solar wind. Typically, the solar wind ion content is about 96% protons and 4% alpha particles, and the alpha particles increase the mass density of the solar wind by 16%. The momentum flux directly affects the location of both the magnetopause and bow shock. Another set of parameters not available in the real-time data stream are the electron moments. The electron temperature is used in the calculation of the upstream magnetosonic Mach number, which determines the location of the bow shock. Lastly, the actual flow direction of the solar wind is not available in the real-time data sets, which is important for the orientation of the bow shock and magnetopause with respect to the Sun-Earth line. Nevertheless, the real-time parameters obtained by the ACE spacecraft are extremely useful for making reasonably accurate predictions of the bow shock and magnetopause locations.

3. RATIONALE

The empirical models of the bow shock and magnetopause are not based on self-consistent physical principles of the entire system in the same way as magnetohydrodynamic computer models. Rather, the empirical models are typically constructed from actual observations of boundary crossings. Thus, much of the physics that controls the location of the magnetopause and bow shock is inherently contained within the models. However, some of this physics can be lost through bad assumptions or an incomplete physical understanding, which leads to inadequate choices of functional forms for the boundary shapes or for parameterization of the controlling factors. An important deficiency in the use of static empirical models for dynamic purposes is that any understanding of the dynamics of the solar wind and magnetosphere has been lost in the construction of the static models. For some physical processes, this is not thought to be a problem. Changes in the solar wind dynamic pressure or magnetosonic Mach number should change the location of the magnetopause and bow shock, respectively, as these boundaries try to maintain an equilibrium state (of course, some oscillations of these boundaries about a new equilibrium location are possible). Some processes, however, operate on a longer time scale or in an integrative manner, and thus are not handled well by static models in a dynamic setting. An important example of this is the magnetospheric erosion due to the magnetic reconnection process. Empirical magnetopause models are parameterized by the instantaneous (convected) value of the z-component of the IMF in GSM coordinates. However, the erosion process of the magnetosphere continues for as long as the IMF is in a particular orientation (e.g., near the subsolar region when the z-component is negative (*Aubry et al.*, 1970)), though there may be a point at which subsolar magnetospheric erosion is balanced by the convection of flux back to the dayside magnetosphere. Nevertheless, the use of empirical models in a dynamic fashion is thought to be reasonable enough to give quick though crude estimates of the bow shock and magnetopause locations for forecasting purposes.

4. PROCEDURE AND COMPLICATIONS

The first step in using the real-time ACE observations is to convect the solar wind time series from the observation location to the near-Earth region. The simplest way to do this is to assume that each element in the solar wind time series represents a uniform planar front, and that the

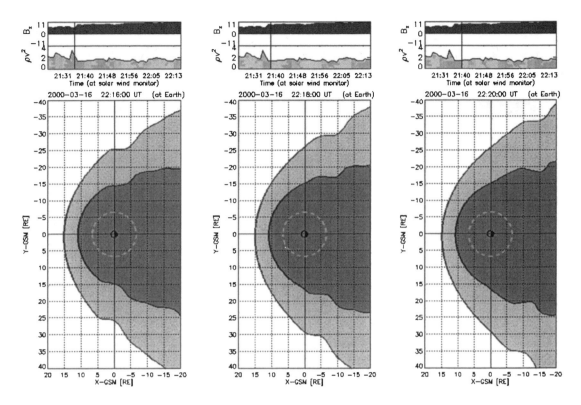

Figure 1. Successive time intervals of the magnetopause (*Petrinec and Russell*, 1996 model) and bow shock (*Farris and Russell*, 1994 and *Farris et al.* 1991 models) taken from a movie placed on the web which used ACE real-time observations as input.

solar wind flows along the aberrated Sun-Earth direction. However, it is understood that there could be significant error in this assumption at times, as shock wave fronts and other discontinuities in the solar wind could result in the arrival of solar wind plasma either earlier or later than the estimated convection time. Also, there is significant uncertainty regarding how much the solar wind is slowed down as it flows through the bow shock and around the magnetopause. In addition, there is also the difficulty in determining how to treat elements of the solar wind time series if they overtake one another by the time they reach the neighborhood of the Earth.

The next step is to calculate the three-dimensional boundary locations for the bow shock and magnetopause for each element of the time series. Once these shapes are calculated, the boundary shapes are reduced to two-dimensions by restricting to the plane of interest (e.g., the equatorial plane). For a given time, each convected front from the time series has a location, and the boundary shape is assumed to be most relevant at this location. The different shapes are then 'stitched' together by linear interpolation from one front location to the next. The resulting shapes

are no longer easily described by simple analytic functions. As one steps forward in time, a series of resulting shapes for the magnetopause and bow shock can be generated. An examination of a sequence of these images as a movie gives the impression of dynamics, as features appear to move in an anti-sunward direction as the sequence progresses. At the time of this writing, such movies are being created and shown on the web at http://muir.spasci.com/DynMod/. The bow shock subsolar location is from *Farris and Russell* [1994] and the bow shock shape is from *Farris et al.*, [1991]. The magnetopause model is from *Petrinec and Russell* [1996]. These movies use the ACE real-time observations to create a forecast of the bow shock and magnetopause locations, and are created every 5-minutes. A few frames from a movie which was displayed on the web site are shown in Figure 1. The solar wind dynamic pressure and z-component of the IMF are shown in the top two panels of each frame. A vertical black line marks the time at the ACE spacecraft. The lower panels show the shapes of the bow shock and magnetopause in the equatorial plane deduced from convection of the ACE time series, for the times at Earth. One can see that perturbations from an ideal conic section appear to move downstream in the successive panels.

5. FEATURES AND PHENOMENA NOT YET INCLUDED IN THE REAL-TIME MODELING EFFORTS

There are many elements which have not been included in the real-time dynamic modeling of the Earth's magnetopause and bow shock based on the static models. Some have been neglected because of computational complexity or extremely infrequent occurrence, while others are not included because it is not clear how they should be incorporated. These are described in detail below.

5.1 Bow Shock

5.1.1 Extreme Conditions

The empirical bow shock shape is typically determined for high magnetosonic Mach numbers (5 or larger), since this is the usual condition of the solar wind. There are very few spacecraft crossings of the bow shock at low Mach numbers, and so it is difficult to empirically determine the bow shock location and shape under these conditions (*Fairfield*, 1971; *Cairns et al.*, 1995). It is expected that the shape of the bow shock becomes more flared as the magnetosonic Mach number decreases towards unity, since the asymptotic Mach cone angle is inversely dependent upon the Mach number (cf., *Bennett et al.*, 1997).

5.1.2 Asymmetries

Because the Earth's bow shock is a magnetosonic fast mode shock, it possibly is not cylindrically symmetric far downstream. The asymmetry of the Mach cone angle is dependent upon the angle between the upstream velocity and IMF vectors. This has been explored in a theoretical context by *Spreiter and Stahara*, 1985; *Khurana et al.*, 1994; *Petrinec and Russell*, 1997 and *Bennett et al.*, 1997.

5.1.3 Polytropic Index

The polytropic index (γ) of the solar wind is not known with certainty. This parameter is used in the calculation of the upstream magnetosonic Mach number (M_{ms}), and is thus used in the determination of the bow shock location. It is usually assumed to be 5/3, though this value may not always be a constant in the solar wind.

5.1.4 Shock Reformation

Under conditions wherein the IMF is quasi-parallel to the normal direction of the shock surface and the Mach number of the solar wind is sufficiently large, then plasma waves upstream of the bow shock can steepen and eventually lead to the local reformation of the bow shock wave (*Scholer*, 1993; *Krauss-Varban and Omidi* [1993] and references therein). These dynamic processes have not been included in any empirical bow shock model to date.

5.1.5 Switch-on Shocks

Under conditions of low solar wind magnetosonic Mach number ($1 < M_{ms}^2 < 1+\gamma(1-\beta)/(\gamma-1)$), plasma β; ($\beta < 2/\gamma$, where β is the ratio of thermal to magnetic pressure), and specific IMF orientation with respect to the bow shock ($\theta_{B-v} = 0°$), part of the shock surface will include a 'switch-on shock' (in the region where $\theta_{B-v} = 0°$). It is expected that in this regime, a region of the bow shock will include intermediate shocks and discontinuities which branch into the magnetosheath (see the recent MHD simulations of *De Sterck and Poedts* [1999] for details). Such conditions, while very interesting, are very rarely encountered in the solar wind and so are not expected to play a significant role in space weather.

5.2 Magnetopause

5.2.1 Asymmetries

When performing fits of magnetopause crossings, modelers typically use a functional form which is axisymmetric about the aberrated Sun-Earth axis. However, it has long been thought that the magnetopause has large-scale indentations near the cusp regions (cf., *Spreiter and Briggs*, 1962), and that the locations of these regions depends upon the tilt angle of the Earth's dipole field. The influence of the magnetopause shape with the dipole tilt angle has been statistically demonstrated with spacecraft crossings (*Petrinec and Russell*, 1995). However, it is not clear if the dipole tilt angle causes actual indentations of the magnetopause at the cusps or merely changes the overall shape of the magnetopause in the noon-midnight meridian plane as compared with the equatorial plane (*Sibeck et al.*, 1991; *Zhou and Russell*, 1997). There are several reasons why the cusp regions are not well understood. First, most spacecraft orbital planes which cross the magnetopause lie close to the ecliptic plane, so the higher latitude magnetopause (where the cusps regions are located) is not often sampled. Second, a surface which includes the cusps can be very complicated, and is probably not well modeled with a simple conic section (cf., *Choe et al.*, 1973). Third, spacecraft magnetopause data sets contain a limited number of magnetopause crossings. When one attempts to parameterize the magnetopause according to solar wind conditions, trying to perform fits to a complicated function which varies with dipole tilt angle introduces more free parameters, and larger associated uncertainties.

In addition to the cusp regions, the magnetotail may also be asymmetric in cross section. Empirical studies have found asymmetries in the magnetotail radius as a function of dipole tilt angle (*Hammond et al.*, 1994) and substorm phase (*Nakamura et al.*, 1997).

5.2.2 Solar Wind History

The geophysical boundary locations and shapes are also believed to be influenced by the integrative effects of the recent history of the solar wind. This is perhaps most important with respect to magnetospheric erosion due to reconnection processes (see Section 3 and the discussion in *Shue et al.*, 2000*b*). Thus far, the history of the solar wind has been difficult to incorporate into the empirical models.

5.2.3 Instabilities

The magnetosheath supports a considerable number of plasma instabilities. Some of these instabilities are of large spatial scale, and can cause waves to propagate along the magnetopause (*Song*, 1994). Some oscillations of the magnetopause can be caused by pressure changes in the solar wind (*Sibeck*, 1990), and the median amplitude of oscillation of the magnetopause is about 0.5 R_E (*Song et al.*, 1988). Empirical studies have also found that magnetopause oscillations are dawn-dusk asymmetric, and are believed to be cause by convected instabilities originating in the foreshock region upstream of the bow shock (*Russell et al.*, 1997).

6. PREDICTIVE ACCURACY AND ASSESSMENT OF MODELS

There has not been much work done in assessing the predictive accuracy of empirical models. Many tests are typically not conducted for the normal situation. The comparisons are often performed to judge which model performs best when extrapolated to extreme conditions. An example

comparison of magnetopause models for a storm interval is described by *Shue et al.*, [1998; 2000*a*]. A recent statistical study by *Shue et al.*, 2000*b* examined many cases of geosynchronous magnetopause crossings.

In order to determine if the models accurately predict the locations of the bow shock and magnetopause, it is important to develop objective, quantitative assessment criteria. It is possible to obtain predict orbit information from spacecraft along with real-time solar wind observations to forecast when and where a given spacecraft will pass through either the magnetopause or bow shock. However, it is usually not possible to obtain plasma or magnetic field parameters in real-time, to determine (either manually or in an automated manner) if the spacecraft actually crossed a boundary. Thus, although it is possible to do an assessment of model predictive accuracy, it cannot be performed in real-time.

A reasonable and simple recipe for assessing the predictive accuracy of the models is as follows: P = 100/N×Σ('pre = obs'), where P is the prediction percentage. For each element in an equally spaced time series, 'pre' equals 'i' (inside) or 'o' (outside), depending upon whether the predicted location of the spacecraft is internal or external to the boundary of interest. Similarly, 'obs' equals 'i' or 'o', for the observation. If 'pre' equals 'obs', then the Boolean operation 'pre = obs' evaluates to 1; otherwise it evaluates to 0. N is the total number of intervals. Such assessment values are to be evaluated as a function of position (e.g., bins of 1 cubic Earth radius), and perhaps also as a function of solar wind conditions. Spatial bins far from the nominal location of a studied boundary will rarely observe the boundary, and so the prediction accuracy will be near 100%, but will be relatively meaningless. Bins closer to the nominal location of the boundary will likely be less accurate but more meaningful. Contingency tables can be constructed for each bin, which will be used to assess the forecast performance (as performed for geosynchronous crossings by *Shue et al.*, 2000*b*). Such information will indicate where the models do well, and where they are most in need of improvement.

7. SUMMARY AND FUTURE WORK

The use of empirical models for the purposes of forecasting or nowcasting has the advantage of being less computationally demanding than the magnetohydrodynamic models. However, the static empirical models may at times not be very accurate when used to predict the actual locations of the dynamic bow shock and magnetopause. It is important to develop objective assessment criteria to determine which empirical models are most accurate, and where weaknesses in the models exist. Work is underway to conduct such tests. In addition, a more interactive web site is under development, which will allow users the capability to specify past time periods in addition to the forecasts now available. In addition, users will have more flexibility in their choice of solar wind monitor for input data, viewing perspectives, addition of spacecraft orbits, and choice of empirical models for the bow shock and magnetopause. Important parameters such as boundary standoff distances and distances of spacecraft from boundaries are also planned to be available, based upon the user's selections. This project is expected to result in a useful tool for the magnetospheric community and for studies of space weather in general.

Acknowledgments. This effort was supported under the Lockheed Martin Space Systems Company independent research and development program.

REFERENCES

Aubry, M. P., C. T. Russell, and M. G. Kivelson, Inward motion of the magnetopause before a substorm, *J. Geophys. Res., 75*, 7018-7031, 1970.

Bennett, L., M. G. Kivelson, K. K. Khurana, L. A. Frank, and W. R. Paterson, A model of the Earth's distant bow shock, *J. Geophys. Res., 102*, 26927-26941, 1997.

Cairns, I. H., D. H. Fairfield, R. R. Anderson, V. E. H. Carlton, K. I. Paularena, and A. J. Lazarus, Unusual locations of Earth's bow shock on September 24 - 25, 1987: Mach number effects, *J. Geophys. Res., 100*, 47-62, 1995.

Choe, J. Y., D. B. Beard, and E. C. Sullivan, Precise calculation of the magnetosphere surface for a tilted dipole, *Planet. Space Sci., 21*, 485-498, 1973.

De Sterck, H. and S. Poedts, Field-aligned magneto-hydrodynamic bow shock flows in the switch-on regime - parameter study of the flow around a cylinder and results for the axi-symmetric flow over a sphere, *Astron. Astrophys., 343*, 641-649, 1999.

Fairfield, D. H., Average and unusual locations of the Earth's magnetopause and bow shock, *J. Geophys. Res., 76*, 6700-6716, 1971.

Farris, M. H., S. M. Petrinec, and C. T. Russell, The thickness of the magnetosheath: constraints on the polytropic index, *Geophys. Res. Letts., 18*, 1821-1824, 1991.

Farris, M. H. and C. T. Russell, Determining the standoff distance of the bow shock: Mach number dependence and use of models, *J. Geophys. Res., 99*, 17681-17689, 1994.

Formisano, V., Orientation and shape of the Earth's bow shock in three dimensions, *Planet. Space Sci., 27*, 1151-1161, 1979.

Formisano, V., V. Domingo, and K. P. Wenzel, The three-dimensional shape of the magnetopause, *Planet. Space Sci., 27*, 1137-1149, 1979.

Hammond, C. M., M. G. Kivelson, and R. J. Walker, Imaging the effect of dipole tilt on magnetotail boundaries, *J. Geophys. Res., 99*, 6079-6092, 1994.

Holzer R. E. and J. A. Slavin, Magnetic flux transfer associated with expansions and contractions of the dayside magnetosphere, *J. Geophys. Res., 83*, 3831-3839, 1978.

Howe, H. C., Jr. and J. H. Binsack, Explorer 33 and 35 plasma observations of magnetosheath flow, *J. Geophys. Res., 77*, 3334-3344, 1972.

Kawano, H., S. M. Petrinec, C. T. Russell, and T. Higuchi, Magnetopause shape determinations from measured position and estimated flaring angle, *J. Geophys. Res., 104*, 247-261, 1999.

Khurana, K. K., M. G. Kivelson, A variable cross-section model of the bow shock of Venus, *J. Geophys. Res., 99*, 8505-8512, 1994.

Krauss-Varban, D. and N. Omidi, Propagation characteristics of waves upstream and downstream of quasi-parallel shocks, *Geophys. Res. Letts., 20*, 1007-1010, 1993.

Kuznetsov, S. N. and A. V. Suvorova, An empirical model of the magnetopause for broad ranges of solar wind pressure and B_z IMF, in *Polar Cap Boundary Phenomena*, edited by J. Moen et al., pp. 51-61, Kluwer Academic Publishers, Netherlands, 1998.

Nakamura, R., S. Kokubun, T. Mukai, and T. Yamamoto, Changes in the distant tail configuration during geomagnetic storms, *J. Geophys. Res., 102*, 9587-9601, 1997.

Nemecek, Z. and J. Safrankova, The Earth's bow shock and magnetopause position as a result of the solar wind-magnetosphere interaction, *J. Atmos. Terr. Phys., 53*, 1049-1054, 1991.

Peredo, M., J. A. Slavin, E. Mazur, and S. A. Curtis, Three-dimensional position and shape of the bow shock and their variation with Alfvenic, sonic and magnetosonic Mach numbers and interplanetary magnetic field orientation, *J. Geophys. Res., 100*, 7907-7916, 1995.

Petrinec, S. M. and C. T. Russell, External and internal influences on the size of the dayside terrestrial magnetosphere, *Geophys. Res. Letts., 20*, 339-342, 1993.

Petrinec, S. M. and C. T. Russell, An examination of the effect of dipole tilt angle and cusp regions on the shape of the dayside magnetopause, *J. Geophys. Res., 100*, 9559-9566, 1995.

Petrinec, S. M. and C. T. Russell, Near-Earth magnetotail shape and size as determined from the magnetopause flaring angle, *J. Geophys. Res., 101*, 137-152, 1996.

Petrinec, S. M. and C. T. Russell, Hydrodynamic and MHD equations across the bow shock and along the surfaces of planetary obstacles, *Space Sci. Rev., 79*, 757-791, 1997.

Roelof, E. C. and D. G. Sibeck, Magnetopause shape as a bivariate function of interplanetary magnetic field B_z and solar wind dynamic pressure, *J. Geophys. Res., 98*, 21421-21450, 1993.

Russell, C. T., S. M. Petrinec, T. L. Zhang, P. Song, and H.

Kawano, The effect of foreshock on the motion of the dayside magnetopause, *Geophys. Res. Letts., 24,* 1439-1442, 1997.

Scholer, M., Upstream waves, shocklets, short large-amplitude magnetic structures and the cyclic behavior of oblique quasi-parallel collisionless shocks, *J. Geophys. Res., 98,* 47-57, 1993.

Shue, J.-H., J. K. Chao, H. C. Fu, C. T. Russell, P. Song, K. K. Khurana, and H. J. Singer, A new functional form to study the solar wind control of the magnetopause size and shape, *J. Geophys. Res., 102,* 9497-9511, 1997.

Shue, J.-H., P. Song, C. T. Russell, J. T. Steinberg, J. K. Chao, G. Zastenker, O. L. Vaisberg, S. Kokubun, H. J. Singer, T. R. Detman, and H. Kawano, Magnetopause location under extreme solar wind conditions, J. Geophys. Res., 103, 17691-17700, 1998.

Shue, J.-H., C. T. Russell, and P. Song, Shape of the low-latitude magnetopause: Comparison of models, *Adv. Space Res., 25, 7/8,* 1471-1484, 2000a.

Shue, J.-H., P. Song, C. T. Russell, J. K. Chao, and Y.-H. Yang, Toward predicting the position of the magnetopause within geosynchronous orbit, *J. Geophys. Res., 105,* 2641-2656, 2000b.

Sibeck, D. G., A model for the transient magnetospheric response to sudden solar wind dynamic pressure variations, *J. Geophys. Res., 95,* 3755-3771, 1990.

Sibeck, D. G., R. E. Lopez, and E. C. Roelof, Solar wind control of the magnetopause shape, location, and motion, *J. Geophys. Res., 96,* 5489-5495, 1991.

Slavin, J. A. and R. E. Holzer, Solar wind flow about the terrestrial planets 1. Modeling bow shock position and shape, *J. Geophys. Res., 86,* 11401-11418, 1981.

Song, P., R. C. Elphic, and C. T. Russell, ISEE 1 & 2 observations of the oscillating magnetopause, Geophys. Res. Letts., 15, 744-747, 1988.

Song, P., Observations of waves at the dayside magnetopause, in *Solar Wind Sources of Magnetospheric Ultra-Low-Frequency Waves, Geophysical Monograph 81,* edited by edited by M. J. Engebretson, K. Takahashi, and M. Scholer, American Geophysical Union, Washington, D.C., pp. 159-171, 1994.

Spreiter, J. R. and B. R. Briggs, Theoretical determination of the form of the boundary of the solar corpuscular stream produced by interaction with the magnetic dipole field of the Earth, *J. Geophys. Res., 67,* 37-51, 1962.

Spreiter, J. R. and S. S. Stahara, Magnetohydrodynamic and gasdynamic theories for planetary bow waves, in *Collisionless Shocks in the Heliosphere: Review of Current Research, Geophysical Monograph 35,* edited by B.T. Tsurutani and R. G. Stone, American Geophysical Union, Washington, D.C., pp. 85-107, 1985.

Zhou, X. W. and C. T. Russell, The location of the high-latitude polar cusp and the shape of the surrounding magnetopause, *J. Geophys. Res., 102,* 105-110, 1997.

_____

Lockheed Martin Advanced Technology Center, B/255, O/L9-42, 3251 Hanover St., Palo Alto, CA 94304-1181, USA

Modeling Inner Magnetospheric Electrodynamics

F. R. Toffoletto, R. W. Spiro and R. A. Wolf

Rice University

J. Birn

Los Alamos National Laboratories

M. Hesse

Goddard Space Flight Center

We describe a model of inner magnetospheric electrodynamics that couples the Rice Convection Model (RCM) to an equilibrium magnetic field model. The equilibrium model is a modified version of the Hesse-Birn [1993] magnetofriction code, adapted for use in the inner magnetosphere. Previous versions of the RCM, which used observation-based or theoretical magnetic field models, were flexible and convenient but lacked theoretical consistency. The coupled code uses the pressure distribution computed from the RCM to modify the magnetic field. We present results using the coupled code to model a substorm growth phase. Under conditions of steady sunward convection the computed inner plasma sheet magnetic field becomes increasingly stretched, the current-sheet thins, and a B_z minimum forms at around $x = -15\ R_E$. In addition, region-1 currents form poleward of the traditional region-2 currents found in the RCM. As the stress in the configuration continues to increase, the numerical method eventually fails. Nature presumably relieves the stress with a substorm expansion phase onset.

1. INTRODUCTION

Theoretical understanding of Earth's inner and middle magnetosphere requires a computational framework that can self-consistently compute the time-dependent electric and magnetic fields, electric currents, and particle distributions. Past computational efforts have fallen short of that goal for one reason or another. Global MHD models (e.g., *Fedder et al.*, 1995; *Raeder et al.*, 1995, *Gombosi et al.*, 1998) solve the complete MHD equations beyond about 3 R_E geocentric distance, but their representations of the inner magnetosphere are unreliable owing to wide grid spacing and to the neglect of energy-dependent and charge dependent particle transport by gradient/curvature drift.

Other modeling efforts have aimed at precise representation of particle dynamics (and sometimes electric fields) within an assumed magnetic field. One set of computational models concentrates on accurate representation of loss and pitch-angle distributions [e.g., *Fok et al.*, 1999] in assumed electric and magnetic-field configurations. Other modeling efforts have treated electric and magnetic fields self-consistently with fluid treatment of the particles, while using approximations that limit them to the magnetotail. Examples of this approach are MHD magnetotail simulations [e.g., *Birn et al.*, 1994] and 2D equilibrium calculations of Erickson [1985, 1992] and Hau [1991].

So-called "convection models", which are based on a formulation originally introduced by Vasyliunas [1970], compute potential electric fields and particle distributions in careful detail, including ionosphere-magnetosphere

Space Weather
Geophysical Monograph 125

coupling and transport of magnetospheric particles by gradient/curvature drift. A single fluid approach to this problem has been taken by Senior and Blanc [1984] and Peymirat and Richmond [1993]. Arguably the most advanced convection model is the Rice Convection Model (RCM) [e.g., *Harel et al.*, 1981a; *Erickson et al.*, 1991; *Wolf et al.*, 1991]. The RCM models the inner and middle magnetosphere and its coupling to the ionosphere. It uses a many-fluid formalism to describe adiabatically drifting isotropic particle distributions within a pre-computed time-dependent magnetic field and associated induction electric fields. The model has been used to simulate several specific, well-studied magnetospheric events [*Chen et al.*, 1982; *Harel et al.*, 1981b; *Karty et al.*, 1982; *Spiro et al.*, 1981; *Spiro and Wolf*, 1984; *Wolf et al.*, 1982].

Erickson and Wolf [1980] demonstrated that typical observation-based magnetic-field models are inconsistent with lossless adiabatic convection in the Earth's plasma sheet. The assumption of lossless convection in such magnetic field models leads to unrealistically high pressures in the inner plasma sheet. (See also Schindler and Birn [1982] for a theoretical discussion). This phenomenon came to be known as the pressure balance inconsistency (PBI). The RCM is particularly sensitive to this effect, because it assumes sunward adiabatic convection and normally uses observation-based magnetic-field models as input. Precipitation loss has been shown to be incapable of resolving the PBI [*Erickson*, 1992]. During periods of sufficiently weak convection, Kivelson and Spence [1988] found that the effects of gradient and curvature drifts could be large enough to negate the PBI. For stronger convection and particularly substorm conditions, the resolution of the PBI is apparently that Nature violates the assumption of lossless adiabatic convection, by releasing plasma down the tail in the form of plasmoids and possibly also by other mechanisms.

One clear implication of the departure from force balance, in a plasma that evolves according to the RCM, is that the magnetic field should be modified self-consistently with the plasma distribution. The present paper represents an effort to construct a full model of the inner and middle magnetosphere, one that treats magnetic fields self-consistently with potential electric fields and particle drifts.

2. MODEL DESCRIPTION

2.1. The MagnetoFriction (MF) Code

The computation of three-dimensional magnetospheric equilibria has long been a goal in magnetospheric physics (For a review see Voigt [1986].) The essential aim is to compute solutions to the static force balance condition

$$\vec{j} \times \vec{B} = \nabla p \qquad (1)$$

that include the dipole region. For simplicity, we assume that the pressure (p) is isotropic everywhere, which is a reasonable assumption for the plasma sheet but not nearer the Earth. Early work was restricted to 2-dimensional solutions of the Grad-Shafranov equation [*Voigt*, 1986; *Hau*, 1991]. Cheng [1995] has developed a sophisticated and accurate numerical scheme for computing 3D magnetic field configurations that satisfy equilibrium condition (1) for an assumed distribution of pressure. However, no procedure has been developed yet for coupling the Cheng [1995] algorithm to a code (like the RCM) that allows calculation of the time evolution of the pressure distribution due to magnetospheric convection. Chodura and Schlüter [1981] suggested using a simple iterative procedure to find magnetostatic equilibria. This algorithm relates the net imbalance in (1) to a weighted pseudo-velocity which is used to modify the magnetic field and pressure distribution using Faraday's law and an appropriately chosen energy equation. While highly flexible equilibrium solutions to the magnetostatic equilibrium problem exist for the tail [*Birn*, 1987], numerical modeling is required to expand the calculation to regions where the dipole field is important.

A three-dimensional magnetospheric tail equilibrium that includes the transition to the dipolar field was calculated for the first time by Hesse and Birn [1993], using the MF method to relax the empirical Tsyganenko [1989] field model for $x < -5\ R_E$ into a force-balanced equilibrium. Improvements to the friction method allowed Hesse [1995] to model the region from $x = -2\ R_E$ to $x = -62\ R_E$. We have further extended this method to the region from $x = 10\ R_E$ to $x = -60\ R_E$ with an inner boundary at $1\ R_E$.

The MF code uses the standard MHD equations, except that the momentum equation includes artificial friction (α) and viscosity (v) terms to iterate towards a solution to equation (1). The basic equations are:

$$\frac{\partial \rho}{\partial t} + \nabla \cdot (\rho \vec{v}) = 0 \qquad (2)$$

$$\frac{\partial \rho \vec{v}}{\partial t} = \vec{j} \times \vec{B} - \nabla \cdot \left(p\vec{\mathbf{I}} + \rho \vec{v}\vec{v} \right) - \alpha \rho \vec{v} + v \nabla^2 \vec{v} \quad (3)$$

$$\frac{\partial p}{\partial t} = -\nabla \cdot [\vec{v} p] - (\gamma - 1) p \nabla \cdot \vec{v} \qquad (4)$$

$$\frac{\partial \vec{B}}{\partial t} = \nabla \times (\vec{v} \times \vec{B}) \qquad (5)$$

$$\nabla \times \vec{B} = \mu_0 \vec{j} \qquad (6)$$

The density ρ is an artificial density that is set to keep the fast-mode speed constant and is initialized with the value

$$\rho(t = 0) = \frac{\gamma p + B^2/\mu_0}{v_f^2} \qquad (7)$$

where v_f is the fast mode speed assumed to have the constant value v_f=1500 km/s. The next-to-last and last terms in equation (3) represent the friction that is employed to term that parameterizes the flux limiting routine in the time integration algorithm. Since the desired solution to

(1) does not depend on α, ν or ρ, they represent a set of convenient parameters that can be adjusted for speedier calculation of the equilibrium. (See Hesse and Birn [1993] for more details.)

Modifications beyond the original method of Hesse and Birn [1993] include: (a) subtracting the dipole field when the current $\vec{j}$ in equation (6) is computed, and (b) normalizing the force imbalance as defined by a parameter $\vec{F}^{(n)}$ (at iteration n) given by

$$\vec{F}^{(n)} = \frac{\left(\vec{j} \times \vec{B} - \nabla p\right)^{(n)}}{\rho(t=0)} \quad (8)$$

We define the average normalized force imbalance $N^{(n)}$ at iteration step n as

$$N^{(n)} \equiv \frac{\langle|\vec{F}^{(n)}|\rangle}{\langle|\vec{F}^{(1)}|\rangle} \quad (9)$$

where angled brackets denote an integral average over the computational volume. The energy equation (4), which is written in a form that does not involve the (artificial) density, guarantees that the pressure follows an adiabatic law as the system moves toward equilibrium. The parameter γ is set equal to 5/3, except in setting up the initial condition magnetic field model, where a higher γ is used to maintain stability and speed convergence. The value of γ =5/3 is chosen for consistency with the RCM (See equation (18)).

The computations were performed on a rectilinear, nonuniformly spaced grid, with grid spacing set to about 0.03 R_E near the Earth and near the equatorial plane and to about 1 R_E at the edge of the simulation box. The initial magnetic-field configuration was a Tsyganenko [1989] empirical model for $Kp = 2$. The initial pressure distribution was taken from an empirical summary of nightside equatorial observations [Lui et al., 1994], extended uniformly along field lines (as required by the model assumptions) and assumed (rather arbitrarily) to exist at other local times as well. The normal component of magnetic field at the boundary of the calculation was adjusted to ensure a solenoidal magnetic field. The numerical method is that described by Hesse and Birn [1992] and consists of a modified leap-frog scheme with a flux limiter based on the method discussed by Book et al. [1975]. The top panel of Figure 1 shows how the normalized force imbalance ($N^{(n)}$) varies as a function of time. (A unit of time is defined as the travel time of an Alfvén wave of speed 1000 km/s through a typical plasma sheet thickness (2 R_E), which is approximately 13 seconds [Hesse and Birn, 1992].) The middle panel in Figure 1 is the friction (α) term as a function of time. The bottom panel of Figure 1 shows the total kinetic energy of the system. The sharp drops occur where the velocity has been set to zero because the algorithm detected a local maximum in the kinetic energy. This is essentially the ballistic method described in Hesse and Birn [1992]. In

order to speed convergence, the frictional coefficient $\alpha^{(n)}$ is increased when the force imbalance grows and reduced when it decreases, according to the criteria

$$\alpha^{(n+1)} = \begin{cases} 2\alpha^{(n)} & \text{if } N^{(n+1)} \geq N^{(n)} \\ \frac{1}{2}\alpha^{(n)} & \text{if } N^{(n+1)} < N^{(n)} \end{cases} \quad (10)$$

This algorithm is motivated by the observed fact that smaller/larger values of $\alpha^{(n)}$ increased/decreased the convergence rate while decreasing/increasing the numerical stability of the code.

2.2. The Rice Convection Model (RCM)

The RCM computes drifts, currents and electric field in the inner magnetosphere in a manner first outlined by Vasyliunas [1970]. The model assumes an isotropic pitch angle distribution. Specifically, we assume that pitch angle scattering keeps the distribution isotropic without changing particle energies; drift-shell splitting effects are therefore not considered. (A more complete description of the basic equations and formulation of the RCM has been given by Harel et al. [1981a].) The equations describing the evolution of the particle distribution are

$$\left[\frac{\partial}{\partial t} + \vec{v}_s \cdot \nabla\right]\eta_s = 0 \quad (11)$$

$$\vec{v}_s = \frac{1}{B^2}\left[\frac{\lambda_s}{q_s}\vec{B} \times \nabla V^{-2/3} + \vec{E} \times \vec{B}\right] \quad (12)$$

where η_s is the number of particles per unit magnetic flux for species s characterized by energy invariant λ_s and charge q_s, and

$$V \equiv \int_s^N \frac{d\ell}{B} \quad (13)$$

is the volume of a flux tube with unit magnetic flux that is computed from the southern (S) to the northern (N) ionospheres. The kinetic energy of an individual particle is then

$$W_s = \lambda_s V^{-2/3} \quad (14)$$

In equation (12), $\vec{E}$ is the electric field (which also implicitly includes the inductive electric field) and is given by

$$\vec{E} = -\nabla(\Phi_{ionosphere} + \Phi_{corotation}) - \vec{v}_{inductive} \times \vec{B} \quad (15)$$

Here $\Phi_{ionosphere}$ is the ionospheric potential in a frame that rotates with the Earth, and $\Phi_{corotation}$ converts the potential to a non-rotating frame. $\Phi_{ionosphere}$ is computed by solving the ionospheric current conservation equation

$$\nabla \cdot (\ddot{\Sigma} \cdot \nabla \Phi_{ionosphere}) = j_{\parallel} \qquad (16)$$

where $\ddot{\Sigma}$ is the height integrated conductivity tensor and $j_{\parallel}$ is the field-aligned current computed from the Vasyliunas [1970] equation

$$j_{\parallel} = \frac{\hat{b}_i}{2} \cdot \nabla_i V \times \nabla_i p \qquad (17)$$

where the subscript i refers to computations and values in the ionosphere and $\hat{b}_i$ is a unit vector in the direction of the magnetic field. The total isotropic particle pressure is given by

$$p = \frac{2}{3} \sum_s \left[\eta_s \mid \lambda_s \mid V^{-\frac{5}{3}} \right] \qquad (18)$$

Note that since λ_s is independent of space and time, the RCM calculation of η_s leads to direct specification of pV^γ. Use of $\gamma=5/3$ in the MF code ensures approximate conservation of pV^γ as the configuration relaxes to equilibrium.

The flux tube content (η_s) is obtained by subdividing a distribution function into energy channels and doing an integration within each energy interval

$$\eta_s = \frac{4\pi\sqrt{2}}{m_k^{\frac{3}{2}}} \int_{\lambda_{min}(s)}^{\lambda_{max}(s)} \sqrt{\lambda} f_s(T_k, \lambda_s) d\lambda \qquad (19)$$

$f_s(T_k, \lambda_s)$ is the distribution function, subscript k corresponds to chemical species and the integration limits are defined as

$$\lambda_{max}(s) = \begin{cases} \lambda_s + \frac{1}{2}(\lambda_s - \lambda_{s-1}) & \text{for the highest } s \\ \frac{1}{2}(\lambda_s + \lambda_{s+1}) \end{cases} \qquad (20)$$

$$\lambda_{min}(s) = \begin{cases} \frac{1}{2}(\lambda_{s-1} + \lambda_s) \\ 0 & \text{for the lowest } s \end{cases} \qquad (21)$$

2.3. Conversion between the RCM and the MF code

Conversion of plasma pressure information from the MF code to multi-channel species information (η_s) for the RCM requires assumptions about the plasma composition

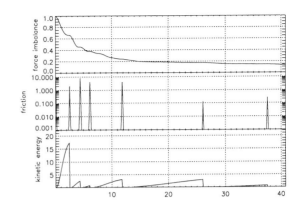

Figure 1. The force imbalance, frictional term, and kinetic energy as a function of normalized time during an iteration of the MF code.

and temperature. We assume that the initial plasma sheet temperatures are given by

$$kT_{ion} + kT_{electron} = \frac{2}{3}(4KeV)V^{-\frac{2}{3}} \qquad (22)$$

where the flux tube volume is in units of R_E/nT. The number density is inferred by assuming that the pressure p is given by

$$p = n(kT_{ion} + kT_{electron}) \qquad (23)$$

The initial and boundary-condition pressures are computed from the MF code. Combining equations (21) and (22) then gives an expression for the number density

$$n = \frac{3}{2} \frac{p}{(4KeV)V^{-\frac{2}{3}}} \qquad (24)$$

and temperature

$$T_{ion} = \frac{2(4KeV)V^{-\frac{2}{3}}}{3k(1 + \Gamma)} \qquad (25)$$

where

$$\Gamma \equiv \frac{T_{electron}}{T_{ion}} = \frac{1}{7.8} \qquad (26)$$

from the empirical result of Baumjohann et al. [1989]. For the results presented in this paper we computed the flux tube content (η_s) by assuming a Maxwellian distribution

$$f_s = n_k \left(\frac{m_k}{2\pi k T_k} \right)^{\frac{3}{2}} e^{-\left(\lambda_s V^{-\frac{2}{3}} / kT_k \right)} \qquad (27)$$

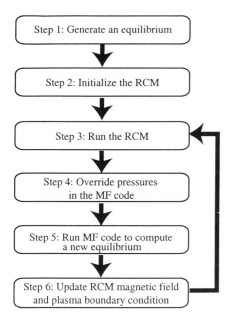

Step 1: Generate an equilibrium

↓

Step 2: Initialize the RCM

↓

Step 3: Run the RCM

↓

Step 4: Override pressures
in the MF code

↓

Step 5: Run MF code to compute
a new equilibrium

↓

Step 6: Update RCM magnetic field
and plasma boundary condition

Figure 2. Flow chart that illustrates the algorithm the couples the RCM with the MF code. See text in section 2.4 for details.

and using equations (19)-(21). In equation (27), the subscript k refers to either ions or electrons. In this paper $n_k = n_{ion} = n_{electron}$ and the initial and boundary-condition temperature are obtained from equations (22) and (26). (Note that the number density (24) is the physical number density and is not related to the artificial number density used in the MF relaxation code.)

2.4. Coupling the RCM and the MF code

The basic algorithm used to couple the RCM with the MF code is outlined in Figure 2. The MF algorithm provides the RCM with the magnetic field structure and the initial and boundary plasma distributions (Step 1 in Figure 2). In practice, the RCM is initialized using the pressure distribution and global magnetic field from a converged MF equilibrium run distribution (Step 2). (The procedure for calculating the initial equilibrium is described in section 2.1.) Equations (18) and (19) are used to translate between pressure (MF formalism) and η_S (RCM formalism). The RCM is then advanced for 5 minutes using an assumed cross-polar-cap potential drop of 100 kV and a sinusoidal distribution of potential around the polar cap boundary (Step 3). Then, the inner magnetosphere pressure distribution within the MF code is replaced by RCM-computed values and the MF code relaxes to a new equilibrium (Steps 4 and 5). This procedure assumes constant pressure along field lines. The timescale for RCM-MF code data interchange was chosen long enough so that the MF code responds to the force imbalance and computes noticeably different new equilibria but short enough to not effect a drastic change that would cause the MF code to become unstable. The resulting magnetic field is returned to the RCM along with

the η_S distribution at the RCM boundary (to represent flux tubes drifting into the RCM modeling region from the tail) (Step 6). The RCM then evolves the electric field and plasma distribution for another 5 minutes (Step 3).

After the pressure has been updated in the MF code (Step 4), but before a new equilibrium field is computed, we change the boundary condition on the normal component of the magnetic field at the tailward boundary (B_n), to represent the effect of dayside merging increasing the tail-lobe field. We change B_n in a way that maintains one-dimensional force balance (in z) with the value of pV^γ specified by the RCM; to change B_n and keep $\nabla \cdot B = 0$, we add a magnetic field that has the same form as the tail field used in the Toffoletto and Hill [1989] magnetic field model. This additional field allows the lobe field to adjust approximately, but self-consistently, to the changing pressure in the plasma sheet. If we do not change B_n, the field lines make a sharp turn next to the tailward boundary, obviously an artifact of the boundary condition.

2.5 Model Results

For the run presented here, we represented the plasma distribution using 20 ion and electron energy invariant species λ_S. (Earlier versions of this code used only a single species plasma [Toffoletto et al., 1996].) For simplicity, we have used a grid-based version of the RCM, which stores η_S values at grid points; this approach tends to be diffusive as compared to the edge-based version of the RCM, where η_S values are kept as individual contours that are moved as the computation proceeds.

In runs with static input magnetic fields, the RCM typically produces field-aligned currents of region-2 sense as a result of the inner edge of the plasma sheet being closer to Earth on the night side than on the day side. These currents have the effect of shielding the inner magnetosphere from the convection electric field.

The contour plots of Plate 1a show RCM-computed results displayed in the equatorial plane at the start of the run: the eight panels represent (across from left to right, then down) electric potential, field-aligned current, $V^{-2/3}$, pressure, and four energy invariant channels of flux tube content (η_S) with the lines representing the flow streamlines for that energy and species. Plate 1b shows the same quantities after 3000 seconds.

When the magnetic field is allowed to change in response to the plasma, as in the case of the growth phase modeled here, stretching of the tail field moves equatorial mapping points of a given ionospheric latitude outward. This field line stretching is shown by the blue curve in the potential plots of Plate 1a and 1b, corresponding to a line of constant ionospheric latitude mapped to the equatorial plane. The changing magnetic field results in a reduced shielding of the electric field, as is evident in the potential plot of Plate 1b.

Plate 1b indicates formation of region-1 sense currents tailward of the region-2 currents. (Yellow/blue colored contours correspond to currents flowing into/out of the ionosphere.) RCM runs with persistent region-1 currents

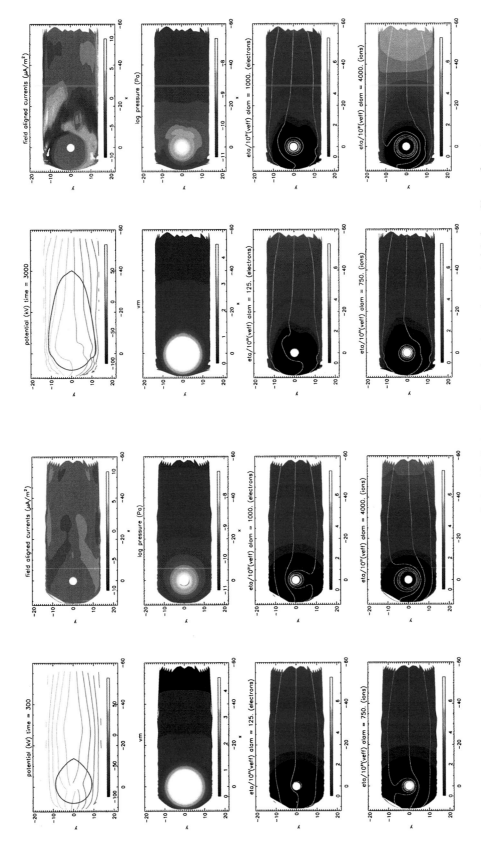

Plate 1. RCM-computed quantities as calculated in the ionosphere, projected onto the equatorial plane. Shown from top left to right, then down are: potential and contour of constant ionospheric latitude (blue curve); field aligned currents into the ionosphere; $V^{-2/3}$; log pressure; and flux tube content (η_s) for two electron and two ion channels. The lines in the (η_s) plots are the flow streamlines for that species. Plate 1a is at t=300 seconds and Plate 1b at t=3000 sec.

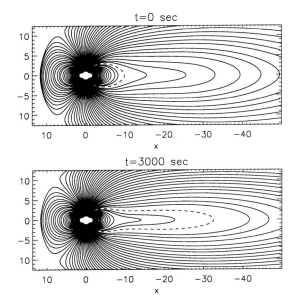

Figure 3. Computed field line plots, in the noon-midnight plane. The dashed line corresponds to field lines with ionospheric footprint at latitude 68°.

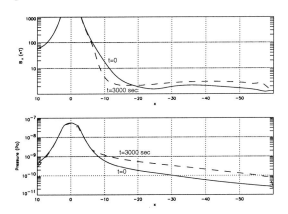

Figure 4. Pressure and z-component of magnetic field in the noon-midnight plane.

have previously been produced in only a few cases (*Karty et al.*, 1982; *Yang et al.*, 1994) with boundary conditions carefully designed to guarantee that pV^γ takes higher values on the flanks than near local midnight. The runs presented here with a self-consistent magnetic field suggest a different way in which region-1 currents can be generated in the plasma sheet. Heavily loaded flux tubes from the tail drift sunward on the flanks, dodging around the Earth. The plasma flow chokes near midnight where an eastward induction electric field almost cancels the westward potential field. The fact that heavily loaded flux tubes make more rapid sunward progress on the flanks leads to a minimum in pV^γ as a function of y near local midnight and to currents of region-1 sense.

Figure 3 shows the change in the magnetic field during the same interval. Figure 3a shows field lines for the initial equilibrated Tsyganenko-type magnetic field model;

Figure 3b shows the result after 3000 seconds. The magnetic field stretching in the inner tail results from the system trying to maintain pressure balance, given the pV^γ distribution dictated by the RCM. Figure 4 shows plots of B_z and p along the x-axis (the Earth-Sun line). After 3000 seconds a weak minimum in B_z exists at around $x = -15$ R_E. (Note that the Tsyganenko 1989 model already had a weak minimum in B_z at around $x = -25$ R_E which subsequently moved closer to the Earth as the computation proceeded.)

The code has been run for longer intervals with the result that the tail field strengthens and the magnetic field in the inner plasma sheet gets more and more stressed until finally the numerical code breaks. This happens when the current sheet gets too thin for the MF-code grid, and the grid spacing in the RCM gets too wide. Nature presumably solves the problem by creating a substorm and thereby violates the adiabatic-drift laws. These results are consistent with the 2D calculations of Erickson [1992], Hau [1991] and Pritchett and Coroniti [1996], where the Earthward convection of plasma resulted in stretching of tail field lines in the inner region of the tail, while field lines that map to the distant plasma sheet become increasingly dipolar.

3. SUMMARY AND CONCLUSION

We have presented results of a simulation that couples the Rice Convection Model to a Magneto-Friction equilibrium solver, which adjusts the magnetic field configuration for consistency with pressures computed by the RCM. To simulate a substorm growth phase, we enforced strong convection on the system but did not allow violation of the adiabatic drift laws. Some results are:

1. The magnetic field stretches in the inner plasma sheet, as the system tries to maintain pressure balance in the face of strong convection, which tries to push high pV^γ flux tubes into the inner plasma sheet.
2. Region-1 currents flow between the central plasma sheet and ionosphere as a result of choking near local midnight.
3. The numerical code eventually breaks as the inner plasma sheet becomes increasingly stressed. Nature's reaction to the same stress is apparently the substorm expansion phase.

Acknowledgments. This work was supported by NSF GEM grants ATM-9625182 and ATM-99000983, Los Alamos Computer Science Institute contract 03891-99-23 and NASA SECTP Grant NAG5-8136. The authors thank Tom Hill and both referees for numerous constructive comments.

REFERENCES

Baumjohann, W., G. Paschmann, and C.A. Cattell, Average plasma properties in the central plasma sheet, *J. Geophys. Res.*, *94*, 6597-6606, 1989.

Birn, J., Magnetotail equilibrium theory: The general three-dimensional solution, *J. Geophys. Res.*, 92, 11101-11108, 1987.

Birn, J., K. Schindler, and M. Hesse, Magnetotail dynamics: MHD simulations of driven and spontaneous dynamic changes, in Substorms 2: Proceedings of the Second International Conference on Substorms, Fairbanks, Alaska,

March 7-11, 1994, ed. J. R. Kan, J. D. Craven and S.-I. Akasofu, University of Alaska, Fairbanks, Fairbanks, Alaska, 135-141, 1994.

Book, D.L., J. P. Boris, and K. Hain, Flux-corrected transport, II, Generalization of the Method, *J. Comput. Phys.*, 18, 248, 1975.

Chen, C.-K., R. A. Wolf, M. Harel, and J. L. Karty, Theoretical magnetograms based on quantitative simulation of a magnetospheric substorm, *J. Geophys. Res.*, 87, 6137, 1982.

Cheng, C. Z., Three-dimensional magnetospheric equilibrium with isotropic pressure, *Geophys. Res. Lett.*, 22, 2401-2404, 1995.

Chodura, R, and A Schlüter, A 3D code for MHD equilibrium and stability, *J. Comput. Phys.* 41, 68, 1981.

Erickson, G. M. and R. A. Wolf, Is steady convection possible in the Earth's magnetotail?, *Geophys. Res. Lett.*, 7, 897, 1980.

Erickson, G. M., Modeling of Plasma Sheet Convection: Implications for Substorms, Ph.D., thesis, Rice University, Houston, Texas, 1985.

Erickson, G. M., A quasi-static magnetospheric convection model in two-dimensions, *J. Geophys. Res.*, 97, 6505-6522, 1992.

Erickson, G. M., R. W. Spiro and R. A. Wolf, The physics of the Harang discontinuity, *J. Geophys. Res.*, 96, 1633-1645, 1991.

Fedder, J. A., S. P. Slinker, J. G. Lyon, and R. D. Elphinstone, Global numerical simulation of the growth phase and the expansion onset for a substorm observed by Viking, *J. Geophys. Res.*, 100, 19083-19093, 1995.

Fok, M. -C., T. E. Moore, D. C. Delcourt, Modeling of plasma sheet and ring current in substorms, *J. Geophys. Res.*, 104, 14557-14569, 1999.

Gombosi, T.I., D. L. DeZeeuw, C. P. T. Groth, K. G. Powell, and P. Song, The length of the magnetotail for northward IMF: Results of 3D MHD simulations, in *Physics of Space Plasmas*, edited by T. Chang, pp. 121-128, MIT Center for Theoretical Geo/Cosmo Plasma Physics, 1998.

Harel, M., R. A. Wolf, P. H. Reiff, R. W. Spiro, W. J. Burke, F. J. Rich and M. Smiddy, Quantitative simulation of a magnetospheric substorm 1, model logic and overview, *J. Geophys. Res.*, 86, 2217-2241, 1981a.

Harel, M., R. A. Wolf, R. W. Spiro, P. H. Reiff, C.-K. Chen, W. J. Burke, F. J. Rich and M. Smiddy, Quantitative simulation of a magnetospheric substorm 2, comparison with observations, *J. Geophys. Res.*, 86, 2242-2260, 1981b.

Hau, L, -N, Effects of steady state adiabatic convection in the configuration of the near-Earth plasma sheet, 2, *J. Geophys. Res.*, 96, 5591, 1991.

Hesse, M. and J. Birn, Three-dimensional MHD modeling of magnetotail dynamics for different polytropic indices, *J. Geophys. Res.*, 97, 3965, 1992.

Hesse, M. and J. Birn, Three-dimensional magnetotail equilibria by numerical relaxation techniques, *J. Geophys. Res.*, 98, 3973, 1993.

Hesse, M., Numerical Modeling of Magnetotail Equilibria, presented at the IUGG XXI General Assembly, Boulder Co., July, 1995.

Karty, J. L., C.-K. Chen, R. A. Wolf, M. Harel and R. W. Spiro, Modeling of high-latitude currents in a substorm, *J. Geophys. Res.*, 87, 777, 1982.

Kivelson, M.G., and H.E. Spence, On the possibility of quasi-static convection in the quiet magnetotail, *Geophys. Res. Lett.*, 15, 1541-1544, 1988.

Lui, A.T.Y., H.E. Spence, and D. P. Stern, Empirical modeling of the quiet time nightside magnetosphere, *J. Geophys. Res.*, 99, 151, 1994.

Peymirat, C., and A. D. Richmond, Modeling the ion loss effect on the generation of region 2 field-aligned currents via

equivalent magnetospheric conductances, *J. Geophys. Res.*, 98, 15467, 1993.

Pritchett, P. L., and F. V. Coroniti, Formation of thin current sheets during convection, *J. Geophys. Res.*, 100, 23551, 1996.

Raeder, J., R. J. Walker, and M. Ashour-Abdalla, The structure of the distant geomagnetic tail during long periods of northward IMF, *Geophys. Res. Lett.*, 22, 349-352, 1995.

Schindler, K. and J. Birn, Self-consistent theory of time-dependent convection in the Earth's magnetotail, *J. Geophys. Res.*, 87, 2263, 1982.

Senior, C., and M. Blanc, On the control of magnetospheric convection by the spatial distribution of ionospheric conductivities, *J. Geophys. Res.*, 89, 261, 1984.

Spiro, R. W., M. Harel, R. A. Wolf and P. H. Reiff, Quantitative simulation of a magnetospheric substorm, 3, plasmaspheric electric fields and evolution of the plasmapause, *J. Geophys. Res.*, 86, 2261-2272, 1981.

Spiro, R. W., and R. A. Wolf, Electrodynamics of convection in the inner magnetosphere, in ed., *Magnetospheric Currents*, edited by T. A. Potemra, AGU, Washington, DC, 247, 1984.

Toffoletto, F.R., and T.W. Hill, Mapping of the solar wind electric field to the Earth's polar caps, *J. Geophys. Res.*, 94, 329-347, 1989.

Toffoletto, F.R., R.W. Spiro, R.A. Wolf, M. Hesse, and J. Birn, Self-consistent modeling of inner magnetospheric convection, in *Third International Conference on Substorms (ICS-3)*, edited by E.J. Rolfe, and B. Kaldeich, pp. 223-230, ESA Publications Division, Noordwijk, The Netherlands, 1996.

Tsyganenko, N. A., A magnetospheric magnetic field model with a warped tail current sheet, *Planet. Space Sci.*, 37, 5, 1989.

Vasyliunas, V. M., Mathematical Models of Magnetospheric Convection and its coupling to the ionosphere, in *Particles and Fields in the Magnetosphere*, edited by B. M. McCormac, p. 60, D. Reidel, Hingham, Mass., 1970.

Voigt, G.-H., Magnetospheric equilibrium configuration and slow adiabatic convection, in *Solar Wind-Magnetosphere Coupling*, edited by Y. Kamide and J. A. Slavin, pp. 233-273, Terra Sci. Publ. Co., Tokyo, 1986.

Wolf, R.A., M. Harel, R.W. Spiro, G.-H. Voigt, P.H. Reiff, and C.K. Chen, Computer simulation of inner magnetospheric dynamics for the magnetic storm of July 29, 1977, *J. Geophys. Res.*, 87, 5949-5962, 1982.

Wolf, R.A., R.W. Spiro, and F.J. Rich, Extension of the Rice Convection Model into the high-latitude ionosphere, *J. Atm. Terrest. Phys.*, 53, 817-829, 1991.

Yang, Y. S., R. W. Spiro and R. A. Wolf, Generation of region-1 current by magnetospheric pressure gradients, *J. Geophys. Res.*, 99, 223-234, 1994.

F. R. Toffoletto, Physics and Astronomy Dept., Rice University MS108, P.O. Box 1892, Houston, TX 77251-1892. (e-mail: toffo@rice.edu)

R. W. Spiro, Physics and Astronomy Dept., Rice University MS108, P.O. Box 1892, Houston, TX 77251-1892. (e-mail: spiro@rice.edu)

R. A. Wolf, Physics and Astronomy Dept., Rice University MS108, P.O. Box 1892, Houston, TX 77251-1892. (e-mail: wolf@alfven.rice.edu)

J. Birn, MS D466, Group NIS-1, Los Alamos National Lab, Los Alamos, NM, 87545. (email: jbirn@lanl.gov)

M. Hesse, Code 696, NASA-Goddard Space Flight Center, Greenbelt, MD 20770. (email: hesse@gsfc.nasa.gov)

Empirical Magnetic Field Models for the Space Weather Program

N. A. Tsyganenko [1]

NASA GSFC, Greenbelt, Maryland

A brief review is presented of the recent progress and the current state of the data-based modeling of the magnetospheric magnetic field. Combining the wealth of the observational data with flexible and realistic models advances our understanding of the dynamics of Earth's magnetic environment. The empirical approach to the modeling not only makes it possible to quantitatively represent the variable magnetosphere, but helps derive from data valuable information on its response to variations in interplanetary conditions. The cornerstones of the empirical modeling are (1) large sets of magnetic field and plasma data, (2) mathematical methods, allowing one to flexibly represent the B-field on a global scale, and (3) parameterization of magnetospheric field sources by the solar wind state variables and ground activity indices. This paper overviews most recent accomplishments made along these lines and discusses ongoing efforts to improve the data-based models; in particular, replicating the highly variable configuration of the magnetotail, introducing more flexible and realistic ring current model, and taking into account the variable shape and size of the magnetopause. Of special importance is the need to develop an improved representation of the high-latitude magnetosphere, including the dynamical large-scale Birkeland currents. Most of these problems can now be tackled, owing to recently devised powerful mathematical techniques, and a large amount of data obtained during several decades of spaceflight.

1. INTRODUCTION

The modeling of the global geomagnetic field has a unique place in space-weather studies. The magnetic field underlies all processes in the near-Earth environment: it links the interplanetary medium with the ionosphere, guides energetic charged particles, channels low-frequency electromagnetic waves, confines the radiation belts, directs electric currents, and stores huge amounts of energy, intermittently dissipated in the course of magnetospheric disturbances. If we know the field configuration, we can compare observations in different regions by mapping them along the magnetic field lines. Empirical models are needed to extract full information on the magnetospheric structure and its response to the solar wind conditions from observations. They serve as our main guide to magnetospheric structure, and their role has been justly compared to that of maps in the exploration of a new country.

Creating an empirical model involves three essentially different tasks. First, one needs to compile large sets of space magnetometer data and tag them by the concurrent information on the solar wind state and by suitable indices of the ground geomagnetic activity. Second, one has to develop flexible and physically sensible mathematical representation of the B-field on a global scale, in which contributions from individual field sources are given by separate terms. Third, a meaningful choice should be made of the input parameters, with which the amplitude and geometry of the principal

---

[1] Also at Raytheon Information, Technology, and Scientific Services (ITSS) Corporation, Lanham, Maryland.

Space Weather
Geophysical Monograph 125
Copyright 2001 by the American Geophysical Union

magnetospheric current systems are to be related. Substantial progress has taken place recently along the above lines [*Tsyganenko*, 1995, 1998b, 2000b; *Ostapenko and Maltsev*, 1997]. Instead of a crude binning into several intervals of the ground activity levels, adopted in early work [*Mead and Fairfield*, 1975; *Tsyganenko and Usmanov*, 1982; *Tsyganenko*, 1987, 1989a] the latest model [*Tsyganenko*, 1996; referred below T96] features: (1) a continuous dependence on the solar wind pressure, Dst-index, and interplanetary magnetic field (IMF) components, (2) an explicitly defined magnetopause with realistic shape and size and controlled by the solar wind, (3) interconnection with the IMF, modeling the open magnetosphere, (4) Birkeland currents, parameterized by the solar wind pressure and IMF.

Significant further advance is expected with abundant data from new missions and recently developed new methods, with which we intend to model the observed polar cusp structure, the variable shape of the magnetopause, and a realistic ring current and magnetotail. After that, we still need better modeling of the high-latitude magnetosphere, including the dynamical large-scale Birkeland currents. These and other problems are briefly overviewed in this paper.

2. THREE "PILLARS" OF THE DATA-BASED MAGNETOSPHERE MODELING

Unlike the main geomagnetic field on the Earth's surface, the distant magnetic field varies constantly due to changing conditions in the solar wind and internal magnetospheric instabilities. Quantitative models should be able to replicate essential features of the response of the magnetospheric configuration to the variable external input. The magnetospheric response to the solar wind and IMF conditions is complicated by "memory" effects, so each measurement inside the magnetosphere should be provided with information not only on the current state of the interplanetary medium, but also with data of the preceding time interval, of an hour or more. Another input is ground-based information, in particular the geomagnetic indices. While the Kp- and AE-indices seem less suited for calibrating models, the Dst index, in spite of its drawbacks [*Campbell*, 1996], can serve as a good measure of the overall strength of the near-Earth electric currents, and its inclusion in parametric relations improves the agreement between observed and predicted fields [*Tsyganenko*, 1996, 2000a].

Magnetospheric and solar wind observations, complemented by concurrent ground-based data, are the first "pillar" of the data-based modeling. Over the past three decades, a vast amount of such data was collected by many spacecraft at different locations, seasons, solar cycle stages, and disturbance levels. *Mead and Fairfield* [1975] compiled the first set of distant magnetic field data, taken by four IMP spacecraft during 1966-1972, and used it to create an empirical model,

binned by Kp-index. *Tsyganenko and Usmanov* [1982] added HEOS-1 and -2 data to the set of Mead and Fairfield and developed a more realistic model with an explicitly defined ring current and a tail current sheet. The dataset was further extended by Tsyganenko and Malkov [see *Peredo et al.*, 1993] who added ISEE 1/2 data from 1977-81, while Fairfield independently added HEOS observations and additional IMP-6 data to the original Mead-Fairfield data base. Editing of those data and merging them into one large database resulted in a set described by *Fairfield et al.* [1994]. It was used in the derivation of the T96 global field model, which not only represented average static configurations, but also revealed the response of individual field sources to changes in the external conditions.

As large as that dataset was, it became clear that much more data were needed, covering the full range of all variables – not only the {X,Y,Z} coordinates, but also the added dimensions of the geodipole tilt angle, the geomagnetic activity level (e.g., Dst-index), solar wind pressure, and IMF components. In this respect, the existing dataset still had many gaps. In particular, the great majority of the data came from quiet and moderately disturbed periods, while unusual conditions in the solar wind (most important for the space weather) were significantly underrepresented. The new observations during the last decade filled numerous gaps in the coverage. In particular, Geotail spacecraft provided excellent mapping of the tail plasma sheet, especially in its near-Earth part, where most interesting space weather phenomena take place. POLAR spacecraft, owing to its highly inclined orbit, greatly improved the sampling at high latitudes, including the polar cusps and the very important region of Birkeland currents. These new data made it possible to quantitatively study the variable structure of the tail current sheet and polar cusps, and stimulated the development of new modeling methods for representing all those features [e.g., *Tsyganenko et al.*, 1998; *Tsyganenko*, 1998, 2000b,c; *Tsyganenko and Russell*, 1999].

The second pillar of a magnetospheric model is its mathematical "frame." Close to Earth ($R \leq 6R_E$), the internal field dominates and its mathematical representation by spherical harmonics, established long ago by Gauss, is simple and straightforward. In contrast, the external part of the magnetic field is highly variable, non-potential, and has a complex structure, dictated by the magnetospheric plasma. Its relative contribution rapidly increases beyond $R \sim 6R_E$ and is dominant at $R \geq 8 - 10R_E$. Accurate, physically sensible, mathematically flexible, and reasonably simple representation of that part of the field is the major challenge.

The recent progress in that area is covered in part in earlier reviews [*Tsyganenko*, 1998b, 2000b]. In general, the approach is to represent the external field by a sum of modules, each representing an individual source, with its own geometry and its own response to external factors and the Earth's dipole tilt. For example, the ring current residing in the inner magnetosphere varies relatively slowly and follows the orientation of the geodipole axis, while the magnetopause and the associated Chapman-Ferraro currents respond more rapidly to variations in the interplanetary medium, and their position is controlled mainly by the solar wind flow.

Modeling the magnetopause and its contribution to the total field has a special place in data-based modeling, since the confining effect of the boundary currents should take into account every source of the magnetospheric field and should also allow a controlled interconnection of the geomagnetic and interplanetary fields. The T96 model uses an analytical magnetopause, based on that of *Sibeck et al.* [1991], obtained by fitting an ellipsoid of revolution to a set of direct crossings. The distant tailward magnetopause in that model was represented by a cylinder, smoothly joining the ellipsoidal surface in the distant tail. *Shue et al.* [1997, 1998] suggested another approximation for the boundary and fitted it to a different set of crossings. Using the hemi-ellipsoidal magnetopause, however, has an important advantage: it allows a simple class of deformations, making it possible to easily reproduce a re-configuration of the cross-tail electric current without violating the shielding of the total field. It also allows one to replicate the realistic shape of the model magnetopause near the polar cusps as a simple modification of one of the ellipsoidal coordinates [*Tsyganenko, 2000b*].

The magnetopause magnetic field can be obtained using a straightforward numerical procedure, which minimizes the rms magnetic flux across the boundary by varying parameters of analytical potential fields, representing the effect of the Chapman-Ferraro currents inside the magnetosphere. That method was first suggested by *Schulz and McNab* [1987] and proved simple and fast tool for shielding all principal sources of the magnetospheric field [*Tsyganenko, 1995, 1996; Tsyganenko and Stern, 1996*].

The cross-tail current and Region 1 Birkeland current have a common feature that makes their global representation somewhat tricky. Namely, both currents intersect the magnetopause and close either on the boundary or in the magnetosheath. The actual path of their closure current cannot be determined on the basis of the internal magnetospheric measurements; moreover, even if it were known, no simple solution for the magnetic field could have been found: the only choice would be to resort to a numerical integration. Fortunately, there is no need to know the actual closure currents: given the boundary condition upon the total normal component B_n, it only suffices to correctly specify the current density **j** *inside* the magnetosphere in order to obtain a unique solution. The closure currents outside the magnetopause can, in fact, be defined in any arbitrary way: for example, we can extend the cross-tail current to infinity in the $\pm y$-direction, as was done in the T87 model [*Tsyganenko, 1987*], or close the current flow lines as circles (T96). This allows one to use

simple analytical current sheets or disks [*Tsyganenko and Peredo, 1994*] and hence dramatically simplifies the model.

The third pillar, connecting the data with the mathematical frame of a model, requires algorithms that relate the strength and spatial configuration of individual field sources with the input parameters, including the state of the solar wind and IMF. It is relatively easy to define that relationship for the magnetopause source: the dominant parameter in that case is the solar wind pressure and, in a crude approximation, the magnetopause shrinks and expands self-similarly, with the scaling factor $[p/\langle p \rangle]^\beta$, where $\beta \sim 1/6$. In fact, the exact self-similarity does not hold, since the magnetopause shape also changes in response to the varying IMF; however, that effect is more difficult to replicate in the models, than the self-similar compression/expansion used in the T96 model.

Much less is known on the actual response of other current systems, especially of the Birkeland current systems and of the magnetotail. *Iijima and Potemra* [1982] correlated the densities of Region 1 Birkeland currents with a variety of interplanetary quantities and found that $\gamma = p^{1/2} B_t \sin(\theta/2)$ provided the highest correlation. The parameter γ was adopted in the T96 model for calibrating the overall strength of the Region 1 current, on a tacit assumption that the total current is roughly proportional to its density, as estimated by Iijima and Potemra. More specifically, the total current was assumed as a linear function of γ, whose slope and intercept, among other model parameters, were fitted by least squares to data.

The same parameter γ was used in a linear form, relating the strength of the cross-tail current and the solar wind state. That expression also included a term containing $\sqrt{p}$, since the ram pressure of the solar wind strongly affects the tail lobe field [e.g., *Fairfield and Jones, 1996*] and, hence, the cross-tail current.

The strength of the ring current in the T96 model was parameterized by a linear function of the Dst-index and $\sqrt{p}$. However, possible change of the ring current radius with growing Dst was neglected. As a result, the model overestimates the magnetic moment of the ring current (and hence its overall contribution to the total field) during major storms.

In spite of the limitations of its approach, large gaps and relatively low resolution of the data, the T96 model not only yields "visually reasonable" magnetospheric configurations, but also in many cases accurately predicts spatial/temporal variations of the field, obtained in independent observations. Figure 1 shows three magnetic field components, observed by POLAR on August 1, 1998. Not only the general features, but also fine details (e.g., between 9 and 19 UT) are reproduced fairly well. The agreement worsens towards the end of the day, after a disturbed period between 13 and 21 UT, and the model also misses a crossing of the field-aligned current between 4 and 5 UT. During that period the IMF was mostly northward and, hence, $\gamma \approx 0$. Due to the simple lin-

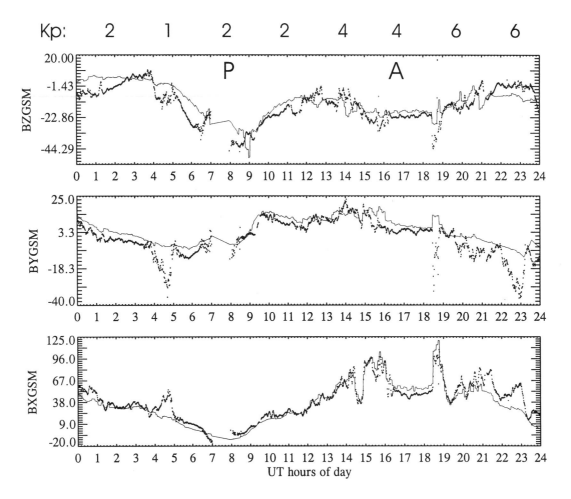

Figure 1. Comparison of the components of external field (geodipole contribution subtracted), observed by POLAR on 08/01/1998 (thick trace), with those predicted by the T96 model (thin trace). Kp-index values are shown above the plot. Letters P and A mark the times of Polar perigee and apogee, respectively.

ear parameterization of the Birkeland currents by γ, adopted in T96, their magnitude was grossly underestimated in the model. The generally insufficient magnitude of the model field-aligned currents, found in many instances, may be due to the virtual absence of the high-latitude observations in the modeling data set.

Figure 2 shows a similar plot, corresponding to an extremely disturbed day (August 27, 1998), with the Kp index ranging between 8 and 6. While the agreement in B_x component is quite good, in particular between 6 and 18 UT, the model grossly overestimates the depression in B_z inside the ring current. In this case, again, the discrepancy was caused by lack of data in the inner magnetosphere and an inflexible model ring current.

3. NEW FRONTIERS

As demonstrated above, efforts to further improve data-based magnetosphere models should address three principal

areas: (i) rebuilding and extending the database, (ii) advanced mathematical representation of external field sources, and (iii) establishing optimal algorithms or relationships, quantifying the response of the magnetospheric currents to the solar wind input and/or the ground indices. In this section we concentrate on the last two aspects: how to improve a model's flexibility and how to best parameterize a model.

Magnetospheric currents have a complex and dynamic geometry. The magnetopause shrinks, expands, and erodes, with varying degrees of connection to the IMF. The tail current sheet warps, bends, and twists, in response to the geodipole wobbling and to the IMF. The tail current density undergoes dramatic redistribution as the tail stretches and rebounds. In addition, the substorms are accompanied by local disruptions of the tail current and its transient diversion into field-aligned currents. The injection of freshly accelerated particles into the inner magnetosphere results in formation of the storm-time ring current, while a significant part of the tail magnetic flux and plasma sheet particles are ejected tailward

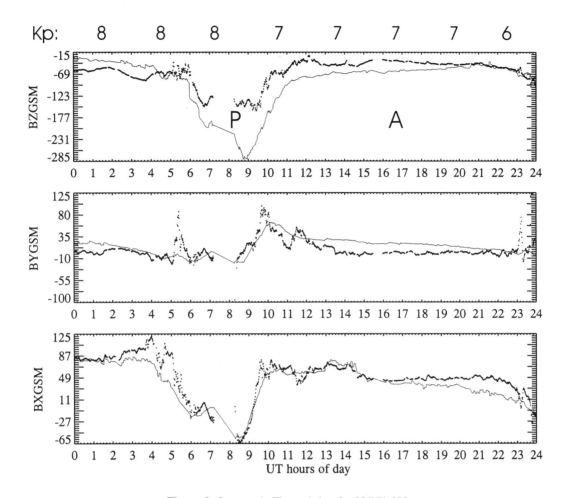

Figure 2. Same as in Figure 1, but for 08/27/1998.

as plasmoids. Large-scale Birkeland currents vary with the IMF and with the substorm cycle – and these are only average changes of the structure. To understand the physics (as well as to achieve concise models), the mathematical representations must resolve each of these variations in a meaningful way.

As recently demonstrated [*Tsyganenko, 1998a*], much flexibility can be added to the existing models by applying a divergence-conserving deformation of the magnetic field, whose basics were outlined by *Stern* [1987]. Using that approach, an economical method to warp and twist the tail current sheet was devised, consistent with the observed shapes of the plasma sheet and with the magnetopause response to the Earth's dipole tilt [*Tsyganenko et al.,* 1998]. Another simple deformation was suggested for the modeling of the structure of polar cusps, observed by POLAR [*Tsyganenko and Russell,* 1999].

In this paper, two other interesting uses of the deformation technique will be briefly described. The first one makes it possible to devise models with a variable profile of the cross-tail current density, in which the position of the inner edge

of the current sheet and its tailward extent are controlled by a single parameter κ, and there is no need to recalculate the shielding currents. The essence of the method is to transform the shielded tail magnetic field vector $\mathbf{B}_T$ to the coordinates σ and τ, in which the T96 model magnetopause is specified as a surface $\sigma = \sigma_0$. In the front part of the magnetosphere ($\tau > 0$), σ and τ are the ellipsoidal coordinates [see *Tsyganenko,* 1989b, 1995, 2000b, for details], while in the rear part ($\tau < 0$) they convert to the cylindrical ones. Then we apply a simple stretch transformation $\tau \rightarrow \tau' = \tau'(\tau)$ and, since $\nabla\sigma \perp \nabla\tau$ and the coordinate σ remains intact, the deformation does not violate the shielding, so that the normal component of the deformed field remains zero. Figure 3 shows the lines of a shielded tail field in the noon-midnight meridian plane for two values of the stretch coefficient κ. The deformation conserves $\nabla \cdot \mathbf{B} = 0$, although it does not conserve $\nabla \times \mathbf{B}$, that is, gives rise to unphysical currents. In our case, however, owing to the smoothness of the transformation, the artificial current density remains relatively small even for large values of the stretch parameter. The above deformation can serve as a promising new tool for improving the accuracy of the

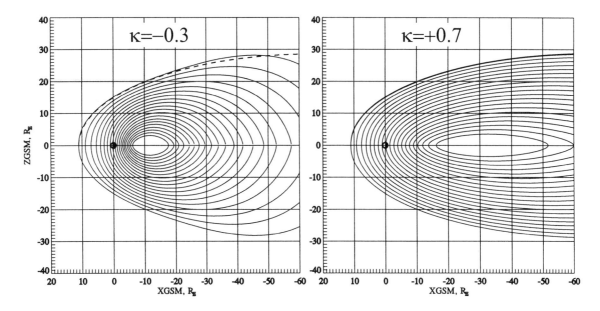

Figure 3. Illustrating the shift of the cross-tail current along the X-axis by using a stretch in ellipsoidal coordinates. The magnetic field lines of the shielded cross-tail current are shown in the noon-midnight meridian plane, and the electric current is directed out of the page.

models in a most important region on the nightside, at the interface of the quasi-dipolar and tail-like field lines. Note that in the T96 model the tail field was represented by a linear combination of only two terms, each with a fixed spatial distribution of the electric current.

In the second example the deformation method is applied to the field of a realistic ring current [*Tsyganenko*, 2000c]. In that work, a quantitative model was developed of the inner magnetospheric magnetic field, combining the effects of the azimuthally asymmetric ring current with those of field-aligned currents, caused by azimuthal variation of the plasma pressure. The axisymmetric part of the model ring current was derived from average profiles of the particle pressure and anisotropy, observed by AMPTE/CCE spacecraft, and was analytically represented by a vector potential, fitted by least squares to one derived by the Biot-Savart integral. The goal of the work was a realistic and computationally efficient global description of the ring current field, so it was important to obtain mathematically simple approximations. A solution for the axially symmetric part of the ring current was found as the field of a spread-out double current loop [*Tsyganenko*, 1998b], deformed in the space $\{\alpha, \gamma\}$ of orthogonal dipolar coordinates, where $\alpha = (x^2 + y^2)/r^3$ and $\gamma = z/r^3$. The difference between the model approximations and the original Biot-Savart field did not exceed a fraction of percent, with fairly simple analytical deformation functions $\alpha' = \alpha'(\alpha, \gamma)$ and $\gamma' = \gamma'(\alpha, \gamma)$. Further details are beyond the scope of this paper, but Figure 4 shows the lines of equal azimuthal current density in the model, in which a small region of the inner eastward current flows inside a much larger and radi-

ally extended westward current, visibly concentrated near the equatorial plane because of the particle anisotropy.

As said above, the main task in parameterizing a model is to find a functional relation or an algorithm, relating the strength of the magnetospheric field sources to the state of the interplanetary medium and/or to the routinely monitored ground-based (or magnetospheric) parameters, including their previous history. The parameterization assumed in the T96 model was by no means optimal, since the adopted relations were based mainly on simple a priori considerations, rather than on thorough correlation studies. That gap was partially bridged in our recent work [*Tsyganenko*, 2000a], in which a variety of functions was studied, describing the tail current response to

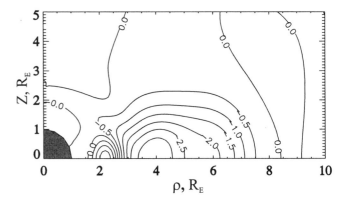

Figure 4. Distribution of the volume electric current density in the ring current model, based on the observed profiles of particle pressure and anisotropy [*Tsyganenko*, 2000c].

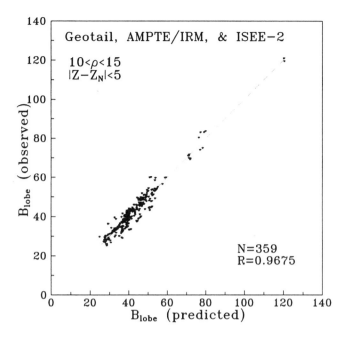

Figure 5. Scatter plot of the observed tail lobe field against that predicted by a linear regression relation, for $10 \leq \rho \leq 15$ [*Tsyganenko, 2000a*].

the solar wind state, and a linear filter technique was used to study the role of previous conditions on the current tail field strength. The study covered a wide range of tailward distance between 10 and 60 R_E and was based on a large set of tail lobe magnetic field data of Geotail (1993–97), AMPTE/IRM (1985–86), and ISEE-2 (1978–80). The tailward variation of the lobe field and its response to the solar wind and IMF conditions were studied using a regression relationship, including various combinations of the interplanetary quantities, measured by WIND and IMP 8. The regression relation included three terms, representing contributions from the Dst-index, solar wind ram pressure P_d, and IMF, using not only the concurrent values of the last two parameters, but also 12 previous values of their 5-min averages.

A detailed search was made for the best combination of the solar wind pressure- and IMF-related parameters, providing the highest value of the multiple correlation coefficient R between the predicted and observed lobe field. It was found that the near-tail field becomes significantly less sensitive to the solar wind pressure for large values of P_d, so that it is better described by a logarithmic law, rather than by a power law. Among various IMF-related regression functions, g, the best results for the near tail were obtained with $g = V h(B_\perp) \sin^3(\theta/2)$, where the function h behaves as $B_\perp^2 = B_y^2 + B_z^2$ for commonly observed values of the IMF, but gradually transforms into a linear dependence for $B_\perp \geq 40 \, nT$. The *Dst* term was found to yield a significant contribution to the regression relation in the nearest bin

of the distance (10–15R_E), but its contribution rapidly fell off at larger distances. Overall, the optimal values of the regression coefficients yielded a very good fit in the near tail, especially in the nearest distance bin, where the correlation coefficient reached $R \approx 0.97$. Figure 5 shows the scatter plot of the observed against predicted values of the lobe field in the distance bin 10–15R_E.

With regard to the time lag effects, in the nearest interval of the distance the best fit linear filter function for the pressure-dependent term was found to rapidly increase with growing time lag, suggesting a significant average delay between the changes of the solar wind pressure and the reaction of the lobe field.

4. CONCLUDING COMMENTS

Significant progress has been made in the development of empirical magnetosphere field models. As shown here, even transient variations of the field components can be reproduced fairly accurately in many cases, even though the model is based on data from many periods in the past, with different interplanetary conditions. The recent accomplishments in the field of the modeling methods, as well as an ample inflow of the newly obtained data allow the development of new better models, which will advance the magnetic mapping of geospace. Future models will include a dynamical parameterization of the field sources, taking into account the effects of a time-delayed response. Finally, it is expected that the empirical modeling will play a crucial role in the future multi-spacecraft missions, making it possible to reconstruct instantaneous magnetospheric configurations from simultaneous field measurements at many different locations [*Tsyganenko, 1998c*].

Acknowledgments. I gratefully acknowledge David Stern for his many thoughtful comments on the manuscript. This work is supported by NASA grants NAS5-32350 and NAS5-32993, and NSF Magnetospheric Physics Program grant ATM-9819873.

REFERENCES

Campbell, W. H., Geomagnetic storms, the Dst ring-current myth and lognormal distribution, *J. Atmos. Terr. Phys., 58,* 1171, 1996.

Fairfield, D. H., N. A. Tsyganenko, A. V. Usmanov, M. V. Malkov, A large magnetosphere magnetic field database, *J. Geophys. Res., 99,* 11,319, 1994.

Fairfield, D. H., and J. Jones, Variability of the tail lobe field strength, *J. Geophys. Res., 101,* 7785, 1996.

Iijima, T., and T. A. Potemra, The relationship between interplanetary quantities and Birkeland current densities, *Geophys. Res. Lett., 9,* 442, 1982.

Mead, G.D., and D.H. Fairfield, A quantitative magnetospheric model derived from spacecraft magnetometer data, *J. Geophys. Res., 80,* 523, 1975.

Ostapenko, A. A., and Y. P. Maltsev, Relation of the magnetic field in the magnetosphere to the geomagnetic and solar wind activity, *J. Geophys. Res., 102,* 17,467, 1997.

Peredo, M., D. P. Stern, and N. A. Tsyganenko, Are existing magnetospheric models excessively stretched ?, *J. Geophys. Res., 980,* 15,343, 1993.

Schulz, M., and M. McNab, Source-surface model of the magnetosphere, *Geophys. Res. Lett., 14,* 182, 1987.

Shue, J.-H., J. K. Chao, H. C. Fu, C. T. Russell, P. Song, K. K. Khurana, and H. J. Singer, A new functional form to study the solar wind control of the magnetopause size and shape, *J. Geophys. Res., 102,* 9497, 1997.

Shue, J.-H., P. Song, C. T. Russell, J. T. Steinberg, J. K. Chao, G. Zastenker, O. L. Vaisberg, S. Kokubun, H. J. Singer, T. R. Detman, and H. Kawano, Magnetopause location under extreme solar wind conditions, *J. Geophys. Res., 103,* 17,691, 1998.

Sibeck, D. G., R. E. Lopez, and E. C. Roelof, Solar wind control of the magnetopause shape, location, and motion, *J. Geophys. Res., 96,* 5489, 1991.

Stern, D. P., Tail modeling in a stretched magnetosphere, 1, Methods and transformations, *J. Geophys. Res., 92,* 4437, 1987.

Tsyganenko, N. A., Global quantitative models of the geomagnetic field in the cislunar magnetosphere for different disturbance levels, *Planet. Space Sci., 35,* 1347, 1987.

Tsyganenko, N. A., A magnetospheric magnetic field model with a warped tail current sheet, *Planet. Space Sci., 37,* 5, 1989a.

Tsyganenko, N.A., A solution of the Chapman-Ferraro problem for an ellipsoidal magnetopause, *Planet.Space Sci., 37,* 1037, 1989b.

Tsyganenko, N. A., Modeling the Earth's magnetospheric magnetic field confined within a realistic magnetopause, *J. Geophys. Res., 100,* 5599, 1995.

Tsyganenko, N. A., Effects of the solar wind conditions on the global magnetospheric configuration as deduced from databased field models, *Eur. Space Agency Spec. Publ., ESA SP-389,* 181, 1996.

Tsyganenko, N. A., Modeling of twisted/warped magnetospheric configurations using the general deformation method, *J. Geophys. Res., 103,* 23551, 1998a.

Tsyganenko, N. A., Data-based models of the global geospace magnetic field: Challenges and prospects of the ISTP era, in *Geospace Mass and Energy Flow: Results From the International Solar-Terrestrial Physics Program, Geophys. Monogr. Ser., v. 104,* edited by J. L. Horwitz, D. L. Gallagher, and W. K. Peterson, p. 371, AGU, Washington, D. C., 1998b.

Tsyganenko, N. A., Toward real-time magnetospheric mapping based on multi-probe space magnetometer data, in *Science Closure and Enabling Technologies for Constellation Class Missions,* edited by V. Angelopoulos and P. V. Panetta, pp. 84-90, U. C. Berkeley, Berkeley, California, 1998c.

Tsyganenko, N. A., Solar wind control of the tail lobe magnetic field as deduced from Geotail, AMPTE/IRM, and ISEE 2 data, *J. Geophys. Res., 105,* 5517, 2000a.

Tsyganenko, N. A., Recent progress in the data-based modeling of magnetospheric currents, in: *Magnetospheric Current Systems, Geophys. Monogr. Ser., vol. 118,* edited by S.-I. Ohtani, R.-I. Fujii, R. Lysak, and M. Hesse, p.61–70, AGU, Washington, D. C., 2000b.

Tsyganenko, N. A., Modeling the inner magnetosphere: The asymmetric ring current and region 2 Birkeland currents revisited, *J. Geophys. Res., 105,* in press, 2000c.

Tsyganenko, N. A., and A. V. Usmanov, Determination of the magnetospheric current system parameters and development of experimental geomagnetic field models based on data from IMP and HEOS satellites, *Planet. Space Sci., 30,* 985, 1982.

Tsyganenko, N. A., and M. Peredo, Analytical models of the magnetic field of disk-shaped current sheets, *J. Geophys. Res., , 99,* 199, 1994.

Tsyganenko, N. A., and Stern, D. P., Modeling the global magnetic field of the large-scale Birkeland current systems, *J. Geophys. Res., 101,* 27,187, 1996.

Tsyganenko, N. A. and C. T. Russell, Magnetic signatures of the distant polar cusps: Observations by Polar and quantitative modeling, *J. Geophys. Res., 104,* 24,939, 1999.

Tsyganenko, N. A., S. B. P. Karlsson, S. Kokubun, T. Yamamoto, A. J. Lazarus, K. W. Ogilvie, C. T. Russell, Global configuration of the magnetotail current sheet as derived from Geotail, Wind, IMP 8 and ISEE 1/2 data, *J. Geophys. Res., 103,* 6827, 1998.

N. A. Tsyganenko, Raytheon ITSS/Code 690.2, NASA Goddard SFC, Greenbelt, MD 20771. (e-mail: ys2nt@lepvx3.gsfc.nasa.gov)

Dynamic Radiation Belt Modeling at the Air Force Research Laboratory

J. M. Albert

Institute for Scientific Research, Boston College, Boston, Massachusetts

D. H. Brautigam, R. V. Hilmer, and G. P. Ginet

Air Force Research Laboratory, Space Vehicles Directorate, Hanscom Air Force Base, Massachusetts

Despite its well-known limitations, diffusion remains an important concept for modeling radiation belt energization and transport. Recent work at AFRL has applied radial diffusion to observations of: (a) $1 - 10$ MeV equatorially mirroring protons at $1.2 < L < 3$, before and after the March 1991 magnetic storm; (b) ~ 1 MeV electrons at $3.5 < L < 6$ following the 9 October 1990 storm; (c) $1.6 - 3.2$ MeV electrons at $L = 4.2$ between December 1994 and September 1996, following enhancements of > 2 MeV electrons at geosynchronous orbit associated with high speed solar wind streams. In general, encouraging results were obtained, but only after implementing: (1) non-steady state initial conditions; (2) time-dependent boundary conditions; (3) realistic magnetic field models; (4) tweaked, scaled, fit, activity-dependent, or enhanced diffusion coefficients. This paper reviews these various modeling efforts, including the modifications that they required, their successes and failures, and lessons learned.

1. INTRODUCTION

In the textbook view [*Roederer*, 1970; *Lyons and Williams*, 1984; *Schulz*, 1991; *Kivelson and Russell*, 1995], the long-term structure of the radiation belts is determined by steady state solutions of the "radial diffusion equation," which also gives the expected rate of relaxation from nonsteady configurations. This equation describes the effect of small, frequent, uncorrelated electric and magnetic field fluctuations on the timescale of the drift period, which break a particle's third adiabatic invariant in a way describable by diffusion. The equation must be augmented by various sources and

sinks appropriate to the type of particle; in particular, electrons are often subject to breaking of the first and second invariant by whistler hiss, which may be modeled by pitch angle (and energy) diffusion.

More recently, an increased appreciation of the role of rare, drastic events has developed [*Li et al.*, 1993; *Hudson et al.*, 1995]. The convective and/or coherent character of these events is often incompatible with diffusion. Nevertheless, diffusion is suited to describing particle behavior during moderate activity, and in the aftermath of abrupt disturbances, especially with enhanced transport coefficients that reflect the disturbed magnetospheric conditions. Also, a statistical picture of large disturbances may lend itself to diffusion in the long run [*Lyons and Schulz*, 1989; *Riley and Wolf*, 1992]. Furthermore, diffusion provides an underlying and significant minimum level of transport that must be considered. This paper surveys several recent efforts at the

Space Weather
Geophysical Monograph 125

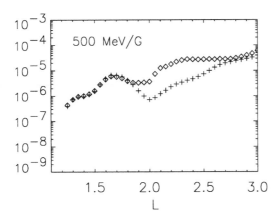 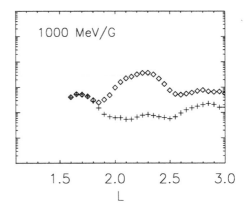

Figure 1. High energy proton phase space density, f (s^3/km^6) vs L, as observed by CRRES. Shown are time averages of data before (pluses) and after (diamonds) the March 24, 1991 magnetic storm.

Air Force Research Laboratory to model radiation belt dynamics using diffusion during different types of geomagnetic conditions.

2. CRRES/PROTEL OBSERVATIONS OF HIGH ENERGY PROTONS

The Proton Telescope (PROTEL) instrument on the CRRES satellite observed protons in the energy range of approximately 1–100 MeV. Phase space density $f(L)$ at constant first adiabatic invariant μ is shown in Figure 1, time averaged over two periods separated by the large magnetic storm of March 24, 1991 [*Albert et al.*, 1998]. Configurations both before and after the storm deviate substantially from the monotonically increasing profiles $f(L)$ expected from steady state radial diffusion. This is to be expected following the large, disruptive March 1991 storm, which was accompanied by particle injection deep into the magnetosphere. It is also likely that geomagnetic activity in the very active year 1989 is at least partially responsible for the prestorm phase space density deficit, which is of comparable relative magnitude.

Albert and Ginet [1998] observed that particle fluxes exhibited slow but systematic evolution during the long periods of relative quiet between the drastic disruptions. They investigated whether conventional diffusion theory was operative and could account for the observed rates of change. The "radial diffusion equation" for protons includes the effects of Coulomb collisions, charge exchange, and the CRAND source as well as radial diffusion:

$$\frac{\partial f}{\partial t} = L^2 \frac{\partial}{\partial L}\left(\frac{D_{LL}}{L^2}\frac{\partial f}{\partial L}\right) + \frac{G(L)}{\mu^{1/2}}\frac{\partial f}{\partial \mu} - \Lambda f + S, \quad (1)$$

where $D_{LL} = D_{LL}^{(m)} + D_{LL}^{(e)}$. The right-hand side of this equation was evaluated using the time-averaged values of f shown in Figure 1, and the left-hand side was evaluated using time series from the corresponding periods. The radial diffusion coefficients used were $D_{LL}^{(m)} = 7 \times 10^{-9}L^{10}$ with $D_{LL}^{(e)} = 2 \times 10^{-5}L^{10}\mu^{-2}$ for the prestorm period, and $D_{LL}^{(e)} = 5 \times 10^{-5}L^{10}\mu^{-2}$ for the poststorm period, where D_{LL} has units of day^{-1} and μ (given relativistically by $p^2/2m_0B$) is in MeV/G. These are very close to the standard values of *Schulz* [1991]. As suggested by Figure 1, before the March 1991 storm the greatest rates of change of f occur for $\mu = 500$; these observed and calculated values are compared in Figure 2. Figure 1 also shows that after the storm the $\mu = 500$ profile had become nearer to a relaxed configuration but the $\mu = 1000$ profile was now very disturbed, and should exhibit large rates of change. These observed and calculated values are also compared in Figure 2. In both cases, good agreement is found.

3. OUTER ZONE ELECTRONS DURING THE MODERATE STORM OF OCTOBER 9, 1990

Radial diffusion was used to model the behavior of outer zone electrons during the October 9, 1990 magnetic storm [*Brautigam and Albert*, 2000]. During this moderate storm, Dst reached a minimum of -133 nT. A careful attempt was made to account for adiabatic effects by binning the data according to the three dynamical adiabatic invariants [*Roederer*, 1970; *Schulz*, 1991], which were calculated using the Kp-dependent (hence time-dependent) Tsyganenko 1989 magnetic field model. The second adiabatic invariant was fixed at $J_2 = 1.78 \times 10^{-16}$ g(cm/s)R_E; the corresponding equatorial pitch angles range from 25° to 60°.

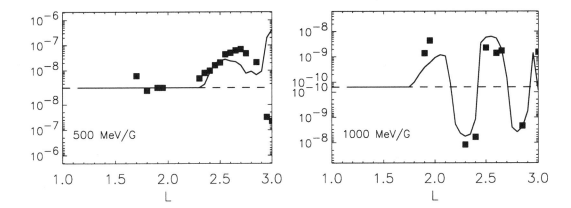

Figure 2. Rates of change, $\partial f/\partial t$, observed by CRRES (filled squares) and calculated from the radial diffusion equation (solid curve), in units $(s^3/km^6)/day$. Both positive and negative values are shown on log scales. The $\mu = 500$ values are for before the March 24, 1991 magnetic storm and the $\mu = 1000$ values are for after the storm.

The solid curves in Figure 3 shows the time history of electron phase space density f at $L = 4.5$ for fixed μ (200 and 1000 MeV/G), as observed by the CRRES/MEA detector. Here L is inversely proportional to the third adiabatic invariant, not the more customary McIlwain quantity. For both values of μ, f decreases during the main phase of the storm but recovers afterward, to a level higher than the initial one.

The radial diffusion simulation used time-dependent outer boundary conditions. To circumvent the problem of incomplete data coverage by CRRES during the storm, the CRRES prestorm boundary fluxes were used, scaled by stormtime LANL/SOPA geosynchronous data. The electric and magnetic radial diffusion coefficients were modeled in a Kp-dependent way, by fitting to previously reported values over the range Kp = 1–6 [*Lyons and Thorne*, 1973; *Lyons and Schulz*, 1989; *Lanzerotti and Morgan*, 1973; *Lanzerotti et al.*, 1978]. This re-

sulted in $D_{LL}^{(m)} = 10^{0.506Kp-9.325}L^{10}$, and the usual expression

$$D_{LL}^{(e)} = \frac{1}{4}\left(\frac{cE}{B_0}\right)^2 \frac{T}{1 + (\omega_D T/2)^2}L^6 \qquad (2)$$

with $E = 0.1 + 0.26(Kp - 1)$ mV/m, where B_0 is the geodipole field strength at the Earth's surface (in gauss), ω_D is the electron drift frequency, and the electric fluctuation autocorrelation time T is taken to be 45 minutes.

Figure 3 shows that the simulation results at $L = 4.5$ (dotted curves) decline at roughly the right time, though not as drastically as the observed values. Both simulated profiles also gradually recover to greater than initial levels, on roughly the right timescale. However, while the final $\mu = 200$ value matches the observed value well, the final calculated $\mu = 1000$ level is not nearly as enhanced as the observed one.

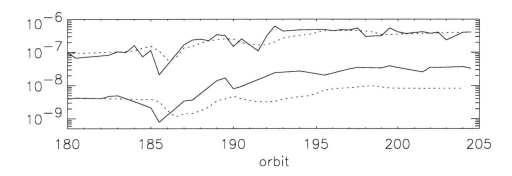

Figure 3. Phase space density (s^3/km^6) at L=4.5 as a function of time as seen by CRRES (solid curve) and simulated (dotted curves). The upper two curves are for $\mu = 200$ MeV/G, and the lower two curves show $\mu = 1000$ MeV/G.

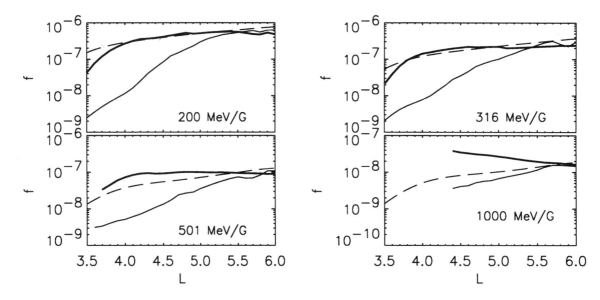

Figure 4. Phase space densities (s³/km⁶) at several values of μ as observed by CRRES at orbit 180 (thin solid curve) and orbit 204 (thick solid curve), and calculated (dashed curve).

Figure 4 shows the initial and final profiles $f(L)$ for different values of μ. It is seen that for $\mu = 200$, the simulated final profile (dashed line) agrees well with the final observed profile (thick solid curve), but that this agreement is only fair for $\mu = 501$. The simulation fails to capture the inward pointing gradient in $f(L)$ observed at $\mu = 1000$, which is suggestive of an inner source, such as local acceleration.

4. MEV ELECTRON ENHANCEMENTS FOLLOWING DROPOUTS CAUSED BY HIGH SPEED SOLAR WIND STREAMS

A study was made of the energetic electron response seen on the equator by GPS (nominally L=4.2) following high speed solar wind streams (HSSWS). Twenty-six of the HSSWS detected by the WIND spacecraft between December 1994 and September 1996 were selected for their large subsequent geosynchronous (nominally L=6.6) electron enhancements. In all 26 cases electron dropouts occured at both geosynchronous and GPS, and in all 26 cases the geosynchronous levels recovered promptly. In 15 of the cases, the flux levels at GPS also recovered from the dropout level, and in 10 of these the final level at GPS was greater than the original level.

Figure 5 shows phase space density (estimated from differential flux) of equatorially mirroring, geosynchronous electrons with $\mu \approx 2100$ MeV/G before, during, and after the passage of a HSSWS. The HSSWS is iden-

tified by the elevated solar wind velocity and pulse in solar wind density. The geomagnetic indices Kp and Dst are shown as well. The geosynchronous phase space density drops shortly after the initiation of the HSSWS but then recovers, to a level exceeding the original level. Also shown is the response of GPS electrons with the same μ, which also decline and then gradually recover to a higher-than-initial level.

Figure 6 shows a different HSSWS event, during which the geosynchronous response is similar. However, the electrons seen by GPS do not show recovery or enhancement following their dropout at around the time of the geosynchronous dropout.

It was found that whether or not the GPS fluxes recovered following a HSSWS-linked dropout is determined by the value of Kp prevalent during the HSSWS. This is quantified by determining a rise rate λ of the GPS levels and an average value of Kp during the rise [*Hilmer et al.*, 2000]. As shown in Figure 7, the GPS recovery rate λ is virtually zero when the averaged Kp is below a threshold of about 3.5, and increases with Kp beyond the threshold.

It is also seen from Figures 5 and 6 that the phase space density at GPS is always below the level at geosynchronous, and that the post-dropout increase at GPS, when it occurs, lags the geosynchronous increase. These observations suggest that the increase seen at GPS is due to the radial transport of particles from geosynchronous, either diffusively or otherwise. To test whether the rise at GPS is consistent with radial diffu-

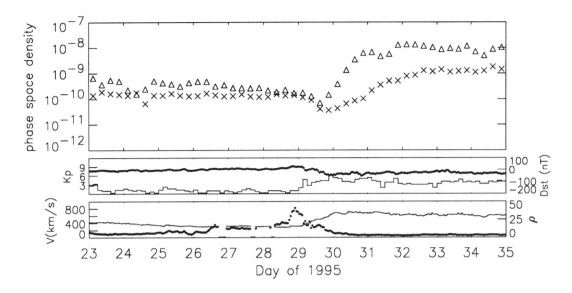

Figure 5. Top: Averaged electron phase space density (s^3/km^6) at geosynchronous (triangles) and GPS (Xs) during a HSSWS. Middle: Kp (solid curve) and Dst (dotted curve). Bottom: solar wind velocity (solid curve) and proton density (dotted curve).

sion, at least locally, a radial profile $\exp(\lambda t + L/L_s)$ was assumed in the vicinity of L=4.2, with $L_s = 0.2$ or 0.4 [*Selesnick et al.*, 1997]. The value of D_{LL} was taken to be Kp-dependent, with the form

$$D_{LL} = \frac{c^2 L^6}{4 R_E^2 B_0^2} P_0(Kp)\nu^{-1.6} \qquad (3)$$

where ν is the drift frequency (in $hour^{-1}$) and P_0 gives the power spectrum of electric field fluctuations as mea-

sured by *Mozer* [1971]. Radial diffusion then implies $\lambda = (1 + 4L_s/L)D_{LL}/L_s^2$. The calculated values of λ are shown in Figure 7, and agree at least qualitatively with the observed rates of change.

5. CONCLUSIONS

Radial diffusion formalism has been shown to be a useful description in a wide range of conditions in the inner magnetosphere. The radial diffusion coefficients

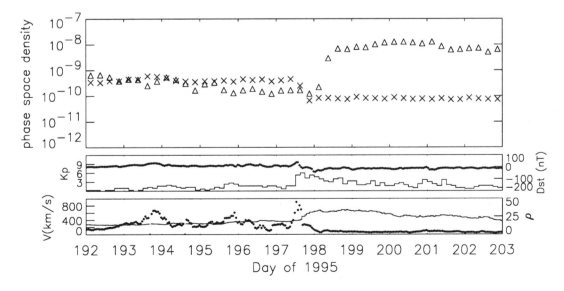

Figure 6. Same as figure 5, for a different HSSWS.

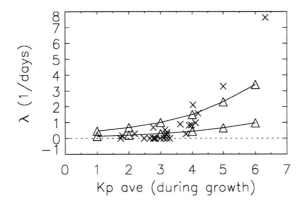

Figure 7. Observed (Xs) and calculated (triangles) growth rate λ vs. Kp. The upper calculated curve used scale length $L_s = 0.2$, and the lower curve used $L_s = 0.4$.

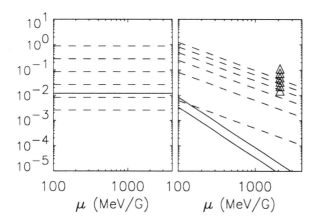

Figure 8. Magnetic (left) and electric (right) radial diffusion coefficients (in days^{-1}) used by *Albert et al.* [1998] (solid curves), *Brautigam and Albert* [2000] (dashed curves), and *Hilmer et al.* [2000] (triangles) at L=4.2 as functions of μ, for $Kp = 1, 2, 3, 4, 5,$ and 6 (bottom to top).

used ranged over wide values, but had the common form $A_m L^{6+2\alpha} \mu^{2-\alpha} + A_e L^{6+2\beta} \mu^{-\beta}$, expressing the effects of magnetic and electric field fluctuations, whose power spectra are assumed to depend on frequency as $\omega^{-\alpha}$ and $\omega^{-\beta}$, respectively [*Claflin and White*, 1974]. *Albert et al.* [1998] and *Brautigam and Albert* [2000] took $\alpha = 2$, $\beta = 2$, while *Hilmer et al.* [2000] took $\beta = 1.6$ and neglected the magnetic term. *Brautigam and Albert* [2000] and *Hilmer et al.* [2000] explicitly parameterized the amplitudes by Kp. Figure 8 shows the values of these expressions evaluated at $L = 4.2$ for different values of μ and Kp. Clearly, more reliable estimates of the field fluctuations under different conditions would be useful. An effort to evaluate the electric field fluctuations from CRRES observations is currently underway [*Bass et al.*, 1998].

The most conspicuous failure of diffusion modeling found was the underprediction by [*Brautigam and Albert*, 2000] of poststorm outer zone electrons with large μ, whose phase space density exhibited an inward-pointing radial gradient. One way this configuration could be produced is through local, cyclotron-resonant heating [*Roth et al.*, 1999; *Summers*, 2000; *Albert*, 2000]. If confirmed, this mechanism could be incorporated naturally into the radial diffusion formalism. In fact, such simulations, with appropriate observations, could help verify or rule out such a mechanism.

Acknowledgment. This work was supported by the Space Vehicles Directorate of the Air Force Research Laboratory and by the Boston College Institute for Scientific Research under USAF contract F19-628-96-C0030.

REFERENCES

Albert, J. M., Gyroresonant interactions of radiation belt particles with a monochromatic electromagnetic wave, *J. Geophys. Res., 105*, 21,191, 2000.

Albert, J. M., G. P. Ginet, and M. S. Gussenhoven, CRRES Observations of radiation belt protons, 1. Data overview and steady state radial diffusion, *J. Geophys. Res., 103*, 9261, 1998.

Albert, J. M., and G. P. Ginet, CRRES Observations of radiation belt protons, 2. Time-dependent radial diffusion, *J. Geophys. Res., 103*, 14,865, 1998.

Bass, J. N., D. H. Brautigam, J. M. Albert, G. P. Ginet, J. R. Wygant, D. E. Rowland, and M. Johnson, Energetic particle radial diffusion coefficients from CRRES electric field data, *EOS Trans., AGU, 79(45)*, F731, 1998.

Brautigam, D. H., and J. M. Albert, Radial diffusion analysis of outer radiation belt electrons during the October 9, 1990, magnetic storm, *J. Geophys. Res., 105*, 291, 2000.

Claflin, E. S., and R. S. White, A study of inner belt protons from 2 to 200 MeV, *J. Geophys. Res., 79*, 959, 1974.

Hilmer, R. V., G. P. Ginet, and T. E. Cayton, Enhancements of equatorial energetic electron fluxes near $L = 4.2$ as a result of high speed solar wind streams, *J. Geophys. Res., 105*, 23,311, 2000.

Hudson, M. K., A. D. Kotelnikov, X. Li, I. Roth, M. Temerin, J. Wygant, J. B. Blake, and M. S. Gussenhoven, Simulation of proton radiation belt formation during the March 24, 1991, SSC, *Geophys. Res. Lett., 22*, 291, 1995.

Kivelson, M. G., and C. T. Russell, *Introduction to Space Physics*, Cambridge University Press, 1995.

Lanzerotti, L. J., and C. G. Morgan, ULF geomagnetic power near L=4, 2. Temporal variation of the radial diffusion coefficient for relativistic electrons, *J. Geophys. Res., 78*, 4600, 1973.

Lanzerotti, L. J., D. C. Webb, and C. W. Arthur, Geomagnetic field fluctuations at synchronous orbit, 2. Radial diffusion, *J. Geophys. Res., 83*, 3866, 1978.

Li, X., I. Roth, M. Temerin, J. R. Wygant, M. K. Hudson, and J. B. Blake, Simulation of the prompt energization and transport of radiation belt particles during the March 24, 1991 SSC, *Geophys. Res. Lett., 20*, 2423, 1993.

Lyons, L. R., and R. M. Thorne, Equilibrium structure of

radiation belt electrons, *J. Geophys. Res., 78,* 2142, 1973.

Lyons, L. R., and D. J. Williams, *Quantitative Aspects of Magnetospheric Physics*, D. Reidel, Boston, 1984.

Lyons, L. R., and M. Schulz, Access of energetic particles to storm time ring current through enhanced radial "diffusion," *J. Geophys. Res., 94,* 5491, 1989.

Mozer, F. S., Power spectra of the magnetospheric electric field, *J. Geophys. Res., 76,* 3651, 1971.

Riley, P., and R. A. Wolf, Comparison of diffusion and drift descriptions of radial transport in the Earth's inner magnetosphere, *J. Geophys. Res., 97,* 16,865, 1992.

Roederer, J. L., *Dynamics of Geomagnetically Trapped Radiation*, Springer-Verlag, New York, 1970.

Roth, I., M. Temerin, and M. K. Hudson, Resonant enhancement of relativistic electron fluxes during geomagnetically active periods, *Ann. Geophysicae, 17,* 631, 1999.

Schulz, M., The magnetosphere, in *Geomagnetism*, vol. 4, edited by J. A. Jacobs, pp. 87-293, Academic, San Diego, Calif., 1991.

Selesnick, R. S., J. B. Blake, W. A. Kolasinski, and T. A. Fritz, A quiescent state of 3 to 8 MeV radiation belt electrons, *Geophys. Res. Lett., 24,* 1343, 1997.

Summers, D., and C. Ma, A model for generating relativistic electrons in the Earth's magnetosphere based on gyroresonant wave-particle interactions, *J. Geophys. Res., 105,* 2625, 2000.

J. M. Albert, Institute for Scientific Research, Boston College, Boston, MA 02135. (e-mail: jay.albert@hanscom.af.mil)

D. H. Brautigam, R. V. Hilmer, and G. P. Ginet, Air Force Research Laboratory/VSBS, 29 Randolph Rd, Hanscom AFB, MA 01731-3010. (e-mail: donald.brautigam@hanscom.af.mil; robert.hilmer@hanscom.af.mil; gregory.ginet@hanscom.af.mil)

Radiation Belt Electron Acceleration by ULF Wave Drift Resonance: Simulation of 1997 and 1998 Storms

Mary K. Hudson, Scot R. Elkington, John G. Lyon, M. Wiltberger and Marc Lessard

Pc 5 ULF waves are seen concurrently with the rise in radiation belt fluxes associated with CME-magnetic cloud events. Four such geomagnetic storm periods in 1997 and 1998, 10-11 January and 14-15 May, 1997; 1-4 May and 26-28 August, 1998, have been simulated with a 3D global MHD code driven by L1-measured solar wind parameters. The field output has been used to advance electron guiding center trajectories in the equatorial plane. The time series has also been analyzed for ULF wave mode structure. Toroidal field line resonances with low azimuthal mode number and frequencies commensurate with the electron drift period are identified in the radial electric field component, along with enhanced power in the poloidal (azimuthal) electric field component. The simulated time scale for inward radial transport of electrons in the several hundred keV to MeV energy range, from geosynchronous orbit to L=3-4, varies with the level of ULF wave power and overall energy input to the magnetosphere. Of the four events studied, the May 1998 storm period was most geoeffective and the January 1997 least so, in terms of simulated radial transport and energization of electrons. This transport rate is consistent with the level of ULF waves excited in the simulations, and the proposed drift resonant acceleration mechanism.

INTRODUCTION

Ultra Low Frequency waves have recently been shown to correlate well with relativistic electron flux enhancements in the MeV energy range, both during recurring high speed stream periods at solar minimum (Rostoker et al., 1998), and for storm periods driven by Coronal Mass Ejections, which occur more frequently at solar maximum (Baker et al., 1998a; 1998b; Hudson et al., 1999). That class of magnetospheric ULF waves which are global oscillations of magnetic and induction electric field in the mHz range can be excited by shear flow of solar wind past the magnetopause, the so-called Kelvin-Helmholz instability (see review by Hughes, 1994). This

mechanism explains a correlation of ULF wave power with solar wind velocity, which must be transferred from a broad frequency spectrum at the magnetopause to interior eigenmodes at standing wave frequencies characteristic of a given field line, a Field Line Resonance. Satellite data show fundamental FLRs on the evening as well as dawnside of the magnetosphere when not sorted by polarization (Lessard et al., 1999, survey of AMPTE IRM data, restricted to L > 6). Fundamental toroidal FLRs with azimuthal magnetic perturbation predominate on the dawnside, while poloidal modes with radial and some compressional perturbation predominate on the duskside, extending well past midnight and around to noon (Anderson, 1994, survey of AMPTE CCE data, L=5-9, see Figures 4 and 12). Fluctuating electric and magnetic fields at frequencies comparable to particle drift frequency lead to radial diffusion and consequent energization for that population moving into a region of

Space Weather
Geophysical Monograph 125

stronger magnetic field while conserving the first adiabatic invariant (Schulz and Lanzerotti, 1974).

This paper presents analysis of four recent storm-time simulations in 1997 and 1998, initiated by a CME-driven change in solar wind input to the magnetosphere. Parameters measured at the WIND spacecraft have been used to drive a 3D global MHD simulation (Goodrich et al., 1998) for the four storms. The field output from these simulations has been analyzed for MHD mode structure (Hudson et al., 2000), and used as input to a guiding center test particle code which advances electron flux in the equatorial plane (Hudson et al., 1999). This paper compares the relative geoeffectiveness of the four storms, as measured by the rate of relativistic electron radial transport and first invariant conserving acceleration. The January 1997 storm had a moderate Dst response (~ -78 nT), averaged horizontal component of the Earth's magnetic field at low latitude, and was characterized by a relatively low solar wind speed of 400-450 km/s, increasing to 400-500 km/s for the May 1997 storm (Dst ~ -115 nT), and almost double (850 km/s at maximum) its nominal value for the May and August 1998 events (Dst $\sim$-216 and -188 nT, respectively). Although the January 1997 storm was noteworthy for the high density ($180\ cm^{-3}$) solar wind impulse early on 11 January, during the northward IMF B_z period of this magnetic cloud passage (Burlaga et al., 1998), the greatest relativistic electron flux increase occurred during the preceeding southward IMF B_z period (Reeves et al., 1998a; 1998b). A growing body of evidence suggests that the solar wind speed (Paulikas and Blake, 1979), in combination with southward IMF B_z (Blake et al., 1997), is the main factor in producing relativistic electron flux enhancement in the magnetosphere. This paper focuses on analysis of the role ULF waves play in mediating the transfer of energy from the solar wind to the electrons.

The time scale for acceleration seen in the four simulation studies, a period of hours to a day, is intermediate between the time scale of days for standard radial diffusion models, confirmed for a relatively quiet period in 1996 using Polar measurements (Selesnick and Blake, 1997), and very rapid injection on an electron drift time scale, as seen for the extreme CME-shock event of March 24, 1991 (Blake et al., 1992; Li et al., 1993), where the interplanetary shock speed was estimated to be as high as 1400 km/s (Shea and Smart, 1993). Both an analytic model (Li et al., 1993; Hudson et al., 1995) and MHD simulation (Hudson et al., 1996; Hudson et al., 1997) have been used to characterize acceleration by the azimuthal electric field associated with this extreme

Storm Sudden Commencement compression of the magnetosphere. An impulsive azimuthal electric field in excess of 100 mV/m was estimated for the dayside outer magnetosphere based on measured fields of 20 mV/m on the nightside within the plasmasphere (Wygant et al., 1994). The SSCs of the four storms simulated here were moderate by comparison.

SIMULATIONS

Plate 1 shows a sequential comparison of the electron integral flux vs. energy and L for the four storms, using an AE8MAX input spectrum (Vette, 1991), which is an empirical average flux vs. energy and L based on solar maximum measurements, otherwise independent of geomagnetic activity level (see Hudson et al., 1999, for simulation of the January 1997 event using as input AE8MIN, average flux model based on solar minimum measurements). Electron guiding center trajectories were advanced in the equatorial plane (restricted to 90° pitch angles) using field output from the Lyon-Fedder-Mobarry 3D global MHD code for the respective storms (Goodrich et al., 1998). Fields are interpolated from the MHD grid to the location of particle guiding centers and a test on first invariant conservation, well maintained for electrons, is imposed. One sees the most rapid inward radial transport and energization, given by first invariant conservation, for the May 1998 storm (7.5 hours shown), and longest time scale for the January 1997 storm (15 hours shown). Inner zone flux changes are not followed because of the L=2.3 inner boundary of the MHD simulation. Electric and magnetic fields are mapped to the ionosphere assuming Hall and Pederson conductivities which evolve in time with magnetic field-aligned currents, simulating auroral ionization (Fedder et al., 1995). No plasmasphere was included for these long time scale runs, which require an evolving plasmasphere, in contrast to simulations of the rapid March 24, 1991 injection event, where the plasmasphere was characterized by initial planetary magnetic activity index Kp (Moore et al., 1987). Outer zone electron fluxes are transported radialy inward by MHD fields to fill the slot region for all four storm comparisons.

While the May 1998 event shows the most rapid radial transport, several factors make this event difficult to compare directly with particle observations, in particular the fact that the 4 May storm occurred during the recovery phase of one initiated on 2 May. Comparing the use of AE8MIN or AE8MAX as a source population was found to make only a small difference in final results. Initiating the simulation with a differential flux spectrum normalized to measurements taken on 1 May from a HEO spacecraft, with 63.4 degree inclination in a 12 hour period orbit, shows only minor variation from the AE8 model (Plate 2). Thus, the variation between storms (Plate 1) due to different MHD field time histories is more striking than the contrast between results obtained using different source populations.

Analysis of the azimuthal mode number at a fixed radial location in each field component indicates that power is concentrated at $m < 5$ (Elkington, 2000), also seen in snapshots of electric field vectors in the equatorial plane (Hudson et al., 1999; 2000). An FFT analysis of 2-3 hour time intervals of the radial and azimuthal electric field components at a fixed longitude determines the frequency spectrum. Plate 3 shows the power spectrum of the radial electric field component at midnight LT for 26 and 27 August 1998, produced for two three-hour intervals. The same toroidal field line resonant mode structure is evident in the radial electric field component as was seen in simulating the 10-11 January 1997 event (Hudson et al., 1999; 2000). A comparison is made in Plate 3a of the radial power profile following arrival of the interplanetary shock at 0640 UT on 26 August, during a period of IMF $B_z > 0$, with increased power at higher frequencies and lower L values seen in Plate 3b, during the $\sim$ 24 hour period of IMF $B_z < 0$ on 27 August. Straight lines plotted indicate electron drift frequency at a specified energy and L value, with positive slope corresponding to constant energy vs. L and negative slope corresponding to constant first invariant. The simulated ULF wave power in the radial and the azimuthal (not shown) electric field component is greater during the negative B_z interval on 27 August than following the interplanetary shock arrival on 26 August, with $B_z > 0$. This result is confirmed by spectral analysis of GOES magnetometer data, plotted in Plate 4 for the August 1998 event. The data are plotted in a geographical coordinate system, with components -E, N and P pointing in the radially outward, eastward and parallel (to the satellite spin axis) directions, respectively. In this coordinate system, the data are approximately radial, azimuthal and compressional in a magnetic coordinate system. No direct comparison can be made between the two 24 hour intervals covering all local times in Plate 4 and the fixed local time analysis of Plate 3, except to note the greater ULF wave activity and power on 27 August during southward IMF B_z. This dependence of ULF wave power on IMF B_z was seen in the GOES magnetetometer data for all four storms.

DISCUSSION

The MHD-test particle simulations confirm the observed correlation between ULF oscillations and inward radial transport and energization of relativistic electrons. The time scale of the transport is consistent with the drift resonant acceleration process proposed by Hudson et al. (1999; 2000) and analyzed in detail by Elkington et al. (1999), who followed a smaller number of test particles in a simplified field model. Acceleration via drift resonance with the radial electric field component comes about because of the non-axisymmetric drift path (Hudson et al., 1999; 2000). An electron with drift frequency $\omega_D = \omega$, the toroidal mode wave frequency, sees a radial electric field oscillating in phase with its drift period, which can lead to continuous acceleration via radial motion in a non-axisymmetric dipole. A similar effect is obtained for the poloidal mode with oscillatory azimuthal electric field in a non-axisymmetric dipole (Elkington, 2000). Resonance with a spectrum of ULF wave frequencies and azimuthal mode numbers is possible, with multiple resonances of finite width in energy and L evident in Plate 3.

To simplify the analysis, Elkington et al. (1999) assumed a tailward displaced dipole magnetic field and a superposed electric field including a uniform dawn-dusk convection field and a summation over m of sinousoidal radial and azimuthal oscilatory components, to simulate poloidal and toroidal modes:

$$B(r,\phi) = \frac{B_0}{r^3} + b_1(1 + b_2 \cos\phi). \quad (1)$$

$$E(r,\phi,t) = E_0(r,\phi) + \sum_{m=0}^{\infty} \delta E_{rm} \sin(m\phi + \omega t + \xi_m) \quad (2)$$

The longest wavelength toroidal mode has a null in oscillatory B_ϕ in the equatorial plane, and maximum E_r, while the corresponding poloidal mode has a null in B_r and maximum in both E_ϕ and compressional B_z components (Cummings et al., 1969). The latter can be neglected at low beta (ratio of kinetic to magnetic pressure) for low azimuthal mode numbers. Bulk energization of an ensemble of 120 particles distributed in drift phase was obtained with a single toroidal mode with frequency corresponding to the initial electron drift frequency. For a range of initial electron energies (40 keV to 5 MeV), resonances corresponding to $(m \pm 1)\omega_D = \omega$ produced energization.

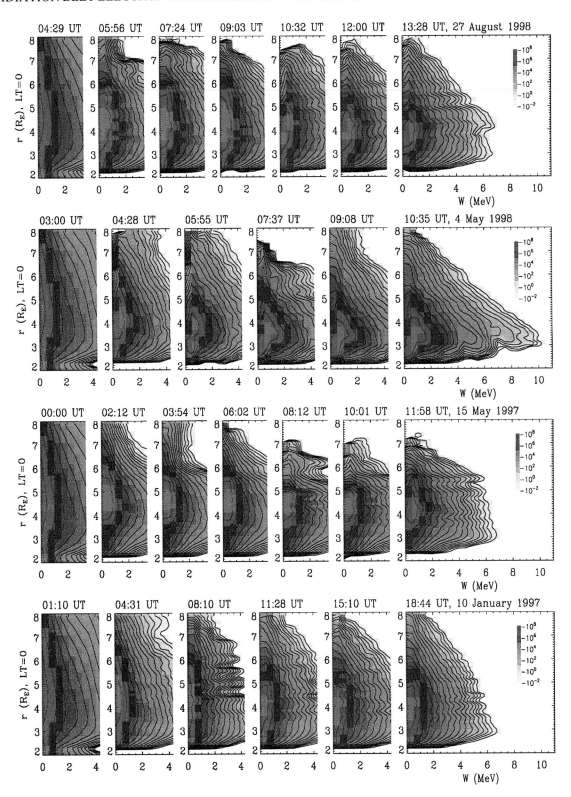

Plate 1. Simulated equatorial plane flux vs. energy and L for four geomagnetic storm events: (a)27 August 1998 and (b)4 May 1998, both at 1.5 hr intervals; (c)15 May 1997 at 2 hr intervals and (d)10 January 1997 at 3 hr intervals. Initialized with AE8MAX source population, advanced in time using equatorial plane fields from 3D MHD simulation driven by WIND input.

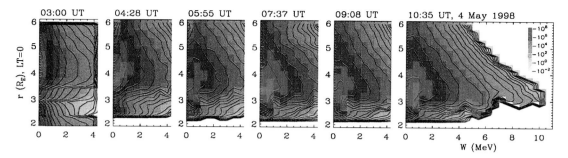

Plate 2. Simulated equatorial plane differential flux vs. energy and L for 4 May 1998, at 1.5 hr intervals, using an electron source population from HEO spacecraft measurements on 1 May 1998 (J. F. Fennell, private communication).

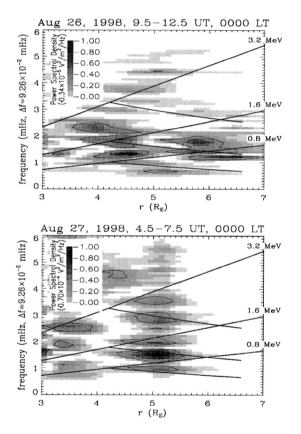

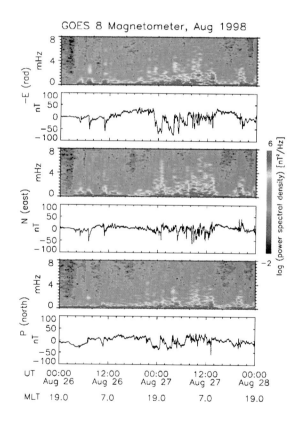

Plate 3. (a)Power spectrum of simulated radial electric field component at midnight ± 1 hour LT for 9.5 to 12.5 UT on 26 Aug 1998; (b)same for 4.5 to 7.5 UT on 27 Aug 1998. Ascending black lines indicate dipole drift frequency as a function of L for electrons with energies indicated; descending lines show same for constant first invariant, same energies. Note power spectral density scale is doubled in (b) relative to (a).

Plate 4. GOES 8 magnetic field data for 26-27 August 1998, plotted from 1-minute averaged key parameter data. Data are plotted in a geographical coordinate system, where the radial (top), eastward (middle) and north (bottom) components are approximately equal to magnetic radial, azimuthal and compressional components; FFT spectra are shown above each time series.

Plate 3 shows the presence of multiple field line resonant frequencies at a given L value. Overlapping resonances result in a continuum of accessible energy-drift phase space, leading to a calculable diffusion time scale of hours to a day (Elkington, 2000), depending on wave power. Poloidal modes (azimuthal electric field perturbation) contribute a comparable amount of acceleration; the non-axisymmetric dipole increases efficiency over standard radial diffusion calculations (Falthammar, 1968). Elkington's calculation assumes wave power consistent with that seen in Plate 3, leading to radial transport and first invariant-conserving energization scaling as $L^{-3/2}$ for relativistic electrons. The convection electric field, which is self-consistently included in the MHD simulation, is also enhanced during periods of increased solar wind coupling. Including a simplified, uniform dawn-dusk E_0 in the model calculation increases the radial distortion of the drift orbit, hence rate of radial transport and energization (Elkington et al., 1999).

Direct comparison with measured electron flux requires careful consideration of the input source population, spacecraft location and detector response (Li et al., 1998; 1999). Increase in the magnitude of D_{st} causes a measurable dropout in geosynchronous electron flux in all four events, due to main phase buildup of the ring current and third invariant conservation (Kim and Chan, 1998). Plots of relative energy flux in four energy channels measured by GPS spacecraft at their equatorial radial crossing point have reproduced the decrease in flux associated with D_{st} (Hudson et al., 1999); for simulated geosynchronous orbit, magnetosheath encounters that coincide with those seen by LANL spacecraft are also reproduced by the MHD-test particle simulations (Hudson et al., 2000), as is the diurnal variation in geosynchronous flux due to the compressed dipole, enhanced during storm periods (Elkington, 2000).

CONCLUSIONS

A comparison of MHD-test particle simulations of four geomagnetic storms driven by CME-generated perturbations of the solar wind has been presented. All four storms were characterized by a buildup of relativistic electron flux in the inner magnetosphere (see ISTP and GEM web sites). The fastest inward radial transport and energization is seen in simulating the May 1998 event, using the AE8MAX initial flux profile for all four storms. The ULF wave power was also greatest for the May 1998 storm. This event proves to be the most complex of the four to analyze because of the

highly compressed state of the magnetosphere (magnetopause compressed into L=4 around 0300 UT on 4 May). We will not attempt a direct comparison with measured particle flux changes, except to note that the range from 12 hours for rise in flux of > 2 MeV electrons at L=4 on 10 January 1997 (Reeves et al., 1998a) to 3 hours for filling of the slot region below 1 MeV on 4 May 1998 (Blake et al., 1998) is consistent with acceleration rates seen in the simulations of these events, at the extremes of the four studied.

The time scale for accleration is hours to roughly a day for all four storms. The stronger the solar wind driver, or solar wind speed as well as prolonged southward IMF B_z, the faster the radial transport and corresponding first-invariant conserving acceleration. Enhanced ULF wave power has the same toroidal mode structure for both the January 1997 and August 1998 simulations, even though solar wind driving conditions were substantially different. There is more simulated power extending to higher frequencies for the May 1998 event (not shown), least for January 1997 (Hudson et al., 1999; 2000), and a comparable amount for May 1997 as for August 1998 (Plate 3). These results are born out in analysis of GOES magnetometer data for the four events.

The global MHD simulations provide an opportunity for ULF wave mode structure analysis not possible with a limited set of spacecraft. Using MHD output fields to advance electron guiding center trajectories provides insight into relative transport rates for different storms. However, the technique is still in its infancy in terms of source population, with work in progress to provide data assmilation (Moorer and Baker, 2000) which would improve upon AE8 or CRRES-based average models (Brautigam and Albert, 2000) as a starting point.

Acknowledgments. We thank J. F. Fennell for providing HEO data, J. B. Blake for discussions of Polar, HEO and Sampex data, and A. A. Chan for discussion and analysis of the acceleration mechanism. This work is supported by NASA grants NAG5-7442, NAG5-8441 and NAG5-7689 to Dartmouth College, along with NSF grant ATM9819927. Use of NPACI and MHPCC is acknowledged.

REFERENCES

Anderson, B. J., An overview of spacecraft observations of 10s to 600s period magnetic pulsations in the Earth's magnetosphere, *Solar Wind Sources of Magnetospheric Ultra-Low-Frequency Waves, Geophysical Monograph 81*, M. J. Engebretson, K. Takahashi and M. Scholer, eds., American Geophysical Union, Washington D.C., 1994.
Brautigam, D. H. and J. M. Albert, Radial diffusion analysis

of outer radiation belt electrons during the October 9, 1990, magnetic storm, *J. Geophys. Res., 105*, 291, 2000.

Baker, D. N., T. I. Pulkkinen, X. Li, S. G. Kanekal, J. B. Blake, R. S. Selesnick, M. G. Henderson, G. D. Reeves, H. E. Spence and G. Rostoker, Coronal mass ejections, magnetic clouds, and relativistic magnetospheric electron events: ISTP, *J. Geophys. Res., 103*, 17279, 1998a.

Baker, D. N., T. I. Pulkkinen, X. Li, S. G. Kanekal, K. W. Ogilvie, R. P. Lepping, J. B. Blake, L. B. Callis, G. Rostoker, H. J. Singer and G. D. Reeves, A strong CME-related magnetic cloud interaction with the Earth's magnetosphere: ISTP observations of rapid relativistic electron acceleration on May 15, 1997, *Geophys. Res. Lett., 25*,2975, 1998b.

Blake, J. B., W. A. Kolasinski, R. W. Filius and E. G. Mullen, Injection of electrons and protons with energies of tens of MeV into L , 3 on March 24, 1991, *Geophys. Res. Lett., 19*, 821, 1992.

Blake, J. B., D. N. Baker, N. Turner, K. W. Ogilvie, and R. P. Lepping, Correlation of changes in the outer-zone relativistic-electron population with upstream solar wind and magnetic field measurements, *Geophys. Res. Lett., 24*, 927, 1997.

Blake, J. B., M. D. Looper, J. E. Mazur, D. N. Baker, X. Li, M. K. Hudson, S. G. Kanekal, R. A. Mewaldt, Multi-satellite observations of the injection of energetic radiation belt particles during May 1998, *EOS Trans. Am. Geophys. Union, 79*, F740, 1998.

Burlaga, L., R. Fitzenreiter, R. Lepping, K. Ogilvie, A. Szabo, A. Lazarus, J. Steinberg, G. Gloeckler, R. Howard, D. Michels, C. Farrugia, R. P. Lin, and D. E. Larson, A magnetic cloud containing prominence material: January, 1997, *J. Geophys. Res., 103*, 277, 1998.

Cummings, W. D., R. J. O'Sullivan and P. J. Coleman, Standing Alfven waves in the magnetosphere, *J. Geophys. Res., 74*, 778, 1969.

Elkington, S. R., M. K. Hudson and A. A. Chan, Acceleration of relativistic electrons via drift-resonant interaction with toroidal-mode Pc-5 ULF oscillations, *Geophys. Res. Lett.,26*, 3273, 1999.

Elkington, S. R., Simulating electron radiation belt dynamics, PhD thesis, Dartmouth College, 2000.

Fedder, J. A., S. P. Slinker, J. G. Lyon and R. D. Elphinstone, Global numerical simulation of the growth phase and the expansion onset for a substorm observed by Viking, *J. Geophys. Res., 100*, 19083, 1995.

Falthammar, C.-G., Radial diffusion by violation of the third adiabatic invariant, in *Earth's Particles and Fields*, ed. by B. M McCormac, p. 157, NATO Advanced Study Institute, Reinhold Book Corp., New York, 1968.

Goldstein, J. Goldstein, M. K. Hudson and W. Lotko, Possible evidence of damped cavity mode oscillations stimulated by the January, 1997 magnetic cloud event, *J. Geophys. Res., 26*, 3589, 1999.

Goodrich, C. C., M. W. Wiltberger, R. E. Lopez, K. Papadopoulos and J. G. Lyon, An overview of the impact of the January 10 - 11, 1997 magnetic cloud on the magnetosphere via global MHD simulation, *Geophys. Res. Lett., 25*, 2537, 1998.

Hudson, M. K., A. D. Kotelnikov, X. Li, I. Roth, M. Temerin, J. Wygant, J. B. Blake, and M. S. Gussenhoven,

Simulation of proton radiation belt formation during the March 24, 1991, SSC, *Geophys. Res. Lett., 22*, 291, 1995.

Hudson, M. K., S. R. Elkington, J. G. Lyon, V. A. Marchenko, I. Roth, M. Temerin and M. S. Gussenhoven, MHD/particle simulations of radiation belt formation during a storm sudden commencement, in *Radiation Belts: Models and Standards*, edited by J. F. Lemaire, D. Heynderickx, and D. N. Baker, *Geophys. Momogr. Ser.*, vol. 97, p. 57, AGU, Washington, D. C., 1996.

Hudson, M. K., S. R. Elkington, J. G. Lyon, C. C. Goodrich and T. J. Rosenberg, Simulation of radiation belt dynamics driven by solar wind variations, *Sun-Earth Plasma Connections, Geophysical Monograph 109*, J. L. Burch, R. L. Carovillano and S. K. Antiochos, eds., American Geophysical Union, Washington D.C., 1999.

Hudson, M. K., S. R. Elkington, J. G. Lyon and C. C. Goodrich, Increase in relativistic electron flux in the inner magnetosphere: ULF wave mode structure, *Adv. Space Res., 25*, 2327, 2000.

Hughes, W. J, Magnerospheric ULF waves: A tutorial overview, *Solar Wind Sources of Magnetospheric Ultra-Low Frequency Waves, Geophysical Monograph 81*, M. J. Engebretson, K. Takahashi and M. Scholer, eds., American Geophysical Union, Washington D.C., 1994.

Lessard, M. R., M. K. Hudson and H. Luhr, A statistical study of Pc3-Pc5 magnetic pulsations observed by AMPTE/Ion Release Module satellite, *J. Geophys. Res., 104*, 4523, 1999.

Kim, H.-J. and A. A. Chan, Fully adiabatic changes in storm time relativistic electron fluxes, *J. Geophys. Res., 102*, 22107, 1997.

Li, X., I. Roth, M. Temerin, J. R. Wygant, M. K. Hudson and J. B. Blake, Simulation of the prompt energization and transport of radiation belt particles during the March 24, 1991, SSC, *Geophys. Res. Lett., 20*, 2423, 1993.

Li, X., D. N. Baker, M. Temerin, T. Cayton, G. D. Reeves, T. Araki, H. Singer, D. Larson, R. P. Lin and S. G. Kanekal, Energetic electron injections into the inner magnetosphere during the January 10-11, 1997, magnetic storm, *Geophys. Res. Lett., 25*, 2561, 1998.

Li, X., D.N. Baker, M. Temerin, T.E. Cayton, G.D. Reeves, R.S. Selesnick, J.B. Blake, G. Lu, S.G. Kanekal and H.J. Singer, Rapid enhancement of the relativistic electrons deep in the magnetosphere during the May 15, 1997, magnetic storm, *J. Geophys. Res., 104*, 4467, 1999.

Moore, T. E., D. L. Gallagher, J. L. Horwitz and R. H. Comfort, MHD wave breaking in the outer plasmasphere, *Geophys. Res. Lett., 14*, 1007, 1987.

Moorer, D.F. and D.N. Baker, Quantitative mapping and forecasting of high energy electron flux in the outer radiation belt, *EOS Trans. Am. Geophys. Union, 80*, 845, 1999.

Paulikas, G. A., and J B. Blake, Effects of the solar wind on magnetospheric dynamics: Energetic electrons at the synchronous orbit, *Quantitative Modeling of Magnetospheric Processes, 21*, Geophys. Monograph Series, 1979.

Rostoker, G., S. Skone and D. N. Baker, On the origin of relativistic electrons in the magnetosphere associated with some goemagnetic storms, *Geophys. Res. Lett., 25*, 3701, 1998.

Reeves, G. D., R. H. W. Friedel, R. D. Belian, M. M. Meier,

M. G. Henderson, T. Onsager, H. J. Singer, D. N. Baker and X. Li, The relativistic electron response at geosynchronous orbit during the January 10, 1997, magnetic storm, *J. Geophys. Res., 103* 17559, 1998a.

Reeves, G. D., D. N. Baker, R. D. Belian, J. B. Blake, T. E. Cayton, J. F. Fennell, R. H. W. Friedel, M. M. Meier, R. S. Selesnick and H. E. Spence, The global response of relativistic radiation belt electrons to the January 1997 magnetic cloud, *Geophys. Res. Lett., 25,* 3265, 1998b.

Schulz M. and L. J. Lanzerotti, *Particle Diffusion in the Radiation Belts*, Springer-Verlag, Berlin, 1974.

Selesnick, R. S. and J. B. Blake, Dynamics of the outer radiation belt, *Geophys. Res. Lett., 24,* 1347, 1997.

Selesnick, R. S. and J. B. Blake, On the source location of radiation belt relativistic electrons, *J. Geophys. Res., 105,* 2607, 2000.

Shea, M. A. and D. F. Smart, March 1991 solar terrestrial phenomena and related technological consequences, in Proc. Int. Conf. Cosmic Rays, 23rd, 1993.

Southwood, D. J., Some features of field line resonances in the magnetosphere, *Planet. Space Sci., 36,* 503, 1974.

Vette, J. I., The AE8 trapped electron model environment, NSSDC/WDC-A-R&S 91-24, 1991.

Wygant, J. R., F. Mozer, M. Temerin, J. B. Blake, N. Maynard, H. Singer and M. Smiddy, Large amplitude electric and magnetic field signatures in the inner magnetosphere during injection of 15 MeV electron drift echoes, *Geophys. Res. Lett., 21,* 1739, 1994.

M. K. Hudson, S. R. Elkington, J. G. Lyon and M. Wiltberger, Physics and Astronomy Dept., Dartmouth College, Hanover, NH 03755. (email: mary.hudson@dartmouth.edu)

M. Lessard, Thayer School of Engineering, Dartmouth College, Hanover, NH 03755

Modeling the Transport of Energetic Particles in the Magnetosphere with Salammbô

D. Boscher, S. Bourdarie

ONERA/DESP, Toulouse, FRANCE

Transport of energetic particles in the trapping region is related to perturbations of the magnetic field, due to those of magnetospheric currents, from the magnetopause, the tail or the ring current. Particles are transported by adiabatic transport along drift shells without field variations or with slow magnetic field perturbations, and by non adiabatic transport across drift shells when perturbations are fast enough. Examples of these different modes of transport are given as well as modeling capabilities. All these transport modes have important implications on the possibility to nowcast or forecast radiation belt particles. Nowcasting is more related to the first type of transport (adiabatic transport) as it induces short term effects. In the forecast, non adiabatic transport has the most important role, as it is long term great variations of the fluxes which are important to predict.

1. INTRODUCTION

In the magnetosphere, the Lorentz force $\vec{F} = q.(\vec{E} + \vec{v} \wedge \vec{B})$ acts on charged particles, with charge q and velocity v, in a field which has two components, an electric field E and a magnetic field B. With the hypotheses that the particle is energetic (high kinetic energy) and located in the inner magnetosphere (where the magnetic field magnitude is high), the electric field term can often be neglected, and the equation is easier to solve. Then, in a static dipole field, the particle motion is the combination of the gyration around the field line, the bounce along the field line, and the drift across field lines due to magnetic field gradient and curvature. The center of the gyration orbit, also called the guiding center, is located on a surface enclosing the dipole center, the drift shell [*Roederer, 1970*]. Two particles with the same guiding center, and no velocity component along the field line, but with different kinetic energy, follow the same drift shell, which can be defined only with the magnetic field.

From the particle point of view, each of these motions can be related to an adiabatic invariant, the "relativistic" magnetic moment M, the second invariant J, and the third invariant J_3. In the dipole field (first order of the Earth magnetic field), the third invariant associated with the drift can be simply related to a more physical parameter, the McIlwain L parameter, which represents the apex of the field line, in Earth radius units. But in a more complicated field (if other internal field terms, or an external magnetic field due to the interaction of the solar wind with the internal field, are added), the McIlwain L parameter is no longer valid (it is only a parameter which describes the first order of the particle drift motion). Nevertheless, in this field, the notion of adiabatic invariant theory is very useful: particles are well organized according to these invariants. And even if the magnetic field is varying, particles conserve their invariants as long as the variations of the field are slower than the particle drift period. For faster variations, particles can conserve one or two adiabatic invari-

Space Weather
Geophysical Monograph 125
Copyright 2001 by the American Geophysical Union

ants depending on the variation time scale of the field compared to the gyration, bounce or drift period.

The transport of energetic particles in the magnetosphere is mainly driven by three different processes, from the simplest to the most complex :
- adiabatic transport along drift shells, without modification of the magnetic field,
- adiabatic transport along drift shells, with slow variations of the magnetic field,
- non-adiabatic transport across drift shells, when the variations of the magnetic field are faster than the particle drift period.

We will discuss these three transport processes, and show examples with modeling as well as with measurements.

2. ADIABATIC TRANSPORT WITHOUT MODIFICATION OF THE FIELD

If the field is constant in time, particles follow their drift shell, the three adiabatic invariants are conserved, as well as the energy and particle flux. This transport seems simple: it corresponds to the particle trapping in the radiation belts. Nevertheless, it can be more complex for high energy protons due to their large Larmor radius, or when the magnetopause is modified.

In a real field, with internal and external contributions, one very important notion is the last closed drift shell (LCDS) [Roederer, 1970]. A drift shell, determined by the field, can cross the magnetopause. All particles below this shell will follow closed trajectories (they are trapped), while particles on shells above this limit will follow open trajectories (they are quasi trapped, which means that they can be trapped for duration less than their drift periods). Particles on shells above this limit coming from outside the magnetosphere can cross it, while particles trapped on this shell easily escape the magnetosphere. This last closed drift shell depends only on the field (it does not depend at first order on the particle energy), but the duration for a particle to cross or escape the magnetosphere depends on its energy.

Protons with very high energy (greater than 1 MeV) have large Larmor radii (around 5000 km for a 10 MeV proton near synchronous orbit). As these radii are not negligible as compared to the curvature of the field lines, particle motions are not so simple as gyration, bounce and drift motion. Particles coming from outside the magnetosphere can reach locations lower than the LCDS. The higher the energy, the lower the location they can reach. An example of the entrance of high energy protons inside the magnetosphere is shown

on Figure 1. On the bottom panel, protons greater than 10 MeV as measured by IMP8 satellite are plotted as a function of time from March 2 to 10, 1993. Two Solar Energetic Particle events (SEP's) are seen and they are able to cross the magnetopause. This is clear at synchronous orbit (middle panel), as well as at low Earth orbit (top panel), where they reach L values as low as L = 4.5. The entrance of high energy particles was calculated using particle tracing codes from different authors [Smart and Shea, 1980; Nymmik, 1998]. The location reached by the particle depends on magnetospheric shielding (the magnetic cut-off).

Another effect of adiabatic transport is particle drift loss. Suppose that the field is not changed at all and the magnetosphere is compressed by a solar wind pressure increase. Then the magnetopause can move inward

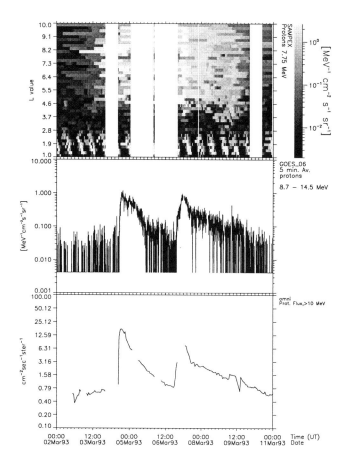

Figure 1. Proton measurements in the 10 MeV range highlighting Solar Energetic Particle events during the period 2-11 March 1993: from bottom to top: IMP8 in the solar wind, GOES at synchronous orbit and SAMPEX on a Low Earth Orbit.

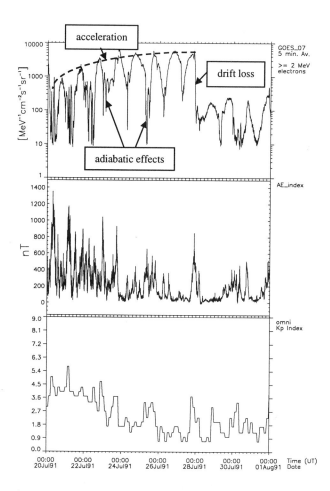

Figure 2. Example of electron transport phenomena during the periood 20 July - 01 August, 1991. From bottom to top: the magnetospheric index Kp, the auroral index AE, and measurements of electrons greater than 2 MeV on GOES 07.

and the old last closed drift shell becomes an open shell: a new LCDS appears. Particles previously trapped in the region between the old and new LCDS cross the magnetopause and are lost from the trapped particle population point of view. This phenomenon is not easy to isolate, because it is always related to a modification of the field. But, combined with the adiabatic motion of drift shells (see Section 3), it is a frequent phenomenon during intense storms seen on high energy protons [see *McIlwain, 1966*], as well as electrons. An example of this loss is shown in Figure 2. The bottom and middle panels show the magnetospheric activity index Kp and the auroral index AE, and the top panel shows GOES-7 synchronous measurements (electrons greater than 2 MeV) during 12 days in 1991. It is clear that losses occur during the July 27, 1991 little magnetic storm.

3. ADIABATIC TRANSPORT WITH SLOW MODIFICATION OF THE FIELD

As the magnetic field is varying slowly (slower than the particle drift period), the three adiabatic invariants are conserved. But due to the field modifications, the drift shells are moving, and the particle energy and flux are changed. This effect is reversible: as the magnetic field comes back to its initial value, particles return to their initial location and energy. As this effect depends on the ratio between the magnetic field variation time scale and the drift period, it depends on the particle energy, therefore it is easier to observe this effect for high energy particles (with faster drift periods). It can be observed for slow variations of the magnetic field, related to either magnetopause or ring current variations. Good examples, well correlated to the ring current variations, were noticed by *McIlwain* [1966] for high energy protons observed by Explorer 26. As the drift period of electrons at synchronous orbit can be fast (around 1000 s for 1 MeV), it is easy to see this effect with these particles. Examples of this adiabatic transport can be seen on Figure 2 several times, as indicated. Sometimes, this effect is combined with the magnetopause compression to give loss of particles as explained in Section 2. As the ring current grows, the local magnetic field inside decreases and drift shells expand in the magnetosphere. If at the same time, the magnetopause is compressed, a lot of particles can be flushed out from the radiation belts. This appears frequently in the magnetosphere, and the effect depends on the strength of the storm: the losses during the March 24th, 1991 storm (Dst around -300 nT) were observed down to a L value around 2.5 (clearly seen on 1 MeV protons, see *A. Korth*, [1996]), while during the March 1989 event, with a Dst around -600 nT, the losses were seen down to L = 1.7.

Modeling the adiabatic transport of particles is not easy, due to uncertainties of the magnetic field. Particle tracing code can be used but, if adiabatic effects occur, it is sufficient to organize particle data with the three adiabatic invariants and then calculate L^* according to the Roederer's definition [*Roederer*, 1970]. This was done by *Kim and Chan* [1997], *Selesnick and Blake* [2000], and by *Brautigam and Albert* [2000], using different magnetic field models. The conclusion of all these studies is that during a particular active period, no magnetic field model can be used to precisely determine the particle transport and their drift shells. Maybe the best, as pointed out by *Selesnick and Blake* [2000] is to use a simple Tsyganenko 89 [*Tsyganenko*, 1989] model.

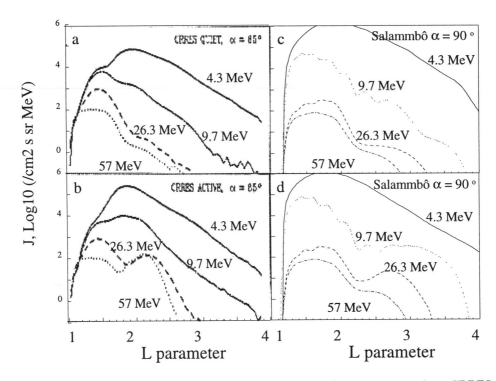

Figure 3. Comparison of L distributions of high energy protons: a) measurements from CRRES PRO-TEL before the March 1991 event (from *Gussenhoven et al.*, [1993]), b) after the event, c) modeling results from Salammbô before the event, d) after the event.

4. NON ADIABATIC TRANSPORT OF PARTICLES

4.1. Radial diffusion

The simplest non adiabatic transport of energetic particles is the well know radial diffusion. It is due to any convection electric field, as well as any perturbation of the magnetic field faster than the particle drift period. It is related to electric fields, ULF waves, magnetosphere shock crossing, dipolarisation or fast magnetopause current variations. In all these processes, particles try to conserve their third adiabatic invariant, but they cannot manage in it: the third invariant is violated (if the perturbations are faster, two or three adiabatic invariants can be violated). Particle trajectory can be determined when fields are varying, but the number of parameters needed to perform such a calculation is very important (the fields and their variations are needed everywhere along the trajectory). In this field, the works of *Hudson et al.* [1997] for the March 1991 storm must be noticed. It shows that the immediate effects of this storm may be explained by the travelling of the solar wind shock through the magnetosphere. Comparison with CRRES measurements show reasonable agreement in modeling the high energy particles.

When uncertainties are important on fields and their variations, it is sometimes better to use radial diffusion equation modeling. In these calculations, the local magnetic field values and perturbations are not needed exactly but rather an average over the drift shells is sufficient. The global results are generally in good agreement with observations, though radial diffusion coefficients are not accurately well known. Moreover, it is easier to take internal losses, friction or pitch angle diffusion into account in these calculations.

For protons, the first works in the whole magnetosphere were made by *Spjeldvik* [1977], but the first version calculating the full distribution was made by *Beutier et al.* [1995a] by the code named Salammbô. In the new version of this code [*Vacaresse et al.*, 1999], radial diffusion is assumed magnetic activity dependent. Moreover, to reproduce the effects of events as the March 1991 storm, magnetospheric shielding was introduced as well as an external source due to Solar Energetic Particles (SEPs). In Figure 3, a comparison is shown between CRRES measurements (a and b, from *Gussenhoven et al.*, [1993]) and Salammbô calculation (c and d, *Vacaresse et al.*, [1999]) for equatorially mirroring particles. The top panels (a and b) represents the fluxes for different energies before the March 1991 event, and the

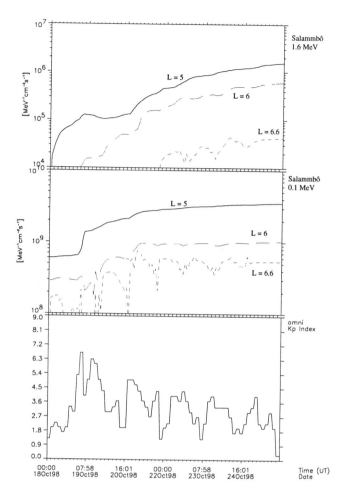

Figure 4. Electron results from Salammbô during an active period from 18 October to 26 October 1998: magnetospheric index Kp (bottom panel), 100 keV flux for 3 L values (middle panel), 1.6 MeV flux for 3 L values (top panel).

present in the magnetosphere. A 3D simulation (2D in space: L and the pitch angles, and 1D for energy, equivalent to 3D in the phase space) is needed to accurately model the transport of electrons, as pitch angle diffusion can interact with radial diffusion in non linear ways. The first calculations in this field were performed by *Lyons and Thorne* [1973]. But at that time, the effects of the waves were understood only as loss process acting on electrons inside the plasmasphere. The first real 3D simulation was performed by *Beutier and Boscher* [1995b]. In the new version [*Boscher et al.*, 2000b], radial diffusion is varying in time according to the magnetospheric activity index Kp, and the waves interacting with the electrons are only in the plasmasphere, but changing with magnetic activity through Kp. Up to around 400 keV, electrons are transported by radial diffusion, and Figure 4 shows a result of that transport during a period from October 18 to October 26, 1998. The bottom panel shows the magnetospheric activity index Kp: the storm begins on October 18th and reaches Kp = 7-, then it decreases slowly with alternate quiet and active periods. The Salammbô results for 100 keV are plotted in the middle panel for different L values in the outer belt. At L = 5, the model gives an increase during the main phase, and the slot region is filled with particles, due to the increase of radial diffusion. But the flux continues to grow as active periods appear and as the boundary condition at L = 7, deduced from synchronous measurements is also increased.

4.2. Recirculation

High energy electrons (in the MeV range) are created during magnetic storm periods. A process which can explain the source of these electrons is the small scale recirculation process [*Boscher et al.*, 2000a]. Particles which undergo pitch angle and radial diffusion at the same time are transported along loops as shown in Figure 5 because diffusion has no privileged direction. This process is not too different from the one discussed by *Fujimoto and Nishida*[1990], but the process described here is made along small loops located near the magnetic equator where electrons will experience gradual acceleration. As a consequence, the acceleration must appear near the equator and near the plasmapause where gradients of the distribution function are highest and even can be reversed.

In measurements, it is not obvious to clearly observe this effect for different reasons:
- the plasmapause is time varying and moreover local time dependent,

bottom panels those after this event. Though the code results do not reproduce accurately the measurements, the calculated effects of the March 1991 event is not too far from the reality: it creates a second belt essentially in the 30 MeV range. This calculation and the one using particle tracing code (*Hudson et al.*, [1997]) suggests that the physics of the interaction between the SEP's and the magnetosphere is understood, though some parameters in the modeling have to be improved. It also must be noticed that the CRRES quiet model is mainly due to the effect of the March 1989 event, which induced losses in the radiation belt down to L = 1.7 (see on the a panel).

Modeling the electron belt with a classical diffusion code is more complex due to the pitch angle diffusion which arises from the interaction between particles and waves

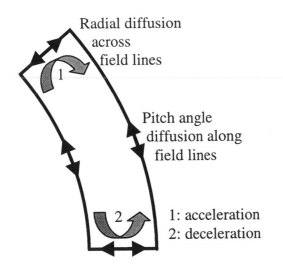

Figure 5. Sketch of the small scale recirculation loop, electron acceleration mechanism.

- measurements have to be done with the same satellite, to avoid intercalibration problems, with good energy resolution, and unidirectional fluxes are needed to avoid shell splitting effects.

First examples of such radial distribution function profiles are now published [*Selesnick and Blake,* 2000, *Brautigam and Albert,* 2000] during magnetic active periods. It seems to show that the source of energetic electrons is located in the inner magnetosphere, and close to the plasmapause.

When radial diffusion and wave-particle interactions act together like in the Salammbô code [*Boscher et al.,* 2000b], it is then possible to accelerate electrons by an order of magnitude in energy. Figure 4 shows such acceleration of electrons during an active period. The 1.6 MeV flux is shown on the Figure on the top panel, and we observe an increase that is slower than for lower energy. During this period, the acceleration region is moved slowly from low L values to higher Ls, following the plasmapause.

5. NOWCASTING AND FORECASTING THE RADIATION BELTS

Radiation belt particles are a major issue in space weather for spacecraft operations. They contribute to dose for solar cells, thermal coatings and on board electronics, they produce backgrounds in detectors and amplifiers, and they decrease the transparency of optics. But the most important issues in space weather are to prevent Single Event Effects (SEE's) and Electrostatic Discharges (ESD's) which endanger the life of satellites.

Radiation belt particles can also be of importance for extra vehicular activities from board the Space Shuttle and the International Space Station. But nowcasting and forecasting the radiation belt environment is not a easy task.

As seen previously, there are two different effects of the radiation belt particle transport:
- short term effects, the drift loss (I) and adiabatic effects (II) (time scale on the order of hours),
- long term effects, related to the non adiabatic transport and acceleration of particles (III) (time scale on the order of days).

These effects are particularly clear in Figure 2 (top panel). The global increase of fluxes corresponds to the gradual acceleration of particles in this region during the first part of the period, or rather a slow diffusion of particles accelerated near the plasmapause. Short term effects are visible particularly as rapid flux dropouts when the satellite is on the night-side of the magnetosphere, and as increases when the satellite is on the day-side, and can be of the same origin. Finally particle loss is clearly seen on the graph as a global decrease by a factor around 40 that is well correlated with a little storm.

The prediction of short term effects necessarily needs a good magnetic field model to accurately calculate the particle drift shells. The other limiting factor is that the magnetic field is never exactly identical everywhere for two active periods, therefore a nowcast/forecast of the magnetic field must be needed for the prediction of energetic particle fluxes. Moreover, to know when losses may appear, it can be necessary to have a good knowledge of the magnetopause dynamics, unless this feature is included in the magnetic field model.

On the other hand, modeling long term effects can be simpler. A model like Salammbô is able to take non adiabatic effects (and acceleration) into account, but parameters used in the code have to be improved. In particular, one key problem is the time dependent radial diffusion coefficients. Nevertheless, using the same input parameter (the magnetospheric activity index Kp) for different periods of magnetic activity, relatively good results were obtained, for protons [*Boscher et al.,* 1998; *Vacaresse et al.,* 2000] as well as for electrons [*Boscher and Bourdarie,* 1998,; *Boscher et al.,* 2000b]. Moreover, in the Salammbô code, simple wave and plasmapause models were used but a reliable model (empirical, magnetic activity dependent) is needed for these two components of the space environment. Finally, to improve radiation belt environment models, accurate measurements of particles are needed, from satellites located on strategic orbits, like a Geostationary transfer orbit.

Finally, to do nowcasting of the radiation belts, one possibility is interpolation between measurements, as it was performed by *Reeves et al.* [1998] for one energy range for the January 1997 event. But it may not be possible unless there will be a constellation of satellites with accurate measurements (good energy resolution and pitch angle distribution) and without intercalibration problems. Another possibility is using a model like Salammbô with input measurements. But short term effects are not so easy to perform unless the magnetic field is well know. As for forecasting, it seems from Salammbô results that it could be done in the future. But once more, the short term effects will be harder to predict.

Acknowledgments. Figures 1, 2 and 4 were made using the PAPCO data analysis package supported by MPAE Lindau through the Max-Planck Gesellschaft, DARA,grant 50 OC 95010, and as part of the HYDRA NASA funding under grant number NAG 5 2231. The authors wish to thank R. Friedel and T. Cayton for providing GPS data and NASA for providing data under ISTP program.

REFERENCES

Beutier T., D. Boscher, M. France, Salammbô: A three-dimensional simulation of the proton radiation belt, *J. Geophys. Res., 100*, 17181, 1995a.

Beutier, T., and D. Boscher, A three-dimensional analysis of the electron radiation belt by the Salammbô Code, *J. Geophys. Res., 100*, 14853, 1995b.

Boscher D., S. Bourdarie, R. Friedel, A. Korth, Long term dynamic radiation belt model for low energy protons, *Geophys. Res. Lett., 25-22*, 4129-4132, 1998.

Boscher, D., S. Bourdarie, Physical modeling of the outer belt high energy electrons, in Workshop on Space Weather *ESA WPP-155*, 411, Noordwijk (Netherlands), Nov 1998.

Boscher, D., S. Bourdarie, R. Thorne, B. Abel, Influence of the wave characteristics on the electron radiation belt distribution, Adv. Space Res., *26*, 163, 2000a.

Boscher, D., S. Bourdarie, R. M. Thorne, B. Abel, Toward nowcasting of the electron radiation belt, SREW conference, to be published in AGU monograph, 2000b.

Brautigam, D., J. Albert, Radial diffusion analysis of outer radiation belt electrons during the October 9, 1990, magnetic storm, *J. Geophys. Res., 105*, 291, 2000.

Fujimoto, M., and A. Nishida, Energization and anisotropization of energetic electrons in the Earth's radiation belts by the recirculation process, *J. Geophys. Res., 95*, 4265, 1990.

Gussenhoven, M. S., E. G. Mullen, D. Brautigam, Improved understanding of the Earth's radiation belts from the CRRES satellite, IEEE Trans. Nucl. Sci., *43*, 353, 1996.

Hudson, M., S. Elkington, J. Lyon, V. Marchenko, I. Roth, M. Temerin, J. Blake, M. Gussenhoven, J. Wygant, Simulations of radiation belt formation during storm sudden commencements, *J. Geophys. Res., 102*, 14087, 1997.

Kim, H.-J., A. Chan, Fully adiabatic changes in storm time relativistic electron fluxes, *J. Geophys. Res., 102*, 22107, 1997.

Korth, A., private communication, 1996.

Lyons, L. R., and R. M. Thorne, Equilibrium structure of radiation belt electrons, *J. Geophys. Res., 73*, 2142, 1973.

McIlwain C. E., Ring current effects on trapped particles, *J. Geophys. Res., 71*, 3623-3633, 1966.

Nymmik, R.A., Predicting the solar and galactic cosmic ray fluxes influencing the upper atmosphere in dependence with solar activity level, 32th COSPAR, Nagoya, Japan, July 1998.

Reeves, G., R. Friedel, R. Hayes, Maps could provide space weather forecasts for the inner magnetosphere, *Eos Trans. AGU, 79*, 613, 1998.

Roederer, J.G., Dynamics of geomagnetically trapped radiation, ed. *Springer-Verlag*, Berlin, Germany, 1970.

Selesnick, R., J.B. Blake, On the source location of radiation belt relativistic electrons, *J. Geophys. Res., 105*, 2607, 2000.

Shea, M., D. Smart, A word grid of calculated cosmic ray cutoff vertical rigidities for 1980.0, 18th International cosmic ray conference, 3:413, 1983.

Spjeldvik, W., Equilibrium structure of equatorially mirroring radiation belt protons, *J. Geophys. Res., 82*, 2801, 1977.

Tsyganenko, N., Magnetospheric magnetic field model with a warped tail current sheet, Planet. Space Sci., *54*, 75, 1989.

Vacaresse, A., D. Boscher, S. Bourdarie, M. Blanc, J.-A. Sauvaud, Modeling the high energy proton belt, *J. Geophys. Res., 104*, 28601, 1999.

D. Boscher and S. Bourdarie, ONERA/DESP, 2, avenue E. Belin, BP425, 31055 Toulouse Cedex 4, France. (e-mail Daniel.Boscher@onecert.fr)

The Search for Predictable Features of Relativistic Electron Events: Results from the GEM Storms Campaign

G. D. Reeves, K. L. McAdams, R. H. W. Friedel, and T. E. Cayton

Los Alamos National Laboratory, Los Alamos, New Mexico

We have examined the relativistic electron events associated with four geomagnetic storms driven by CME-produced magnetic clouds in order to discover some of the underlying consistent behavior that is common to these events. The geomagnetic conditions for the four events showed considerable variation as did the fluxes at fixed energies and fixed *L*-shell. By combining data from nine different satellites and by selecting specific parameters we found several characteristics of the relativistic electron events that were consistent from storm to storm. Our analysis focused on geosynchronous orbit and on $L\approx4.2$ which is near the heart of the outer electron belt. We examined a large number of parameters and combinations of parameters and found that among the most consistent sets were (1) the temporal behavior of the electron fluxes at $L\approx4.2$ and $L\approx6.6$, (2) the temporal evolution of the spectra at $L\approx4.2$ and $L\approx6.6$, and (3) the gradients of the phase space densities between $L\approx4.2$ and $L\approx6.6$. The characteristics of those parameters may be common to all relativistic electron events and if so that they hold important clues to the physical processes operating during these events. The fact that they are common to these four storms already provides an important framework for evaluating the success of various predictive space weather models that are being developed and tested against these four events.

1. INTRODUCTION

A common feature of geomagnetic storms, and one with important space weather implications, is the enhancement of electrons with relativistic energies in the Earth's outer radiation belts. Electrons with sufficiently high energies can penetrate spacecraft and instrument shielding causing damage to materials and circuits. The amount of energy required depends on the type and thickness of the shielding, but most studies consider electrons with energies greater than 2 MeV, which will penetrate approximately 300 mils of aluminum.

Relativistic electrons can cause damage in a number of ways, but the most common concerns are damage to materials through total radiation dose and damage to electronics through deep dielectric charging (also referred to as bulk charging). When electrons penetrate the protective layer of shielding they will impact other materials. If those materials are insulators then charge will become embedded in the material. If the rate of charge deposition is higher than the rate of charge leakage, then bulk charge will build up and will eventually discharge. The results can be minor, such as spurious signals in coaxial cables, or dramatic, such as the failure of critical electrical components. Deep dielectric charging from relativistic electron enhancements is a well-known hazard that is typically mitigated through prudent spacecraft design. Nevertheless, the risk is probabilistic and difficult to quantify, even in very cautious designs, and a number of serious spacecraft failures have been at least cir-

Space Weather
Geophysical Monograph 125
Copyright 2001 by the American Geophysical Union

cumstantially linked to enhanced relativistic electron fluxes [*Baker et al.,* 1998].

The risk of deep dielectric charging is proportional to the flux of relativistic electrons and to the hardness of the spectrum. Both are known to vary on a wide range of time scales ranging from minutes to a solar cycle. Relativistic electron enhancements are most commonly observed in the approach to solar minimum when high-speed solar wind streams from coronal holes buffet the Earth's magnetosphere. In fact the strong correlation between solar wind velocity and relativistic electron fluxes remains one of the best predictors of relativistic electron events [*Paulikas & Blake,* 1979; *Baker et al.,* 1990; *Belian et al.,* 1996]. When coronal holes are particularly long-lived, periodic enhancements of the relativistic electron fluxes can be observed every Carrington rotation - sometimes for over half a year.

If coronal holes were the only solar driver for relativistic electron events we would have a firm basis for predictions, but they are not. Relativistic electron events are also commonly caused by more random and impulsive solar activity. *Blake et al.,* [1992] and *Li et al.,* [1993] have documented the extremely sudden (several minute) enhancement of relativistic electrons to energies greater than 25 MeV by the passage of an interplanetary shock in March 1991. Relativistic electron enhancements are also commonly observed in association with Coronal Mass Ejections (CMEs) with and without associated shocks [e.g. *Baker et al.,* 1998; *Knipp et al.,* 1998; *Reeves et al.,* 1998a].

While there is a good association between solar wind drivers and the relativistic electron response, the relationship is not quantitatively predictable. Not all CMEs produce relativistic electron enhancements. Likewise *Paulikas & Blake's* [1979] plot of solar wind velocity vs. relativistic electron flux showed significant scatter until 27-day averages were taken. *Blake et al.,* [1997] suggested that the amount of southward IMF B_Z might play a role in producing a "seed population" of $\approx$100 keV electrons for subsequent acceleration. However, southward IMF B_Z and high solar wind velocity are exactly the ingredients necessary for building up the storm-time ring current, but *Reeves* [1998] showed that, while 90% of geomagnetic storms in 1994 were accompanied by relativistic electron enhancements, the correlation between their peak amplitudes was less than about 60%. While there is a strong association between solar wind drivers and the ring current and radiation belt responses the weaker quantitative correlations invite further study.

Because there is considerable variety in the solar wind conditions that drive relativistic electron enhancements there is also considerable variety from event to event. This can sometimes make the job of modeling a particular event in all its particulars seem rather daunting. In this study we take a different approach which is to try to determine the fundamental underlying conditions that are common among several events in order to provide a baseline for predictive models and for greater physical understanding.

2. THE GEM CAMPAIGN STORMS

The Geospace Environment Modeling (GEM) program is sponsored by the US National Science Foundation and has, as one of its goals, improved observational, theoretical, and predictive understanding of the dynamics of the inner magnetosphere during geomagnetic storms. To this end the GEM Inner Magnetosphere/Storms campaign selected three geomagnetic storms for intensive study. These storms took place in May 1997, September 1998, and October 1998. In this study we also compare the January 1997 storm which has been extensively documented in the scientific literature [e.g. *Reeves et al.,* 1998a].

Figure 1 shows the geophysical conditions for each of these four events. From top to bottom, the panels show the *Dst* index, the *Kp* index, the IMF B_Z component, the solar wind ion density, and the solar wind velocity. For the four storms the minimum *Dst* ranged from -85 to -228, and maximum solar wind velocity ranged from 500 km/s in May 1997 to 930 km/s in September 1998 while the other parameters also showed considerable variation. All four storms in this study were driven by CME-produced magnetic clouds.

Plate 1 shows the relativistic electron fluxes throughout the outer radiation belts for the four storms. These plots show the flux of electrons with energies greater than approximately 2 MeV as a function of *L*-shell with a 3-hour time resolution (where "*L*", also known as the McIlwain parameter, is approximately equal to the equatorial radial distance for a given magnetic flux tube in units of Earth radii). The high time-resolution for these events, which is essential for resolving the storm-time dynamics, was obtained by combining the data from nine separate satellites: POLAR, GOES-8, GOES-9, GPS NS-24, GPS NS-33, 1990-095, 1991-080, 1994-084, and SAMPEX [*McAdams et al.,* 2001]. Plate 1 shows that both the intensity and the temporal behavior of the >2 MeV electron fluxes varies from event to event. Although it is not shown in the figure, those differences are apparent at all measured energies.

By combining data from a large number of satellites we can essentially specify the relativistic electron fluxes as a function of time, energy, *L*-shell, and local time to produce what is essentially a "data synthesis model" for each event [e.g. *Reeves et al.,* 1998b]. These models are extremely useful for detailed analysis of a particular event and how it may have affected a particular satellite which may or may

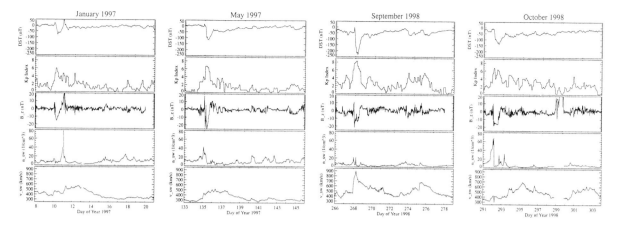

January 1997   May 1997   September 1998   October 1998

Figure 1. Geophysical conditions for the four storms. For each event we plot, from top to bottom, the *Dst* index, the *Kp* index, the IMF B_Z component, the solar wind density, and the solar wind velocity.

not have been equipped with environmental monitoring instruments. However, production of these data synthesis models is labor intensive and does not, by itself, provide any physical insight into the acceleration and transport processes.

3. COMMON ELEMENTS OF THE GEM STORMS

Using the specification of the four storms provided by all nine satellites, we searched for common elements of the storms which could provide a basis for inter-comparison and for developing physical models. We examined a large number of parameters and combinations of parameters and found that among the most consistent sets were (1) the temporal behavior of the electron fluxes at $L \approx 4.2$ and $L \approx 6.6$, (2) the temporal evolution of the spectra at $L \approx 4.2$ and $L \approx 6.6$, and (3) the gradients of the phase space densities between $L \approx 4.2$ and $L \approx 6.6$.

3.1 Temporal Behavior of Fluxes at L=4.2 vs. L=6.6

A useful rule of thumb for spacecraft operators has been that the relativistic electron fluxes peak 3-4 days after the onset of a geomagnetic storm. This time delay was thought to provide ample lead time for any needed change in operations. The time delay was also one of the primary observational motivations for the proposition that a slow "recirculation" process was responsible for the relativistic electron acceleration [*Nishida* 1976; *Fujimoto & Nishida* 1990].

One of the interesting observations from study of the January 1997 storm was that the relativistic electron fluxes at $L \approx 4.2$ increased much more quickly, reaching a peak within 12 hours after they began to rise. *Reeves et al.*, [1998a] concluded that the acceleration of the relativistic electrons in the heart of the radiation belts was relatively rapid (less than 1 day) and that the peak observed 3-4 days following the storm onset was a characteristic primarily of

the outer edge of the trapping region which includes $L \approx 6.6$. They attributed the delay at geosynchronous orbit to a redistribution of particles in the magnetosphere - perhaps outward radial diffusion.

We see from Plate 2 that this behavior is indeed common to all four storms. In each event there is an initial decrease in the relativistic electron fluxes which is attributable to an adiabatic response to the ring current - the so-called *Dst* effect [*Kim & Chan*, 1997]. Following the dip there is a very rapid increase in the fluxes measured at $L \approx 4.2$. In all cases the rise to maximum at $L \approx 4.2$ is on the order of a half a day. After reaching maximum levels the fluxes at $L \approx 4.2$ remain nearly constant or decrease slightly. The October 1998 storm is different in two respects. As with the other storms the fluxes at $L \approx 4.2$ increase rapidly following the main-phase *Dst* decrease, but they also continue to rise slowly over the following several days. Careful examination of the *L*-shell profiles show that this is because the radial position of the flux peak (in *L*) moved outward in time during the October storm causing the fluxes at the fixed position of $L \approx 4.2$ to increase. We suspect that this may be related to the continued activity reflected in the *Kp* index (Figure 1).

The other difference is that the >2 MeV electron fluxes were already elevated before the storm, so this event did not produce an overall increase in the fluxes at $L \approx 4.2$. In fact, we note that new activity can actually produce a sustained decrease in the relativistic electron fluxes even at $L \approx 4.2$. This is seen, for example, on Day 274 of the September 1998 event when a new interval of moderate activity (IMF B_Z south, $Kp > 4$, and $Dst < -85$ nT) appeared to "reset" the fluxes at $L \approx 4.2$ to a lower level.

3.2 Spectral Behavior at L=4.2 vs. L=6.6

While the temporal evolution of the fluxes at $L \approx 4.2$ and $L \approx 6.6$ exhibit some important common characteristics, the precise nature of those characteristics is a function of en-

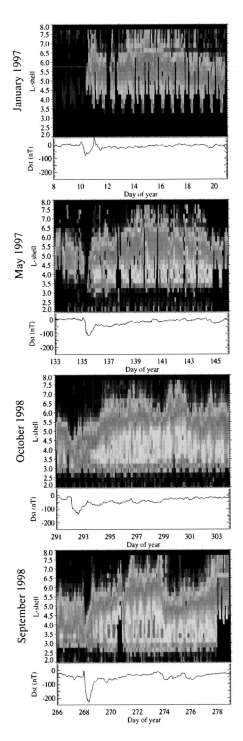

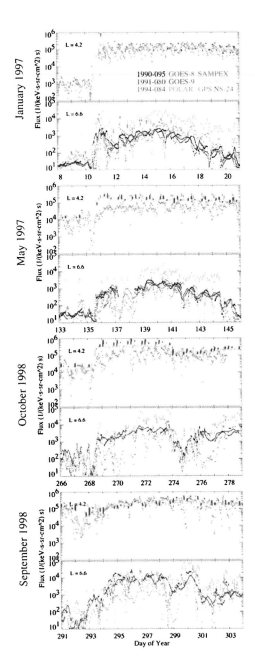

Plate 2. The flux of >2 MeV electrons measured near L=4.2 near the heart of the outer electron belt and at L=6.6, geosynchronous orbit. The fluxes measured by each satellite are color coded by spacecraft.

Plate 1. The flux of >2 MeV electrons as a function of time and L-shell. The color scale ranges in log10(flux) from 2 (black) to 6 (red). The Dst index for each event is shown below for reference.

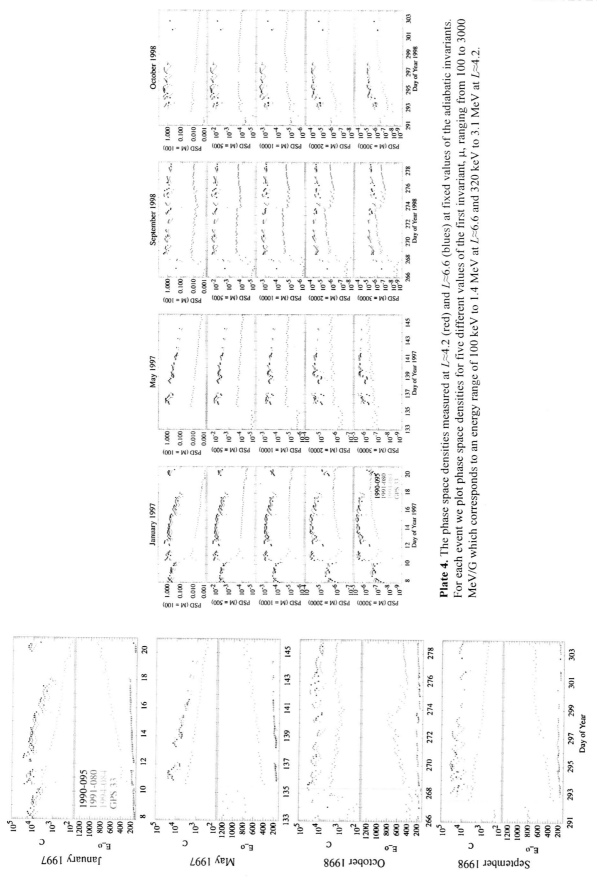

Plate 4. The phase space densities measured at $L \approx 4.2$ (red) and $L \approx 6.6$ (blues) at fixed values of the adiabatic invariants. For each event we plot phase space densities for five different values of the first invariant, μ, ranging from 100 to 3000 MeV/G which corresponds to an energy range of 100 keV to 1.4 MeV at $L \approx 6.6$ and 320 keV to 3.1 MeV at $L \approx 4.2$.

Plate 3. Parameters of an exponential fit to the spectra measured at $L \approx 4.2$ (red) and $L \approx 6.6$ (blues). C represents the "density" of electrons above 200 keV and E_0 is the spectral slope or "hardness".

ergy, and fluxes at different energies can appear to exhibit quite different temporal behavior. However, when one examines the characteristics of the spectra themselves, a satisfying consistency emerges.

McAdams et al., [2001] fit the fluxes measured by GPS at $L\approx4.2$ and the fluxes measured by the LANL geosynchronous satellites at $L\approx6.6$ to a simple exponential, $j(E) = C \exp(-E/E_0)$. Plate 3 shows the temporal profiles of C, which is the number of electrons with energy greater than 200 keV (a "density") and E_0, which is the spectral slope (or "temperature").

One interesting feature is that C and E_0 change much more abruptly than the >2 MeV fluxes. On this time scale the change appears nearly instantaneous. (When no points are plotted the fluxes are too low to obtain a meaningful spectral fit.)

Subsequent to the sudden change in spectrum the "density" begins to decrease at both L-shells (C decreases) and the spectrum at $L\approx4.2$ gets harder (E_0 increases). The changes at $L\approx4.2$ are consistent with a sudden "injection" of particles and a subsequent loss of the low-energy component of the spectrum (probably due to pitch angle scattering). At geosynchronous orbit the increase in "density" is nearly simultaneous with that seen at $L\approx4.2$ and the decrease over time is at roughly the same rate (with some interesting differences). The spectral slope at geosynchronous seems to stay roughly constant. It actually peaks and decreases roughly in concert with the >2 MeV fluxes, but the changes are small compared to the changes at $L\approx4.2$.

3.3 Phase Space Densities at L=4.2 *vs.* L=6.6

Theoretical studies and physical models of the relativistic electron belts are almost always cast in terms of phase space densities at fixed values of the adiabatic invariants while spacecraft observations are almost always at fixed energies. Closer connection between the models and observations can be obtained if measured fluxes can be converted to phase space density. This can be done if certain assumptions are made. Here we restrict our analysis to locations where the satellites are near the magnetic equator (geosynchronous and $L\approx4.2$ for GPS). We use a model magnetic field to calculate the first invariant, we assume that the second invariant is strictly zero (e.g. 90° pitch angles), and we assume that the magnetic field is time stationary [*McAdams et al.,* 2001].

The assumption that the magnetic field is time stationary during the storms is the most important assumption for the interpretation of the results. In particular it means that the adiabatic "*Dst* Effect" has not been removed. However, without complete knowledge of the actual, global, time-dependent magnetic field any other assumption would be equally model-dependent.

The resulting phase space densities for five values of μ, the first invariant, are plotted in Plate 4. The range of μ corresponds to electron energies of 100 keV to 1.4 MeV at geosynchronous and 320 keV to 3.1 MeV at $L\approx4.2$. We use the spectral fits shown in Plate 3 to derive the fluxes used to calculate the phase space densities at appropriate values of μ.

Several interesting and consistent features are apparent in the phase space density profiles. First we note that for the range of μ values plotted, and for all four storms, the phase space densities at $L\approx6.6$ are always greater than those at $L\approx4.2$ indicating that there is a sufficiently large source population at high L-shells that diffusive inward transport is a viable mechanism.

Second, we find that there are distinct differences at high and low values of μ, particularly at $L\approx4.2$. At low μ the phase space density decreases in the ten days following the storm, which is indicative of the loss of low-energy particles through pitch angle scattering. Interestingly, at low values of μ the gradient in phase space density remains nearly constant over this relatively long time period.

Instead of decreasing over time, the phase space densities at higher μ for $L\approx4.2$ systematically increase over time. At $L\approx6.6$ the phase space densities do not show as big a difference with increasing μ as they do at $L\approx4.2$, although, they can also show a gradual increase for some period of time. In the several days following the storm, though, the gradients at high μ decrease until, after about ten days the phase space densities are nearly equal.

While the physical processes responsible for these phase space density profiles are still a matter of current research, it is clear that there is a degree of consistency among these four storms which any successful model of relativistic electron events would need to reproduce.

4. CONCLUSIONS

We have examined the relativistic electron events associated with four geomagnetic storms driven by CME-produced magnetic clouds in order to discover some of the underlying consistent behavior that is common to these events. The geomagnetic conditions for the four events showed considerable variation. The *Dst* index ranged from -85 to -228 nT and maximum solar wind velocity ranged from 500 km/s to 930 km/s. When plotted as fluxes at fixed energies and fixed L-shells, the relativistic electron response in the outer radiation belts also varies considerably from storm to storm.

By combining data from nine different satellites and by selecting specific parameters, we found several characteristics of the relativistic electron events that were consistent from storm to storm. Our analysis focused on geosynchronous orbit and on $L \approx 4.2$, which is near the heart of the outer electron belt and, importantly, where the GPS satellites cross the magnetic equator.

We found that the fluxes of >2 MeV electrons at $L \approx 4.2$ and $L \approx 6.6$ for the three GEM storms were quite similar to the behavior described by *Reeves et al.,* [1998a] for the January 1997 storm. The >2 MeV fluxes at $L \approx 4.2$ increase quite rapidly (on the order of 12 hours) and then remain nearly constant over the next ten days. The well-known 3-4 day delay between storm onset and peak relativistic electron fluxes appears to be characteristic of the outer edges of the electron belts, near geosynchronous orbit.

The relativistic electron spectra also had common characteristics among the four storms. The change in spectra at both L-shells was even more abrupt than the change in fluxes. In particular the "density" of electrons with energies above 200 keV increase sharply and then decrease monotonically over time. The rate of decrease at $L \approx 4.2$ and $L \approx 6.6$ was nearly the same. Although the spectral slope at $L \approx 6.6$ changes over time the changes are small compared to those at $L \approx 4.2$ and do not show the same consistent hardening over time.

We also used the spectral fits and an assumed magnetic field to calculate phase space densities which also showed a more consistent behavior than the fluxes at fixed energy. We found that for all values of μ considered here the phase space densities at geosynchronous were higher than at $L \approx 4.2$, but that the gradient decreased with increasing μ. At $L \approx 4.2$ we found that the phase space densities at low μ decreased slowly over time while at high μ they increased. The differences between low and high μ at $L \approx 6.6$ were not as strong so, while at low μ the phase space density gradient remained nearly constant, at high μ the gradient decreased over time and nearly disappeared after ten days.

We suspect that these features may be common to all relativistic electron events and, if so, they hold important clues to the physical processes operating during these events. The fact that they are common to these four storms already provides an important framework for evaluating the success of various predictive space weather models that are being developed and tested against these events.

Acknowledgments. We would like to thank R. D. Belian, R. A. Christensen, S. Kanekal, J. B. Blake, R. Selesnick, W. C. Feldman, and J. F. Fennell for their help in acquiring and processing the data used in this study. We also thank the US Department of Energy Office of Basic Energy Science (OBES) for support of this study.

REFERENCES

Baker, D. N., R. L. McPherron, T. E. Cayton, and R. W. Klebesadel, Linear prediction filter analysis of relativistic electron properties at 6.6 RE, *J. Geophys. Res., 95,* 15,133, 1990.

Baker, D. N., T. Pulkkinen, X. Li, S. G. Kanekal, J. B.. Blake, R. S. Selesnick, M. G. Henderson, G. D. Reeves, H. E. Spence, and G. Rostoker, Coronal mass ejections, magnetic clouds, and relativistic magnetospheric electron events: ISTP, *J. Geophys. Res., 103,* 17,279, 1998.

Belian, R. D., T. E. Cayton, R. A. Christensen, J. C. Ingraham, M. M. Meier, G. D. Reeves, and A. J. Lazarus, Relativistic electrons in the outer-zone: An 11-year cycle; their relation to the solar wind, in *Workshop on the Earth's Trapped Particle Environment,* edited by G. D. Reeves, AIP Press, New York, 13-18, 1996.

Blake, J. B., D. N. Baker, N. Turner, K. W. Ogilvie, and R. P. Lepping, Correlation of changes in the outer-zone relativistic electron population with upstream solar wind and magnetic field measurements, *Geophys. Res. Lett., 24,* 927-929, 1997.

Blake, J. B., W. A. Kolasinski, R. W. Fillius, and E. G. Mullen, Injection of electrons and protons with energies of tens of MeV into $L<3$ on 24 March 1991., *Geophys. Res. Lett., 19,* 821, 1992.

Fujimoto, M., and A. Nishida, Energization and isotropization of energetic electrons in the Earth's radiation belt by the recirculation process, *J. Geophys. Res., 95,* 4625, 1990.

Kim, H.-J., and A. A. Chan, Fully-adiabatic changes in storm-time relativistic electron fluxes, *J. Geophys. Res., 102,* 22,107-22,116, 1997.

Knipp, D. J., N. Crooker, M. Engebretson, X. Li, A. H. McAllister, T. Mukai, S. Kokubun, G. D. Reeves, T. Obara, A. Weatherwax, and B. A. Emery, An overview of the early November 1993 geomagnetic storm, *J. Geophys. Res., 103,* 26,197-26,220, 1998.

Li, X., R. Roth, M. Temerin, J. R. Wygant, M. K. Hudson, and J. B. Blake, Simulation of the prompt energization and transport of radiation belt particles during the March 24, 1991, SSC, *Geophys. Res. Lett., 20,* 2423, 1993.

McAdams, K. L., G. D. Reeves, R. H. W. Friedel, and T. E. Cayton, Mulit-satellite comparisons of the radiation belt response to the GEM magnetic storms, *J. Geophys. Res., in press,* 2001.

Nishida, A., Outward diffusion of energetic particles from the Jovian radiation belt, *J. Geophys. Res., 81,* 1771, 1976.

Paulikas, G. A., and J. B. Blake, Effects of the solar wind on magnetospheric dynamics: Energetic electrons at the synchronous orbit, in *Quantitative Modeling of Magnetospheric Processes, Geophys. Monogr., Amer. Geophys. Un., 21,* 180, 1979.

Reeves, G. D., D. N. Baker, R. D. Belian, J. B. Blake, T. E. Cayton, J. F. Fennell, R. H. W. Friedel, M. G. Henderson, S. Kanekal, X. Li, M. M. Meier, T. Onsager, R. S. Selesnick, and H. E. Spence, The Global Response of Relativistic Radiation Belt Electrons to the January 1997 Magnetic Cloud, *Geophys. Res. Lett., 17,* 3265-3268, 1998a.

Reeves, G. D., R. H. W. Friedel, and R. Hayes, Maps could provide space weather forecasts for the inner magnetosphere, *EOS Trans. AGU, 79,* 613, 1998b.

Reeves, G. D., Relativistic electrons and magnetic storms: 1992-1995, *Geophys. Res. Lett., 25,* 1817, 1998.

T. E. Cayton, R. H. W. Friedel, M. G. Henderson, K. L. McAdams and G. D. Reeves, Los Alamos National Laboratory, NIS-1 MS D-466, Los Alamos, NM 87545. (reeves@lanl.gov).

Forecasting Kilovolt Electrons

R. A. Wolf, R. W. Spiro, T. W. Garner, and F. R. Toffoletto

Dept. of Physics and Astronomy, Rice University, Houston, Texas, U.S.A.

The Magnetospheric Specification Model (MSM) represents the state-of-the-art in operational specification of fluxes of magnetospheric electrons in the 1-100 keV range. When driven by upstream solar-wind data, it can provide very short-term forecasts. Partly theoretical and partly empirical, the MSM calculates average particle drifts using data-driven models for the electric and magnetic fields and for initial and boundary fluxes. For the geosynchronous-orbit region, where it has been thoroughly tested, the MSM represents general trends in a reasonable way, but the accuracy of its predicted flux for a given spacecraft at a given instant is less impressive. The RMS error in the $\log_{10}(\text{flux})$ is about 0.45. However, it will probably not be easy to make significant global improvement in that level of accuracy. One barrier to achieving high accuracy is our incomplete understanding of the physics of magnetospheric substorms, which are the main cause of dramatic increases in near-geosynchronous kilovolt electrons. For the immediate future, the most promising approach to improving accuracy is the assimilation of real-time particle-flux data from spacecraft near geosynchronous orbit. The first step along that line showed a potential for modest improvement in accuracy, within about 2 hours local time of the data-providing spacecraft. In the long run, we can hope for models that represent the essential physics of substorm-associated particle injections. Recent runs with the Rice Convection Model coupled with a friction-code equilibrium solver provide a different view of substorms that emphasizes aspects that are crucial for particle injection into the inner magnetosphere. In this view, a substorm is the physical process that violates the adiabatic drift laws to produce flux tubes with lower values of $pV^{5/3}$ than typical middle-plasma-sheet flux tubes. These reduced plasma population flux tubes are the primary ones injected into the inner magnetosphere during substorms.

INTRODUCTION

Electrons in the 1-100 keV energy range are a major cause of surface charging problems for spacecraft in the geosynchronous-orbit region [*e.g., Lanzerotti et al.*, 1998]. The Magnetospheric Specification Model (MSM) [*Tascione et al.*, 1988; *Bales et al.*, 1993; *Lambour*, 1994; *Wolf et al.*, 1997] was developed in the late 1980s and early 1990s to provide operational nowcasts of these electron fluxes.

Dramatic variations in the flux of geosynchronous electrons in the 1-100 keV range mainly result from substorms, which are primarily driven by the solar wind with a time delay of the order of half an hour. Until reliable solar-wind forecasts become available, detailed prediction of kilovolt

Space Weather
Geophysical Monograph 125
Copyright 2001 by the American Geophysical Union

313

electron fluxes is limited by the time for the solar wind to travel from an upstream solar-wind monitor to Earth's vicinity, resulting in a maximum lead-time of about one hour. From a magnetospheric modeling point of view, such short-term forecasts use essentially the same computational machinery as nowcasts. While this paper concentrates on nowcasting, the types of models described could, in principle, provide very-short-term forecasts when suitably modified for incorporation of real-time data from upstream solar wind monitors. Of course, accurate short-term forecasts will require a substantial improvement over present capabilities: at present, we cannot accurately predict the onset time, intensity, or location of a substorm from solar-wind data.

We begin with a brief description of how the MSM works and of its extensive performance testing. We move on to discuss data assimilation, which is the most likely initial route to improved performance. We conclude with speculation about the possibilities for advanced models that incorporate more sophisticated theoretical understanding.

THE MAGNETOSPHERIC SPECIFICATION MODEL

The MSM was developed for the U. S. Air Force with the primary objective of specifying the flux of kilovolt electrons using established physics and real-time input data. It is now run routinely by the NOAA Space Environment Center, and its nowcasts are available on the web (http://sec.noaa.gov/prc/msm/index.html).

The MSM, an offspring of the Rice Convection Model (RCM), combines established theoretical formulas with real-time data. The theoretical formulation includes gradient, curvature, and $\mathbf{E} \times \mathbf{B}$ drift, and application of Liouville's theorem, suitably adjusted for estimated loss of plasma by electron precipitation and ion charge exchange. A key approximation, made for the sake of computational efficiency, is the assumption that the particle pitch-angle distributions are kept isotropic by some mechanism (*e.g.*, slow-moving electrostatic structures) that randomizes pitch angles without changing particle energies. The MSM treats the magnetospheric plasma distribution using a multi-fluid approach with each fluid being characterized by a particle species (electron, H^+, or O^+) and a value of the energy invariant λ, appropriate for an isotropic distribution, which is related to the kinetic energy E by

$$E = \lambda \left(\int \frac{ds}{B} \right)^{-2/3}$$

where the term in the parentheses is the volume of a magnetic flux tube with unit magnetic flux. The quantity λ

can be shown to be constant along a particle drift path [*Harel et al.*, 1981].

The MSM assumes that kilovolt H^+ and O^+ ions are lost by charge exchange with the cold neutral gas, using charge-exchange lifetimes from a pre-computed lookup table [*J. Bishop*, private communication]. The principal electron loss rate is assumed to be precipitation into the loss cone. For electron precipitation loss rates, the MSM uses a combination of formulae, including one based on *Schumaker et al.* [1989] for the plasma sheet, one based on the work of *Lyons et al.* [1972] for electrons above 40 keV in the plasmasphere, and various simple interpolation and extrapolation schemes to cover large gaps between published estimates. Procedures have been, to some extent, optimized to improve agreement with observed electron fluxes [*Lambour*, 1994].

The MSM utilizes the magnetic field model of *Hilmer and Voigt* [1995], driven by three physical input parameters: the solar-wind ρv^2, which determines the magnetopause standoff distance; the magnetic storm index Dst, which determines the strength of the ring-current; and the auroral boundary index (the low-latitude edge of auroral electron precipitation projected to local midnight) [*Gussenhoven et al.*, 1983], which is used to estimate the amount of magnetic flux in the tail. In operation, the MSM interpolates magnetic field information among approximately 1000 pre-computed configurations for a canonical set of the input parameters. To model a given time period, magnetic field configurations are constructed at 15-minute intervals by interpolation of the pre-computed models, using input values appropriate to each interval.

While the Rice Convection Model computational machinery is based on a self-consistent calculation of the electric field and plasma distribution in the inner magnetosphere, the MSM relies on an experiential data-driven analytic formulation of the global electric field. Inputs for the time-dependent potential electric field model are: (1) the cross-polar-cap potential drop, which governs the overall strength of convection; (2) the auroral boundary index and its time derivative, which govern the spatial extent of the auroral flow pattern and the degree of penetration to low ionospheric latitudes; and (3) the polar-cap pattern type, which is based on IMF B_y, if available. In the polar cap, the MSM electric field is a scaled version of the *Heppner-Maynard* [1987] model, as converted to analytic formulae by *Rich and Maynard* [1989]. Penetration of convection to the subauroral ionosphere is designed to imitate the Rice Convection Model results of *Spiro et al.* [1988]. The expression for the potential pattern in the auroral-zone is designed to exhibit conventional sunward flow and to merge smoothly with the Heppner-Maynard-based polar cap and the RCM-based subauroral formulae. In computing particle drifts, the MSM implicitly takes account of the induction electric field corresponding to the changing magnetic field described above.

The particle data used to set MSM's initial- and boundary-condition fluxes are assembled from statistical studies [*e.g.*, *Huang and Frank*, 1989; *Garrett et al.*, 1981], where available, as well as various published event studies. The fluxes are characterized by particle species (electrons, H^+, O^+), energy, *L*-value, and *Kp* index. *Lambour* [1994] gives a more complete description of the procedure.

The MSM has been tested at Rice [*Lambour*, 1994] and, much more extensively, by *Hilmer and Ginet* [2000], who used a large data set from the Charge Control System on a Defense Satellite Communications System spacecraft. A sample result of their testing is shown in Figure 1. In general, the MSM gets basic trends approximately correct, but is unimpressive on details. Most notably, geosynchronous energetic electrons sometimes exhibit dramatic flux dropouts [*Thomsen et al.*, 1994], which the MSM usually either underestimates or misses entirely. More rarely, it misses or greatly underestimates a peak flux. The overall RMS error in $\log_{10}$(flux) was approximately 0.45 in the Hilmer-Ginet tests.

DATA ASSIMILATION

The MSM was designed in 1987. The Magnetospheric Specification and Forecast Model, which included machinery for running from solar wind data, was completed in 1994. These codes no longer represent new technology, nor is their performance overwhelmingly impressive. Nonetheless, it will not be easy to design an operational algorithm that will offer dramatically better performance. One factor that severely limits accuracy is the fact that the primary dynamic process that causes abrupt increases in geosynchronous electron fluxes is the magnetospheric substorm. The substorm was not understood when the MSM was designed in the late eighties, and it is still not understood today, though substantial progress has been made.

The most promising near-term avenue toward improved nowcasting of geosynchronous-region keV electrons lies in the intelligent use of data. The most obvious data that the MSM does not use, but could, are the measured geosynchronous electron fluxes, which are available in near-real time. In the structure of the MSM, those data constitute a natural output of the program, not a natural input. Use of geosynchronous particle data entails using observations to constrain theoretically computed fluxes.

One obvious problem with using real-time-measured particle fluxes to correct the model is that there are only a few

observing spacecraft in the vastness of the magnetosphere. It is not clear how far from the point of observation the data can constrain the model realistically. A second problem is that when the MSM gets particle fluxes dramatically wrong, it is often because the electric and magnetic field models are wrong. We have not yet developed automatic algorithms that can objectively constrain electric- and magnetic-field models based on geosynchronous particle fluxes.

However, a straightforward first attempt at using geosynchronous particle-flux data to constrain the MSM has been made by *Garner et al.* [1999]. It utilizes the "direct insertion" approach by overriding model-computed electron fluxes within the time loop of the model. The constraint is applied at 15-minute intervals for a region around the spacecraft. The size of the constrained region has been determined by testing and optimization. Because the data is assimilated within the time loop of the MSM, which

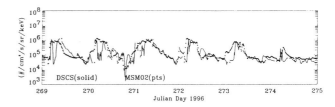
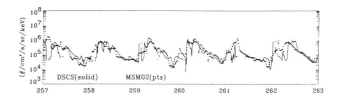
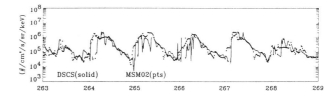
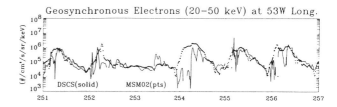

Figure 1. Comparison between MSM predictions (points) and DSCS spacecraft measurements (solid curve) of geosynchronous electrons for 25 days in 1996. From *Hilmer and Ginet* [2000].

follows particle drifts, the constraint propagates through the system with the drifting particles. (See Plate 1.)

Figure 2 shows the result of a test of the data assimilation procedure. Data were obtained from the MPA instrument, developed by Los Alamos National Laboratory, on one geosynchronous satellite. To test the efficacy of the data assimilation procedure, we compared MSM-predicted 40-keV electron fluxes with measurements from a similar Los Alamos instrument on a spacecraft that was located about 30° east of the first one. Data assimilation clearly improved accuracy, but undramatically. The unconstrained MSM run missed several flux spikes. (The MSM was first intentionally run with only partial inputs to provide ample room for improvement by the data assimilation procedure.) When run using the data assimilation procedure, the MSM results showed some of those spikes, though it underpredicted the peak fluxes. Some conclusions from similar tests were:

1. When run for quiet geomagnetic conditions, the data assimilation was effective in correcting the MSM's quiet, steady flux level, which was based on a statistical model, to what was actually observed on the day in question.

2. In tests for a magnetic storm using geosynchronous spacecraft separated by about 120° longitude, the data assimilation procedure had no significant effect. The problem was that electrons from the data-providing spacecraft rarely drifted close to the test spacecraft. (See Figure 3.)

A joint Rice/Los Alamos effort is currently underway to try to develop more sophisticated data assimilation schemes, for example to use the geosynchronous particle data to correct the MSM's electric field model. One promising approach involves using the upper cutoff energy for plasma-sheet electrons observed at geosynchronous orbit as an indicator of the strength of the electric field in that region. Another possible approach centers on running MSM several times for each hour, assuming different electric field strengths, then choosing the electric field that fits the observed fluxes best.

Of course, the effectiveness of data assimilation increases in direct proportion to the number of spacecraft that supply data. Although the test illustrated in Plate 1 and Figure 3 assimilated data from only one spacecraft, several geosynchronous spacecraft are typically available, and substantial fleets of small craft are currently under discussion.

The effectiveness of the data assimilation for inner-magnetospheric particles increases with increasing particle energy, because the constraint propagates at the drift velocity. Suppose that a spacecraft is in a highly elliptic orbit

and that it passes $L = L_0$ and local time $\phi = \varphi_0$ at time t_0, measuring fluxes. If that measurement is approximately valid for time Δt, then use of Liouville's theorem in the code propagates the measured fluxes to a longitude range of 360° ($\Delta t/t_{drift}$). That covers a whole drift shell if $\Delta t > t_{drift}$. For substorm-associated 10-50 keV particles, Δt cannot be more than ~0.5 hours, and the drift period is 2-10 hours. Thus the drift spreads the information only about 18-90° of longitude. Drift shells are also much more variable for keV particles than for MeV particles. Several spacecraft in petal orbits will provide a better picture of MeV particles than they will for keV particles.

COMMENTS ON POSSIBILITIES FOR DRAMATICALLY IMPROVED THEORETICAL MODELS

For computational modeling of the inner magnetosphere, the Holy Grail has been the assembly of computational machinery for solving a complete and defensible set of theoretical equations. An adequate large-scale theoretical model of the physical system must solve a set of equations something like the following:

1. An equation that calculates the distribution of magnetospheric particles by requiring the distribution function to be constant along a drift path, except for the effects of sources, losses, and diffusion. The drift path is defined by conservation of the first two adiabatic invariants, or, for the case on an isotropic distribution, by the single energy invariant λ.

2. An equation for calculating Birkeland current down into the ionosphere from magnetospheric pressure gradients, using *Vasyliunas'* [1970] equation or an acceptable alternate.

3. An equation that conserves current in the ionosphere and allows calculation of the potential electric field, given an appropriate model of ionospheric conductance, the distribution of Birkeland current, and appropriate boundary conditions.

4. An algorithm for calculating the magnetic field so that it is in approximate force balance with the model plasma and consistent with Maxwell's equations.

We are now beginning to construct solutions to such a set of coupled equations. *Toffoletto et al.* [1996; 2000a,b] have coupled the RCM to a friction-code equilibrium solver. The RCM solves equations representing conditions 1, 2, and 3, though assuming isotropic pitch angles and neglecting loss. The friction code maintains approximate force balance (condition 4) out to approximately 40 R_E in the tail. *Toffoletto et al.* [2000a] have started from a

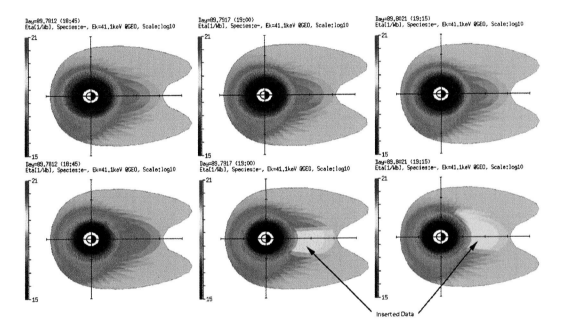

Plate 1. Sample comparison of MSM predicted distributions of ~41 keV electrons at geosynchronous orbit without data ingestion (top row) and with data ingestion (bottom). The plots show the $\log_{10}(\eta)$, where η is the number of particles per unit magnetic flux in the relevant MSM invariant energy range; η is proportional to the distribution function. From *Garner et al.* [1999].

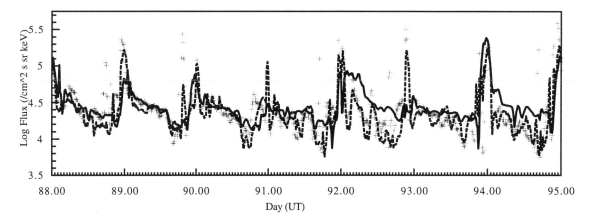

Figure 2. Comparison of observational data (crosses), MSM without data ingestion (solid curve) and MSM with data ingestion (dashed curve). The ordinate represents the log base 10 of the flux in units of cm^{-2} s^{-1} sr^{-1} keV^{-1}. From *Garner et al.* [1999].

magnetosphere based on a *Tsyganenko* [1989] magnetic field, but relaxed to approximate force equilibrium with an average observed pressure distribution. When this configuration is subjected to strong convection (100 kV across the tail), it evolves in time. The tail lobe field gets stronger and stronger as field lines in the inner plasma sheet become more and more stretched, and as the current sheet thins. (See Figure 3 of *Toffoletto et al.* [2000a].) The magnetic configuration strongly resembles the growth phase of a magnetospheric substorm. However, the time evolution does not extend beyond the growth phase. The system evolves to a state in which the inner plasma sheet consists of flux tubes that have values of pV^γ that are much too large to take dipolar form and become part of the inner magnetosphere. (Here, p is the pressure, V is the flux tube volume, and $\gamma = 5/3$.) Thus solving the complete set of conditions does not, by itself, produce an adequate picture of inner-magnetospheric dynamics during the expansion phase of a substorm. Enforcing those four conditions throughout the inner and middle plasma sheet in fact seems to preclude significant particle injections into the inner magnetosphere.

From an inner-magnetospheric point of view, the essential role of the expansion phase of the magnetospheric substorm is violation of the adiabatic drift law that p_sV^γ be constant along a drift path in the isotropic case. Here, p_s is the partial pressure due to particles of energy invariant λ_s. The expansion phase dramatically reduces p_sV^γ on certain inner-plasma-sheet flux tubes. Magnetic reconnection in the middle plasma sheet reduces p_sV^γ on closed field lines by releasing much of the particle energy into tailward-moving plasmoids. Cross-tail current disruption would also reduce p_sV^γ on an additional set of inner-plasma sheet flux tubes: field lines slip on the plasma and

dipolarize, which represents a drastic reduction in the flux tube volume V.

Toffoletto et al. [2000b] carried out two simulations in which p_sV^γ-values were reduced on selected inner-plasma-sheet flux tubes, in patterns specifically designed to mimic the near-Earth X-line and cross-tail-current disruption pictures of substorms. The magnetic configuration changes corresponding to the near-Earth X-line case are shown in Figure 4. While these runs need to be carried further in order to examine the properties of the resulting particle injections, it is already clear that the two assumptions imply very different patterns of ionospheric electric field [*Toffoletto et al.*, 2000b].

In this view of the substorm expansion phase, the essence of the process is the pattern of non-adiabatic reduction of p_sV^γ, or, more generally, on the distribution function $f(\lambda, x)$ or $f(\mu, J, \mathbf{x})$. It is specifically the particles that experienced the violent non-adiabaticity that become part of the storm-time ring current. The creation of flux tubes with different values of p_sV^γ also gives rise to crucial inter-

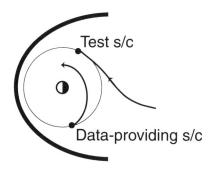

Figure 3. Cartoon illustrating path of electrons forward in time from a data-providing spacecraft and backward in time from the test spacecraft, for active times.

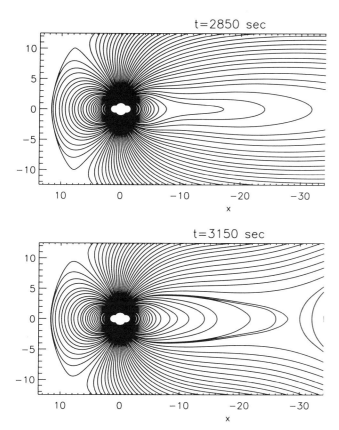

t=2850 sec

t=3150 sec

Figure 4. Change of field-line configuration, in the noon-midnight meridian plane, at the onset of the substorm expansion phase. The upper diagram, labeled t = 2850 sec, represents the situation at the end of the growth phase, while the lower diagram, labeled t = 3150 sec, represents the situation just after the beginning of the expansion phase.

change motions, with the smallest-$p_s V^\gamma$ tubes forced earthward.

The MSM bypasses these problems in a simplistic fashion. It has little difficulty compressing high-$p_s V^\gamma$ flux tubes into the inner magnetosphere since its magnetic field is not in force balance with the plasma distribution (condition 4). Interchange motion does not occur, because the model doesn't enforce conditions 2 or 3. Operationally, the MSM avoids prediction of excessively high fluxes in the inner magnetosphere by simply placing a ceiling on its calculated electron fluxes.

In order to move beyond MSM to an accurate first-principles calculation of kilovolt electron fluxes, we will have to either understand or be able to parameterize the non-adiabatic processes that constitute the essence of the magnetospheric substorm. The crucial parameter from the viewpoint of the inner magnetosphere is the distribution function $f(\lambda, x)$ or $f(\mu, J, \mathbf{x})$ within plasma flux tubes that have been subject to the non-adiabaticity.

Acknowledgments. The authors are grateful to the referees for very helpful comments on the manuscript. This research was supported by NASA under the Sun-Earth Connection Theory Program, grant NAG5-8136. The data-assimilation work was supported by the Defense Threat Reduction Agency, Dulles, VA.

REFERENCES

Bales, B., J. Freeman, B. Hausman, R. Hilmer, R. Lambour, A. Nagai, R. Spiro, G.-H. Voigt, R. Wolf, W. F. Denig, D. Hardy, M. Heinemann, N. Maynard, F. Rich, R. D. Belian, and T. Cayton, Status of the development of the Magnetospheric Specification and Forecast Model, in *Solar-Terrestrial Predictions-IV: Proceedings of a Workshop at Ottawa, Canada, May 18-22, 1992*, edited by J. Hruska, M. A. Shea, D. F. Smart, and G. Heckman, pp. 467-478, NOAA, Environmental Res. Labs, Boulder, 1993.

Garner, T. W., R. A. Wolf, R. W. Spiro, and M. F. Thomsen, First attempt at assimilating data to constrain a magnetospheric model, *J. Geophys. Res., 104* (A11), 25145-25152, 1999.

Garrett, H. B., D. C. Schwank, and S. E. DeForest, A statistical analysis of the low-energy geosynchronous plasma environment-1. electrons, *Planet. Space Sci., 29*, 1021-1044, 1981.

Gussenhoven, M. S., D. A. Hardy, and N. Heinemann, Systematics of the equatorward diffuse auroral boundary, *J. Geophys. Res., 88*, 5692-5708, 1983.

Harel, M., R. A. Wolf, P. H. Reiff, R. W. Spiro, W. J. Burke, F.. J. Rich, and M. Smiddy, Quantitative simulation of a magnetospheric substorm, 1. Model logic and overview, *J. Geophys. Res., 86*, 2217-2241, 1981.

Heppner, J. P., and N. C. Maynard, Empirical high-latitude electric field models, *J. Geophys. Res., 92*, 4467-4489, 1987.

Hilmer, R. V., and G. P. Ginet, A Magnetospheric Specification Model Validation Study: Geosynchronous electrons, *J. Atm. Solar-Terrest. Phys.* to be published in *J. Atmos. Solar Terrest. Phys.*, 2000.

Hilmer, R. V., and G.-H. Voigt, A magnetospheric magnetic field model with flexible current systems driven by independent physical parameters, *J. Geophys. Res., 100* (A4), 5613-5626, 1995.

Huang, C. Y., and L. A. Frank, A statistical study of the central plasma sheet: Implications for substorm models, *Geophys. Res. Lett., 13*, 652-655, 1986.

Lambour, R. L., Calibration of the Rice Magnetospheric Specification and Forecast Model for the Inner Magnetosphere, Ph. D. thesis, Rice University, Houston, TX, 1994.

Lanzerotti, L. J., K. LaFleur, C. G. Maclennan, and D. W. Maurer, Geosynchronous spacecraft charging in January 1997, *Geophys. Res. Lett., 25* (15), 2967-2970, 1998.

Lyons, L. R., R. M. Thorne, and C. F. Kennel, Pitch angle diffusion of radiation belt electrons within the plasmasphere, *J. Geophys. Res., 77*, 3455, 1972.

Rich, F. J., and N. C. Maynard, Consequences of using simple analytic functions for the high-latitude convection electric field, *J. Geophys. Res., 94*, 3687-3701, 1989.

Schumaker, T. L., M. S. Gussenhoven, D. A. Hardy, and R. L. Carovillano, The relationship between diffuse auroral and plasma sheet electron distributions near local midnight, *J. Geophys. Res., 94*, 10061-10078, 1989.

Spiro, R. W., R. A. Wolf, and B. G. Fejer, Penetration of high-latitude-electric-field effects to low latitudes during SUNDIAL 1984, *Ann. Geophys., 6*, 39-50, 1988.

Tascione, T. F., H. W. Kroehl, R. Creiger, J. W. Freeman, Jr., R. A. Wolf, R. W. Spiro, R. V. Hilmer, J. W. Shade, and B. A. Hausman, New ionospheric and magnetospheric specification models, *Rad. Sci., 23*, 211-222, 1988.

Thomsen, M. F., S. J. Bame, D. J. McComas, M. B. Moldwin, and K. R. Moore, The magnetospheric lobe at geosynchronous orbit, *J. Geophys. Res., 99*, 17283-17293, 1994.

Toffoletto, F. R., R. W. Spiro, R. A. Wolf, M. Hesse, and J. Birn, Self-consistent modeling of inner magnetospheric convection, in *Proc. Third International Conference on Substorms (ICS-3), ESA SP-389*, edited by E. J. Rolfe, and B. Kaldeich, pp. 223-230, ESA Publications Division, Noordwijk, The Netherlands, 1996.

Toffoletto, F. R., J. Birn, M. Hesse, R. W. Spiro, and R. A. Wolf, Modeling inner magnetospheric electrodynamics, in *Proceedings of the Chapman Conference on Space Weather*, edited by P. Song, G. L. Siscoe, and H. J. Singer, Am. Geophys. Un., Washington, D. C., 2000*a*.

Toffoletto, F. R., R. W. Spiro, R. A. Wolf, J. Birn, and M. Hesse, Computer experiments on substorm growth and expansion, submitted to *Proceedings of ICS-5*, 2000*b*.

Vasyliunas, V. M., Mathematical models of magnetospheric convection and its coupling to the ionosphere, in *Particles and Fields in the Magnetosphere*, edited by B. M. McCormac, pp. 60-71, D. Reidel, Hingham, MA, 1970.

Tsyganenko, N. A., A magnetospheric magnetic field model with a warped tail current sheet, *Planet. Space Sci., 37*, 5-20, 1989.

Wolf, R. A., J. W. Freeman, Jr., B. A. Hausman, R. W. Spiro, R. V. Hilmer, and R. L. Lambour, Modeling convection effects in magnetic storms, in *Magnetic Storms*, edited by B. T. Tsurutani, J. K. Arballo, W. D. Gonzalez, and Y. Kamide, pp. 161-172, Am. Geophys. Un., Washington, D. C., 1997.

T. W. Garner, Dept. of Physics and Astronomy, Rice University, MS 108, P. O. Box 1892, Houston, TX 77251

R. W. Spiro, Dept. of Physics and Astronomy, Rice University, MS 108, P. O. Box 1892, Houston, TX 77251

F. R. Toffoletto, Dept. of Physics and Astronomy, Rice University, MS 108, P. O. Box 1892, Houston, TX 77251

R. A. Wolf, Dept. of Physics and Astronomy, Rice University, MS 108, P. O. Box 1892, Houston, TX 77251

Specification of Energetic Magnetospheric Electrons

D. F. Moorer

Department of Aerospace Engineering Sciences/Laboratory for Atmospheric and Space Physics,
University of Colorado, Boulder, CO

D. N. Baker

Department of Astrophysical, Planetary, and Atmospheric Sciences/Laboratory for Atmospheric and Space Physics,
University of Colorado, Boulder, CO

In order to make prompt and accurate assessments of whether enhancements in near-Earth energetic particles are related to space system operational problems, it is necessary to know well the present (and recent past) space radiation environment. Ideally, this means that one should be able to specify the temporal behavior of electrons and ions at all relevant altitudes, latitudes, and local times over the entire energy range of interest to space system operators. Through the use of data from a variety of scientific and operational spacecraft, it has been possible in recent times to develop "dynamic" radiation belt models. Our present work in this regard uses a type of data assimilation technique. With such modeling, we are generally able to achieve high accuracies ($\geq 75\%$) of energetic electron flux specification throughout the outer electron radiation belt ($2 \leq L \leq 7$). Future work employing these methods should allow accurate, reliable specification for magnetospheric ion populations as well.

1. INTRODUCTION

In January 1997, the TELSTAR 401 satellite owned by AT&T failed during a severe geomagnetic storm. A suspected cause was deep dielectric charging from relativistic electrons [Reeves et al., 1998]. In May of 1998, the Pan Am Sat's Galaxy 4 satellite became inoperable under similar conditions for possibly the same reason [Baker et al., 1998b]. In April 2000, Brazilsat was lost after another intense geomagnetic storm. The combined cost of placing spacecraft such as these into orbit is estimated to approach 750 million dollars [e.g., Baker et al., 1998b]. The commercial cost in terms of lost business due to communication satellite downtime, of course, pushes this figure much

higher. Indeed, despite improvements in spacecraft design, relativistic electrons continue to be a real and costly threat to all satellites operating in the outer radiation belt [Gussenhoven et al., 1987; Baker et al., 1994].

Figure 1 shows graphically the increasing number of insured spacecraft in near-Earth orbit and the number of serious anomalies suffered by spacecraft over the last nine years [Kunstadter, 1999; see also Baker, 2000]. High energy electrons are an important component of the environment causing such anomalies. Electron fluxes can vary in intensity by orders of magnitude in less than an hour [Li et al., 1998]. They can degrade satellite performance in a number of ways including microstructural damage, background noise in detectors, errors in digital circuits, electrostatic charge buildup in insulators, and spacecraft failure [Garrett, 1980; Liemohn, 1984, Baker et al, 1994]. Therefore, radiation belt modeling is crucially important in establishing the space environment to help avoid premature satellite failure, to improve satellite manufacturing, and to contain launch costs. Indeed, there continues to be a press-

Space Weather
Geophysical Monograph 125

Risks to Human Technology

■ Number of Serious Anomalies
▨ Insured Spacecraft in Low-Earth Orbit
▨ Insured Spacecraft in Geostationary Orbit

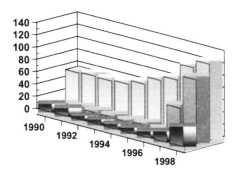

Figure 1. The numbers of insured spacecraft in both geostationary orbit and low-Earth orbit continue to rise as well as the numbers of serious anomalies [from Kunstadter, 1999].

ing need for an easy-to-use, easy-to-access, dynamic, global standard radiation belt model [Rodgers, 1996]. This will become especially urgent in the near future during and after solar maximum when there are usually a greater number of severe geomagnetic storms. In recognition of these needs and to alleviate the associated negative effects on terrestrial and space systems, the charter for the recently established National Space Weather Program (NSWP) has charged researchers to "specify and *forecast* ions and trapped electrons from one to twelve R_E" [NSWP, 1996].

Current models are static or quasi-static in nature. Plate 1 shows a comparison between a one-year (1998) average electron spectrum for a Los Alamos geostationary spacecraft and two current empirical models, AE-8 and CRRESELE. AE-8 estimates the flux to be too high on average and the CRRESELE model, though somewhat more realistic in their range, do not have coverage for the entire portion of the spectrum needed for environmental studies. Moreover, as we will show here, such static models as AE-8 fail to capture time variations of fluxes throughout the outer zone.

In the research reported here, we make use of the large, accurate satellite data bases currently available. Our purpose is to develop a system that inter-calibrates electron flux data from several currently-operating spacecraft. Our method incorporates those data by data assimilation into a baseline radiation belt model and produces a more accurate specification of electron flux for operational use.

The technique employed here is a variation of four-dimensional data assimilation, a technique that has long been used in meteorology to improve model forecasting performance [Ghil and Malanotte-Rizzoli, 1991]. Similar to the benefits it has provided meteorological forecasters, data assimilation appears to improve empirical modeling

of the radiation belts in that it helps to capture the dynamics of electron flux. In doing so, it takes a step toward a fully dynamic empirical model. The initial validation results demonstrate that the four-dimensional data assimilation technique is useful for specifying with great accuracy the high-energy electron flux throughout the outer radiation belt.

2. THE MODELING SYSTEM

The purposes of our model are threefold: 1) to improve empirical modeling of electron flux in the outer radiation belt; 2) to improve *ex post facto* satellite anomaly analysis; and 3) to take a step toward a fully dynamic empirical model.

2.1. Baseline model description

Any data assimilation technique incorporates data into, and adjusts, a baseline model. The purpose of this baseline model is to provide a generalized starting point for the assimilation process. The spacecraft whose data form the baseline model is the Combined Release and Radiation Effects Satellite (CRRES), part of the SPACERAD project of the United States Air Force [Johnson and Kierein, 1992]. For 14 months from July 1990 to September 1991 (during the last solar maximum), it flew in a geosynchronous transfer orbit with an inclination of 18°, a perigee of 350 kilometers and an apogee of 33,000 kilometers. The CRRES instrument collecting electron data was the High Energy Electron Fluxmeter (HEEF), designed to measure flux of 1-10 MeV electrons in ten differential flux channels.

The baseline model used here is the CRRESELE (CRRES electron) Science Module of the PL-GEOSpace Space Environment Model based on HEEF data [Brautigam and Bell, 1995]. The CRRESELE module was chosen because its output was easily accessible and because its flux accuracy is very good across the portion of the energy spectrum with which its instruments dealt. In creating the CRRESELE utility, the 0.512 second count rates from each of 10 HEEF energy channels were binned by L and pitch angle. For a given particle energy and L, the bin average pitch angle distributions were mapped to the magnetic equator using the International Geomagnetic Reference Field (IGRF85) internal magnetic field model and the Olson-Pfitzer static external magnetic field model. The result is a set of 8 omni-directional flux maps binned in L shell and B/Bo for a given energy. The CRRESELE utility uses these models to calculate electron omni-directional fluxes for 10 energy intervals (0.5 to 6.6 MeV) for each of six geomagnetic activity levels. These levels are based on the Ap15 index (mean of Ap recorded for 15 days prior to the day in question) [Mayaud, 1980]. The input parameters for the CRRESELE utility, then, are magnetic field model, energy channel, and geomagnetic activity.

We chose 18 variants from the CRRESELE utility: 3 energy channels under 6 different geomagnetic conditions. We chose a three-dimensional geocentric Cartesian grid with the x-axis through Greenwich meridian and the equator, z-axis through the north rotational pole, and y-axis completing the right-hand Cartesian system. This grid encloses the outer belt torus with x- and y-coordinates extending + 7 R_E and z-coordinates + 3 R_E. Spacing for the grid is 0.4 R_E in each direction (2551 km3). The electron flux is specified for each grid location based on energy channel and geomagnetic conditions.

2.2. Model boundaries, limitations and initialization

Although this model will accept a request to determine electron flux at any energy, it is optimized for, those electron energies most likely to cause deep dielectric charging [Vampola, 1987]. It is most accurate at those orbits where sensing platforms currently provide data. Linearity is assumed when adjusting grid flux values up or down by a multiplier. This is allowed because the baseline model takes into account most of the spectral and intensity variations before the data assimilation process begins.

We have conducted data assimilation only within the radiation belt outer torus from 3 to 7 R_E at the equator and 3 R_E in the z-direction in a geocentric Cartesian coordinate system. This allows analysis of the outer belt from about L=3 to L=6.8. CRRES recorded data only for 14 months during solar maximum. Initial validation demonstrates that CRRESELE spectral and intensity baselines support data assimilation even at times outside solar maximum although exact differences between solar maximum and minimum have not been determined.

There are five input parameters: year, date, time and location of the target spacecraft, and energy range. The year, date, and time allow the program to select the appropriate satellite data files that will provide a means to adjust the model.

2.3. Model calibration technique.

A software program has been developed that contrasts observed and CRRES-specified electron flux. In addition to contrasting these two, it also intercalibrates electron flux data from various sensors, accounts for local time variations, accounts for differences among spacecraft sensor accuracies, and determines more accurately the state of the outer radiation belt. All of these corrections contribute to output accuracy. To hopefully improve upon the Ap15 method of geomagnetic condition used by CRRESELE, a new high-resolution geomagnetic condition is determined using GPS spacecraft to accurately locate the most intense "heart" of the outer belt. Once this position is determined correctly, electron fluxes may be adjusted as described next.

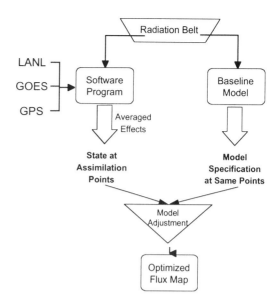

Figure 2. Diagram of model process. On the left side of this diagram, data from various charged particle sensing satellites are intercalibrated and incorporated into a software program. Given these in-situ observations, one knows the state of outer belt electrons at certain points in time and space. On the right side of this diagram, the baseline model also predicts the state at those same points. A comparison is made and an adjustor determined. It adjusts the baseline model producing a new, more accurate picture of the outer belt. At this point, one re-enters the adjusted model to determine the electron flux at any point.

The baseline model electron fluxes are adjusted in magnitude according to the difference between all observed and CRRES-specified fluxes by assigning a multiplier to each point in the outer belt torus. This becomes the optimized flux map. The software program then reenters this new flux map, finds the correct spacecraft location, records and reports the electron flux there. It is assumed that information on any disturbances to the outer belt from some nominal calm state will be transmitted throughout the belt in a relatively short time [e.g., Schulz and Lanzerotti, 1974]. Therefore, hourly averaged *in situ* observations at all points in the outer belt capture the trend of radiation belt conditions. See Figure 2.

3. MODEL COMPARISON

Plate 2 is a 2-year plot of geostationary flux (shown in blue) for electrons, $315 \leq E \leq 500$ keV. The red line is the flux predicted by the AE-8 model at geostationary orbit for this same energy range. AE-8 is a widely used model and is the world standard against which all other models are compared. As can easily be seen, however, AE-8 consistently overestimates the true flux for this period.

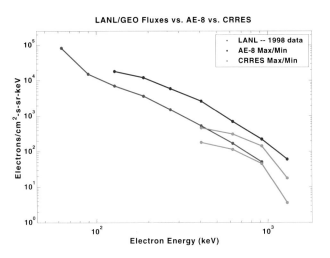

Plate 1. Comparison among yearly geostationary average fluxes, AE-8, and CRRESELE model output.

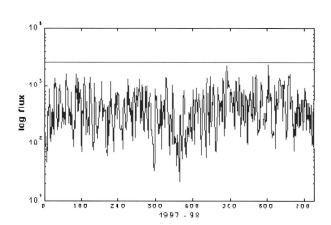

Plate 2. Comparison of electron flux experienced by a geostationary spacecraft with AE-8 MAX/MIN.

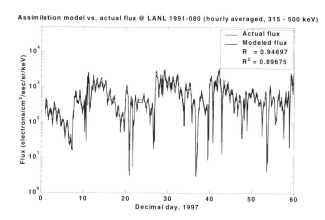

Plate 3. Sixty-day plot of model output versus actual flux, 315-500 keV for a geostationary spacecraft.

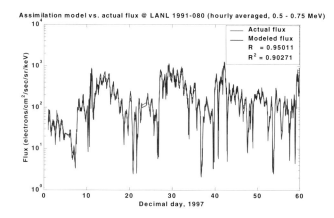

Plate 4. Sixty-day plot of model output versus actual flux, 500-750 keV for a geostationary spacecraft.

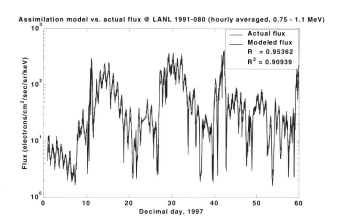

Plate 5. Sixty-day plot of model output versus actual flux, 0.75-1.1 MeV for a geostationary spacecraft.

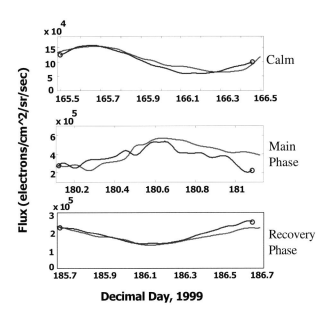

Plate 6. Forecast results for calm conditions and storm main and recovery phases. Red is actual flux; blue is predicted flux.

Over sample test periods, both AE-8 MAX and MIN had an average error of about 1500%. In similar tests, there was ~300% error even for the more dynamic CRRESELE module.

4. VALIDATION RESULTS

Plates 3, 4, and 5 display validation results for our four-dimensional data assimilation approach. These figures represent sixty-day (calm and storm period) "out-of-sample" validation tests of model efficiency. That is, data for the target spacecraft were removed from the model. The model was then asked to calculate the electron flux at the target spacecraft's location. For these tests, there were five spacecraft providing data to the data assimilation process: two Los Alamos spacecraft: LANL 1990-095 and LANL 1994-084; and three GPS spacecraft: NS 24, NS 33, NS 39. The three energy ranges tested: 315-500 keV, 500-750 keV, and 0.75-1.1 MeV. In all tests, the correlation coefficient is near 0.95 meaning that ~90% of the variance is being captured by the model. The assimilation method replicated observations well except at the lowest observed electron intensities.

5. FORECASTING

The accurate specification of outer radiation belt electron flux is an interesting and valuable endeavor. The success of such an effort holds great promise for *ex post facto* satellite anomaly analysis, radiation belt flux mapping, and satellite design and operation [Roederer, 1996; Wrenn and Sims, 1996]. The real value of any model is its ability to forecast, that is, to provide a picture of a system's future state given a set of input parameters [Sinha and Kuszta, 1983]. Meteorologists have sought to provide predictions of terrestrial weather and, indeed, it is the need to forecast that has driven progress in meteorological forecasting [Ghil et al., 1991].

Fung [1996] presented the idea of parameterizing the state of the magnetosphere such that an historical analog could be found for present conditions which could then be used for specification and forecast. That technique established these magnetospheric driver parameters: interplanetary magnetic field vector, the solar wind pressure, and solar F10.7 radio flux. The methodology presented here is a simpler adaptation of that technique using only solar wind speed as the driver. The electron energy range tested is > 600 keV. Data from the GOES > 600 keV electron channel are easily available and will be used as the means for assessing the outer belt state [Onsager et al., 1996]. There are three steps to the forecast model: 1) assessing the state of the radiation belt; 2) assessing the magnitude of the solar wind speed; 3) locating the best match from the historical electron flux and solar wind speed databases (see

Figure 3). To assess the trend and state of outer radiation belt electron flux, we first gather data from the assimilating spacecraft for some hours prior to the start-time of the forecast. This captures both the trend and the state of the outer belt and becomes the hourly flux sequence that is to be matched in the historical electron flux database.

This state of the outer belt is compared to all hourly averaged data from 1995 to present. A correlation coefficient is calculated that records the magnitude of difference between the present and past sequence. Those thirty sequences best matching the magnitude and trend (highest correlation coefficient) of the target sample become a new subset of data. We use the calendar dates for this subset in the next step.

We next examine solar wind sequences. The daily average solar wind speed preceding the target time of interest is collected. This sequence is compared to the daily average for the dates of the subset chosen above and a correlation coefficient for each historical sequence is calculated. From this list, the dates of the three best solar wind daily average matches are chosen.

Lastly, we reenter the historical database, find each of those dates, and record the electron flux history that occurred during the 48 hours following that point in time. These histories are averaged and become the forecast for our present time and conditions.

6. FORECAST RESULTS

No information is provided to the analog model past the point where the forecast is to begin. To examine the robustness of this analog forecasting process, a test period of time is examined that includes a geomagnetic storm. Day 165 through day 190 of 1999 is the test period chosen (See Figure 4). GOES 8 is the test vehicle and, as stated previously, 0.6 MeV integral flux is chosen as the test energy. During this 25-day period, analog forecast accuracy is examined at three times: calm conditions prior to the storm, storm main phase, and during storm recovery.

A sample of calm condition forecast is shown in Plate 6. Accuracy during this period is excellent. The forecast begins at noon on Day 165 and proceeding for 24 hours. The analog matches almost exactly the flux experienced by GOES 8. It was thought that the most exacting test would come during the severe temporal flux gradients apparent during storm main phase. Indeed, flux experienced by GOES 8 did undergo a strong enhancement during this time and the model was not able to capture it to the same accuracy as during calm conditions. However, the trend was captured and important information about maximum flux levels reached during this storm is obtained. The storm recovery phase is important because flux levels remain high even after all geomagnetic indices have returned to normal. These electron flux levels can create a hazardous charging environment for geostationary spacecraft. The period examined was six days after storm onset. Accuracies during this time were exceptional.

7. SUMMARY AND DISCUSSION

Modeling of radiation belt electron flux with data assimilation seems to be increasingly practical due to the continued growth of computing power and the availability of near-continuous data streams. Continuous data assimilation schemes capable of effectively analyzing this huge satellite database are needed for dynamic analysis of the radiation belts. Work described in this paper has tested a data assimilation strategy to simulate the electron flux dynamics experienced by a geostationary spacecraft during a sixty-day period. Data were assimilated into the CRRESELE module baseline model. One of the main objectives of this study was to determine whether an adaptation of data assimilation is effective in capturing electron flux to a finer resolution than has previously been achieved with empirical models. It appears that this technique holds some important potential for capturing outer belt electron dynamics. Forecasting is the second of two tasks presented in the National Space Weather Program's objective to "specify and forecast ions and electrons from 1 to 12 R_E." [NSWP, 1996] It is apparent from the results presented here that analog forecasting begins to meet this objective and can be a valuable tool on both the scientific and operational sides of space physics.

The study of today's space weather problems and the challenges to spacecraft design and operation demand higher-frequency and finer resolution models to aid in ex post facto anomaly analysis and to open new doors in radiation belt research. Economic constraints effectively limit

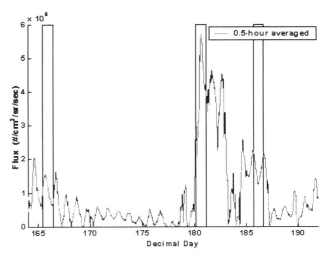

Figure 4. The three green vertical bars represent the three 24-hour forecast test times. The first (left) is a calm period, the middle is during storm main phase, the last (right) is during storm recovery phase.

the number and spatial resolution of *in situ* observations. Data sets are always incomplete and generally distributed non-uniformly in space and time. Because of such limitations and the complex relationships among the physical principles involved, a dynamic approach to data analysis is required. A new technique was used here to generate complete, dynamically consistent data sets and images of outer belt electrons. The results look very promising.

Acknowledgments. This work was supported by grants from the NASA/ Goddard Space Flight Center Graduate Student Researcher Program and the National Science Foundation. Additionally, data assimilation is, by its nature, extremely data intensive. Therefore, for providing data and advise on its use, the authors thank colleagues at the Los Alamos National Laboratory, NOAA Space Environment Center, National Space Science Data Center, Boston University, and the Air Force Research Laboratory.

REFERENCES

Acuña, M.H., et al., The Global Geospace Science Program and its investigations, *Space Science Review*, 71, 5, 1995.
Anthes, R.A., Data Assimilation and initialization of hurricane prediction models, *J. Atmos. Sci.*, 31, 702-719, 1974.
Baker, D.N., J.B. Blake, R.W. Klebesadel, and P.R. Higbie, Highly relativistic electrons in the Earth's magnetosphere, 1, Lifetimes and temporal history, 1979-1984, *J. Geophys. Res.*, 91, 4265-4276, 1986.
Baker, D.N., et al., Satellite anomalies linked to electron increase in the magnetosphere, *Eos Trans.* AGU, 75, 401, 1994.
Baker, et al., Coronal mass ejections, magnetic clouds, and relativistic magnetospheric electron events: ISTP, *J. Geophys. Res.*, 103, A8, 1998a.
Baker, D.N., et al., Disturbed Space Environment May Have

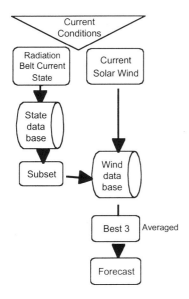

Figure 3. Analog forecast algorithm.

Been Related to Pager Satellite Failure, *EOS Trans*. AGU, 79, 477, 1998b.

Brautigam, D.H. and Bell, J.T., *CRRESELE Documentation*, Phillips Laboratory, Environmental Research Papers, No. 1178, 1-16, 1995.

Burlaga, L.F., et al., Magnetic loop behind an interplanetary shock: Voyager, Helios, and IMP 8 observations, *J. Geophys. Res.*, 86,6673, 1981.

Charney, J.G., et al., Use of incomplete historical data to infer the present state of the atmosphere, *J. Atmos. Sci.*, 26, 1969.

DiMego, G.J., The National Meteorological Center regional analysis system, *Mon. Wea. Rev.*, 100, 1988.

Fung, S.F., Recent Development in the NASA Trapped Radiation Models, in *Radiation Belts Models and Standards*, Geophys. Mo Monogr. Ser., edited by J.F. Lemaire, D. Heynderickx, and D.N. Baker, AGU Washington DC, 1996.

Garrett, H.B., *Spacecraft Charging: A Review, in Space Systems and Their Interactions With Earth's Space Environment*, H.B. Garrett and C.P. Pike, eds., American Institute of Aeronautics and Astronautics, New York, 1980.

Ghil, M. and Malanotte-Rizzoli, P., Data Assimilation in Meteorology and Oceanography, *Advances in Geophysics*, 33, 141, 1991.

Gussenhoven, M.S., et al., New low-altitude dose measurements, *IEEE Trans. Nuc. Sci.*, NS-34, 676, 1987.

Johnson, M.H. and Kierein, H., Combined Release and Radiation Effects Satellite (CRRES), *Journal of Spacecraft and Rockets*, 9, 4, 1992.

Kunstadter,C.T.W., "Were we crying wolf?, Insurance implications of the Leonids and other space phenomena", AIAA Leonids Conference, Los Angeles, CA, May 1999.

Lepping, R.P., et al., The WIND magnetic field investigation, *Space Science Review*, 71, 207-229, 1995.

Li, X., et al., Energetic electron injections into the inner magnetosphere during the Jan. 10-11, 1997 magnetic storm, *Geophys. Res. Letters*, 25, 14, 1998.

Liemohn, H., Single Event Upset of Spacecraft Electrons, Proceedings of Spacecraft Anomalies Conference, NOAA, 1984.

Mayaud, P.N., *Derivation, Meaning and Use of Geomagnetic Indices*, Geophys. Monogr., 22, AGU, Washington, D.C., 1980.

Murray-Smith, D.J., *Continuous System Simulation*, Chapman & Hall, New York, 1997.

National Space Weather Program Strategic Plan (FCM-P30-1995, Office of the Federal Coordinator For Meteorological Services and Supporting Research, Silver Spring, Md., 1996).

National Space Weather Program Implementation Plan: Capabilities, Goals, and Strategy, 1996.

Ogilvie, K.W., et al., SWE, A comprehensive plasma instrument for the WIND spacecraft, *Space Science Review*, 71, 55-77, 1995.

Onsager, T.G., R. Grubb, J. Kunches, L. Matheson, D. Speich, R. Zwickl, and H. Sauer, "Operational Uses of the GOES Energetic Particle Detectors," in *GOES-8 and Beyond*, Edward R. Washwell, Editor, Proc. SPIE 2812, 281-290, 1996.

Paulikas, G.A., and J.B. Blake, Effects of the solar wind on magnetospheric dynamics: Energetic electrons at the synchronous orbit, in *Quantitative Modeling of Magnetospheric Processes*, Geophys. Monogr. Ser., vol. 21, edited by W.P. Olson, AGU, Washington, DC, 1979.

Rodgers, D.J., Empirical Radiation Belt Models, *Radiation Belts: Models and Standards*, Geophysical Monograph 97, American Geophysical Union, 1996.

Reeves, G.D., et al., The relativistic electron response at geosynchronous orbit during the January 1997 magnetic storm, *J. Geophys. Res.*, 103, 17559, 1998.

Richmond, A.D., Assimilative Mapping of Ionospheric Electrodynamics, *Adv. Space Res.*, 12, 6, 1992.

Roederer, J.G., Introduction to Trapped Particle Flux Mapping, in *Radiation Belts Models and Standards*, Geophys. Mo Monogr. Ser., edited by J.F. Lemaire, D. Heynderickx, and D.N. Baker, AGU Washington DC, 1996.

Schulz, M. and L.J. Lanzerotti, *Particle Diffusion in the Radiation Belts*, Springer-Verlag, New York, 1974.

Sinha, N.K., and B. Kuszta. *Modeling and Identification of Dynamic Systems,* Van Nostrand Reinhold Company, 1983.

Stauffer, D.R. and N.L. Seaman, Multiscale Four-Dimensional Data Assimilation, *Journal of Applied Meteorology*, 33, 416, 1994.

Stauffer, D.R. and N.L. Seaman, Use of Four-dimensional Data Assimilation in a Limited-Area Mesoscale Model. Part I: Experiments with Synoptic-Scale Data, *Monthly Weather Review,* 118, 1250, 1990.

Tascione, T., *Space Weather Requirements: Commercial Priorities*, Geospace Environment Modeling Conference, June, 1998.

Vampola, A.L., The aerospace environment at high altitudes and its implications for spacecraft charging and communications, *J. Electrost.*, 20, 21, 1987.

Vette, J.I., *The AE8 Trapped Electron Model Environment*, NSSDC/WDC-A-R&S, 91-24, 1991.

Wrenn, G.L., and A.J. Sims, Internal Charging in the Outer Zone and Operational Anomalies, in *Radiation Belts Models and Standards*, Geophys. Monogr. Ser., edited by J.F. Lemaire, D. Heynderickx, and D.N. Baker, AGU Washington DC, 1996.

D. F. Moorer, Department of Aerospace Engineering Sciences/Laboratory for Atmospheric and Space Physics, University of Colorado, Boulder, CO 80303.

D. N. Baker, Department of Astrophysical, Planetary, and Atmospheric Sciences/Laboratory for Atmospheric and Space Physics, University of Colorado, Boulder, CO 80303.

Predicting Geomagnetic Storms as a Space Weather Project

Syun-Ichi Akasofu

International Arctic Research Center, University of Alaska Fairbanks

To be successful, space weather researchers need to establish a new discipline by synthesizing and integrating the traditional four major disciplines: solar physics, interplanetary physics, magnetospheric physics and ionospheric physics (aeronomy). Although much progress within the four major disciplines and their sub-disciplines has been made in the past and more progress is still needed, such efforts <u>alone</u> cannot accomplish the task of forecasting space weather and predicting geomagnetic storms. This paper describes an example of the integration efforts - entirely physically-based. It is not intended to be a review. Elements that need to be integrated include modeling of the background solar wind flow; parameterizing of solar events on the source surface; modeling of the propagation of the shock waves; detecting of the shock waves by IPS, comets and other methods; characterizing geomagnetic storms by the ε parameter (or any others); estimating the Dst and AE indices from the ε parameter (or any others); simulating ionospheric effects; examining effects on power line systems and oil pipeline systems, etc. It is emphasized that in this whole series of research, one of the major missing links is related to our present inability to predict the IMF polar angle θ as a function of time. A concerted effort among the four disciplines is needed in making this prediction possible.

1. INTRODUCTION

A study of the solar-terrestrial relationship has developed into four major disciplines: solar physics, interplanetary physics, magnetospheric physics and ionospheric physics (aeronomy). Researchers have made considerable progress within each field of study during the 20th century, although there are many challenging problems left unsolved. Meanwhile, there has been much discussion about space weather research during the last few years. It is important to realize that the term "weather" in this context implies that the goal of space weather research should be to forecast space weather and, at least, predict geomagnetic storms in terms of the two geomagnetic indices Dst and AE as a function of time.

To be successful in this particular effort, space weather researchers need to establish a new discipline that synthesizes and integrates the four major disciplines and their sub-disciplines (Akasofu, 1996). In this paper, a research scheme needed for success in predicting geomagnetic storms is described by presenting an example of this integration process. The needed efforts are:

1. Modeling of the background solar wind flow
2. Parameterizing solar events on the source surface
3. Modeling the propagation of shock waves (including the simulation of past events)
4. Detecting shock waves by IPS, comets and others
5. Estimating the velocity, density, and IMF at the earth
6. Characterizing geomagnetic storms

Space Weather
Geophysical Monograph 125

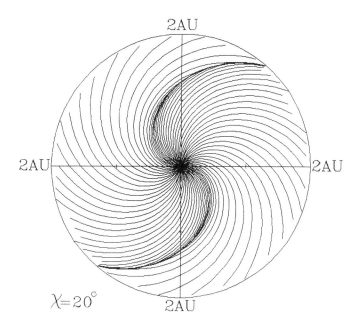

Figure 1. The spiral structure of the interplanetary magnetic field lines on the equatorial plane. The magnetic equator on the source surface is assumed to be sinusoidal, the amplitude being 20° in latitude (see the top diagram of Figure 3a).

7. Predicting the size of the auroral oval
8. Examining effects on powerlines and oil pipeline systems.

Unfortunately, there have been a relatively small number of papers which dealt with the entire subject in the past (Akasofu and Fry, 1986; Dryer, 1994). One of the purposes of this paper is to emphasize the need for the synthesis approach in accomplishing our goal of space weather research; it is not intended to be a review. The latest improvement of the method described here is presented in a companion paper (Fry et al., 2000)

2.1. Modeling the Background Solar Wind Flow

The solar wind exhibits considerable variations even without any specific solar events, such as solar flares, CMEs, and sudden filament disappearances. This is particularly the case when long-lasting coronal holes, high-speed streams flowing from them, and the interplanetary sector boundary structures corotating with the sun are present. Since any effects of specific solar events propagate into the existing structures and interact with them, it is important to devise, first of all, a simple way to model them.

Conditions on the source surface, an imaginary spherical surface of 2.5 solar radii, are important in modeling interplanetary conditions. One of the most important aspects

on the source surface is the magnetic equator (or the so-called "neutral line"). The axis of the dipolar field on the source surface rotates from 0° to 180° (or from 180° to 0°) during the sun's 11-year cycle variations (Saito et al., 1989). We found that we can reproduce much of the main feature of solar wind variations during the sunspot minimum period at the earth or at any point to about a distance of 2 au by assuming that the solar wind speed is minimum at the sinusoidal magnetic equator and increases toward higher latitudes (see the top diagram of Figure 2a).

As the sun and its source surface rotate every 25 days, a fixed point in space (not on the source surface) at a distance of 2.5 solar radii scans horizontally the velocity field from the solar longitude 360° to 0° in one solar rotation along a heliographic latitude line (e.g., 0° at the June and December solstices). Solar wind particles leave radially from this particular point one by one with different velocities as the sun rotates; the point depicts a sinusoidal variation of the speed of solar wind particles during one rotation.

Subsequent changes of the radial speed of individual particles can be modeled by adopting a kinematic solution in our method (Hakamada and Akasofu, 1982). A faster flow of particles interact with a slower flow of particles to form a shock wave and a reverse shock. Thus, by integrating the velocity as a function of time graphically, one can determine the distance traveled by individual particles. This method has been carefully calibrated on the basis of MHD solutions by Olmsted and Akasofu (1985), Sun et al. (1985), and Olmsted and Akasofu (1986).

A magnetic field line originating from the source surface can be traced by following particles leaving a particular point on the source surface (not a fixed point in space). The resulting interplanetary magnetic field structure is the familiar Parker spiral, together with the corotating interaction region produced by the formation of the shock wave structure. Figure 1 shows such an example. The computed velocity (V), density (n) and IMF magnitude B agrees reasonably well with the observed ones.

This method of using the sinusoidal curve was used in our early study, but what is really needed in modeling the solar wind pattern in interplanetary space is the distribution of the solar wind speed and the magnetic field magnitude on the source surface. Fortunately, there have been a number of studies in this particular aspect (cf. Hoeksema et al., 1983; Wang et al., 2000).

2.2. Parameterizing Solar Events

In our scheme, a solar event is represented by a high-speed source area on the source surface, which is super-

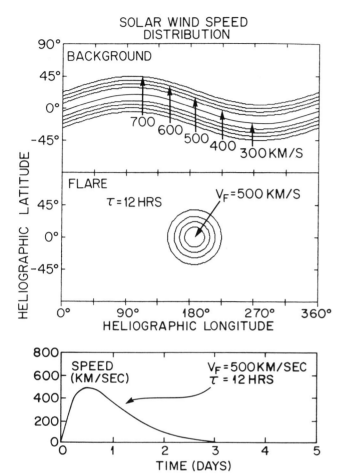

Figure 2a. From the top: The background flow speed of the solar wind on the source surface (the solar longitude-latitude map); the solar wind speed is minimum (300 km/sec) at the magnetic equator and increases toward higher latitudes (Akasofu and Fry, 1986). The middle diagram represents a solar event; a higher speed from a circular area is added to the background flow; the speed at the center of the circular area reaches $V_F = 500$ km/sec: The bottom diagram shows how the speed at the center of the circular area in the middle diagram varies in time; it reaches $V_F = 500$ km/sec 30 minutes after the onset and decreases slowly (expressed by the parameter $\tau_F = 12$ hrs in this particular case).

posed on the background structure described in the previous section. The source area is represented by a circular area (or an elliptical area); the speed is highest at the center and has a Gaussian distribution. The speed at the center varies in time in a characteristic way, which is parameterized by $\tau_F(V_F(t/\tau_F) \bullet e^{(1-t/\tau_F)}$. Thus, a solar event is parameterized by the maximum speed V_F at the center at the peak of the event, the area size σ_F and the time variations τ_F, together with longitude λ_F and latitude ϕ_F and the

start time T_F of the event (Akasofu and Fry, 1986). Figure 2a shows graphically the adopted parameters, $\lambda_F = 180°$, $\phi_F = 0°$, $V_F = 500$ km/sec, $\sigma = 30°$, and $\tau_F = 12$ hrs on the source surface for a hypothetical event. It is assumed that the event takes place on the center of the disk of the sun at 12 UT on December 8. Therefore, in this particular case, the center of the sun, the location of the solar event and the earth are almost on the same solar radial line at the onset of the solar event. If we consider CMEs or sudden filament disappearances, another set of parameters may be needed to describe them. Obviously, the same is the case for an MHD approach.

2.3. Modeling the Propagation of Shock Waves

Figure 2b shows the propagation of the resulting shock wave in the equatorial plane at 0, 6, 12, and 18 UT on December 10, namely 36, 42, 48, and 54 hours after the hypothetical event on December 8, respectively. The earth's location is indicated by a star mark. Since the event is assumed to occur on the center of the disk, the

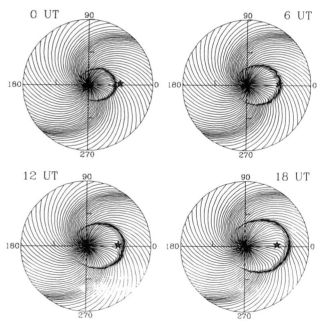

Figure 2b. The propagating shock wave generated by a hypothetical growth of the high speed flow (the middle diagram in Figure 3a) on the equatorial plane of the circular area of radius 2au. The sun is located at the center; the earth's location is indicated by a star. The shock wave can be recognized by distorted IMF field lines. The resulting solar wind variations at the time of the passage of the shock wave at the earth are illustrated in Figure 4.

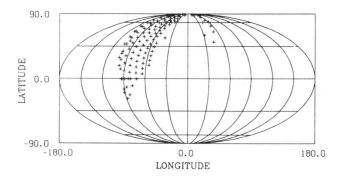

Figure 3a. The projection of the 3-D shock front (Figure 3a) on the sky map at 16 UT on September 24, 1978. The center of the map is the direction toward the sun. The shock front is located in the northwestern part of the sky.

center of the shock wave is propagating approximately along the sun-earth line.

2.4. Detecting Shock Waves by IPS

In predicting a geomagnetic storm after a specific solar event, it is desirable to detect the advancing shocks midway between the sun and the earth. A space probe at the midpoint is ideal, but is practically unavailable. For this reason, we have searched for other methods. One of them is to use interplanetary scintillation (IPS). In order to demonstrate this method, we show here a study of an event in September, 1978 (Akasofu and Lee, 1989, 1990). We constructed successive 3-D surfaces of the shock wave and projected them onto the so-called 'sky map', a map of the sky centered at the direction of the sun (Figure 3a). The available IPS observation during the event showed an intense IPS area in the upper left of the sky (Figure 3b), in agreement with the projection (Hewish et al., 1985). Such an observation assures that we chose the necessary parameters of the solar event on the source surface reasonably well before the arrival of the shock wave

2.5. Estimating the Velocity, Density, and IMF at the Earth

Our modeling enables us to predict the solar wind quantities, such as velocity, density, and IMF at any location within a distance of 2 au. Figure 4 shows these quantities at the earth for the hypothetical solar event discussed in Section 2.3 (see also Section 2.1). A similar estimation can be made at the L5 point or any other points (for example, the L3 and L4 points, which will be occupied by the approved STEREO mission) in the inner interplanetary space. Actually, our method could reproduce reasona-

bly well the observed shock wave at 7 au, the location of Pioneer 11 (Akasofu et al., 1985).

2.6. Characterizing Geomagnetic Storms

The prediction of a geomagnetic storm requires the prediction of the Dst and AE indices as a function of time.

The next step is to identify the expression for the electric power that generates the storm components and the resulting storm fields. One example is given by Perreault and Akasofu (1978), Akasofu (1981), and Pudovkin and Semenov (1986):

$$\varepsilon \text{ (megawatts)} = 20 \text{ V (km/sec) } B^2 \text{ (nT) } \sin^4 (\theta/2) \quad (1)$$

where θ denotes the IMF polar angle.

The first important test of adopting ε (megawatts = MW) in predicting geomagnetic storms is whether or not ε can characterize the variety of geomagnetic storms and whether or not we can infer the two geomagnetic indices AE(t) and Dst(t) as a function of time from ε(t).

Figure 5 shows, from the top, ε, calculated Dst, calculated AE and the observed AE for the March, 1973 storm. Knowing that Dst is proportioned to the total kinetic energy of the ring current particles (Dessler and Parker, 1959), the calculated Dst can be obtained by

$$\frac{dDst}{dt} = \alpha\varepsilon - \frac{Dst}{\tau_R} \quad (2)$$

It is assumed that 70% of the power is dissipated in the ring current, so that α is 0.7; τ_R is the lifetime of ring current particles (~ 7 hrs or less). In terms of ε, the inten-

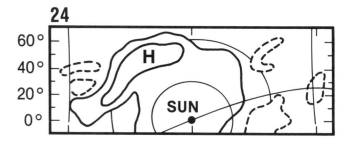

Figure 3b. The observed interplanetary scintillation (IPS) on the sky map at 16 UT on September 24. A high IPS region is seen in the northwestern part of the sky map at about the location where the shock wave was located (Hewish et al., 1985).

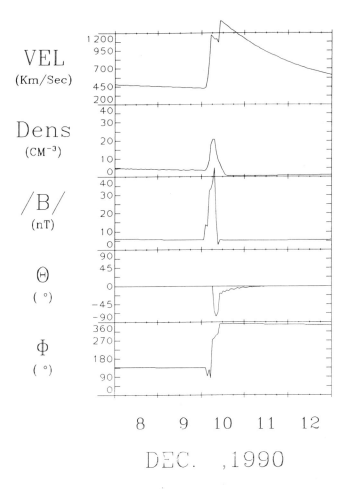

Figure 4. The computed solar wind velocity, density, IMF magnitude B, IMF THETA and PHI angles for the hypothetical event illustrated in Figure 2b.

sity of geomagnetic storms may be roughly classified as follows:

weak storms	$\varepsilon \sim 0.25$ MW (e.g. V = 500 km/sec, B = 5 nT)
moderate storms	$\varepsilon \sim 1.4$ MW (e.g. V = 700 km/sec, B = 10 nT)
very intense storms	$\varepsilon > 8.0$ MW (e.g. V = 1000 km/sec, B = 20 nT)

There is so far no theoretical study which can relate ε to the AE index. This is because the magnetosphere responds to an increased ε in two ways, the directly driven component and the unloading component. The directly driven component correlates fairly well with ε, but the unloading component does not (Sun and Akasofu, 2000) and thus cannot be predicted at this time. The AE index includes both components. Therefore, the empirical relationship between ε and AE has to be established.

$$AE \ (nT) = -300 \ (\log \varepsilon)^2 + 11700 \ \log \varepsilon - 113200. \qquad (3)$$

It can be seen from Figure 5 that the Dst variations computed on the basis of ε can reproduce fairly well both the observed characteristics of the storm and its time variations. However, it is obvious that the empirical relationship between ε and AE should be improved. Figure 6 shows ε computed for the hypothetical event discussed in Section 2.3 and 2.5. Both the AE (CAE) and Dst (CDST) indices are also computed. The quantity, Φ_{pc}, will be discussed in Section 2.7. The size of the auroral oval can be predicted in a similar empirical way (cf. Hardy et al., 1984; Newell et al., 1996; Sotirelis et al., 1998).

2.7. Predicting Ionospheric Effects

The ε parameter has a good correlation with the polar cap potential (Reiff et al., 1987):

$$\Phi_{pc} = (0.93 \ \varepsilon - 3.19)^{1/2}$$

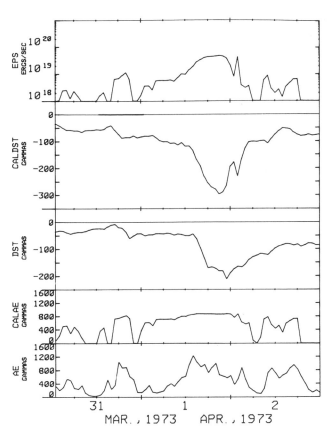

Figure 5. From the top, the ε parameter computed on the basis of the solar wind observations, the calculated Dst index (CAL, Dst), observed Dst index, calculated AE inde (CAL, AE) and observed AE index for the March 1973 storm.

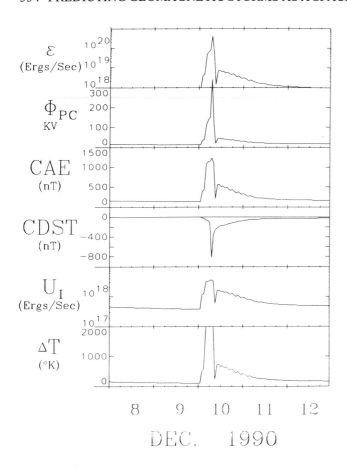

Figure 6. The computed ε parameter, polar cap potential Φ_{pc}, calculated AE (CAE) and Dst (CDst) indices for the hypothetical event illustrated in Figure 4.

It is this potential drop that drives a flow of ionospheric plasma from the dayside hemisphere into the polar cap region. For details, see Maurits and Watkins (1996) and Maurits et al. (2000).

2.8. *Effects on Power Transmission Lines and Oil/Gas Pipelines*

Although it has been emphasized by the space science community that solar events could cause serious problems on power transmission lines, there have so far been only a few studies to examine how geomagnetic storms can actually affect power transmission lines. Akasofu and Merritt (1979) and Akasofu and Aspnes (1982) demonstrated that changing magnetic fields produced by the auroral electrojet induce electric currents in the neutral line of a three-phase transmission line and that such extra currents are converted into pulses signals in the circuit breaker system.

For the trans-Alaska oil pipeline, we found a rather simple formula; the current I in the pipe is given by I =

V/Ω, where V is the induced voltage (~ 1 volt/m x 1000 km) for moderate auroral activity (the total length of the pipe ~ 1000 km) and Ω the total resistance of the pipe. Thus, I ~ 100 amperes for V = 1000 volt and Ω = 10 ohms (E. Wescott, private communication). Leaking currents from the pipe to the ground causes corrosion of the pipe. Our finding is now used to monitor the corrosion effect of the trans-Alaska oil pipeline.

3. MISSING LINKS

It is quite obvious that we cannot succeed in predicting geomagnetic storms until we can find a way to predict θ(t). Thus, as emphasized earlier (Akasofu, 1996), the prediction of θ(t) after a solar event is most crucial in predicting geomagnetic storms.

Recently, IMF changes associated with solar events are discussed in terms of magnetic "clouds", magnetic "flux ropes", "loops", "tubes", etc. (cf. Lepping et al., 1990; Lepping et al., 1995; Farrugia et al., 1995; Chen et al., 1997; Marubashi, 1997; Gonzalez et al., 1998, Mulligan et al., 1998; Fenrich and Luhmann, 1998; Lu et al., 1998; Osherovich et al., 1999; Rust, 1999; Moldwin et al., 2000). Perhaps some efforts are needed to standardize such terms. For example, if the structures are detached magnetically from the sun, they may be called "clouds" (Simnett et al., 1997), while the structures are magnetically anchored to the sun, they may be called "loops". For the rigorous MHD (2 1/2 D and 3 D) simulations for this particular phenomenon, see Detman et al. (1991), Vandas et al. (1996, 1998, 2000).

Three important issues in this regard are how these structures are related to (i) CMEs, (ii) whether or not the expanding CME constitute the so-called "driver gas" for the shock wave, and (iii) if it would be at all possible to predict changes of the IMF polar angle θ as a function of time (cf. Tsurutani et al., 1988).

Further, there occur often very sharp changes (in time) of the azimuth angle of the IMF during major geomagnetic storms, suggesting sometimes a large-scale movement of the current sheet. Figure 7 shows how a shock wave associated with a solar event in the southern hemisphere can push up the current sheet, so that the earth's position with respect to the current sheet (above before the passage of the shock) can change (below after the passage). Some of the changes of θ may be related to flappings of the current sheet. Further, the passage of the shock wave can increase the magnitude of the IMF (cf. Tsurutani et al., 1988), so that ε can be increased if θ > 90°; it may be noted that Smith et al. (1998) confirmed numerically (via MHD) their results.

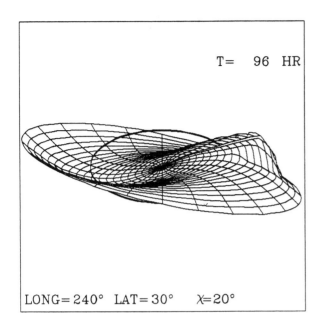

Figure 7. Deformation of the solar current sheet by a solar event in the southern hemisphere before (OHR) and 96 hours (96 HR) after the onset.

In examining the variability of the solar wind velocity V, the IMF magnitude B and the IMF polar angle θ, the most variable quantity is θ. In order for ε to be greater than 1 MW, for V = 500 - 1000 km/sec and B ~ 10 nT, it is necessary for θ to be greater than ~ 90°. This situation is generally called the "southward turning of the IMF".

In this regard, it is likely that intense solar events produce a high value of V ~ 500 - 1000 km/sec and B > 10 nT. However, if θ happens to be 0° or very small, ε cannot reach 1 MW. Some early attempts to relate θ to magnetic features in the photosphere have failed (Tang et al., 1985). More efforts are needed for this particular task along the line studied by Detman et al. (1996, 1998, 2000) and others.

4. SUMMARY

As a summary, Figure 8 shows the geomagnetic storm prediction scheme presented in this paper in a block diagram form. It is satisfying to see many solar physicists and magnetospheric physicists starting to work together to study jointly on space weather by learning the others'

Geomagnetic Storm Prediction Scheme

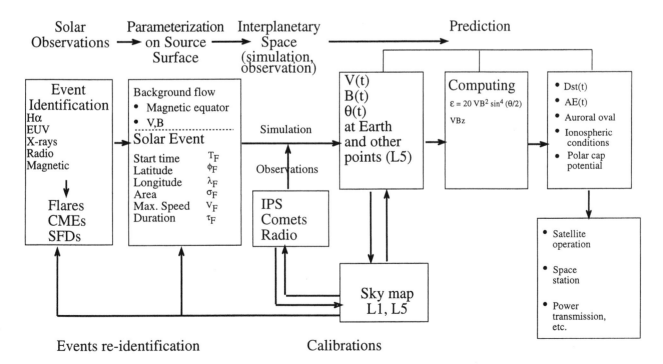

Figure 8. The physically-based geomagnetic storm prediction scheme presented in this paper in a block-diagram form.

discipline (e.g. National Space Weather Program Strategic Plan; National Space Weather Program Implementation Plan). However, much more concerted efforts are needed for the success in the prediction of geomagnetic storms. In particular, we should focus our effort in predicting IMF θ(t); this requires the knowledge of the interplanetary magnetic structure associated with the clouds and loops.

Acknowledgements. The author would like to thank C.D. Fry, C.S. Deehr, M. Dryer, K. Hakamada, Z. Smith, L. Snyder, and W. Sun for their decades of discussion on the subject.

REFERENCES

Akasofu, S.-I., New scheme provides a first step toward geomagnetic storm prediction, *EOS, 77*, 25, 1996.

Akasofu, S.-I. and L.-H. Lee, Modeling of a series of interplanetary disturbance events in September, 1978, *Planet. Space Sci., 38*, 575, 1990.

Akasofu, S.-I. and L.-H. Lee, Modeling of an interplanetary disturbance event tracked by the interplanetary scintillation method, *Planet. Space Sci., 37*, 73, 1989.

Akasofu, S.-I., C. Olmsted, T. Saito, and T. Oki, Quantitative forecasting of the 27-day recurrent magnetic activity, *Planet. Space Sci., 36*, 1133, 1988.

Akasofu, S.-I. and C.F. Fry, A first generation numerical geomagnetic storm prediction scheme, *Planet. Space Sci., 34*, 77, 1986.

Akasofu, S.-I., W. Fillius, W. Sun, C. Fry, and M. Dryer, A simulation of two major events in the heliosphere during the present sunspot cycle, *J. Geophys. Res., 90*, 8193, 1985.

Akasofu, S.-I. and J.D. Aspnes, Auroral effects on power transmission line systems, *Nature, 295*, 136, 1982.

Akasofu, S.-I., Energy coupling between the solar wind and the magnetosphere, *Space Sci. Rev., 28*, 121, 1981.

Akasofu, S.-I., Prediction of development of geomagnetic storms using the solar wind-magnetosphere energy coupling function ε, *Planet. Space Sci., 29*, 1151, 1981.

Akasofu, S.-I. and R.P. Merritt, Electric currents in power transmission lines induced by auroral activity, *Nature, 279*, 308, 1979.

Baker, D.N., T.I. Pulkkinen, S. Li, S.G. Kanekal, K.W. Ogilvie, R.P. Lepping, J.B. Blake, L.B. Callis, G. Rostoker, H.J. Singer, and G.D. Reeves, A strong CME-related magnetic cloud interaction with the Earth's magnetosphere: ISTP observations of rapid relativistic electron acceleration on May 15, 1997, *Geophys. Res. Lett., 25*, 2975, 1998.

Bothmer, V. and D.M. Rust, The field configuration of magnetic clouds and the solar cycle, *Coronal Mass Ejection, Geophys. Monograph 99*, American Geophys. Union, 1997.

Burke, W.J., T.L. Fehringer, D.R. Weimer, C.Y. Huang, M.S. Gussanhoven, F.J. Rich, and L.C. Gentile, Observed and predicted potential distributions during the October 1995 magnetic cloud passage, *Geophys. Res. Lett., 25*, 3023, 1998.

Burlaga, L.F., K.W. Behannon, and L.W. Klein, Composure streams, magnetic clouds and major magnetic storms, *J. Geophys. Res., 92*, 5725, 1987.

Chen, J., R.A. Howard, G.E. Bruckner, II, S.E. Paswaters, O.C. St. Cyr, R. Schwenn, P. Lamy, and G.M. Sinnet, Evidence of an erupting magnetic flux rope: LASCO coronal mass ejection of 1997 April 13, *Ap. J., 490*, L191, 1997.

Dessler, A.J. and E.N. Parker, Hydromagnetic theory of geomagnetic storms, *J. Geophys. Res., 64*, 2239, 1959.

Detman, T.R., M. Dryer, T. Yeh, S.M. Han, S.T. Wu, and D.J. McComac, A time-dependent, 3-D MHD numerical study of interplanetary magnetic draping around plasmoids in the solar wind, *J. Geophys. Res., 96*, 9531, 1991.

Dryer, M., Interplanetary studies: Propagation of disturbances between the sun and the magnetosphere, *Space Sci. Rev., 67*, 363, 1994.

Farrugia, C.J., V.A. Osherovich, and L.F. Burlaga, The magnetic flux rope versus the spheromak as models for interplanetary magnetic clouds, *J. Geophys. Res., 100*, 12,293, 1995.

Feldstein, Y.I. and G.V. Starkov, Dynamics of auroral belt and polar geomagnetic disturbances, *Planet. Space Sci., 15*, 209, 1967.

Fenrich, F.R. and J.G. Luhmann, Geomagnetic response to magnetic clouds of different polarity, *J. Geophys. Res., 25*, 2999, 1998.

Fry, C.D., W. Sun, C.S. Deehr, M. Dryer, and Z. Smith, S.-I. Akasofu, M. Tokomaru, and M., Koijima, Improvements to the HAF solar wind model for space weather predictions, *J. Geophys. Res.*, 2000.

Gonzalez, W.D., A.L. Clua de Gonzalez, A. Dal Lago, B.T. Tsurutani, J.K. Arballo, G.T. Lakhina, B. Buti, C.M. Ho, S.-T. Wu, Magnetic cloud field infusities and solar wind velocities, *Geophys. Res. Lett., 25*, 963, 1998.

Hakamada, K. and S.-I. Akasofu, Simulation of three-dimensional solar wind disturbances and resulting geomagnetic storms, *Space Sci. Rev., 31*, 3, 1982.

Hardy, D.A., L.K. Schmidt, M.S. Gussenhoven, F.J. Marshall, H.,C. Yeh, T.L. Schumaker, A. Hubard, J. Pantazis, Precipitating electron and ion detector (SSJ/4) for the block 5D/Flights 6-10 DMSP satellites: Calibration and data presentation, Rep. AFGL-TR-84-0317, Air Force Geophys. Lab. Hanscom AFB, Hanscom, Mass. 1984.

Hewish, A., S.J. Tappin, and G.R. Faggen, Origin of strong interplanetary shocks, *Nature, 314*, 137, 1985.

Hoeksema, J.T., J.M. Wilcox, and P.H. Scherrer, The structure of the heliospheric current sheet: 1978-1982, *J. Geophys. Res., 88*, 9910, 1983.

Lepping, R.P., J.A. Jones, and L.F. Burlaga, Magnetic field structure of interplanetary magnetic clouds at 1 au, *J. Geophys. Res., 95*, 11,957, 1990.

Lepping, R.P., L.F. Burlaga, A. Szabo, K.W. Ogilvie, W.H. Mish, D. Varsiliadis, A.J. Lazarus, J.T. Steinberg, C.J. Farrugia, L. Janoo, and F. Mariani, The wind magnetic cloud and events of October 18-20, 1995: Interplanetary properties act as triggers for geomagnetic activity, *J. Geophys. Res., 102*, 14,049, 1997.

Lu, G., D.N. Baker, R.L. McPherron, C.J. Farrugia, D. Lummerzheim, J.M. Ruohoniemi, F.J. Rich, D.S. Evans, R.P. Lepping, M. Brittnacher, .X. Li, R. Greenwald, G. Sofko, J. Villain, M. Lester, J. Thayer, T. Moretto, D. Milling, O. Troshichev, A. Zaitzev, V. Odintzov, G. Makarov, and K. Hayashi, Global energy deposition during the January 1997 magnetic cloud event, *J. Geophys. Res., 103*, 11,685, 1998.

Marubushi, K., Interplanetary magnetic flux ropes and solar filaments, *Coronal Mass Ejection, Geophys. Monograph 99*, American Geophys. Union, 1997.

Maunder, E.W., Magnetic disturbances, 1882 to 1903, as recorded at the Royal Observatory, Greenwich, and their association with sun-spots, *Mar. Not. Roy. Astron. Soc., 65*, 2, 1905.

Maurits, S.A. and B.J. Watkins, UAF Eulerian model of the polar ionosphere, Solar-Terrestrial Energy Program: Handbook of Ionospheric Models, SCOSTEP, p. 95, 1996.

Maurits, S.A., J. McAllister, and B.J. Watkins, WWW-based visualization of the real time run of a space weather forecasting model in preparation, 2000

Moldwin, M.B., S. Ford, R. Lepping, J. Slavin, and A. Szabo, Small-scale magnetic flux ropes in the solar wind, *Geophys., Res. Lett., 27*, 57, 2000.

Mulligan, T., C.T. Russell, and J.G. Luhman, Solar cycle evolution of the structure of magnetic clouds in the inner heliosphere, *Geophys. Res. Lett., 25*, 2959, 1998.

Newell, P.T., Y.-I. Feldstein, Y.I. Galpainod, C.-I. Meng, Morphology of nightside precipitation, *J. Geophys. Res., 101*, 10, 737, 1996.

Olmsted, C. and S.-I. Akasofu, One-dimensional kinematics of particle stream flow with application to solar wind simulation, *Planet. Space Sci., 33*, 831, 1985.

Olmsted, C. and S.-I. Akasofu, A method for determining corotating interaction regions in the equatorial plane of the heliosphere from solar wind velocity data, *J. Geophys. Res., 91*, 13, 689, 1986.

Osherovich, V.A., J. Fainberg, and K.G. Stone, Multi-tube model for interplanetary magnetic clouds, *Geophys. Res. Lett., 26, 401*, 1999.

Perreault, P. and S.-I. Akasofu, A study of geomagnetic storms, *Geophys. J.R. Astr. Soc., 54*, 547, 1978.

Pudovkin, M.I. and V.S. Semenov, Implications of the stagnation line model for energy input through the dayside magnetopause, *Geophys. Res. Lett., 13*, 213, 1986.

Reiff, P.H., R.W. Spiro, and T.W. Hill, Dependence of polar cap potential drop on interplanetary parameters, *J. Geophys. Res., 86*, 7639, 1987.

Rust, D.M., Magnetic helicity in solar filaments and coronal mass ejection, *Magnetic Helicity in Space and Laboratory Plasmas, Geophys. Monograph, 111*, AGU, 1999.

Saito, T., T. Oki, C. Olmsted, and S.-I. Akasofu, A representation oft he magnetic neutral line on the solar source surface in terms of the sun's axial dipole at the center and two equatorial dipoles in the photosphere, *J. Geophys. Res., 97*, 14, 993, 1989.

Shepherd, S.G., R.A. Greenwald, and J.M. Ruohoniemi, A possible explanation for rapid, large-scale ionospheric responses to southward turnings of the IMF, *Geophys. Res. Lett., 26*, 3197, 1999.

Simnett, G.M., S.J. Tappin, S.P. Plunkett, D.K. Bedford, C.J. Eyles, O.C. St. Cyr, R.A. Howard, G.E. Brueckner, D.J. Michels, J.D. Moses, D. Socker, K.P. Dere, C.M. Korendyke, S.E. Paswaters, D. Wang, R. Schwenn, P. Lamy, A. Llebaria, and M.V. Bout, LASCO observations of disconnected magnetic structures out to beyond 28 solar radii during coronal mass ejections, *Solar Phys., 175*, 685, 1997.

Sotirelis, T., P.T. Newell, and C.-I. Meng, Shape of the open-closed boundary of the polar cap as determined from observations of precipitating particles by up to four DMSP satellites, *J. Geophys. Res., 103*, 399, 1998.

Srivastara, N., W.D. Gonzalez, A.L.C. Conzalez, and S. Masuda, On the solar origins of intense geomagnetic storms observed during 6-11 March 1993, *Solar Phys., 183*, 419, 1998.

Sun, W. and S.-I. Akasofu, On the formation of the storm-time ring current belt, *J. Geophys. Res., 105*, 5411, 2000.

Sun, W., S.-I. Akasofu, Z.K. Smith, and M. Dryer, Calibration of the kinematic method of studying solar wind disturbances on the basis of a one-dimensional MHD solution and a simulation study of the heliosphere disturbances between 22 November and 6 December 1977, *Planet. Space Sci., 33*, 933, 1985.

Tang, F., S.-I. Akasofu, E. Smith, and B. Tsurutani, Magnetic fields on the sun and the north-south component of transient variations of the interplanetary magnetic field at 1 au, *J. Geophys. Res., 90*, 2703, 1985.

Tsurutani, B.T., W.D. Gonzalez, F. Tang, S.-I. Akasofu, and E.J. Smith, Origin of interplanetary southward magnetic fields responsible for major magnetic storms near solar maximum (1978-79), *J. Geophys. Res., 93*, 8519, 1988.

Vandas, M., S. Fischer, M. Dryer, Z. Smith, and T. Detman, Self-consistent simulation of cylindrical magnetic cloud propagation in the heliosphere with its axis both perpendicular to, and lying within, the ecliptic plane, *Adv. Space Res. 174/5*, 327 , 1996.

Vandas, M., S. Fischer, M. Dryer, Z. Smith, and T. Detman, Propagation of a spheromak: 2. Three-dimensional structure of a spheromak, *J. Geophys., Res., 103*(A10), 23717, 1998.

Vandas, M. and D. Odstrcil, Magnetic cloud evolution: A comparison of analytical and numerical solutions, *J. Geophys. Res., 105*(A6), 12605.

Wang, Y.-M., R.R. Sheeley Jr., and J. Lean, Understanding the evolution of the sun's open magnetic flux, *Geophys. Res. Lett., 27*, 621, 2000.

Syun-Ichi Akasofu, 930 Koyukuk Drive, P.O. Box 757340, Fairbanks, AK 99775-7340.

Predicting Geomagnetic Activity: The D_{st} Index

Robert L. McPherron and Paul O'Brien

Institute of Geophysics and Planetary Physics, University of California Los Angeles, Los Angeles, California

Geomagnetic activity is usually characterized by magnetic indices. Most indices have long records that allow statistical studies of the causes of activity and of related phenomena. Correlations between indices and possible drivers provide the basis for empirical prediction. Here we examine solar wind control of D_{st}, an index that is thought to be linearly proportional to the total energy in the terrestrial ring current. We use linear prediction filtering, a technique in which an autoregressive (AR) filter maps past values of the index to the next value, and a moving average (MA) filter maps current and past values of the solar wind input to the next value of the index. These ARMA filters may be determined from historical records by least square optimization. Nonlinear systems can be approximated in a piecewise fashion by localizing the filter. We do this by using narrow bins of the solar wind electric field; *VBs*. Our model utilizes 37 years of hourly observations to estimate the coefficients representing the quiet time ring current, the solar wind dynamic pressure, the ring current decay rate, and the rate of ring current injection in a simple differential equation. We find that pressure and decay coefficients are fit by exponential functions of *VBs*, decreasing as *VBs* increases, but ring current injection is a linear function of *VBs*. Integration of our model using observations of the solar wind and analytic fits to the coefficients produces a time series that contains 76% of the variance in the original data. The prediction residuals have a Gaussian distribution with zero mean and rms error of 10.6 nT.

1. INTRODUCTION

Space weather consists of a variety of phenomena driven by the solar wind such as substorms, magnetic storms, acceleration of relativistic electrons, and ULF waves. In this paper we report an empirical study of magnetic storms as characterized by the D_{st} index. Magnetic storms occur when the number and energy of positive ions and electrons drifting in the outer radiation belts increase significantly. Since electrons and protons drift in opposite directions they produce a ring current around the earth. The direction of this current is westward causing a decrease in the surface field. The D_{st} index is a measure of the total energy of these drifting particles. D_{st} is obtained by finding the instantaneous average of the deviations from a quiet day in the horizontal component of the magnetic field at a number of low latitude magnetic observatories.

A magnetic storm typically consists of three phases. The initial phase is a result of an increase in solar wind dynamic pressure. This increase presses the magnetopause current closer to the earth causing a positive perturbation in H. The main phase is a consequence of a southward turning of the interplanetary magnetic field (IMF). When the IMF turns southward magnetic reconnection occurs on the dayside allowing a fraction of the solar wind electric field to penetrate the magnetosphere [*Reiff and Luhmann*, 1986]. This field transports ions from the tail to the inner magnetosphere where they are trapped in the ring current, causing the D_{st} index to become more negative. The recovery phase is a consequence of the IMF turning northward shutting off the magnetospheric electric field. Particle injection decreases while the drifting ions charge exchange with atmospheric neutral particles losing their energy and thereby decreasing the strength of the ring current.

The purpose of this paper is to illustrate how the strength of the ring current as measured by the D_{st} index can be predicted by the method of local linear filters. To do this we utilize 37 years of solar wind and geomagnetic data to produce filters for a variety of states of the magnetosphere. We demonstrate that the coefficients of these models can be represented by analytic functions of a single variable, the solar wind electric field, VBs. We then show that these functions may be used to make multi-step predictions of the index from observations of the solar wind. We evaluate the quality of these predictions showing that they generally provide accurate predictions.

2. PREDICTION FILTERS

A linear prediction filter is written in the following way.

$$O(t) = \sum_{i=1}^{N} a_i O(t - i\Delta t) + \sum_{j=0}^{M} b_j I(t - j\Delta t) \quad (1)$$

The output of a system at the next time step is the sum of two parts. The first part is a weighted sum of previous values of the output. This self-prediction in called auto regression and it represents internal dynamics of the system. The second part is a weighted sum of the current and past values of the input. This part represents the external dynamics. Together the autoregressive (AR) filter coefficients, a_i, and the moving average (MA) coefficients, b_j, constitute an autoregressive moving average (ARMA) filter. With a single set of filter coefficients the representation is completely linear. Such filters can approximate even nonlinear systems, but the more nonlinear the system, the lower the accuracy of the predictions.

ARMA filters are actually discrete representations of differential equations [*Klimas et al., 1998*]. Representation of the relation between the input and output of a causal system by a linear prediction filter is equivalent to describing it by a differential equation. Integration of this equation from a known initial condition with a measured input as driver is how prediction is accomplished. Note that the current output cannot be calculated until the current input is measured. If for example the time step is one day the output for the day cannot be calculated until the end of the day. Only if the actual input can be measured well in advance of its arrival, or the delay in the system response is long compared to the time step, can this be considered a "prediction" technique.

A representation of the behavior of the ring current in terms of linear prediction filters is motivated by a consideration of the physical processes that produce the surface magnetic fields. According to the D-P-S relation, D_{st} is directly proportional to the total energy in the drifting ring current particles [*Dessler and Parker, 1959; Sckopke, 1966*]. This implies that the D_{st} (a negative quantity) becomes more negative when energy is injected and less negative when energy is lost. It has been found that injection is proportional to the solar wind electric field and decay is proportional to the strength of D_{st}. *Burton et al.* [1975] expressed these facts with the following equation.

$$\frac{dD_{st}^{*}}{dt} = Q(t) - \frac{D_{st}^{*}}{\tau} \quad (2)$$

The quantity D_{st}* is the component of the measured D_{st} index that is caused by a symmetric ring current. However it is well known that measured D_{st} is a superposition of the effects of the ring current, the disturbed magnetopause current, and the quiet time ring current present when the baselines for the magnetic perturbations are determined. These facts are summarized by the relation

$$D_{st}^{*} = D_{st} - b\sqrt{p} - c \quad (3)$$

Here p is the dynamic pressure of the solar wind, b is the constant of proportionality relating changes in D_{st} to changes in pressure, and c is the effect of the quiet time magnetopause. If we substitute this relation into equation 2, approximate the time derivatives with first differences, and rearrange terms we obtain

$$\Delta D_{st} = \left(\frac{-\Delta t}{\tau}\right) D_{st} + b\left(\Delta\sqrt{p}\right) + \left(\frac{b\Delta t}{\tau}\right)\sqrt{p} + \left\{Q(t) + \frac{c\Delta t}{\tau}\right\} \quad (4)$$

The dependent variable in this difference equation is ΔD_{st} while D_{st}, $\sqrt{p}$, $\Delta\sqrt{p}$, and $Q(t)$ are independent variables. The coefficients of the equation are τ, b, c, and whatever constants are required to describe the input function $Q(t)$. Measurements at a specific time define one instance of the relation between the dependent and independent variables. Taking many successive measurements we obtain an over-determined set of equations for the unknown coefficients. *Burton et al.* [1975] assumed that all of these coefficients were constants independent of the state of the system and determined each constant separately by a different method. Their most important result was that $Q(t)$ was a linear function of the solar wind electric field, i.e. $Q(t) = \alpha VBs$. Below we assume that injection is an arbitrary function of the solar wind, i.e. $Q(t) = A(VBs)$. We then show that this is a linear function except for details near $VBs = 0$.

The work of *Klimas et al.* [1998] and *Vassiliadis et al.* [1999a] makes quite different assumptions about both the form of the differential equation and the constancy of the coefficients. Instead of using a single feedback term they use two (a_1 and a_2). This is equivalent to a second order differential equation where a_1 can be interpreted as the damping constant and a_2 as the resonant frequency of the system. They also assume that the coefficients depend on the state of the system. The two papers differ in how they localize the ARMA filters, but both assume that the state of the system depends on three variables: D_{st}, dD_{st}/dt, and VBs. *Klimas et al.* [1998] neglect the pressure correction to D_{st}. *Vassiliadis et al.* [1999b] makes this correction prior to calculating the state-dependent filter coefficients obtaining coefficients b and c that change with season. The authors justify the use of the second order equation by the observation that the D_{st} index appears to oscillate after pressure pulses and during the main phase decrease.

In this paper we continue to use the Burton equation (2) because we see no compelling theoretical reason to believe that the symmetric ring current should display oscillatory behavior. Also, how well the Burton equation with state dependent coefficients is able to explain the behavior of the ring current has not been established. Finally, our use of hourly averages precludes observation of any oscillations shorter than two-hour period, and this is the order of the oscillations reported by *Vassiliadis et al.* [1999a]. Our work also differs from that of Klimas and Vassiliadis because we assume that the system's state is defined only by the solar wind electric field, *VBs*.

A justification for our view is provided by our previous work described in *O'Brien and McPherron* [2000a]. In this work we used a new statistical procedure that does not involve linear filters to study the validity of the Burton equation. The entire 37-year history of solar wind measurements was examined creating a sequence of joint probability distributions corresponding to fixed bins of *VBs*. Each distribution shows the probability of a particular ΔD_{st} given a specific D_{st} with *VBs* essentially constant. In each distribution we fit the median ΔD_{st} versus D_{st} to a polynomial in D_{st}. The fits are almost precisely straight lines. The slope and intercepts of these lines depend on *VBs* allowing us to obtain analytic fits to the parameters of the Burton equation. The absence of any quadratic or higher terms in the polynomial fits implies that ΔD_{st} is a linear function of D_{st} throughout the D_{st} - ΔD_{st} phase space. This is precisely the relation implied by the Burton equation!

3. ANALYSIS METHOD

Justified by our previous experience we continue our examination of the Burton equation with linear filters as expressed in (4). However, we now assume that the unknown parameters in this equation are functions only of *VBs*. We thus divide the 37-year history of solar wind observations into subsets characterized by specific intervals of *VBs*. For each bin τ, b, c, and $A(VBs)$ are assumed constant. This fact complicates the solution of (4) because the coefficients are then constrained by linear relations between them, e.g. the coefficient for D_{st} is fixed, while other coefficients are related to this one. If we ignore this fact for a moment (4) can be written in the form

$$\Delta D_{st} = \alpha D_{st} + \beta \Delta \sqrt{p} + \gamma \sqrt{p} + \delta \qquad (5)$$

In this equation D_{st} and $\sqrt{p}$ are scalar time series with one-hour time resolution. The first differences of these [ΔD_{st} and $\Delta \sqrt{p}$] are easily calculated with due consideration for flags denoting missing records. Equation 5 with fixed coefficients applies to every unflagged hourly record in each subset of the data. This collection of records defines an over-determined set of equations for the unknown coefficients. Unfortunately, the coefficients are not independent because the unknown τ relates them. However, if we fix τ, which is possible since *VBs* is fixed, then the coefficients are related by linear constraints. We can use either constrained least square or nonlinear optimization techniques to obtain the solution for the unknown parameters. The two methods give identical results. It should be emphasized, however, that the parameters cannot be treated as independent in a standard least square analysis. The constraints produce a significant correction.

3.1. Solution for Southward IMF

To define the filter coefficients for southward IMF we must define a set of *VBs* bins. If these are too narrow then a stable solution cannot be obtained. We find by experience that 100 or more records are needed to accurately define the coefficients for a specific *VBs* bin. However, the extreme values of *VBs* that create large storms are so rare that it is impossible to obtain this many records for values of *VBs* exceeding 10 mV/m. We have used a set of bins of increasing width as the magnitude of *VBs* increases to partially compensate for the decreasing probability of a given *VBs*. Even so, the extreme bins are quite wide and the calculated model coefficients are highly variable and somewhat suspect. In spite of this problem the coefficients display a systematic dependence on *VBs*.

Figure 1 presents four panels displaying graphs of various parameters as a function of *VBs*. The top left panel contains the rate of ring current injection as a function of *VBs*. The rate is clearly a linear function of *VBs* as shown by the straight line fit. The slope of this line, (4.65 nT/hr)/(mV/m), is close to the value obtained in previous studies. The top right panel presents the ring current decay rate versus *VBs*. This rate decreases exponentially from about 15 hours with northward IMF to approximately 5 hours when *VBs* is greater than 10 mV/m. An analytic fit to this curve is shown in the graph. The form of this function was derived from physical arguments by *O'Brien and McPherron* [2000a], but the parameters were determined by nonlinear inversion from the data plotted here. The bottom left panel shows the pressure constant versus *VBs*. This function also decreases exponentially from its value for northward field (>8 $nT/\sqrt(nP)$) to a value close to 2. An arbitrary function has been fit to the data with the results shown in the graph. The fourth panel shows the number of records used to determine the coefficients for each *VBs* bin. For the most extreme bin near 20 mV/m there were only 10 occurrences in the 37 years of data.

3.2. Calculation of the Ring Current Injection Function

Previous work suggests that the rate of injection into the ring current and its rate of decay are controlled by solar wind *VBs* [*Burton et al.*, 1975; *O'Brien and McPherron*, 2000a]. This assumption allows us to combine the injection function $Q(t) = A(VBs)$ and the quiet magnetopause correction $(c\Delta t/\tau)$ into a single term in the regression equation (4). We then sort our data into bins of constant *VBs* and use the data from each bin to determine the coefficients of the regression equation. The results plotted in Figure 1 indicate that the offset term, $\delta(VBs)$, in (5) is nearly a linear function of *VBs*. The injection function, *A(VBs)* can be calculated from the offset, δ, using the equation

$$A(VB_s) = \delta(VB_s) - c\frac{\Delta t}{\tau(VB_s)} \qquad (6)$$

If we assume that there is no ring current injection for northward IMF, i.e. *A(VBs=0)* = 0, then the constant *c* in (6) can be determined from our regression coefficients. We find $c = \tau(0)\delta(0)/\Delta t = -4.6$ *nT*. Substituting this value into (6) gives us an equation for *A(VBs)*, for any value of *VBs*. Because of the nonlinear dependence of τ on *VBs* this equation is clearly nonlinear. However, note that the magnitude of the magnetopause correction term is always less than 1.0 while the offset ranges from 0 to 100. Thus the offset is important only near *VBs* = 0. Elsewhere *A(VBs)* is well approximated by the linear relation shown in the fig-

ure. Despite this, any small departure from linearity near zero is important. A linear fit to *A(VBs)* intercepts the *VBs* axis at 0.53 (mV/m). Using the linear fit for smaller values of *VBs* implies *VBs* causes a loss of energy from the ring current. Thus, as did *Burton et al.* [1975] we introduce a cutoff electric field *Ec* = -0.53 mV/m so that for weaker fields we set *A(VBs)* = 0. A better approximation would be to use a polynomial fit in the interval 0.0 to 1.0 mV/m.

3.3. Quality of the D$_{st}$ Predictions

Solutions for the ARMA filter coefficients may be used to model the data that defined them. The predicted changes for each *VBs* bin are given by $\Delta \vec{D}_{st} = (X)\vec{c}$ where *(X)* is the matrix with columns containing the independent variables in (5), and $\vec{c}$ is the vector of coefficients dependent on *VBs*. Using the collection of solutions calculated for all *VBs* bins we can generate a time series of ΔD_{st} that may be compared to the observed ΔD_{st}. This change may then be added to the current value of the observed D_{st} to predict the next D_{st}. Note that this prediction utilizes the observed D_{st} and solar wind parameters for each hour. This procedure is referred to as one-step prediction.

A scatter plot of the predicted ΔD_{st} versus observed D_{st} shows a correlation of only 0.606. This average translates to a prediction efficiency of only 37%. This means that most of the variance in ΔD_{st} is not predictable! None-the-less, when the predicted change is added to the current measured D_{st}, the predicted D_{st} for the next hour agrees with the observed with a prediction efficiency of order 99%. The reason for this is that most of the predicted D_{st} consists of the currently observed D_{st} so that the next D_{st} is very close to that observed. This one-step prediction accuracy has little meaning since in an operational situation measured D_{st} values are not available. Operationally we must perform multi-step predictions as discussed in the next section.

3.4. Multi-step Prediction of D$_{st}$ Using Analytic Fits to Model Coefficients

A multi-step prediction is achieved by integrating the regression equation (5) from a known initial condition. First expand ΔD_{st} to obtain

$$D_{st}(n+1) = (1+\alpha)D_{st}(n) + \beta\Delta\sqrt{p} + \gamma\sqrt{p} + \delta \qquad (7)$$

If we start with a known value of D_{st}, now available within about 24 hours, we can integrate forward in time to the current time. In this integration we "feedback" the previous prediction to calculate the next ΔD_{st} that is added

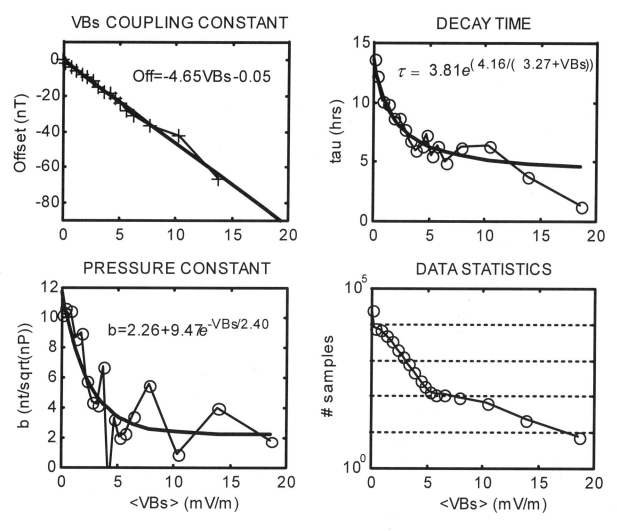

Figure 1. Graphs of the coefficients of the ΔD_{st} prediction equation (4) versus *VBs*. Parametric fits to the relations are shown in each panel. Clockwise from the upper left the panels present the offset (δ in equation 5), the decay time τ, the number of hourly records used to make the fits in each *VBs* bin, and the pressure constant b.

to the previous prediction to obtain the next D_{st}. This is referred to as multi-step prediction and provides a much more stringent test of the model. Once the integration reaches the current time it can be advanced only as fast as data on the input are acquired. In this integration we use the analytic fits to the various coefficients rather than the tabular results of modeling. This assures more continuous change in the coefficients as the state of the solar wind changes.

Figure 2 presents illustrations of the results from this integration. Each panel contains a short segment of continuous solar wind data that has been integrated starting from a measured value of D_{st}. Unfortunately, the main problem in predicting geomagnetic activity is the presence of frequent gaps in the solar wind input data. To perform the integra-

tion across such gaps it is necessary to assume the behavior of the solar wind. These assumptions are almost always wrong and introduce large errors in the predicted time series. It takes many hours for the prediction to converge back to the correct solution. To avoid this problem here we have scanned the entire 37-year record identifying all intervals with 12 or more hours of continuous data. For each of these intervals we start the integration with the known value of D_{st} and integrate to the end of the interval using feedback of the prediction. This technique allows us to determine the best possible prediction efficiency for the method. The four examples were selected by the requirement that the interval contain more than 100 hours of data, and the prediction efficiency for these intervals exceeds 89%. Thus the four intervals are the best possible illustra-

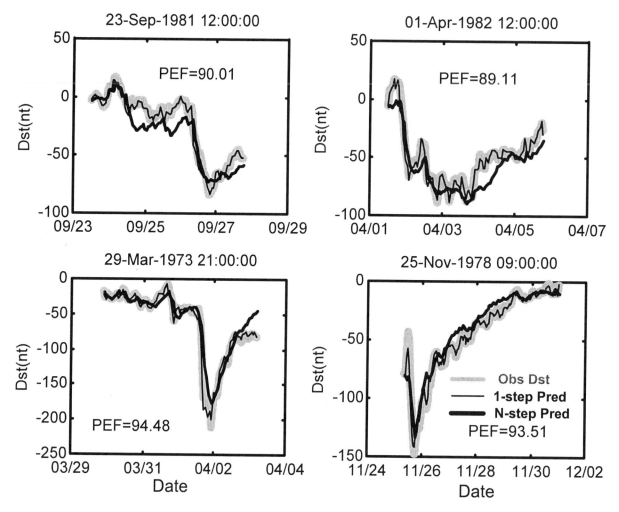

Figure 2. A comparison of the waveforms of the modeled and observed changes in D_{st}. The four longest intervals of contiguous data with prediction efficiencies above 89% were selected for the illustration.

tions of the quality of the predictions. Three curves are plotted in each panel: the observed D_{st}, the 1-step prediction, and the multi-step prediction. The 1-step prediction is indistinguishable from the original. The heavy line shows the multi-step prediction is also very close to the observed index for these intervals.

The average correlation for the entire data set is 0.872 corresponding to a prediction efficiency of 76%. The rms error of the prediction relative to the data is only 10.6 nT! This is close to the expected error in D_{st} produced by a variety of problems in its generation. It is also better than the second order fits obtained by *Vassiliadis et al.* [1999a] assuming that the coefficients depend on D_{st}, ΔD_{st}, and *VBs*. However, this comparison is somewhat misleading since they use 5-minute data that contains considerable more variance than does hourly data.

4. DISCUSSION AND CONCLUSIONS

In this paper we have developed an algorithm for predicting the hourly D_{st} index from upstream observations of the solar wind. Our technique is an extension of the work of *Burton et al.* [1975] who derived a simple first order differential equation for D_{st} based on physical principles. This equation equates the rate of change of pressure corrected D_{st} to the difference between injection by the solar wind and decay by charge exchange. *Burton et al.* [1975] assumed that all coefficients in the differential equation were constant with time and state of the magnetosphere, and then determined their values empirically from small datasets of 2.5-minute data. They demonstrated that an integral of the equation from a known starting value, driven by measured inputs, provided a good fit to the data and could be used as a forecast tool provided real time observations are available.

It is extremely difficult to compare our results to those obtained by earlier workers. Some have used higher time resolution, most have used smaller datasets, often authors consider only a fixed point in the solar cycle, in some cases they have used a second order differential equation (more coefficients in the filters), and different methods of localization. Others have used alternative solar wind coupling functions. Pressure correction are often ignored or not determined self-consistently with estimates of the ARMA coefficients. A comparison of our prediction efficiency (~76%), or rms error (~11 nT), with values quoted in previous papers suggests our method is equal or superior to others. However, comparison with 5-minute D_{st} predictions are misleading because these data have higher variance and are therefore more difficult to predict. On the other hand, other models use more parameters than does ours so that should improve the quality of their fits.

Our results show that both the ring current decay time and the dynamic pressure coefficients depend on the strength of the solar wind electric field. The dependence of τ on VBs is consistent with previous reports that claim that it is much shorter during the main phase than in the recovery phase. *O'Brien and McPherron* [2000a] have suggested this variation is a consequence of changes in location of the inner boundary of convection and the charge exchange decay rate relevant to this location as a function of VBs.

The exponential decrease of b with VBs may have a simple explanation. As VBs increases a stronger tail current is closer to the Earth. An increase in dynamic pressure will increase the tail field thus increasing the tail current. The effect of this current on the Earth is opposite to that of the magnetopause current and so partially compensates for the increase caused by an increase in the magnetopause current. Let us assume that the pressure contribution to D_{st} is the difference between the effects of the magnetopause and magnetotail, and that the effect of the magnetopause does not change with VBs.

$$\Delta D_{st} = \Delta Magnetopause - \Delta Magnetotail$$
$$\Delta D_{st} = b_{MP}\sqrt{p_{dyn}} - b_{Tail}(VBs)\sqrt{p_{dyn}}$$
$$= (b_{MP} - b_{Tail}(VBs))\sqrt{p_{dyn}} = b(VBs)\sqrt{p_{dyn}}$$

Our value for $b(VBs)$ during northward field is smaller than the value expected theoretically (8 rather 16 $nT/\sqrt{(nP)}$) suggesting that at quiet times the contribution to D_{st} from the tail is half that expected from the magnetopause. When VBs reaches -20 mV/m, $b(VBs)$ appears to be about 2. If this is the correct explanation of the apparent change in b

then the tail current must affect D_{st} far more than previously realized.

In conclusion we believe that we have improved our ability to predict the hourly D_{st} index from upstream data. The algorithm is simple, involving only three parametric relations for the coefficient dependence on the solar wind electric field. The method requires continuous tracking of the solar wind and that the data be transferred to the Earth in real time so as to provide sufficient time delay to allow a calculation of the next hourly value prior to the end of the hour. A version of this algorithm has been implemented at www.igpp.ucla.edu/swcgag/. A predicted D_{st} index is updated hourly and past values are compared to the Kyoto Sym-H index for validation. A brief description of the real time algorithm will appear in *O'Brien and McPherron* [2000b].

Acknowledgments. The authors would like to acknowledge frequent helpful discussions of the D_{st} prediction problem with A. Klimas, D. Vassiliadis and T. Detman. This work has been supported by grants from the NSF under the space weather prediction program [NSF ATM 96-13667 & NSF ATM 99-72069].

REFERENCES

Burton, R.K., R.L. McPherron, and C.T. Russell, An empirical relationship between interplanetary conditions and D_{st}, *J. Geophys. Res.*, *80*(31), 4204-4214, 1975.
Dessler, A.J., and E.N. Parker, Hydromagnetic theory of magnetic storms, *J. Geophys. Res.*, *64*(12), 2239-2259, 1959.
Klimas, A.J., D. Vassiliadis, and D.N. Baker, D_{st} index prediction using data-derived analogues of the magnetospheric dynamics, *J. Geophys. Res.*, *103*(A9), 20,435-20,447, 1998.
O'Brien, T.P., and R.L. McPherron, An empirical phase-space analysis of ring current dynamics: solar wind control of injection and decay, *J. Geophys. Res.*, *105*(A4), 7707-7719, 2000a.
O'Brien, T.P., and R.L. McPherron, Forecasting the ring current index D_{st} in real time, *JASTP*, in press, 2000b.
Reiff, P.H., and J.G. Luhmann, Solar wind control of the polar-cap voltage, *Solar Wind-Magnetosphere Coupling*, 1986.
Sckopke, N., A general relation between the energy of trapped particles and the disturbance field over the earth,, *J. Geophys. Res.*, *71*(13), 3125-3130, 1966.
Vassiliadis, D., A.J. Klimas, and D.N. Baker, Models of D_{st} geomagnetic activity and of its coupling to solar wind parameters, *Phys. Chem. Earth (C)*, *24*(1-3), 107-112, 1999a.
Vassiliadis, D., A.J. Klimas, J.A. Valdivia, and D.N. Baker, The D_{st} geomagnetic response as a function of storm phase and amplitude and the solar wind electric field, *J. Geophys. Res.*, *104*(A11), 24,957-24,976, 1999b.

R. L. McPherron and Paul O'Brien, Institute of Geophysics and Planetary Physics, University of California Los Angeles, Los Angeles, CA 90095-1567 (e-mail: rmcpherron@igpp.ucla.edu; tpoiii@igpp.ucla.edu)

Space Weather Effects on Power Systems

Geomagnetic Laboratory, Geological Survey of Canada, Ottawa, Canada

Space weather disturbances cause geomagnetic field variations that induce electric currents into power transmission systems on the ground. These geomagnetically induced currents (GIC) flow to ground through the windings of power transformers where they produce extra magnetic flux that can saturate the transformer core. This leads to transformer heating, increased power demand, and ac harmonic generation, which can interfere with power system operation. This paper examines the magnetic disturbances on March 24, 1940, February 11, 1958, August 4, 1972, and March 13, 1989 that were responsible for the most significant power system effects. The blackout of the Hydro-Québec system on March 13, 1989 was due to an enhancement of a westward substorm electrojet resulting from loading and unloading of energy in the magnetosphere. Power system effects, including transformer overheating, later on March 13 can be attributed to an eastward convection electrojet caused by the 'directly-driven' flow of energy from the solar wind. Power system problems during the earlier disturbances are also shown to be caused by rapid changes of the convection electrojets. This shows that the convection current systems, as well as substorm currents, need to be included when predicting space weather effects on power systems.

1. INTRODUCTION

The first magnetic storm that had a noticeable effect on power systems occurred on March 24, 1940 (Davidson, 1940). Power systems were again affected during the magnetic storm of February 10, 1958; however, it was not until 1967 that detailed investigations began (Slothower and Albertson, 1967). In the following years an extensive investigation was made by Alberston and co-workers who showed how widespread was the occurrence of geomagnetically induced currents (GIC) and the range of effects they could have on power systems (Albertson et al, 1973, 1974). Some of their recording systems were still deployed during the major magnetic disturbance on August 4, 1972 and this became one of the best documented GIC events (Albertson and Thorson, 1974).

During solar cycle 21 there were no major effects on power systems due to magnetic disturbances. However, a number of studies gathered more information about the processes involved and how to model them. In Finland, Pirjola and co-workers made a long series of GIC recordings and developed techniques for calculating the electric fields and GIC produced in a power system during geomagnetic disturbances (Pirjola, 1985; Lehtinen and Pirjola, 1985; Pirjola and Lehtinen, 1985). In North America, Bolduc and Aubin (1978) showed how to calculate the transformer saturation produced by GIC, Boteler et al (1989) reported observations of the increased ac harmonics that result from saturation, and Albertson et al (1981) and Kappenman et al (1981) studied how these geomagnetic effects influenced power system operation.

In spite of the aforementioned studies, the arrival of the magnetic storm on March 13, 1989, during the up-swing of solar cycle 22, and its effects on power systems came as a surprise. The storm was one of the largest recorded since observations began in the 1840s and produced widespread technological effects (Allen et al, 1989). Power systems in

Space Weather
Geophysical Monograph 125

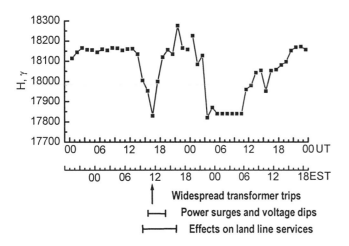

Figure 1. The March 24, 1940 magnetic disturbance (as shown by hourly means from Cheltenham) plus the times of power system problems.

North America and Europe experienced relay trips, voltage drops, and transformer heating. The most significant effect was a blackout of the Hydro-Québec power system.

Space weather affects power systems because geomagnetic field variations induce electric currents into the power transmission lines. These GIC flow to and from ground through the windings of power transformers and cause partial saturation of the transformer core. This disrupts the ac operation of the transformer causing extra heating that can damage winding insulation, increased reactive power demand leading to a drop in system voltage, and higher levels of ac harmonics which can trigger tripping of protective relays. In extreme cases, the combination of all these things can have serious effects on system stability and lead to a power blackout such as occurred on the Hydro-Québec system.

There are a number of possibilities for preventing these space weather effects. Some transformers (3-phase 3-legged core type) are less susceptible to saturation from GIC. However changing existing transformers to this type is uneconomic and for handling high power levels a 3-phase transformer in one unit becomes impractically large and sets of 3 single-phase transformers are used instead. An alternative approach involves blocking the flow of GIC. Hydro-Québec have placed blocking capacitors in their power transmission lines, and Kappenman et al (1991) have developed a suitable device for insertion in transformer neutral-ground connections. These remedies are not considered economic for all systems and many power system operators rely on advance warning of magnetic storms to implement operating strategies designed to reduce system vulnerability.

Several studies have been made to assess the geomagnetic hazard to particular power systems (Mäkinen, 1993; Boteler et al, 1997). Such work can determine, in general statistical terms,

the size of GIC that can be expected during different levels of geomagnetic activity. However, in real-time forecasting we are trying to predict the GIC and power system response that will be produced by specific events. To help in this endeavour it is worth looking at exactly what characteristics of past disturbances produced significant effects on power systems. This paper presents a summary of the power system effects and examines their cause for the four magnetic disturbances that have had the biggest impact on power systems: March 24, 1940; February 10, 1958; August 4, 1972; and March 13, 1989. Identifying the magnetospheric and ionospheric current systems responsible for the critical geomagnetic field variations should help in forecasting future space weather effects on power systems.

2. MARCH 24, 1940

On Easter Sunday 1940 a magnetic storm produced widespread effects on power systems and communication systems. Germaine (1940) reports that effects on land line services occurred between 10.00 and 16.00 Eastern standard time (EST). Davidson (1940) provides detailed accounts of the power system problems, including the following items:
Minneapolis area:
10.45 am to 1.45 pm EST power system disturbances
11.50 am EST most severe power surges
    Central Maine:
10.50 am to 2.00 pm EST numerous voltage dips
11.48 am EST two transformer banks tripped out
    Eastern Pennsylvania:
11.48 am EST reactive power surges of 20% and
    two 75,000 KVA transformer banks tripped
    Chats Falls, Ontario:
11.48 am EST four transformers tripped out
These reports show that the peak of the disturbance occurred just before local noon in eastern North America.

The original magnetic observatory recordings of the March 24, 1940 magnetic storm have been lost or were off scale. However some information about the disturbance can still be obtained from the archived hourly mean values. Figure 1 shows the hourly mean values from Cheltenham magnetic observatory (Geograph. Lat. 38.7 Long. 283.2) on the east coast of the United States. This shows that the power system problems occurred at the time of large negative change in the northward magnetic field. Such a change would be produced by enhancement of an overhead westward electric current.

3. FEBRUARY 10-11, 1958

A major magnetic storm occurred on February 10-11, 1958 and produced effects on a number of power systems in North America. Slothower and Albertson (1967) report large reactive

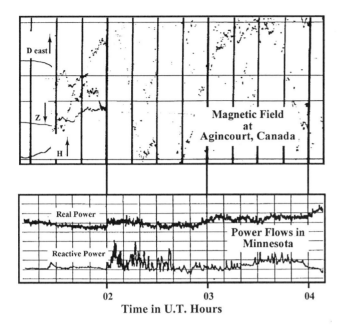

Figure 2. Magnetogram from Agincourt and reactive power flow in Minnesota showing the increase in reactive power flow at 02.00 UT concides with an increase in magnetic activity.

power flows seen on the Northern States Power Company lines at Minnesota. In Ontario transformers at Port Arthur and Raynor Generating Station were simultaneously tripped by differential relay operation (Acres, 1975). Figure 2 shows that the start of increased reactive power flow and the transformer trips coincided with a sudden increase in magnetic activity at 02.00 UT recorded at the Agincourt Magnetic Observatory near Toronto. A sudden jump in the magnetic field was also recorded on the rapid-run magnetogram at Fredericksburg (Winckler et al, 1959).

The Fredericksburg magnetogram shows a positive excursion in H at the time of the power system problems. Agincourt magnetic recordings were off scale for the worst part of the storm, however notes of the daily extremes made by the observatory operators (Ross and Evans, 1962) record that a maximum positive excursion of 949 nT occurred at 02.53 UT on February 11. Both the Fredericksburg and Agincourt magnetic excursions indicate that an eastward electrojet was responsible for the magnetic disturbance that caused the power system effects in Minnesota and Ontario.

4. AUGUST 4, 1972

On August 4, 1972 a magnetic storm produced widespread effects on power systems in the United States and Canada (Albertson and Thorson, 1974; Acres, 1975). These included tripping of transformers and capacitor banks, increased reactive power demand and voltage drops. The most pronounced effects

began at 22.42 UT. At this time there was also an outage of the L4 communication cable system in the mid-western United States. An investigation by Anderson et al (1974) found that the system outage coincided with a particularly rapid change of the magnetic field. The disturbance was centred over western Canada with a peak rate of change of magnetic field intensity of 2200nT/min. The disturbance extended down over the mid-western United States, and the rate of change of the magnetic field at the cable location was estimated to be 700 nT/min.

Satellite observations showed that at the time of the power system and cable disturbances there was a severe compression of the magnetopause and Anderson et al (1974) concluded that currents on the magnetopause were responsible for the magnetic field variations that caused the problems. However, recent model calculations by Boteler and Jansen van Beek (1999) have shown that the observed magnetic disturbance was too localised to have been caused by magnetopause currents. Contour plots of the disturbance are instead consistent with an ionospheric current as the source. Equivalent current plots (Figure 3) derived from the observed magnetic field variations show that a rapid intensification of an eastward electrojet was responsible for the magnetic disturbance and the power system and cable problems.

5. MARCH 13, 1989

On March 13, 1989 power systems experienced one of the largest magnetic storms ever recorded. The resulting geomagnetically induced currents caused widespread problems.

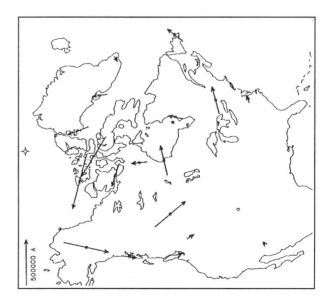

Figure 3. Equivalent current vectors derived from ground magnetic field hourly mean values centred at 22.30 UT, Aug 4, 1972. The length of the vectors indicates the size of the currents.

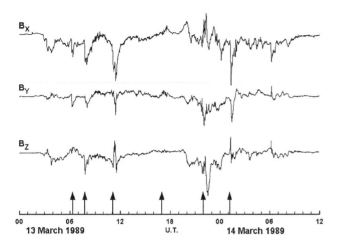

Figure 4. Magnetic variations on March 13, 1989 at Ottawa and the times of power system problems.

In North America these effects occurred at six times during the storm as shown in Figure 4 (Boteler and Jansen van Beek, 1993). In addition, damage due to transformer heating was detected after the storm.

The most significant effect produced by the March 13, disturbance was the Hydro-Québec blackout. Technical descriptions of the power system problems are given by Czech et al (1992) and Blais and Metsa (1993). The system collapse at 07.45 UT coincided with the onset of a magnetic substorm associated with the rapid increase of a westward electrojet.

Widespread power system problems also occurred later in the storm at 21.58 UT. Equivalent current plots derived from the magnetic observatory recordings show that there was a strong eastward electrojet extending across North America at this time (Figure 5).

On March 14 the Meadow Brook 500/138 kV power transformer on the Allegheny power system was removed from service because of evidence of heating (Gattens et al, 1989). Inspection found 4 areas of discolored paint on Phases 1 and 2 coils on both HV and LV sides. Calculations showed that total saturation of the core would produce a temperature of 400°C in part of the transformer tank. Gattens et al estimated that GIC of 80 A would have been necessary to produce the damage that was found.

Seven days after the storm, routine tests on the Pennsylvania, New Jersey, Maryland (PJM) system found indications of transformer overheating at the Salem nuclear power station (Balma, 1992). Further tests later in March confirmed this evidence and the transformers were removed from service. Subsequent inspection showed damage in the A phase and C phase transformers of Salem unit 1 and the transformers had to be replaced. In phase B, the damage was not as severe. However, in September 1989 evidence was found of heating on the phase B transformer of Salem unit 2 and this transformer was removed from service and had to be replaced (Balma, 1992).

Because the transformer damage was only discovered after the storm it is not possible to unequivocally identify which phase of the disturbance was the cause. However, data from the Fredericksburg magnetic observatory, near the Allegheny and PJM systems, can be used to show when these systems experienced the largest disturbance. Figure 6 shows the largest disturbance was a positive excursion in B_x at approximately 22.00 UT (17.00 EST) which coincides with the power system problems mentioned earlier. This suggests the transformer damage on the Allegheny and PJM systems was caused by the eastward electrojet that occurred in the evening sector on March 13.

6. DISCUSSION

The ionospheric currents responsible for the magnetic disturbances that affect power systems at mid to high latitudes are associated with two different processes in the magnetosphere (see Rostoker, 1991; McPherron, 1995). Eastward and westward convection electrojets in the evening and morning sectors are part of a two cell current circulation in the polar cap and auroral zone. This results from convection of magnetic field lines within the magnetosphere which is directly driven by coupling of energy from the solar wind. In the midnight sector a westward substorm electrojet occurs as a result of disruption of a cross-tail current and its diversion

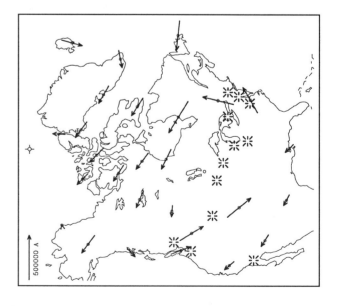

Figure 5. Equivalent current vectors derived from ground magnetic field observations at 21.58 UT, March 13, 1989.

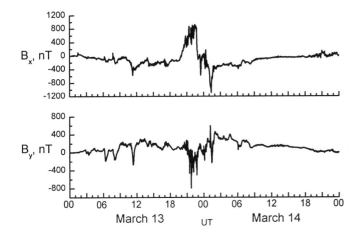

Figure 6. Magnetic variations on March 13, 1989 recorded at Fredericksburg, showing the disturbance at 22.00 UT experienced by the Allegheny and PJM systems.

through field-aligned currents into the ionosphere. This is part of a sequence involving loading of energy into the tail of the magnetosphere and its subsequent unloading into the auroral ionosphere. It has generally been thought that the directly-driven convection electrojets vary slowly, and that only the loading and unloading of energy leading to the substorm electrojet would cause the rapid magnetic field variations that produce power system problems.

Reviewing the disturbances considered in this paper can give a guide to the current systems responsible for major power system effects. The Hydro-Québec blackout on March 13, 1989, was caused by the rapid intensification of a nightside westward electrojet and represents a good example of problems due to the substorm process of loading and unloading of energy in the magnetosphere. However, power system effects later in the day were caused by an eastward electrojet which is part of the directly-driven convection current system. This indicates that the convection current system, as well as the substorm current system, can vary fast enough to cause power system problems.

Of the earlier disturbances, both the power system effects and the L4 cable outage on August 4, 1972 and the power system effects on February 10, 1958 have been shown to coincide with eastward electrojets. The eastward electrojet is unambiguously identified with the convection current system so the cause of the disturbances in these two cases is clear.

The power system problems in eastern North America on March 24, 1940 are associated with a westward ionospheric current which can be produced by either the substorm or convection systems. In this case the location of the disturbance near local noon excludes the substorm current system as the cause. If the disturbance is due to the convection current system there should be a simultaneous increase in the eastward

electrojet in the evening sector. At the time of the disturbance observatories in Europe were in the evening sector and show a positive change in the northwards magnetic field indicative of a strong eastward electrojet. These observations are consistent with the two-cell convection current system with an eastward electrojet in the evening sector over Europe and a westward electrojet in the morning and extending round to noon over North America.

The events presented here are not claimed to be a complete list of space weather disturbances affecting power systems. Also, this analysis has concentrated on effects to power systems in North America and a similar analysis needs to be done to trace the cause of power system effects in Europe and other regions. No conclusions can therefore be drawn about the relative importance of the substorm or convection current systems in causing GIC problems. However the results presented here show that the convection current system can vary fast enough to cause significant GIC effects. Thus both the substorm and convection current systems need to be considered when trying to predict space weather effects on power systems.

7. CONCLUSIONS

Space weather effects on power systems have been reported for the last sixty years. Particularly significant effects were observed in North America during major disturbances on March 24, 1940, February 10, 1958, August 4, 1972, and March 13, 1989. The effects range from relay trips and voltage dips to a widespread blackout and transformer damage.

The March 13, 1989 power system effects can be linked to two different ionospheric current systems. The Hydro-Québec blackout and effects on other power systems at 07.45 UT were caused by the sudden enhancement of a westward substorm electrojet. In contrast, power system effects at 21.58 UT were produced by an eastward convection electrojet. Transformer damage, discovered later, was likely also caused by this eastward electrojet.

Power system effects during the August 4, 1972 and February 10, 1958 disturbances were produced by rapid changes of an eastward electrojet produced by a sudden increase in magnetospheric convection.

On March 24, 1940 power system problems in North America occurred just before local noon and were associated with a westward ionospheric current. This coincided with an eastward electrojet in the evening sector which is indicative of an enhanced convection current system.

Both the loading and unloading substorm process and the directly driven convection process can produce sudden changes of the ionospheric currents and the large magnetic field changes that cause power system problems.

Acknowledgements. This work was funded by the Geological Survey of Canada and Ontario Hydro. I am grateful to Dr L. Trichtchenko, G. Jansen van Beek, and R. Libbey for help with the preparation of this paper.

REFERENCES

Acres Consulting Services Ltd, Study of the disruption of electric power systems by magnetic storms, *Earth Phys. Branch Open File 77-19*, Dept. of Energy, Mines, Resources, Ottawa, 1975.

Albertson, V.D. and J.M. Thorson, Power system disturbances during a K-8 geomagnetic storm: August 4, 1972, *IEEE Trans. Power App. & Sys.*, vol. PAS-93, 1025, 1974.

Albertson, V.D., J.M. Thorson, R.E. Clayton, and S.C. Tripathy, Solar-induced-currents in power systems: cause and effects, *IEEE Trans. Power App. & Sys.*, PAS-22, 471, 1973.

Albertson, V.D., J.M. Thorson, and S.A. Miske, The effects of geomagnetic storms on electrical power systems, *IEEE Trans. Power App. & Sys.*, PAS-93, 1031, 1974.

Albertson, V.D., J.G. Kappenman, N. Mohan, and G.A. Skarbakka, Load-flow studies in the presence of geomagnetically induced currents, *IEEE Trans. Power App. & Sys.*, PAS-100, 594, 1981.

Allen, J., L. Frank, H. Sauer, and P. Reiff, Effects of the March 1989 solar activity, *EOS Trans. AGU*, 70, 1479, 1989.

Anderson, C.W., L.J. Lanzerotti, and C.G. Maclennan, Outage of the L-4 system and the geomagnetic disturbances of August 4, 1972, *Bell Syst. Tech. J.*, 53, 1817, 1974.

Balma, P.M., Geomagnetic effects on a bank of single phase generator step-up transformers, *Proc. Geomagnetically Induced Currents Conference, Millbrae, California, Nov 8-10, 1989*, EPRI Report TR-100450, 20-1, 1992.

Blais, G. and P. Metsa, Operating the Hydro-Québec grid under magnetic storm conditions since the storm of 13 March 1989, Proc. *Solar-Terrestrial Predictions Workshop, May 18-22, 1992, Ottawa*, vol 1, 108, 1993

Bolduc, L. and J. Aubin, Effects of direct currents in power transformers, *Electric Power Systems Research,* 291, 1978.

Boteler, D.H., R.M. Shier, T. Watanabe, and R.E. Horita, Effects of geomagnetically induced currents in the B.C. Hydro 500 kV system, *IEEE Trans. Power Delivery*, 4, 818, 1989.

Boteler, D.H. and G. Jansen van Beek, Mapping the March 13, 1989, magnetic disturbance and its consequences across North America, *Proceedings, Solar-Terrestrial Predictions Workshop, May 18-22, 1992, Ottawa*, vol 3, 57, 1993.

Boteler, D.H. and G. Jansen van Beek, August 4, 1972, Revisited: A new look at the geomagnetic disturbance that caused the L4 cable system outage, *Geophys. Res. Lett.*, 26, 577, 1999.

Boteler, D.H., S. Boutilier, A.K. Wong, Q. Bui-Van, L. Hajagos, D. Swatek, R. Leonard, B. Hughes, I.J. Ferguson, and H.D. Odwar, Geomagnetically Induced Currents: Geomagnetic Hazard Assessment, Phase II, *Geological Survey of Canada*, Open File No 3420, 1997.

Czech, P., S. Chano, H. Huynh, and A. Dutil, The Hydro-Québec system blackout of 13 March 1989: System response to geomagnetic disturbance, Proc. *Geomagnetically Induced Currents Conference, Nov 8-10, 1989, Millbrae, California,* EPRI Report TR-100450, 1992.

Davidson, W.F., The magnetic storm of March 24, 1940 - effects in the power system, *Edison Electric Institute Bulletin*, July 1940.

Gattens, P.R., R.M. Waggel, R. Girgis, and R. Nevins, Investigation of transformer overheating due to solar magnetic disturbances, *Effects of Solar-Geomagnetic Disturbances on Power Systems, Special Panel Session Report,* pp 29-32, IEEE PES Summer Meeting, Long Beach, California, July 12, 1989

Germaine, L.W., The magnetic storm of March 24, 1940 - effects in the communication system, *Edison Electric Institute*, July 1940.

Kappenman, J.G., V.D. Albertson, and N. Mohan, Current transformer and relay performance in the presence of geomagnetically induced currents, *IEEE Trans. Power App. & Sys.*, PAS-100, 1078, 1981.

Kappenman, J.G., S.R. Norr, G.A. Sweezy, D.L. Carlson, V.D. Albertson, J.E. Harder, and B.L. Damsky, GIC mitigation: a neutral blocking/bypass device to prevent the flow of GIC in power systems, *IEEE Trans. Power Delivery*, vol. 6, no 3, 1271, 1991.

Lehtinen, M. and R. Pirjola, Currents produced in earthed conductor networks by geomagnetically induced electric fields, *Annales Geophysicae*, 3, 479, 1985.

Mäkinen, T., Geomagnetically induced currents in the Finnish power transmission system, *Geophysical publications No 32*, Finnish Meteorological Institute, Helsinki, 101pp, 1993.

McPherron, R.L., Magnetospheric dynamics, *Introduction to Space Physics*, ed. Kivelson, M.G. and Russell, C.T., Cambridge Univ. Press, 400, 1995.

Pirjola, R., On currents induced in power transmission systems during geomagnetic variations, *IEEE Trans. Power App. & Sys.*, PAS-104, 2825, 1985.

Pirjola, R. and M. Lehtinen, Currents produced in the Finnish 400 kV power transmission grid and in the Finnish natural gas pipeline by geomagnetically induced electric fields, *Annales Geophysicae*, 3, 485, 1985.

Ross, W.E. and A.E. Evans, Records of observations at the Agincourt magnetic observatory 1957 and 1958, *Publications of the Dominion Observatory, Ottawa*, vol XXIII, no 3, 1962.

Rostoker, G., Some observational constraints for substorm models, *Magnetospheric Substorms*, ed. J.R. Kan, T.A. Potemra, S. Kokubun, T. Iijima, AGU Monograph 64, 61, 1991.

Slothower, J.C. and V.D. Albertson, The effects of solar magnetic activity on electric power systems, *J. Minn. Academy of Science*, 34, 94, 1967.

Winckler, J.R., L. Peterson, R. Hoffman, and R. Arnoldy, Auroral X-rays, cosmic rays, and related phenomena during the storm of February 10-11, 1958, *J. Geophys. Res.*, 64, 597, 1959.

D. H. Boteler, Geomagnetic Laboratory, Geological Survey of Canada, 7 Observatory Cres., Ottawa, Ontario K1A 0Y3, Canada.

Advanced Geomagnetic Storm Forecasting
for the Electric Power Industry

John Kappenman

Metatech Corporation, Duluth, Minnesota

Geomagnetic disturbances may impact the operational reliability of electric power systems. Solar Cycle 22 (the most recent solar cycle extending from 1986-1996) demonstrated to the power industry the need to take into consideration the potential impacts of geomagnetic storms. Experience gained from the unprecedented scale of these recent storm events provides compelling evidence of a general increase in electric power system susceptibility. Important infrastructure advances have recently been put in place that provide solar wind data. This new data source along with numeric model advances allows the capability for predictive forecasts of severe storm conditions, which can be used by impacted power system operators to better prepare for and manage storm impacts.

1. POWER GRIDS AND SPACE WEATHER BACKGROUND

Society reliance on electricity for meeting essential needs has steadily increased for many years. This unique energy service requires coordination of electrical supply, demand, and delivery—all occurring at the same instant. Geomagnetic storms can disrupt these complex power grids. The interaction of local geomagnetic field disturbances with terrestrial systems can result in the flow of induced currents on these systems.

Geomagnetically-induced currents (GIC) can flow through the power system, entering and exiting the many grounding points on a transmission network. GICs are produced when shocks resulting from sudden and severe magnetic storms subject portions of the Earth's surface to fluctuations in the planet's normally quiescent magnetic field.

These fluctuations induce electric fields in the Earth that cause currents to flow in the Earth and in electrical conductors on the surface, resulting in GIC's that flow through transformers, power system lines, and grounding points. Only a few amps are needed to disrupt transformer operation, but over 300 amps have been measured in the grounding connections of transformers in affected areas [*Kappenman,* 1990]. Unlike threats due to ordinary weather, space weather can readily create large-scale problems because the footprint of a storm can extend across a continent. As a result, simultaneous widespread stress occurs across a power grid to a level where widespread failures and even regional blackouts may occur. Systems in the upper latitudes of the Northern Hemisphere are at increased risk because auroral activity and its effects center on the magnetic poles. North America is particularly exposed to these storm events because the Earth's magnetic north pole tilts toward this region and therefore the dense critical power grid infrastructure across the continent is in closer proximity to intense auroral current expansions. Figure 1 illustrates the geographical extent of three levels of the rate of change of the ground-level geomagnetic disturbance, which drives GIC, for the large magnetic storm of March 13, 1989.

Space Weather
Geophysical Monograph 125

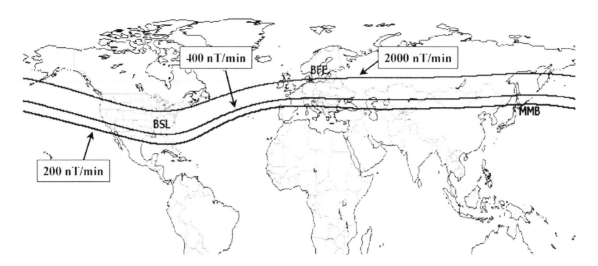

Figure 1. Exposed Regions of the Northern Hemisphere. The footprints of a superstorm can be extensive; the above diagram shows regions of the Northern Hemisphere that can be exposed to intense storm activity such as the Great Geomagnetic Storm of March 1989. For perspective, the level of storm severity that precipitated the Hydro Quebec collapse (~400 nT/min) was observed at locations as far south as Bay St Louis and would map along geomagnetic latitudes for different time onsets around the world, which would encompass most of North America and Europe. Much stronger intensities were observed at more northerly locations, these intensities are approximately 5 times more severe (2000 nT/min) than the levels that triggered the Hydro Quebec collapse. Levels at one-half the intensity of those that triggered the Hydro Quebec collapse (200 nT/min) have also shown capable of causing power reliability problems which for this storm extended to even lower latitudes

2. LESSONS LEARNED FROM SOLAR CYCLE 22

The events that led to the collapse of the Hydro Quebec system in the early morning hours of March 13, 1989 illustrate the challenges that lie ahead in managing this risk [*NERC*, 1989]. At 2:42am LT, (7:42 UT), all operations were normal; at that time a large impulse in the Earth's magnetic field erupted along the US/Canada border. Within a matter of seconds, the voltage on the network began to sag as the storm developed; automatic compensating devices rapidly turned themselves "ON" to correct this voltage problem. However, these automatic devices were themselves vulnerable to the storm and all 7 of these critical compensators failed within less than a minute. The failure of the compensators led to a voltage collapse and complete blackout of the second largest utility grid in North America. All together, the chain of events from start to complete province-wide blackout took an elapsed time of only 90 seconds. The rapid manifestation of storm events and their impacts on the Hydro Quebec system allowed no-time to even assess, let alone provide meaningful human intervention. (See Figure 2 – Four Minutes of a Superstorm, March 13, 1989).

The rest of the North American system also reeled from this storm. Over the course of the next 24 hours, five more large disturbances propagated across the continent, the only difference being that they extended much further south and

very nearly toppled power systems from the Midwest to the Mid-Atlantic Regions of the US. The NERC (North American Electric Reliability Council) in their post analysis attributed over 200 significant anomalies across the continent to this one storm. In spite of the large number of significant events that were observed, it is now recognized that North America was very lucky that day [*NERC*, 1989]. This same storm produced dB/dt fluctuations twice as intense over the lower Baltic, than any that were experienced in North America. Over the next 5 years, smaller storms demonstrated time and again for the power industry that significant impacts could be triggered at even lower storm levels [*NERC*, 1991, *IEEE*, 1996].

For perspective, the limited climatologic data available suggests that storms of even larger intensity and with a larger planetary footprint are possible than the one that occurred in March 1989. Also, the power industry realizes that its vulnerability continues to incrementally grow over time [*Kappenman*, 2000]. As a result, the challenges of this solar cycle may be even greater than those posed 10 years ago.

3. FORECASTING FOR RISK MANAGEMENT

In the long run, improvements in forecasts will play an important role in managing these risks. Utility companies concerned about storms, rely on contingency strategies for

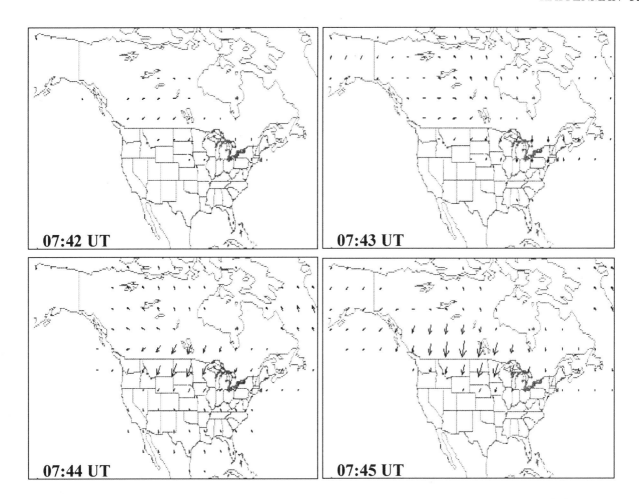

Figure 2. Four minutes of a superstorm. Space weather conditions capable of threatening power system reliability can rapidly evolve. The system operators at Hydro Quebec and other power system operators across North America faced such conditions during the March 13, 1989 superstorm. The above figure shows the rapid development and movement of a large geomagnetic field disturbance between the times 7:42-7:45 UT (2:42-2:45 EST) on March 13, 1989. Depicted by vectors are the direction and intensity of the dB/dt of the horizontal geomagnetic fields in the region at each time. The disturbance of the magnetic field began intensifying over eastern US-Canada border and then rapidly intensified while moving to the west across North America over the span of a few minutes. From calm conditions, the Hydro Quebec system collapsed in just 90 seconds during this disturbance. The magnetic field disturbances observed at the ground are caused by large electrojet current variations that interact with the Earth's magnetic field. The intensities ranged from 400 nT/min at Ottawa at 7:44UT to over 892 nT/min at Glen Lea. The large electrojet current variations that created these magnetic field disturbances moved from eastern Canada to Alaska in less than 8 minutes, a velocity of approximately 10 to 15km/sec.

weathering severe magnetic disturbances. Choosing the best contingency procedures depends on being able to predict storm severity and how the storm will affect the local system. NASA and NOAA have put in place important real-time solar wind monitoring with the ACE Satellite, which has been operational since January 1998. This capability allows, for the first time, the ability to accurately predict the occurrence of large threatening storms with enough lead-time to take meaningful well-informed actions to better prepare the power grid for the impact of the storm.

The first of these advanced forecast systems has been developed by Metatech and is now operational at National Grid Company, the utility that provides service to England and Wales. National Grid is the largest electric transmission operating company in the world and is providing the vital proving ground for this new approach in Space Weather forecasting for the power industry. This forecasting capability has been operational at National Grid since May of 1999 as a Phase 1 system and in a fully complete mode of operation since January 31, 2000.

Methods of classifying geomagnetic storm activity in the past have typically used two letter indices (for example K1 to K9 for the smallest to largest geomagnetic storm) to rank the severity of the storm over broad three-hour time

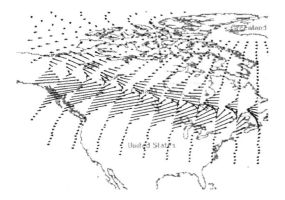

Figure 3. Electrojet conditions are forecast. The above shows an example North American display of the electrojet current pattern. This model uses real-time solar wind data from the NASA ACE satellite to forecast the expected storm conditions. The model provides important details on the storm intensity, the equatorward boundaries and temporal variations of the electrojet currents in the ionosphere that are used to estimate ground-level magnetic field disturbances and impacts to critical infrastructures. The estimate is made nominally 45 minutes in advance, which allows important lead-time for impacted systems. Because large-scale changes can occur rapidly, this model recalculates the electrojet current continuously at a one-minute cadence.

windows and planetary or large region locations. Space Weather is a very complex, detailed, and dynamic process that is ever-changing over the course of the storm. Just as the diversity of terrestrial weather impacts to critical operational infrastructure (such as rain/snow, thunderstorms, heat/cold, hurricanes, etc.) cannot be adequately classified by 3-hour, 2-letter, planetary indices; neither should the inherently dynamic impacts of space weather remain with this outdated classification approach. The operation of critical infrastructures such as power grids is a continuous minute-by-minute coordinated and supervised operation. Thus, the forecasting capability for geomagnetic activity also needs to provide continuous updates of the rapidly changing space environment conditions to best meet the operational needs of power systems in managing this storm risk.

4. DEFINING NEW STANDARDS IN FORECAST CAPABILITY

The advanced forecast/nowcast system provides minute by minute continuous high-resolution, accurate specification of geomagnetic storm conditions which depict not only important intensifications during a storm, but also the region-specific locations of major storm activity, with a typical 45 minute lead-time [*Kappenman,* 2000]. The "lead-time" provided by the warning system is vital as storm on-

sets can develop suddenly and some power grid operational changes can not otherwise be implemented in time to address the paramount priority of power system reliability.

In addition to a detailed specification or forecast of the environment, it is necessary to provide a detailed assessment of the potential impact upon operations of a power network due to a storm. This requires the needed utilization of an end-to-end set of models to discretely extend from solar wind or locally observed geomagnetic field conditions to derive a forecast or nowcast set of impacts upon the operation of the client infrastructure. These measurements and models increase the capability to predict not only on a global scale, but also more importantly for concerned transmission grid operators, to provide a projection of region and time-specific meso-scale processes of concern.

In predictive forecasts of impending geomagnetic storm conditions, solar wind conditions are used to derive expected electrojet current conditions as depicted in Figure 3. The electrojet model determines the characteristics of the current flowing in the electrojet approximately 100 km above the Earth's surface. This model predicts the ionospheric Hall and Pedersen currents flowing in the auroral zone approximately 30 to 45 minutes ahead in time using the real-time solar wind data. The model also needs to take into consideration ionospheric composition conditions (adjustments in conductivity, current intensity, and equatorward boundaries of the current systems) during the course of the storm as well in order to provide sufficiently accurate estimates of eastward and westward current conditions. To compute the coupling of the electromagnetic fields produced by the electrojet to a fixed power grid, it is necessary to translate the fields produced by these moving currents to calculate the local magnetic field in the regions of interest to the client. The derivation of the geomagnetic field disturbances is performed by two-dimensional calculations of a gaussian spatial profile for an electrojet over a layered (with depth) ground conductivity profile.

Given either the predicted (forecast) or observed (nowcast) magnetic fields in the region of interest, it is necessary to compute the electric fields at the same locations. These electric fields couple to the high-voltage power lines and thereby induce the quasi-dc currents that flow in the network. As is well understood, the conductivity of the Earth itself is crucial in determining the electric fields produced by a given magnetic field. Lower conductivities (higher resistivities) produce larger electric fields from a given magnetic field. Deep earth conductivity models that typically range from depths of 40 to 700 km are necessary due to the low-frequency content of the magnetic fields and their associated skin-depth. The ground conductivity in the upper 40-km of the Earth varies considerably based on the specific Earth location due to the geology and stability of the rock in the crust. In most cases, the dominant factor in a conductivity profile is the variation with depth, although there are cases where abrupt horizontal variations in conductivity may be important [*Kappenman*, 1997].

It is straightforward to compute the coupling to each line in the complex power grid. In this calculation, the orientation of the electric field vectors is applied throughout the client network. This is done by interpolation and analytic integration, so that the net voltage (electric field times distance) induced in each line segment can be determined rapidly. This allows the computation of the GIC flow in every line and transformer in the power network. Once the GIC is determined throughout the network and in each transformer, additional calculations can be made to provide power system operating staff with precise system impact estimates such as system and regional reactive power demands, numbers of transformers in saturation, and other important system impact visualizations.

The advanced and comprehensive forecast technologies now employed provide power system operators with a clear and to-the-point summary of space weather conditions, similar to the way that ordinary weather information is presented, such as radar and satellite imagery. As depicted in Figure 3, this system assesses the environment created by the storm in detail [*Tascione*, 1994]. An even more important feature of this forecast system is provided by the ability to assess the storm's impact on the power grid. A detailed model of the power grid is calculated to provide a quantifiable assessment for the system operator of the potential impact on a short-term basis to enable well-informed judgements about appropriate operational response measures [*Kappenman*, 2000]. Figure 4 shows a sample output from the detailed power grid model display.

Space Weather forecasting is more difficult than ordinary weather forecasting. Because of the inherent dynamic nature of storms, forecasting models like this need to be continuously updated. Magnetospheric, ionospheric and power grid models all need to be continuously recalculated by data provided by a solar wind monitoring satellite located one-million miles upstream of the Earth. This data provides the lead-time, because it typically takes another 45 minutes to one-hour for the solar wind to arrive at the Earth. The solar wind data updates are then provided to the models for recalculation of space weather forecasts on a continuous one-minute cadence [*Kappenman*, 1997, *Kappenman*, 2000].

The enormous US power network is controlled regionally by more than 100 separate control centers that coordinate responsibilities jointly for the impacts upon real-time network operations. Other power networks around the world use similar regional control center approaches for

providing the 24-hour continuously supervised operation of their networks. Not one of these large power system control centers would do without continuous high quality weather data in managing the operation of their systems. The same paradigm needs to be developed and adopted by the power industry for the threats posed by space weather.

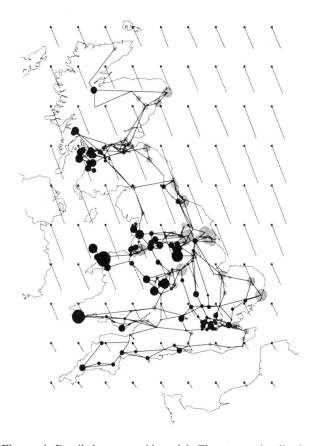

Figure 4. Detailed power grid model. The storm visualization shown above is designed to provide a clear and concise picture of the location and intensity of storm impacts across the transmission network. In this example the storm conditions are displayed for National Grid over England and Scotland. The 400kV and 275kV transmission system is displayed with small circles indicating the magnitude (circle size varies) and polarity (circle black-gray changes) of the GIC at each transformer. Also shown are the vector icons of the magnitude and direction of the local magnetic field during the storm, which is responsible for the GIC flows. Text and graphic summaries can also be provided on system or region reactive power demands, numbers of transformers in saturation and other important system impact details. This calculation is made in a forecast and nowcast mode and both are recalculated on a one-minute cadence. The forecast mode utilizes data from the Electrojet model output as shown in Figure 3 and effectively provides a nominal 45-minute warning of storm impacts. The nowcast mode utilizes locally sensed magnetic field data and provides a system-wide assessment of current conditions.

REFERENCES

IEEE Working Group K-11 Report, The effects of GIC on protective relaying, *IEEE Transactions on Power Delivery*, 11, No.2, 725-739, April 1996.

Kappenman, J.G., V. D. Albertson, Bracing for the geomagnetic storms, *IEEE Spectrum Magazine*, 27-33, March 1990.

Kappenman J. G., W. A. Radasky, J. L. Gilbert, I. A. Erinmez, Advanced Geomagnetic Storm Forecasting: A Risk Management Tool for Electric Power Operations, Submitted for *IEEE Plasma Society Special Publication on Space Plasmas and Space Weather*, 2114-2121, December 2000.

Kappenman J. G., L. J. Zanetti, and W. A. Radasky, Space weather from the users perspective: geomagnetic storm forecasts and the power industry, *EOS, Transactions American Geophysics Union*, 78, No. 4, pp 37, 41, 44-45, January 28, 1997.

NERC Disturbance Analysis Working Group Report, The 1989 System Disturbances, March 13, 1989 geomagnetic disturbance, pp. 8-9, 36-60, 1990.

NERC Disturbance Analysis Working Group Report, *The 1991 System Disturbances, Geomagnetic disturbances in 1991*, 50-57, 1991.

Tascione T. F., *Introduction to the Space Environment*, 2nd Edition, pp 48-50, pp. 104-111, Kreiger Publishing Co, 1994.

John G. Kappenman, Metatech Corporation, 5 W. First Street, Suite 301, Duluth, MN 55802, USA. (email: JKappenma@aol.com).

Ionospheric Climatology and Weather Disturbances: A Tutorial

R. W. Schunk

Center for Atmospheric and Space Sciences, Utah State University, Logan, Utah

The ionosphere can have a detrimental effect on both civilian and military systems, including OTH radars, HF communications, surveillance, and navigation systems that use GPS satellites. In an attempt to mitigate the adverse effects, specification and forecast models are being developed that are based on sophisticated data assimilation techniques. However, the model development represents a significant challenge because the ionosphere is known to display a marked variation with altitude, latitude, longitude, universal time, season, solar cycle, and geomagnetic activity. This variation results from the couplings, time delays, and feedback mechanisms that are inherent in the system as well as from the effects of solar, interplanetary, magnetospheric, and mesospheric processes. The various processes act in concert to define both a background state (climatology) and a disturbed state (weather), which are described in this tutorial. First, a brief description of the background ionospheric state and the physical processes that are responsible for establishing this state is given. The tutorial then focuses on a discussion of weather disturbances and features, the causes of weather, and the status of weather modeling.

1. INTRODUCTION

Since its discovery in 1901 by Marconi, the ionosphere has been studied extensively with the aid of balloons, rockets, satellites, and a wide range of ground-based instruments (radars, magnetometers, and optical devices). In addition, there have been literally hundreds of model simulations conducted for nearly all solar cycle, seasonal, and geomagnetic conditions. Based on these studies, it is now well known that the primary source of ionization and energy for the ionosphere is solar EUV and x-ray radiation, but magnetospheric electric fields, particle precipitation, and field-aligned currents can have a significant effect at high and middle latitudes. Also,

gravity waves and tides propagating upwards from the lower atmosphere can appreciably affect the ionosphere.

The various chemical, electrodynamical, radiative, and transport processes acting on and within the ionosphere act in concert to determine the density, composition, and temperature structure of the ionosphere. However, the Earth's strong internal magnetic field also plays an important role, and its effect is different at high, middle, and low latitudes. Despite the fact that different physical processes operate in the different latitudinal domains and that the effect of the geomagnetic field is different, the ionosphere tends to separate into layers at all latitudes (Figure 1). The D region (60-100 km) is controlled by chemical processes and the dominant species are negative and positive molecular ions, water cluster ions, and molecular neutrals. Like the D region, the E region (100-150 km) is controlled by chemical processes and the dominant species are molecular ions (NO^+, O_2^+, N_2^+) and neutrals (N_2, O_2). In the F_1 and F_2 regions (150-1000 km) both transport and chemical processes are important and

Space Weather
Geophysical Monograph 125

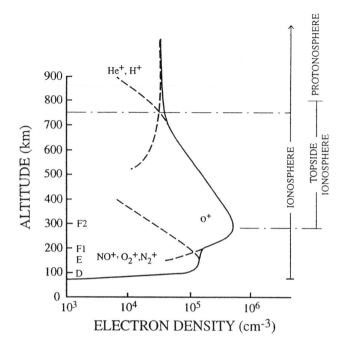

Figure 1. Schematic diagram of the different regions in the Earth's ionosphere. From *Banks et al.* [1976].

the dominant ion and neutral species are O^+ and O, respectively. The F_2 region is where the maximum plasma density, N_mF_2, occurs and it can vary by more than a factor of 10^3 over the globe at a given time. At higher altitudes, the light ion H^+ dominates (≥ 1000 km).

To lowest order, only a few chemical reactions need to be included in a mathematical description of the ionosphere. At F region altitudes, O^+ is created via photoionization of neutral O, $O + h\nu \rightarrow O^+ + e$, and is lost in reactions with N_2 and O_2; $O^+ + N_2 \rightarrow NO^+ + N$ and $O^+ + O_2 \rightarrow O_2^+ + O$. The molecular ions are then lost via recombination with electrons; $NO^+ + e \rightarrow N + O$ and $O_2^+ + e \rightarrow O + O$. Above the F region, the main source of H^+ is the accidentally resonant charge exchange reaction $O^+ + H \Leftrightarrow H^+ + O$.

Because the processes acting within and on the ionosphere display characteristic features and trends, the ionosphere also displays "characteristic" or "average" features, and these correspond to the climatology of the system. However, superimposed on the background ionosphere is a wide range of weather features due to ionospheric disturbances. As a consequence, the ionosphere can vary appreciably from hour to hour and from day to day at any location. The weather features occur because the external processes acting on the ionosphere can be localized, spatially structured, and unsteady. They also occur because there are time delays

with respect to when changes in magnetospheric processes affect the ionosphere and then the neutral atmosphere. In this tutorial, a brief description is given of both the climatology of the ionosphere and some selected weather features.

2. HIGH LATITUDES

At high latitudes, the geomagnetic field lines at ionospheric altitudes are nearly vertical and extend into deep space. As a consequence, magnetospheric electric fields and energetic particles can penetrate to ionospheric altitudes. The effect of the magnetospheric electric fields is to cause the ionosphere to $\mathbf{E} \times \mathbf{B}$ drift (convect) across the polar region at altitudes above about 150 km. When the interplanetary magnetic field (IMF) is southward, the magnetosphere tries to impose a 2-cell convection pattern on the ionosphere, with antisunward flow across the polar cap and return flow at lower latitudes. However, the ionosphere also has a tendency to corotate with the Earth. When the magnetospheric and corotational electric fields are combined, the resulting $\mathbf{E} \times \mathbf{B}$ drift pattern takes the form shown in Figure 2. On the sunlit side of the terminator, the plasma density is elevated due to photoionization. As the plasma convects across the terminator into the dark polar cap, the plasma density

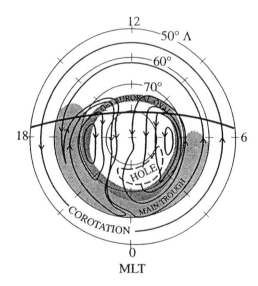

Figure 2. Schematic diagram showing the characteristic ionospheric features that can occur in the high-latitude F region. The solid curves with arrows show the streamlines of the plasma in a quasi-inertial magnetic reference frame. The solid curved line shows the terminator. Also shown are the polar hole, the auroral oval, and the main electron density trough. From *Brinton et al.* [1978].

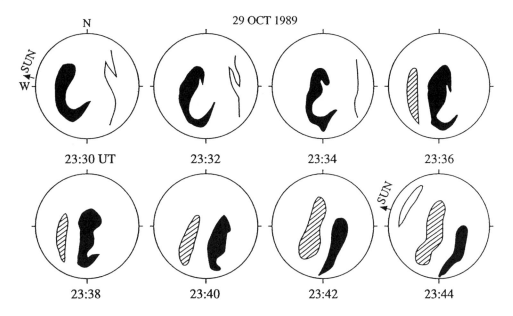

Figure 3. Propagating plasma patches observed at Qaanaaq, Greenland, on 29 October 1989. The patches are elongated in the dawn-dusk direction and are propagating in an antisunward direction. From *Fukui et al.* [1994].

decays due to chemical reactions with the neutral gas. If the antisunward convection speed is large, the plasma can move great distances before it decays appreciably, and the net result is that a "tongue of ionization" extends across the polar cap. On the other hand, if the antisunward convection speed is slow, the plasma will have sufficient time to decay to low values and a "polar hole" can form just poleward of the nocturnal auroral oval. When the plasma enters the auroral oval, the density is enhanced due to impact ionization from precipitating energetic electrons. After exiting the nightside oval, the plasma convects around the polar cap toward sunlight both on the dawn and dusk sides. On the dusk side, the sunward flow imposed by the magnetosphere opposes the antisunward flow associated with the corotational electric field and a low-speed or "stagnation" region occurs. The plasma that enters this low-speed region can decay to very low values due to the long residence time in darkness. The plasma eventually moves out of the low-speed region and then corotates with the Earth, and this leads to the formation of the "main electron density trough."

The simple 2-cell convection pattern described above only occurs when the IMF is southward ($B_z < 0$). In this case, the strength of the electric field in the polar cap, which determines the antisunward convection speed of the plasma, varies with the dynamic pressure of the solar wind. However, the electric field can be distributed either symmetrically or nonuniformly between the two convection cells, depending on the IMF dawn-dusk

component. Also, when the IMF is northward, multicell or turbulent convection patterns can exist. Associated with the different convection patterns are auroral precipitation patterns, which can contain discrete auroral arcs in addition to diffuse precipitation. In general, the magnetospheric convection and precipitation patterns can be spatially structured and unsteady, particularly during geomagnetic storms and substorms. They can also vary continuously from one convection pattern type to another throughout the day, and all of these effects account for the weather features and disturbances seen at high latitudes. A few examples of weather features are given in what follows. Further examples can be found, for example, in *Schunk and Nagy* [2000].

Propagating plasma patches are a manifestation of weather at high latitudes. They are regions of enhanced plasma density that generally appear when the IMF turns southward. They seem to be created either in the dayside auroral oval near noon or just equatorward of the dayside oval. Once formed, they convect in an antisunward direction across the polar cap at the prevailing convection speed. The patch densities are a factor of 2-10 greater than background densities and their horizontal dimensions vary from 200-1000 km. The plasma patches can be nearly circular or elongated in the direction perpendicular to their direction of propagation. Typically, intermediate-scale irregularities (1-10 km) and scintillations are associated with the patches. Figure 3 shows an example of elongated plasma patches. The figure corresponds to a digitization of

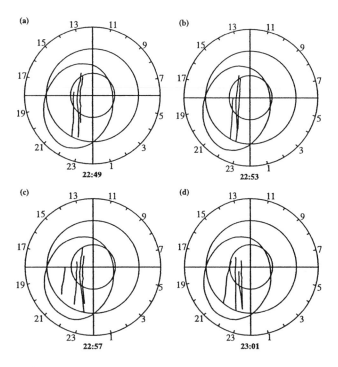

Figure 4. Multiple sun-aligned polar cap arcs observed at Qaanaaq, Greenland, on 19 February 1989. A corrected geomagnetic coordinate system is used and the oval-shaped curves indicate the polar cap boundary. From *Valladares et al.* [1994].

a sequence of all-sky photographs taken at 2-minute intervals at Qaanaaq, Greenland, on 29 October 1989 [*Fukui et al.*, 1994]. The Sun's direction is shown by an arrow on the first and last photographs. At 23:30 UT, a patch that is extended (~1000 km) in the dawn-dusk direction appears and then moves in an antisunward direction across Greenland. Within 14 minutes there are three propagating plasma patches in the all-sky camera's field-of-view. The spacing between the patches and the width of the patches are about 200 km.

Another weather feature that is prevalent in the polar cap is sun-aligned arcs. These arcs are observed as discrete 6300 Å emission structures in the polar cap. The arcs appear when the IMF is near zero or northward and are a result of electron precipitation. The arcs are relatively narrow ($\leq$ 300 km), but are extended along the noon-midnight direction (1000-3000 km). The characteristic energy of the electron precipitation varies from 300 eV to about 5 keV and the energy flux varies from 0.1 to 5 erg cm^{-2} s^{-1}. A single sun-aligned arc or multiple sun-aligned arcs can appear, and then they drift toward either the dawn or dusk side of the polar cap at speeds up to a few hundred meters per second. Figure 4

shows an example of the temporal evolution of multiple sun-aligned arcs. The arcs were observed at Qaanaaq, Greenland, on 19 February 1989, and the figure corresponds to reconstructions of 6300 Å all-sky images. Initially, three arcs were visible when the measurements began (22:49 UT), but 8 minutes later a fourth arc appeared and it then drifted toward the other arcs.

The frequent occurrence of propagating plasma patches and sun-aligned arcs, coupled with geomagnetic storm and substorm dynamics, imply that the plasma density in the polar cap should display a large hour-by-hour variation when the ionosphere is viewed from a single ground station. Figure 5 shows the variation of the critical frequency f_oF_2, measured by a digisonde located at Qaanaaq, Greenland, over a 3-day winter period (17-19 January 1989). In this figure, f_oF_2 is plotted at 5-minute intervals over a 24-hour period, but the data from all three days are plotted on the same 24-hour time axis. Although a diurnal trend can be detected, the hour-by-hour and day-by-day variability in f_oF_2 is large in winter. Since the F-region peak density, N_mF_2, varies as the square of f_oF_2, its hour-by-hour variation is larger than that shown in Figure 5. In summer, the N_mF_2 variability is much smaller than that in winter, because the bulk of the polar cap is sunlit.

3. MID-LATITUDES

At mid-latitudes, the ionosphere is not appreciably affected by magnetospheric electric fields and tends to

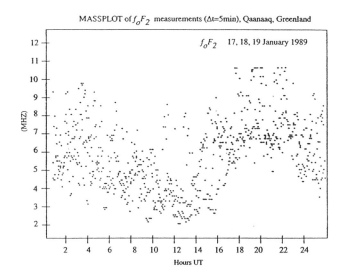

Figure 5. Mesurements of f_oF_2 at Qaanaaq, Greenland, plotted every 5 minutes over a 24-hour period. Three days (17-19 January 1989) of data are plotted on the same time axis. From *Schunk and Sojka* [1996].

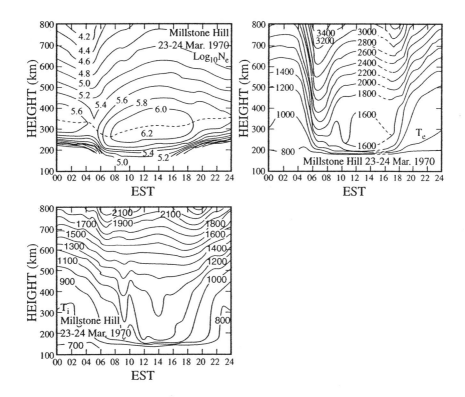

Figure 6. Diurnal variation of electron density (top left), electron temperature (top right), and ion temperature (bottom) measured by the Millstone Hill incoherent scatter radar on 23-24 March 1970. From *Roble et al.* [1978].

corotate with the Earth. However, the geomagnetic field, which is inclined to the vertical, has an important effect on the plasma transport processes. On the dayside, the plasma produced at F-region altitudes by photoionization can diffuse both upwards and downwards along **B**. The plasma that diffuses downwards is lost in chemical reactions with the neutral atmosphere. The plasma that diffuses upwards can escape the topside ionosphere, flow along the dipolar field lines, and then enter the conjugate ionosphere. The plasma that flows upwards can also charge exchange with neutral hydrogen ($O^+ + H \Leftrightarrow H^+ + O$), and the resulting H^+ ions that are created during the daytime at high altitudes can return to the ionosphere at night. This latter transport process acts to maintain the nocturnal ionosphere. Another important transport process involves the neutral wind. On the dayside, there is a component of the neutral wind that blows away from the subsolar point toward the poles and it drives the ionization down the geomagnetic field lines, where it can be lost more rapidly in chemical reactions with a more dense neutral atmosphere. On the nightside, the meridional (north-south) wind blows from the poles toward the equator, and the ionization is driven up the geomagnetic field lines to altitudes where the neutral atmosphere is less

dense and the loss rates are lower. This is another mechanism that acts to maintain the nocturnal ionosphere.

The chemical and transport processes act in concert to establish the dynamical behavior of the mid-latitude ionosphere. Despite the fact that the different processes are highly variable, the mid-latitude ionosphere displays characteristic features and trends that correspond to the climatology of the region. There are distinct solar cycle, seasonal, and diurnal trends. The diurnal variation of the mid-latitude ionosphere is shown in Figure 6. This figure shows contours of n_e, T_e, and T_i as a function of altitude and Eastern Standard Time (EST) as measured by the Millstone Hill incoherent scatter radar on 23-24 March 1970 [*Roble et al.*, 1978]. At sunrise, the electron density begins to increase rapidly due to photoionization. After this initial sunrise increase, n_e displays a slow increase throughout the day, and then it decays at sunset when the photoionization source disappears. At night, the ionization decay is controlled by plasma transport processes. The equatorward neutral wind lifts the F-layer to an altitude where the decay rate is smaller, which acts to maintain the nocturnal ionosphere. The F-layer is also maintained by a downward flow of plasma from the overlying plasmasphere.

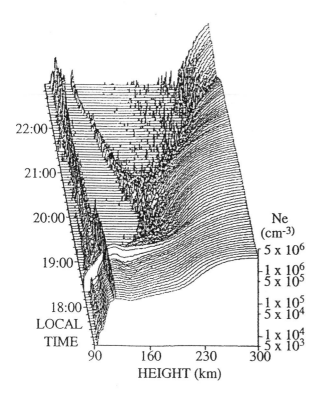

Figure 7. Altitude profiles of electron density at different times between 17:00 and 22:00 local time showing sporadic E and descending intermediate layers. The measurements were made with the Arecibo incoherent scatter radar on 7 May 1983. From *Riggin et al.* [1986].

At sunrise, T_e increases rapidly, with a time constant of the order of seconds, because of photoelectron heating. However, as the electron density slowly builds up, T_e decreases (between 07-10 EST), because of the increasing heat capacity of the electron gas and the stronger coupling to the relatively cold ions. From about 10 to 16 EST, T_e does not vary appreciably, and then T_e decreases at sunset when the photoelectron heat source disappears. At night, T_e remains elevated above T_n because of a downward flow of energy from the plasmasphere, which produces the positive gradient in the nocturnal T_e profile at altitudes above 200 km. With regard to T_i, its durinal variation is controlled by the neutral gas below about 400 km (T_i follows T_n) and by the electron gas above 400 km (T_i follows T_e).

Superimposed on the mid-latitude climatology are several weather disturbances. During geomagnetic storms and substorms, the convection and precipitation patterns expand, and the electric fields and particle precipitation become more intense. These changes in the energy deposition at high latitudes then affect the mid-latitude ionosphere. When a storm or substorm commences, gravity waves can be excited at high latitudes and their subsequent propagation to lower latitudes leads to "traveling ionospheric disturbances." Also, the expansion of the 2-cell convection pattern to lower latitudes results in the transport of high-density plasma in the mid-latitude/afternoon sector in both a sunward and poleward direction. This creates a latitudinally narrow region of "storm enhanced densities" [*Foster*, 1993].

Two other important mid-latitude weather features are sporadic E layers and descending intermediate layers. Sporadic E layers are ionization enhancements in the E region at altitudes between 90-120 km. The sporadic E layers are relatively narrow (0.6-2 km) and are composed primarily of metallic (Fe^+, Mg^+) ions, which are produced during meteor oblation. Typically, the density in sporadic E layers is about a factor of 10 greater than background densities, and after formation the layers tend to descend at a slow speed (0.6-4 m/s). Figure 7 shows an example of a sporadic E layer, as measured by the incoherent scatter radar at Arecibo. During the early evening (17:10-19:10 AST), a sporadic E layer was present at 116 km with a peak electron density of about 5×10^5 cm^{-3}. After sunset (18:10 AST), the sporadic E layer descended to 114 km and its peak density decreased to about 10^4 cm^{-3}. Subsequently, the layer continued to descend and at 21:48 AST the layer descended to 105 km.

Like sporadic E layers, intermediate layers are ionization enhancements. However, in contrast to sporadic E layers, intermediate layers are relatively broad (10-20 km wide), are composed of molecular ions (NO^+, O_2^+), and occur in the altitude range of from 120 to 180 km. They frequently appear at night in the valley between the E and F regions, but they can also occur during the day. They tend to form on the bottomside of the F-layer and then slowly descend throughout the night. Like sporadic E layers, intermediate layers are formed by a wind shear, whereby the vertical wind component changes direction with altitude. The layers form at the altitudes where the vertical wind converges, and the slow downward drift of the layers results from the downward motion of the wind convergence nodes. Figure 7 shows a distinct intermediate layer that was measured above Arecibo on 7 May 1986. The intermediate layer appeared at about 20:30 AST and then descended from 160 to 120 km during the night.

4. LOW LATITUDE

At low latitudes, the geomagnetic field is nearly horizontal and this introduces some important transport

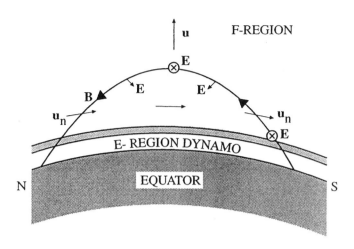

Figure 8. Neutral winds in the low-latitude E region generate dynamo electric fields as the ions are dragged across **B**. The dynamo electric fields are transmitted along **B** to F region altitudes, where they induce **E×B** plasma drifts. Meridional neutral winds induce field-aligned plasma drifts at F region altitudes. The corotational electric field causes the plasma to **E×B** drift toward the east with the corotation speed.

effects. At F-region altitudes and above, the plasma is constrained to move along **B** like beads on a string, and therefore, a meridional wind can very effectively induce an interhemispheric plasma flow (Figure 8). Also, at E-region altitudes, neutral wind-induced ionospheric currents act to generate zonal electric fields and these electric fields are transmitted to F-region altitudes along the highly conducting magnetic field lines. The electric fields are eastward during the day, which induce upward **E × B** drifts, and westward at night, which leads to downward **E × B** drifts. In addition to the interhemispheric flow and the vertical **E × B** drift, the low-latitude ionosphere also has a tendency to corotate with the Earth.

As with the mid-latitude ionosphere, the low-latitude ionosphere generally has higher plasma densities on the dayside than on the nightside. However, the low-latitude ionosphere also has a unique and persistent density feature known as the "Appleton anomaly", which arises as follows. During the day, the **E × B** plasma drift is upward near the magnetic equator. The plasma lifted in this way then diffuses down the **B** field lines and away from the equator due to gravity. This combination of upward **E × B** drift and downward diffusion produces a fountainlike pattern of plasma motion, which is called the "equatorial fountain." The downward plasma diffusion acts to produce density enhancements on both sides of the magnetic equator, and this feature is called the Appleton anomaly (Figure 9). The figure shows two ionization

crests, one on each side of the magnetic equator, that extend from the afternoon to well into the evening. Note the strong longitudinal variation of the Appleton anomaly, which results from the offset of the geomagnetic and rotational axes.

As with mid and high latitudes, the low latitude domain experiences both daily and hourly variations as well as severe weather disturbances. The daily and hourly variations are due to variations in the neutral wind and the zonal electric fields. Figure 10 shows electron density profiles obtained from GPS/MET satellite occultations near the magnetic equator. The n_e profiles were obtained within ± 5° of the magnetic equator at dusk (between 2000 and 2200 local time). Near the F-region peak, where the profiles are reliable, there is more than an order of magnitude variation in the electron density.

With regard to severe weather, perhaps the most important phenomenon is equatorial spread F. Plasma instabilities lead to density irregularities in the F region and they appear as spread F echoes (Figure 11). The scale size of the density irregularities ranges from a few centimeters to a few hundred kilometers. At night, fully developed spread F is characterized by plasma bubbles, which are vertically elongated wedges of depleted plasma that drift upward from beneath the bottomside F layer to altitudes as high as 1500 km. The individual magnetic flux tubes in a vertical wedge are typically depleted along their entire north-south extent. When bubbles form, they drift upward with speeds that vary from 100 m/s to 5 km/s. The electron density in the bubbles can be more than two orders of magnitude lower than background densities.

Severe spread F typically occurs near dusk (18:00-22:00 local time). As noted earlier, the zonal electric field is eastward during the day, which induces an upward **E × B** drift. At dusk, this upward **E × B** drift is frequently enhanced (Figure 12) and the F layer rises as the ionosphere corotates into darkness. In the absence of sunlight, the lower ionosphere rapidly decays and a steep vertical density gradient develops on the bottomside of the raised F layer. This configuration is conducive for a Rayleigh-Taylor instability, in which a heavy fluid is situated on top of a light fluid. This phenomenon is shown schematically in Figure 13, which displays the evolution of equatorial spread F that was observed on 14-15 March 1983, with simultaneous measurements from multiple instruments [*Argo and Kelley*, 1986]. Near the dusk terminator the F-layer was observed to rise and then descend. During its descent, spread F occurred, and subsequently, plasma bubbles formed on the bottomside of the F-layer. The bubbles drifted upward as the entire disturbed region corotated toward midnight. After

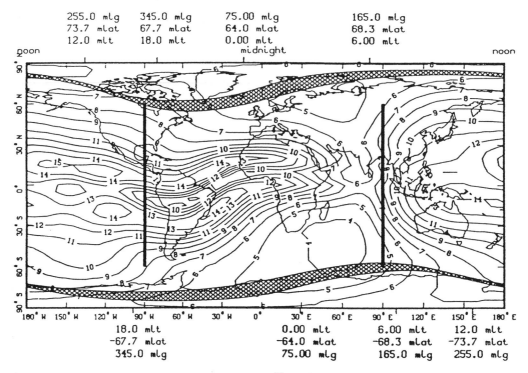

Figure 9. Contours of the critical F2-layer frequency ($f_oF_2 \sim n_e^{1/2}$) in MHz obtained from the IRI empirical model of the ionosphere. The contours are plotted versus geographic latitude and longitude. The conditions are for solar maximum, September equinox, and 0 UT. The vertical bars show the sunset/sunrise terminators and the shaded regions show the auroral ovals. From *Szuszczewicz* [1989].

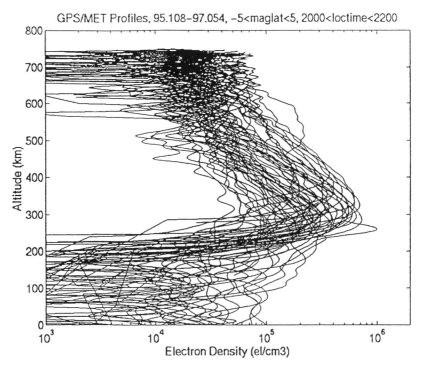

Figure 10. Electron density profiles obtained from GPS/MET satellite occulations. The profiles were obtained within $\pm 5°$ of the magnetic equator and between 2000-2200 local time. (Courtesy of B. Schreiner.)

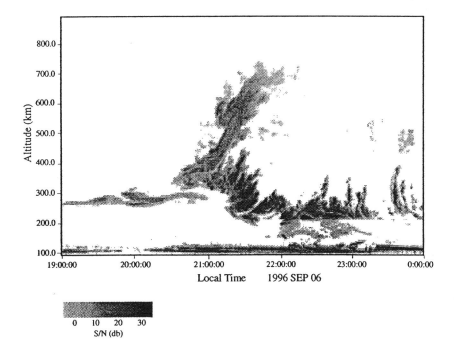

Figure 11. Spread F event measured by the JULIA coherent scatter radar on 6 September 1996. Displayed is a range-time-intensity plot of coherent backscatter signal-to-noise ratios. From *Hysell and Burcham* [1998].

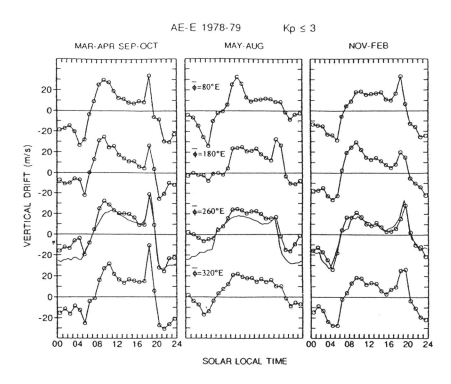

Figure 12. Empirical model values of vertical plasma drifts in four longitude sectors and for three seasons. The values correspond to low magnetic activity and moderate to high solar activity. Also shown are the seasonal Jicamarca drift patterns for similar solar flux and geomagnetic conditions. From *Fejer et al.* (1995).

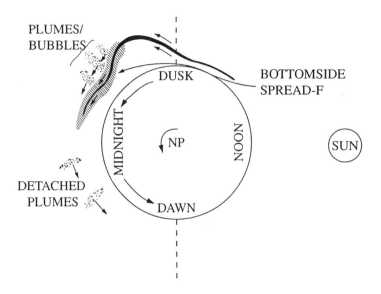

Figure 13. Schematic diagram showing the evolution of equatorial spread F and plasma bubbles that is consistent with simultaneous measurements from multiple instruments. From *Argo and Kelley* [1986].

midnight, the spread F disturbance ceased, but the bubbles (detached plumes) persisted.

5. CONCLUSION

At the present time, the climatology of the ionosphere at high, middle, and low latitudes has been established. Likewise, the physical and chemical processes that act on and within the ionosphere-thermosphere system are known. The challenge for the future is to model the large daily and hourly variations of the ionosphere as well as the severe weather disturbances. This challenge is currently being attacked by the experimentalists via long-duration campaigns involving multiple instruments making simultaneous ionospheric measurements. On the theoretical side, coupled physical models, multi-grid models, and data assimilation models are being employed in attempts to more rigorously describe the ionosphere and its drivers. A significant advance in ionospheric specification and forecasting can be anticipated during the coming decade.

Acknowledgement. This research was supported by NASA grant NAG5-8227 and NSF grant ATM-9612638 to Utah State University.

REFERENCES

Argo, P. E. and M. C. Kelley, Digital ionosonde observations during equatorial spread F, *J. Geophys. Res., 91*, 5539, 1986.
Banks, P. M., R. W. Schunk, and W. J. Raitt, The topside ionosphere: A region of dynamic transition, *Ann. Rev. Earth Planet. Sci., 4*, 381-440, 1976.
Brinton, H. C., J. M. Grebowsky, and L. H. Brace, The high-latitude winter F region at 300 km: Thermal plasma observations from AE-C, *J. Geophys. Res., 83*, 4767, 1978.
Fejer, B. G., et al., Global equatorial ionospheric vertical plasma drifts measured by the AE-E satellite, *J. Geophys. Res., 100*, 5769-5776, 1995.
Foster, J. C., Storm-time plasma transport at middle and low latitudes, *J. Geophys. Res., 98*, 1675-1689, 1993.
Fukui, K., J. Buchau, and C. E. Valladares, Convection of polar cap patches observed at Qaanaaq, Greenland, during the winter of 1989-1990, *Radio Sci., 29*, 231-248, 1994.
Hysell, D. L. and J. D. Burcham, JULIA radar studies of equatorial spread F, *J. Geophys. Res., 103*, 29155, 1998.
Riggin, D. et. al., Radar studies of long-wavelength waves associated with mid-latitude sporadic E layers, *J. Geophys. Res., 91*, 8011-8024, 1986.
Roble, R. G., et. al., The calculated and observed ionospheric properties during Atmosphere Explorer-C satellite crossing over Millstone Hill, *J. Atmos. Terr. Phys.*, 40, 21, 1978.
Schunk, R. W. and A. F. Nagy, *Ionospheres*, Cambridge University Press, Cambridge, UK, 2000.
Schunk, R. W. and J. J. Sojka, Ionosphere-thermosphere space weather issues, *J. Atmos. Terr. Phys., 58*, 1527-1574, 1996.
Szuszczewicz, E. P., *Proceedings of the Solar-Terrestrial Predictions Workshop*, Leura, Australia, 1989.
Valladares, C., H. C. Carlson, and F. Fukui, Interplanetary magnetic field dependency of stable sun-aligned polar cap arcs, *J. Geophys. Res., 99*, 6247-6272, 1994.

Robert W. Schunk, Center for Atmospheric and Space Sciences, Utah State University, Logan, UT 84322-4405.

On Forecasting Thermospheric and Ionospheric Disturbances in Space Weather Events

R. G. Roble

High Altitude Observatory, National Center for Atmospheric Research, Boulder, Colorado

It is well known that solar EUV radiation and auroral heat and momentum sources have a significant effect on thermospheric and ionospheric structure and dynamics. Upper atmosphere general circulation models using these forcings have been reasonably successful in simulating the thermosphere and ionosphere responses for a number of geophysical event studies. These models can be used as forecast models of thermospheric and ionospheric structure and dynamics by using predicted inputs of solar EUV and UV radiation, auroral hemispheric power of precipitating particles, cross-polar cap potential drop and ion convection patterns. It is also necessary to have a satisfactory initial state to start the simulation. The NCAR TIE-GCM that simulates the thermosphere and ionosphere between 95 and 800 km altitude is used to show the sensitivity of the thermosphere and ionosphere to space weather events.

1. INTRODUCTION

There have been several different upper atmosphere 3-D models that have been developed over the years to study processes in the thermosphere and ionosphere and especially their dynamic response to solar and auroral variability [*Mayr and Volland*, 1966; *Fuller-Rowell and Rees*, 1980; *Dickinson et al.*, 1981; *Mikkelsen and Larsen*, 1993; *Namgaladze et al.*, 1990]. These models have been used for thermospheric research and in the analysis of satellite and ground-based data for many years. In this brief paper, I will present results from one of these models the 'NCAR TIE-GCM' to illustrate the feasibility of using such a model in a time-dependent simulation of space weather using realistic solar and auroral forcings.

2. TIE-GCM

The National Center for Atmospheric Research (NCAR) Thermosphere-Ionosphere-Electrodynamics General Circulation Model (TIE-GCM) that extends between 95 and 800 km has been used by many investigators to simulate the time-dependent response of the thermosphere and ionosphere system to variations in solar EUV and UV radiation, auroral particle precipitation and ion convection during geomagnetic storms and substorms. It incorporates all of the features developed in previous general circulation models of the upper atmosphere, the TGCM [*Dickinson et al.*, 1981, 1984], the TIGCM [*Roble et al.*, 1988], and it has the main features of the TIE-GCM that have been described by [*Richmond et al.*, 1992] but several changes to that model have been made as described in the numbered paragraphs below. The changes are to better represent aeronomical processes that in turn give a better comparison of model simulations with data. The T refers to thermosphere, I ionosphere, and E electrodynamics in each of the GCM characterizations described above. The model does not extend into the mesosphere and

Space Weather
Geophysical Monograph 125

TIE-GCM

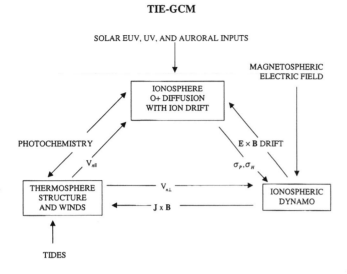

Figure 1. Schematic of coupled physical and chemical processes in the TIE-GCM.

thus must represent lower atmospheric variability by specifying boundary conditions at 95 km. Currently the model only includes a zonally averaged seasonal variation and Hough Mode tidal forcings for the diurnal (1,1) tide and the semi-diurnal (2-2 through 2-6) tidal components [*Forbes et al.*, 1993]. The aeronomic processes in the model have been described by *Roble et al.* [1987] and *Roble* [1995].

The TIE-GCM includes self-consistent ionospheric electrodynamics, that is, a calculation of the electric fields and currents generated by the ionospheric dynamo, and consideration of their effects on the neutral and ionospheric dynamics. It is used for studies that focus on the thermosphere and its coupling with the ionosphere and magnetosphere. The thermosphere/ionosphere/electrodynamic interactions are shown schematically in Figure 1. These interactions are calculated at every grid point and time step in the model. The essential differences between the model that was described in those previous papers and the model used for these simulations include the following:

1. The neutral chemical reaction rates for the aeronomic scheme described by *Roble* [1995] have been updated to be consisted with the JPL-97 compilation [*De-More et al.*, 1997]. The ion chemistry is consistent with *Buonsanto et al.* [1995].

2. The CO_2 infrared cooling parameterization has been updated to include the model of *Fomichev et al.* [1999] to account for a variable CO_2 mixing ratio that is important for non-local thermodynamic equilibrium (non-LTE) processes in the upper mesosphere and lower thermosphere. All calculations assume an $O-CO_2$ vibra-

tional relaxation rate of $3 \times 10^{-12} cm^{-3} s^{-1}$ that seems to work reasonably well on all of the terrestrial planetary thermospheres [*Bougher et al.*, 1998].

3. The tidal forcing at the lower boundary near 95 km, that represents diurnal and semi-diurnal components excited in the lower atmosphere, has been specified using results from the Global Scale Wave Model (GSWM) [*Hagan*, 1996]. The amplitudes and phases of the propagating diurnal (1,1) and semi-diurnal tide (2-2 through 2-6) were obtained from the GSWM at 95 km altitude for various seasons and used as lower boundary amplitudes and phases in the TIE-GCM. They, however, have been adjusted to be consistent with the UARS measurements of winds by *McLandress et al.* [1996].

4. Solar Ionization rates are calculated using the EUVAC solar flux model and absorption cross-sections from *Richards et al.* [1994]. Solar photodissociation rates for thermospheric processes have been determined using, in part, the parameterizations given in *Brasseur and Solomon* [1986].

The output fields calculated by the TIE-GCM are given in Appendix A.

3. TIE-GCM CALCULATED STRUCTURE FOR EQUINOX CONDITIONS

The TIE-GCM has been used to calculate the global circulation, temperature and compositional structure for equinox and solstice conditions. A diurnally reproducible solution for equinox conditions during solar cycle maximum is shown in Figure 2. The upper panel shows the global temperature variation as well as vectors of neutral wind velocity at an altitude near 350 km. There is a diurnal temperature variation of about 500K and the winds have high velocity in the polar region driven by ion convection. At low latitudes the winds are greater at night when the ion drag is relatively small. The lower panel shows the ionospheric dynamo calculation of electric potential and vectors of ion drift. At high latitudes, the ions follow the two cell pattern of ionospheric convection with drift speed on the order of 500-800 m/s. At low latitudes the ion drifts are small during the daytime when the E-region is electrically conducting but drift with velocities approaching the neutral wind velocities at low latitudes at night. These drifts are driven by the F-region dynamo that operates after the E-region recombines at night. These and other electrodynamic couplings between the thermosphere/ionosphere have been discussed in detail by *Richmond and Roble* [1997].

TIE-GCM
350 km, 0 UT, Equinox, Solar Maximum

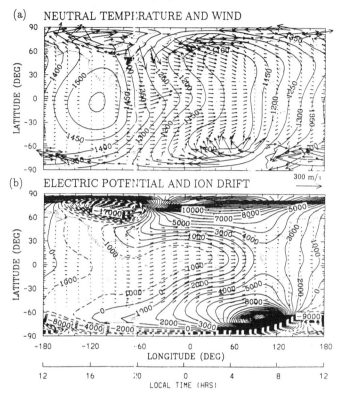

Figure 2. (a) Calculated neutral gas temperature and vectors of neutral wind at 350 km altitude for equinox solar cycle maximum conditions and (b) calculated electric potential (volts) and vectors of ion drift.

4. SIMULATING SPACE WEATHER FOR MARCH 15 - MAY 15, 1998

The TIE-GCM is used to calculate the thermosphere and ionosphere response to solar and auroral forcing during the period March 15 through May 15, 1998. The solar F10.7 variation during the period is shown in Figure 3 and it clearly illustrates a 27 day solar rotation period on a background characteristic of near solar minimum conditions. The solar F10.7 variation is used to drive the EUVAC solar spectral irradiance model of *Richards et al.* [1994] that covers the spectral range from 0.1 to 200. nm. This model is used to calculate the solar ionization, photodissociation and heating rates as described by *Roble et al.* [1987] and *Roble* [1995].

The auroral model used in the TIE-GCM has been described by *Roble and Ridley* [1987]. It includes an ion convection model and aurora particle precipitation model and both are driven by parameterizations related to the 3 hr geomagnetic index Kp. The variation of Kp

used in the simulation is shown in Figure 4. The parameterizations based on Kp are derived from many space weather studies using satellite data and the results of AMIE simulations. The main driving parameters include the cross-polar cap potential drop, the configuration of the convection pattern and its expansion with increased geomagnetic activity. It also includes an analytic auroral oval as described by *Roble and Ridley* [1987] with characteristic auroral particle mean energy and energy flux related to variations in Kp derived from satellite particle precipitation studies.

Similar to the procedure used in space weather event studies, the TIE-GCM is run for a geomagnetic quiet period with constant solar and auroral forcings until a diurnally reproducible steady state solution is obtained. From the steady state solution the TIE-GCM is run then forward in time using the time-dependent geophysical indicies to drive the solar spectral irradiance and auroral models. The TIE-GCM time step is 3 minutes and the solar spectral irradiance is linearly interpolated between daily values and the Kp parameterizations are linearly varied in response to the 3 hr Kp values. The simulation described below saves histories hourly through the period. The results, however, are displayed as daily zonal averages as a satellite in a sun synchronous orbit at 12 Local Time would see the thermosphere and ionosphere properties. This is just

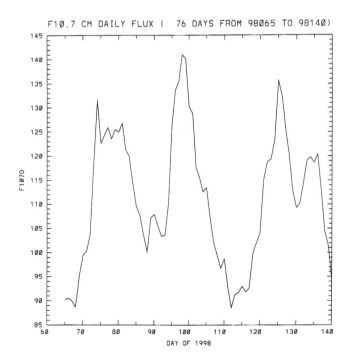

Figure 3. Solar F10.7 variation during days 60-140 in 1998.

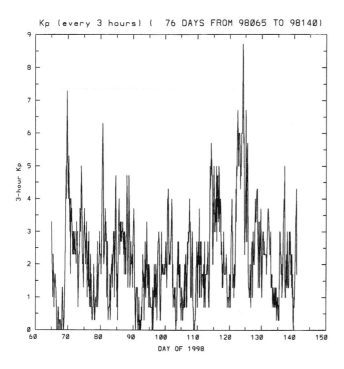

Figure 4. Geomagnetic index Kp variation during days 60-140 in 1998.

for presentation of results in this paper, but the global data comprising the figures are available on an hourly basis for other types of space weather uses and displays.

The calculated neutral gas temperatures at two altitudes (300 km and 150 km) are shown in Figure 5. The solar F10.7 peaks near days 80, 100 and 125 and there are corresponding increases in thermospheric temperature on those days. The Kp index peaks near day 125 and the are major changes to temperature between days 122 and 126. The variability in these figures are caused by variations in the solar EUV ionizing radiation and the Joule heating and particle precipitation in the aurora.

The atomic oxygen number density and the O/O_2 ratio variations during the period are shown in Figures 6 and 7 respectively. There is a slow seasonal variation as well as enhancements during the peaks in solar F10.7 and Kp, similar to the daily temperature variations shown in Figure 5. The O/O_2 ratio is important for predicting density variations detected by remote UV sensing instruments.

The calculated electron density variations at the same two heights are shown in Figure 8. Since the presentation is along a constant height and constant solar local time (12 hr), the variability shows up as variations to

the otherwise straight lines that would be generated in the absence of any solar or auroral variability.

These figures illustrate the space weather effects on thermospheric and ionospheric variability.

5. DEVELOPMENT OF SPACE WEATHER MODELS FOR OPERATION PURPOSES

The above simulation illustrates the variability of various thermospheric and ionospheric properties associated with space weather. The existing space weather models can be used in either the nowcast or forecast

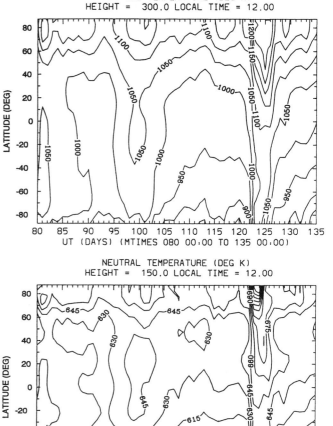

Figure 5. Calculated zonal average temperature (K) as sampled daily by a sun synchronous satellite at 12LT at two altitudes (a) 300 km and (b) 150 km.

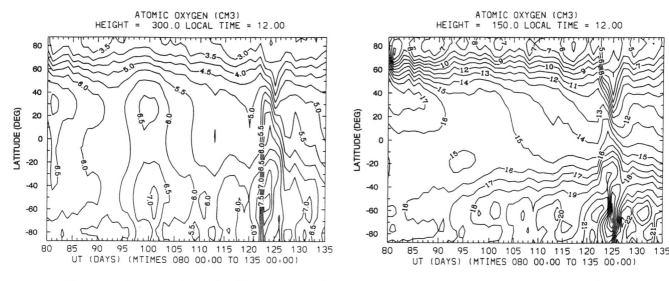

Figure 6. Same caption as for Figure 5 except for atomic oxygen number density. (a) Particles/cm³ in units of 10^8, (b) particles/cm³ in units of 10^9.

modes as well as for previous event studies once the appropriate solar and auroral inputs can be specified.

Nowcasting uses previous and instantaneous solar and geophysical indices or AMIE data to drive the TIE-GCM.

- Model solar EUV and UV spectral irradiance inputs are specified by instantaneous solar indicies and model ion convection and aurora particle precipitation parameterizations are driven by specified instantaneous geomagnetic index or solar wind properties.

- Assimilative Mapping of Ionospheric Electrodynamics (AMIE) can be used with ground-based and satellite data to derive instantaneous time-dependent auroral inputs for the TIE-GCM.

Forecasting uses predicted solar and auroral indicies to integrate the model forward in time for future forecasts and they can be updated as often as desired.

- With solar and auroral indicies predicted for various days in advance the TIE-GCM can be run to give, say, 1, 3, 5, or 10 day forecasts of space

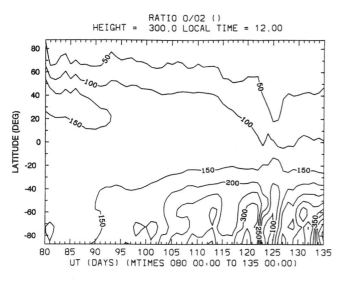

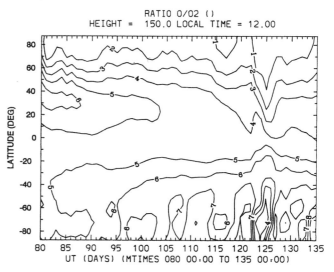

Figure 7. Same caption as for Figure 5 except for atomic oxygen to molecular oxygen number density ratio.

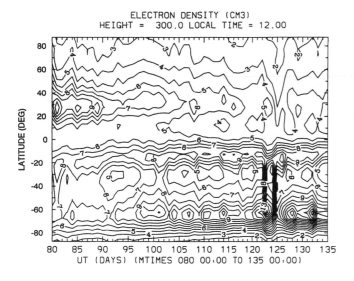
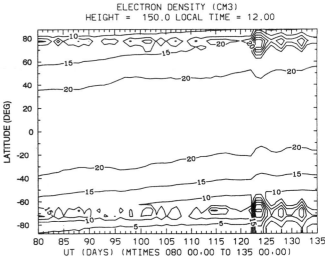

Figure 8. Same caption as for Figure 5 except for electron number density. (a) Particles/cm^3 in units of 10^5, (b) particles/cm^3 in units of 10^4.

weather, much as the weather forecasters do for predicting the future troposphere and stratosphere weather. The TIE-GCM is numerically fast so the predictions can be updated daily or as needed.

6. FUTURE SPACE WEATHER MODEL DEVELOPMENT

- A major problem that has not yet been addressed for space weather studies is data assimilation for numerical initial conditions. There is a strong need to develop such techniques so that the model can be run forward smoothly without introducing transients.

- There is also a need for continued aeronomical research and data analysis to upgrade many aspects of the aeronomy in the TIE-GCM, such as the rate coefficients for CO_2 and NO vibrational exchange with O.

- There is a need to improve boundary conditions to better represent magnetosphere/ionosphere energy and plasma exchange.

- There is a need to improve the lower boundary of the TIE-GCM to represent coupling with the lower atmosphere. The recent development of the TIME-GCM (M indicating Mesosphere) that in-

Table A1. TIE-GCM Output Fields

Field	Symbol
Neutral gas temperature	Tn
Neutral winds (zonal, meridional and vertical)	Un, Vn, Wn
Neutral composition (major species)	N_2, O_2, O
Neutral composition (minor species)	NO, N(^{4}S), N(^{2}D), O(^{1}D), O(^{1}S), He, and Ar
Neutral density	Rho (gm/cm^3)
Electron density	Ne
Ion Composition	NO$^+$, O$_2^+$, O$^+$(^{4}S), O$^+$(^{2}D), O$^+$(^{2}P), N$^+$, H$^+$
Electron and Ion temperature	Te, Ti
Electric Fields and Currents	E, J
Electrical conductivity	local and height integrated Pedersen, Hall and Parallel conductivity
Airglow emissions	a large number of UV and visible airglows

cludes the mesosphere and lower thermosphere down to 10 mb or 30 km is a first step in that direction [*Roble and Ridley*, 1994]. Ultimately, a model of the whole atmosphere is needed to adequately represent coupling with the lower atmosphere [*Roble*, 2000].

- It is important to continually validate model performance by comparisons with observations. There is also a need to develop skill tests to provide an objective assessment of the various models' ability to specify space weather events accurately.

- Satellite Observations – There are many satellite projects that have a need for a comprehensive model that predict dynamics, temperature, compositional and airglow structure. These include the military and commercial satellites, NASA satellites such as TIMED, ISTP, UARS, as well as various ground-based programs such as CEDAR, GEM, and Space Weather.

APPENDIX A: TIE-GCM OUTPUT FIELDS

The TIE-GCM output includes a global description of the following fields at every time step.

REFERENCES

Brasseur, G., and S. Solomon, *Aeronomy of the Middle Atmosphere*, 452 pp., D. Reidel Publishing Co., second edition, Dordrecht, Holland, 1986.

Buonsanto, M. J., P. G. Richards, W. K. Tobiska, S. C. Solomon, Y.-K. Tung, and J. A. Fennelly, Ionospheric electron densities calculated using different EUV flux models and cross-sections: Comparison with radar data, *J. Geophys. Res., 100*, 14569-14580, 1995.

DeMore, W. B., et al., Chemical kinetics and photochemical data for use in stratospheric modeling, Evaluation Number 10, *JPL-Publication 94-1*, NASA-JPL, Pasadena, CA, 1994.

Dickinson, R. E., E. C. Ridley, and R. G. Roble, A three-dimensional general circulation model of the thermosphere, *J. Geophys. Res., 86*, 1499-1512, 1981.

Dickinson, R. E., E. C. Ridley, and R. G. Roble, Thermospheric general circulation with coupled dynamics and composition, *J. Atmos. Sci., 41*, 205-219, 1984.

Flemming, E. L., S. Chandra, M. R. Schoeberll, and J. J. Barnett, Monthly mean global climatology of temperature, wind, geopotential height, and pressure for 0–120 km, *NASA Tech. Memo TM-100697*, 85 pp., Goddard Space Flight Center, Greenbelt, Maryland, 1988.

Fomichev, V. I., J.-P. Blanchet, and D. S. Turner, Matrix parameterization of the $15\mu m$ CO_2 band cooling in the middle and upper atmosphere for variable CO_2 concentration, *J. Geophys. Res.*, in press, 1999.

Forbes, J. M., R. G. Roble, and C. G. Fesen, Acceleration, heating and compositional mixing of the thermosphere due to upward-propagating tides, *J. Geophys. Res., 98*, 311-321, 1993.

Fuller-Rowell, T. J., and D. Rees, A three-dimensional time-dependent global model of the thermosphere, *J. Atmos. Sci., 37*, 2545-2567, 1980.

Hagan, M. E., Comparative effects of migrating solar sources on tidal signatures in the middle and upper atmosphere, *J. Geophys. Res., 101*, 21,213-21,222, 1996.

McLandress, C., G. G. Shepherd, B. H. Solheim, M. D. Burrage, P. B. Hays, and W. R. Skinner, Combined mesosphere/thermosphere winds using WINDII and HRDI data from the Upper Atmosphere Research Satellite, *J. Geophys. Res., 101*, 10441-10453, 1996.

Mikkelsen, I. S., and M. F. Larsen, Comparisons of spectral thermospheric general circulation model simulations and E and F region chemical release wind observations, *J. Geophys. Res., 98*, 3693-3709, 1993.

Namgaladze, A. A., Yu. N. Koren'kov, V. V. Klimenko, I. V. Karpov, F. S. Bessarb, V. A. Surotkin, T. A. Glushcenko, and N. M. Naumova, A global numerical model of the thermosphere, ionosphere, and protonosphere, *Geomagnetism and Aeronomy, 30*, 515-521, 1990.

Richards, P. G., J. A. Fennelly, and D. G. Torr, EUVAC: A solar EUV flux model for aeronomical calculations, *J. Geophys. Res., 99*, 8981-8992, 1994.

Richmond, A. D., E. C. Ridley, and R. G. Roble, A thermosphere/ionosphere general circulation model with coupled electrodynamics, *Geophys. Res. Lett., 19*, 601-604, 1992.

Richmond, A. D., and R. G. Roble, Electrodynamic coupling effects in the thermosphere/ionosphere system, *Adv. Space Res., 20*, 1115-1124, 1997.

Roble, R. G., and E. C. Ridley, An auroral model for the NCAR thermosphere general circulation model (TGCM), *Annales. Geophysicae, 5A*, (6), 369-382, 1987.

Roble, R. G., E. C. Ridley, A. D. Richmond, and R. E. Dickinson, A coupled thermosphere/ionosphere general circulation model, *Geophys. Res. Lett., 15*, 1325-1328, 1988.

Roble, R. G., and E. C. Ridley, A thermosphere-ionosphere-mesosphere-electrodynamics general circulation model (TIME-GCM): Equinox solar cycle minimum simulations (30–500 km), *Geophys. Res. Lett., 21*, 417-420, 1994.

Roble, R. G., E. C. Ridley, and R. E. Dickinson, On the global mean structure of the thermosphere, *J. Geophys. Res., 92*, 8745-8758, 1987.

Roble, R. G., Energetics of the mesosphere and thermosphere, in *The Upper Mesosphere and Lower Thermosphere: A Review of Experiment and Theory, Geophys. Mono., 87*, 1-21, 1995.

Roble, R. G., On the feasibility of developing a global atmospheric model extending from the ground to the exosphere, in *Coupling of Processes Across the Stratopause, Geophys. Mono., 92*, in press, 2000.

R. G. Roble, High Altitude Observatory, National Center for Atmospheric Research, 3450 Mitchell Lane, Boulder, CO 80301. (e-mail: roble@ucar.edu)

Geomagnetic Storm Simulation With a Coupled Magnetosphere–Ionosphere–Thermosphere Model

Joachim Raeder

Institute of Geophysics and Planetary Physics, University of California, Los Angeles, California

Yongli Wang

Department of Earth and Space Sciences, University of California, Los Angeles, California

Timothy J. Fuller-Rowell

NOAA Space Environment Center, Boulder, Colorado
CIRES, University of Colorado

We present the first global, self-consistent, fully electrically coupled magnetosphere–ionosphere–thermosphere model, based on the UCLA magnetosphere–ionosphere model and the NOAA Coupled Thermosphere Ionosphere Model (CTIM). Initial results from this coupled model for the January 10, 1997 geomagnetic storm event are encouraging. In particular, the model produces a much more realistic electrodynamic and ionospheric response as compared with the previous magnetosphere model that relied on parameterizations for the ionospheric conductance. This is attributed to the much more realistic conductance calculations provided by CTIM. Like in previous studies with the magnetosphere model, the cross-polar-cap potential is too high. Examining the cause will require further investigation.

1. INTRODUCTION

Contemporary terrestrial weather forecasts rely on a combination of dense observation networks and sophisticated numerical models which project the current weather conditions into the future. Similarly, space weather forecasting will require large scale numerical models of Earth's space environment as a key operational element, along with sufficient timely observations to provide model input and to ini-

tialize model fields. Besides their utility for space weather forecasting, such models are also essential tools to understand the environment and the plasma physical processes in it, because *in situ* measurements are often too sparse to enable a unique interpretation of the data, and because the complexity of the space environment limits our theoretical understanding. By comparing predictions with data, large scale models also serve to test our knowledge about the prevailing processes, i.e., successful model predictions suggest that the assumptions underlying the model are correct, whereas model failures point to deficiencies of our understanding. Models have also become an increasingly important tools to analyze and interpret experimental data, by putting *in situ* measurements from a single (or at most a few) spacecraft into

Space Weather
Geophysical Monograph 125

perspective and thus extending the "view" of the observations. Thus, progress is often made by combining data analysis with global modeling, since both are essentially complementary. In the foreseeable future tens to hundreds of spacecraft might provide data, in which case models will play a crucial role in assimilating these measurements in order to provide global synoptic maps of Earth's space environment.

No global comprehensive model of Earth's space environment exists today. However, regional models have been developed that treat limited regions or processes. For example, global MHD models cover the outer magnetosphere [*Lyon et al.*, 1998; *Ogino*, 1986; *Winglee and Menietti*, 1998; *Gombosi et al.*, 1998; *Tanaka*, 1995; *Janhunen et al.*, 1995; *Raeder*, 1999], but omit the particle drift physics of the inner magnetosphere within a few R_E from Earth as well as the plasma and neutral constituents of the ionosphere–thermosphere system. Some of these models [*Lyon et al.*, 1998; *Tanaka*, 1995; *Janhunen et al.*, 1995; *Raeder*, 1999] include an ionosphere submodel that solves a potential equation to close the field aligned currents originating in the magnetosphere. The ionospheric conductance in these submodels is either taken to be constant or derived by using empirical models for EUV ionization and parameterizations for electron precipitation in the auroral zone [e.g. *Slinker et al.*, 1998; *Raeder*, 1999]. However, such submodels do not fully represent the ionosphere–thermosphere system, but approximate only one aspect of it, namely the closure of field aligned currents, and rely on a number of approximations and parameterizations.

On the other hand, fully dynamical models of the ionosphere–thermosphere system exist, for example the Thermosphere Ionosphere Mesosphere Electrodynamics General Circulation Model (TIME-GCM) [*Roble and Ridley*, 1994] or the NOAA Coupled Thermosphere Ionosphere Model (CTIM) [*Fuller-Rowell et al.*, 1996]. These models depend on magnetospheric input, for example the electric field and particle precipitation. For the lack of direct observations of these quantities with sufficient resolution, they are usually taken from empirical (climatological) models, parameterizations, or data-assimilative models, for example AMIE [*Richmond and Kamide*, 1988].

Clearly, comprehensive models of Earth's space environment require the coupling of magnetosphere models with ionosphere–thermosphere models because these types of models are complementary. The global magnetosphere models lack the physical first-principle calculations at the ionospheric end which can be provided by an ionosphere–thermosphere (IT) model; vice versa, IT models require input that magnetosphere models can provide, and at the same time provide input to magnetosphere models.

In this paper we present the first attempt of producing such a coupled model consisting of the UCLA global magnetosphere model with the NOAA CTIM model. In the following sections we briefly describe the two models, discuss the coupling issues, and present some initial results from the coupled model.

2. THE UCLA GLOBAL MAGNETOSPHERE MODEL

The UCLA global magnetosphere model solves the MHD equations in a large volume surrounding Earth such that the entire interaction region between the solar wind and the magnetosphere is included. Specifically, the simulation domain comprises the bow shock, magnetopause, and the magnetotail up to several hundred R_E from Earth. Thus the model input is given by the solar wind plasma and IMF (Interplanetary Magnetic Field), which for most studies (and space weather applications) is taken from measurements of a solar wind monitor such as WIND or ACE. This model has been developed and continually improved over the many years and now goes well beyond a "three-dimensional global MHD simulation model". Besides numerically solving the MHD equations with high spatial resolution, the model also includes ionospheric processes and their electrodynamic coupling with the magnetosphere. The coupling between the magnetosphere and the ionosphere is an essential part of the model because the ionosphere controls to a large part magnetospheric convection, by providing the resistive closure of the field aligned currents that are generated from the interaction of the solar wind with the magnetosphere [*Raeder et al.*, 1998]. Processes that occur in the near-Earth region on polar cap and auroral field lines and that are inherently kinetic have been parameterized in the model using empirical relationships [*Raeder et al.*, 2000]. These processes include the field aligned potential drops that are associated with upward field aligned currents and related electron precipitation, and the diffuse electron precipitation that is caused by pitch angle scattering of plasma sheet electrons [*Lyons et al.*, 1979; *Weimer et al.*, 1987; *Kennel and Petschek*, 1966]. In previous simulations the electron precipitation parameters were used to determine the ionospheric Pedersen and Hall conductances using the empirical *Robinson et al.* [1987] formulae, while the new model uses CTIM conductances (see below). The model uses the conductances and the field aligned currents to solve the ionospheric potential equation. The embedded ionosphere model yields many ionospheric quantities that are observable from the ground and low Earth orbiting satellites. The primary quantities are the field aligned currents, the Hall and Pedersen conductance, and the electric potential. From these, other related quantities are derived, such as the total and equivalent ionospheric current, dissipation rates, and ground magnetic perturbations. In particular, the ground magnetic perturbation can be computed at any point of the auroral zone and the polar cap, as well as related geomagnetic indices like the AE, AU, and AL indices. The availability of the synthetic magnetograms and indices allows for direct comparisons with ground data [*Raeder et al.*, 2000] and are of great significance for space weather.

The numerical grid is rectangular and nonuniform with the highest spatial resolution (about 0.3 R_E) near Earth and in the tail plasma sheet. It extends about 20 R_E in the sunward direction, 300 R_E in the tailward direction and ± 40 R_E in the Y and Z directions. The gas-dynamic part of the MHD equations is spatially differed by using a technique in which fourth-order fluxes are hybridized with first-order (Rusanov) fluxes [*Harten and Zwas*, 1972; *Hirsch*, 1990]. The magnetic induction equation is treated somewhat differently [*Evans and Hawley*, 1988] in order to conserve $\nabla \cdot \mathbf{B} = 0$ exactly. The time stepping scheme for all variables consists of a low-order predictor with a time-centered corrector, which is accurate to the second order in time. Thus, the numerical scheme is flux-limited, i.e., it produces diffusion only to the extent needed at shocks and discontinuities. In regions where all variables vary smoothly, the gas dynamic variables are computed with fourth-order accuracy, and the magnetic field with second-order accuracy. The outer boundary conditions are fixed at the given solar wind values on the upstream side. At the other boundaries we apply open, i.e., zero normal derivative, boundary conditions. For optimal performance the code is parallelized for state-of-the-art massively parallel computers (IBM-SP2, SGI-O2000, Beowulf Clusters) using domain decomposition [*Fox et al.*, 1988] and MPI message passing. Real time operation with a grid of about 10^6 cells takes about 40 processors of either of these computers. A more detailed description of the model can be found in [*Raeder*, 1999; *Raeder et al.*, 2000].

3. THE NOAA COUPLED IONOSPHERE THERMOSPHERE MODEL

CTIM is a global multi-fluid model of the thermosphere–ionosphere system with a long heritage [see *Fuller-Rowell et al.*, 1996, and references therein]. CTIM solves both neutral and ion fluid equations self-consistently from 80 to 500 km for the neutral atmosphere and from 80 to 10,000 km for the ionosphere on a spherical grid with 2° latitude resolution and 18° longitude resolution. The thermosphere part solves the continuity equation, horizontal momentum equation, energy equation, and composition equations for the major species O, O_2, and N_2 on 15 pressure levels. The ionosphere model part solves the continuity equations, ion temperature equation, vertical diffusion equations, and horizontal transport for H^+ and O^+, while chemical equilib-

rium is assumed for N_2^+, O_2^+, NO^+, and N^+. The horizontal ion motion is governed by the magnetospheric electric field. The coupled model includes about 30 different chemical and photo-chemical reactions between the species. Compared to the magnetosphere, the CTIM timescales are relatively long, allowing for numerical timesteps of the order of 1 minute. Consequently, CTIM is computationally very efficient and runs considerably faster than real-time (>10 times) on a single CPU.

CTIM's primary inputs are the solar UV and EUV fluxes (parameterized by the solar 10.7 cm radio flux), the tidal modes (forcing from below), auroral precipitation, and the magnetospheric electric field, each of which is usually taken from parameterized empirical models.

CTIM provides several outputs that are of prime importance for space weather, for example global two- and three-dimensional ionosphere and thermosphere state fields, like electron density, neutral density, neutral wind, chemical composition, NmF2, hmF2, and total electron content (TEC). Specifically, the electron parameters are important because they strongly affect HF communication and navigation systems, and the neutral densities are important because they determine the drag on LEO satellites and space debris.

A more thorough description of CTIM, including the detailed equations, reaction rates, and examples can be found in [*Fuller-Rowell et al.*, 1996].

4. COUPLING ISSUES

Figure 1 shows schematically how the magnetosphere model and CTIM are coupled. The "MI coupling module" and the "ionosphere potential solver" are part of the magnetosphere model. Specifically, the MI coupling module maps field aligned currents from the magnetosphere to the ionosphere and computes electron precipitation fluxes (F_E) and mean energies (E_0). In the opposite direction, it maps the ionospheric electric field to the inner boundary of the magnetosphere model (which is a sphere of about 3-4 R_E radius centered on Earth) where it is used as an MHD boundary condition [*Raeder*, 1999; *Raeder et al.*, 2000]. The potential solver solves the ionosphere potential equation

$$\nabla \cdot \Sigma \cdot \nabla \Phi = -j_\parallel \sin I + j_{\parallel,d}$$

on the surface of a sphere with 1.015 R_E radius. Here, Φ is the ionosphere potential, Σ is the conductance tensor, $j_\parallel$ is the magnetospheric field aligned current (FAC), $j_{\parallel,d}$ is a parallel current arising from the ionospheric dynamo, and I is the magnetic field inclination in the ionosphere. The ionospheric dynamo current $j_{\parallel,d}$ arises from the ion-neutral drag in the ionosphere, in which the electric field $\mathbf{E}'$ in the reference frame of the neutrals is given by $\mathbf{E}' = \mathbf{E} + \mathbf{U} \times \mathbf{B}$

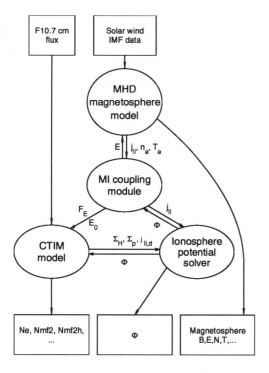

Figure 1. Schematic of the coupled magnetosphere–ionosphere–thermosphere model. The arrows represent the data flow.

[*Kelley*, 1989] where **U** is the velocity of the neutrals. From $\nabla \cdot \mathbf{J} = 0$ and $\mathbf{J} = \sigma \cdot \mathbf{E}'$ follows by integration over z:

$$j_{\parallel,d} = -\nabla \cdot \int_z \sigma \cdot (\mathbf{U} \times \mathbf{B})\,dz$$

where σ is the conductivity tensor and **B** is the magnetic field.

In coupling the magnetosphere model with CTIM, the MI coupling module provides the electron precipitation parameters and the magnetospheric electric field. In turn, CTIM provides the ionospheric conductance and $j_{\parallel,d}$ to the potential solver. Thus, as far as the magnetosphere model is concerned, we replace empirical conductance calculations [*Robinson et al.*, 1987] with first-principle calculations and also account for the ionospheric dynamo effect. The latter effect is probably of minor importance in most situations, but may become significant during storm recovery [*Richmond and Roble*, 1987]. With this coupling, CTIM is also driven with more realistic magnetospheric inputs and depends on fewer empirical parameters.

Because the ionosphere–thermosphere time scales are significantly slower than the magnetospheric time scales the computational coupling between the models can be relatively loose. Typical numerical timesteps for the MHD model are about 0.2 to 0.5 seconds, whereas a typical CTIM

timestep takes about 60 seconds. Thus CTIM is only invoked every 60 seconds, and for the MHD timesteps between CTIM invocations the parameters Σ and $j_{\parallel,d}$ are held constant in the magnetosphere model. Moreover, CTIM runs on a separate computational node asynchronously from the MHD calculations, i.e., after CTIM receives parameters from the magnetosphere model, it returns immediately Σ and $j_{\parallel,d}$ from its last step. This makes the calculations much more efficient because the magnetosphere model needs not to wait for CTIM to finish. Thus the CTIM calculation lags the magnetosphere calculations by 60 seconds. However, this time is at least partially compensated for by the instantaneous mapping of quantities between the MHD boundary and the ionosphere, which in reality takes a finite amount of time.

5. MODEL COMPARISONS FOR THE JANUARY 10/11, 1997 GEOMAGNETIC STORM

The magnetic storm of January 10/11, 1997 has been discussed in substantial detail in the literature [see for example: *Fox et al.*, 1998; *Goodrich et al.*, 1998; *Spann et al.*, 1998], thus we refer the reader to these papers regarding the specifics of this event. Briefly, a magnetic cloud encountered Earth's magnetosphere at 0441 UT on 1/10/1997, following a storm sudden commencement at 0050 UT on the same day. The IMF B_z associated with the cloud turned initially southward and stayed southward for the next 12 hours. Associated with the southward IMF was strong geomagnetic activity. The Canopus electrojet CL index [*Rostoker et al.*, 1995; *Goodrich et al.*, 1998] reached values of -1800 nT, and the Polar UVI imager recorded strong auroral emissions throughout this period [*Spann et al.*, 1998]. *Lu et al.* [1999] presented a study of this storm in which they used AMIE [*Richmond and Kamide*, 1988] and DMSP data to estimate ionospheric power inputs and the cross polar cap potential. In the following we use results from this study to compare three different simulation runs with data and with each other.

In run 1 we used the simplest ionosphere model possible, namely a flat ionospheric conductance of 5 S. With a flat conductance model we do not expect significant perturbations of the ground magnetic field. As shown by *Fukushima* [1969, 1975, see also *Kamide et al.* 1981] the ground magnetic effects of the field-aligned and the poloidal ionospheric currents cancel. Because there is no toroidal ionospheric current for a flat conductance model, only minor ground perturbations are expected. These arise from the deviations of our model from the idealizations in Fukushima's theorem, namely a radial magnetic field and an infinite ionosphere.

Results from this run are shown in Figure 2. Here, we focus on the first 10 hours of the cloud event during which

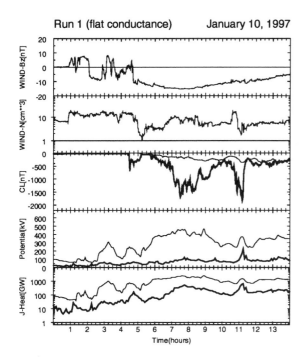

Figure 2. Simulation results for run 1 (flat ionospheric conductance) for January 10, 1997. From top to bottom: IMF B_z; solar wind number density; Canopus CL index; cross polar cap potential; and Joule heating rate. In the lower 3 panels thin lines are for the model results, thick lines are for the data.

most of the auroral activity occurred. The figure shows, from top to bottom, the IMF B_z and the solar wind number density measured by Wind for reference (not time shifted), the Canopus CL index, the cross-polar-cap potential in the northern hemisphere, and the Joule heating rate in the northern hemisphere. Thick lines show the data (courtesy of G. Lu, NCAR/HAO), and thin lines show the model results. As expected, there is virtually no geomagnetic activity in the model, i.e., the modeled CL index never drops below -300 nT, despite the fact that the Canopus values reach -1800 nT. The modeled polar cap potential is very high compared to the AMIE result and consequently the Joule heating rate is almost an order of magnitude larger than the AMIE estimate. In view of earlier results [*Raeder et al.*, 1998] this is also expected and is discussed in more detail below.

Figure 3 shows the results from run 2, where we used a parameterized conductance model [see *Raeder et al.*, 1998, 2000, for details] in lieu of CTIM. The modeled CL index is considerably improved over the case with flat conductance and shows activity levels that are comparable to the observed ones. However, the model also shows several activations that are not real. The potential values are at times somewhat lower compared to run 1, but at other times even higher and reach occasionally 600 kV, which is several times

the AMIE estimate. Similarly, the model overestimates the Joule heating rate by about one order of magnitude most of the time.

Results from the fully coupled model are shown in Figure 4. It is immediately evident that the coupled model reproduces the electrojet activity (as measured by CL) more faithfully than the simpler models. The model potential is still considerably higher than the AMIE estimate, however, the differences in the Joule heating rate are less for this model than for the other two models. In the fourth panel of Figure 4 we also show the integrated electron precipitation energy flux over the northern hemisphere and compare it with

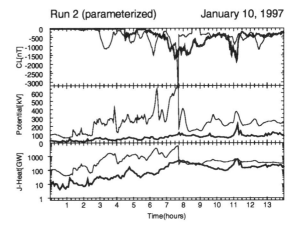

Figure 3. Like Figure 2, but for run 2.

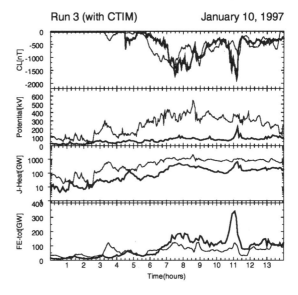

Figure 4. Like Figure 3, but for run 3. In addition, the fourth panel shows the energy flux of precipitating electrons in the northern hemisphere.

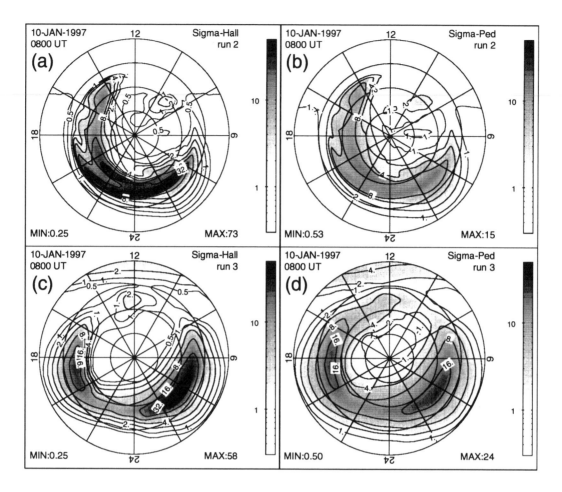

Figure 5. Conductance patterns in the northern polar cap between 50° magnetic latitude and the geomagnetic pole at 0800 UT on January 10, 1997: a) Hall conductance for run 2, b) Pedersen conductance for run 2, c) Hall conductance for run 3, d) Pedersen conductance for run 3. Conductance values are given in Siemens and contour-lines are spaced by a factor of 2.

earlier experimental estimates [*Lu et al.*, 1998]. The energy input produced by the model is slightly less than the data, but generally agrees fairly well with the observations, except during the activation around 1100 UT.

6. DISCUSSION AND SUMMARY

Clearly, the coupled model produces a significantly better response than either run 1 or run 2. The main reason appears to be that the self-consistently and from first principles computed ionospheric conductance distribution is far more realistic than either a flat conductance model or simple parameterization. The flat conductance model eliminates any asymmetries and the effects of Hall conductance (uniform Hall conductance falls out of the potential equation), and thus is not expected to yield a realistic response to magnetospheric input.

Figure 5 shows the comparison of the Pedersen (left column) and Hall (right column) conductance from run 2 (parameterized conductances, top row) and from run 3 (CTIM conductances, bottom row). These snapshots were taken at 0800 UT, that is during a period of strong geomagnetic activity. The differences between the two conductance models are significant. Generally, the parameterized model produces a larger Hall conductance but a smaller Pedersen conductance. The empirical model also skews the conductance pattern towards the evening sector and produces very low conductance values in the morning sector, that is between 0600 and 1200 magnetic local time (MLT). Furthermore, the empirical model has very large north-south gradients, in particular in the nightside and in the Hall conductance. CTIM, on the other hand produces much weaker gradients, and in that model moderately large (a few Siemens) conductance values are also seen at lower latitudes (below 60°) in the nightside.

The large nightside conductance gradient and the morningside conductance gap appear to be the main reason for the excessive potential values and their erratic jumps in run 2. The strong gradient must be associated with a strong field-aligned current, a strong ionospheric electric field, or both. If the magnetosphere does not supply an adequate field-aligned current, a strong electric field across the gradient inevitably develops. The magnetosphere responds to that electric field and eventually sets up a field-aligned current that reduces this electric field (note that, if the magnetospheric response were of the opposite sense an instability would develop). However, the magnetospheric response is not necessarily a perfect match nor does it occur instantaneously. Thus, the system switches erratically between different states. Although such processes may indeed occur in nature – evident by the rapidly time-varying aurora – the resolution of our model is by far not sufficient to capture these processes at their correct scales. On the other hand, the CTIM conductance is much smoother and apparently much closer to reality than the empirical model. Although it still produces a far too large cross-polar-cap potential the ground perturbations match exceptionally well with the observations. This suggests that the Hall currents in the model are roughly correct but that the Pedersen currents deviate from reality to a larger extent. Unfortunately, neither the ionospheric currents, nor the ionospheric conductances can be measured directly, and thus we have to rely on indirect constraints, like ground magnetic perturbations, to assess and improve the model.

Despite the improvements brought by CTIM the high cross-polar-cap potential in the model remains a critical issue. Such discrepancy has been found in earlier studies [*Raeder et al.*, 1998] and is also evident in other models [*Fedder et al.*, 1998; *Hill and Toffoletto*, 1998; *Peroomian et al.*, 1998]. Possible causes are discussed in [*Raeder et al.*, 1998] and include reconnection rates in the MHD model that are too high, and the lack of adequate region-2 currents in the model. Based on the results presented here we can now almost certainly rule out inaccuracies of the ionospheric conductance. More detailed comparisons with data, for example with directly measured FACs, are required to find the cause of this discrepancy.

In summary, we have presented the first global, self-consistent, fully electrically coupled, magnetosphere–ionosphere–thermosphere model. Initial results from this model for the January 10, 1997 geomagnetic storm are encouraging, and yield in particular a much more realistic ionospheric response as compared with the previous magnetosphere model that relied on parameterizations for the ionospheric conductance.

Acknowledgments.
We thank Gang Lu (NCAR/HAO) for providing the AMIE results for this event. The research at UCLA was supported by NSF grants ATM-97-13449 and ATM-98-01937. YW was supported by LLNL grant ICSR-99-008. Computations were performed on the IBM-SP2 of the San Diego Supercomputer Center and the SGI-Origin2000 at the National Center for Supercomputer Applications. IGPP publication 5477. The Editor would like to thank the reviewers of this manuscript.

REFERENCES

Evans, C. R., and J. F. Hawley, Simulation of magnetohydrodynamic flows: A constrained transport method, *Astrophys. J.*, *332*, 659, 1988.
Fedder, J. A., S. P. Slinker, and J. G. Lyon, A comparison of global numerical simulation results to data for the January 27-28, 1992, geospace environment modeling challenge event, *J. Geophys. Res.*, *103*, 14,799, 1998.
Fox, G. C., M. A. Johnson, G. A. Lyzenga, S. W. Otto, J. K. Salmon, and D. W. Walker, *Solving Problems on Concurrent Processors*, Prentice Hall, Englewood Cliffs, N.J., 1988.
Fox, N. J., M. Peredo, and B. J. Thompson, Cradle to grave tracking of the January 6-11, 1997 Sun-Earth connection event, *Geophys. Res. Lett.*, *25*, 2461, 1998.
Fukushima, N., Equivalence in ground magnetic effect of Chapman-Vestine's and Birkeland-Alfven's current systems for polar magnetic storms, *Rep. Ionos. Space Res. Jpn.*, *22*, 219, 1969.
Fukushima, N., Generalized theorem for no ground magnetic effect of vertical currents connected with Pedersen currents in the uniform conducting ionosphere, *Rep. Ionos. Space Res. Jpn.*, *30*, 35, 1976.
Fuller-Rowell, T. J., D. Rees, S. Quegan, R. J. Moffett, M. V. Codrescu, and G. H. Millward, A coupled thermosphere-ionosphere model (CTIM), in *STEP Report*, edited by R. W. Schunk, p. 217, Scientific Committee on Solar Terrestrial Physics (SCOSTEP), NOAA/NGDC, Boulder, Colorado, 1996.
Gombosi, T. I., D. L. DeZeeuw, C. P. T. Groth, K. G. Powell, and P. Song, The length of the magnetotail for northward IMF: Results of 3D MHD simulations, in *Phys. Space Plasmas (1998)*, edited by T. Chang and J. R. Jasperse, vol. 15, p. 121, Cambridge, Mass., 1998.
Goodrich, C. C., J. G. Lyon, M. Wiltberger, R. E. Lopez, and K. Papadopoulos, An overview of the impact of the January 10-11, 1997, magnetic cloud on the magnetosphere via global MHD simulation, *Geophys. Res. Lett.*, *25*, 2537, 1998.
Harten, A., and G. Zwas, Self-adjusting hybrid schemes for shock computations, *J. Comput. Phys.*, *9*, 568, 1972.
Hill, T. W., and F. R. Toffoletto, Comparison of empirical and theoretical polar cap convection patterns for the January 1992 GEM interval, *J. Geophys. Res.*, *103*, 14,811, 1998.
Hirsch, C., *Numerical Computation of Internal and External Flow*, vol. II, John Wiley, New York, 1990.
Janhunen, P., T. I. Pulkkinen, and K. Kauristie, Auroral fading in ionosphere-magnetosphere coupling model: Implications for possible mechanisms, *Geophys. Res. Lett.*, *22*, 2049, 1995.
Kamide, Y., A. D. Richmond, and S. Matsushita, Estimation of ionospheric electric fields, ionospheric currents, and field-aligned currents from ground magnetic records, *J. Geophys. Res.*, *86*, 801, 1981.
Kelley, M. C., *The Earth's Ionosphere*, Academic Press, New York, 1989.

Kennel, C. F., and H. E. Petschek, Limit on stably trapped particle fluxes, *J. Geophys. Res.*, *71*, 1, 1966.

Lu, G., et al., Global energy deposition during the January 1997 magnetic cloud event, *J. Geophys. Res.*, *103*, 11,685, 1998.

Lyon, J. G., R. E. Lopez, C. C. Goodrich, M. Wiltberger, and K. Papadopoulos, Simulation of the March 9, 1995, substorm: Auroral brightening and the onset of lobe reconnection, *Geophys. Res. Lett.*, *25*, 3039, 1998.

Lyons, L. R., D. Evans, and R. Lundin, An observed relation between magnetic field aligned electric fields and downward electron energy fluxes in the vicinity of auroral forms, *J. Geophys. Res.*, *84*, 457, 1979.

Ogino, T., A three dimensional MHD simulation of the interaction of the solar wind with the Earth's magnetosphere: The generation of field aligned currents, *J. Geophys. Res.*, *91*, 6791, 1986.

Peroomian, V., L. R. Lyons, M. Shultz, and D. C. Pidmore-Brown, Comparison of assimilative mapping and source surface model results for magnetospheric events of January 27 to 28, 1992, *J. Geophys. Res.*, *103*, 14,819, 1998.

Raeder, J., Modeling the magnetosphere for northward interplanetary magnetic field: Effects of electrical resistivity, *J. Geophys. Res.*, *104*, 17,357, 1999.

Raeder, J., J. Berchem, and M. Ashour-Abdalla, The Geospace Environment Modeling grand challenge: Results from a Global Geospace Circulation Model, *J. Geophys. Res.*, *103*, 14,787, 1998.

Raeder, J., et al., Global simulation of the geospace environment modeling substorm challenge event, *J. Geophys. Res.*, *in press*, 2000.

Richmond, A. D., and Y. Kamide, Mapping electrodynamic features of the high latitude ionosphere from localized observations, *J. Geophys. Res.*, *93*, 5741, 1988.

Richmond, A. D., and R. G. Roble, Electrodynamic effects of thermospheric winds for the NCAR thermospheric general circulation model, *J. Geophys. Res.*, *92*, 12,365, 1987.

Robinson, R. M., R. R. Vondrak, K. Miller, T. Dabbs, and D. Hardy, On calculating ionospheric conductances from the flux and energy of precipitating electrons, *J. Geophys. Res.*, *92*, 2565, 1987.

Roble, R. G., and E. C. Ridley, A Thermosphere - Ionosphere - Mesosphere - Electrodynamics General Circulation Model (TIME-GCM): Equinox solar cycle minimum simulations (30-500 km), *Geophys. Res. Lett.*, *21*, 417, 1994.

Rostoker, G., J. C. Samson, F. Creutzberg, T. J. Hughes, D. R. McDiarmid, A. G. McNamara, A. Vallace-Jones, D. D. Wallis, and L. L. Cogger, CANOPUS - A ground based instrument array for remote sensing the high latitude ionosphere during the ISTP/GGS program, *Space Sci. Rev.*, *71*, 743, 1995.

Slinker, S. P., J. A. Fedder, J. Chen, and J. G. Lyon, Global MHD simulation of the magnetosphere and ionosphere for 1930-2330 ut on November 3, 1993, *J. Geophys. Res.*, *103*, 26,243, 1998.

Spann, J. F., M. Brittnacher, R. Elsen, G. A. Germany, and G. K. Parks, Initial response and complex polar cap structures of the aurora in response to the January 10, 1997 magnetic cloud, *Geophys. Res. Lett.*, *25*, 2577, 1998.

Tanaka, T., Generation mechanisms for magnetosphere-ionosphere current systems deduced from a three-dimensional MHD simulation of the solar wind-magnetosphere-ionosphere coupling processes, *J. Geophys. Res.*, *100*, 12,057, 1995.

Weimer, D. R., D. A. Gurnett, C. K. Goertz, J. D. Menietti, J. L. Burch, and M. Sugiura, The current - voltage relationship in auroral current sheets, *J. Geophys. Res.*, *92*, 187, 1987.

Winglee, R. M., and J. D. Menietti, Auroral activity associated with pressure pulses and substorms: A comparison between global fluid modeling and Viking UV imaging, *J. Geophys. Res.*, *103*, 9189, 1998.

Forecasting ionospheric electric fields: An interplanetary coupling perspective

Nelson C. Maynard

Mission Research Corporation, Nashua, New Hampshire

William J. Burke

Air Force Research Laboratory, Space Vehicles Directorate, Hanscom AFB, Massachusetts

Understanding of the temporal and spatial development of interplanetary-driven ionospheric convection is evolving rapidly. Detailed comparisons of the IMF measurements by Wind with ionospheric electric field observations by sounding rockets and the Polar spacecraft conclusively demonstrate that signals can interact with the magnetosphere on significantly longer or shorter than nominal advection lag time scales. The antiparallel merging hypothesis of Crooker provides a unifying perspective for interpreting measurements. The timing and location of interactions depend on IMF B_X in such a way that merging can occur simultaneously in the northern and southern hemispheres with interplanetary features that are significantly separated in time on a single streamline. The measurements also show that small-scale variations of the interplanetary electric field couple to the magnetosphere and ionosphere more directly than has been previously reported. These variations are plausible sources of commonly seen temporal and spatial deviations from statistical dayside convection patterns. To quantitatively apply this new understanding to improve space weather forecasting requires multiple satellites in the solar wind.

1. INTRODUCTION

Electric fields produce significant space weather effects on the propagation of transionospheric electromagnetic signals and, through Joule heating, on satellite drag. Predictions of these effects require a thorough grasp of electrical interactions between the interplanetary medium and the Earth's magnetosphere-ionosphere (M-I) system. In some average sense the interplanetary magnetic field (IMF) components in the

$Y - Z$ plane affect the sizes, shapes and intensities of ionospheric convection (potential) patterns [*Heppner and Maynard*, 1987. *Weimer* [2001] has developed detailed statistical models for high-latitude potential patterns that use time-averaged values of IMF clock angles and magnitudes, as well as the solar wind densities and velocities as input parameters. Data from monitors upstream in the solar wind allow these models to make one-half- to one-hour predictions of global convection patterns. Experience teaches that significant variability in the intensities and distributions of convective electric fields cannot be accounted for by the statistical parameters. The purpose of this paper is to summarize results from two recent sounding rocket experiments that suggest (1) the dayside ionosphere responds on a much finer scale to interplanetary variations than previously

Space Weather
Geophysical Monograph 125

thought, and (2) the timing of near-cusp auroral events depends on all three IMF components [*Maynard et al.*, 2000a]. To help understand the significance of our new results we begin with a brief overview of electric field driven plasma flows in the high-latitude M-I system.

Cowley and Lockwood [1992] developed a conceptually simple picture of how merging between the IMF and earth-bound magnetic field lines adjacent to the dayside cusps drive ionospheric plasma convection. Magnetic tension forces associated with the polarity of IMF B_Y initially cause plasma to move toward the east or west across the cusp before turning into the polar cap [*Heppner and Maynard*, 1987]. Flows continue as newly opened flux tubes are dragged tailward by the solar wind as it propagates a few tens of R_E beyond the Earth. The addition of new open flux to the polar caps expands the adiaroic portion of the open-closed boundaries, thereby communicating the flow to the whole pattern. As used by *Siscoe and Huang* [1985], the term "adiaroic" describes a boundary across which there is no flux transfer. After 5 to 10 min, the "new" open flux becomes "old." While it continues to move tailward as more open flux is added through the cusp, its influence on the flow pattern diminishes. Reconnection of open flux in the magnetotail produces analogous affects on nightside convection patterns that also depend on the IMF orientation, but with a further 30 to 60 min delay.

Originally magnetic merging was assumed to occur along the subsolar magnetopause between antiparallel components of the IMF and the Earth's dipole field [e. g. *Sonnerup*, 1974]. *Crooker* [1979] suggested that many observations were easier to explain if merging were confined to regions on the dayside magnetopause where internal and external magnetic field vectors were completely antiparallel. A consequence of the antiparallel merging hypothesis is that in cases where the IMF has a significant B_Y component the merging line splits into two segments located near the two dayside cusps [*Crooker*, 1985]. For IMF $B_Y > 0$ the merging-line segments are on the evening side of the northern hemisphere cusp and the morning side of the southern hemisphere cusp. The opposite spatial relations between the cusps and the merging-line segments maintain when IMF $B_Y < 0$. While questions relating to the dominance of antiparallel or component merging are not yet resolved, recent sounding rocket results support the antiparallel hypothesis [*Maynard et al.*, 2000a, 2000b].

Two independent sets of observations bear on the problem. *Ridley et al.*, [1997, 1998] investigated responses to changes in the IMF by examining differences from base convection patterns previously esablished us-

ing the AMIE technique [*Richmond and Kamide*, 1988]. They found that difference patterns acquired characteristic shapes that depended on the orientation of the new IMF conditions and appeared over most of the polar cap region simultaneously. The difference-pattern magnitude increased linearly with time, maintaining both its position and shape. The communication lag between the time when changed interplanetary conditions first contacted the magnetosphere and difference patterns were established in the ionosphere was 8.4 (± 8.2) min. The large variation indicates that interaction processes are still not well defined. The average time for reconfiguration after the change started was 12 min. The second set of response time measurements utilized Super-DARN radar capabilities. *Ruohoniemi and Greenwald* [1998] and *Shepherd et al.*, [1999] showed that changes in convection patterns appeared nearly simultaneously at all local times after abrupt switches in the IMF direction. *Shepherd et al.*, [1999] suggested that draping of the IMF around the magnetopause delays the onset of interactions until they can happen simultaneously along the whole merging line. Their experimental results were viewed as being inconsistent with models requiring single-point reconnection that spreads antisunward at a few km/s.

2. FOUNDATIONS FOR A NEW DIRECTION

On December 2 and 3, 1997, two sounding rockets were launched to the geographic west from the SvalRak range at Ny-Ålesund (78.92° N, 11.95° E). The experimental payloads of both rockets included energetic particle spectrometers to measure the energy distributions of downcoming electrons and ions, double-probe antennas to measure electric field vectors and fluxgate magnetometers to detect signatures of field-aligned currents and propagating magnetohydrodynamic waves [*Maynard et al.*, 2000a]. Supporting ground-based measurements of auroral luminosities at 557.7 and 630.0 nm were made by meridian scanning photometers and all-sky imagers. Measurements from the SuperDARN radar in Finland and from the ion drift meter on a Defense Meteorological Satellite Program (DMSP) satellite provided useful synoptic views of the prevailing large-scale convection patterns. Both launches occurred near magnetic noon while B_X was the dominant IMF component. On December 2, IMF $B_Y < 0$ and B_Z was mostly northward. At the time of the December 3 flight the polarities of all three components were reversed but their magnitudes were comparable to those of the first flight. The antiparallel merging hypothesis required that on December 2 the evening/morning convection cell would be driven by merging in the southern/northern hemisphere. The converse relationships prevailed at the time of the December 3 flight.

Most previous research emphasized the roles of the Y and Z components of the IMF in determining the magnitude and timing of the interactions. However, IMF B_X also can be important. *Maynard et al.*, [2000a] compared data from the first sounding rocket from Svalbard with IMF and solar wind velocity measurements from the Wind and IMP 8 spacecraft and with ground-based optical and radar measurements to help distinguish spatial and temporal variations. The rocket's westward trajectory carried it toward auroral forms associated with morningside boundary layers. The rich set of vector dc electric and magnetic fields and energetic particle fluxes gathered by the rocket revealed a complex electrodynamic picture of the cusp/boundary layer region. Four factors were important in separating temporal and spatial effects: (1) near the December solstice the Earth's north magnetic pole tilts away from the Sun; (2) at the UT of launch the dipole axis was rotated toward dawn; (3) B_X was the dominant IMF component; and (4) the variability of interplanetary driving was low. The first three factors affect how the dayside magnetopause presents itself to the oncoming solar wind and thus the locations of the merging sites. The fact that the interplanetary variability during the flight was low allowed us to use small but distinctive changes in the IMF as timing markers for electric field variability in the dayside ionosphere. No signatures of dayside merging at a northern hemisphere site were detected by either the rocket or ground sensors. From an interpretive point of view, the key observations were of electric field variations in the interplanetary medium that correlated directly with those observed by the sounding rocket. However, the rocket detected the signals about 10 min earlier than expected with lag times estimated for simple advection between Wind and the Earth.

The effective interplanetary electric field (IEF) was calculated using the formula $E_{KL} = V_X B_{YZ} \sin^2(\theta/2)$ where θ is the IMF clock angle [*Kan and Lee*, 1979]. This formula was initially derived by *Sonnerup* [1974] for the maximum rate of component merging in the subsolar region. The correlated rocket and Wind measurements required that planes of constant IEF phase be tilted toward the Earth, and that first contact with the magnetopause occurred in the southern hemisphere. Using correlations of electric field measurements at the rocket with those at the Wind and IMP-8 satellites *Maynard et al.*, [2000a] estimated the polar and azimuthal tilt angles of IEF phase fronts as they propagated toward the Earth in the solar wind. Figure 1

schematically shows the deduced merging site in the southern hemisphere and the two tilt angles of the constant-phase planes. The interaction at the measured correlation time was only possible with merging in the southern hemisphere. Consequently, the observed northern hemisphere convection pattern was stirred in part by merging of the IMF with closed field lines near the poleward edge of the southern hemisphere cusp thereby adding open flux to the northern polar cap. Subsequent motions of adiaroic polar cap boundaries were detected in the rocket electric field measurements, [*Siscoe and Huang*, 1985]. The observations indicate that IMF B_X significantly affected the location and timing of merging interactions. Implicit in our placing of the merging site is the antiparallel merging hypothesis [*Crooker*, 1979].

On the following day the second rocket flight was launched under southward IMF conditions. As mentioned above, the rocket was launched to the west while Svalbard was near magnetic noon, and the signs of all three IMF components were reversed from their values on December 2. Background electric fields measured by

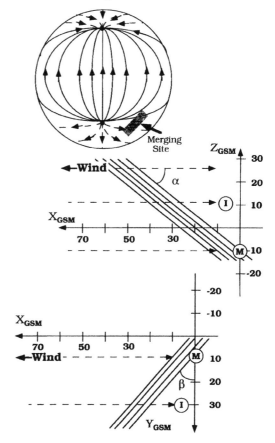

Figure 1. Schematic representation of merging site and depiction of phase plane tilts from *Maynard et al.*, [2000a].

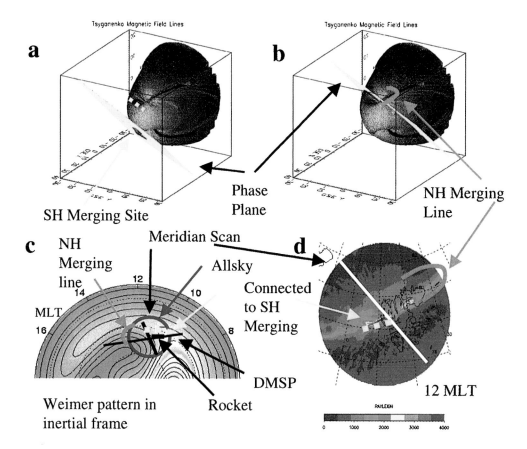

Plate 1. Cartoon illustrating the various merging sites in the magnetosphere and ionosphere and their relationship to the aurora from *Maynard et al.*, [2000b]. The top two panels shows a 3-D representation of the magnetopause. The yellow phase plane is shown abutting the Southern-Hemisphere-merging site in the left panel and encountering the northern-hemisphere cusp region in the right panel. Newly opened field lines in the Southern Hemisphere would pass by the dotted yellow region in the Northern Hemisphere in the top left. The orange line in the top right schematically delineates possible Northern Hemisphere merging sites, applying the antiparallel criterion. The bottom two panels translate the merging sites to the ionospheric configuration of convection and aurora.

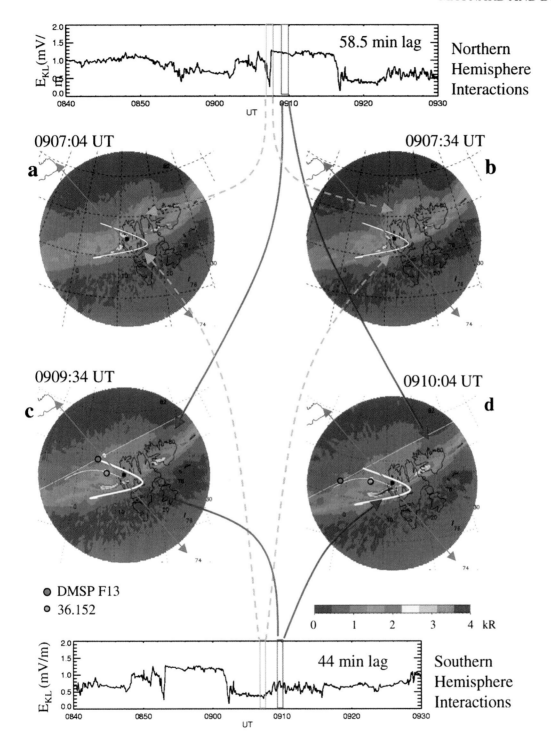

Plate 2. All-sky images of 630.0 emissions over Ny-Ålesund acquired at (a) 0907:04, (b) 0907:34 UT, (c) 0909:34, and (d) 0910:04 UT projected to 200 km in altitude. E_{KL} for the two different lags determined to be the best for the northern (southern) hemisphere correlations are shown in the top (bottom) data traces, with the image times keyed by the arrows to the respective images. The open curve in each image approximates the region controlled by southern hemisphere merging. Auroral enhancements in (to the right of) this region are in response to increases in E_{KL} in the bottom (top) trace. For reference, the DMSP F13 orbit and the rocket trajectory mapped to 200 km altitude have been superposed on the image (adapted from *Maynard et al.*, [2000b]).

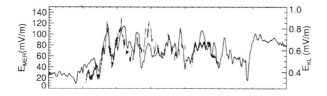

Figure 2. Electric field data from the second rocket flight, launched at 0906 UT on December 3, 1997, compared with E_{KL} measured by Wind. From *Maynard et al.,* [2000b].

the rocket and a DMSP satellite indicate that the rocket entered the prenoon convection cell. Again a detailed correlation was found between electric field variations detected at different times by the rocket, Wind, and IMP 8. Figure 2 shows the correlation between E_{KL} and the meridional component of the electric field detected at the location of the rocket. The correlation coefficient between the two data sets was 0.79. Again the lag time for signal detection in the ionosphere was much less than that required for normal advection between Wind and the Earth [*Maynard et al.*, 2000b]. The combined rocket, Wind and IMP 8 data also required that the constant IEF phase fronts be similarly tilted and that first contact occured at the magnetopause on the morningside of the southern hemisphere cusp.

Auroral acivity in the vicinity of the rocket occurred in the prenoon convection cell and had to be driven at a southern hemisphere merging line. *Maynard et al.,* [2000b] used this inference to distinguish afternoon and morning convection cell portions of all-sky images that were driven by processes in the northern and southern hemispheres, respectively. This allowed a separation of spatial and temporal effects. The key to understanding the data was applying the antiparallel merging hypothesis of *Crooker* [1979].

Plate 1 shows a cartoon from *Maynard et al.,* [2000b]. The top two plots show how a tilted phase front first impacts the magnetosphere on the dawn side of the southern cusp, and later interacts near the northern cusp. The magnetosphere shape was determined with the Tsyganenko-96 magnetic field model *Tsyganenko and Stern,* 1996]. The subsolar magnetopause coincides with the last closed field line surface. Open field lines behind the cusp provide the full outline of the magnetopause. If component merging had been dominant, interactions would have occurred near the equator and affected both hemispheres at about the same time. If antiparallel merging dominated, the northern hemisphere should respond much later than the southern hemisphere. Thus, it is possible to have the two hemispheres reacting to different segments of the solar wind data stream associated with different lag times. This was in fact the case. Newly opened field lines at the

southern hemisphere merging point drape over the edge of the northern hemisphere in the region of the yellow dots. The orange line in the top right plot represents possible antiparallel sites for positive B_Y and smaller B_Z of both polarities. The bottom left plot of Plate 1 shows the approximate field of view of the all-sky image overlaid onto a *Weimer* [2001] convection pattern displayed in inertial coordinates [*Maynard et al.*, 1995]. The approximate ionospheric footprints of the merging sites are represented by the yellow dots and the orange line. The westward rocket trajectory is indicated by a heavy black line. The bottom right plot overlays the same possible merging sites on an all-sky image. The white line shows the direction of the meridian scanning photometer measurements taken at the same time.

Plate 2c-d shows two all-sky images acquired during the time of the second rocket flight when the correlation with Wind IEF data was established. The approximate area of emissions originating from merging in the southern hemisphere is represented by an open-ended hook. The size and position of the hook were approximated from the locations of bright auroral emissions observed during the rocket flight. The cusp is clearly bifurcated in the left image. This same break point is evident at other times, depending on the relative activity occurring to the east of the Ny-Ålesund observatory. During this interval the intensities of both E_{KL} and auroral emissions within the hook increased. The rocket trajectory, represented by a light line on the figure, passed just poleward of the most intense emissions. Auroral emissions from east of the hook originate from northern hemisphere merging events. The traces at the bottom (top) display E_{KL} properly lagged for merging in the southern (northern) hemisphere. The difference in time between interactions with the same E_{KL} features in the two hemispheres was ~14.5 min.

The lag time for northern hemisphere interactions in the afternoon convection cell was determined using auroral signatures as fiduciaries that are characteristic of three brief intervals when IMF B_Z turned northward or approached zero. All-sky images acquired during one of the northward IMF intervals are shown in Plate 2a-b. Auroral forms characteristic of northward IMF [*Sandholt et al.*, 1998] were observed at the poleward edge of the cusp region (noted by the top orange dashed arrows). They appeared at that location, rather than originating to the east and propagating westward across the cusp, clearly identifying their origin. Attention is also directed to the markedly lower level of auroral emissions from the region controlled by southern hemisphere merging (inside of the hook). This is consistent with the low levels of E_{KL} at the times keyed by the two bottom dashed arrows.

Returning to Plate 2c-d, we note that much weaker auroral emissions emanated from the region at the end of the top set of red arrows, referring to northern hemisphere merging. Also there is a general lack of auroral emissions to the east, even though E_{KL} in the top trace is large. The antiparallel criterion explains this lack of aurora to the east. At the appropriate interaction time IMF B_X had decreased, tilting the IMF more toward the vertical. Since the phase front had already passed the subsolar magnetopause, there were no possible antiparallel sites for the new orientation. Thus, merging and the auroral emissions turned off even though the IEF intensity was high. Using the antiparallel hypothesis *Maynard et al.*, [2000b] harmonized 40 min of all-sky and meridian scanning photometer data with the same time lags.

3. CONCLUSIONS

In summary, the data presented here have opened new avenues for understanding interplanetary coupling to the M-I system. They show that interaction timing depends strongly on IMF B_X. This probabaly accounts for some of the variability in the initial response times reported by *Ridley et al.*, [1998]. The orientation and propagation of the phase front is a three-dimensional problem that cannot ignore B_X. Applying the antiparallel merging hypothesis harmonizes extremely complex data sets. Small details in the IEF have counterparts in the ionospheric electric field, confirming that the ionosphere responds directly to interplanetary driving. Northern hemisphere convection is in part driven at merging sites on the magnetopause near the northern and southern hemisphere cusps. As a result, it becomes possible to separate spatial from temporal variability derived from two hemispheric sources, which respond to different segments of the solar wind data stream. We believe that resultant source-bifurcation of the cusp is a common feature that arises in direct consequence to antiparallel merging.

The interactions differ according to the presentation of the magnetopause to the solar wind and the IMF. The rocket measurements, especially with northward IMF, support the concept of expanding/contracting adiaroic boundaries for the polar cap. In the vicinity of the cusp, under B_Y dominant conditions with southward IMF, coupling to the solar wind results in direct tugging of the field lines as open flux is carried tailward. Even though flux tubes opened in the north and in the south are tugged in the same direction, there is a natural break between the two. Inertial coordinates logically separate the two cells. It thus follows that the small convection cell is driven from the opposite hemisphere. How long

the IMF directly couples to the ionosphere through the lobes remains to be determined. *Cowley and Lockwood* [1992] postulate that after ∼10 min the newly opened flux no longer acts as a driver of convection patterns. *Farrugia et al.*, [2000] also found that small variations in the IMF coupled to particle precipitation, field-aligned currents, and ionospheric currents in the closed field line region that maps to a sunward flowing mixing layer near the flanks of the magnetosphere. In this case, however, the lag was longer than the normal advection time, indicating that the different merging locations and physical processes affect coupling.

4. FORECASTING APPLICATIONS

Since the M-I system responds in detail to interplanetary drivers, a model could in principle be constructed for forecasting if we could determine how the IEF engages the magnetopause under a wider variety of conditions. This is a long-term goal toward which we have only taken a few steps. As a whole the ionosphere serves as an integrator that feeds back actively into the magnetosphere. How coupling modifies the driving must be understood before detailed modelling becomes practical.

In the near term, predictions of convection patterns using the statistical models driven by data from an upstream monitor could logically be improved by treating the propagation as a 3-D problem. The influence of B_X on the timing and driving rate can be factored in. The distance that the upstream monitor is away from the Sun-Earth line in both Y and Z, as well as the tilt of the phase front, must be considered when determining the timing of the interaction at the magnetopause. We also need to understand the differences between large changes in the IMF and shocks versus the small-scale variations discussed above.

There is growing evidence that through optical techniques we will be able to predict large geomagnetic events with a day or more warning [*Fox et al.*, 1998]. However, for detecting the relatively small interplanetary variations needed to predict substorm occurrences, we will be limited to the shorter time scales required for information to propagate from monitors near the first libration point (L_1) to the Earth. The coherence of interplanetary parameters between the halo orbit of ISEE 3 to the Earth ranged from good to poor [*Russell et al.*, 1980]. The correlation between L_1 and Earth improved if the distance between the Sun-Earth line and the observing spacecraft ($d_\perp$) was less than a few tens of R_E [*Crooker et al.*, 1982; *Lyons et al.*, 1997] and if interplanetary variability was relatively high. Working within the $d_\perp$ limitation, our finding that surfaces of constant IEF phase are tilted with respect to the Sun-Earth line appears to introduce an insuperable error of $\sim \pm 10$ min for predicting when an interplanetary structure will reach the magnetopause. With a single satellite near L_1, it is impossible to determine the tilt angles. However, since three points determines a plane, with two near-Earth satellites such as IMP 8 and Geotail in the solar wind, the two tilt angles can be specified using standard correlation techniques [*Russell et al.*, 1980]. We are encouraged by the examples presented by *Russell et al.*, [1980] indicating that once established, high correlations (constant tilt angles?) lasted for hours. In fact, *Maynard et al.*, [2000b] were able to show agreement between IEF and dayside auroral variations by assuming that the tilt angles remained constant during the 50 min period shown in Plate 2.

Acknowledgments.
We are grateful to our many coinvestigators, the NASA Wallops support team, the Norwegian SvalRak range crew, and many others whose help made the rocket flights possible. They are further enumerated and acknowledged in the *Journal of Geophysical Research* papers. Drs. Dan Ober and Dan Weimer assisted in the preparation of the figures. This work was supported by NASA under contract NAS5-96034, the Air Force Office of Scientific Research under task 2311PL013, and the Air Force Research Laboratory under contract F19628-98-C-0061.
The Editor would like to thank the reviewer of this manuscript.

REFERENCES

Cowley, S. W. H., and M. Lockwood, Excitation and decay of solar wind-driven flows in the magnetosphere-ionosphere system, *Ann. Geophysicae, 10,* 103, 1992.

Crooker, N. U., Dayside merging and cusp geometry, *J. Geophys. Res., 84,* 951, 1979.

Crooker, N. U., Split separator line merging model of the dayside magnetopause, *J. Geophys. Res., 90,* 12,104, 1985.

Crooker, N. U., G. L. Siscoe, C. T. Russell, and E. J. Smith, Factors controlling the degree of correlation between ISEE 1 and ISEE 3 interplanetary magnetic field measurements, *J. Geophys. Res., 87,* 2224, 1982.

Farrugia, C. J., P. E. Sandholt, N. C. Maynard, W. J. Burke, J. D. Scudder, D. M. Ober, J. Moen, and C. T. Russell, Pulsating mid morning auroral arcs, filamentation of a mixing layer in a flank boundary layer, and ULF waves observed during a Polar-Svalbard conjunction, *J. Geophys. Res.,* in press, 2000.

Fox, N. J., M. Peredo, and B. J. Thompson, Cradle to grave tracking of the January 6-11, 1997 Sun-Earth connection event, *Geophys. Res. Lett., 25,* 2461, 1998.

Heppner, J. P., and N. C. Maynard, Empirical high latitude ionospheric electric field models, *J. Geophys. Res., 92,* 4467, 1987.

Kan, J. R., and L. C. Lee, Energy coupling function and solar wind magnetosphere dynamo, *Geophys. Res. Lett., 6,* 577, 1979.

Lyons, L. R., G. T. Blanchard, J. C. Samson, R. P. Lepping, T. Yamamoto, and T. Moretto, Coordinated observations demonstrating external substorm triggering, *J. Geophys. Res., 102,* 27,039, 1997.

Maynard, N. C., W. F. Denig, and W. J. Burke, Mapping ionospheric convection patterns to the magnetosphere, *J. Geophys. Res., 100,* 1713, 1995.

Maynard, N. C., et al., Driving dayside convection with northward IMF: Observations by a sounding rocket launched from Svalbard, *J. Geophys. Res., 105,* 5245, 2000a.

Maynard, N. C., W. J. Burke, R. F. Pfaff, P. E. Sandholt, J. Moen, D. M. Ober, D. R. Weimer, A. Egeland, and M. Lester, Observations of simultaneous effects of merging in both hemispheres, submitted to *J. Geophys. Res.,* 2000b.

Richmond, A. D., and Y. Kamide, Mapping electrodynamic features of the high latitude ionosphere from localized observations: Technique, *J. Geophys. Res., 93,* 5741, 1988.

Ridley, A. J., G. Lu, C. R. Clauer, and V. O. Papitashvili, Ionospheric convection during nonsteady interplanetary magnetic field conditions, *J. Geophys. Res., 102,* 14,563, 1997.

Ridley, A. J., G. Lu, C. R. Clauer, and V. O. Papitashvili, A statistical study of the ionospheric convection response to changing interplanetary magnetic field conditions using the assimilative mapping of ionospheric electrodynamics technique, *J. Geophys. Res., 103,* 4023, 1998.

Ruohoniemi, J. M., and R. A. Greenwald, The response of high-latitude convection to a sudden southward IMF turning, *Geophys. Res. Lett., 25,* 2913, 1998.

Russell, C. T., G. L. Siscoe, and E. J. Smith, Comparison of ISEE-1 and -3 interplanetary magnetic field measurements, *Geophys. Res. Lett., 7,* 381, 1980.

Sandholt, P. E., C. J. Farrugia, J. Moen, and S. W. H. Cowley, Dayside auroral configurations: Responses to southward and northward rotations of the interplanetary magnetic field, *J. Geophys. Res., 103,* 20,279, 1998.

Shepherd, S. G., R. A. Greenwald, and J. M. Ruohoniemi, A possible explanation for rapid, large-scale ionospheric responses to southward turnings of the IMF, *Geophys. Res. Lett., 26,* 3197, 1999.

Siscoe, G. L., and T. S. Huang, Polar cap inflation and deflation, *J. Geophys. Res., 90,* 543, 1985.

Sonnerup, B. U Ö., Magnetopause reconnection rate, *J. Geophys. Res., 79,* 1546, 1974.

Tsyganenko, N.A., and D. P. Stern, Modeling the global magnetic field of the large-scale Birkeland current systems, *J. Geophys. Res., 101,* 27187, 1996.

Weimer, D. R., An improved model of ionospheric electric potentials including substorm perturbations and application to the GEM November 24, 1996 event, *J. Geophys. Res., , 106], 407,* 2001.

N. C. Maynard, Mission Research Corporation, One Tara Blvd.; Ste. 302, Nashua, NH 03054. (nmaynard@mrcnh.com)

W. J. Burke, Air Force Research Laboratory, Space Vehicles Directorate, 29 Randolph Road, Hanscom AFB, MA 01731. (william.burke2@hanscom.af.mil)

Capturing the Storm-Time F-Region Ionospheric Response in an Empirical Model

T.J. Fuller-Rowell, M.V. Codrescu, and E.A. Araujo-Pradere

CIRES, University of Colorado and NOAA, Space Environment Center, Boulder, Colorado

Understanding the response of the thermosphere and ionosphere to geomagnetic disturbances has reached a point where some of the F-region ionospheric characteristics can be captured in an empirical model. The integrated effect of magnetospheric energy injection during a storm disturbs the thermospheric temperature, global circulation, and neutral composition. The composition modifications accumulate over tens of hours, have a particular seasonal/latitude structure, and are very slow to recover. The long-lived reaction of the neutral upper atmosphere produces an ionospheric response that is consistent from storm to storm, and so enables the characteristics to be captured by an empirical model. The goal of this first empirical storm-time model is to establish a correction to the F-region peak density, or critical frequency, as a function of season and latitude for any given a_p time history of a storm. Guided by the knowledge gained from previous data analysis, and from simulations with a physically-based model, observations of the F-region peak density from available sites and from many storms have been sorted by latitude and season, and by the magnitude of the storm. A coherent picture begins to emerge particularly in the summer and equinox mid-latitudes. Several features are still unable to be included in the empirical model, although they are clearly important, and can be simulated in physical models. These include the local-time dependence, and the dynamic response to transient large-scale gravity waves. The latter in particular requires accurate knowledge of the spatial and temporal variation of the geomagnetic sources.

1. INTRODUCTION

The maturity of an area of science can often be reflected in the ability to capture the physics within an empirical model. The knowledge of the physics enables the appropriate choice of sorting parameters. As this maturity grows so does the accuracy with which the empirical models are able to reproduce observations. For the thermosphere, the neutral component of the upper atmosphere, the Mass Spectrometer and Incoherent Scatter (MSIS) empirical model [*Hedin et al.*, 1987] is the undisputed standard for specifying neutral temperature and composition around the globe as a function of season, latitude, longitude, Universal Time, solar and geomagnetic activity. The model has evolved from the early versions in the 1970s to the current version of the 1990s, by including more data and by incorporating new physical understanding of the controlling parameters. The International Reference Ionosphere (IRI) can be regarded as the equivalent standard for the ionosphere [*Bilitza*, 1990]. The global distribution of ionospheric parameters can be specified as a function of a similar set of sorting variables. The one exception is that, currently, the IRI has no geomagnetic activity dependence. The reason is simple; the characteristics of the ionospheric response to storms had not been understood to a level where the development of an empirical model was feasible.

Physical understanding of the ionospheric response to storms has now sufficiently matured to enable the first rudimentary characterization of the storm-time response of the peak F-region ionospheric density to be made as a function of season and latitude. This step has been made possible by a combination of analyses of extensive networks of ionospheric data and detailed simulations using coupled thermosphere ionosphere models.

Space Weather
Geophysical Monograph 125

Geomagnetic storms result when high-speed plasma injected into the solar wind from coronal mass ejections or coronal holes impinges upon Earth's geomagnetic field. If the arriving solar wind plasma has a southward magnetic field, energy is coupled efficiently into Earth's magnetosphere and upper atmosphere. The magnitude of the ensuing geomagnetic storm has come to be defined by the strength of the low latitude magnetic index, D_{st}, which is a measure of the magnetospheric ring current. Although the ring current is not the main driver of the upper atmosphere, the sources for both are related; D_{st} can reflect the level of magnetospheric energy input to the upper atmosphere.

A geomagnetic storm, from the perspective of the upper atmosphere, is a period of intense energy input from the magnetosphere for a period of several hours to days. The manifestations of a storm are well defined and include the following effects. Auroral electron precipitation increases in magnitude and expands to lower latitudes than normal, and these particles heat and ionize the gas and increase the conductivity of the atmosphere. The magnetospheric electric field mapped to the atmosphere intensifies and expands in concert with the aurora, and combines with the increased conductivity to produce large enhancements of Joule heating, which is the dominant atmospheric energy source during a storm. Joule heating can increase from tens of gigawatts during quiet times, to hundreds of gigawatts during severe geomagnetic disturbances. The combined input can dump thousands of Terajoules of energy into the upper atmosphere during the course of a storm, and can raise the temperature of the gas by hundreds of degrees Kelvin. Thermal expansion of the atmosphere raises neutral density and can have significant effects on satellite drag. Ionospheric ions drift in response to the electric field and, by colliding with the atmosphere, can drive winds in excess of 1 km/s at high latitudes. Many interesting science problems center around these high latitudes processes, and the combined energy source is the driver of the global storm-time changes in the upper atmosphere.

The energy injection drives global thermospheric storm winds and composition changes, and many of the ionospheric changes at midlatitude can be understood as a response to these thermospheric perturbations. Wind surges propagate from high latitudes around the globe and transport plasma along magnetic field lines to regions of altered chemical composition, changing ion recombination rates. Some of the increases in the peak F region electron density (NmF_2) and total electron content (TEC) result from these "traveling atmospheric disturbances." The divergent nature of the wind causes upwelling of molecular rich thermospheric gas from lower altitudes. These regions of enhanced molecular species at F region altitudes, or composition "bulges, " can be transported by the pre-existing background wind fields and by the storm winds [*Fuller-Rowell et al.*, 1996]; the changed composition again feeds back to the ionosphere. The regions of upwelling (increases of N_2 and O_2) cause the ionosphere to decay faster and create negative phases of ionospheric storms.

The dynamical changes are complicated because they are driven by the highly variable magnetospheric energy sources. The wind surges propagate and interact around the globe and often appear as a random mixture of waves. Exactly where a composition bulge will be created is also difficult to determine; composition changes are created by persistent divergence of the wind field, in areas of significant energy injection (mainly Joule heating). Accurate knowledge of the spatial and temporal distribution of the magnetospheric sources is required to predict where and when these composition changes will manifest themselves.

Much of the interest in understanding the response of the upper atmosphere to geomagnetic storms has stemmed from the need to predict the ionospheric response. The need arises for practical reasons; the requirement for ground-to-ground communication via the ionosphere using HF radio propagation and from ground-to-satellite through the ionosphere at higher frequencies. The parameters that have received the most attention are NmF_2, or the critical frequency (foF_2), which is related to the maximum usable frequency (MUF) for oblique propagation of radio waves, and TEC, which is significant for the phase delay of high frequency ground-to-satellite navigation signals.

PHYSICAL UNDERSTANDING

In spite of the complexities in the observed response of the thermosphere and ionosphere to a geomagnetic storm, systematic features are apparent. The breakthrough in understanding of the storm-time ionosphere has come from analysis of extensive ionospheric observations and from interpretation of the data by physical models. Rodger et al. [1989] showed that the response of the ionosphere reveals both seasonal and local-time (LT) dependencies during a geomagnetic storm. They demonstrated that at a southern magnetic mid-latitude station, a consistent local time signature in the ratio of disturbed to quiet NmF_2 existed throughout the year, with a minimum in the morning hours around 06 LT and a maximum in the evening hours around 18 LT (see Figure 1). The local-time "AC" variation was superimposed on a "DC" shift of the mean level that varied with season, being most positive in winter (May-July) and most negative in summer (October-February). The data supported the widely held belief that positive phases " of storms (increases in electron density) are more likely in winter mid-latitudes, and "negative phases " of storms (decreases in electron density) are more likely in summer. Field and Rishbeth [1997] showed that these same characteristics are true for other longitude sectors. Rodger et al [1989] stressed the point that individual storms show large deviations from the average behavior.

A cause of the LT variation was suggested by Fuller-Rowell et al. [1994], who extended the theory of Prölss [1993]. Prölss suggested that negative storm effects are due to regions in which the neutral gas composition is changed — the ratio of molecular gas concentration (N_2 + O_2) to the atomic oxy-

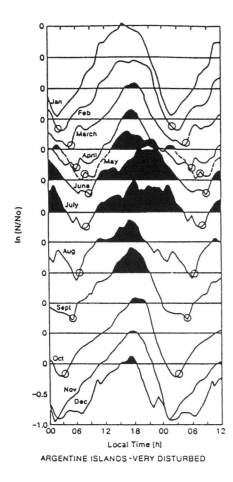

Figure 1. The average seasonal and local time variations in ln(N/N0), at Argentine Islands (650 S) for 1971-1981, taken from Rodger et al. [1989]

gen concentration is increased. Such a region, which we call a "composition bulge" because it represents a region of increased mean mass, is originally produced through heating and upwelling of air by the magnetospheric energy inputs at auroral latitudes. The bulge can be moved to middle latitudes by the nightside equatorward winds and brought onto the dayside as Earth rotates.

Fuller-Rowell et al. [1994] made computational simulations of this process for storms at equinox. They showed that once the composition bulge is created it can be transported by a wind field, either that of the background quiet-day thermospheric circulation, or of the storm circulation driven by the high latitude heat input. They attributed the local time AC effect derived by Rodger et al. [1989] to an oscillation in latitude of the composition bulge in response to the diurnally varying winds. Skoblin and Förster [1993] also showed a case where steep gradients in thermospheric composition could be advected by the meridional wind.

The work of Fuller-Rowell et al. [1996] extended this idea to suggest an explanation of the seasonal variations. Nu-

merical computations suggest that the prevailing summer-to-winter circulation at solstice transports the molecular rich gas to mid- and low-latitudes in the summer hemisphere over the day or two following the storm (see Figure 2a). In the winter hemisphere, poleward winds restrict the equatorward movement of composition (see Figure 2b). The altered neutral-chemical environment in summer subsequently depletes the F-region mid-latitude ionosphere to produce a negative phase. In winter mid-latitudes a decrease in molecular species, associated with downwelling, persists and produces a positive phase. The seasonal migration of the bulge is superimposed on the diurnal oscillation.

This seasonal dependence of the response of the neutral atmosphere composition as a function of season is depicted in Figure 2. The top panel shows summer, the middle is winter conditions, and the lower panel is the equinox case. Each one shows a snap-shot of the northern hemisphere from 10° latitude to the pole, at about 300 km altitude, 24 hours after a 12 hour storm. The figure is from numerical simulations using the coupled thermosphere ionosphere model (CTIM; *Fuller-Rowell et al.*, 1996). In the upper panel the "bulge" of increased mean molecular mass (equivalent to an increase in the N_2/O ratio) has been transported by the wind field to low latitudes. In the middlepanel, in the winter solstice, the composition bulge has been constrained to high latitudes. The lower panel for equinox is the intermediate case.

Fuller-Rowell et al. [2000] demonstrated that numerical simulations are able to capture the seasonal variation in the ionospheric response to storms and indicated clear "scientific success." Plate 1 shows a numerical simulation of a 25-day period during November and December of 1997, for sites in the winter hemisphere (Rome) and the summer hemisphere (Grahamstown, SA). The top traces for each site show the hourly F-region peak critical frequency (foF_2), compared with the monthly mean. Immediately below are shown the ratio between the hourly value and the monthly mean, illustrating the increase or decrease during the geomagnetic disturbance on days 326 and 327. An increase occurs in the winter hemisphere at Rome, and a decrease occurs in the summer hemisphere at Grahamstown. The results from a numerical simulation using the CTIM model where variability of the high-latitude electric field was included [*Codrescu et al.*, 2000], show good agreement in both cases. The lower panel in Plate 1 show the variation in the geomagnetic index a_p during the interval.

An important point to note is that the physical understanding that has emerged does not translate automatically to operational value. In contrast to the "visual" success of the model, Fuller-Rowell et al. [2000] showed that detailed quantitative comparisons of the physical model with data, which are necessary for space weather applications, were less impressive. The accuracy, or value, of the model was quantified by evaluating the daily standard deviation, the root-mean-square error, and the correlation coefficient between the data and model predictions. During the storm periods, the RMS error of the model improved on IRI only slightly, and there were

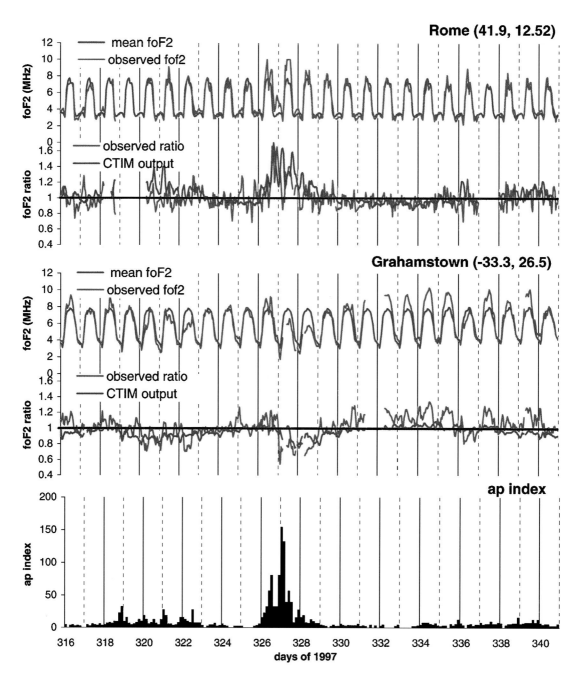

Plate 1. Numerical simulation of a period during November and December of 1997 for Rome in the winter hemisphere and Grahamstown in the summer hemisphere. The top traces for each site show a comparison of the observed foF_2 and the monthly mean. Immediately below each is a comparison between the observed and CTIM modeled ratio of storm to quiet foF_2 values. The lower panel shows the variations in the geomagnetic index a_p.

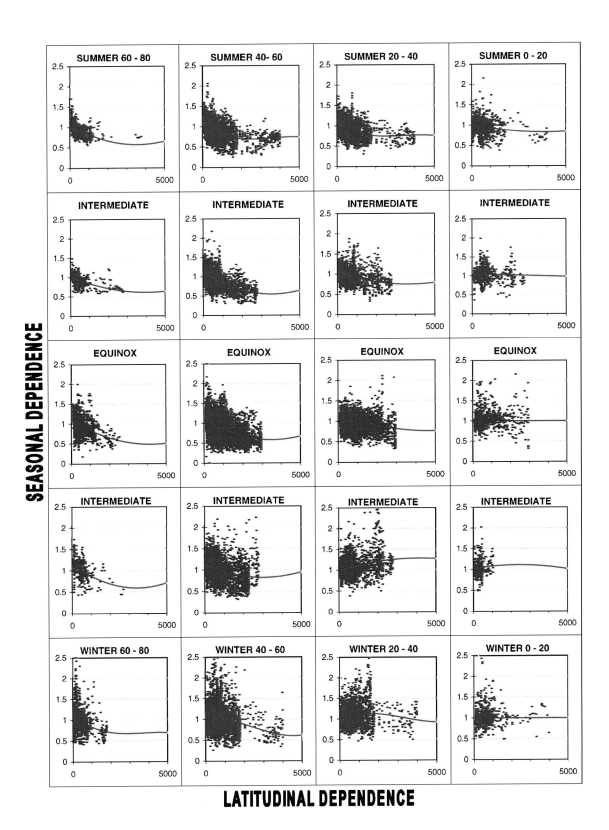

Plate 2. The seasonal/latitude response of the storm-to-quiet foF_2 as a function of filtered a_p, $X(t_0)$.

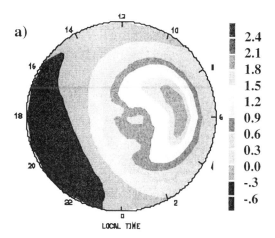

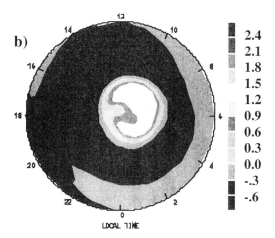

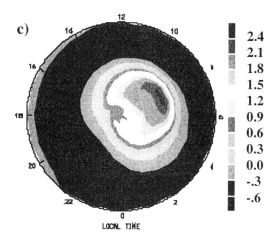

Figure 2. The equatorial extent of the "composition bulge" at 1200 UT, for equivalent storms in the northern hemisphere for a) summer, b) winter, and c) equinox, from $10°$ latitude to the pole.

occasionally false-alarms. Using unsmoothed data over the full interval, the correlation coefficients between the model and data were low, between 0.3 and 0.4. Isolating the storm intervals increased the correlation to between 0.43 and 0.56, and by smoothing the data the values increased up to 0.65. The study illustrated the substantial difference between scientific success and a demonstration of value for space weather applications.

EMPIRICAL STORM-TIME CORRECTION MODEL

The physical understanding that has emerged from the numerical simulations and the data analysis has set the stage for developing an empirical storm-time ionospheric correction model. Specifically, we would like to be able to correct the monthly median, or climatological value (International Reference Ionosphere; IRI, for instance), of the F-region density during a geomagnetic storm, as a function of season and latitude. One advantage of the empirical approach is that data is used to establish the correct relationship between the parameters and the ionosphere, providing the correct sorting parameters are selected. The physical model can be used to determine the qualitative relationship, but we do not have to rely on the model to provide the quantitative dependence. The question is, do we have the knowledge and understanding to include a geomagnetic dependence in the IRI, that demonstrates a real improvement? The goal of this first empirical storm-time model is to establish a correction to the F-region peak density, or critical frequency, as a function of season and latitude for any given a_p time history of a storm.

The key to developing a successful algorithm is to determine the relationship between the parameter representing the magnitude of the geomagnetic forcing and the ionospheric response (the function $F(\tau)$ in equation 1). The knowledge gained from the past data analysis and numerical simulations suggests an ionospheric storm-time correction algorithm of the form:

$$\Phi = \{a_0 + a_1 X(t_0) + a_2 X^2(t_0) + a_3 X^3(t_0)\}$$

$$\{1 + a_4 \sin(LT + \alpha)\} \qquad (1)$$

where $X(t_0) = \int F(\tau) a_p(t_0 - \tau) d\tau$

The target parameter Φ, is the value used to scale the quiet-time ionospheric F-region critical frequency, as a function of latitude, day-of-year, and local time. $X(t_0)$ is a new index representing the integrated effect of geomagnetic activity over the previous day or two. The index is calculated using a filter function, $F(\tau)$, that weights the time history of a_p. which is described further below. "a", "a_1", "a_2", and "a_3" are coefficients for the polynomial fit to the non-linear relationship between the filtered a_p and the ionospheric response, which are dependent on magnetic latitude and season, and "a_4" and "α" are the amplitude and phase of the local-time dependence.

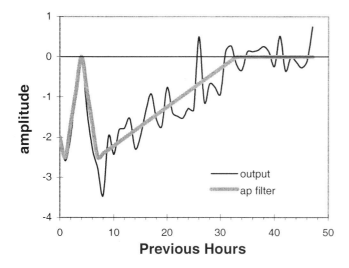

Figure 3. Filter shape, $F(\tau)$, representing the relationship between the geomagnetic index a_p and the ionospheric storm-time response at summer mid-latitudes.

The optimum shape and length of the filter $F(\tau)$ defining the relationship between the time history of a_p and the ionospheric response at mid-latitude (the impulse response function) was determined by a linear regression technique [*Detman and Vassiliadis*, 1997]. Ideally, the function should be derived for each latitude and season, but with the limited data volume this was not possible. Data from summer mid-latitudes were used to develop the filter, by minimizing the mean square error between the filter input (a_p) and filter output (the ionospheric ratios). Figure 3 shows the shape of the unsmoothed filter (dashed line), together with a smoothed profile (solid line) that was adopted for the algorithm. During the early time history of a_p, 0 to 7 hours into the past, the filter shape is quite variable probably due to the ionospheric effects of gravity wave propagation and penetration electric fields. The predominant negative response is due to the development of the composition changes, as described earlier, and the slow recovery over the next day or so, reflects the time-scale for molecular diffusion that gradually returns the neutral atmosphere to its pre-storm configuration.

The theory described previously suggests that the ionospheric response will depend strongly on both season and latitude. Using the new index $X(t_0)$, ionosonde data from 75 ground-based stations covering latitudes from 79°S to 83°N for a number of storms over the last twenty years have been sorted as a function of latitude and season (see Araujo-Pradere et al. 2000a and b, for further details). The results of this analysis are shown in Plate 2. Each panel shows a scatter plot of the storm-to-quiet foF_2 ratio as a function of the filtered a_p index, $X(t_0)$. This has been done for four magnetic latitude bands 0-20°, 20-40°, 40-60°, and 60-80°, and five seasonal bands centered on the summer and winter solstices, equinox, and the two intermediate times. The data show a consistent negative response in summer mid-latitudes, while in the win-

ter hemisphere the response is not so well defined, showing a boundary around 40°. The consistent response in summer is due to the prevailing summer-to-winter circulation. In the winter hemisphere, theory suggested a boundary exists in the prevailing circulation and in the composition response. Such a boundary also exists in the sorted data producing a negative phase in latitudes greater than 40°, while in lowest latitudes a decrease in molecular species, associated with downwelling, persists and produce the characteristic positive storm.

Another important difference between summer and winter hemispheres is the variability in both sets of data. Summer hemisphere and equinox mid latitudes shown a very coherent behavior, with a narrow band of variability around the fit following the negative phase, while winter hemisphere shows the highest dispersion around the fit. In each panel a polynomial cubic fit to the data has been determined to provide the set of coefficients "a_0", "a_1", "a_2", and "a_3" required in Equation 1.

TESTING THE EMPIRICAL MODEL

The empirical storm-time correction model has been tested on a twenty-five day interval towards the end of 1997, between November 12 and December 6. During this period, a significant disturbance occurred on November 22/23. Figure 4 illustrates the response of the ionosphere to the event and the prediction of the empirical model at two sites, Rome in the northern winter midlatitudes (top panel), and Grahamstown, SA, in the southern summer midlatitudes (lower panel). The filtered a_p index, using the function in Figure 3, shows a large increase on the two storm days. Both panels show the ionospheric response depicted as the storm-to-quiet ratio of foF_2. At Rome, the ionospheric response is positive, consistent with expectations in winter midlatitudes. At Grahamstown, the ionospheric F region decreases, again consistent with expectations in summer midlatitudes. In both cases, the empirical model captures the direction of the change, and the magnitude is particularly good in summer.

OUTSTANDING ISSUES

At this stage in the development of the empirical algorithm, we have not included the local time dependence represented by coefficient a_4 in Equation 1. The analysis by Rodger et al. [1989] showed a strong local time signature with a variation of about 40% in NmF_2, but we have been unable to show such a strong dependence in the present analysis.

At the present time it is also not possible to capture the transient effect during the early phase of a geomagnetic storm. The numerical simulation and observations clearly show the propagation of large-scale gravity waves that are known to have a significant influence on the ionosphere at mid latitudes. At present their ionospheric consequences are represented by the structure in the new index based on the time filtered geomagnetic index, represented in Figure 3.

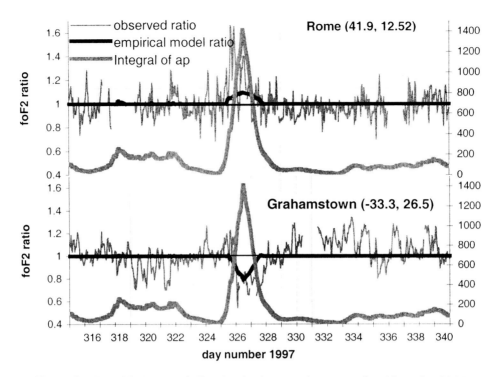

Figure 4. Observed and empirical model foF_2 ratios for the event that occurred on November 22/23 at two sites, Rome in the northern hemisphere (winter) and Grahamstown in the southern hemisphere (summer).

The response at low latitudes is also poorly represented by the empirical model. This is due to our current rudimentary understanding of the temporal evolution of the ionosphere at low latitude during geomagnetic storms. Due to the nature of the magnetic field configuration, the low latitude ionosphere is strongly influenced by electric fields. Changes in the low-latitude electric field can arise from the penetration of magnetospheric fields and through the development of polarization fields from the changing thermospheric wind fields. The effect on the ionosphere at low latitudes has yet to be understood. When the response of the equatorial region has reached the same level of maturity as at mid-latitudes, then maybe a relatively simple algorithm that captures the response will become apparent.

CONCLUSION

Our understanding and knowledge of the upper atmosphere and ionosphere response to an increase in geomagnetic activity has matured considerably over the last ten years. An empirical storm-time ionospheric correction model has been developed that shows promise for improving IRI. The algorithm is designed to scale climatological F-region peak critical frequencies, foF_2, or electron densities, NmF_2, and possibly total electron content, TEC, every hour through a geomagnetic event. This model is to be included in a new version of IRI in an effort to parallel the development of MSIS by including a dependence on geomagnetic activity.

The prospect for improvement of the model in the future is uncertain. The most promising area is to redo the analysis of the local-time variation. The analysis may require the amplitude of the local-time response to be normalized by the mean response in a given latitude and seasonal bin (as given in Plate 2). Work in this direction appears to be showing promise, which will hopefully further improve the fit to the data, and reduce the residual error.

Another avenue to pursue is the use of the state-space reconstruction method, to separate the response characteristics as a function of the phase of the disturbance. This approach may be a way of capturing the affect of the propagating wave features or penetrating electric field, but will have to await further analysis. The index a_p was chosen to capture the time-history of the geomagnetic events in order to maximize the use of the model by the community, for operational use, and to enable the correction model to be implemented in the next generation IRI. The limitations of a single index are obvious, but the availability of a simple alternative, suitable for implementation in an empirical model, is not clear. The time-history of solar wind parameters is one possible option to replace a_p in the future.

REFERENCES

Araujo-Pradere, E.A., T.J. Fuller-Rowell, and M.V. Codrescu, An ionospheric storm-time correction model. *Radio Science, submitted,* 2000a.

Araujo-Pradere, E.A. and T.J. Fuller-Rowell, A model of the perturbed ionosphere using auroral power as the input. *Geof. Int., 39*, 1, 2000b.

Bilitza, D., International reference ionosphere 1990, National Space Science Data Center, WDC A for Rockets and Satellites, Greenbelt, MD, 1990.

Codrescu, M.V., T.J. Fuller-Rowell, J.C. Foster, J.M. Holt, S.J. Cariglia, Electric field variability associated with Millstone Hill electric field model, *J. Geophys. Res., 105*, 5265-5273, 2000.

Detman, T.R. and D. Vassiliadis, Review of techniques for magnetic storm forecasting, *AGU Monograph, 98*, 253-266, 1997.

Field, P. and H. Rishbeth, The response of the ionospheric F2-layer to geomagnetic activity: an analysis of worldwide data, *J. Atmos. Solar-Terr. Phys., 59*, 163-180, 1997.

Fuller-Rowell, T.J., M.V. Codrescu, R.J. Moffett, and S. Quegan, Response of the thermosphere and ionosphere to geomagnetic storms, *J. Geophys. Res., 99*, 3893-3914, 1994.

Fuller-Rowell, T.J., M.V. Codrescu, R.J. Moffett, and S. Quegan, On the seasonal response of the thermosphere and ionosphere to geomagnetic storms, *J. Geophys. Res., 101*, 2343-2353 1996.

Fuller-Rowell, T.J., E.A. Araujo-Pradere, and M.V. Codrescu, An empirical ionospheric storm-time correction model. *Adv. Space Res., 25, 1*, 139-146, 2000.

Hedin, A.E., MSIS-86 thermospheric model. *J. Geophys. Res., 92*, 4649, 1987.

Prölss, G.W., Common origins of positive ionospheric storms at middle latitudes and the geomagnetic effect at low latitudes, *J. Geophys. Res., 98*, 5981-5991, 1993.

Rodger, A.S., G.L. Wrenn, and H. Rishbeth, Geomagnetic storms in the Antarctic F region, II, physical interpretation, *J. Atmos. Terr. Phys., 51*, 851-866, 1989.

Skoblin, M.G., and M. Forster, An Alternative Explanation of Ionization Depletions in the Winter Night-time Storm Perturbed F2-Layer, *Ann. Geophys., 11*, 1026-1032, 1993.

T.J. Fuller-Rowell, M.V. Codrescu and E.A. Araujo-Pradere CIRES, University of Colorado and NOAA Space Environment Center, 325 Broadway, Boulder, CO 80303, USA.

Ionospheric Response for the Sept. 24-25, 1998 Magnetic Cloud Event

R. M. Winglee, D. Chua, M. Brittnacher, G. K. Parks

Geophysics Program, University of Washington, Seattle

Heavy ionospheric ions play a crucial part in the overall space weather environment, particularly in the generation of high energy particle populations during disturbed periods. Multi-fluid global simulations are used to quantify the ionospheric outflows during the Sept 24-25, 1998 magnetic cloud event. The model results show that jumps in dynamic pressure associated with structures within the magnetic cloud lead to a temporary broadening of the cusp/cleft region and enhancements in the ionospheric outflows both in latitude and local time in association with dayside brightenings seen in Polar UVI observations. However, the inferred power in UVI is several times greater than that in the model outflows. After about 15 mins the cusp/cleft region in the model decreases to the nominal size of about 1-2 hrs in local time and a few degrees in latitude in conjunction with an overall decline in UVI intensity, and the energy fluxes are comparable. During the period of B_y dominated IMF, the cleft ion fountain is seen in the model to reach 50-70 R_E into the tail. For southward IMF period, the cleft ion fountain is restricted to less than about 20 R_E into the tail, but convection around the flanks still leads to significant ionospheric mass loading in the deep tail.

1. INTRODUCTION

Intense ionospheric outflows have been observed for more than two decades by polar orbiting spacecraft (for reviews see *Moore* [1991], *Moore and Delcourt* [1995] and *Yau and André* [1997]). During storms the oxygen concentration in the ring current can be comparable to if not dominate that of H^+ [*Krimigis et al.*, 1985; *Hamilton et al.*, 1988; *Roeder et al.*, 1996] while in the deep tail the oxygen concentration can reach a few percent[*Seki et al.*, 1996, 1998]. The magnetic cloud event of September 24-25, 1998 is a good example of the importance of how heavy ionospheric ions produce mass loading of the magnetosphere and the possible genera-

tion of energetic particles that modify the space weather conditions during disturbed times.

This event was seen by Polar to produce a direct and prompt response in the polar ionosphere generating enhanced flows of heavy ionospheric ions at high altitudes [*Moore et al.*, 1999]. The outflows during the event were sufficient to produce a mantle that was dominated by heavy O^+ ionospheric ions. The Fast Auroral SnapshoT (FAST) explorer also observed enhanced ionospheric outflows during the event [*Strangeway et al.*, 2000]. These observations indicate that the ionospheric outflows were enhanced for several hours after the passing of the shock front. The peak particle flux was shown to be correlated with the peak Poynting flux into the auroral region. Recently, *Wygant et al.* [2000] has also demonstrated a direct correlation between downward Poynting flux and UVI intensifications.

Cladis et al. [2000] has noted that the velocity profile of the ionospheric ions is consistent with that expected from centrifugal acceleration as field lines are convected

Space Weather
Geophysical Monograph 125

over the polar cap. The total amount of oxygen outflows produced is dependent on the solar wind conditions and on the concentration of oxygen at the lower end of the field line, which in term is dependent on the solar EUV photon flux and magnetospheric activity [*Yau et al.*, 1988; *Yau and André*, 1997, *Craven et al.*, 1997]. Centrifugal acceleration was originally proposed in the 1980's to explain the presence of enhanced heavy ionospheric fluxes at high altitudes and their potential for the mass loading of the near-earth plasma sheet [*Cladis*, 1986; *Horwitz*, 1987; *Winglee*, 1998, 2000]].

In this paper data from Polar UVI are used to map the energy deposition into the auroral oval. Multi-fluid simulations driven by the observed solar wind conditions are then used to map the expected ionospheric outflow of H^+ and O^+ in relation to the UVI brightenings. The intercomparison of the energy input and output from the auroral regions enables the identification of potential sources on global scales, and to quantify the response of these sources to changes in the solar wind conditions. The global modeling then allows the ionospheric outflows to be self-consistently mapped into the magnetosphere, including the deep tail. The model outflows provide a minimum estimate since they only include centrifugal acceleration but not wave-particle interactions that would be associated with the above increases in Poynting flux.

2. MULTI-FLUID SIMULATIONS

The details for the numerical scheme and boundary conditions are given in *Winglee* [1998, 2000]. The equations are essentially the standard multi-fluid equations for a plasma [e.g., *Parks*, 1991] with the ion and electron equations kept separate unlike MHD which combines the equations for a single fluid description. The grid spacing increases from 0.4 R_E in the dayside and mid-tail regions, to about 3 R_E in the distant tail at $x \simeq -200R_E$ (GSM) and at the flanks at $\pm 60R_E$. The solar wind boundary is at $x = 35R_E$.

The inner radius of the simulations is set to 3 R_E which is typical of global simulation models. However, at 3 R_E the effective gravitational force on oxygen would be negligible. Therefore in order to ensure that the oxygen ions experience essentially the same average gravitational force within 1 to 2 R_E of the inner boundary, the gravity term $g = GM_E/R_E^2$ in the above equations is set at 30 m/s^2 as opposed to its actual value of 10 m/s^2, i.e. g is set so that the plasma in the simulations experience a potential well comparable to the one that they would encounter if the particles

started as the ionosphere proper. At higher altitudes convective or centrifugal processes dominate any effects that the enhanced gravitation field might have on the outflowing ions. Test runs with this value show that the outflow of oxygen is inhibited relative to the H^+ outflow when there is only weak forcing by the solar wind.

The initial profile is assumed to decrease as R^{-6}. When ionospheric oxygen is assumed to be present it is given the same density profile with a relative concentration of 5%. This fixed profile while not very realistic allows the outflow physics and the solar wind controlling influences to be properly isolated. Dynamic ionospheric composition [*Craven et al.*, 1997; *Gallagher et al.*, 1998] are now just becoming available. Future models need to incorporate dynamic changes in the ionospheric composition, but quantitative models for such changes as a function of the solar wind conditions presently do not exist.

The initial pressure along each flux tube is assumed constant (i.e. set by mapping the simulation grid point along the dipole field line to the equator). The equatorial bulk temperature is set to about 60 eV and in the polar cap it is less that 0.3 eV. The total ion outflow seen in the simulations is about 1×10^{26} ions/s for the northward interplanetary magnetic field (IMF) case increasing to 4×10^{26} ions/s for southward IMF and about 50% smaller in the presence of the gravity field imposed. These numbers are of the order of those estimated by *Chappell et al.* [1987]. More recent estimates for the total outflow by *Yau and André* [1997] indicate that the H^+ is limited to about 10^{26} ions/s while the O^+ outflow rate ranges from about 10^{25} ions/s during quiet times to a few times 10^{26} ions/s during active times. Thus the values assumed here while slightly high for average values are appropriate for the active period being treated here.

The system is driven by the solar wind conditions (Figure 1) observed by the Wind spacecraft which was approximately 185 R_e upstream. Prior to the arrival of the magnetic cloud, the solar wind conditions are close to the nominal with a density of about 7 cm^{-3} and a speed of about 420 km/s. With the arrival of the shock at Wind at about 2325 UT the density jumps by a factor of 2, and the speed increases to about 600 km/s, producing an increase of nearly a factor of four in the solar wind dynamic pressure. At 600 km/s the propagation time to Earth is about 33 min. As the event proceeds the wind speed gradually increases to about 800 km/s. There are two additional large jumps in the dynamic pressure seen at Wind at about 0003 and

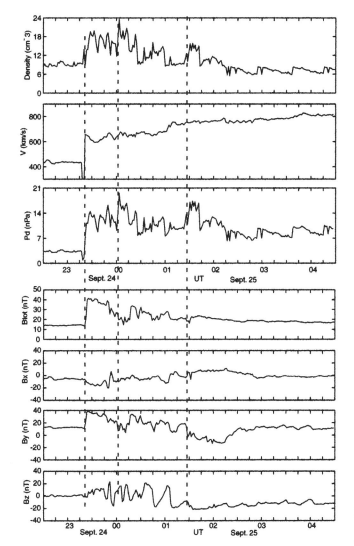

Figure 1. Solar wind conditions for the Sept. 24-25, 1998 magnetic cloud event as observed by Wind which was approximately 185 R_E upstream. The propagation time to Earth at a speed of 600 km/s is about 33 min. The solar wind density, speed and dynamic pressure are shown in the top panel with the IMF (in GSM) are shown in the lower panels. The dashed lines indicated the arrival of the cloud and the subsequent arrival to two large pressure jumps within the cloud.

0127 UT. These pressure jumps can be used to mark timing for changes in the magnetospheric/ionospheric response.

The IMF for this period is shown in the lower part of Figure 1. Prior to the arrival of the magnetic cloud, the IMF is dominated by a strong (10 nT) duskward field. With the arrival of the magnetic cloud, IMF B_y remains the dominant component with its magnitude increasing by nearly a factor of 3. It is not until about 0120 UT at

Wind that the IMF becomes predominantly southward and remains that way for several hours. This change in the IMF conditions will be shown to produce a change in the apparent source of the ionospheric outflows, and their distribution in the tail.

3. ENERGY AND MASS FLOWS INTO AND OUT OF THE AURORAL OVAL

The response of the aurora to the arrival of the magnetic cloud event as observed by the Ultra-Violet Imager (UVI) on Polar is shown in Plate 1. The arrival of the cloud is seen at 2346 UT as an intensification of the dayside emissions as well as an increase in pre-existing emissions in the nightside oval. The post-noon emissions then intensify and extend down towards the terminator and the nightside emissions towards higher latitudes. The strongest emissions are seen for the period between 2352 and 0004 UT. While there are some subsequent local intensifications, large-scale brightenings of the auroral oval are not seen again until about 0155 UT (not shown).

In order to determine the sources and sinks for the production on these auroral emissions, we examined the inflow and outflow of the different ion species through a surface set at 6 R_E. The height of 6 R_E is chosen because ionospheric ions crossing this surface have the potential for propagating into the plasma sheet and deep tail as opposed to being confined to just the near-Earth region. At the same time, the energy influx from solar wind ions provides a simple diagnostic for the size and position of the cusp/cleft region.

Plate 2 shows the energy influx of solar wind ions derived from the model that crosses the 6 R_E surface and is mapped to the ionosphere. It is seen that the initial pressure pulse causes the appearance of a very wide cusp/cleft region (Plate 2b), with a latitudinal extent of about 10° and about 5 hrs in local time. The cusp proper is represented by the hook feature at about 80° and its sweeping around from the dusk side to the dawn side is produced by the presence of the large positive B_y in the IMF. Within the next 30 minutes (Plates 2c-2d) the size of the cusp/cleft is seen to decrease to the more typical size of a few degrees in latitude and about 1-2 hrs in local time. This reduction in time occurs during a period when the dynamic pressure and IMF are approximately constant.

Note that the peak power in the solar wind ions can be comparable to that seen in the UVI images. The cusp proper as identified by the model does not correspond to any intense emissions seen in UVI since it consists primarily of ion downflow as opposed to elec-

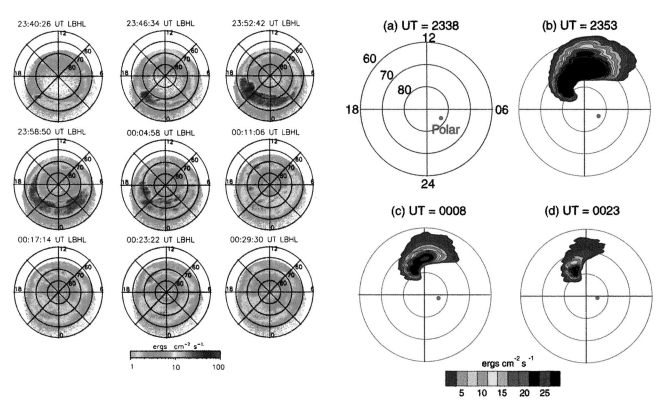

Plate 1. Polar UVI images in the LBHL band showing the energy deposition into the northern auroral oval for a 50 minute period covering the arrival of the cloud (seen at 2346 UT), and the subsequent brightening and expansion of the oval and eventual recovery.

Plate 2. Downward energy flux in solar wind ions crossing the surface at 6 R_E and mapped down to auroral altitudes. This diagnostic provides a simple means to identify the size of the cusp/cleft region. Prior to the arrival of the cloud, the energy flux is less than 2 erg cm^{-2} s^{-1} and no contours are shown. With the arrival of the cloud the energy flux increases to about 25 erg cm^{-2} s^{-1} over an extended region in latitude and local time. The high latitude extension is the cusp and is not seen in the UVI images, but the equatorward extension which would correspond to the cleft corresponds approximately with the UVI images in both power an extent. With the response of the ionosphere and magnetosphere, the size of the cusp cleft in latitude is seen to decrease.

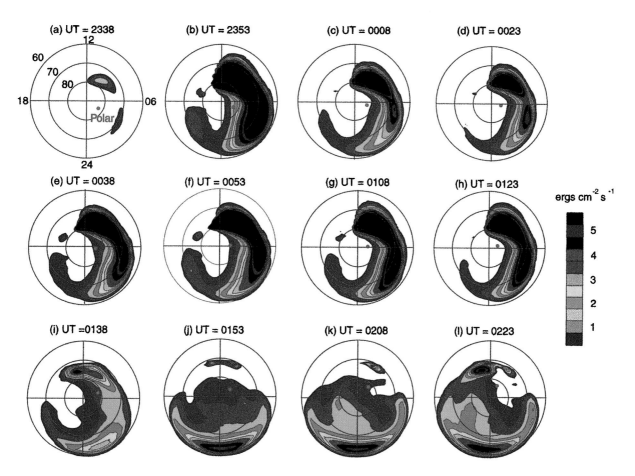

Plate 3. Outward energy flux of ionospheric ions crossing the 6 R_E surface. This energy flux is mapped back to auroral heights for comparison with the energy flux seen by UVI. Note the scale is 1/5 of that in Plate 2. These outflows are driven by centrifugal acceleration, which is enhanced by pressure pulses. The outflows are predominantly on the dawn side for positive B_y IMF and in the nightside for southward IMF.

tron downflow. Instead it is the equatorward extension of the downflow that overlaps with the near-noon emissions in the UVI images.

The corresponding energy flux carried by ionospheric ions crossing the 6 R_E surface is shown in Plate 3. For comparison with the UVI data, the energy flux is mapped back down to the ionosphere to give a lower estimate for the power in the ionospheric outflows. The format is the same as in Plate 3 except the energy scale is only a quarter of that in Plate 3 and the period has been extended to cover the first 3 hrs of the event. The arrival of the magnetic cloud (Plate 3b) is seen to produce substantial increases in the outflow of ionospheric ions across the dawnside of the auroral oval and into the dawnside polar cap. This outward flux is produced in the simulations by centrifugal acceleration (cf. *Cladis et al.* [2000]) in which increases in the solar wind speed, density and IMF lead to enhancements in the cross-polar cap potential. This enhanced potential in turns causes the auroral field lines to experience faster convection so that plasma at low altitudes is centrifugally accelerated outwards. Each of the pressure pulses produces increased outflows which then subside over a period of about half an hour (e.g. Plates 3b-3e).

Note that while the power in the outflows on the dawnside are comparable to the power in UVI emissions, the model outflows on the dusk side are very much smaller. This difference could be suggestive that in this case wave-particle heating associated with increased downward Poynting flux over the region is an important factor for generating ionospheric outflows as suggested by *Strangeway et al.* [2000] and *Wygant et al.* [2000].

At 0138 UT (Plate 3i), there is a marked change in the pattern. Prior to this time the outflows are predominantly on the dawnside and this feature is correlated with the period of dominant B_y in the IMF. At $\sim$ 0110 UT at Wind the IMF turns predominantly southward. This change causes the outflows to be primarily in the nightside region centered around midnight local time. These nightside outflows overlay the nightside brightenings seen by UVI at this time.

The total energy input into the ionosphere as determined by UVI is shown in Figure 2a. The period is limited to 0120 UT because beyond this time UVI does not see the full auroral oval. It is seen that with the arrival of the magnetic cloud there is an increase in the power into the oval of the order of 200 GW extending over a period of about 15 min, after which the power decays with an e-folding period of about 23 minutes.

The corresponding total energy in the ionospheric outflows (15 min sampling only) crossing the 6 R_E surface are shown in Figure 2b. The dashed line in Figure 2a shows a comparison of the power in the outflows over the period in which UVI can calculate the total power in the aurora. It is seen that while the total power in the outflows increases with the arrival of the magnetic cloud and subsequent pulses, the ionospheric outflows in the simulations can only account for 40% of the total energy input during the first part of the magnetic cloud event.

This discrepancy is probably due to the fact that the simulation model assumes fixed ionospheric conditions. As such the model only incorporates outflows driven by centrifugal acceleration. The model does not allow changes in the ionospheric ion composition and corresponding outflow due to heating from precipitating energetic particles. Such effects are probably very important, particularly in the pre-midnight sector where the model predicts only limited outflows whereas the UVI imaging indicates large energy deposition. *Moore et al.* [1999] indicate that during the event, ionospheric heating was such that O^+ is the dominant ion species, as opposed to the 5% relative concentration assumed in the present simulations.

A second factor that may contribute to the discrepancy is that the energy is being deposited at low altitudes and ions accelerated outwards by such energy deposition need not reach high altitudes but instead may be confined to the inner magnetosphere. This latter effect produces the modulation in the fraction of energy carried by the O^+ which is shown in Figure 2c. If the energy fraction approaches 46%, the oxygen ions have the same velocity of the ions for a relative concentration of 5%. Due to gravity, the actual fraction reaching the $6R_E$ surface is lower than the ionospheric concentration. Prior to the arrival of the magnetic cloud event, the IMF is still moderately strong with a 15 nT B_y component so that even at this early stage O^+ ions are carrying nearly 38% of the outward ion energy flux. With the arrival of the magnetic cloud, the enhanced convection drives the oxygen energy flux up to its limit of 44% for fixed ionospheric boundary conditions.

The mapping of the oxygen outflows into the magnetotail is shown in Plate 4 for the period covering the initial cloud interaction during the period when the IMF was predominantly duskward. Prior to the arrival of the pressure pulse, the oxygen density is primarily localized to the near-Earth region within ± 15 R_E. The high latitude extensions of the oxygen density arise from the cleft ion fountain and because of the positive IMF B_y resides primarily on the northern dawn and southern dusk sections. The plasma sheet lies orthogonal to this plane which is seen in the tail views as the flat expanse

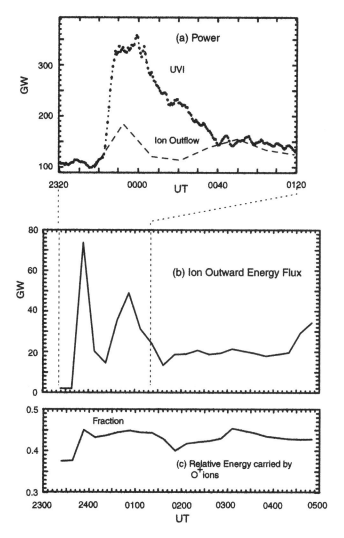

be considered small, O^+ ions at this level would still be significant in providing several tens of percent in the total mass and energy to the tail, and would actually dominate the high energy portions of the particle distributions.

Plate 5 shows the continuation of the oxygen outflows out to 0323 UT. It is seen that the deep tail cleft ion fountain is a sustained feature (Plate 5a) provided that the solar wind speed remains high and the IMF is dominated by B_y. With the southward turning of the IMF (Plates 5b and 5c) the cleft ion fountain becomes more aligned with the north/south direction and is also more restricted to the near-Earth region entering the plasma sheet at 20 - 30 R_E down the tail.

The strong outflows near midnight indicated by the polar plots of Figure 2 are seen in Plate 5 to be restricted to populating the near-Earth region ($< 20\ R_E$). These ions are then convected to the dawn and dusk flanks where they are then convected into the deep tail. This convection pattern produces the broader contribution in y direction, but still allows mass loading of the deep tail by ionospheric ions.

4. SUMMARY AND CONCLUSIONS

The presence of different particle populations within the magnetosphere provide important clues as to the mechanism and processes controlling the energy and mass transport during periods of magnetospheric activity. The magnetic cloud events provide an important means to assess the sources and driving forces for ionospheric outflows due to forcing by the solar wind since they provide long periods of almost constant IMF. The magnetic cloud event of Sept. 24-25, 1998 represents just such an opportunity. The leading edge of the event was dominated by a strong B_y component while the second part was dominated by southward IMF, and it was well observed by the Polar spacecraft which indicated elevated fluxes of oxygen being accelerated out of the ionosphere.

The simulation results show that the initial arrival of the cloud event, which is marked by a large increase (10-15 nPa) in solar wind dynamic pressure, leads to a broadening of the cusp/cleft region to nearly 10 degrees in latitude and 5 hrs in local time. As the magnetosphere is able to respond, the cusp/cleft is seen to close up to more typical values of a few degrees in latitude and about 1 hr in local time. The power carried by the solar wind ions at this time is comparable to that seen in the dayside auroral emissions as observed by UVI on polar.

Figure 2. Comparison of the total energy input into the auroral oval as determine by Polar UVI and the power output carried by ionospheric outflows as determined by the global model. While the initial pressure pulses produce substantial outflows, the power inflow at least initially dominates the energy of the outward flowing ionospheric ions, but as the ionosphere and magnetosphere respond, the two are seem to converge. While O^+ is only a small percentage of the total density, it is makes a significant contribution to the total energy flux carried by the ionospheric ions.

running from the northern dusk side diagonally to the southern dawnside.

With the passing of the cloud, the oxygen densities are seen be pulled into the mid-tail (Plate 4b) and eventually into the deep tail (Plate 4c). This is true for both the oxygen in the plasma sheet as well as for the cleft ion-fountain. The oxygen is able to move quickly down the tail as it is approximately co-streaming with the light H^+ ions. While the density of the 0.02 cm^{-3} may

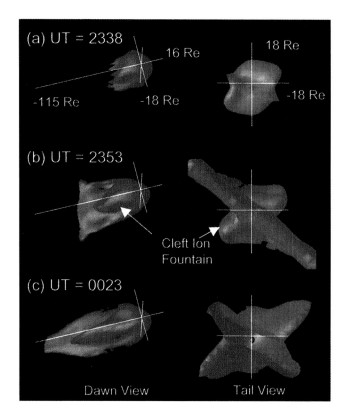

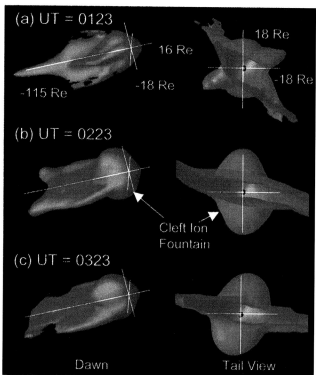

Plate 4. Evolution of the oxygen density as seen by the changes in the density isosurface of $0.02\ \text{cm}^{-3}$. Left hand side shows a view from the northern dawnside while the right hand side shows the view from the tail. The cleft ion fountain is seen as the high latitude feature extending from the northern (southern) dawn (dusk) side and entering the plasma sheet that is seen as the flat expanse rising from the southern dawnside to the northern dusk side. The tilting of the plasma sheet and cleft ion fountain is due to the positive B_y in the IMF.

Plate 5. As in Plate 5, but for later times during the event. Due to the change to an IMF dominated by southward field the cleft ion fountain becomes localized to the noon-midnight meridian and to the near-earth region. The amount of oxygen in the plasma sheet is seen to expand in y and still maintains a significant presence in the deep tail.

The presence of the large B_y in the IMF causes the cleft to drape around the dawnside. This draping coincides with the position of enhanced ionospheric outflows driven by centrifugal acceleration. This acceleration for the first half of the event occurs through enhanced convection velocities driven by increases in the B_y component of the IMF, as well as by increases in the solar wind dynamic pressure. The latter effect provides only short-lived (15-30 min) enhancements in the outflows while the IMF dependence leads to a sustained cleft ion fountain that is present essentially the whole duration of the strong B_y IMF. The cleft ion fountain for this case is seen to feed energetic O^+ into the deep tail at 50 - 70 R_E.

For the southward IMF portion of the event, the strongest ionospheric ion outflows are restricted to the midnight sector. Under these conditions most of the ions enter the current sheet in the near-Earth region (within $20-30 R_e$). These ions when they enter the current sheet in the nightside are then convected towards the dayside along the flanks. These particles then are convected down the tail along the flanks and back into the center of the current sheet, so that the deep tail is still heavily populated by energetic ions of ionospheric ions.

The energy carried by the above ionospheric outflows is shown to be comparable to that calculated from the auroral emissions as seen by UVI. The main exception arises during transients associated with pressure jumps, which do not allow sufficient time for the ionosphere/magnetosphere to respond to the new solar wind conditions. In the present case, between about 40 to 50% of this energy is carried by oxygen ions when their concentration is held fixed at only 5%. This concentration is actually smaller than observed by *Moore et al.* [1999] but already indicates the significance of such heavy ions in the mass loading and energy transport throughout the magnetosphere.

Acknowledgment. This work was supported by NSF Grant ATM-9731951 and by NASA grants NAG5-6244, NAG5-8089 to the Univ. of Washington. The simulations were supported by the Cray T-90 at the San Diego Supercomputing Center which is supported by NSF.

REFERENCES

André, and A. Yau, Theories and observations of ion energization and outflow in the high latitude magnetosphere, *Space Sci. Rev., 80*, 27, 1997.

Chappell, C. R., T. E. Moore, and J. H. Waite, Jr., The ionosphere as a fully adequate source of the earth's magnetosphere, *J. Geophys. Res., 92*, 5896, 1987.

Cladis, J. B., Parallel acceleration and transport of ions from polar ionosphere to plasma sheet, *Geophys. Res. Lett., 13*, 893, 1986.

Cladis, J. B., H. L. Collin, O. W. Lennartsson, T. E. Moore, W. K. Peterson, and C. T. Russell, Observations of centrifugal acceleration during compression of magnetosphere, *Geophys. Res. Lett., 27*, 915, 2000.

Craven, J. D., D. L. Gallagher, and R. H. Comfort, Relative concentration of He^+ in the inner magnetosphere as observed by the DE 1 retarding ion mass spectrometer, *J. Geophys. Res., 102*, 2279, 1997.

Gallagher, D. L., P. D. Craven, and R. H. Comfort, A simple model of magnetospheric trough total density, *J. Geophys. Res., 103*, 9293, 1998.

Hamilton, D. C., et al. Ring current development during the great geomagnetic storm of February 1986, *J. Geophys. Res., 93*, 14343, 1988.

Horwitz, J. L, Core plasma in the magnetosphere, *Rev. Geophys. Spc. Phys., 25*, 579, 1987.

Krimigis, S. M., et al., Magnetic storm of September 4, 1984: a synthesis of ring current spectra and energy densities measured with AMPTE/CCE *Geophys. Res. Lett., 12*, 329, 1985.

Moore, T. E., Origins of magnetospheric plasma, *Rev. Geophys., 29*, 1039, 1991.

Moore, T. E., and D. C. Delcourt, 'The Geopause', *Rev. Geophys., 33*, 175, 1995.

Moore, T. E., et al., Ionospheric mass ejection in response to a CME, *Geophys. Res. Lett., 26*, 2339, 1999.

Parks, G. K, *Physics of Space Plasmas* Addison-Wesley, Redwood City, California, 1991.

Roeder, J L., et al., CRRES observations of the composition of the ring-current ion populations *Adv. Space Res., 17*, 17, 1996.

Seki, K., et al., Coexistence of Earth-origin O^+ and solar wind-origin H^+/He^{++} in the distant magnetotail, *Geophys. Res. Lett., 23*, 985, 1996.

Seki, K., et al., Statistical properties and possible supply mechanisms of tailward cold O^+ beams in the lobe/mantle regions, *J. Geophys. Res., 103*, 4477, 1998.

Strangeway, R. J., et al., Cusp field-aligned currents and outflows, *J. Geophys. Res.,,* in press, 2000.

Winglee, R. M., Multi-fluid simulations of the magnetosphere: The identification of the geopause and its variation with IMF, *Geophys. Res. Lett., 25*, 4441, 1998.

Winglee, R. M., Mapping of ionospheric outflows into the magnetosphere for varying IMF conditions, *J. Atmos., Solar Terrestrial Physics, 62*, 527, 2000.

Wygant, J., et al., Polar comparisons of intense electric fields and Poynting flux near and within the plasmasheet-tail lobe boundary to UVI images: An energy soruce for the aurora, *J. Geophys. Res.,* in press 2000.

Yau, A. W., and M. André, Source of ion outflow in the high latitude ionosphere, *Space Sci. Rev., 80*, 1, 1997.

Yau, A. W., W. K. Peterson, and E. G. Shelley, Qunatitative parameterization of energetic ionospheric ion outflows, in *Modeling Magnetospheric Plasma*, edited by T. E. Moore, and J. H. Waite, Jr., Geophys. Monogr. 44, p. 211, 1988.

M. Brittnacher, D. Chua, G. K. Parks, R. M.Winglee, Geophysics Program, University of Washington, Seattle WA 98195-1650. (e-mail: winglee@geophys.washington.edu)

FAST Observations of Ion Outflow Associated with Magnetic Storms

J. P. McFadden[1], Y. K. Tung[1], C. W. Carlson[1], R. J. Strangeway[2], E. Moebius[3], and L. M. Kistler[3]

New observations from the FAST mission are used to study the relationship between the auroral ionospheric ion outflow and the injection of ring current plasma during magnetic storms. One set of observations follows the injection of the low energy tail of the ring current (<25 keV) at low latitudes (L-shell<4). These ions are found to be primarily of ionospheric origin and to evolve on the two hour time scale of the spacecraft orbit indicating a relatively brief injection. Higher latitude observations show intense auroral outflows are observed in both the cusp and near midnight during magnetic storms. Cusp outflows that are energetic enough to contribute hot plasma to the plasmasheet are likely lost out the magnetotail whereas nightside auroral outflows place ionospheric ions directly onto newly reconnected plasmasheet field lines. These observations suggest how substorms can play a role in the development of magnetic storms. Electrons accelerated during substorms produce waves that heat the ionospheric plasma producing intense conic outflows into the plasmasheet. The higher density plasmasheet ions are subsequently convected to the inner magnetosphere by the large convection electric fields during the magnetic storm. Without the large convection fields, the substorm outflows have no effect on Dst since these ions never reach the inner L-shells. Without substorm outflows, the plasmasheet density is relatively modest so that convection to the inner L-shells produces only small drops in Dst. In this model, it is through ionospheric ion outflows that substorms affect the development of magnetic storms.

1. INTRODUCTION

The sources of plasmasheet and ring current plasma are an important element in understanding the dynamics of the Earth's magnetosphere. Models that attempt to predict the "space weather" of the near Earth environment must be able to account for relative abundances of ionospheric and solar wind plasma in the magnetosphere. To distinguish these sources of plasma, one generally looks for O^+ as a proxy for an ionospheric source. When the composition is primarily H^+, it is assumed that the source is the solar wind even though the ionosphere can be a major source of H^+. *Hamilton et al.* [1988] found that the storm time ring current was primarily H^+ for moderate magnetic storms, but for the one great magnetic storm observed by AMPTE, the ring current was dominated by O^+ during the storm's maximum phase. These observations, made during solar minimum, were extended by *Daglis* [1997] who used CRRES data to show that the large magnetic storms during solar maximum were also dominated by O^+. This supports the hypothesis of *Hamilton et al.* [1988] that the initial fast Dst recovery during great magnetic storms

[1] Space Sciences Lab., University of California, Berkeley, CA
[2] IGPP, University of California, Los Angeles, CA
[3] Space Science Center, University of NH, Durham, NH

Space Weather
Geophysical Monograph 125

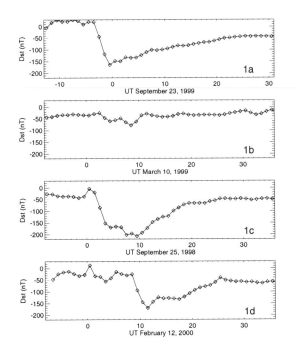

Figure 1. Dst for 99-9-23, 99-3-10, 98-9-25, and 00-2-12.

can be explained in terms of the rapid loss of energetic (∼100 keV) O^+ which has a shorter charge exchange lifetime. More recent observations point to a highly asymmetric ring current [*Grafe*, 1999; *Greenspan and Hamilton*, 2000] during the main phase of storms which is explained by drift-loss ring current models [*Kozyra et al.*, 1998; *Liemohn et al.*, 1999]. In these models both dayside outflows and charge exchange play roles in the evolution of the ring current.

Since one of the primary space weather perturbations is due to the ring current, it is important to identify the source and energization of ring current plasma. Injections of ions at geosynchronous altitudes were observed on early satellite missions [*Arnoldy and Chan*, 1969; *De-Forest and McIlwain*, 1971] and explained in terms of a substorm injection boundary [*Mauk and McIlwain*, 1974]. The relationship between geosynchronous ion injections associated with substorm dipolarizations and storm related ring current ion injections has been a subject of debate [see *McPherron*, 1997, and references therein] including arguments that magnetic storms are the result of a succession of substorm ion injections. It has also been argued [*Kamide et al.*, 1997] that substorm injections are not fundamental to ring current formation since they are limited to the outer L-shells. To affect Dst during magnetic storms, ion injections must place significant plasma on the inner L-shells (L<4).

Ring current injections appear to originate in the plasmasheet as shown by particle tracing through model

electric and magnetic fields [*Chen et al.*, 1994]. *Lennartsson and Sharp* [1982] and *Young et al.*, [1982] found that the plasmasheet (L>5) was typically dominated by H^+ except during disturbed times when O^+ could be significant. During the early phase of magnetic storms, the O^+/H^+ ratios were the highest and the mean plasma energies were elevated. These observations and simulations suggest that an enhanced outflow of ionospheric plasma into the plasmasheet precedes large magnetic storms and that this plasma subsequently convects or is injected into the inner magnetosphere by enhanced solar wind - magnetospheric coupling. This plasma, after betatron heating and perhaps additional nonadiabatic heating, eventually becomes the storm time ring current. The two primary sources of ionospheric outflow into the plasmasheet are the polar cusp and near midnight oval. In this paper we investigate ion injections into the inner magnetosphere and ion outflows from the cusp and midnight oval.

2. LOW LATITUDE OBSERVATIONS

We begin with an example of low latitude ion injection associated with the magnetic storm on September 22-23, 1999. Figure 1a shows the provisional Dst reached a minimum -164 during a three hour drop then gradually recovered over the next two days. Prior to the storm, the Wind spacecraft showed a southward turning of a rather large >20 nT interplanetary magnetic field at ∼20:35 UT on September 22, which continued for about 3 hours. This field should have reached Earth ∼20:50 UT and appears to have triggered a substorm with AE reaching ∼900 thirty minutes later. AE eventually exceeded 1500 several times over the course of the next two hours.

Plate 1 shows a series of three low latitude FAST passes (Orbits 12222, 12223, and 12224) through magnetic noon during this magnetic storm. For each orbit, the H^+, He^+, O^+, and total ion energy flux are plotted from top to bottom. Prior to orbit 12222, FAST detected no significant low latitude ions for several days. The series of plots show an injection of ions that peaks in the middle panel at 2:30 UT and begins to fade on the following orbit. Subsequent orbits show the ions at ILAT<50° disappear after a few orbits, however a more tenuous flux of O^+ ions at 50°<ILAT<60° remain discernible for two days. The primary features of the plot are the rather abrupt energy-latitude injection boundary, the penetration of these ions to at least L=2, and the dominance of O^+ in the injection. Ion injections of this type are observed for all drops in Dst when FAST low latitude data are available.

The sharp upper cutoff of the ion flux between 5 and 10 keV marks the energy boundary where $E \times B$ and ∇B drifts oppose each other along the particle trajectories, resulting in longer drift times and often significant charge exchange losses before the ions reach the dayside. Below this energy cutoff, ions $E \times B$ drift eastward and above this cutoff the ions ∇B drift westward around the Earth. The increase in background counts near the end time of the plots, in particular the ESA plots, is due to penetrating radiation from the inner radiation belt.

We now examine another ion injection associated with a rather weak (Dst=-78) magnetic storm on March 10, 1999 (Figure 1b). The provisional Dst for early March shows this to be an active period after the recovery from a magnetic storm on March 1, with Dst fluctuating between -20 and -50 from March 2 to March 9. Low latitude ions are observed on most of these days, although they are generally weak and remain at L>3. Wind data show a fluctuating solar wind magnetic field (~5-10 nT) with short periods of southward B_z but no extended period with strong southward B_z. The AE index is also generally high, exceeding 500 one or more times each day during this period, indicating an average of several substorms per day. At the end of the day on March 9, B_z turns southward and remains southward until about 5 UT on March 10. B_z then turns northward for ~2.5 hours, then southward for another 4.5 hours, before turning northward again for an extended period (~7 hours). The magnitude of B was slightly higher (B~10 nT) on March 10 during the periods with extended B_z southward. These last two southward turnings are associated with two periods of enhanced AE index (1:40-4:30 and 7:30-10:00 UT) and low latitude ion injections observed by FAST.

Plate 2 shows the low latitude ion injections associated with the drop in Dst on March 10 (Figure 1b), with measurements made at 7-8 MLT. The low latitude ions on orbit 10083 are the product of two ion injections. The first ion injection was observed to peak near orbit 10081 (~5 UT). Those ions between 9:52 and 10:01 UT in Plate 2a are the remnants of this earlier injection, similar to the ions in Plate 1c. The more intense ions between 9:45 and 9:52 UT in Plate 2a were the second ion injection which developed into the more equatorward ion flux seen between 12:04 and 12:14 UT in Plate 2b. By orbit 10085 (Plate 2c) no ions at L<3 (ILAT< 55°) were observed and a gradual recovery of the plasmasphere's density was measured by FAST. (Note: The enhanced counts in the bottom panel of Plate 2c at 14:15 UT are primarily due to penetrating energetic electrons.)

The provisional Dst shows two minima on March 10 (Figure 1b), at 4-5 and 8-9 UT, which appear to be associated with the two ion injections seen in Plate 2. Since the low latitude ions were observed at a local time (~7:30 MLT) far from the injection region near midnight, an ~1 hour delay must be expected between the injection and arrival of convecting ions at FAST (~9:45 UT in Plate 2a). The estimated injection time (~8:45) is consistent with the minimum Dst value at 8-9 UT. This Dst minimum is lower than the minimum at 4-5 UT, which is consistent with two ion injections being present at 8-9 UT as seen in Plate 2a.

We hypothesize that enhanced auroral activity or substorms, as indicated by AE, are responsible for the initial ion outflow into the plasmasheet, and that subsequent convection injected these ions into the inner magnetosphere. It also appears that the first ion injection convected out of the inner magnetosphere shortly after 10 UT since Dst increased from -62 to -35 by 10-11 UT. Similar changes in Dst are observed with the drift-loss ring current model [Kozyra et al., 1998; Liemohn et al., 1999]. Since Dst remained in the -30 to -40 range for most of the day, one might anticipate that the second ion injection remained within the inner magnetosphere when it decoupled from the sunward flows. From the decrease in the southward component of the IMF B_z at ~10 UT, the eventual northward turning of B_z at ~11:30 UT, the drop in AE index to below 500 at ~10 UT, and drop in AE below 200 at ~11:45 UT, the decoupling likely occurred between 10 and 12 UT, with the more probable time closer to 10 UT. Since FAST detected low latitude ions at L<3 at ~12 UT (Plate 2b) and observed their subsequent disappearance on the following orbits, it appears that co-rotation shifted these ions to later local times. Weaker fluxes of ions at L<3 were observed 24 hours later as one would expect if the ions were trapped on co-rotational orbits. The abrupt disappearance of these ions in Plate 2c strongly suggests that these low latitude injections are localized in both space and time. In the following sections we will look closely at auroral ion outflows in an attempt to investigate whether auroral activity can account for these low latitude ions.

3. CUSP OUTFLOW

The polar cusp has long been known to be a source of ionospheric outflow [Lockwood, 1985; Moore et al., 1996]. Yau et al. [1985] presented statistics of ion outflow which showed the largest average outflows were in the vicinity of the cusp. However statistical studies are not necessarily representative of the large outflows observed during active periods associated with magnetic

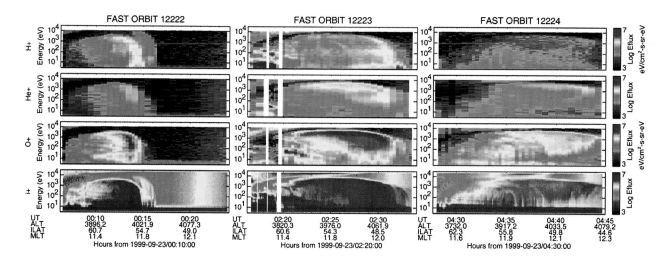

Plate 1. Plots of low latitude ion injections during FAST orbits 12222,12223,12224 on 99-9-23. This injection is characteristic of many injections observed by FAST during magnetic storms. The ion injection is primarily O$^+$ and evolves on two hour time scales. Penetrating radiation contaminates the ESA data (bottom plot) near 45° invariant latitude (ILAT).

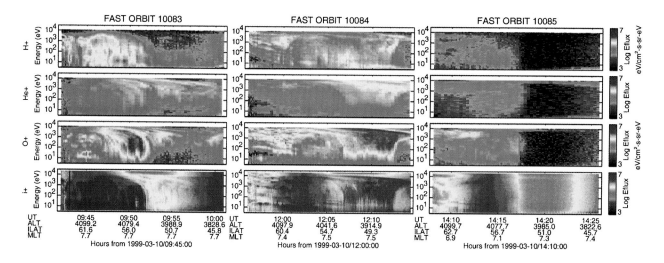

Plate 2. Plots of low latitude ion injections during FAST orbits 10083,10084,10085 on 99-3-10. Orbit 10083 has both a new injection (> 54° ILAT) and the remains of an earlier injection (< 54° ILAT). Orbit 10084 shows the evolution of the new injection from orbit 10083, and orbit 10085 shows the ions at ILAT< 55° have disappeared due to co-rotation.

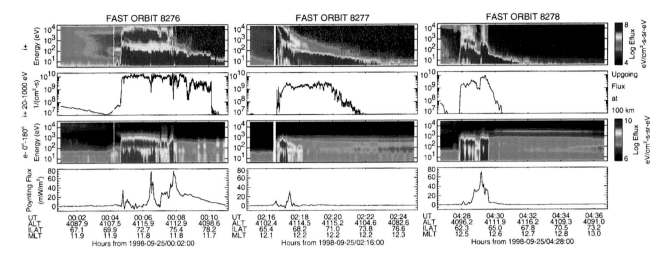

Plate 3. Ion spectrogram, upgoing ion flux, electron spectrogram and Poynting flux in the cusp for FAST orbits 8276, 8277, 8278 on 98-9-25. Conic outflows from the cusp during this magnetic storm were some of the largest observed during the FAST mission.

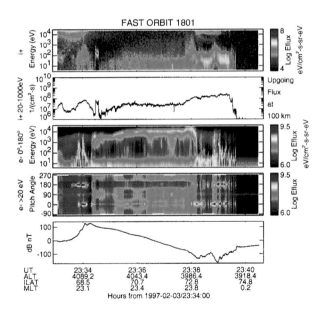

Plate 4. Ions, upgoing ion flux, electrons, electron pitch angle distribution, and magnetic field minus model field for a midnight crossing of the auroral oval. Plot shows conic outflows from the downward (∼23:34 UT) and upward (23:34:20-23:38:10 UT) current regions are much smaller than polar cap boundary conic outflows (23:38:10-23:39:30).

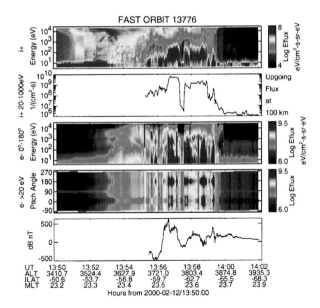

Plate 5. A polar cap boundary (PCB) conic outflow during the magnetic storm on 00-2-12. Format is the same as Plate 4. The figure shows PCB conic outflows (second panel) during magnetic storms can be as large as those observed from the cusp (Plate 3).

storms. Instead we examine an individual magnetic storm and look for the source of ionospheric outflow in the cusp and compare timing of this outflow with the drops in Dst. We begin with a summary of the cusp outflows observed by FAST for the September 24, 1998 magnetic storm.

This magnetic storm is associated with a CME that reached the magnetopause at ~23:45 UT on September 24. Using observations from the Polar spacecraft, *Moore et al.* [1999] showed energetic ion outflows were observed in the northern cusp region for several hours after the arrival of the CME, with enhanced outflows continuing during the Polar southern auroral pass seven hours later. Plate 3 shows the FAST northern auroral crossings on the first three orbits during the September 25, 1998 magnetic storm. Orbit 8276 shows an enhanced ion outflow in the cusp region between 00:05 and 00:10 UT that exceeds 10^{10}/cm^2-s when projected to 100 km. On orbit 8277, cusp outflows are reduced by nearly an order of magnitude to $\sim 2 \times 10^9$/cm^2-s, but return to nearly 10^{10}/cm^2-s on orbit 8278. *Strangeway et al.* [2000] show the outflow is correlated with the Poynting flux (bottom panels).

The extended period of strong ion outflow makes the September 25, 1998 cusp event a prime candidate for supplying ionospheric O$^+$ to the plasmasheet and eventually to the storm ring current. In addition, the high O$^+$ outflow energies reported by *Moore et al.* [1999] and observed by TIMAS were more than adequate for O$^+$ escape into the mantle and eventually into the plasmasheet. The peak of the magnetic storm occurred about 10 hours after the cusp ion outflows began which is adequate time for cusp plasma to convect across the pole and into the plasmasheet, and return to the nightside ring current.

What is not so clear is whether cusp ion outflows can account for the initial drop in Dst. A typical convection time for cusp field lines to move across the polar cap at hundreds of meters per second [*Dungey*, 1965] and return to the inner magnetosphere is ~4 hours. During active periods the cross polar cap potential can increase so the convection time may be reduced to as little as 2-3 hours. Dst had dropped to -85 at 2-3 UT and continued to -152 at 3-4 UT on September 25. Thus the initial phase of the magnetic storm could be due to cusp ion outflow being recycled to the inner magnetosphere. However, *Moore et al.* [1999] show that the cusp ion outflow was relatively high in velocity (~100 km/s) so that these ions would be far down the tail (>50 R_e) by the time the field line convected to the nightside oval. If a near Earth neutral line is providing reconnection in the tail, as expected during active periods, then much

of the cusp outflow would be lost down the tail.

Another possibility is that a high density plasma was already present in the plasmasheet, or injected shortly after the arrival of the CME, and was subsequently convected into the inner magnetosphere causing the initial drop in Dst. FAST data show no atypical large cusp or large near-midnight outflows on orbits prior to the CME arrival. However, the northern border of the polar cap at 00:25 UT was located at ILAT~82° suggesting that a major substorm had occurred prior to the FAST pass. An examination of Polar UVI images shows that a major substorm began shortly after the arrival of the CME, with all of the nightside oval brightened, and with the near midnight brightening extending to ILAT> 80° (private communication, George Parks). AE rose from a little over 500 to nearly 2000 at the arrival of the CME indicating the CME probably triggered this intense substorm. As we will show below, intense outflows are observed from the nightside oval during substorms. Thus ion outflow associated with the substorm may have provided the initial plasma that formed the ring current during the early phase of this magnetic storm.

4. ION OUTFLOWS NEAR MIDNIGHT

Near midnight ion outflows fall into 3 categories: outflow associated with inverted-V electrons or upward current regions, outflow associated with upgoing electron beams or downward current regions, and outflow associated with counterstreaming electrons near the polar cap boundary. Plate 4 shows a FAST pass near midnight that contains all three outflows. Starting at the top, the panels show ion energy spectra, upgoing ion number flux between 20 and 1000 eV, electron spectra, electron pitch angles, and eastward magnetic perturbation field. A positive (negative) slope in the perturbation field denotes a downward (upward) east-west aligned current sheet. The period from 23:34:20 to 23:38:15 UT is an example of an upward current region where the low altitude ion outflows are typically low energy (<30 eV) conics that form more energetic ion beams as they pass through the inverted-V acceleration region. (An ion beam appears briefly in the top panel at 23:34:30 UT.) These conics are associated with electromagnetic ion cyclotron (EMIC) waves and typical fluxes are < 10^8/cm^2-s when mapped to 100 km.

Ion outflow in the downward current region, 23:33:25-23:34:20 UT, consists of more energetic conics [*Carlson et al.*, 1998] trapped below a parallel electric field [*Gorney et al.*, 1985], with heating provided by broadband electric low frequency (BBELF) waves and/or electron solitary waves [*Ergun et al.*, 1998]. Although these ions constitute the highest energy conics, their fluxes are modest and similar to upward current region fluxes.

The region of ion outflow near the polar cap, 23:38:15 to 23:39:10, is associated with counterstreaming electrons and contains highly structured, filamentary, upward and downward currents (bottom panel). Velocity dispersed downgoing ions (although not clear in Plate 4) are often observed in this region showing that these outflows occur on newly reconnected field lines [Zelenyi et al., 1990]. Polar cap boundary (PCB) ion conics constitute the most intense ion outflows observed during the FAST campaign, with typical fluxes of 10^8-10^9/cm^2-s. The association of these outflows with BBELF and counterstreaming electrons suggests that these waves provide the ion heating and that the waves are generated by the counterstreaming electrons.

Using data from the FAST satellite, Tung et al. [2000] examined PCB ion conics and found a correlation of these outflows with substorms. Thus substorms represent a mechanism for injecting a large flux of ionospheric plasma into the plasmasheet. The observed outflows were primarily light ions, H$^+$ and He$^+$, rather than O$^+$. Total ion outflows above 20 eV were conservatively estimated (using a 1-2 hour local time extent) to be between 10^{23}/s and 10^{24}/s. These estimated outflows represent less than 1% of the average auroral outflows presented by Yau et al. [1985], and thus may seem unimportant to the total ion outflow budget. However, the DE-1 observations were taken near solar maximum and included non-winter hemisphere outflows, both of which can greatly enhance total outflow. The Tung et al. [2000] study took place in the winter hemisphere during solar minimum, and thus may not be representative of PCB conic outflow during more active or in sunlit conditions. Below we will show an example of a much larger PCB conic outflow associated with a magnetic storm where the conics are primarily O$^+$.

Plate 5 shows a PCB conic outflow event in the southern hemisphere during the magnetic storm on February 12, 2000, with the same format as Plate 4. This particular outflow occurred in the middle of the magnetic storm (Figure 1d) and during a period of enhanced AE (>1000). Plate 5 shows what may have been the beginning of a poleward expansion phase of a substorm since the aurora is rather far south (ILAT< 65°) but with fairly large energy flux ($\sim$ 10 ergs/cm^2-s). On the following orbit ($\sim$16 UT), the polar cap boundary was located at a more nominal ILAT$\sim$69° and the aurora was much weaker (<1 erg/cm^2-s). Counterstreaming electrons and the associated PCB ion conics were observed over 6° of invariant latitude. The PCB conic outflow in Plate 5 appears to encompass most of the width of the oval, with only diffuse aurora appearing to the south. Unlike the outflows described by Tung et al. [2000], this outflow was dominated by O$^+$. Low latitude ions observed by FAST were also primarily O$^+$. Since this outflow occurred in the summer hemisphere on the rise to solar max, enhanced solar output and sunlight conditions may explain the enhanced O$^+$.

Similar PCB conic outflows (10^9-10^{10}/cm^2-s) were observed during the two previous FAST orbits. On orbit 13774, which skimmed along the auroral oval, PCB conic outflows were measured over a wide local time (1 to 8 MLT) demonstrating that during active periods these outflows are not necessarily restricted to midnight. The largest outflows (> 10^{10}/cm^2-s) were also observed on this orbit ($\sim$10 UT) and may have provided the ionospheric plasma to the plasmasheet that subsequently formed the ring current, producing the peak in the magnetic storm (Figure 1d). Since this outflow was over a large local time range and with fluxes that were 1-2 orders of magnitude larger than those studied by Tung et al. [2000], it is likely that the PCB outflows dominated the total ion outflow.

As seen in Figure 1d, the provisional Dst for the February 12, 2000 magnetic storm begins in hour 9-10 UT with a drop from -25 to -94. The provisional Dst had a minimum of -169 in hour 11-12 UT, recovered to -124 in hour 13-14 UT, then remained near this value for about 5 additional hours before beginning a rather rapid decay to greater than -60 the next day. If one tries to associate these drops in Dst with ion outflows observed on FAST and with enhanced AE, one can hypothesize the following.

We begin with the initial Dst drop at 3-4 UT on February 12. Three periods of enhanced AE were observed prior to the drop in Dst. No FAST auroral data were available for the first two periods, however FAST did measure near midnight outflows of $\sim$$10^9$/cm^2-s at $\sim$03:18 UT as AE declined to $\sim$500. At this time the polar cap boundary was located at ILAT$\sim$74.5° near midnight and the PCB electron precipitation was weak ($\sim$1 erg/cm^2-s). These observations, when combined with a declining AE, indicate the 3:18 UT PCB outflow occurred at the end of a substorm expansion. If similar outflows occurred during the first two periods of enhanced AE, these ions would be convected into the inner magnetosphere during the third period of enhanced AE and may be responsible for the short drop in Dst in hour 3-4 UT on February 12.

We now examine the main phase of the magnetic storm. A fourth period of enhanced (>1000) AE occurred at 9:50-10:10 UT and is likely related to the strong southward B$_z$ that began shortly before 8 UT.

FAST observed extremely large PCB O^+ conic outflows ($\sim 10^{10}$/cm^2-s) at $\sim$9:45 UT over an extended local time (1-8 MLT). This large outflow appears $\sim$1 hour before the minimum Dst. The IMF had a strong southward component until $\sim$11:30 UT and should have strongly driven magnetospheric convection. This suggests that the largest PCB conic outflow into the plasmasheet was likely convected to the inner magnetosphere and formed the ring current responsible for the minimum in Dst at 11-12 UT. AE continued to be relatively active and the following two FAST orbits measured additional intense PCB outflows (see Plate 5). These outflows provided additional ions to the plasmasheet which, when convected to the inner magnetosphere, may have kept Dst relatively low until $\sim$19 UT. Although the above associations cannot prove that PCB outflows provide the ionospheric plasma that produced this magnetic storm, they provide strong circumstantial evidence that substorm outflows are a primary driver of Dst.

5. CONCLUSION

Models that attempt to predict the space weather associated with magnetic storms must be able to account for the relative abundance of ionospheric and solar wind plasma in the inner magnetosphere. Observations from FAST show that the low energy portion of the storm injected ring current is primarily of ionospheric origin and related to auroral activity as indicated by AE. Intense ion conic outflows from the cusp and a near midnight polar cap boundary (PCB) were examined for two storms. The cusp outflows were likely ejected out the tail whereas the substorm related PCB outflows contribute substantial ionospheric plasma to the plasmasheet. These observations suggest how substorms can play a role in the development of magnetic storms. Electrons accelerated near the polar cap boundary during substorms generate waves that heat the ionospheric plasma producing intense conic outflows into the plasmasheet. The higher density plasmasheet ions are subsequently convected to the inner magnetosphere by the large convection electric fields during the magnetic storm. Without the large convection fields associated with magnetic storms, the substorm outflows have no effect on Dst since these ions never reach the inner L-shells. Without substorm outflows, the plasmasheet density is relatively modest so that convection to the inner L-shells produces only small drops in Dst. In this model, it is through ionospheric ion outflows that substorms affect the development of magnetic storms.

Acknowledgments. The analysis of FAST data was supported by NASA grant NAG-3596. Dst values were obtained from the web site http://swdcdb.kugi.kyoto-u.ac.jp and we thank T. Kamei and M. Sugiura for preparation of these quantities.

REFERENCES

Arnoldy, R. L., and K. W. Chan, Particle Substorms observed at the geostationary orbit, *J. Geophys. Res., 74,* 5019, 1969.

Carlson, C. W., et al., FAST observations in the downward auroral current region: energetic upgoing electron beams, parallel potential drops, and ion heating, *Geophys. Res. Lett., 25,* 2017, 1998.

Chen, M. W., L. R. Lyons, and M. Schulz, Simulations of phase space distributions of storm time proton ring current, *J. Geophys. Res., 99,* 5745, 1994.

Daglis, I. A., The role of magnetosphere-ionosphere coupling in magnetic storm dynamics, in Magnetic Storms, *Geophys. Mono. 98,* 107, 1997.

DeForest, S. E., and C. E. McIlwain, Plasma clouds in the magnetosphere, *J. Geophys. Res., 76,* 3587, 1971.

Dungey, J. W., The length of the magnetospheric tail, *J. Geophys. Res., 70,* 1753, 1965.

Ergun, R. E., et al., FAST satellite observations of large-amplitude solitary waves, *Geophys. Res. Lett., 25,* 2041, 1998.

Gorney, D. J., Y. T. Chiu, and D. R. Croley, Jr., Trapping of ion conics by downward parallel electric fields, *J. Geophys. Res., 90,* 4205, 1985.

Grafe, A., Are our ideas about Dst correct?, *Ann. Geophys., 17,* 1, 1999.

Greenspan, M. E., and D. C. Hamilton, A test of the Dessler-Parker-Sckopke relation during magnetic storms, *J. Geophys. Res., 105,* 5419, 2000.

Hamilton, D. C., et al., Ring current development during the great geomagnetic storm of February 1986, *J. Geophys. Res., 93,* 14,343, 1988.

Kamide, Y., et al., Magnetic storms: current understanding and outstanding questions, in "Magnetic Storms", *Geophys. Mono. 98,* 1, 1997.

Kozyra, J. U., et al., Effects of a high-density plasma sheet on ring current development during November 2-6, 1993, magnetic storm, *J. Geophys. Res., 103,* 26,285, 1998.

Lennartsson, W., R. D. Sharp, A comparison of the 0.1-17 keV/e ion composition in the near equatorial magnetosphere between quiet and disturbed conditions, *J. Geophys. Res., 87,* 6109, 1982.

Liemohn, M. W., et al., Analysis of early phase ring current recovery mechanisms during geomagnetic storms, *Geophys. Res. Lett., 26,* 2845, 1999.

Lockwood, M., et al., The cleft ion fountain, *J. Geophys. Res., 90,* 9736, 1985.

Mauk, B. H., and C. E. McIlwain, Correlation of Kp with the substorm-injected plasma boundary, *J. Geophys. Res., 79,* 3193, 1974.

McPherron, R. L., The role of substorms in the generation of magnetic storms, in "Magnetic Storms", *Geophys. Mono. 98,* 131, 1997.

Moore, T. E., et al., The cleft ion plasma environment at low solar activity, *Geophys. Res. Lett., 23,* 1877, 1996.

Moore, T. E., et al., Ionospheric mass ejection in response to a CME, *Geophys. Res. Lett., 26,* 2339, 1999.

Strangeway, R. J., et al., Cusp field-aligned currents and ion outflows, *J. Geophys. Res., 105,* 21129, 2000.

Tung, Y.-K., et al., Auroral polar cap boundary ion outflow observed on FAST, *J. Geophys. Res.,* in press, 2000.

Yau, A. W., et al., Energetic auroral and polar ion outflow at DE 1 altitudes: magnitude, composition, magnetic activity dependence, and long-term variations, *J. Geophys. Res., 90,* 8417, 1985.

Young, D. T., H. Balsiger, and J. Geiss, Correlations of magnetospheric ion composition with geomagnetic and solar activity, *J. Geophys. Res., 87,* 9077, 1982.

Zelenyi, L. M., R. A. Kovrazkhin, and J. M. Bosqued, Velocity-dispersed ion beams in the nightside auroral zone: AUREOL 3 observations, *J. Geophys. Res., 95,* 12119, 1990.

J. P. McFadden, Y. K. Tung, and C. W. Carlson, Space Sciences Laboratory, University of California, Berkeley, CA 94720. (e-mail: mcfadden@ssl.berkeley.edu)

E. Moebius and L. M. Kistler, Space Science Center, University of New Hampshire, Durham, NH 03824.

R. J. Strangeway, Institute of Geophysics and Planetary Physics, University of California, Los Angeles, CA 90095.

Specification and Forecasting of Outages on Satellite Communication and Navigation Systems

S. Basu and K. M. Groves

Space Vehicles Directorate, Air Force Research Laboratory, 29 Randolph Road, Hanscom AFB, MA 01731

The ionized upper atmosphere often develops electron density irregularities which cause amplitude and phase scintillations of satellite signals. Scintillations, when intense, can cause outages in satellite communication systems and in Global Positioning System (GPS) for navigation. The performance of these systems can be improved by providing global specification and forecast of scintillation. A scintillation specification system, Scintillation Network Decision Aid (SCINDA), has been developed for the South American sector. An equatorial satellite, Communication Navigation Outage Forecasting System (C/NOFS), equipped with suitable sensors, has been planned for the specification and forecast of equatorial scintillation.

INTRODUCTION

The earth's ionized upper atmosphere often becomes turbulent and develops irregularities of electron density. These irregularities scatter radio waves from satellites in the frequency range of 100 MHz – 4 GHz (Basu et al., 1988; Aarons, 1993; Aarons and Basu, 1994). In the presence of a relative motion between the satellite, the ionosphere and the receiver, the received signal exhibits temporal fluctuations of intensity and phase, called scintillations. Intensity scintillations cause signals to fade below the average level. When the depth of fading exceeds the fade margin of a receiver, the signal becomes buried in noise and signal loss and cycle slips are encountered. Phase scintillations induce frequency shift and, when this shift exceeds the phase lock loop bandwidth, the signal is lost and the receiver spends valuable time to reacquire the signal. Overall, in the presence of scintillations, the performance of communication and navigation systems is degraded. When loss of signal occurs in a satellite based communication or navigation system due to scintillation, the problem is often attributed to an interference or to a failure of equipment at the transmitting or the receiving end, or to the satellite itself. If forecasts of scintillation can be provided, the users will not waste their resources and may instead evolve alternate strategies.

Scintillations are strong at high latitudes, weak at middle latitudes and intense in the equatorial region (Basu et al., 1988). Scintillation at all latitudes attains its maximum value during the solar maximum period when the F-region ionization density increases and the irregularities occur in a background of enhanced ionization density. As such, scintillation effects, expected during the upcoming solar maximum period between 2001-2004, are of concern to systems engineers. Scintillation during the solar minimum period is much reduced in magnitude mainly because of decreased background ionization density. The scintillation climatology is well-known but the space-time variability of scintillation and its trigger mechanisms remain unresolved (Basu et al., 1996).

An empirical scintillation model, WBMOD, is available (Secan et al., 1995, 1997). The name of the model was derived from the Wideband satellite which provided the multi-frequency scintillation data input to the model. Later, the model was upgraded in the equatorial region with extensive geostationary satellite scintillation database of the

Space Weather
Geophysical Monograph 125

"WORST CASE" FADING DEPTHS AT L-BAND

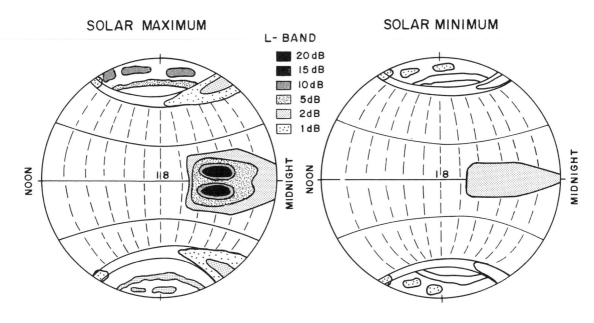

Figure 1. Schematic of the global morphology of scintillations at L-band frequencies during the solar maximum (left panel) and solar minimum (right panel) conditions.

Air Force Research Laboratory at Hanscom AFB, MA. The high latitude upgrade of the model was achieved by using the HiLat and Polar Bear satellite scintillation data. Since scintillation exhibits extreme variability in space and time, such climatological models are useful only for planning purposes. For the support of space based communication and navigation systems, we require weather models, i.e., real-time specification and forecast of scintillation. Such weather models provide, at least, the situational awareness to the user who, under certain circumstances, can switch to other satellites for a better satellite link performance.

This paper provides an overview of scintillation at high and low latitudes. The issues related to the specification and forecast of polar scintillation are discussed. A scintillation specification system, developed for the South American sector, is outlined (Groves et al., 1997). It also describes an equatorial satellite, called the Communication/Navigation Outage Forecasting System, C/NOFS, which has been planned for specifying and forecasting scintillation in the equatorial region.

RESULTS AND DISCUSSIONS

Figure 1 shows an updated schematic of the global distribution of worst case scintillation at L-band frequencies during the solar maximum and the minimum period, originally published by Basu et al., 1988. The left hand panel

represents the solar maximum condition when scintillation attains its maximum value. The magnetic north and south poles are at the top and the bottom and the magnetic equator is in the middle. The noon meridian is on the left, midnight is on the right and 18 LT is in the middle. It may be noted that L-band scintillation is most intense in the equatorial region, moderate at high latitudes and generally absent at middle latitudes.

At high latitudes, scintillations are found to be associated with large scale plasma structures. It has been established that for the interplanetary magnetic field component B_z southward, large scale plasma structures, patches and blobs, are observed. The polar cap patches are convected through the dayside auroral oval into the polar cap and then exit into the nightside auroral oval to form auroral blobs (Weber et al., 1984; Tsunoda, 1988; Carlson, 1994; Rodger et al., 1994; Basu and Valladares, 1999; Pedersen et al., 2000). The association of polar cap patches with intermediate scale irregularities (tens of km to tens of meters), responsible for intense scintillations, has been examined in the framework of observations and modeling (Basu et al., 1995; Basu and Valladares, 1999). When the B_z component of the interplanetary magnetic field (IMF) is northward, ordered reversals of horizontal plasma flow occur in the polar cap. A soft particle precipitation along the flow reversal forms long sun-aligned polar cap arcs (Kelley, pg. 364, 1989). The arcs become associated with small-scale

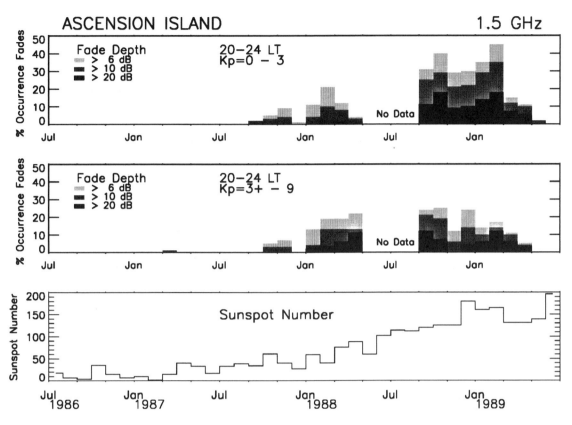

Figure 2. Variation of the occurrence of 1.5 GHz scintillation with the sunspot number observed at Ascension Island near the crest of the equatorial anomaly during the pre-midnight period. The top panel shows the occurrence statistics for magnetically quiet (K_p = 0 - 3) conditions, the middle panel for disturbed ($K_p = 3^+$ - 9) conditions and the bottom panel shows the variation of the sunspot number.

irregularities of electron density due to plasma instabilities in the velocity shear region (Keskinen et al., 1988). Scintillations associated with polar cap patches are most intense owing to the high ionization density of patches and sun-aligned arcs, associated with low ionization density, cause only weak scintillations. Polar scintillation is most conspicuous in winter when the solar ionizing radiation is not available to smooth out the irregularities.

In the equatorial region, at the time of sunset, the magnetic field line integrated ionospheric conductivity changes rather abruptly across the sunset line or the terminator. The zonal neutral wind and the terminator induced conductivity gradient interact and drive the ionosphere unstable (Kelley, pg. 121, 1989). Large scale plasma bubbles are formed in the bottomside of the ionosphere which rise to great heights and become populated by electron density irregularities (Bernhardt et al., 2000). The tens of meter to kilometer scale irregularities, distributed over a wide range of altitudes, cause intense L-band scintillation. Being associated with relatively smaller scale irregularities, L-band scintillation decays shortly after midnight whereas 250 MHz scintillation may persist till sunrise.

Figure 2 shows the occurrence statistics of 1.5 GHz scintillation at Ascension Island during premidnight hours (20-24 LT). This station is located under the crest of the equatorial anomaly in F-region ionization, also known as the Appleton anomaly, where the ionospheric turbulence is encountered in an environment of high background ionization density. The top two panels correspond to magnetically quiet and active periods respectively and the bottom panel shows the variation of the sunspot number. There is no data during June-August, 1988, but past data indicate that during these months scintillation occurrence is minimum in the American-Atlantic longitude sector. During the solar maximum period, L-band scintillations can indeed be intense causing signals to fade below 20 dB. It is interesting to note that scintillations in the equatorial region are much enhanced during magnetically quiet period (top panel) as compared to the magnetically disturbed period (middle panel). This may be due to the fact that quiet-time scintillations follow a definite occurrence pattern, starting after sunset and decaying after midnight, whereas the disturbed period scintillation is distributed in time being dictated by the onset time of magnetic storms. The generation

CHILE: 1 OCTOBER 1994

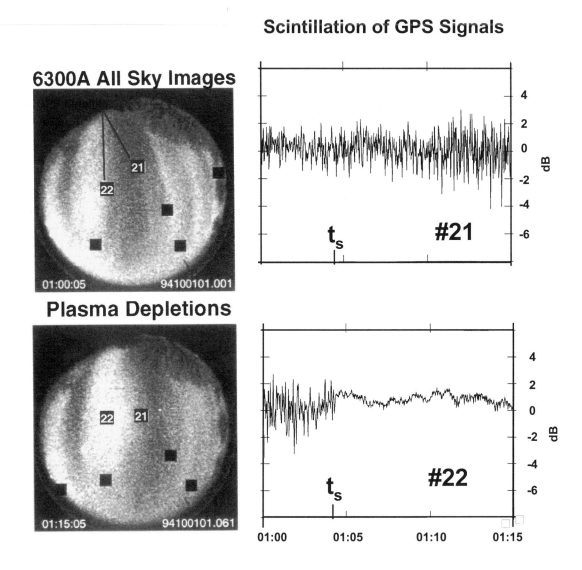

Figure 3. The left panels show the 630.0 nm all sky images of plasma depletions at 0100 UT and 0115 UT obtained at Chile near the crest of the equatorial anomaly on 1 October 1994. The numbers 21 and 22 represent 250 km intersections of raypaths of two specific GPS satellites, while the black squares denote additional GPS satellite intersection points. The right hand panels show the scintillation of GPS satellites (PRN 21 and 22) during 0100 – 0115 UT, the tic marks in the ordinate represent 1 dB intervals.

mechanisms of equatorial scintillation have not been identified but it seems that it is internally driven and cannot be predicted by following the trail of energy from the sun.

Figure 3 shows the results of simultaneous GPS scintillation and 630.0 nm all sky imager observations near the equatorial anomaly region in Chile (Weber et al., 1996). The left hand panels illustrate two 630.0 nm images at 15 minute intervals. The north-south elongated dark bands represent the footprints of plasma bubbles at 250 km altitude. The positions marked 21 and 22 indicate the intersections of the ray paths from the ground-based GPS receiver to GPS satellites with PRN numbers 21 and 22. Each GPS satellite is identified by the specific pseudorandom noise (PRN) code it transmits. The right hand panels indicate that both satellites scintillate when both raypaths intersect the dark band (see top left image). A few minutes later, at time t=t$_s$, owing to the eastward motion of plasma bubbles as well as the satellite motion, the satellite with PRN 22 emerges from the scintillating region. The other satellite continues to encounter scintillations as its raypath remains within the dark band. It should be pointed out that only weak scintillations of GPS signals, corresponding to a signal excursion of only 5 dB, were observed during this solar minimum period. During the solar maximum period, scintillation of GPS signals at this location exceeds 20 dB. It should also be mentioned that measurements with GPS satellites offer a unique opportunity to perform multi-point scintillation observations from one station.

Figure 4 illustrates four azimuth-elevation plots of the trajectories of GPS satellites at one hour intervals, as viewed from Antofagasta, Chile, on 26 October 1999. The concentric circles indicate elevation angles at 30^0 intervals, the outermost circle indicating an elevation angle of 0^0 and the centre of the circles corresponds to the zenith. The azimuth is reckoned with reference to the four cardinal directions, namely, north (N), east (E), south (S) and west (W) as indicated in the diagram. The trajectories of GPS satellites identified by their PRN numbers are shown. The varying diameters of circles along the satellite trajectories indicate the variation of the intensity scintillation index, S_4, which is defined as the ratio of the standard deviation of signal intensity and the average signal intensity. It may be noted that S_4 indices as high as 0.9, corresponding to signal fluctuations exceeding 20 dB, were measured in 1999 with the solar maximum period approaching. Scintillation indices are more than a factor five higher than that in Figure 3. Further, at this location, which is equatorward of the anomaly crest, the east-west widths of depletions also increase as the solar maximum is approached. Thus a larger number of GPS satellites will simultaneously suffer strong scintillations in the equatorial region during the solar maximum period.

In the Introduction, it has been mentioned that a global climatological model of scintillation, WBMOD, is available. The input to this empirical model corresponds to the sunspot number, magnetic index, day number, universal time, the locations of the satellite and the receiver and the signal frequency. The model output specifies the magnitudes of phase and intensity scintillation and their temporal structures as defined by their spectra. The model is climatological in nature and is useful for planning purposes. It is, however, of limited value to many users in view of the extreme day-to-day variability of scintillation, particularly in the equatorial region (Basu et al., 1996). Since the cause of this variability is not yet resolved, a physics based predic-

tion model is not possible. This has led to the development of a real-time local area scintillation specification system based on two station measurements in the South American sector known as SCINDA, Scintillation Network Decision Aid (Groves et al., 1997). The magnitude of scintillation and the zonal irregularity drift is measured by using spaced receiver scintillation measurements. Such measurements are made at 250 MHz with two geostationary satellites in the east and the west of two stations, one near the magnetic equator (Ancon, Peru) and the other at about 10° magnetic latitude (Antofagasta, Chile) at about the same longitude. The data is brought to the user by the internet and the data drives a scintillation model which considers the upwelling and the zonal motion of irregularities as well as their mapping along the magnetic field to produce three dimensional scintillation structures. Plate 1 shows such structures mapped by the SCINDA system, and the colors, green, yellow and red indicate the increasing levels of scintillation.

The SCINDA system is currently incorporating multi-point GPS L-band scintillation measurements at the two stations. It also utilizes the DMSP (Defense Meterological Satellite Program) satellite in-situ measurements of the latitude variation of electron density around the magnetic equator near the longitude of the SCINDA stations. By using the pre-sunset DMSP measurements of the nature of latitude variation of electron density around the magnetic equator, which is dictated by the zonal electric field, it has been possible to predict nighttime scintillation two hours in advance of its onset (Basu et al., 2000).

In the near future, a space-based system, capable of specifying and forecasting equatorial scintillation, is expected. Figure 5 shows the conceptual schematic of this satellite, called the Communication/Navigation Outage Forecasting System (C/NOFS) (Bernhardt et al., 2000). The satellite will be launched in an elliptical orbit 400 km x 700 km with an orbital inclination of 13°. The satellite in-situ sensors will include a Langmuir probe, digital ion drift meter, vector electric field and neutral wind sensors. It will also incorporate a tri-frequency beacon (150, 400, 1067 MHz) and a GPS occultation receiver. These sensors are expected to forecast the occurrence of scintillation by probing the destabilizing and stabilizing forces in the ionosphere, namely the pre-reversal enhancement of zonal electric field from the electric field, and zonal/meridional neutral wind measurements. Real-time data driven electron density models will be derived and validated against electron density profiles obtained by the GPS occultation sensor. Radio wave scattering theory will be combined with electron density irregularity data and electron density profiles to specify amplitude and phase scintillation and their

Antofagasta - 27 October 1999

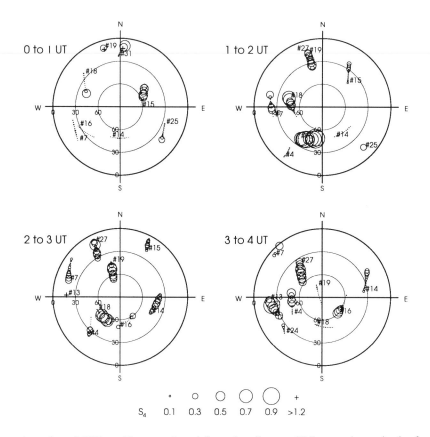

Figure 4. The trajectories of GPS satellites, as viewed from Antofagasta, Chile, are shown in the four sets altitude-azimuth plots at hourly intervals. The four concentric circles indicate elevation angles at 30^0 intervals, the outermost circle corresponding to 0^0 elevation angle and the centre of the circles indicating the zenith. The circles around the GPS satellite trajectories indicate the S_4 index of scintillation according to the legend at the bottom of the left panel.

temporal structures. Scintillation specifications will be validated by the measurements of scintillation at three beacon frequencies. The space-based C/NOFS and the ground based SCINDA will be able to specify, forecast and validate scintillation products for the users.

SUMMARY

During the solar maximum period, ionospheric scintillation can affect many space based communication and navigation systems in the VHF/UHF range. The global climatology of scintillation in the VHF/UHF range of frequencies is fairly well established and it can be used in the planning of new communication links. However, in view of the extreme variability of scintillation in space and time, the climate models cannot support operational systems that require real-time scintillation specification.

At high latitudes, during solar maximum conditions, scintillation caused by polar cap patches is of concern to

many systems. The large scale polar cap patches have been modeled in the framework of convection appropriate for southward IMF. For the development of high latitude scintillation specification and forecast system, we need to determine the trajectories of patches for varying IMF and the saturation amplitude of irregularities need to be established from the standpoint of plasma instabilities.

In the equatorial region, scintillation specification system is a reality. Systems, such as SCINDA, could be developed primarily because the large scale irregularity motion is ordered and the irregularities causing scintillations can be mapped between the northern and southern crests of the equatorial anomaly. Although theoretical studies, intensive observations and refined modeling have continued for many decades, the forecasting of equatorial scintillation remains a challenging task. It is hoped that satellites, such as C/NOFS, will be able to track simultaneously the stabilizing and the destabilizing forces and thereby advance our capability to forecast equatorial scintillation.

Scintillation Network Decision Aid
(SCINDA)

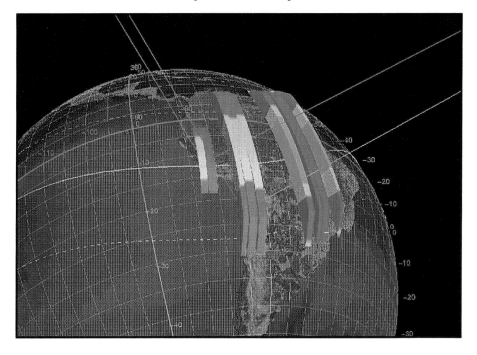

Plate 1. Illustrates the three-dimensional irregularity structures mapped by the Scintillation Network Decision Aid, SCINDA, in South America, which is based on an irregularity model and 250 MHz scintillation measurements made at Ancon, Peru, and Antofagasta, Chile.

C/NOFS

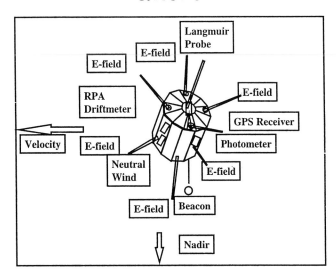

Figure 5. Schematic of the Communication/Navigation Outage Forecasting System (C/NOFS), a proposed equatorial satellite with its suite of sensors.

Acknowledgments. The work at Air Force Research Laboratory was partially supported by the Air Force Office of Scientific Research task 2310G9.

REFERENCES

Aarons, J., The longitudinal morphology of equatorial F-layer irregularities relevant to their occurrence, *Space Sci. Rev., 63,* 209, 1993.

Aarons, J., and S. Basu, Ionospheric amplitude and phase fluctuations at the GPS frequencies. In: Proceedings of ION GPS-94, Arlington, VA, pp. 1569, 1994.

Basu, S., E. MacKenzie, and Su. Basu, Ionospheric constraints on VHF/UHF communication links during solar maximum and minimum periods, *Radio Sci., 23,* 363, 1988.

Basu, S., Su. Basu, J.J. Sojka, R.W. Schunk, and E. MacKenzie, Macroscale modeling and mesoscale observations of plasma density structures in the polar cap, *Geophys. Res. Lett., 22,* 881, 1995.

Basu, S., E. Kudeki, Su. Basu, C.E. Valladares, E.J. Weber, H.P. Zengingonul, S. Bhattacharyya, R. Sheehan, J.W. Meriwether, M.A. Biondi, H. Kuenzler, and J. Espinoza, Scintillations, plasma drifts, and neutral winds in the equatorial ionosphere after sunset, *J. Geophys. Res., 101,* 26795, 1996.

Basu, Su., and C.E. Valladares, Global aspects of plasma structures, *J. Atmos. Solar Terr. Phys., 61,* 127, 1999.

Basu, S., K.M. Groves, Su. Basu, and P.J. Sultan, Specification and forecasting of scintillations in communication and navigation systems: current status and future plans, *J. Atmos. Solar Terr. Phys.,* 2000 (in press).

Bernhardt, P.A., J.D. Huba, C.A. Selcher, K.F. Dymond, G.R. Carruthers, G. Bust, C. Rocken, and T.L. Beach, New systems for space-based monitoring of ionospheric irregularities and radio wave scintillation, this issue, 2000.

Carlson, H.C., Jr., The dark polar ionosphere: Progress and future challenges, *Radio Sci., 29,* 157, 1994.

Groves, K.M., S. Basu, E.J. Weber, M. Smitham, H. Kuenzler, C.E. Valladares, R. Sheehan, E. MacKenzie, J.A. Secan, P. Ning, W.J. McNeill, D.W. Moonan, and M.J. Kendra, Equatorial scintillation and systems support, *Radio Sci., 32,* 2407, 1997.

Kelley, M.C., The Earth's Ionosphere, Academic Press, San Diego, CA, 1989.

Keskinen, M.J., H.G. Mitchell, J.A. Fedder, P. Satyanarayana, S.T. Zalesak, and J.D. Huba, Nonlinear evolution of the Kelvin-Helmholtz instability in the high latitude ionosphere, *J. Geophys. Res., 93,* 137, 1988.

Pedersen, T.R., B.G. Fejer, R.A. Doe, and E.J. Weber, An incoherent scatter radar technique for determining two-dimensional horizontal ionization structure in polar cap F region patches, *J. Geophys. Res., 105,* 10,637, 2000.

Rodger, A.S., M. Pinnock, J.R. Dudney, K.B. Baker, and R.A. Greenwald, A new mechanism for polar patch formation. *J. Geophys. Res., 99,* 6425, 1994.

Secan, J.A., R.M. Bussey, E.J. Fremouw, and S. Basu, An improved model of equatorial scintillation, *Radio Sci., 30,* 607, 1995.

Secan, J.A., R.M. Bussey, E.J. Fremouw, and S. Basu, High latitude upgrade to the Wideband ionospheric scintillation model, *Radio Sci., 32,* 1567, 1997.

Tsunoda, R.T., High latitude F region irregularities: a review and synthesis, *Rev. Geophys., 26,* 719, 1988.

Weber, E.J., J. Buchau, J.G. Moore, J.R. Sharber, R.C. Livingston, J.D. Winningham, and B.W. Reinisch, F-layer ionization patches in the polar cap, *J. Geophys. Res., 89,* 1683, 1984.

Weber, E.J., S. Basu, G. Bishop, T. Bullett, H. Kuenzler, P. Ning, P. Sultan, C.E. Valladares., and J. Araye, Equatorial plasma depletion precursor signatures and onset observed at 11° south of the magnetic equator, *J. Geophys. Res., 101,* 26,829, 1996.

S. Basu and K.M Groves, Air Force Research Laboratory, 29 Randolph Road, Hanscom AFB, MA 01731 (e-mail: santimay@aol.com; Keith.Groves@hanscom.af.mil).

New Systems for Space Based Monitoring
of Ionospheric Irregularities and Radio Wave Scintillations

P. A. Bernhardt[1], J. D. Huba[1], C. A. Selcher[2], K. F. Dymond[3], G. R. Carruthers[3], G. Bust[4], C. Rocken[5], T. L. Beach[6]

The ionosphere has long been known to be the primary source of amplitude and phase fluctuations for VHF, UHF, and L-Band radio waves. Monitoring ionospheric irregularities that affect radio propagation is currently being implemented using *in situ*, radio and optical sensors on satellites in low-earth orbit (LEO). The remote sensing instruments provide observations of extreme ultra-violet (EUV) emissions or total electron content (TEC) to reconstruct images of electron densities. The GPS MET satellite has provided a global description of the ionosphere using GPS occultation from space. The ARGOS satellite launched in February 1999 has produced new images of the ionosphere using both EUV limb scanning and computerized ionospheric tomography (CIT) techniques. Recent measurements from the ARGOS satellite have provided data for improved modeling of both global scale and small-scale structures in the ionosphere. Future satellite sensors will be launched with EUV limb scanners, GPS occultation receivers, radio beacons for CIT, and beacon receivers for global mapping of ionospheric scintillations. These satellites will be placed in a variety of orbit inclinations to cover the equatorial, mid-latitude, and polar ionospheres.

I. INTRODUCTION

The earth's upper atmosphere contains partially ionized plasma that is constantly changing under the influence of

[1]Plasma Physics Division, Naval Research Laboratory, Washington, DC 20375

[2]Information Technology Division, Naval Research Laboratory, Washington, DC 20375

[3]Space Science Division, Naval Research Laboratory, Washington, DC 20375

[4]Applied Research Laboratory, University of Texas at Austin, Austin, TX 78713

[5]GPS Science and Technology Program, UCAR, Boulder, CO 80307

[6]Space Vehicles Directorate, Air Force Research Laboratory, Hanscom AFB, MA 01731

Space Weather
Geophysical Monograph 125

solar extreme ultraviolet (EUV) radiation, recombination chemistry, neutral winds, and electric fields [Rishbeth and Garriott, 1969; Kelley 1989]. The ionosphere extends from 50 km to above 1000-km altitude with variations in ion species balanced by equal densities of electrons. At altitudes below 150 km, the ions are primarily molecular. The peak densities are found in the F-layer near 300 to 400 km where atomic oxygen is the primary ion species. Above 1500-km altitude, the plasmasphere is composed of atomic hydrogen and helium ions along with an equal number of electrons to maintain neutrality. The atmospheric region known as the ionosphere is both important and complex. The ionosphere affects terrestrial radio signals. Satellite to ground links are affected by electron density irregularities that can degrade received signal strength. Communication, radar, and navigation systems often rely on predictions and measurements of ionospheric propagation conditions. Over-the-horizon radars require accurate models of the ionosphere to determine target locations.

A major task for space-weather is the now-casting and forecasting of ionospheric conditions that affect radio wave propagation. Now-casting identifies existing ionospheric conditions that affect radio systems. Forecasting provides a prediction of an ionospheric disturbance that will impact radio communications, navigation, geolocation, surveillance, etc. Forecasting can be as simple as following the motion a known disturbed region in the ionosphere or as complex as modeling the effects of a solar disturbance as it propagates toward the earth to directly or indirectly affect the ionosphere.

II. THE EFFECTS OF IONOSPHERIC IRREGULARITIES ON RADIO SYSTEMS

There are eight major ionospheric propagation effects on satellite-to-ground and ground-to-ground radio links [Davies, 1990]. These are phase fluctuations, amplitude fluctuations, absorption, frequency shifts, Faraday rotation, group delay, scattering, and multipath. The significance of each effect depends on the specific application, system design, and frequency of use. Absorption in the D-region has the strongest effect on medium-wave frequencies (0.3 to 3 MHz). High-frequency waves (3 to 30 MHz) are adversely affected by large and rapid amplitude and phase fluctuations. Very high-frequency waves (30 to 300 MHz) suffer from the effects of Faraday rotation fading and amplitude scintillations from F-region ionospheric irregularities. Ultra high-frequencies (300 MHz to 3 GHz) are still affected in the auroral zones and the equatorial region at certain local times and seasons when F-region irregularities are severe [Basu, Mackenzie, and Basu, 1988; Groves et al., 1997].

As an example, consider the propagation of a spherical wave generated from a source above the ionosphere. If the wave encounters irregularities in the ionosphere, the phase front of the wave will become distorted. Plate 1 shows a calculated phase front disturbance just below a model equatorial irregularity. Such ionospheric "bubbles" are found in the equatorial ionosphere for a few hours after sunset [Kelley, 1989]. The electron densities in the bubble are calculated using the computer model developed at NRL by Zalesak, S.T., S.L. Ossakow, and P.K. Chaturvedi [1982]. In this model the peak density is 7.8×10^5 cm^{-3} and the contours are linearly spaced to zero density below and above the peak at 435 km altitude. The densities are calculated on a rectangular grid with spacing of 5 km.

Scintillations are produced by diffraction from irregularities with scales sizes less than Fresnel zone $L_F = (\lambda z)^{1/2}$ which is on the order of 500 meters with radio wavelengths $l \sim 0.7$ meter and ionospheric distances $z \sim 350$ km. The effects of scintillations are reproduced in the electron density model by adding irregularities with a spatial frequency ($k = 2\pi f_s$) of k^{-p} using the spectral index value of

$p = 3.5$ [Yeh and Liu, 1982; Aarons, 1982] in the range 0.0004 m$^{-1} < f_s < 0.1$ m^{-1}. The irregularities structures are imposed on the electron density bubble in proportion to the magnitude of the local density gradient. The resulting density irregularity has a resolution of 10 meters.

The primary effect of electron density in the ionosphere is to increase the phase path along the propagation direction. The differential phase between 150 and 400 MHz transmissions are shown in the upper right panel of Plate 1. Except for 2π phase ambiguities, the phase is proportional to the variation of the total electron content shown in the last panel of Plate 1. The jagged sides of one the TEC curve are due to the imposition of the small-scale irregularity spectrum.

Figure 1 shows the calculated phase disturbance near 400 km altitude at VHF (150 MHz), UHF (400 MHz), and L-Band (1067 MHz), respectively. The magnitude of the phase disturbance is inversely proportional to frequency but the phase fronts are similarly shaped. The NRL Scintillation and Diffraction (SCINDIF) model applies scalar diffraction theory [Goodman, 1968] to these waves to yield amplitude and phase disturbances that have strong frequency dependence. The amplitude scintillations vary from 35 dB at VHF to less than 1 dB at L-Band. The last row of panels in Figure 1 shows the spatial variations of the phase fluctuations. Only the VHF phase is noticeably affected by the irregularity. In this case, errors in ground TEC determined from VHF/UHF phase measurements would be in error directly below the irregularity but TEC obtained from UHF/L-Band differential phase would be acceptable. Plate 1 and Figure 1 illustrate that accurate prediction of adverse affects that the ionosphere has on communications, navigation and radar systems requires accurate measurements of electron densities.

This paper focuses on space-based methods for detecting regions of the ionosphere that will impact satellite-to-earth radio links. Once these regions are located, they can be used to validate the accuracy of forecasts and models. If the drift velocity of the disturbances is known, near-term forecasts can provide a warning of impending affects on radio systems.

Nearly all of our knowledge of the ionosphere has been obtained by (1) remote sensing with radio waves [Davies, 1990; Hunsucker, 1991], (2) remote sensing with optical emissions [Chamberlain, 1961; Meier, 1991], and (3) *in situ* measurements of particles and fields [Kelley, 1989; Pfaff, Borovsky, and Young, 1998]. Each technique can be applied using ground-based instruments, space-borne sensors, or ground-to-space systems. Ground-based radio sensors of the ionosphere include the sweep-frequency pulsed radar device (ionosonde), the coherent or incoherent scatter radars, and relative ionospheric opacity meter

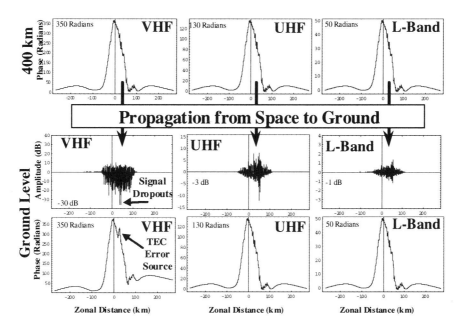

Figure 1. Computed ground scintillations from the localized equatorial irregularity.

(riometer) [Kelley 1989, Davies, 1990; Hunsucker, 1991]. Ground based optical sensors include the Fabry-Perot interferometers, an all-sky imager, as well as the optical spectrometer and photometer. Ground sensors provide the time-history of the ionosphere near the site where the sensor is located.

Space-based plasma monitors may use *in situ* probes to measure electron density, electron temperature, ion density, average ion temperature, ion-molecular weight, plasma irregularities, and ion drift velocity vectors [Kramer, 1996]. These data are acquired at the altitude of the orbit. The Defense Meteorological Satellite Program (DMSP) currently is employing the Special Sensor Ionospheric Plasma Drift/Scintillation Monitor to provide these measurements from the Block 5 Series of satellites in the 811-853 km altitude range. The Fast Auroral Snapshot Explorer (FAST) satellite provides *in situ* measurements of rapidly varying electric and magnetic fields and the flow of electrons and ions in the auroral regions. FAST is in a 350 km by 4200 km orbit with an inclination of 83 degrees to focus on high latitude particles and fields associated with the aurora. The San Marco satellite was designed to study the structure, dynamics, and aeronomy of the equatorial thermosphere. Ionospheric irregularities, ion drifts, and electric fields were measured using electric field and ion velocity instruments in a 262 km by 629 km orbit with an inclination of 2.9 degrees.

Both ground-based and *in situ* measurements are restricted in their region of observation. Radio and optical remote sensors on satellites can be designed to cover large volumes of observations. These sensors measure the integrated electron or ion densities along straight-line paths. Electron densities are obtained using discrete-inverse or tomographic reconstruction techniques. One example of this type of sensor is the GPS/MET experiment launched in April 1995. By receiving signals from Global Positioning System (GPS) Satellites, profiles of the ionosphere as each satellite sets or rises behind the Earth's ionosphere [Yunck, Liu, and Ware, 2000]. Complementary geometry has been demonstrated using transmissions from the TRANSIT and other beacons in LEO to ground receivers [Austin, Franke and Liu, 1988]. The analysis of TEC data obtained with this geometry is called Computerized Ionospheric Tomography (CIT). The Scintillation and TEC Receiver in Space (SCITRIS) sensor is being developed at NRL for flight near 400 km altitude to provide direct measurements of satellite transmissions to obtain TEC and scintillation data from different geometries than available from the ground.

Finally, it has been known for the past 40 years that the upper atmosphere radiates in the far ultraviolet (FUV). Sounding rockets measurements have determined that electron density profiles can be obtained by observing emissions from atomic oxygen produced by radiative recombination between atomic oxygen ions and electrons [Meier, 1991]. The reconstruction of ionospheric densities from the full-scan, EUV instruments has been described by Kamalabadi et al. [1999] and Bernhardt et al. [1998]. The current and future use of these remote sensing techniques provides the basis for this paper.

III. SPACE-BASED SENSORS OF IONOSPHERIC IRREGULARITIES

Several new sensor suites have been or will be available for measurements of ionospheric irregularities. The Advanced Research and Global Observation Satellite (ARGOS) is providing comprehensive data on the upper atmosphere since February 1999. New satellites in this category are the Communication Navigation Outage Forecasting System (C/NOFS) from the Air Force, the University of Surrey PICOsat sponsored by the United States Department of Defense Space Test Program, and the Constellation Observing System for Meteorology, Ionosphere, and Climate (COSMIC or ROCSAT-3) a collaborative Taiwan-US experiment. Finally, a space-based sensor called Scintillation and TEC Receiver in Space (SCITRIS) is being developed to provide global, real-time maps of ionospheric irregularities and ground scintillations.

A. ARGOS

The Advanced Research and Global Observation Satellite (ARGOS) was launched into a sun-synchronous, polar orbit on 23 February 1999. The three-ton ARGOS satellite contained a total of 9 science and technology experiments sponsored by the Department of Defense. Three ARGOS instruments illustrated were developed at the Naval Research Laboratory to provide direct observations of the upper atmosphere. First, the High Resolution Airglow/Aurora Spectroscopy (HIRAAS) payload consists of three high-spectral-resolution ultraviolet-spectrographs designed to measure the naturally occurring thermospheric and ionospheric airglow. Extreme ultraviolet (EUV) imaging of the ionosphere is accomplished with scans of the Earth's limb to cover from 50 to 750 km in altitude. EUV photons are collected by HIRAAS in a rectangular aperture that has 5 km vertical by 120 km horizontal resolution. Second, the Global Imaging Monitor of the Ionosphere (GIMI) instrument uses far ultraviolet emissions (FUV) to image the earth's ionosphere and the upper atmosphere. One of the GIMI charge coupled device (CCD) sensor covers the 131 to 160 nm with a 6° by 9° field of view. Each pixel in the GIMI imager views a 3 x 3 km portion of the ionosphere at the tangent altitude. Third, the Coherent Radio Tomography (CERTO) beacon radiates two frequencies at 150 and 400 MHz to a linear array of ground receivers to provide images of the electron densities in the ionosphere using tomographic techniques. There is no instrumental limit to the resolution of the TEC measurements but Fresnel diffraction limits tomographic reconstructions to about 1 km. Irregularities on the order of 0.5 km or less can be detected by observing rapid amplitude and phase fluctuations in the received beacon signals.

A good example of the global-scale, space-based observations provided by ARGOS comes from the HIRAAS observations of the ionosphere. The optical determination of F-layer electron densities uses extreme ultraviolet (EUV) radiation associated with O$^+$-ions in the ionosphere [Dymond, et al., 1997]. Radiative recombination proceeds by the reaction

$$O^+ + e^- \rightarrow O^* \rightarrow O + hv(91.1nm)$$
$$rate: \alpha = 4.4x10^{-13} cm^3 / s \tag{1}$$

where α is calculated for a 0.1 eV electron temperature by Julienne et al. [1974]. The measured intensity of the 91.1 nm radiation is proportional to the integral of the product of the O$^+$-ion and electron densities. In the F-region, atomic oxygen ion and electron concentrations are nearly equal so the 91.1 nm intensities are approximately given by

$$I_{91.1} = \alpha \int_S [O^+][e^-]ds \cong \alpha \int_S n_e^2 ds \quad photons/cm^2/s \tag{2}$$

where the integration path is along the line-of-sight S of the detector, and I$_{91.1}$ is the intensity from the integrated volume emission. Observations of 91.1 nm emissions are most easily accomplished at night when other line emissions excited by solar illumination are not present. The nadir 91.1 nm emission from 600 km altitude is typically 35 Rayleighs [Meier, 1991].

During daytime, the brightest feature in the ultraviolet (UV) airglow is the 83.4 nm emission. The dominant source of this emission is photoionization of O by the reaction

$$O + hv(SolarEUV) \rightarrow O^+(^4P) + e^-$$
$$\rightarrow O^+(^4S^o) + e^- + hv(83.4nm) \tag{3}$$

which occurs below 200 km altitude. The diffuse 83.4 nm line radiation illuminates the F-region O$^+$-ions from below. Multiple scattering causes the O$^+$-ions in the daytime F-region to become visible in the 83.4 emission. The volume excitation rate for this scattering is described by Meier [1991]. The 83.4 nm intensity is given by

$$I_{83.4} = \int_S J([O^+])ds \tag{4}$$

where the volume emission rate J([O$^+$]) is a complex function of O$^+$ density that includes the radiative transport

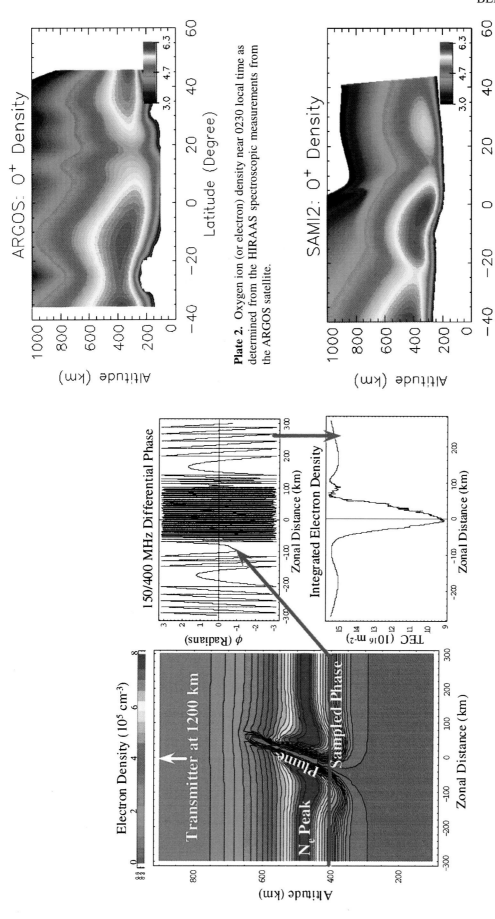

Plate 2. Oxygen ion (or electron) density near 0230 local time as determined from the HIRAAS spectroscopic measurements from the ARGOS satellite.

Plate 3. Simulation of the night-time, low- and mid-latitude ionosphere with the SAMI2 model. The conditions for the model match the environment that produced the data in Figure 3.

Plate 1. Equatorial irregularity phase disturbance.

of 83.4 nm photons, solar EUV source, and O density profile. A method for retrieving the O^+ number density from satellite limb scans of 83.4 nm is described by Picone et al. [1997].

The results from the HIRAAS instrument on the ARGOS satellite are compared with the SAMI2 model developed at NRL [Huba et al., 2000a,b]. Plate 2 is a colored contour plot of the logarithm of the O^+ density as a function of geographic latitude and altitude. The contours are of $\log_{10}[n(O^+)]$ in units cm^{-3}. The O^+ density is reconstructed from the measured UV emission from the 91.1 nm emission using discrete inverse theory techniques. The position and time of the satellite for geomagnetic latitude (θ_g) of 45° was 285° E and 07:38 UT, and at θ_g = -40° was 265 E and 08:03 UT. The dominant features in Plate 2 are the following. The O^+ density has maxima at near θ_g = -40°, -15°, and 40° at altitudes of ~ 400 km, 350 km, and 300 km, respectively. There is a weak latitudinal minima at θ_g = -30° and a stronger and broader minima at θ_g = 20°. The topside above 600 km has latitudinal structure that is evident in the transition from the green to light blue contour levels.

The simulation results use a mesh of 201 grid points for each flux tube. In the altitude range 150 - 3000 km 120 flux tubes are used, in the range 3000 - 10000 km 24 flux tubes are used. The inputs to simulation are the following parameters: year - 1999, day - 328 (Nov 24), longitude - 275 E, Ap - 21, F10.7 - 181, and F10.7A - 181. The actual satellite trajectory was from a longitude 285 E to 265 E. Plate 3 is a colored contour plot of the logarithm of the O^+ density as a function of geographic latitude versus altitude at time 0200 LT. The contours and units are the same as in Plate 2. The comparison between Plates 3 and 4 is excellent. The general morphology of the ionosphere is captured by SAMI2: the shape and magnitude of the contours. The biggest discrepancy between the ARGOS reconstructed density and SAMI2 is the position and intensity of the higher latitude (θ_g 40°) density maximum. SAMI2 has a weaker density maximum that is located at a lower latitude (θ_g ~ 25°). Interestingly, SAMI2 does capture some of the latitudinal structure at high altitude (above 600 km).

For comparison, Plate 4 illustrates a colored contour plot of the logarithm of the electron density as a function of geographic latitude versus altitude at time 0200 LT using the IRI (International Reference Ionosphere) model. The contours and units are the same as in Plate 4. The IRI model does not capture the detailed morphology of the ionosphere very well for this event. It has significantly higher electron densities over a broader altitude range than the Argos data or SAMI2 results at latitudes less than ~ 5°. There is very little latitudinal structure at higher altitudes

(above 600 km); also, the electron density is much larger than the ARGOS reconstruction or SAMI2 results at these higher altitudes. The one area where the IRI model is closer to the observational results than SAMI2 is at latitudes above ~ 20°. The IRI electron density is larger and extends to higher latitudes compared to the SAMI2 results.

Higher horizontal resolution than provided by HIRAAS limb scanner can be obtained using the sensors on the GIMI or CERTO sensors on ARGOS. GIMI, with a spatial resolution of 3 km, is sensitive to FUV radiation from atomic oxygen at 135.6 nm. The source of this excited state is the radiative recombination given in equation (1). ARGOS is in a polar orbit and GIMI is oriented to view along the wake direction of the orbit. Images of FUV emissions are recorded looking tangent to the magnetic field near the equator. Consequently, field aligned irregularities at low-latitude show up well in the optical emissions.

A sample of the GIMI data near the geomagnetic equator is illustrated in Figure 2. The F-layer is laminar except for tilted black stripes at two locations in the image. These stripes are equatorial bubbles that were simulated in Plate 1. Currently, the GIMI images are qualitative and an absolute EUV intensity is not available. Calibration of the GIMI data is currently in process using stellar intensities. Even without calibration, the GIMI position data yields the locations of equatorial bubbles that can be a space-based proxy for scintillations at ground sites located below the irregularities.

GIMI can detect the location of plasma bubbles from space but the FUV sensor does not have enough resolution to resolve kilometer scale irregularities that are responsible for scintillation. The CERTO radio beacon on ARGOS has designed to provide scintillation measurements at 150.012 MHz and 400.032 MHz. In addition, differential phase measurements are made with the beacon to provide TEC data along straight-line paths between the beacon and ground receivers.

The CERTO measurements are limited to 500 m because of Fresnel diffraction effects. These measurements augment the larger scale measurements provided by HIRASS and GIMI. In the absence of the EUV observations, the TEC measurements can be used to modify existing models to provide finer scale structure. Plate 5 shows an image of horizontal structures that would not be easily obtained using limb scanning technique. These data were based on TEC observations at two receivers located at 11° and 18° Latitude. The IRI model was used to provide the background ionosphere. The TEC data from the receivers are processed with an algebraic reconstruction technique to yield the tomographic image [Bernhardt et al., 2000]. The main features in the observations are a trough

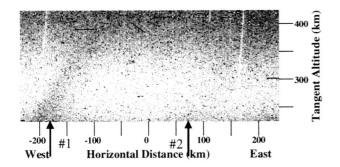

Figure 2. ARGOS/GIMI Image of Equatorial Irregularities 0212 Local Time, 334 Longitude,31 July 1999, 0352 GMT. The wavelength of the observations is 135.6 nm. Tilted density depressions occur at the arrows marked #1 and #2.

near 11° latitude and modulation of the electron density by acoustic gravity waves to the right of the trough. Finally, electron densities provided by the HIRAAS instruments and the CIT technique using CERTO provide cross-calibration for the instruments

B. C/NOFS

The Communication/Navigation Outage Forecasting System (C/NOFS) will be launched in 2003. C/NOFS is being planned by the U.S Air Force to predict ionospheric disturbances near the equator that have the potential to disrupt military space systems used for communications and navigation. C/NOFS is being managed at the Air Force Research Laboratory at Hanscom Air Force Base, Mass and is being executed as a part of the DoD Space Test Program at Kirtland AFB in New Mexico.

The basis for C/NOFS is that the conditions for the development and initiation of equatorial irregularities, such as shown in Plate 1, can be measured before the irregularities form [Kelley, 1989]. Destabilizing factors such as gravity, electric fields, vertical density gradients, and zonal neutral winds compete with stabilizing factors such as meridional neutral winds, E-region densities, and ion-electron recombination for the generation of equatorial bubble structures. C/NOFS will attempt to measure these factors for use in irregularity predictions.

After launch, the first year of C/NOFS operations will be primarily for basic research to validate models that use space-based measurements to predict ionospheric scintillations. The primary *in situ* instruments are the Planar Langmuir Probe (PLP) for plasma density, the Vector Electric Field Instrument (VEFI) for electric fields related to plasma drifts and irregularity development, the Digital Ion Drift Meter (DIDM) for measurements of plasma drift velocities. The electron density profiles will be provided by the CNOFS Occultation Receiver for Ionospheric Sensing and Specification (CORISS) using

transmissions from GPS satellites. The Coherent Electromagnetic Radio Tomography (CERTO) beacon will provide direct measurements of scintillations at ground receivers and two-dimensional images of electron densities by employing reconstructions with computerized ionospheric tomography (CIT). Starting in the latter part of the first year of operation and continuing for another two years, the C/NOFS mission will focus on applications of the prediction techniques.

The orbit inclination of C/NOFS will be about 13 degrees and the orbit altitude will range from 400 to 700 km. This will permit sampling the equatorial ionosphere above and below the peak of the F-layer. The CNOFS satellite will revisit the same portion of the ionosphere every 90 to 100 minutes to validate forecasts or to observe the progress of known scintillation regions. Further, because equatorial irregularities extend up to 2000 km north and south of the equator along magnetic field lines, the CNOFS satellite can effectively sample a wide band of critical magnetic latitudes at every longitude.

D. PICOsat

The name PICOsat was developed from the first initials of the four instruments on board the satellite. The two ionospheric sensors are the Ionospheric Occultation eXperiment (IOX) and the CERTO beacon. Bernhardt et al. [1998] have shown that the images from computerized ionospheric tomography can be improved by using both beacon to ground and radio occultation measurements. The inclination of PICOsat will be near 55° to match the inclination of GPS satellites. This provides the maximum number of occultation opportunities. This inclination also provides a unique scan geometry for radiotomography this is halfway between the polar obits of ARGOS and the equatorial orbit of CNOFS and STRV-1d.

D. COSMIC

The Constellation Observing System for Meteorology, Ionosphere, and Climate (COSMIC) uses multi-point sensors that employ line-of-sight measurements for studying the Earth's atmosphere. The three sensors on each COSMIC satellite are (1) GPS occultation receiver, (2) Tiny Ionospheric Photometer (TIP) and (3) the Tri-Band Beacon (TBB). These instruments are complementary in their observations by using both radio and optical measurements. The six COSMIC satellites will be placed in planes separated by 60 degrees at 800-km operational altitude and 72 degrees inclination. The COSMIC system has been described in detail by Rocken et al. [2000]. The GPS occultation receiver will make over 3000 globally distributed occultations per day for ionospheric sensing. These occultations will be used to provide 3-dimensional

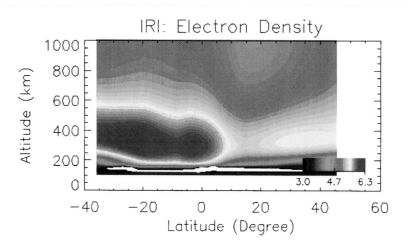

Plate 4. Night-time electron densities provided by the International Reference Ionosphere in an attempt to match to the observations shown in Figure 3.

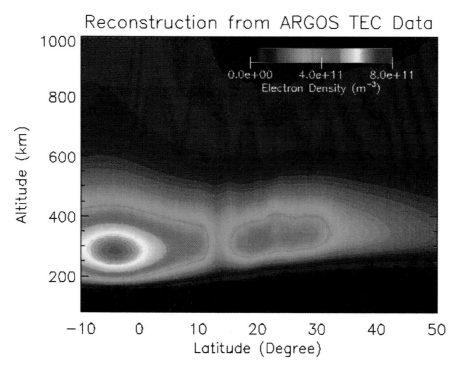

Plate 5. Mid-latitude ionospheric irregularities obtained with the CERTO beacon and two ground receivers.

images of electron density and irregularities in the ionosphere [Hajj, 2000]. The Tiny Ionospheric Photometer will be a down looking ultraviolet radiometer for measuring horizontal electron density gradients in the night-time F-region. The TIP data will combined with the GPS receiver data to yield maps of the two dimensional structure of the ionosphere [Dymond, et al., 2000]. The Tri-Band-Beacon (TBB) radiates phase coherent signals to an array ground receivers for studies of the ionosphere. Computerized Ionospheric Tomography (CIT) of the ionosphere will be obtained using total electron content (TEC) data from both the TBB ground receivers and the GPS receivers [Bernhardt et al., 2000]. The three instruments on COSMIC will work together to produce high-resolution, global images of ionospheric irregularities.

E. SCITRIS

The Naval Research Laboratory is developing the Scintillation and TEC Receiver in Space (SCITRIS) sensor to provide measurements of the complex amplitude data from radio waves passing through the ionosphere. This data will be processed to yield TEC along nearly horizontal paths using paths between satellites in low-earth-orbit (LEO). This data will provide the missing horizontal paths needed for radio tomography and will be used for global scintillation estimates. With the geometry illustrated in Plate 1, SCITRIS can make direct measurements of the phase front of VHF/UHF/L-Band radio waves before they have become distorted by diffraction effects. Based on those measurements and diffraction calculations using the SCINDIF code, estimates will be available for ground based scintillations at any desired frequency. For these observations, the radio beacon transmissions must originate from above the ionosphere and the receiver must be located below the peak of the F-layer. There are more than twenty 150- and 400-MHz radio sources between 800 and 1000 km altitude from the OSCAR, COSMOS, TSIKADA, NADEZDA, and RADCAL programs. The receiver orbit near 400 km can be provided by the Space Shuttle launch SPARTAN or by the International Space Station. SCITRIS is a candidate for flight in the 2003 to 2010 time period.

IV. CONCLUSIONS

The complementary nature of the radio and optical remote sensing is currently being fully exploited for ionospheric observations. Radio beacons in low-earth-orbit provide VHF, UHF and L-Band transmissions to arrays of ground receivers for TEC and scintillation measurements. GPS receivers record L-band transmissions from the constellation of GPS satellites around the globe to provide electron density profiles and L-Band scintillations. Optical spectrometers, photometers and imagers provide a global record of electron and ion densities in the ionosphere. Each type of measurements has inherent strengths; and all types of sensors are required to provide a complete picture of the ionosphere. The horizontal viewing instruments such as the GPS occultation receiver and the EUF limb scanners, are primarily used to produce ionosphere profiles. The oblique and vertical-looking instruments such as the radio beacon transmissions to ground stations and the down-looking photometer primarily provide horizontal structure of the ionosphere. Finally, the redundancy of multiple sensors can provide cross-calibrations.

Acknowledgments. The authors are grateful for discussions with Paul Straus of Aerospace Corporation, Robert Meier at NRL and Robert McCoy at ONR. This research was sponsored by the Office of Naval Research, the Air Force Office of Scientific Research and the COSMIC Program Office at the University Corporation for Atmospheric Research.

REFERENCES

Aarons, J., Global Morphology of Ionospheric Scintillations, *Proceedings of the IEEE*, *70*, 360-378, 1982.

Austen, R.J., S.J. Franke, and C.H. Liu, Ionospheric imaging using computerized tomography, *Radio Sci.*, *23*, 299-307, 1988

Basu S., E. Mackenzie, and Su. Basu, Ionospheric constraints on VHF/UHF communication links during solar maximum and minimum periods, *Radio Sci.*, *23*, 363, 1988.

Bernhardt, P.A., R.P. McCoy, K.F. Dymond, J.M. Picone, R.R. Meier, F. Kamalabadi, D.M. Cotton, S. Charkrabarti, T.A. Cook, J.S. Vickers, A.W. Stephan, L. Kersley, S.E. Pryse, I.K. Walker, C.N. Mitchell, P.R. Straus, H. Ha, C. Biswas, G.S Bust, G.R. Kronschnabl, and T.D. Raymund, Two-dimensional mapping of the plasma density in the upper atmosphere with computerized ionospheric tomography (CIT), *Physics of Plasmas*, *Vol.5.*, No. 5, pp. 2010-2021, 1998.

Bernhardt, P.A., C. A. Selcher, S. Basu, Gary Bust, and S.C. Reising, Atmospheric Studies with the Tri-Band Beacon Instrument on the COSMIC Constellation, *TAO, 11*, 291-312, 2000.

Chamberlain, J.W., *Physics of the Aurora and Airglow*, American Geophysical Union, 1995.

Davies, D., *Ionospheric Radio*, IEE Electromagnetic Waves Series 31, Peter Peregrinus Ld., Exeter, England, 1990.

Dymond K.F., S.E.Thonnard, R.P. McCoy, R.J. Thomas, An optical remote sensing technique for determining nighttime F region electron density, *Radio Sci.*, *32*, 1985-1996, 1997.

Dymond, K.F., J.B. Nee, and R.J. Thomas, The tiny ionospheric photometer: An instrument for measuring ionospheric gradients for the COSMIC constellation, *TAO, 11*, 273-290, 2000.

Goodman, J.W., *Introduction to Fourier Optics*, McGraw-Hill, New York, 1968.

Groves, K.M., S. Basu, E.J. Weber, M. Smitham, H. Kuenzler, C.D. Valladares, R. Sheehan, E. MacKenzie, J.A. Secan, P. Ning, W.J. McHeill, D.W. Moonan, and M.J. Kendra, Equatorial scintillation and systems support, *Radio Sci.*, *32*, 2047-2064, 1997.

Hajj, G.A., L.C. Lee, X. Pi, L.J. Romans, W.S. Schreiner, P. R. Straus, C. Wang, COSMIC GPS ionospheric sensing and space weather, *TAO*, *11*, 235-272, 2000.

Huba, J.D., G. Joyce, and J.A. Fedder, The formation of an electron hole in the topside equatorial ionosphere, *Geophys. Res. Lett.*, *27*, 181, 2000a.

Huba, J.D., G. Joyce, and J.A. Fedder, SAMI2 (Sami2 is Another Model of the Ionosphere): A new low-latitude ionosphere model, *J. Geophys. Res.*, *105*, 23035-23053, 2000b.

Hunsucker, R. D., *Radio techniques for probing the terrestrial ionosphere*, Springer-Verlag, New York, 1991.

Julienne, P.S., J. Davis, and E. Oran, Oxygen recombination in the tropical nightglow, *J. Geophys. Res.*, *79*, 2540-2551, 1974.

Kelley, M.C., *The Earth's Ionosphere*, pp. 261-310, Academic Press, 1989.

Kramer, H.J., *Observation of the Earth and Its Environment*, 3rd Edition, Springer-Verlag, New York, 1996.

Meier, R.R., Ultraviolet spectroscopy and remote sensing of the upper atmosphere, *Space Sci. Rev.*, *58*, 1-186, 1991.

Newton, R.R., The Navy Navigation Satellite System, **Space Research VII**, pp. 735-763, North-Holland, 1976.

Picone, J.M., R.R. Meier, O.A. Kelley, K.F. Dymond, R.J. Thomas, D.J. Malendez-Alvira, and R.P. McCoy, Investigation of ionospheric O^+ remote sensing using 834 A airglow, *J. Geophys. Res.*, *102*, 2441-2456, 1997.

Pfaff, R.F., J.E. Borovsky, and D.T. Young, Measurement Techniques in Space Plasmas: Particles (Vol 102) and Fields (Vol. 103), *Geophysics Monograph*, American Geophysical Union, 1998.

Rishbeth, H. and O.K Garriott, *Introduction to Ionospheric Physics*, pp. 126-159, Academic Press, New York 1969.

Rocken, C., Y-H. Kuo, W.S. Schreiner, D. Hunt, S. Sokolovskiy, and C. McCormick, COSMIC System Description, *TAO*, *11*, 21-52, 2000.

Yeh, K.C, and C.H. Liu, Radio Wave Scintillations in the Ionosphere, *Proceedings of the IEEE*, *70*, 324-360, 1982.

Yunck, T.P., C.H. Liu, and R. Ware, A history of GPS sounding, *TAO*, *11*, 1-20, 2000.

Zalesak, S.T., S.L. Ossakow, and P.K. Chaturvedi, Nonlinear equatorial spread F: The effects of neutral winds and background Pedersen conductivity, *J. Geophys. Res.*, *87*, 151-166, 1982

T.L. Beach, Space Vehicles Directorate, Air Force Research Laboratory, Hanscom AFB, MA 01731

P.A. Bernhardt, Code 6794, Naval Research Laboratory, Washington, DC 20375

G Bust, Applied Research Laboratory, University of Texas at Austin, Austin, TX 78713

K.F. Dymond, Code 7620, Naval Research Laboratory, Washington, DC 20375.

J.D. Huba, Code 6794, Naval Research Laboratory, Washington, DC 20375

C. Rocken, GPS Science and Technology Program, UCAR, Boulder, CO 80301

C.A. Selcher, Code 5551, Naval Research Laboratory, Washington, DC 20375

C. Rocken, UCAR, P.O. Box 3000, Boulder, CO 80307